Acta Numerica 1994

Acta Numerica 1994

Published by the Press Syndicate of the University of Cambridge
The Pitt Building, Trumpington Street, Cambridge CB2 1RP
40 West 20th Street, New York, NY 10011-4211, USA
10 Stamford Road, Oakleigh, Melbourne 3166, Australia

First published 1994

Printed in the United States of America

Library of Congress cataloging in publication data available

A catalogue record for this book is available from the British Library

ISBN 0-521-461812 hardback

Contents

Formalization and computational aspects of image analysis 1
Luis Alvarez and Jean Michel Morel

Domain decomposition algorithms 61
Tony F. Chan and Tarek P. Mathew

Aspects of the numerical analysis of neural networks 145
S.W. Ellacott

A review of pseudospectral methods for solving partial differential equations 203
Bengt Fornberg and David M. Sloan

Exact and approximate controllability for distributed parameter systems 269
R. Glowinski and J.L. Lions

On the numerical evaluation of electrostatic fields in composite materials 379
Leslie Greengard and Monique Moura

Numerical geometry of surfaces 411
Malcolm Sabin

Numerical analysis of dynamical systems 467
Andrew Stuart

Acta Numerica (1994), *pp.* 1–59

Formalization and computational aspects of image analysis

Luis Alvarez
Departamento de Informatica y Sistemas
Universidad de Las Palmas
Campus de Tafira, 35017 Las Palmas, Spain

Jean Michel Morel
C.E.R.E.M.A.D.E.
Université Paris IX Dauphine
75775 Paris cedex 16, France

In this article we shall present a unified and axiomatized view of several theories and algorithms of image multiscale analysis (and low level vision) which have been developed in the past twenty years. We shall show that under reasonable invariance and assumptions, all image (and shape) analyses can be reduced to a single partial differential equation. In the same way, movie analysis leads to a single parabolic differential equation. We discuss some applications to image segmentation and movie restoration. The experiments show how accurate and invariant the numerical schemes must be and we compare several (old and new) algorithms by discussing how well they match the axiomatic invariance requirements.

CONTENTS

1 Introduction 2
2 Image multiscale analysis 6
3 Axiomatization of image multiscale analyses and classification of the main models 12
4 Shape multiscale analyses 16
5 Relation between image and shape multiscale analyses 20
6 Multiscale segmentation 22
7 An example: texture discrimination 26
8 Movies multiscale analysis 30

9 Invariance and stability requirements for numerical schemes of the fundamental equation. 32
10 Finite difference schemes for the AMSS model 36
11 Morphological (set evolution) schemes 42
12 Conclusions 50
Appendix A. The 'fundamental theorem' of image analysis 50
Appendix B. Proof of the scale normalization lemma 51
Appendix C. Classification of shape multiscale analyses 53
References 55

1. Introduction

1.1. What will be done, what not, and why?

Before starting with what will be the main object of this survey – image multiscale analysis – we intend to give a very brief account of what image processing is and the choices we have made about what should be developed here and what should be omitted. Image processing may be viewed as a long list of techniques for capturing, transmitting, and extracting information from digital images, in close relation with what is assumed to be relevant to human perception. Here is, accordingly, the list of subjects treated in a classical manual of image processing: *Visual perception, Digitization, Compression, Enhancement, Restoration, Reconstruction, Matching, Segmentation, (Semantic) Representation* (Rosenfeld and Kak, 1982). This defines image processing as a somewhat abstract theory. Other manuals focus on practical applications to perception-based control of robots (Horn, 1986) vision theory (Marr, 1982), while many other monographs treat a single technological application: radar vision, microscopy, satellite imaging, compression standards, character recognition, etc. These involve specific mathematical techniques which will not be presented here. Indeed, whenever some *a priori* (statistical, structural) knowledge about the processed image is at hand, the process must be adapted accordingly and (for instance) the use of stochastic filtering techniques, is justified, but specific.

The IEEE monographs and proceedings give a good account of what is being done in image processing and one can get a rather complete view of the image analysis subject by reading the proceedings of the biannual ICCV (International Conference on Computer Vision). Now, whatever the envisaged applications are, the nine items from Kak and Rosenfeld are a reliable common denominator. They represent what everybody should know before starting any application dealing with images or a formalized theory of vision. From the nine terms quoted above, only the second (digitization) and the last (semantic representation) fall outside our field because the first relates to the engineering of captors and the last to artificial intelligence and structured programming. Starting with the mathematical classification of the subjects, let us say that compression and reconstruction rely on sharp mathematical techniques related to harmonic analysis. Indeed, the main step of compression–reconstruction devices is a decomposition of the image on a well chosen functional orthonormal basis which can be classical (Fourier, Haar, Hadamard) or new: Wavelets (Meyer, Mallat, Daubechies, etc.), Wavelet packages, etc. All these theories, from the mathematical as well as from the numerical viewpoint, are well explained in several recent books (see e.g. Meyer (1992)), accessible to both mathematicians and engineers, and we simply choose not to present them here. However, there is another reason for this: we must distinguish between techniques for *analysing* images (which therefore strongly rely on the geometry of images) and techniques for *storing* them, where there is no need for geometrical invariance in the numerical representation. When one wants to compress, 'tous les coups sont permis'. In addition to this intuitive difference there is a corresponding strong difference in mathematical techniques. As we shall see, image-analysis geometry-preserving techniques must be fully nonlinear (probably one of the first to understand this and draw the mathematical consequences was Matheron (1975)). Therefore, from the initial nine subjects from the Kak–Rosenfeld classification we shall keep only four: *visual perception, enhancement, restoration and segmentation.* (We have also omitted matching because this is a secondary task, only effectuated after some of the four preceding processes have been applied.) Because the four mentioned subjects obey the same geometric requirements, we shall see that they can be treated by a common theory which we shall call *multiscale analysis.* We shall therefore not treat them as primary subjects and they will simply appear as natural consequences or applications of the theory of geometric multiscale analysis.

1.2. Geometric multiscale analysis

This section is devoted to a short overview of what will be covered.

A numerical image can be modelled as a real function $u_0(x)$ defined in $\mathbb{R}^N$ (in practice, $N = 2$ or 3). The main concept of vision theory and image analysis is *multiscale analysis* (or '*scale space*'). Multiscale analysis associates with $u(0) = u_0$ a sequence of simplified (smoothed) images $u(t, x)$ which depend upon an abstract parameter $t > 0$, the *scale.* The image $u(t, x)$ is called *analysis of the image* u_0 *at scale* t. The formalization of *scale space* has received much attention in the past ten years; more than a dozen of theories for image, shape or 'texture' multiscale analysis have been proposed and recent mathematical work has permitted a formalization of the whole field. We shall see that several formal principles (or axioms) are sufficient to characterize and unify these theories and algorithms and to show that some of them simply are equivalent. These principles are *causality* (a concept in vision theory which can be led back to a maximum principle), *Euclidean (and/or affine) invariance*, which means that image analysis does not depend upon the distance and orientation in space of the analysed image, and *morphological invariance* which means that image analysis does not depend upon a contrast change.

The characterization and classification of the numerous theories of image and shape analysis will be obtained by identifying the underlying partial differential equations (which have been more or less implicit in many theories!). The axiomatic characterization leads, as we shall see, to a significant improvement in most proposed algorithms as well as to new ones with more invariance properties. Among the theories which will be axiomatically or numerically tested here, we shall mention

- the *Raw Primal Sketch* by Hildreth and Marr,
- the *Scale Space* by Witkin, Koenderink, etc.,
- the *Intrinsic Heat Equation* by Gage, Hamilton, Grayson, Angenent, etc.,
- the *Motion by Mean Curvature* (Osher, Sethian, Evans, Spruck, Giga, Goto, Barles, Souganidis, etc.),
- the *Entropy Scale Space* by Kimia, Tannenbaum and Zucker,
- the *Texton* theory by Julesz,
- the *Dynamic Shape* by Koenderink and Van Doorn,
- the *Curvature Primal Sketch* by Mackworth and Mocktarian, Asada and Brady, etc.,
- the *Morphologie Mathématique* by Matheron, Serra and the 'Fontainebleau school',
- the *Anisotropic Diffusion* by Perona and Malik,
- the *Affine Scale Space of Curves* by Sapiro and Tannenbaum,
- the *Affine Morphological Scale Space* of images by Alvarez, Guichard, Lions and Morel,
- the *Affine Morphological Galilean Scale Space* of movies by the same authors.

The classification of these multiscale theories will lead us to focus on the only one of them which simultaneously matches all invariance and stability requirements partially satisfied by the others: the Affine Morphological Scale Space (AMSS). This multiscale analysis can be defined by a simple Partial Differential Equation (PDE),

$$\frac{\partial u}{\partial t} = |\mathrm{D}u|(t \cdot \mathrm{div}(\mathrm{D}u/|\mathrm{D}u|))^{1/3}, \quad u(0,x) = u_0(x), \tag{1.1}$$

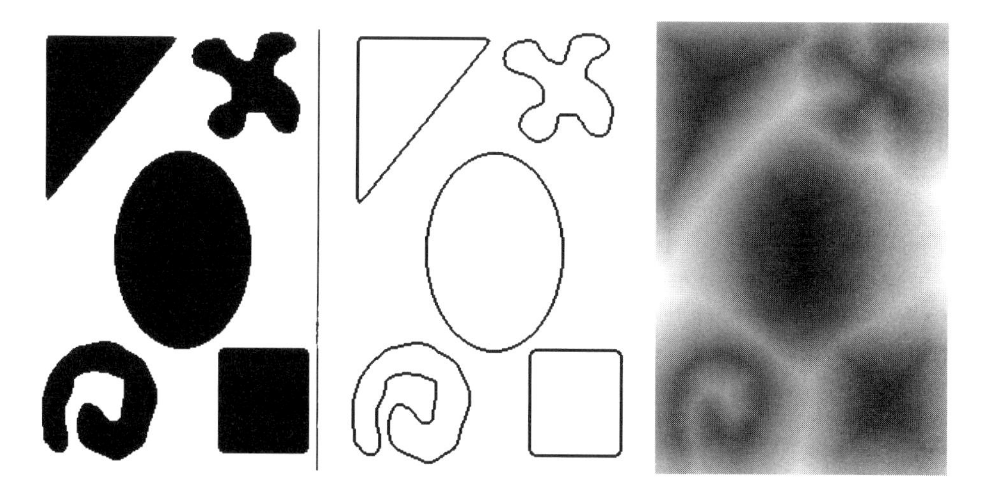

Fig. 1. Three different representation of a 200×300 pixels binary image (from left to right): (a) the classical discrete representation, (b) the level curve representation, (c) the distance function to the level curve. See Section 9 for more details.

where $u(t, x)$ denotes the image analysed at scale t and point x. (This parabolic equation admits a unique 'viscosity solution' in the Crandall–Ishii–Lions (1991) sense.) As we shall see, the equation of the AMSS handles independently all *level sets* of the analysed image and is therefore compatible with the Morphologie Mathématique (which asks for contrast invariance). In addition, contrast invariance means that the boundary of every level set of the image is analysed as a shape and we get a common multiscale analysis for shapes and images.

A multiscale formalization of *image segmentation* can be developed with analogous principles and leads to multiscale segmentation algorithms. Multiscale segmentation associates an initial image u_0 with a sequence $u(t, x)$, $K(t, x)$, where $u(t, x)$ is the image simplified at scale t and $K(t, x)$ the set of boundaries of the homogeneous regions of u_0 at scale t. Thanks to the formalization, many segmentation algorithms can be reduced to one.

As a first application of the axiomatic method, we shall show how both scale space and image segmentation theories lead to texture segmentation algorithms as well as to a rigorous discussion of Julesz' axiomatic theory of texture discrimination. The experimental result of this discussion is unexpected.

We finally devote some pages to the above mentioned Affine Morphological Galilean analysis of movies, with Guichard's (1993) remarkable experimental results in movie denoising. The underlying equation,

$$\frac{\partial u}{\partial t} = (|\nabla u| \operatorname{curv}^{1/3}(u))^{1-q} \left(\left(|\nabla u| \operatorname{sgn}(\operatorname{curv}(u)) \operatorname{accel}(u)\right)^+\right)^q, \quad (1.2)$$

is also a parabolic equation, where curv denotes the nonlinear differential operator computing the curvature of the level lines, and accel represents the 'apparent acceleration' observed at a given space–time point of the movie.

Roughly speaking, this survey has two parts: in the first, (Sections 1 to 8) we develop the above mentioned theories and give comparative numerical results on test images. The second is devoted to a long discussion of old and new algorithms for Mean Curvature Motion (applied to images or shapes) and the AMSS model (equation (1.1)).

2. Image multiscale analysis

2.1. *A short story of the subject*

Computer vision deals with a philosophical, psychological, physiological and technical question which can be stated in a few words: how can the local brightness information arriving at the retina of some individual (or any optical sensor) be transformed into a global percept of the objects surrounding him, including their distance, colour and shape? In the 1960s, this question was translated into a very practical framework with the new possibilities for experimentation offered by digital pictures with computers. The new technology has enabled accurate measurements of human visual performance on digital pictures and the first experiments in 'computer vision'. The joint developments in pschophysics and computer vision have led to a new doctrine: the existence of *low level vision.* The story of the doctrine is well explained in David Marr's book *Vision* and we shall just give a few hints of how this doctrine developed.

On the other hand, several psychophysical experiments due to Bela Julesz and his school proved that the reconstruction of the spatial environment from binocular information was an automated, reflex process, independent of any learning. Julesz also studied the 'preattentive' perception of textures and proved the existence of a process for discriminating textures independently of any *a priori* knowledge. The discrimination process is fast, parallel and Julesz and his school discussed it in *mathematical terms* from statistics and geometry. These experiments, as well as the neurobiological experiments of Hubel and Wiesel, gave proof of the existence, in the first milliseconds of the perception process, of a series of parallel, fast and irreversible operations applied to the retina information and already yielding very rich and useful information to further understanding of the 'image'.

2.2. *The visual pyramid as an algorithm*

We shall call this series of operations the 'visual pyramid'. In mathematical terms, it may be thought of as an algorithm but not in the Turing sense; in the more general sense where we define an algorithm as a black box transforming its input into an output in a deterministic way by a physical process: in any case, this machine is assumed to be physically implemented in the brain. We must distinguish the problem of how this machine works in the case of the brain and what it really does as an information processor (what Chomsky called performance versus competence). Indeed, the second question is simply a mathematical question, while the first one is very relevant in neurobiology. We shall now focus on the mathematical question and treat it in a rather rough way by answering the three questions:

(a) What is the input of the visual pyramid?
(b) What is its output?
(c) What basic principles must obey the visual pyramid if it is considered to be a physical system?

2.3. What is the input of the visual pyramid?

A simple model to discuss image processing is to define an 'image' as a 'brightness' function $u_0(x)$ at each point x of a domain of the plane. This domain, which may be the plane itself, is a model of the retina or any other photosensitive surface. In what follows, we shall take the plane for simplicity. For commodity, in the discussion, we shall always assume that $u_0(x)$ is in the space $\mathcal{F}$ of all continuous real functions $u(x)$ on $\mathbb{R}^N$ such that $\|(1+|x|)^{-N}u(x)\| \leq C$ for some N and C. Of course, the datum of $u_0(x)$ is not absolute in perception theory, but can be considered as the element of an equivalence class. If y is a vector of the plane, the shifted datum $u_0(x-y)$, which is the image shifted by y, is an equivalent datum. In the same way, the change of $u_0(x)$ into $u_0(Rx)$ where R is an isometry of the plane should not change the visual analysis. Finally, we can think of u_0 as belonging to a projective class, that is, as a representative of the class $u_0(Ax)$ where A is any projective map of the plane. Indeed, a plane image can be viewed by an observer from any distance and orientation in space (think of a painting in a gallery). Therefore, the input $u_0(x)$ is assumed to be equivalent to any of its anamorphoses $u_0(Ax)$. We shall assume, in the following, A to be only any affine map, which makes sense when the ratio between the size of the observed objects and the distance to the sensor is small. Last but not least, the observation of $u_0(x)$ does not generally give any reliable information on the number of photons sent by any visible place to the optical sensor. Therefore, the equivalence class in consideration will be $g(u_0(x))$, where g stands for any (unknown) contrast function depending on the sensor. This last assumption, that only isophotes matter, is associated with the 'mathematical morphology' school. So we shall call it the 'morphological' assumption.

To summarize, an image is an equivalence class of functions $u_0(Ax)$ where

A is a translation or an isometry in most classical geometrical models, A is any affine map in the simplified projective model and it is a class of functions $g(u_0(x))$ where g is any continuous nondecreasing function in the morphological model. We can combine these models and consider a morphological projective model, that is, an equivalence class under the action of all gs and As: $g(u_0(Ax))$.

2.4. *What is the* output *of the visual pyramid?*

Starting from the local brightness information, each layer of the visual pyramid is assumed to yield more and more global 'low level' information about the image. This information is assumed to be usable for the geometric reconstruction (stereovision) as well as for 'high level vision', that is, the interpretation of the scene. Whatever its use might be, most models define the basic output as either

- a smoothed image (from which reliable 'features' can be extracted by local and therefore differential operators); or
- a segmentation, that is, either a decomposition of the image domain into homogeneous regions ('strong segmentation'), with boundaries or a set of boundary points or 'edge map'.

In both cases, the output depends on two variables: a variable x which denotes the centre of a spatial neighbourhood and a variable t which can be identified or correlated with

- the 'height' in the visual pyramid (or distance from the first layer: the retina). This distance corresponds to the biological time between the 'arrival' at the retina and the first arrival at a given layer; and
- the degree of globality of the local information in the considered layer, that is, the size of the neighbourhood in the retina which influences what happens at x.

To summarize, the output of the vision pyramid is:

- either a multiscale image $u(t, x)$; or
- or a multiscale 'edge map' $K(t, x)$, where t is a parameter which can be identified with a time of analysis or with a measure of the spatial globality of the information provided by $u(t, x)$. The bottom of the output is the original image $u(0, x) = u_0(x)$.

2.5. *What basic principles must the visual pyramid obey?*

The causality These principles come first from the 'preattentive' assumption that no feedback is allowed, the visual information being processed in parallel through a sequence of filters. This means that what happens at

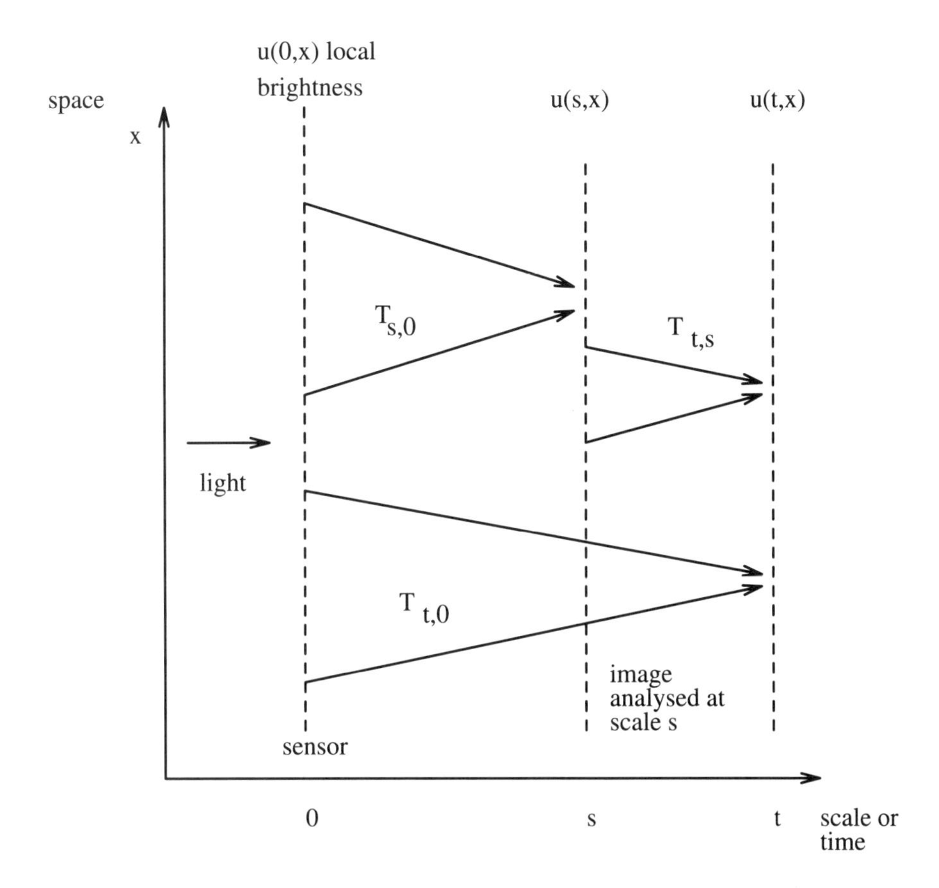

Fig. 2. The scale-space visual pyramid.

higher scales cannot influence what happens at lower scales: the pyramid acts 'from fine to coarse'. Furthermore, there is no time for taking into account at scale t what happens at a significantly smaller scale s. So we assume that the output at scale t can be computed from the output at a scale $t-h$ for very small h. (Take into account the fact that the visual pyramid is a series of filters through which new visual information is constantly being processed. So, to look at what happens at a smaller scale means to 'look into the future', at newly arriving perceptual images.) To formalize this relation, we call $T_t : \mathcal{F} \to \mathcal{F}$ the map which associates an image u_0 with its 'smoothed image' at the scale t, $T_t u_0$. (In the same way, we denote by T_t the map associating a set K of boundaries with a set of boundaries $T_t K$ simplified at scale t.) This mapping is obtained by constructing 'transition filters' which we call $T_{t+h,t} : \mathcal{F} \to \mathcal{F}$ and hence we have the

Pyramidal Structure (Causality 1) $T_{t+h} = T_{t+h,t} T_t, \quad T_0 = \mathrm{Id}$. Furthermore, the operator $T_{t+h,t}$ will always be assumed to act 'locally', that is, to look at a small part of the processed image. In other terms, $(T_{t+h,t}u_0)(x)$ must essentially depend upon the values of $u_0(y)$ when y lies in a small neighbourhood of x.

We shall give two formal versions of this 'locality assumption'. Let us now just give its 'physical' interpretation: if the basic elements of the pyramid are assumed to be 'neurons', this only means that a neuron is primarily influenced by its neighbours. A clear argument for this is time: only neurons which are close can have an influence without transmission delay. Let us finish with an intuitive requirement which is called in image-processing 'causality'. Since the visual pyramid is assumed to yield more and more global information about the image and its features, it is clear that when the scale increases, no new feature should be created by the multiscale analysis: the image and the boundaries at scale $t' > t$ must be simpler than the boundaries at scale t. The causality assumption must of course be formalized. Its formalization has been discussed by Hummel (1986), Koenderink (1984, 1990a), Yuille (1988), Witkin (1983), Perona and Malik (1987) in the framework of image processing, by Kimia *et al.* (1992) in the framework of shape analysis and by Muerle and Allen (1968), Brice and Fennema (1970), Horowitz and Pavlidis (1974) in early works on image segmentation.

The result of the discussion in the case of image processing is that causality must be formalized as *pyramidality* plus a *local comparison principle*: if an image u is locally brighter than another v, then this order must be conserved some time by the analysis (prevalence of local behaviour on global behaviour). In formal terms, it can be expressed as the

Local Comparison Principle (Causality 2) If $u(y) > v(y)$ for y in a neighbourhood of x and $y \neq x$, then for h small enough,

$$(T_{t+h,t}u)(x) \geq (T_{t+h,t}v)(x).$$

In the case of edge detection, there are several formalizations, but the simplest states that no new boundary is created when the scale increases, that is, $T_{t'}K$ is contained in T_tK if $t' > t$.

We finally need some assumption stating that a very smooth image must evolve in a smooth way with the multiscale analysis. Somehow, this belongs to the 'causality' galaxy, but we prefer to call it regularity and it clearly corresponds to the assumption of the existence of an infinitesimal generator for the multiscale analysis.

Regularity Let $u(y) = \frac{1}{2}(A(x-y), x-y) + (p, x-y) + c$ be a quadratic form of $\mathbb{R}^N$. There exists a function $F(A, p, x, c, t)$, continuous with respect to A, such that

$$\frac{(T_{t+h,t}u - u)(x)}{h} \to F(A, p, x, c, t), \quad \text{when} \quad h \to 0.$$

Morphological and affine invariance In addition to the causality requirement, we must keep in mind that the pyramid acts on equivalence classes of images of the form $g(u_0(Ax))$, where g is any nondecreasing continuous function and A any isometry (or any affine map) of the plane. Therefore, the output should not depend upon u_0 but on the equivalence class. So the transition operators $T_{t+h,t}$ must somehow commute with the perturbations g and A. In the case of a change in contrast g, this is easily translated into the

Morphological invariance $gT_{t+h,t} = T_{t+h,t}g$, which means that change of contrast and multiscale analysis can be applied in any order. If A is an isometry, the same kind of relation must be true. Denote by Au the function $Au(x) = u(Ax)$. Then we state the

Euclidean invariance $AT_{t+h,t} = T_{t+h,t}A$.

Let us now examine the case of an arbitrary linear map A. The commutation relation cannot be so simple because A can reduce or enlarge the image. (Think of the case where A is a zoom defined by $Au(x) = u(\lambda x)$ for some positive constant λ.) Since the zoom has changed the scale of the image, we can just impose a weak commutation property:

Affine invariance For any A and $t \geq 0$, there exists a C^1 function $t'(t, A) \geq 0$ such that

$$AT_{t'(t,A),t'(s,A)} = T_{t,s}A.$$

Moreover, the function $\phi(t) = (\partial t'/\partial \Lambda)(t, \lambda \,\mathrm{Id})$ is positive for $t > 0$.

This relation means that the result of the multiscale analysis T_t is independent of the size and position in space of the analysed features: an affine map corresponds to the anamorphosis of a plane image when it is presented to the eye at any distance large enough with respect to its size and with an arbitrary orientation in space. (The general visual invariance should be projective, but for small objects at some distance, we shall be contented with the affine invariance.) (See Forsyth *et al.* (1991), Lamdan *et al.* (1988).) The assumption on t', $\phi(t) = (\partial t'/\partial \lambda)(t, \lambda \,\mathrm{Id}) > 0$, can be interpreted by looking at the relation $\lambda(\mathrm{Id})T_{t'} = T_t(\lambda \,\mathrm{Id})$ when λ increases, i.e. when the image is shrunk before analysis by T_t. Then, the corresponding analysis time before shrinking is increased. In more informal terms we can say that the analysis scale increases with the size of the picture.

Let us point out the fact that the affine invariance must be stated in such a general framework because we have, up until now, made no attempt to fix the relation between the abstract 'scale' parameter and the concrete scale understood as having some relation with the size of objects. As for all future results, they will be true whatever the change in abstract scale $T_t \to T_{\sigma(t)}$ provided σ is a smooth increasing function: $\mathbb{R}^+ \to \mathbb{R}^+$. Now, the next lemma will permit a full normalization of the scale, thanks to affine invariance.

Lemma 1 (Normalization of scale.) Assume that $t \to T_t$ is a one-to-one family of operators satisfying pyramidality and affine invariance. Then the function $t'(t,B)$ only depends on t and $|\det B|$: $t'(t,B) = t'(t, |\det B|^{1/2})$ and is increasing with respect to t. Moreover, there exists an increasing differentiable rescaling function $\sigma : [0,\infty] \to [0,\infty]$, such that $t'(t,B) = \sigma^{-1}(\sigma(t)|\det B|^{1/2})$ and if we set $S_t = T_{\sigma^{-1}(t)}$ we have $t'(t,B) = t|\det B|^{1/2}$ for the rescaled analysis.

We shall give a proof of this lemma in Appendix B, because it is of particular relevance in image processing.

To summarize, the multiscale analysis T_t must (or may) satisfy:

- **Causality** $T_{t+h} = T_{t+h,t}T_t$, $T_{t,t} = T_0 = \mathrm{Id}$, and $(T_{t+h,t}u)(x) > (T_{t+h,t}v)(x)$ if $u(y) > v(y)$ for y in a neighbourhood of x and $y \neq x$. In the case of boundary multiscale analysis, this last assumption is replaced by $T_tK \subset T_sK$ if $t > s$.
- **Regularity** Let $u(y) = \frac{1}{2}(A(x-y), x-y) + (p, x-y) + c$ be a quadratic form of $\mathbb{R}^N$. There exists a function $F(A,p,x,c,t)$, continuous with respect to A, such that
$$\frac{(T_{t+h,t}u - u)(x)}{h} \to F(A,p,x,c,t) \quad \text{when} \quad h \to 0.$$
- **Morphological invariance** $gT_{t+h,t} = T_{t+h,t}g$ for any change of contrast g.
- **Euclidean invariance** $AT_{t+h,t} = T_{t+h,t}A$ for any isometry A of $\mathbb{R}^N$.
- **Affine invariance** $AT_{t'+h',t'} = T_{t+h,t}A$ with $t' = |\det A|^{1/2}t$ and $h' = |\det A|^{1/2}h$. Notice that the affine invariance implies the Euclidean invariance.
- **(Optional) linearity** $T_{t+h,t}(au + bv) = aT_{t+h,t}(u) + bT_{t+h,t}(v)$. We add this property because it has been very much in use in computer vision.

There are therefore five main axioms and we shall see that they allow us to classify and characterize the theories of multiscale image and shape processing completely, to unify and to improve several of them.

3. Axiomatization of image multiscale analysis and classification of the main models

Theorem 1 (Koenderink, 1990a; Hummel, 1986; Yuille and Poggio, 1986.)

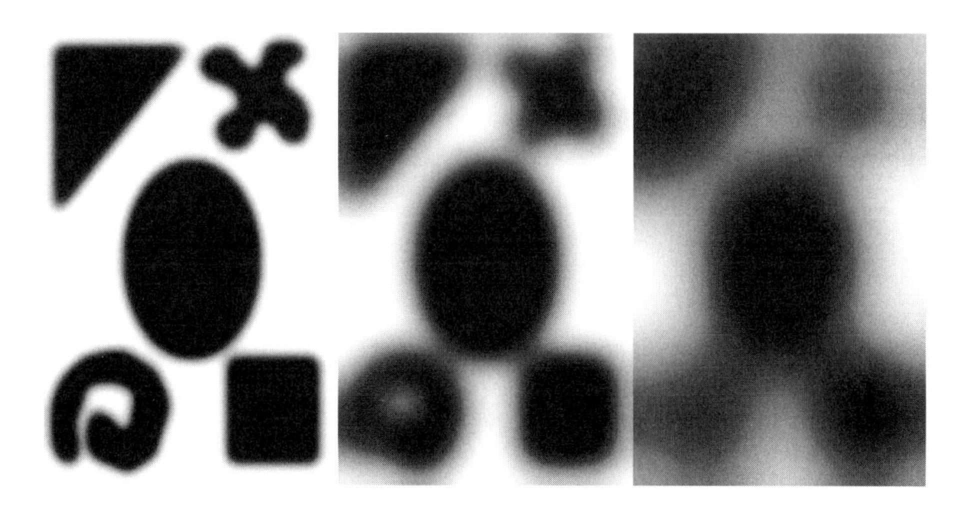

Fig. 3. Linear multiscale analysis (heat equation). Evolution of the original image given in Figure (1a) (from left to right): (a) $t = 10$; (b) $t = 70$; (c) $t = 300$.

If a multiscale analysis is causal, Euclidean invariant and linear, then it obeys (up to a rescaling $t \to \sigma(t)$) the heat equation

$$\frac{\partial u}{\partial t} = \Delta u,$$

(that is, $u(t,x) = (T_t u)(x)$ obeys the heat equation).

This model of multiscale analysis (the 'raw primal sketch') is due (among others) to Hildreth and Marr (see Marr (1982)) and Witkin (the *scale space*) (1983). See also Koenderink (1984) who was the first to state the heat equation explicitly. More recently Lindeberg (1990) studied the associated discrete scale space and Florack *et al.* (1992) showed how to use the heat equation to find corners, T-junctions, etc. by simple differential operators.

What happens if we remove the linearity axiom? As noted in Perona and Malik (1988) in their nonlinear theory of scale space, 'anisotropic diffusion', we can get nonlinear heat equations

$$\frac{\partial u}{\partial t} = F(\mathrm{D}^2 u, \mathrm{D}u, u, x, t). \tag{3.1}$$

The converse implication and a complete study of nonlinear models is given in Alvarez *et al.* (1992a):

Theorem 2 (Fundamental theorem.) If an image multiscale analysis T_t is causal and regular then $u(t,x) = (T_t u)(x)$ is a viscosity solution of (3.1), where the function F, defined in the regularity axiom, is nondecreasing with

Fig. 4. 'Dilation' multiscale analysis. Evolution of the original image given in Figure (1a) (from left to right): (a) $t = 5$; (b) $t = 10$; (c) $t = 15$.

respect to its first argument $\mathrm{D}^2 u$. Conversely, if u_0 is a bounded uniformly continuous image, then equation (3.1) has a unique viscosity solution.

(For a quick proof of this theorem, see Appendix A.)

The particular case of the heat equation corresponds to $F(A, p, c, x, t) = \operatorname{trace}(A)$. As a consequence of this theorem, all multiscale models can be classified and new, more invariant models can be proposed:

Theorem 3 Let $N = 2$. If a multiscale analysis is causal, regular, Euclidean invariant and morphological, then it obeys an equation of the form

$$\frac{\partial u}{\partial t} = |\mathrm{D}u| F(\operatorname{div}(\mathrm{D}u/|\mathrm{D}u|, t), \tag{3.2}$$

where $\operatorname{div}(\mathrm{D}u/|\mathrm{D}u|)(x)$ can be interpreted as the curvature of the level line of the image $u(t, x)$ passing by x and $F(s, t)$ is nondecreasing with respect to the real variable s.

An important particular case is when F is a constant function: if $F = +1$ or $F = -1$, then the equation becomes $\partial u/\partial t = |\mathrm{D}u|$ (resp. $\partial u/\partial t = -|\mathrm{D}u|$), which corresponds to the so-called morphological erosion when the sign is '$-$' and to a morphological dilation when the sign is '$+$' (see Brockett and Maragos (1992)). *Dilation* and *erosion* are the basic operators of the Morphologie Mathématique, founded by Matheron (1975) and his 'Fontainebleau School'. The dilation at scale t is defined by

$$\mathrm{D}_t u_0(x) = \sup_{y \in B(x,t)} u_0(y),$$

where $B(x, t)$ is a set centred at x: the 'structuring element' which is gen-

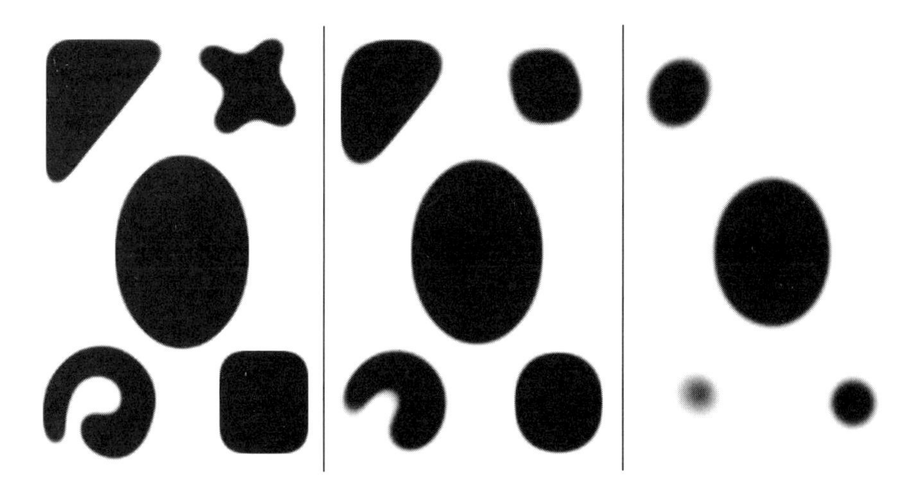

Fig. 5. 'Mean curvature motion' multiscale analysis. Evolution of the original image given in Figure (1a) by using scheme (10.4) (from left to right): (a) $t = 3$; (b) $t = 9$; (c) $t = 14$.

erally a ball with radius t. Assume, for instance, that u_0 is the characteristic function of a set X, then $\mathrm{D}_t u_0$ is the characteristic function of the t-neighbourhood of X. For the erosion, one simply replaces 'sup' by 'inf'. As noted by Michel Rascle, another relevant example satisfying the multiscale morphological axioms and therefore a parabolic PDE is the family of zooms with ratio t,

$$(T_t u_0)(x) = u(t, x) = u_0(tx).$$

Indeed, the preceding theorem applies and it is easily seen that the underlying equation is

$$\frac{\partial u}{\partial t} = \frac{1}{t}(\mathrm{D}u \cdot x).$$

Note that this formulation may be useful, the zooming operators on a digital picture being in no way easy to implement. Now, the preceding examples have been very particular instances of the equation and there are many other possibilities for morphological multiscale filtering! If we set $F(s, t) = s$, we obtain the 'mean curvature equation' (MCM)

$$\frac{\partial u}{\partial t} = t \cdot |\mathrm{D}u| \mathrm{div}(\mathrm{D}u/|\mathrm{D}u|). \tag{3.3}$$

This equation comes from a reformulation by Osher and Sethian (1988) of a differential geometry model studied by Grayson (1987) and Gage and Hamilton (1986). It is also very close to the 'anisotropic diffusion' of Perona and Malik (1987) and to an image restoration equation due to Rudin *et al.*

(1992a):

$$\frac{\partial u}{\partial t} = \mathrm{div}(\mathrm{D}u/|\mathrm{D}u|).$$

Now, the most invariant model is new: it is proved in Alvarez *et al.* (1992a) that

Theorem 4 (AMSS Model.) Let $\mathbb{N} = 2$. There is a single causal, regular, morphological and affine invariant multiscale analysis. Its equation is

$$\frac{\partial u}{\partial t} = |\mathrm{D}u|(t \cdot \mathrm{div}(\mathrm{D}u/|\mathrm{D}u|))^{1/3}. \tag{3.4}$$

We shall better understand this equation in the framework of shape analysis: Sapiro and Tannenbaum (1992a,b), who independently discovered the model as a *shape scale space* have given it a remarquable geometric interpretation in this framework. Of course, fundamental Theorem 2 can be applied in any dimension. Let us just state a last example of scale space in dimension 3 (of particular relevance for medical solid images) (see again Alvarez *et al.* (1992a) and Caselles *et al.* (1993)).

Theorem 5 Let $N = 3$. There is (up to a rescaling) a single causal, regular, affine invariant and morphological multiscale analysis, associated with the equation

$$\frac{\partial u}{\partial t} = |\mathrm{D}u|(G(u)^{+})^{1/4}. \tag{3.5}$$

By $G(u)$ we denote the Gaussian curvature, that is the determinant of $\mathrm{D}^2(u)$ restricted to $\mathrm{D}u^{\perp}$. We shall expand more on this subject when looking for movie analysis equations.

4. Shape multiscale analyses

We could deduce the shape analysis statements from image analysis statements. However, since in this case the axiomatics is particularly simple and intuitive, we shall list well-adapted principles, which are, however, equivalent to the general image analysis principles. For more details, see Kimia *et al.* (1992), Lopez and Morel (1992), Mackworth and Mockhtarian (1992).

We define a shape or ('silhouette') as a closed set X whose boundary is a Jordan curve of $\mathbb{R}^2$. We denote by $T_t(X)$ the *shape analysed at scale t*. X is identified with its characteristic function $X(x) = 1$ if $x \in X$ and 0 else. We call multiscale analysis any family of operators $(T_t)_{t \geq 0}$ acting on shapes and we set $X(t) = T_t(X)$. As before, we shall state *causality* principles, the first of which remains unchanged:

Pyramidal Structure (Causality 1) $T_{t+h} = T_{t+h,t}T_t, \quad T_0 = \mathrm{Id}$. Instead of the local comparison principle, we shall give a very intuitive statement: the shape local inclusion principle.

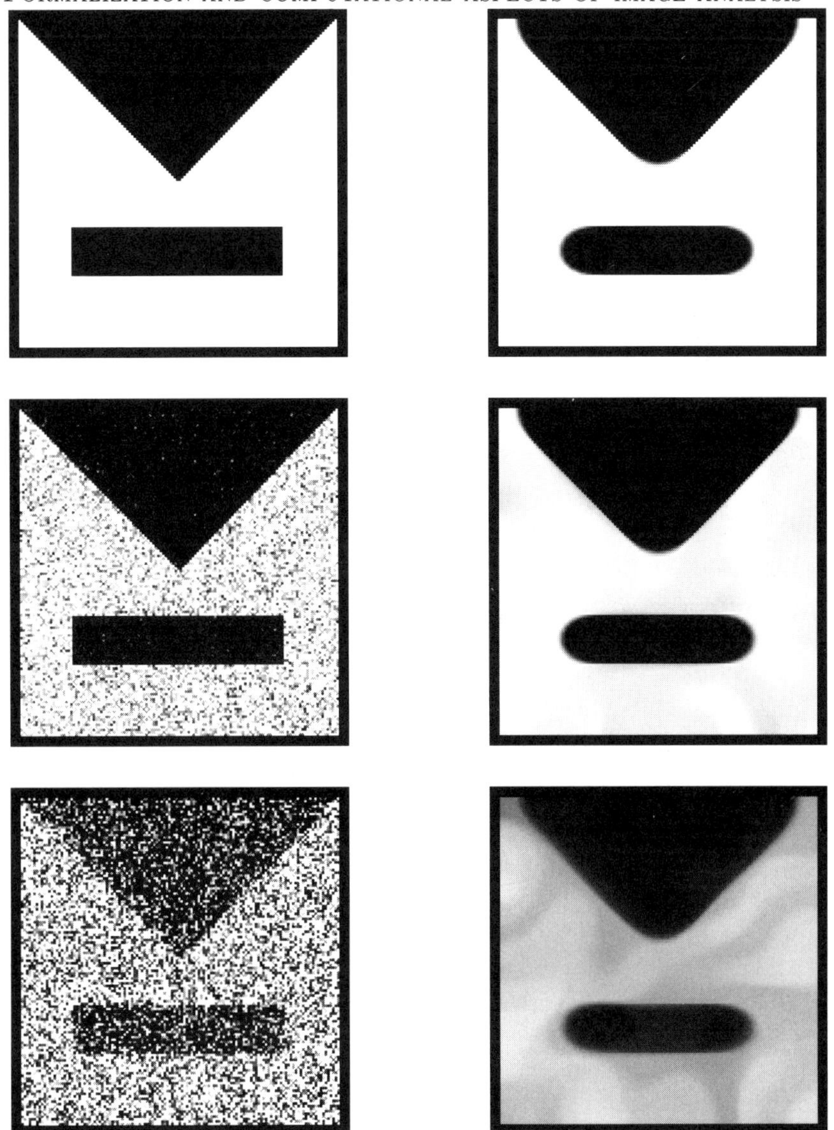

Fig. 6. Affine Invariant Morphological Multiscale Analysis. Authors: L. Alvarez–F. Guichard. A multiscale analysis T_t associates an image $u_0(x)$ with more regular images $u(x,t) = (T_t u_0)(x)$, where t is the scale of analysis. In this experiment, $u(t,x)$ is computed by the AMSS, $\partial u/\partial t = |\mathrm{D}u|(t \cdot \mathrm{curv}(u))^{1/3}$ where $\mathrm{D}u$ is the gradient of u, $\mathrm{curv}(u)(x) = \mathrm{div}(\mathrm{D}u/|\mathrm{D}u|)(x)$ the curvature of the level line of u passing by x. In order to illustrate the use of the AMSS model as a way of keeping only reliable information in image analysis, we display on the left-hand column two increasingly distorted (noisy) versions of an original synthetic image (up-left). The left-down image is obtained from the left-up by giving 70% of its pixels a random value. In the right-hand column, we display the respective analyses at the *same* scale t of these images. Two phenomena are illustrated by comparing the right and left columns: first, the *causality* (a smoothing effect), selective loss of information from fine to coarse; second, the *morphologically invariant* behaviour. Unless the level sets of the triangle and rectangle have quite different mean brightness in the noisy images, their shape is handled by the analysis in the same way. Size of the images: 128×128. CPU time: 21 s.

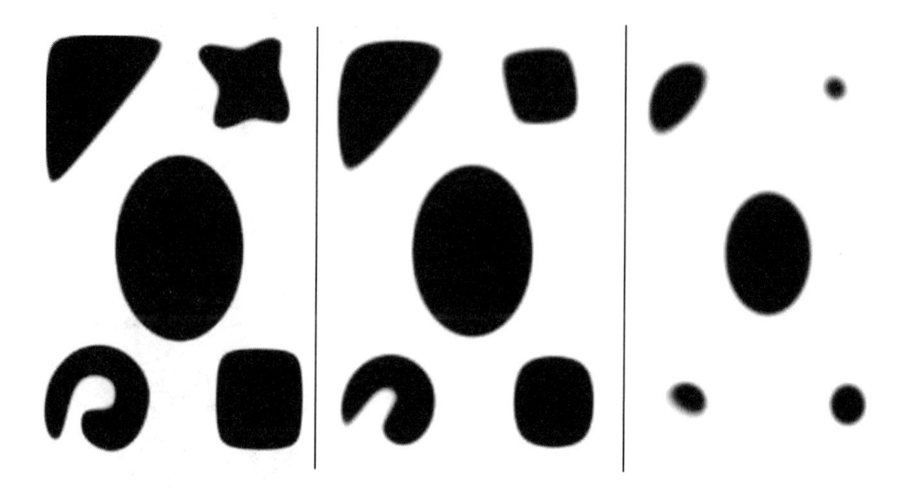

Fig. 7. AMSS multiscale analysis. Evolution of the original image given in Figure 1(a) by using the scheme (10.8) (from left to right): (a) $t = 3$; (b) $t = 9$; (c) $t = 14$.)

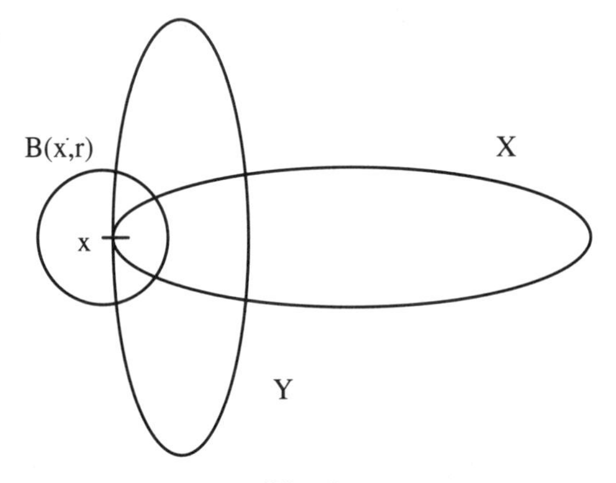

Fig. 8.

Assume that X and Y are two silhouettes and that for some $x \in \partial Y$ and some $r > 0$, one has $X \cap B(x, 2r) \subset Y \cap B(x, 2r)$. Assume further that the inclusion is strict in the sense that ∂X and ∂Y only meet possibly at x. Then we shall say that *the shape X is included in shape Y around x*.

Shape local inclusion If X is included in Y around x, then for h small enough, $T_{t+h,t}(X) \cap B(x,r) \subset T_{t+h,t}(Y) \cap B(x,r)$.

This last axiom implies that the value of $T_{t+h,t}(X)$ for h small, at any point x, is determined by the behaviour of X near x. We are allowed to take r infinite. Therefore the shape local inclusion principle also implies that if a shape is globally contained in another, this order is preserved for every scale (Mackworth and Mokhtarian, 1992).

Both preceding principles allow, as we shall see, the shape analysis to be localized in space and time and we therefore only need to state what the multiscale analysis makes of very simple shapes in order to specify it. So we add a 'basic principle' which will state what happens to disks. As we shall see, disks are somehow a 'basis' for shape analysis because whenever we know how they are analysed, we know what will happen to every other shape.

Basic principle Let $D = D(x, 1/r)$ be a disk with curvature $1/r$ and centre x. Then $T_{t+h,t}(D)$ is a disk with radius $\rho(t, h, 1/r)$ and centre x. Moreover, the function $h \to \rho(t, h, 1/r)$ is differentiable with respect to h at $h = 0$ and the differential is continuous with respect to $1/r$.

The 'basic principle' implies that the multiscale analysis behaves in a smooth and isotropic way. In the following, we set

$$g(t, 1/r) = \frac{\partial \rho}{\partial h}(t, 0, 1/r). \tag{4.1}$$

Note that $g(t, s)$ is defined for $t \geq 0$ and $s \in \mathbb{R}$. Now in order to define $g(t, 0)$ we must assume that lim $g(t, 1/r)$ exists when r tends to $+\infty$ or $-\infty$. The radius r may be positive or negative, according to the orientation of the normal $\vec{n}(x)$. In the case where the curve is a Jordan curve enclosing a set X, we take as a *convention that* $\vec{n}(x)$ is *pointing outside* X and *the curvature is negative if* X *is convex at* x, *positive else.* It may seem natural to assume therefore that $g(t, \kappa)$ is odd with respect to κ. Indeed, this corresponds to the assumption that a black disk on grey background and a white disk behave in the same way.

4.1. The fundamental equation of shape analysis

When a point x belongs to an evolving curve, we denote by $\dot{x}$ the time derivative of x, which is a vector of $\mathbb{R}^2$. By curv(x) we denote the curvature of a curve which is C^2 at x. Recall that the curvature is defined as the inverse of the radius of the osculatory circle to the curve at x. The curvature is zero if the radius is infinite.

Theorem 6

1. Under the three principles (pyramidal, local shape inclusion, 'basic'), the multiscale analysis of shapes is governed by the curvature motion equation

$$\dot{x} = g(t, \text{curv}(x))\vec{n}(x) \tag{4.2}$$

where g is defined by (4.1).

2. If the analysis is affine invariant, then the equation of the multiscale analysis is, up to rescaling,

$$\dot{x} = \gamma(t \cdot \text{curv}(x))\vec{n}(x) \tag{4.3}$$

where γ is defined by $\gamma(x) = a \cdot x^{1/3}$ if $x \geq 0$ and $\gamma(x) = b \cdot x^{1/3}$ if $x \leq 0$ and a, b are two nonnegative values.

3 If we add that $T_t(X^c) = T_t(X)^c$ (reverse contrast invariance) then the function g in (i) is odd and we get

$$\dot{x} = (t \cdot \operatorname{curv}(x))^{1/3}\vec{n}(x). \tag{4.4}$$

We prove this theorem in Appendix C. By the expression 'governed by', we mean that the equation must be satisfied at any point (t, x) of $\partial X(t)$ where the boundary of the silhouette is smooth enough to give a classical sense to both terms of the equation.

Remark It has been proved (Gage and Hamilton, 1986; Grayson, 1987; Angenent, 1989) that equation (4.2) with $g(t, s) = s$ has smooth solutions. This equation has been introduced in picture processing by Kimia *et al.* (1992a), Mackworth and Mockhtarian (1992) and Alvarez *et al.* (1992a) in different contexts. An early version of an algorithm leading to equation (4.2) is given in Koenderink and Van Doorn (1986). See also Yuille (1988). Equation (4.4) has been introduced and axiomatically justified (with a more complicated axiomatic however) in Alvarez *et al.* (1992a). It has also been proposed in Sapiro and Tannenbaum (1992a). The existence and regularity of the solution are proved in Sapiro and Tannebaum (1992b). The axiomatic presentation adopted here follows Cohignac *et al.* (1993a).

5. Relation between image and shape multiscale analyses

In this section, we shall show how the AMSS model (1.1) can be deduced from the shape evolution (4.4). We give in Appendix C proofs of the axiomatic deduction of this last equation, so that our exposition will be rather complete. In addition, the shape multiscale analysis equation has an easy geometric interpretation as an 'intrinsic diffusion' which we shall explain at the end of this section. Since the multiscale analysis satisfies the obvious inclusion principle that

$$\text{If} \quad A \subset B \quad \text{then} \quad T_{t+h,t}(A) \subset T_{t+h,t}(B),$$

we can, as well known in the 'mathematical morphology school' (Matheron 1975, Maragos 1987), associate with a picture u the set of its level sets

$$X_a u = \{(x, y), u(x, y) \geq a\}.$$

Then, assuming that each level set is a union of silhouettes, we can simply define $T_t(u)$ from the multiscale analysis $T_t(X)$ of silhouettes by

Morphological Principle For any a, t, h and

$$u : X_a T_{t+h,t}(u) = T_{t+h,t}(X_a u).$$

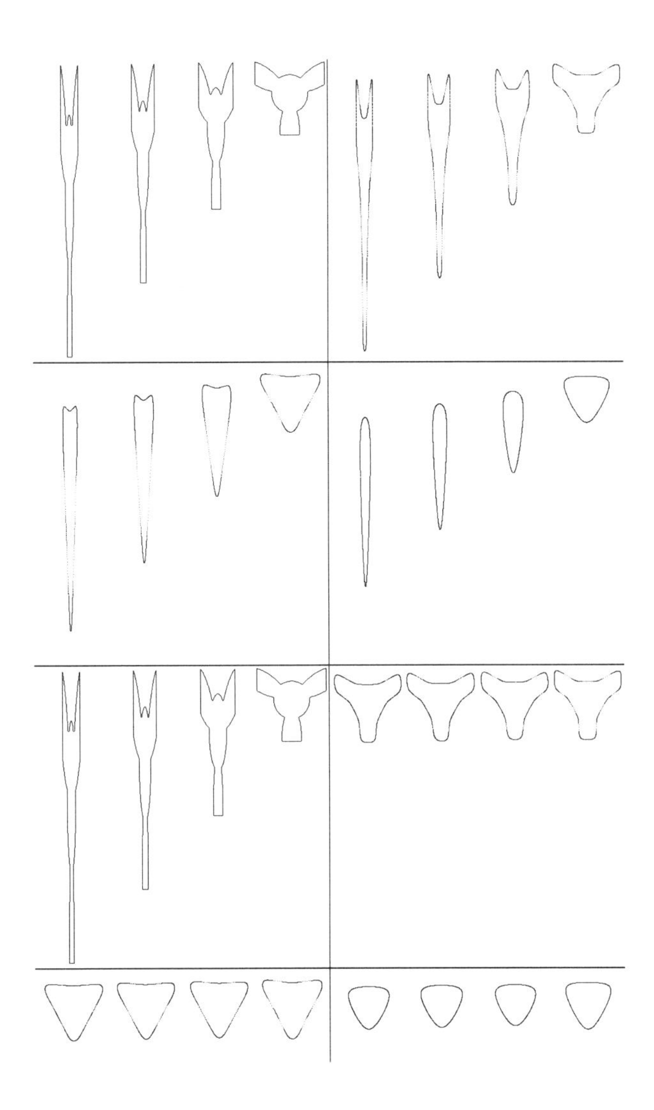

Fig. 9. Affine invariant shape recognition made possible thanks to the AMSS model. Authors: Cohignac–Eve–Guichard–Lopez–Morel. Which numerical implementation can be done using the AMSS model? We wish, after many iterations of the discrete algorithm, to be able to recognize shapes. So the numerical algorithm should strictly have the same invariance properties as the continuous model. In the case of affine invariance, this happens to be a rather new problem in numerical analysis. We display here the results of an algorithm which will be discussed later on. These results illustrate the affine invariance of the analysis. A shape is distorted by affine anamorphoses with respective eigenvalues (1/2,2), (1/3,3) et (1/4,4) (first quadrant, up-left). The quadrants 5, 6, 7, 8 display the multiscale analyses of these shapes at four successive scales. If we apply the inverse anamorphoses to the shapes after analysis, they must be the same (if the algorithm is correct!). This can be checked in quadrants 2, 3, 4.

Theorem 7 Assume that a multiscale image analysis T_t satisfies the morphological principle and that each of the level sets is governed by equation (4.2). Then $T_t u$ satisfies

$$\frac{\partial u}{\partial t} = g(t, \mathrm{curv}(u))|\mathrm{D}u|. \tag{5.1}$$

Proof. Let us prove (3.2) at any (x, a) such that $u(t, x) = a$ and $u(t, y) = a$ implies $\mathrm{D}u(t, y) \neq 0$ and u is C^2 at (t, y). The first condition implies by the implicit function theorem that x belongs to a Jordan curve Γ enclosing two regions Γ^+ and Γ^-, with $u(t, x) < a$ on Γ^- , $u(t, x) > a$ on Γ^+ and $u(t, x) = a$ on Γ. Set $u(t) = u(t, .)$. The relation $X_a T_{t+h,t} u(t) = T_{t+h,t} X_a u(t)$ implies that $T_{t+h,t}\Gamma^+$ is equal to a connected component of $X_a u(t + h)$. Therefore, the point $x(t+h)$ defined by equation (4.2) with initial value $x(t) = x$ belongs to the boundary of $X_a u(t+h)$. So we obtain $u(t+h, x(t+h)) - u(t, x(t)) = 0$. Dividing by h and passing to the limit yields

$$\frac{\partial u}{\partial t}(t, x) + \mathrm{D}u(t, x)\dot{x}(t) = 0$$

and using equation (4.2) and $\vec{n}(x) = -\mathrm{D}u/|\mathrm{D}u|$ we obtain equation (3.2).

The Affine Scale Space (ASS) model (4.4) of Sapiro and Tannenbaum yields a simple geometric interpretation of the AMSS model (1.1). Let us consider two ways of parametrizing a smooth Jordan curve $x(s)$:

- either in an Euclidean-invariant way by imposing $|x_s| = 1$; or
- or in an affine-invariant way by setting $|\det(x_s, x_{ss})| = 1$.

In the second case, we say that the curve has been parametrized by its 'affine length parameter' s. This parametrization is affine covariant because $|\det(Ax_s, Ax_{ss})| = |\det A||\det(x_s, x_{ss})|$ for any affine map A. Sapiro and Tannenbaum (1992b) proved (it is an easy computation) that the ASS model is equivalent to the following 'intrinsic heat equation ':

$$\frac{\partial x(s,t)}{\partial t} = t^{1/3}\frac{\partial^2 x(s,t)}{\partial s^2}, \quad x(0, s) = x_0(s).$$

So the application of the AMSS model to an image u can be interpreted as the affine invariant diffusion of all the level lines (isophotes) of u. Therefore we obtain a nonlinear generalization of the linear classical scale spaces.

6. Multiscale segmentation

Segmentation is acknowledged as the main tool in image interpretation. As we shall see, the segmentation problem is quite well understood in the framework of multiscale analysis. To define this problem in two lines, let us say that the objective of segmentation is to find the homogeneous regions of an image as well as their boundaries. However, the term 'homogeneous' is extremely vague and in order to state what homogeneity is, one has to rely either on perceptual experiments or on axiomatic definitions. In any case, homogeneity may concern clues as different as brightness, colour and texture. In this section, we shall not discuss what these clues are (this will be axiomatically introduced in the next section for textures). We shall assume that the image datum is composed of a finite set of k 'channels', each channel being itself a real image. In the simplest case, there is a single channel, the brightness. In the colour image case, the actual technology (partly based on perceptual criteria) yields three channels, i.e. three images of the same size (red, green, blue). In the case of other clues, like texture elements, the number of channels is unlimited and experiments showed in the next section involve up to forty channels computed from an initial grey level image and with the same size. The segmentation problem assumes that an initial multichannel image u_0, with $u_0(x) \in \mathbb{R}^k$, and an initial boundary map K_0, where K_0 is a subset of the image domain with finite Hausdorff length, are given. The initial boundaries can simply be the boundaries of all pixels. The segmentation process computes a multiscale sequence $(K(t))$ of segmentations, as well as a multiscale sequence of piecewise homogeneous images $u(t)$. $K(t)$ is assumed to be the set of the boundaries of the homogeneous regions of $u(t)$.

We shall give simple multiscale principles which will closely determine the segmentation process. Setting, as usual, $T_t K_0 = K(t)$, we impose

- **Causality** $K(t') \subset K(t)$ if $t < t'$ and $T_{t,s}T_s = T_t, \quad T_0 = \mathrm{Id}$.
- **Euclidean invariance** $T_{t+h,t}R = RT_{t+h,t}$ for every isometry R.
- **Invariance by zooming** Let Z_λ denote the zoom with ration λ. Then

$$T_t Z_\lambda = Z_\lambda T_{\lambda^{1/2}t}.$$

Finally we need an axiom fixing the relations between the boundaries $K(t)$ at scale t and the image $u(t)$. Our choice (see Morel and Solimini (1993)) is to take the simplest principle, which Mumford and Shah (1988) called the 'cartoon principle'.

Cartoon Principle $u(t)$ is locally constant in $\mathbb{R}^2 \setminus K(t)$ and equal to the mean value of u_0 on each connected component of $\mathbb{R}^2 \setminus K(t)$.

The causality and cartoon principles nearly fix the type of segmentation algorithm to be used: it is a *region growing* algorithm. Let us define *region* as every connected component of $\mathbb{R}^2 \setminus K(t)$. Since, for

$$t' > t, \quad K(t') \subset K(t),$$

we deduce that the regions at scale t' are unions of regions at scale t (which is another way of stating the causality axiom). Thus, in order to completely fix the multiscale analysis, we only need a criterion for region 'merging'. Mumford and Shah (1988) proposed the following Euclidean and zoom-invariant criterion: every segmentation is associated with an energy (which somehow measures its complexity),

$$E(u(t), K(t)) = \int_{\mathbb{R}^2 \setminus K(t)} (u(t,x) - u_0(x))^2 \, dx + t \cdot \mathrm{length}(K(t)).$$

Then two regions of the segmentation will be merged at scale t if and only if the energy of the resulting segmentation decreases. The associated 'recursive merging' algorithm is extremely simple. We start with an initial trivial segmentation of the image at scale $t = 0$. In this case, the image simply is (e.g.) divided into 'pixels' (small squares in the actual technology).

Region Growing Variational Algorithm

- Fix u_0, K_0 as the initial trivial segmentation, where K_0 is the union of boundaries of all pixels.
- A scale t being fixed: for every pair of regions, check whether their merging decreases the Mumford–Shah energy and, if so, merge them. Then the new K is obtained by removing the common boundary of both regions and u takes the mean value of u_0 on the union of both regions as the new value.
- Increment the scale t and go back to the preceding step.

Mumford and Shah (1989) proved that a segmentation which is minimal for their energy has a finite number of regions with smooth boundaries. It is, however, impossible to find a minimal segmentation, because the energy is highly not convex. Therefore, it is useful to get information about the segmentations obtained by a concrete algorithm computing local minima. Here is such a theorem, which justifies the use of region growing associated with the Mumford–Shah energy.

Theorem 8 (Koepfler *et al.*, 1994.) Let us say that a segmentation of a bounded vectorial image is *2-normal at scale t* if no pair of regions can be merged without increasing the Mumford–Shah energy. Then the set of 2-normal segmentations is compact for the Hausdorff distance and there is a bound, only depending on t, for the number of regions of a 2-normal segmentation.

We display in the following several examples of segmentations (Figures 10–11). This section has been kept short, as more than a thousand papers have been written on segmentation algorithms. Now, Morel and Solimini (1993), following and updating the terminology of Zucker (1976), have discussed more than ten classes of algorithms in image processing and have shown that they are merely variants of the Mumford–Shah energy minimizing process. Among the many theories which lead to the Mumford–Shah formalization (or variants), let us mention the 'snakes' (Kass *et al.*, 1987), the survey by Haralick and Shapiro (1985), the stochastic segmentation of

Fig. 10. Image segmentation by the Munford–Shah criterion. Authors: Koepfler–Morel. Up-left to right-down: an original satellite image (one single channel), the boundary map when it has been decided to keep 100 regions, the 'cartoon' (i.e. the associated piecewise constant image) and finally the superposition of the original image and the boundary map.

Geman and Geman (1984), the Blake and Zisserman 'weak membrane' model (1987), and the region growing algorithms of Brice and Fennema (1970), Muerle and Allen (1968), Pavlidis (1972), Horowitz and Pavlidis (1974). An affine invariant version of the Mumford–Shah theory has been proposed by Ballester and Gonzalez (1993).

7. An example: texture discrimination

Some theories of low level vision happen to be axiomatized and these can therefore be matched with the mathematical axiomatics which we developed above. This is the case for the Julesz Texton Theory (Julesz, 1981; 1986; Julesz and Kroese, 1988), which attempts to give a formal account of how human perception can discriminate textures in a few milliseconds. In order to understand the intuitive concept of *texture*, it is sufficient to look at the texture pair in Figure 11. The discrimination of both textures is easy for human perception, therefore whatever the computation involved in it is should also be. In order to confirm his texton theory experimentally, Julesz created pairs of simple different shapes, which were taken as building elements for creating two different but undiscriminable textures. This showed that not all *different* texture pairs are discriminable. Look at Figure 12 where a region of '10s' is surrounded by '5s'. Unless these shapes are individually quite distinguishable, they become quite equivalent when they are building elements of a 'texture'.

The axioms of shape analysis can be matched with the (not completely formal) axiomatics proposed by Julesz for his theory of preattentive texture discrimination. Therefore the texton theory can be numerically tested, and the result is quite different from the previous attempts and surprising. Indeed, previous attempts at formalization were based on linear image multiscale filtering followed by nonlinear mechanisms compatible with physiological data. One of the most conclusive in this way is due to Malik and Perona (1991), whose algorithm experimentally matches the human performance in texture discrimination. Now, as we shall see, a faithful implementation of the texton theory yields hyperdiscrimination! Julesz created pairs of textures which are undiscriminable for the preattentive vision but not for the algorithm deduced from his axioms.

In order to give a brief account of the texton theory, let us say that Julesz assumes that in perception only the local means of the curvatures and orientations of the shapes (or 'textons') of level sets are retained. These local means are called *texton densities*. Thus, every image u_0, supposedly containing textures, is associated with a multichannel image $u_i(t,x)$, $i = 1, 2, \dots, k$ where each image $u_i(t,x)$ stands for the density of the texton of type i at scale t. Then a segmentation process, computing 'texture gradients' is effectuated and yields the boundaries of the textures.

Therefore, the Julesz theory can be immediately translated into the multiscale analysis framework developed above. The translation yields a texture segmentation algorithm authorized to check how valid the texton theory is.

Texture Segmentation Algorithm

1 Apply the fundamental equation (AMSS) to the image: it is the only multiscale analysis allowing an invariant computation of multiscale curvatures and orientations.

2 Texton densities can be deduced as local means of curvatures and orientations at several scales.

3 Assuming that the analysed image only contains two texture regions, apply the region growing Mumford–Shah algorithm and stop the segmentation process when only two regions remain. If the regions are those predicted by the theory (and if the contrast of densities between them is strong), there is discrimination.

Let us briefly state how we organize the numerical experimentation of Julesz theory by using the above discussed scale-space algorithms.

7.1. Formalization and computation of textons densities

Let $u_0(x)$ be an image and $u(t, x)$ its multiscale analysis by equation (1.1). Then the curvature and orientation at scale t and location x are given respectively by $\mathrm{curv}(u)(x)$ and $\mathrm{D}u/|\mathrm{D}u|(x)$ which are both well defined and computed by equation (1.1).

Notice that textons can be white on black or white on black, so we define for any $\theta \in [0, \pi]$ and $t > 0$ the 'density on a ball $B(x, \Delta)$ of textons at scale t and orientation θ' by

$$G_{B_\Delta}(x) * (1_{\Theta_{\theta,\delta\theta}}\, \mathrm{curv}^+(u)(t, x))$$

$$G_{B_\Delta}(x) * (1_{\Theta_{\theta,\delta\theta}}\, \mathrm{curv}^-(u)(t, x))$$

where $\Theta_{\theta,\delta\theta}$ is the set of points where $\mathrm{D}u/|\mathrm{D}u| = (\cos\phi, \sin\phi)$ satisfies $|\theta - \phi| \leq \delta\theta$ and $G_{B_\Delta}(x)$ is the isotropic Gaussian with variance Δ. By a^+ we mean $\max(a, 0)$ and by a^-, $\max(-a, 0)$.

Then we get a formal definition of texton densities at each scale and orientation. Based on the obtained texton density channels, a segmentation of the image can be achieved. If the Julesz theory is correct, all the pairs of textures which we discriminate should be clearly discriminated by at least one of the texton densities.

Conversely, any pair of undiscriminable textures should be undiscriminated by all channels.

The basic facts which can be discovered by experimentation (Figures 11 and 12) are following. The textures proposed by Julesz, and most textures proposed in the literature as discriminable are easily discriminated by texton densities even rougher than those described above: Indeed, it is sufficient to compute global curvature densities, that is, we take $\delta\theta = \pi$ and do not

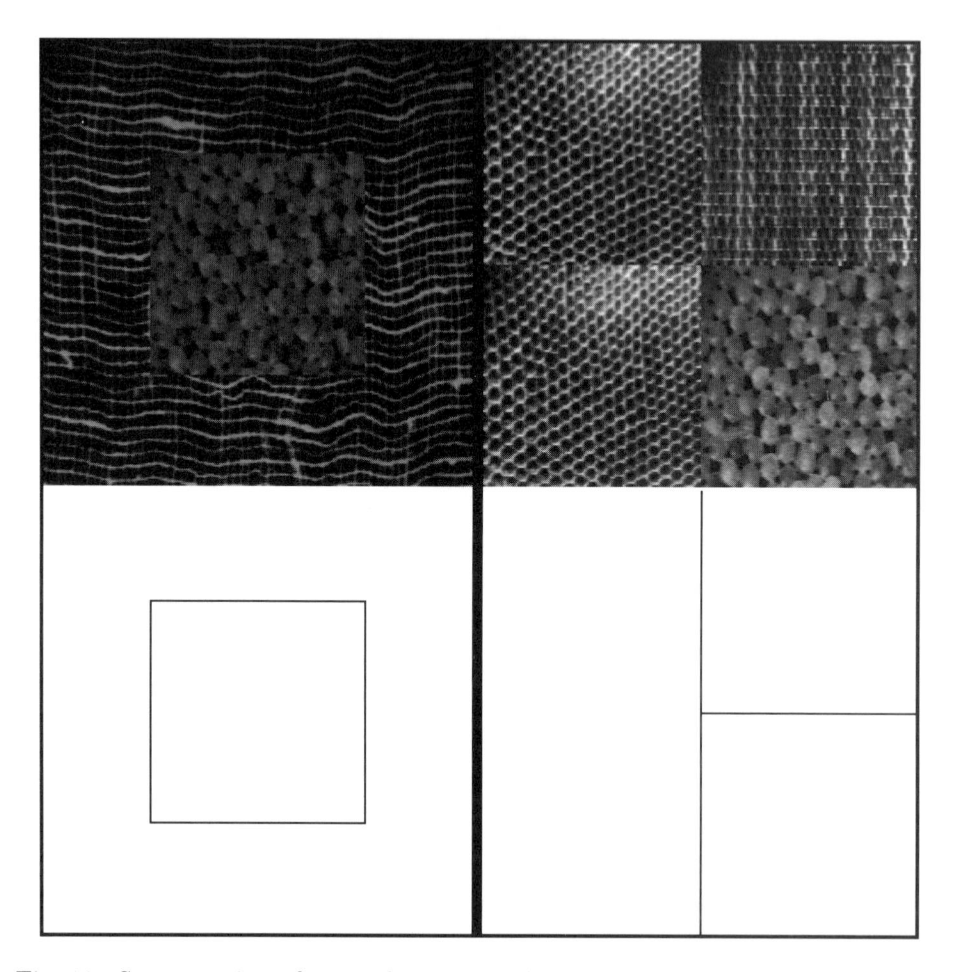

Fig. 11. Segmentation of natural textures. Authors: Koepfler–Lopez–Morel. On the left: two natural textures extracted from the Brodatz collection. On the right, three different Brodatz textures. Below: boundaries computed by the multichannel segmentation algorithm based on the Mumford–Shah functional. The 18 used channels are curvatures at different scales. Note how the same texture is repeated twice on the left of the right-hand image. The algorithm, which only takes into account local mean values of curvature, yields the right segmentation.

take into account orientation. The same process is followed for the Julesz undiscriminable textures!

The computational theory of texture discrimination which we have presented is the most complete attempt at computationally discussing the psychophysical theory of textons. It is, however, not the first successful attempt. The first one, already mentioned, is due to Malik and Perona (1991), who modelled texton density computations by a combination of linear oriented filters and nonlinear mechanisms. The Malik and Perona algorithm

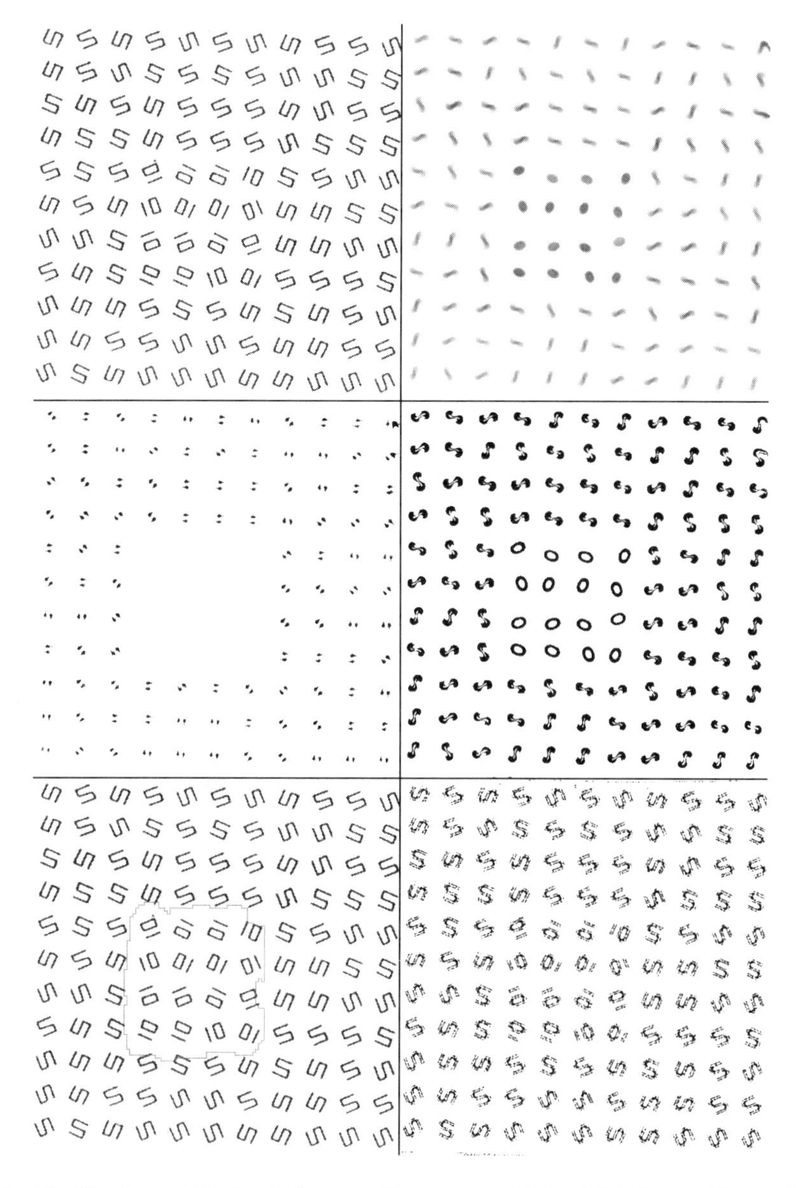

Fig. 12. Verification of Julesz' theory of textons which yields hyperdiscrimination algorithms. Experiment: Lopez–Morel. Julesz created textures which are undiscriminable for the human preattentive vision, as apparently predicted by the texton theory. However, the AMSS model, whose requirements match the axioms of the texton theory, leads to discriminate–undiscriminable texture pairs. Up–down and left–right: (a) a preattentively undiscriminable texture pair, u_0; (b) $u(t) = T_t u_0$: image obtained after application of the multiscale analysis T_t for $t = 7$ pixel units; (c) negative part of the curvature, $\mathrm{curv}(u(t))^-$; (d) positive part $\mathrm{curv}(u(t))^+$; (e) segmentation obtained by the Mumford–Shah piecewise constant model applied to the two-channel image of the curvatures; (f) negative curvature $\mathrm{curv}(u_0)^-$ of u_0. This last image shows that curvature-based discrimination was not possible on the original image. In order to explain the hyperdiscrimination of the AMSS model, the texton theory must be made more precise (Lopez and Morel, 1992).

matches the human preattentive vision and does not exceed it, as the above presented algorithm does. It is, however, easily seen by looking at their algorithm that it essentially computed curvature-based features, with less and less accuracy as the scale increased. The same article presents a very precise account of the previous attempts to make texton theory computational. (See, in particular, Treisman (1985), Voorhees and Poggio (1987), Enns (1986).)

8. Movie multiscale analysis

In this section, we return to scale space theory and look for its adaptation to movies. It is rather easy, since we still stay in the framework of the Fundamental Theorem 2. Indeed, a movie can be modelled as a 3D image, $u(x, \theta)$, where $x \in \mathbb{R}^2$ and $\theta \in \mathbb{R}$ is the time parameter. The statement and justification of the causality axioms remain unchanged, as well as the morphological invariance. Of course, the Euclidean and affine invariance make less sense in 3D, and we shall keep their 2D versions, the 2D-Euclidean invariance and the 2D-affine invariance, where isometries (respectively affine maps) are restricted to the image plane. We, however, add two specific axioms related to motion.

- **Time affine invariance** For any affine time rescaling $A_{a,b} : u(x, \theta) \to u(x, a\theta + b)$, there exist $t'(a, t)$ such that $T_t A_{a,b} = A_{a,b} T_{t'}$.
- **Galilean invariance** Denote by B_v any Galilean motion operator defined by $B_v u(x, \theta) = u(x - v\theta, \theta)$. Then $B_v(T_t u) = T_t(B_v u)$.

The Galilean invariance means that the analysis is invariant under 'travelling', that is motion of the whole picture at constant velocity v does not alter the analysis. In the following, we distinguish the 'spatial gradient' $\nabla u = (u_x, u_y)$, and the space–time gradient, $\mathrm{D}u = (u_x, u_y, u_\theta)$.

Theorem 9 If a multiscale analysis of movies is causal, regular, time-translation invariant, space-Euclidean invariant and morphological, then it is governed by the equation

$$\frac{\partial u}{\partial t} = |\nabla u| \; F(\mathrm{curv}(u), \mathrm{accel}(u), t). \tag{8.1}$$

If, in addition, we assume the analysis to be time-affine and space-affine invariant, the equation is (up to a rescaling)

$$\frac{\partial u}{\partial t} = (|\nabla u| \, \mathrm{curv}^{1/3}(u))^{1-q} \; ((|\nabla u| \mathrm{sgn}(\mathrm{curv}(u)) \, \mathrm{accel}(u))^{+})^{q}, \tag{8.2}$$

for some $q \in [0, 1]$.

(See Alvarez *et al.* 1992a,b.) In the above formulae, we use the convention that the power preserves the sign, that is $a^q = |a|^q \mathrm{sgn}(a)$. Hence, when

$q = \frac{1}{4}$, we obtain the equation

$$\frac{\partial u}{\partial t} = |\mathrm{D}u|(G(u)^+)^{1/4}. \tag{8.3}$$

This equation was mentioned in Section 2 as the only affine invariant morphological scale space in $\mathbb{R}^3$. Of course, this full affine invariance has no meaning for classical movies: what is the meaning of a rotation involving spatial and time variables? Now, in the field of relativity theory, such invariance makes sense because the Lorentz transform is nothing but a spatial-time rotation. In other words, when $q = \frac{1}{4}$, we have an equation which is both Galilean invariant and relativist invariant. We did not explain what the differential operator accel(u) is. (As for curv(u), it is simply the curvature of the spatial level curves, as in Section 2.) We could give the explicit formula for accel(u) in terms of the first and second derivatives of u, but this would prove disastrous, since the formula takes several lines. We shall use two ways to characterize accel and justify its name of 'apparent acceleration'. Let us first explain a classical notion in motion analysis: the apparent velocity. Assuming that a movie displays moving objects, let us call $x(\theta)$ the trajectory of a point of one of these objects. If we make the assumption that the object is *Lambertian*, which means that the light that it is sending to the camera is constant, then we can ensure that $u(x(\theta), \theta) = C$ for some constant C. Differentiating with respect to θ yields $\langle \nabla u, \dot{x}(\theta)\rangle + u_\theta = 0$. So we see that the component of the velocity in the direction of the spatial gradient can be recovered from the partial derivatives of u:

Definition We call *apparent velocity* of a movie $u(x, \theta)$ at point (x, θ) the scalar

$$v_1 = -u_\theta / |\nabla u|.$$

Now we are in a position to justify (and define) accel(u):

Lemma 2 (Interpretation of accel.) Consider a picture in translation motion $u(x, \theta) = w(x - \int_0^\theta \vec{v}(\tau)\,\mathrm{d}\tau)$, where $\vec{v}(\theta)$ is the instantaneous real velocity vector. Then the apparent velocity v_1 is equal to the true velocity in the direction of the gradient, $\langle \vec{v}, \nabla u / |\nabla u|\rangle$. Let $V = (\vec{v}, 1)$ be the real space–time velocity. Then

$$\mathrm{accel}(u) = -\langle \mathrm{D}v_1, V\rangle. \tag{8.4}$$

The first result is easy to check. The second formula can be taken as a definition of accel and shows that in the case of objects in translation motion, accel(u) is the derivative of the apparent velocity in the direction of $-\nabla u$. This is why we call it 'apparent acceleration'. For a proof, see Alvarez *et al.* (1992a,b) or Guichard (1993).

The second way of explaining what accel is consists in using Guichard's (1993) numerical approximant of it, which has proved essential when dealing with actual movies. Normal movie display series of frames where objects may jump more than 3 (and up to 50...) pixels from one frame to the next. Indeed, quick motions may let an object jump from one side of the screen to the other side in a very few frames. So, real movies are dramatically undersampled in time, which does seem to affect the trained zapper, but makes classical (local) numerical schemes impossible. So we define a nonlocal search space for the 'possible velocity vectors',

$$\mathcal{W} = \{\mathrm{V} = (\alpha, \beta) \text{ for all } \alpha \text{ and } \beta \text{ in } \mathbb{R}\}.$$

Theorem 10

$$\begin{aligned} &|\nabla u|(\mathrm{sgn}(\mathrm{curv}(u))\,\mathrm{accel}(u))^{+} \qquad (8.5)\\ = &\min_{w_1, w_2 \in \mathcal{W}} \frac{1}{\Delta\theta^2}(|u(x - w_1, \theta - \Delta\theta) - u(x, \theta)| \\ &+ |u(x + w_2, \theta + \Delta\theta) - u(x, \theta)| + |\langle \nabla u, w_1 - w_2 \rangle|) + o(1). \end{aligned}$$

Interpretation Of course, for numerical experiments, we shall not compute the minimum for all vectors in $\mathcal{W}$, but only for the vectors on the grid. We have two differents parts in the second term: the first part is the variation of the grey level value of the point x, for candidate velocity vectors: w_1 between $\theta - \Delta\theta$ and θ, and w_2 between θ and $\theta + \Delta\theta$. This variation must be as small as possible, because a point is not supposed to change its grey level value during its motion. The second part is nothing but an 'apparent acceleration', that is, the difference between w_1 and w_2 in the direction of the spatial gradient $|\nabla u|$.

Since movie analysis theory is very recent and computationally heavy (but not desperate), it has not yet received technological applications. The most promising applications at hand are detection of hidden trajectories and denoising of movies. Both are done in the same way, by simply applying the equation to a movie at some small scale. As can be deduced from the equation itself, a trajectory will be easily eliminated if either its acceleration is high or the moving object is small (and therefore with high curvature). The multiscale analysis acts like a sieve, by removing first all the most erratic trajectories, and leaving as $t \to +\infty$ only the Galilean trajectories. (Which, by the Galilean invariance principle, remain unaltered.) This phenomenon is illustrated in Figures 13 and 14.

9. Invariance and stability requirements for numerical schemes of the AMSS model

In order to compare schemes, we shall fix in this section a list of formal requirements which derive from the axiomatic analysis developed above. We

Fig. 13. Eliminating parasites in a real movie (Author: F. Guichard). We present (up) a sequence of three images of a real movie where a man is talking and moving his head. Notice that in the second image we have added two black rectangles. In the middle and down, we present two states of the multiscale analysis of the above sequence.

first have the properties which imply convergence of the schemes towards the equations, and second the invariance properties of the equations themselves. All schemes will be defined as the iteration of some nonlinear discrete filter, which we call T. We set $u^n = T^n u_0$, where u_0 is the (discretized) initial image. All schemes T will depend on a scale step Δt and a space step Δx.

We shall consider different discrete representations of u^n (see Figure 1):

- The classical discrete representation on a grid $u^n_{i,j} = u^n(i,j)$, with $1 \leq i \leq N$, $1 \leq j \leq N$. The image is the union of the squares centred

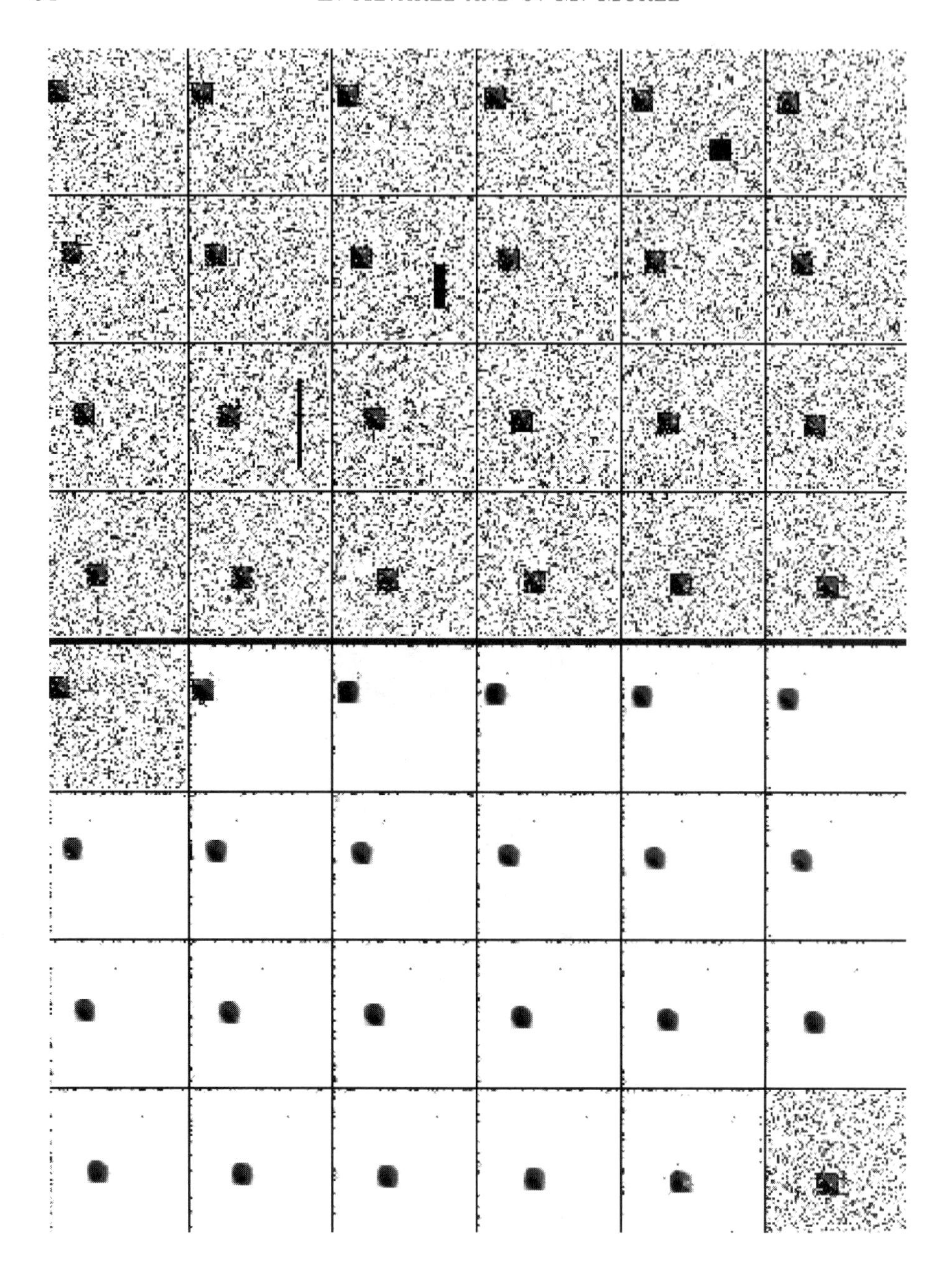

Fig. 14. Movie example. (Author: F. Guichard). Up: original noisy movie of a square and a point in uniform motion, 24 images 64×64. Down: the same movie analysed at a small scale, $q = 0.3$. Spurious trajectories due to noise are eliminated. Only the 'true' trajectories remain: the square and the point.

at the points (i, j), and the brightness in each square is constant: $u(i, j)$. Each one of the squares is called *pixel* (for 'picture element').

- The representation of u as a collection of all its Jordan level curves. In order to define these curves, one considers the level sets $X_a = \{u(x) \geq a\}$, where a admits, for technological reasons, 255 values. Thus each X_a is a finite union of squares in the initial image u_0. By elementary planar topology, its boundary ∂X_a is a finite union of Jordan curves, which we call 'level curves'. Conversely, in this discrete framework, u_0 can easily be reconstructed from all its Jordan level curves. So we call this numerical representation the 'level curve' representation. By the previous sections, we know that it is equivalent to applying AMSS to the initial image or to running ASS on each level curve. But of course, these are very different ways of handling an image in practice.
- We have now to say how we represent each level curve $C(s)$. A first solution is to discretize it as a polygon $C(i), i = 1, \ldots, N$ where the $C(i)$ have real (floating point) values. Then the description of the image will be very precise, with an 'underpixel' accuracy.
- Another solution to represent the level curves is to associate them with their distance function, which is a Lipschitz image. Of course, this representation is heavy since an image can have many level curves (more than pixels) and each becomes associated with a new image. However, the interest in this representation is that, as the distance function is Lipschitz, classical finite difference codes are easy to implement and we avoid the direct programming of a finite element method for curve evolution (Osher and Sethian, 1988). To visualize this representation on the screen, we use the following distance function (see Figure 1).

$$u^n_{i,j} = 128 + \text{dist}((i, j), C) \quad \text{if } (i, j) \text{ is not surrounded by } C(s)$$

and

$$u^n_{i,j} = 128 - \text{dist}((i, j), C) \quad \text{if } (i, j) \text{ is surrounded by } C(s).$$

Let H_λ, R, and A the operators defined by $H_\lambda u(x) = u(\lambda x)$, $Ru(x) = u(Rx)$ where R is an isometry of $\mathbb{R}^2$, $Au(x) = u(Ax)$ where A is a linear application of $\mathbb{R}^2$. Since the schemes to be considered will be very different in structure, we cannot state all following properties in a completely formal way. What they mean in each particular case will be clear in context.

- `[Consistency]` We shall say that a scheme is consistent if the discrete operator T (which only depends upon Δx and Δt) tends to the differential one (the second member of AMSS or ASS, MCM), when the steps Δx and Δt tend to 0 in a suitable way.
- `[Convergence]` $\forall t \; T^n u \to T_t u$ a.e. when Δt, $1/n$ and Δx tend to 0 in a suitable way. (T_t denotes the continuous multiscale analysis under consideration.)

- `[L∞ stability]` Let c, d be real numbers. If $c \leq u(x) \leq d$ then $c \leq Tu(x) \leq d$.
- `[Order Preserving]` If $u(x) \geq v(x)$ for all x in $\mathbb{R}^2$, then $(Tu)(x) \geq (Tv)(x)$.
- `[Morphology]` For all nondecreasing functions h: $Th(u) = h(Tu)$. This means that *the level sets* are handled independently by the discrete operator.
- `[Isom. invariance]` $TR_\theta = R_\theta T$.
- `[Affine invariance]` $TA = AT$ for any affine map with determinant equal to 1.

In the following sections, we define schemes and discuss which of the above properties they satisfy. As must be clear from the above list of requirements, as well as from the axiomatic analysis of the first sections, consistency is by no way a sufficient criterion for a good scheme in image analysis. The algorithms we look for must satisfy as much as possible all causality and invariance properties, and consistency, though necessary, should be a consequence rather than a primary requirement.

We shall not only discuss the AMSS model, but also the Mean Curvature Motion (MCM) equation. As we have shown, the MCM model has all the desirable properties except for full affine invariance. (It is, however, Euclidean and zoom invariant.) Therefore, it can prove quite useful in image processing and shape recognition. In addition, as we already have seen, it has been very much used by the image processing community and there is a rich list of attempts to run MCM (or algorithms which *a posteriori* prove to be equivalent to MCM). This will help us when we come to discuss schemes for the AMSS model. Of course, this last model is still much more puzzling, because unless it is written as a PDE, its behaviour is, because of the affine invariance, highly nonlocal (a circle is equivalent to an ellipse of any eccentricity!).

10. Finite difference schemes for the AMSS model

10.1. Mean curvature motion

We start with the 'MCM' equation given by

$$u_t = |\nabla u| \operatorname{curv}(u) = \frac{u_y^2 u_{xx} - 2u_x u_y u_{xy} + u_x^2 u_{yy}}{u_x^2 + u_y^2}.$$

In order to discretize this equation by finite differences we shall introduce an explicit scheme which uses a fixed stencil of $3 * 3$ points to discretize the differential operators. For simplicity, we assume that the spatial increment Δx is the same in the x- and y-axes. We approach the first derivatives u_x

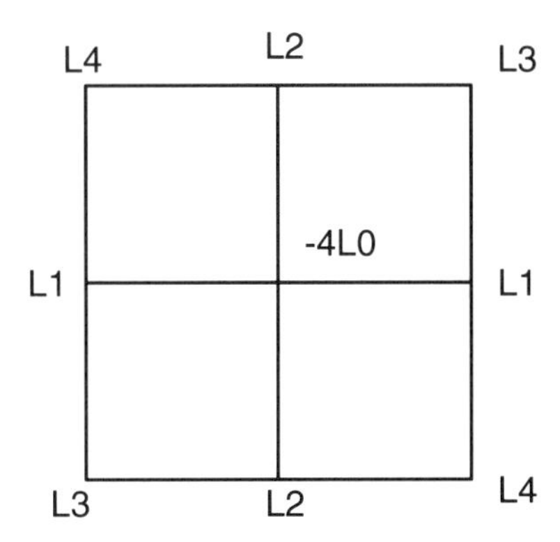

Fig. 15. A 3×3 stencil.

and u_y at a point (i, j) of the lattice by using the following linear scheme:

$$\begin{aligned}
(u_x)_{i,j} &= \frac{2(u_{i+1,j} - u_{i-1,j}) + u_{i+1,j+1} - u_{i-1,j+1} + u_{i+1,j-1} - u_{i-1,j-1}}{4\Delta x} \\
&\quad + \mathcal{O}(\Delta x^2) \\
(u_y)_{i,j} &= \frac{2(u_{i,j+1} - u_{i,j-1}) + u_{i+1,j+1} - u_{i+1,j-1} + u_{i-1,j+1} - u_{i-1,j-1}}{4\Delta x} \\
&\quad + \mathcal{O}(\Delta x^2).
\end{aligned}$$

Denoting by ξ the direction orthogonal to the gradient of u, one easily sees that $|\nabla u| \operatorname{curv}(u)$ is equal to $u_{\xi\xi}$. We have

$$\xi = (-\sin\theta, \cos\theta) = \left(\frac{-u_y}{\sqrt{u_x^2 + u_y^2}}, \frac{u_x}{\sqrt{u_x^2 + u_y^2}} \right),$$

and

$$u_{\xi\xi} = \sin^2\theta \, u_{xx} - 2\sin\theta\cos\theta \, u_{xy} + \cos^2\theta \, u_{yy}. \tag{10.1}$$

We want to write $u_{\xi\xi}$ as a linear combination of the values of u on the fixed stencil 3*3. Of course, the coefficients of the linear combination may depend on ξ. Because the direction of the gradient (and then ξ) is defined modulo π, by symmetry we must assume that the coefficients of points symmetrical with respect to the central point are the same (see Figure 15).

In order to have consistency, we must find $\lambda_0, \lambda_1, \lambda_2, \lambda_3, \lambda_4$, such that

$$\begin{aligned}
(u_{\xi\xi})_{i,j} &= \frac{1}{\Delta x^2}(-4\lambda_0 u_{i,j} + \lambda_1(u_{i+1,j} + u_{i-1,j}) + \lambda_2(u_{i,j+1} + u_{i,j-1}) \\
&\quad + \lambda_3(u_{i-1,j-1} + u_{i+1,j+1}) + \lambda_4(u_{i-1,j+1} + u_{i+1,j-1})) \\
&\quad + \mathcal{O}(\Delta x^2).
\end{aligned} \tag{10.2}$$

We write

$$u_{i+1,j} = u_{i,j} + \Delta x(u_x)_{i,j} + \tfrac{1}{2}\Delta x^2(u_{xx})_{i,j} + \mathcal{O}(\Delta x^3),$$

and the same relation for the other points of the stencil. By feeding (10.2) with these relations and by using relation (10.1), we obtain four relations between our five coefficients

$$\begin{cases} \lambda_1(\theta) = 2\lambda_0(\theta) - \sin^2\theta, \\ \lambda_2(\theta) = 2\lambda_0(\theta) - \cos^2\theta, \\ \lambda_3(\theta) = -\lambda_0(\theta) + 0.5(\sin\theta\cos\theta + 1), \\ \lambda_4(\theta) = -\lambda_0(\theta) + 0.5(-\sin\theta\cos\theta + 1). \end{cases} \tag{10.3}$$

There remains one degree of freedom for our coefficients given by the choice of $\lambda_0(\theta)$. We shall choose $\lambda_0(\theta)$ following the stability and geometric invariance criteria. Denoting by $u^n_{i,j}$ an approximation of $u(i\Delta x, j\Delta x, n\Delta t)$ we can write our explicit scheme as

$$u^{n+1}_{i,j} = u^n_{i,j} + \Delta t(u^n_{\xi\xi})_{i,j}. \tag{10.4}$$

Note that this scheme can be rewritten as

$$u^{n+1}_{i,j} = \sum_{k,l=-1}^{1} \alpha_{k,l} u^n_{i+k,j+l}$$

where $\alpha_{k,l}$ satisfy $\sum_{k,l=-1}^{1} \alpha_{k,l} = 1$.

The following obvious lemma shows a general condition for having [L^∞ stability] in this type of scheme:

Lemma 3 Let a finite difference scheme be given by

$$T(u)_{i,j} = \sum_{k,l=-1}^{1} \alpha_{k,l} u_{i+k,j+l},$$

where $\alpha_{k,l}$ satisfy

$$\sum_{k,l=-1}^{1} \alpha_{k,l} = 1.$$

Then the scheme satisfies [L^∞ stability] if and only if $\alpha_{k,l} \geq 0$ for any k, l.

Proof. If $\alpha_{k,l} \geq 0$ for any k, l, set $\min = \inf_{i,j}\{u_{i,j}\}$, $\max = \sup_{i,j}\{u_{i,j}\}$ and take a point (i, j). Then [L^∞ stability] follows from the inequality:

$$\min = \sum_{k,l=-1}^{1} \alpha_{k,l} \min \leq \sum_{k,l=-1}^{1} \alpha_{k,l} u_{i+k,j+l} = (Tu)_{i,j} \leq \sum_{k,l=-1}^{1} \alpha_{k,l} \max = \max.$$

On the other hand, if there exists $\alpha_{k_0,l_0} < 0$ then choosing u and (i, j) such that $u_{i+k_0,j+l_0} = \min$ and $u_{i+k,j+l} = \max$ for any other k, l, we obtain

$$(Tu)_{i,j} = \sum_{k \neq k_0, l \neq l_0}^{1} \alpha_{k,l} \max + \alpha_{k_0,l_0} \min = \max + \alpha_{k_0,l_0}(\min - \max) > \max$$

and therefore `[L∞ stability]` is violated.

Following this lemma, in order to have `[L∞ stability]` in the scheme (10.4) we must seek for λ_0 such that $\lambda_1, \lambda_2, \lambda_3, \lambda_4 \geq 0$ and $(1 - 4\lambda_0/\Delta x^2) \geq 0$. Unfortunately, because of the relations between our coefficients, it is impossible to obtain these relations, except for particular values of $\theta = (0, \frac{1}{4}\pi, \frac{1}{2}\pi, \ldots)$. Indeed, We remark that for θ in $[0, \frac{1}{4}\pi]$,

$$\lambda_1 \geq \lambda_2 \quad \text{and} \quad \lambda_3 \geq \lambda_4.$$

But

$$\lambda_2(\theta) \geq 0 \Rightarrow \lambda_0(\theta) \geq \tfrac{1}{2}\cos^2(\theta)$$

$$\lambda_4(\theta) \geq 0 \Rightarrow \lambda_0(\theta) \leq \tfrac{1}{2}(1 - \sin\theta\cos\theta).$$

So, we cannot find $\lambda_0(\theta)$ satisfying both inequalities, since

$$\tfrac{1}{2}\cos^2\theta \geq \tfrac{1}{2}(1 - \sin\theta\cos\theta).$$

Then, if we choose $\lambda_0(\theta) \geq \frac{1}{2}\cos^2\theta$ we have $\lambda_4(\theta)$ very negative. If we take $\lambda_0(\theta) \leq \frac{1}{2}(1 - \sin\theta\cos\theta)$ we obtain $\lambda_2(\theta)$ very negative. We prefer to choose λ_0 between both functions, and then to have λ_2 and λ_4 negative, but slightly (see Figure 14).

On the other hand, if we impose on λ_0 the following geometrical requirements:

(i) Invariance by rotation of angle $\frac{1}{2}\pi$

$$\lambda_0(\theta + \tfrac{1}{2}\pi) = \lambda_0(\theta).$$

(ii) Pure diffusion in the case $\theta = 0, \frac{1}{2}\pi, \ldots$

$$\lambda_0(0) = 0.5.$$

This condition implies that $\lambda_2(0) = \lambda_3(0) = \lambda_4(0) = 0$.

(iii) Pure diffusion in the case $\theta = \frac{1}{4}\pi, \frac{3}{4}\pi, \ldots$

$$\lambda_0(\tfrac{1}{4}\pi) = 0.25.$$

This condition implies that $\lambda_1(\frac{1}{4}\pi) = \lambda_2(\frac{1}{4}\pi) = \lambda_4(\frac{1}{4}\pi) = 0$.

(iv) Symmetry with respect to the axes $i + j$ and $i - j$,

$$\lambda_0(\tfrac{1}{2}\pi - \theta) = \lambda_0(\theta).$$

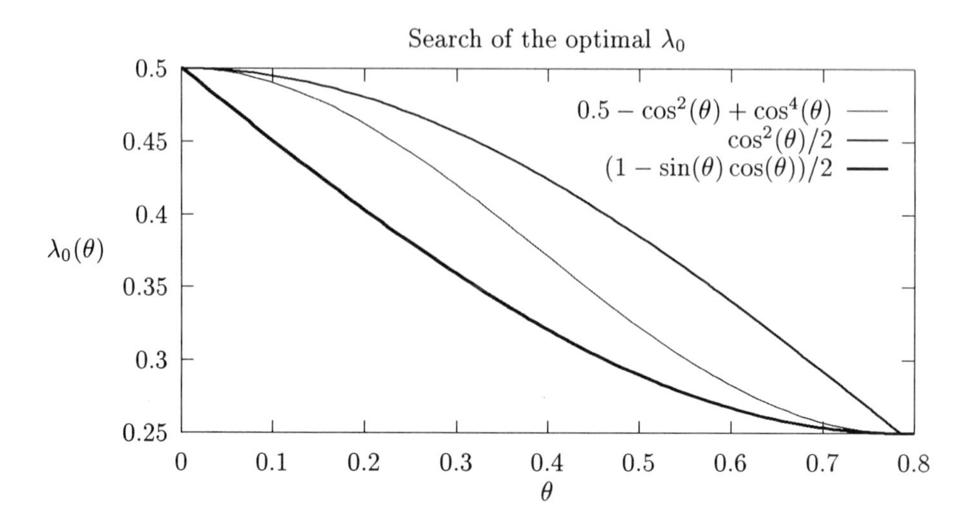

Fig. 16.

We remark that, by the above conditions, it is sufficient to define the function $\lambda_0(\theta)$ in the interval $[0, \frac{1}{4}\pi]$ because it can be extended by periodicity elsewhere.

We have tested two choices for the function $\lambda_0(\theta)$ using the trigonometric polynomials as basis. The first one corresponds to an average of the boundary functions:

$$\lambda_0(\theta) = \tfrac{1}{4}(\cos^2\theta + 1 - \sin\theta\cos\theta). \tag{10.5}$$

As we shall see this choice is well adapted to the 'affine curvature motion' equation. However, if we extend this function by periodicity, the extended function is not smooth at $\frac{1}{4}\pi$. If we seek for a smooth function for $\lambda_0(\theta)$, we must impose $\lambda_0'(0) = \lambda_0'(\frac{1}{4}\pi) = 0$. The simplest polynomial, of degree as small as possible, satisfying the above conditions, and between both boundary functions is

$$\lambda_0(\theta)) = 0.5 - \cos^2\theta\sin^2\theta. \tag{10.6}$$

We deduce the values of the other λs using (10.3). For instance with the above choice of $\lambda_0(\theta)$ we have

$$\begin{cases} \lambda_1(\theta) = \cos^2\theta(\cos^2\theta - \sin^2\theta), \\ \lambda_2(\theta) = \sin^2\theta(\sin^2\theta - \cos^2\theta), \\ \lambda_3(\theta) = \cos^2\theta\sin^2\theta + 0.5\sin\theta\cos\theta, \\ \lambda_4(\theta) = \cos^2\theta\sin^2\theta - 0.5\sin\theta\cos\theta. \end{cases}$$

We have tested this scheme in a workstation and we have noticed that if we impose a 'natural' stability condition such as

$$\frac{\Delta t}{\Delta x^2} \leq \frac{1}{2},$$

then the algorithm has good behaviour and remains stable in the sense that there exists experimentally a (small with respect to 255) $\epsilon > 0$ such that for any $n \in \mathbb{N}$ and (i, j),

$$-\epsilon + \inf_{i,j}\{u^0_{i,j}\} \leq u^n_{i,j} \leq \sup_{i,j}\{u^0_{i,j}\} + \epsilon.$$

10.2. The AMSS model

We will use the ideas developed in the above section. We rewrite the AMSS equation (1.1) as

$$u_t = (|\nabla u|^3 \operatorname{curv}(u))^{1/3} = (u_y^2 u_{xx} - 2u_x u_y u_{xy} + u_x^2 u_{yy})^{1/3}. \tag{10.7}$$

We remark that

$$|\nabla u|^3 \operatorname{curv}(u) = |\nabla u|^2 u_{\xi\xi}$$

where ξ corresponds to the direction orthogonal to the gradient. Therefore, in order to discretize this operator, it is enough to multiply the discretization of $u_{\xi\xi}$ presented in the above section by $|\nabla u|^2$. We choose $\lambda_0(\theta)$ given by (10.5) because it corresponds to a trigonometric polynomial of degree two and then multiplying it by $|\nabla u|^2$ the coefficients

$$\eta_i = |\nabla u|^2 \lambda_i, i = 0, 1, 2, 3, 4,$$

are polynomials of degree two with respect to u_x and u_y. Indeed, we obtain for $\theta \in [0, \frac{1}{4}\pi]$

$$\begin{aligned} (|\nabla u|^2 u_{\xi\xi})_{i,j} \quad = \quad & \frac{1}{\Delta x^2}(-4\eta_0 u_{i,j} + \eta_1(u_{i+1,j} + u_{i-1,j}) + \eta_2(u_{i,j+1} + u_{i,j-1}) \\ & +\eta_3(u_{i-1,j-1} + u_{i+1,j+1}) + \eta_4(u_{i-1,j+1} + u_{i+1,j-1})) \\ & +\mathcal{O}(\Delta x^2), \end{aligned}$$

where $\eta_0, \eta_1, \eta_2, \eta_3, \eta_4$ are given by

$$\begin{cases} \eta_0 = 0.25(2u_x^2 + u_y^2 - u_x u_y), \\ \eta_1 = 0.5(2u_x^2 - u_y^2 - u_x u_y), \\ \eta_2 = 0.5(u_y^2 - u_x u_y), \\ \eta_3 = 0.25(u_y^2 + 3u_x u_y), \\ \eta_4 = 0.25(u_y^2 - u_x u_y). \end{cases}$$

Finally, the finite difference scheme for the AMSS equation is

$$u^{n+1}_{i,j} = u^n_{i,j} + \Delta t(|\nabla u^n|^2 u^n_{\xi\xi})^{1/3}_{i,j}. \tag{10.8}$$

We have tested this algorithm and we have noticed that in this case the condition for the experimental stability (in the sense presented in the above subsection) is

$$\frac{\Delta t}{\Delta x^2} \leq \frac{1}{10}.$$

Remark The finite difference schemes that are presented satisfy `[Consistency]` and we conjecture `[Convergence]`. Morphological invariances are obtained asymptotically by taking a little time step Δt. The experimental results presented in Figures 5 and 7 have been obtained by using these schemes with $\Delta x = 1$ and $\Delta t = 0.1$ in the case of MCM and $\Delta t = 0.01$ in the case of affine curvature motion. One has to take Δt that small because unless experimental stability is achieved with $\Delta t \leq 0.1$, the experimental affine invariance needs $\Delta t < 0.05$.

11. Morphological (set evolution) schemes

11.1. A theoretical scheme

In this subsection, we discuss theoretical (only discretized in time) schemes inspired by the Mathematical Morphology School. These schemes will be, in contrast to the above presented finite difference schemes, fully geometrically invariant and stable. Now, as we shall see in a latter subsection, their implementation on a grid is problematic and they will only be practically implementable by working with the 'Jordan level curve' representation of the images (Section 11.3). Denote by C a set of convex sets which is stable either by isometries or by linear maps. For example, we can take for C the set of disks, ellipses or triangles,.... For all $t \geq 0$, let us set

$$C_t(x) = \{B \in C,\ \text{area}(B) = t^{3/2},\ \text{ and } x \text{ is the barycentre of } B\}.$$

Then we define two operators

$$\begin{aligned} IS_t(u)(x) &= \inf_{B \in C_t(x)} (\sup_{y \in B}(u(y)), \\ SI_t(u)(x) &= \sup_{B \in C_t(x)} (\inf_{y \in B}(u(y)). \end{aligned}$$

For example, if we take for C the set of disks, then IS_t is by definition the 'dilation operator' with radius t, and SI_t is the 'erosion' with radius t, and these are the basic operators of the mathematical morphology discussed in Section 3 (see Serra (1982), Maragos (1987)). If we impose that C is stable by all affine maps (for instance C can be the set of all ellipses), we obtain 'special' morphological operators which clearly are affine invariant, because C_t is unchanged under any affine map with determinant 1.

Consider the schemes

$$\text{(i)} \quad u^{n+1} = IS_{\Delta t}(u^n),$$

$$\text{(ii)} \quad u^{n+1} = SI_{\Delta t}(u^n).$$

It is proved in Guichard *et al.* (1993) that scheme (i) is consistent with the equation

$$\frac{\partial u}{\partial t} = |\mathrm{D}u|(\mathrm{curv}^+(u))^{1/3}$$

while scheme (ii) is consistent with the equation

$$\frac{\partial u}{\partial t} = |\mathrm{D}u|(\mathrm{curv}^-(u))^{1/3}.$$

In order to approximate equation (1), one can alternate affine closing and affine opening and one is led to the (consistent with AMSS) scheme (first announced in Cohignac *et al.* (1993b)):

$$\text{(iii)} \quad \left\{ \begin{array}{rcl} u^{n+1/2}(x) & = & \inf_{B \in C_t(x)} (\sup_{y \in B}(u^n(y))), \\ u^{n+1}(x) & = & \sup_{B \in C_t(x)} (\inf_{y \in B}(u^{n+\frac{1}{2}}(y))). \end{array} \right. \tag{11.1}$$

Let us end with a variant of the schemes (i), (ii) and (iii) which is consistent with MCM. Catté *et al.* (1993) proved that if one takes for C the set of all segments in the plane, then scheme (iii) converges towards the MCM model. We shall explain in the next sections why this type of scheme, though fully invariant in theory, can hardly be implemented on a fixed grid. However, it can, as we shall later see, be well adapted to the 'Jordan level curve' representation of images.

11.2. Iterated median filters and curvature motion

Morphological schemes for MCM have been proposed by Koenderink and Van Doorn (1987) and Merriman *et al.* (1992). They define a weighted median filter which weighs the contribution of points y close to a point x according to a Gaussian law of distance. So the algorithm can be rewritten (when applied to a characteristic function $\tilde{u}^0$ of a level set).

Weighted median filter

1 Let $\tilde{u}^0$ be a binary image:

$$\left\{ \begin{array}{ll} \tilde{u}^0(x) = 1, & \text{if } x \text{ belongs to the level set } E \\ \tilde{u}^0(x) = 0, & \text{else.} \end{array} \right.$$

2 We solve the heat equation, with initial datum $\tilde{u}^n$, for a small time Δt, by a convolution with a Gauss function. We obtain a new function v^n.

3 We set (median filter)

$$\left\{ \begin{array}{ll} \tilde{u}^{n+1}(x) = 1, & \text{if } v^n(x) \geq \frac{1}{2} \\ \tilde{u}^{n+1}(x) = 0, & \text{else .} \end{array} \right.$$

4 We turn back to step 2.

Barles and Georgelin (1992), proved that this theoretical scheme is consistent with and convergent to the MCM evolution of the boundary of E. Earlier works by Yuille (1988) and Mascarenhas (1992) have also proved the consistency. However, we shall see that its implementation on a fixed grid is problematic. A simpler and obvious discrete implementation of the preceding weighted iterated filter on a grid was proposed long ago by the Mathematical Morphology School. If A is a finite set of real numbers, we call $\mathrm{med}(A) = a$ any real number such that

$$\mathrm{Card}\{r \in A, r \geq a\} = \mathrm{Card}\{r \in A, r \leq a\}.$$

Of course, the possible as make an interval and if we want to specify a, we take for a the middle point of this interval. Then a classical morphological filter, with well-known 'denoising' properties is the original 'iterated median filter' (Matheron, 1975; Serra, 1982; Maragos, 1987)

$$\text{(iv)} \qquad u^{n+1}(x) = \mathrm{med}_{y \in B}(u^n(y)).$$

This filter (where B is a circular fixed stencil around x) can be viewed as a simplification of the Gaussian-weighted median filter, where the Gauss function has been replaced by the characteristic function of a disk.

Let us now pass to discrete versions on a grid. Among the schemes discussed in this article, the morphological schemes (i)–(iv) discretized on a fixed grid are the only ones to be both morphological and order preserving (whatever the discretization of sup, inf, med on the grid is.) They are also consistent. Now, they do not have all geometrical properties. They only satisfy the geometrical invariances of the grid and are by no way scale invariant or affine invariant. So in practice they prove to be useless schemes when discretized on a fixed grid: because of the lack of rotational invariance, they make corners appear in the shapes with sides parallel to the principal directions of the stencil. However, since these schemes are the only ones to be fully morphological and monotonic, it is very desirable to have them implemented in one way or the other. There is, however, no cheap solution: either one uses grid refinements or one adapts these theoretical schemes to the Jordan level curve representation (Section 11.3).

Let us say a little more about the consistency of these schemes when discretized on a grid. Unless consistency is proved, in practice they are not: $\mathrm{curv}(u)$ is incorrectly calculated. If we use a fixed stencil, it will simply be computed as equal to zero. Indeed, it is well known in computer graphics that the curvature of a discretized circle is impossible to compute accurately on a fixed window unless this window is very large: if e.g. the radius is equal to 20, the window should have a width of at least 12 pixels to yield a roughly correct guess of the curvature. If the window is, say, 5 pixels wide, the curvature will simply be implicitly computed as equal to 0 by a median filter and so we shall have $u^{n+1} = u^n$, which stops the evolution. In other words, when Δt tends to 0, the scheme tends to an equation

$$\partial u/\partial t = |\mathrm{D}u| F(\mathrm{curv}(u))$$

where F is defined by

$$F(s) = \begin{cases} s, & \text{if } s \geq \mathrm{C}\,, \\ 0, & \text{if } -\mathrm{C} \leq s \leq \mathrm{C}\,, \\ s, & \text{if } s \leq -\mathrm{C}\,, \end{cases}$$

and C is a constant depending on the grid step. So consistency is in practice not true; there is a thresholding effect on the curvature. A striking example of the lack of consistency can be seen in Figure 17 the iterated weighted median filter (with an $8 * 8$ stencil) is applied to both a grey level image (left down) and a binary image (left up) which is one of its level sets. The right-hand images display what happens when $n \to \infty$: the displayed images remain steady under the median filter because the curvature of all level sets is implicitly computed as equal to 0.

11.3. Geometrical curve evolution schemes

Curve sampling problems We have seen in Section 5 MCM and the AMSS can be reformulated as intrinsic heat equations,

$$\dot{C}(t,s) = \partial^2 C(t,s)/\partial s^2$$

where s stands for the length parameter along the curve in the first case and for the 'affine length' parameter in the second. This formulation suggests a computationally cheap implementation of MCM, which was developed by Mackworth and Mocktharian (1992).

Intrinsic heat equation

- Discretize C_0 as a polygon with vertices $C(i), i = 1, \ldots, N$. Parametrize this polygon with length s.
- C_n being given, together with its length parametrization, convolve it with a discretized Gaussian filter with (small) variance δt: $\tilde{C}_{n+1}(i) = (G_{\delta t} * C_n)(i)$.
- Reparametrize $\tilde{C}_{n+1}$ with length, which yields $C_{n+1}(i)$ and go back to the previous step.

This type of algorithm is very accurately scale invariant and isotropic and satisfies, in practice, the *local shape inclusion principle* when δt is chosen small enough (close to the pixel size). The consistency of such a scheme is easily proved but the convergence is, as far as we know, unproved. One could be willing to extend this kind of scheme to the ASS model: it suffices to compute the affine length instead of the Euclidean one in the last step. This is, however, *impossible*. Indeed, it is easily seen that any C^2 curve (for which affine length is well defined) can be approached in C^1 by C^1 curves with affine length tending to zero. This is due to the fact that any straight line has zero affine length. Conversely, every polygon can be approximated by C^2 curves whose affine length tends to zero. So the computation of the affine length makes no computational sense in the irregular context of shapes and images. There has, however, been some attempt to base schemes for the ASS model on discrete versions of affine length: Bruckstein *et al.* (1992) define the affine length of a polygon as the number of vertices and iterate a convolution kernel on it. Therefore, they obtain a Mackworth–Mockhtarian scheme, but the evolution is highly dependent on the sampling of the polygon.

A geometrical scheme In order to avoid the above discussed problems, we shall define an easy to implement version of the alternate scheme 11.1.

Let C^0 be a Jordan curve, X^0 the set enclosed by this Jordan curve, and x a point of $C^0 = \partial X^0$. Let K_t be an affine invariant and translation invariant set of convex sets (e.g. all ellipses,...) with area equal to $t^{3/2}$. Then, we define the *t-affine opening* $O_t X^0$ as the set of the barycentres of the elements of K_t contained in X^0. In the same way, we define the *affine closing* $F_t X^0$ as the complementary of the opening of X^c. And we consider the alternate scheme

$$X^{n+1/2} = O_t X^n,$$

$$X^{n+1} = F_t X^{n+1/2}.$$

This is the set evolution scheme associated with the convergent image evolution scheme 11.1. Indeed, it is easy to check that u^n obeys 11.1 if and only if its level sets obey the preceding scheme.

We shall now define a practical scheme associated with this theoretical scheme. We notice that, in order to define the opening, we can restrict the sets in K_t to be *chord sets*, that is, sets bounded on the one side by the boundary of the set and on the other by a chord. This leads to a computationally easier variant.

If X is a set enclosed by a Jordan curve $C(s)$, let $(C(s_0), C(s_1))$ be a 'chord', that is, a segment joining two points of C with $s_0 < s_1$. We define the Δt-opening as follows. For every s_0, we consider the chords $(C(s_0), C(s))$ and we call $A(s)$ the area of X between the chord and the curve $C(\sigma)$ for $s_0 \leq \sigma \leq s$. The function $A(s)$ is nondecreasing and we take $s_1 = \sup\{s, A(s) \leq \Delta t^{3/2}\}$. We call $CH(s_0)$ the chord set associated with the chord $(C(s_0), C(s_1))$. It has an area less than or equal to $\Delta t^{3/2}$.

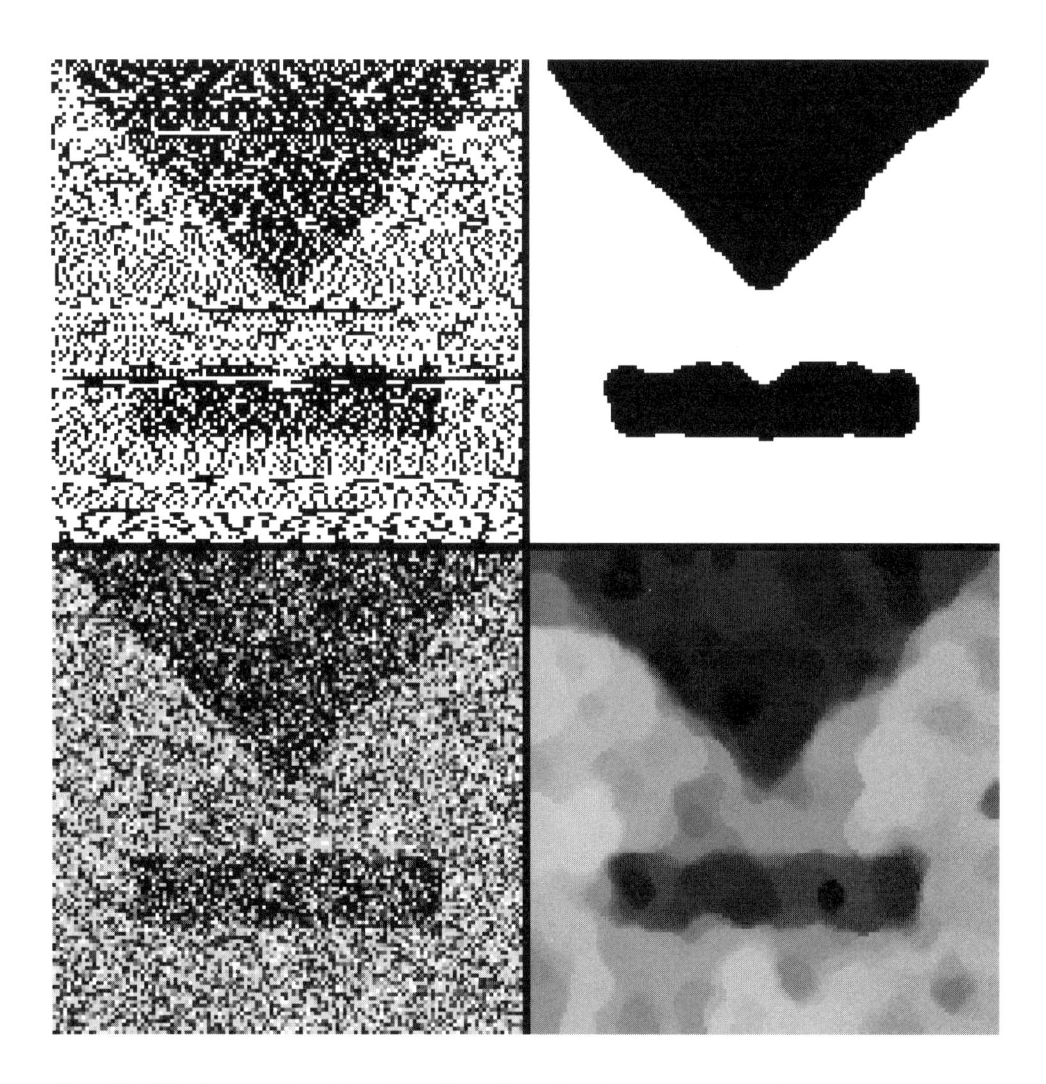

Fig. 17. The iterated weighted median filter (with an $8*8$ stencil) is applied to both a grey level image (left down) and a binary image (left up) which is one of its level sets. The right-hand images display what happens when $n \to \infty$: the displayed images remain steady under the median filter because the curvature of all level sets is implicitly computed as equal to 0.

Then the Δt-opening of X is defined by

$$O_{\Delta t}X = X \setminus (\cup_s CH(s)).$$

In the same way, we can define the closing $F_{\Delta t}X$, by $F_{\Delta t}X = (O_{\Delta t}X^c)^c$. Of course, $O_{\Delta t}X$ may be no Jordan set. Now, if we discretize X and C, we

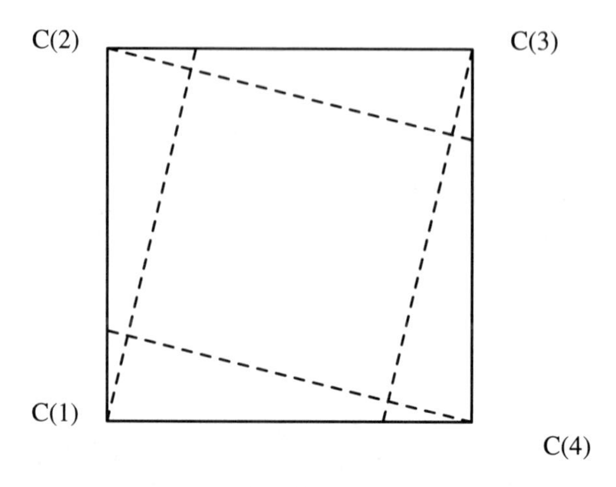

Fig. 18.

can ensure that $O_{\Delta t}X$ remains a finite union of Jordan sets on which the algorithm can be iterated. We now define a practical algorithm based on the same principle.

1 $C(i), \quad i = 1 \cdots N$, is a polygon curve enclosing X.
2 For every i, we find the last j such that the area of the chord set $CH(C(i), C(j))$ is less than $\Delta t^{3/2}$. Then, we define a new vertex $\tilde{C}(i)$, contained in the vertex $(C(j), C(j+1))$. $\tilde{C}(i)$ is chosen so that either $\tilde{C}(i) = C(j)$ or the area of the chord set associated with the chord $(C(i), \tilde{C}(i))$ is $\Delta t^{3/2}$. Denote this chord set by $CH(C(i), \tilde{C}(i))$.

The new polygon $O_{\Delta t}C$ is defined as

$$O_{\Delta t}X = X \setminus (\cup_i CH(C(i), \tilde{C}(i)).$$

It is a polygon or a union of polygons. The algorithm for computing $O_{\Delta t}X$ is complex if Δt is large, but very simple if Δt is of the same order as the distance between two consecutive vertices.

We notice that such a method is 'self-sampling' since the number of vertices cannot increase. The final algorithm consists in alternating $O_{\Delta t}$ and $F_{\Delta t}$ as explained at the beginning of this section.

11.4. Comparison and cross-validation of the schemes

To summarize the above presentation, we have rejected several existing schemes and essentially proposed two for the AMSS model: first a finite difference scheme and second a curve evolution scheme (which can in theory be used for handling images but, because of high computational cost, this has not yet been done). In any case, we can compare both schemes on images which are characteristic functions. The match, however, would not

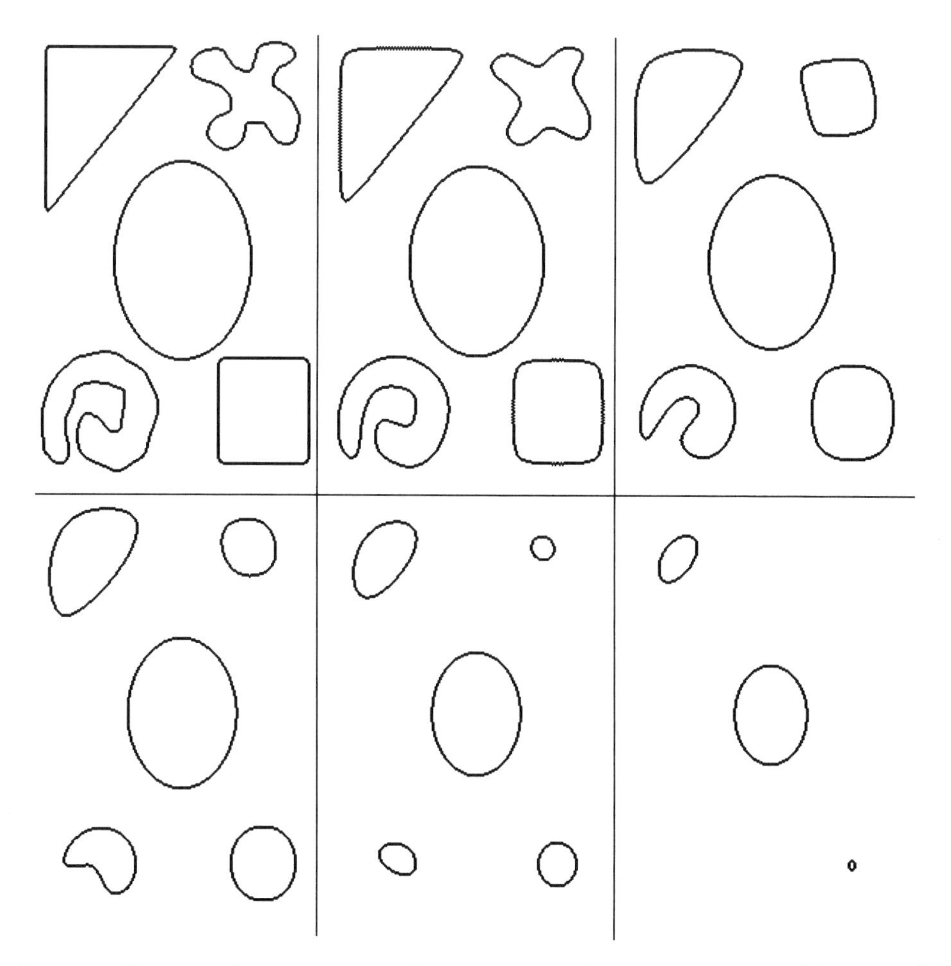

Fig. 19. Curve evolution. From left to right and from up to down: (a) $t = 0$; (b) $t = 9$, $t = 15$, $t = 21$, $t = 25$.

be fair, were the finite difference scheme to be applied to an image which presents abrupt fronts. So, following Cohignac *et al.* (1993a) we used the 'morphologization' of the numerical scheme for comparison. We know that the morphological invariance can be restored by using the 'morphological school' idea of running the algorithm separately on each level set of the image and then reconstructing it. Moreover, a very efficient numerical idea of Osher and Sethian (1988) is to run the algorithm not on a set, but on some Lipschitz function having this set as a level set (see Section 9), therefore allowing good behaviour of numerical schemes by avoiding explicit front tracking (Figure 1(c)). When done in this framework, Cohignac *et al.* (1993a) proved a full cross-validation of the above selected finite difference scheme and curve evolution scheme: complicated objects such as a spiral evolve in exactly the same way when handled with schemes with so different an implementation (see Cohignac *et al.* (1993a) for more details).

12. Conclusions

In this review, we have described with the unified formalism of *multiscale analysis* more than 20 acknowledged theories of image, shape and texture analysis. Following recent mathematical work, we have shown how axiomatic analysis reduced these theories to not more than one model for standard multiscale analysis, the AMSS model, while a large number of segmentation multiscale devices were reduced to a single variational algorithm. The same analysis applied to movies also led to a single model, with applications to motion analysis and movie denoising. As an application of the unified theory, we have shown how it permitted a rigorous axiomatic and experimental discussion of a psychophysical theory such as Julesz texture discrimination theory.

The second part of our review deals with algorithms implementing a highly geometrically invariant equation, the AMSS model, as well as its Euclidean version, MCM. We have discussed several algorithms directly proposed for these equations, as well as procedures which, although not directed towards solving any PDE, happen to be discrete versions of the MCM.

Unless we have considered a wide range of theories and algorithms, the conclusion of the study considerably narrows the idea of what has been aimed and attained by the image analysis research. Indeed, under the variety of methods lies essentially one aim for the multiscale analysis: the computation of *multiscale curvature, multiscale orientation, multiscale affine curvature, and multiscale apparent velocity.* **And nothing else!** This is due to the obvious invariance requirements, which do not leave space for any other differential operator. In the same way, the invariance requirements practically specify the forms the algorithms must take. In any case, image processing has contributed to mathematics by proposing new variational problems and new methods for solving them (for the segmentation problem). In the case of smoothing multiscale analyses, it has brought a new equation, the AMSS model, as well as new variants of schemes for the MC equation.

Appendix A. The 'fundamental theorem' of image analysis

Fundamental Theorem 2 If an image multiscale analysis T_t is causal and regular then $u(t,x) = (T_t u)(x)$ is a viscosity solution of (3.1), where the function F, defined in the regularity axiom, is nondecreasing with respect to its first argument $\mathrm{D}^2 u$. Conversely, if u_0 is a bounded uniformly continuous image, then equation (3.1) has a unique viscosity solution.

Proof. Assume for simplicity that $u(t,x)$ is C^2 in the neighbourhood of (t,x). Then, we have

$$u(t,y) = u(t,x) + (\mathrm{D}u, y-x) + \tfrac{1}{2}\mathrm{D}^2u(y-x,y-x) + o(|y-x|^2).$$

Let $\epsilon > 0$ and Q_ϵ a quadratic form given by

$$Q_\epsilon(y) = u(t,x) + (\mathrm{D}u, y-x) + \tfrac{1}{2}\mathrm{D}^2u(y-x,y-x) + \epsilon \cdot |y-x|^2.$$

Then, in a neighbourhood of (t,x)

$$Q_\epsilon(y) < u(t,y) < Q_\epsilon(y) \qquad \text{for } y \neq x,$$

and by using the causality principle we obtain

$$(T_{t+h,t}Q_{-\epsilon})(x) \leq (T_{t+h,t}u(t))(x) \leq (T_{t+h,t}Q_\epsilon)(x).$$

On the other hand, we also have

$$Q_{-\epsilon}(x) = Q_\epsilon(x) = u(t,x) = (T_{t,t}Q_{-\epsilon})(x) = (T_{t,t}Q_\epsilon)(x).$$

Therefore we deduce from the above relations

$$\begin{aligned}\frac{\partial(T_{t+h,t}Q_{-\epsilon})}{\partial h}(x) &\leq \liminf \frac{T_{t+h,t}u(,x) - T_{t,t}u(x)}{h} \\ &\leq \limsup \frac{T_{t+h,t}u(,x) - T_{t,t}u(x)}{h} \leq \frac{\partial(T_{t+h,t}Q_\epsilon)}{\partial h}(x).\end{aligned}$$

By using the regularity principle and the continuity of the function F, and taking $\epsilon \to 0$ we obtain that $u(t,x)$ satisfies equation (3.1). Finally, in order to obtain that $F(A,p,c,x,t)$ is nondecreasing with respect to A, we notice that if $A \leq B$ then the quadratic forms

$$\begin{aligned}Q_A(y) &= \tfrac{1}{2}(A(y-x),(y-x)) + (p,y-x) + c, \\ Q_B(y) &= \tfrac{1}{2}(B(y-x),(y-x)) + (p,y-x) + c,\end{aligned}$$

satisfy $Q_A(y) \leq Q_B(y)$ and by using an obvious adaptation of the above proof, we obtain $F(A,p,c,x,t) \leq F(B,p,c,x,t)$ if $A \geq B$.

To simplify the exposition, we have showed that equation (3.1) is true in the case where u is a C^2 function. By using the same ideas in the framework of viscosity solutions (see Crandall *et al.* (1991)), it is possible to show that equation (3.1) is true in the sense of viscosity solutions for any $u(t,x)$ uniformly continuous satisfying the causality and regularity principles. The fact that if $u_0(x)$ is a bounded uniformly continuous function, equation (3.1) has a unique viscosity solution is proved in Chen *et al.* (1991), Crandall *et al.* (1991) and Evans and Spruck (1991).

Appendix B. Proof of the scale normalization lemma

Normalization Lemma (Normalization of scale.) Assume that $t \to T_t$ is a one-to-one family of operators satisfying [affine invariance]. Then the function $t'(t,B)$ only depends on t and $|\det B|$: $t'(t,B) = t'(t, |\det B|^{1/2})$ and increases with respect to t. Moreover, there exists an increasing differentiable rescaling function σ: $[0,\infty] \to [0,\infty]$, such that

$$t'(t,B) = \sigma^{-1}(\sigma(t)|\det B|^{1/2})$$

and if we set $S_t = T_{\sigma^{-1}(t)}$ we have $t'(t,B) = t|\det B|^{1/2}$ for the rescaled analysis.

Proof. First we notice that for any linear transforms B and C and any t one has the semigroup property

$$\text{(i)} \qquad t'(t,BC) = t'(t'(t,B),C).$$

Indeed, we have $BCT_{t'(t,BC)} = T_t BC = BT_{t'(t,B)}C = BCT_{t'(t'(t,B),C)}$. The map which associates T_t with t being one to one, this implies the stated relation.

Next, we show that

$$\text{(ii)} \qquad t'(t,A) \text{ increases with respect to } t.$$

Let us prove that $t'(t,A)$ is one to one with respect to t for any A. Indeed, if not, there would be some A and some (s,t) such that $t'(t,A) = t'(s,A)$. Thus $T_t A = AT_{t'(t,A)} = AT_{t'(s,A)} = T_s A$ and therefore $t = s$ because T_t is one to one. Notice that this implies, in particular, that $t'(0,A) = 0$. Therefore, since $t'(t,A)$ is nonnegative (by definition), one to one and continuous with respect to t, we can deduce that it is increasing with respect to t.

Moreover $t'(t,A)$ satisfies

$$\text{(iii)} \qquad t'(t,R) = t \quad \text{for any orthogonal transform } R.$$

Indeed, let R be an orthogonal transform. Then iterating the formula of (i) we have

$$t'(t'(t'(t'(\dots t'(t,R)\dots,R),R),R) = t'(t,R^n).$$

Remark that there is a subsequence of R^n tending to Id. (Indeed, there is a subsequence R^{n_k} which converges to some H, orthogonal, because the orthogonal group is compact. Therefore, the subsequence $R^{n_{k+1}-n_k}$ converges to Id.) Since there exists a subsequence of R^n tending to Id and since t' is continuous we have for this subsequence $\lim t'(t,R^n) = t'(t,\text{Id}) = t$. Assume by contradiction that $t'(t,R) = t''$ with $t'' < t$ then $t'(t'(t,R),R) = t'(t'',R) \leq t'(t,R) = t''$ and by recursion,

$$t'(t,R^n) = t'(t'(t'(t'(\dots t'(t,R)\dots,R),R),R) \leq t'' < t.$$

This is a contradiction. Thus $t'(t,R) \geq t$. We prove the converse inequality in the same way and we obtain $t'(t,R) = t$.

We note that any linear transform B of $\mathbb{R}^2$ can be obtained as a product of orthogonal transforms and of linear transforms of the kind $A(\lambda)$: $(x,y) \to$

$(\lambda x, y)$ where λ is nonnegative. We only need to make a singular value decomposition of B: $B = R_1 \mathrm{D} R_2$, where R_1 and R_2 are both orthogonal transforms and D is a transform of the kind $(x, y) \to (\lambda_1 x, \lambda_2 y)$ where λ_i are non negative. Now, it is clear that D can be decomposed as $\mathrm{D} = A(\lambda_1) R A(\lambda_2) R^{-1}$ where R is the orthogonal transform: $(x, y) \to (-y, x)$. Using (i), $t'(t, R_i) = t$, the singular value decomposition and $A(\lambda_1)A(\lambda_2) = A(\lambda_1\lambda_2)$, we obtain

$$t'(t, B) = t'(t, \lambda_1\lambda_2) = t'(t, |\det B|^{1/2}).$$

Using (i) and (ii), we have

$$\text{(iv)} \qquad t'(t, \lambda\mu) = t'(t'(t, \mu)\lambda)$$

for any positive λ and μ. Differentiating this relation with respect to μ at $\mu = 1$ yields

$$\lambda \frac{\partial t'}{\partial \lambda}(t, \lambda) = \frac{\partial t'}{\partial \lambda}(t, 1) \frac{\partial t'}{\partial t}(t, \lambda). \tag{B.1}$$

Choose σ such that $\phi\sigma' = \sigma$ and set

$$t'(t, \lambda) = G(t, \sigma(t)\lambda),$$

where

$$\sigma(t) = \exp\left(\int_1^t \mathrm{d}s/\phi(s)\right).$$

Then the preceding relation (B.1) yields $\partial G/\partial x(x, y) = 0$. Thus $G(x, y) = \beta(y)$ for some differentiable nondecreasing function β. We obtain that $t'(t, \lambda) = \beta(\sigma(t)\lambda)$. Returning to the definition of $\phi(t)$, we have

$$\phi(t) = \partial t'/\partial\lambda(t, 1) = \partial\beta(\sigma(t)\lambda)/\partial\lambda(t, 1)$$

and

$$\phi(t) = \sigma(t)\beta'(\lambda\sigma(t)) = \phi(t)\sigma'(t)\beta'(\sigma(t)\lambda).$$

Thus the derivative of $\beta(\sigma(t))$ is 1 and integrating this last relation between 0 and t yields $\beta(\sigma(t)) = t + \beta(\sigma(0))$. Using the fact that $t'(0, \lambda) = 0$ (which derives from the injectivity of the T_t), we obtain $\beta(\sigma(0)) = 0$ and therefore $t'(t, \lambda) = \sigma^{-1}(\lambda\sigma(t))$. To finish the proof, we set $S_t = T_{\sigma^{-1}(t)}$ and we prove that the affine invariance is true for S_t with $t'(t, \lambda) = \lambda t$. $S_t B = T_{\sigma^{-1}(t)} B = B T_{t'(\sigma^{-1}(t), \lambda)} = B T_{\sigma^{-1}(\lambda\sigma(\sigma^{-1}(t)))} = B T_{\sigma^{-1}(\lambda t)} = B S_{\lambda t}$.

Appendix C. Classification of shape multiscale analyses

Theorem

(i) Under the three principles (pyramidal, local shape inclusion, 'basic'), the multiscale analysis of shapes is governed by the curvature motion equation

$$\dot{x} = g(t, \mathrm{curv}(x))\vec{n}(x), \tag{C.1}$$

where g is defined by (4.1).

(ii) If the analysis is affine invariant, then the equation of the multiscale analysis is, up to rescaling,

$$\dot{x} = \gamma(t \cdot \mathrm{curv}(x))\vec{n}(x), \tag{C.2}$$

where $\gamma(x) = a \cdot x^{1/3}$ if $x \geq 0$ and $\gamma(x) = b \cdot x^{1/3}$ if $x \leq 0$ and a, b are two nonnegative values.

(iii) If we add that $T_t(X^c) = T_t(X)^c$ [Reverse contrast invariance] then the function g in (i) is odd and we get

$$\dot{x} = (t \cdot \mathrm{curv}(x))^{1/3}\vec{n}(x). \tag{C.3}$$

Proof. (i) Let X be a silhouette and assume that $T_t(X)$ has a boundary which is a C^2 manifold in a neighbourhood of a point x of ∂X. Then it has a curvature κ at point x and we consider a subosculatory and a surosculatory disk, that is, a disk D with curvature $\kappa - \epsilon$ and a disk D$'$ with curvature $\kappa + \epsilon$, both tangent to the silhouette at x. Applying the same two principles as in the lemma, we see that

$$T_{t+h,t}(\mathrm{D}) \cap B(x,r) \subset T_{t+h,t}(X) \cap B(x,r) \subset T_{t+h,t}(\mathrm{D}') \cap B(x,r).$$

Thus, denoting by $x(t+h)$ the point of $\partial T_{t+h}(X)$ such that $x(t+h) - x(t)$ is parallel to $\vec{n}(x)$, we obtain

$$\rho(t,h,\kappa-\epsilon) - \rho(t,0,\kappa-\epsilon) \leq (x(t+h)-x(t))\cdot\vec{n}(x) \leq \rho(t,h,\kappa+\epsilon) - \rho(t,0,\kappa+\epsilon).$$

Dividing by h and passing to the limit when h tends to 0 yields

$$\frac{\partial \rho}{\partial h}(t,0,\kappa-\epsilon) \leq \liminf \frac{x(t+h)-x(t)}{h} \cdot \vec{n}(x),$$

$$\limsup \frac{x(t+h)-x(t)}{h} \cdot \vec{n}(x) \leq \frac{\partial \rho}{\partial h}(t,0,\kappa+\epsilon).$$

We obtain equation (C.2) by passing to the limit when ϵ tends to 0 and using the fact that $\kappa \to \partial\rho/\partial h(t,0,\kappa)$ is continuous.

(ii) After renormalization, we can use the identity

$$T_{t+h,t}\mathrm{D}_\lambda = \mathrm{D}_\lambda T_{(t+h)\lambda,t\lambda},$$

(where $\mathrm{D}_\lambda = \lambda\,\mathrm{Id}$) and so, we can deduce that the function ρ of the basic principle must satisfy $\rho(t,h,\lambda/r) = \lambda^{-1}\rho(\lambda t, \lambda h, 1/r)$ Therefore, we obtain the relation (after differentiation with respect to h at 0)

$$g(t, \lambda s) = g(\lambda t, s),$$

for any $t > 0$, $\lambda > 0$ and $s \in \mathbb{R}$. Changing t in t/λ and taking $\lambda = 1/t$ we get $g(t,s) = g(1,ts) = \beta(ts)$ for the function β defined as $\beta(x) = g(1,x)$.

On the other hand, we can use the identity $T_{t+h,t}A = AT_{t+h,t}$, where A is the linear transform whose determinant is one,

$$(x,y) \to (\lambda x, (1/\lambda)y), \quad \lambda > 0.$$

Let us apply this identity to the unit disk Δ. Look at the point $x_0 = (1,0)$ on the boundary of Δ. Then the velocity of x_0 is $\beta(-t)$, and this velocity is transformed into $\lambda\beta(-t)$. Now, look at $A\Delta$. Since $A\Delta$ is an ellipse with curvature $-\lambda^3$ at point Ax_0, the velocity of Ax_0 is $\beta(-t\cdot\lambda^3)$. Using the first identity, we obtain $\beta(-t.\lambda^3) = \lambda\beta(-t)$. Taking $t = 1$, we get $\beta(x) = b\cdot x^{1/3}$ for $x < 0$ ($b = \beta(-1)$). Now, apply the same technique to Δ^c and we get the result $\beta(x) = a\cdot x^{1/3}$ for $x > 0$ ($a = \beta(1)$).

(iii) With the same technique as above we obtain that the function β is odd. □

REFERENCES

L. Alvarez, F. Guichard, P.L. Lions and J.M. Morel (1992a), 'Axioms and fundamental equations of image processing', Report 9216, C.E.R.E.M.A.D.E.,, Université Paris Dauphine, *Arch. Rat. Mech.*, to appear.

L. Alvarez, F. Guichard, P.L. Lions and J.M. Morel (1992b), 'Axiomatisation et nouveaux opérateurs de la morphologie mathematique', *C.R. Acad. Sci. Paris* **315**, 265–268.

L. Alvarez, P.L. Lions and J.M. Morel (1992c), 'Image selective smoothing and edge detection by nonlinear diffusion (II)', *SIAM J. Numer. Anal.* **29**, 845–866.

L. Alvarez and L. Mazorra (1992), 'Signal and image restoration by using shock filters and anisotropic diffusion', Preprint, Dep. de Inf. U.L.P.G.C. ref:0192, *SIAM J. Numer. Anal.*, to appear.

S. Angenent (1989), 'Parabolic equations for curves on surfaces I, II', University of Wisconsin-Madison Technical Summary Reports, 19, 24.

H. Asada and M. Brady (1986), 'The curvature primal sketch', *IEEE Trans. Patt. Anal. Machine Intell.* **8**(1).

C. Ballester and M. Gonzalez (1993), 'Affine invariant multiscale segmentation by variational method', *Proc. Eighth Workshop on Image and Multidimensional Signal Processing (8–10 September, Cannes)*, IEEE (New York), 220–221.

G. Barles (1985), 'Remarks on a flame propagation model', Technical Report No 464, INRIA Rapports de Recherche.

G. Barles and C. Georgelin (1992), 'A simple proof of convergence for an approximation scheme for computing motions by mean curvature', Preprint.

G. Barles and P.E. Souganidis (1993), 'Convergence of approximation schemes for fully nonlinear second order equation', *Asymp. Anal.*, to appear.

C. Brice and C. Fennema (1970), 'Scene analysis using regions', *Artificial Intelligence* **1**, 205–226.

W. Brockett and P. Maragos (1992), 'Evolution equations for continuous-scale morphology', *ICASSP, San Francisco* 23–26.

A.M. Bruckstein, G. Sapiro and D. Shaked (1992), 'Affine-invariant evolutions of planar polygons', Preprint.

V. Caselles, F. Catté, T. Coll and F. Dibos (1992), 'A geometric model for active contours in image processing', Report 9210, C.E.R.E.M.A.D.E., Université Paris Dauphine (Paris).

F. Catté, F. Dibos and G. Koepfler (1993), 'A morphological approach of mean curvature motion', Report 9310, C.E.R.E.M.A.D.E., Université Paris Dauphine (Paris).

Y-G. Chen, Y. Giga and S. Goto (1989), 'Uniqueness and existence of viscosity solutions of generalized mean curvature flow equations', Preprint, Hokkaido University.

T. Cohignac, F. Eve, F. Guichard, C. Lopez and J. M. Morel (1993a), *Numerical Analysis of the Fundamental Equation of Image Processing*, to appear.

T. Cohignac, C. Lopez and J. M. Morel (1993b), 'Multiscale analysis of shapes, images and textures', *Proc. Eighth Workshop on Image and Multidimensional Signal Processing (8–10 September, Cannes)*, IEEE (New York) 142–143.

M. G. Crandall, H. Ishii and P.L. Lions (1991), 'User's guide to viscosity solution of second order partial differential equation', C.E.R.E.M.A.D.E., Preprint (Paris).

G. Dal Maso, J.M. Morel and S. Solimini (1989), 'Une approche variationnelle en traitement d'images: résultats d'existence et d'approximation', *C.R. Acad. Sci. Paris* **308**, 549–554.

I. Daubechies (1992), 'Ten lectures on wavelets', SIAM (Philadelphia).

F. Dibos and G. Koepfler (1991), 'Propriété de régularité des contours d'une image segmentée', *C.R. Acad. Sci. Paris* **313**, 573–578.

J. Enns (1986), 'Seeing textons in context', *Perception and Psychophysics* **39**(2), 143–147.

L.C. Evans and J. Spruck (1992), 'Motion of level sets by mean curvature I', Preprint.

O. Faugeras (1993), 'A few steps toward a projective scale space analysis', *C.R. Acad. Sci. Paris* to appear.

L. Florack, B. ter Haar Romeny, J.J. Koenderink and M. Viergever (1991), 'General intensity transformations and second order invariants', *Proc. 7th Scandinavian Conference on Image Analysis (Aalborg)* 13–16.

L. Florack, B. ter Haar Romeny, J.J. Koenderink and M. Viergever (1992), 'Scale and the differential structure of images', *Image Vision Computing* **10**.

D. Forsyth, J.L. Mundy and A. Zisserman (1991), 'Invariant descriptors for 3-D object recognition and Pose', *IEEE Trans. Patt. Anal. Machine Intell.* **13**, No.10.

M. Gage and R.S. Hamilton (1986), 'The heat equation shrinking convex plane curves', *J. Diff. Geom.* **23**, 69–96.

S. Geman and D. Geman (1984), 'Stochastic relaxation, Gibbs distributions and the Bayesian restoration of images', *IEEE Patt. Anal. Machine Intell.* **6**.

F. Guichard (1993), 'Multiscale analysis of movies', *Proc. Eighth Workshop on Image and Multidimensional Signal Processing (8–10 September, Cannes)*, IEEE (New York), 236–237.

F. Guichard, J.M. Lasry and J.M. Morel (1993), 'A monotone consistent theoretical scheme for the fundamental equation of image processing', Preprint.

M. Grayson (1987), 'The heat equation shrinks embedded plane curves to round points', *J. Diff. Geom.* **26**, 285–314.

R.M. Haralick and L.G. Shapiro (1985), 'Image segmentation techniques', *Comput. Vision Graph. Image Process.* **29**, 100–132.

B. Horn (1986), *Robot Vision*, MIT (Cambridge, MA).

S. L. Horowitz and T. Pavlidis (1974), 'Picture segmentation by a directed split-and-merge procedure', *Proc. Second Int. Joint Conf. Pattern Recognition* 424-433.

R. Hummel (1986), 'Representations based on zero-crossing in scale-space', *Proc. IEEE Computer Vision and Pattern Recognition Conf.*, 204–209.

B. Julesz (1981), 'Textons, the elements of texture perception, and their interactions', *Nature* **290**.

B. Julesz (1986), 'Texton gradients: the texton theory revisited', *Biol. Cybern.* **54**, 245–251.

B. Julesz and J.R. Bergen (1983), 'Textons, the fundamental elements in preattentive vision and perception of textures', *Bell System Tech. J.* **62**(6), 1619–1645.

B. Julesz and B. Kroese (1988), 'Features and spatial filters', *Nature* **333**, 302–303.

M. Kass, A. Witkin and D. Terzopoulos (1987), 'Snakes: active contour models', *1st Int. Comput. Vis. Conf. IEEE 777*.

B.B. Kimia (1990), 'Toward a computational theory of shape', PhD Dissertation Department of Electrical Engineering, McGill University, Montreal, Canada.

B.B. Kimia, A. Tannenbaum and S.W. Zucker (1992), 'On the evolution of curves via a function of curvature, 1: the classical case', *J. Math. Anal. Appl.* **163**(2).

G. Koepfler, J.M. Morel and S. Solimini (1991), 'Segmentation by minimizing a functional and the "merging" methods', *Proc. 'GRETSI Colloque' (Juan-les-Pins, France)*.

G. Koepfler, C. Lopez and J.M. Morel (1994), 'A multiscale algorithm for image segmentation by variational method', *SIAM J. Numer. Anal.* **31**, to appear.

J.J. Koenderink (1984), 'The structure of images', *Biol. Cybern.* **50**, 363–370.

J.J. Koenderink (1990a), *Solid Shape*, MIT Press (Cambridge, MA).

J.J. Koenderink (1990b), 'The brain, a geometry engine', *Psychol. Res.* **52**, 122–127.

J.J. Koenderink and A.J. van Doorn (1986), 'Dynamic shape', *Biol. Cybern.* **53**, 383–396.

J.J. Koenderink and A.J. van Doorn (1987), 'Representation of local geometry in the visual system', *Biol. Cybern.* **55**, 367–375.

Y. Lamdan, J.T. Schwartz and H.J. Wolfson (1988), 'Object recognition by affine invariant matching', in *Proc. CVPR 88*.

T. Lindeberg (1990), 'Scale-space for discrete signal', *IEEE Trans. Patt. Anal. Machine Intell.* **12**, 234–254.

C. Lopez and J.M. Morel (1992), 'Axiomatisation of shape analysis and application to texture hyperdiscrimination', *Proc. Trento Conf. on Surface Tension and Movement by Mean Curvature*, De Gruyter (Berlin).

A. Mackworth and F. Mokhtarian (1986), 'Scale-based description and recognition of planar curves and two-dimensional shapes', *IEEE Trans. Patt. Anal. Machine Intell.* **8**(1).

A. Mackworth and F. Mokhtarian (1992), 'A theory of multiscale, curvature-based shape representation for planar curves', *IEEE Trans. Patt. Anal. Machine Intell.* **14**, 789–805.

J. Malik and P. Perona (1991), 'Preattentive texture discrimination with early vision mechanisms', *J. Opt. Soc. Am.* A **7**(5), 923–932.

P. Maragos (1987), 'Tutorial on advances in morphological image processing and analysis', *Opt. Engrg* **26**(7).

D. Marr (1976), 'Analyzing natural images: a computational theory of texture vision', *Cold Spring Harbor Symp. on Quantitative Biology, XL* 647–662.

D. Marr (1982), *Vision*, Freeman (San Francisco).

P. Mascarenhas (1992), 'Diffusion generated motion by mean curvature', Preprint.

G. Matheron (1975), *Random Sets and Integral Geometry*, John Wiley (New York).

B. Merriman, J. Bence and S. Osher (1992), 'Diffusion generated motion by mean curvature', CAM Report 92-18, Department of Mathematics, University of California (Los Angeles CA 90024.1555, USA).

Y. Meyer (1992), *Ondelettes et Algorithmes Concurrents*, Hermann (Paris).

J.M. Morel and S. Solimini (1988a), 'Segmentation of images by variational methods: a constructive approach', *Rev. Matematica de la Universidad Complutense de Madrid* Vol. 1 **1,2,3**, 169–182.

J.M. Morel and S. Solimini (1988b), 'Segmentation d'images par méthode variationnelle: une preuve constructive d'existence', *C. R. Acad. Sci. Paris*.

J.M. Morel and S. Solimini (1993), *Variational Methods in Image Segmentation*, Birkhauser (Boston) to appear.

J.L. Muerle and D. C. Allen (1968), 'Experimental evaluation of techniques for automatic segmentation of objects in a complex scene', in *Pictorial Pattern Recognition* (G. C. Cheng *et al.*, eds), Thompson (Washington), 3–13.

D. Mumford and J. Shah (1988), 'Boundary detection by minimizing functionals', *Image Understanding* (S. Ullman and W. Richards, eds).

D. Mumford and J. Shah (1989), 'Optimal Approximations by Piecewise Smooth Functions and Associated Variational Problems', *Commun. Pure Appl. Math.* XLII **4**.

S. Osher and J. Sethian (1988), 'Fronts propagating with curvature dependent speed: algorithms based on the Hamilton–Jacobi formulation', *J. Comput. Phys.* **79**, 12–49.

T. Pavlidis (1972), 'Segmentation of pictures and maps through functional approximation', *Comput. Graph. Image Process.* **1**, 360–372.

T. Pavlidis and Y.T. Liow (1988), 'Integrating region growing and edge detection', *Proc. IEEE Conf. on Comput. Vision Patt. Recognition.*

P. Perona and J. Malik (1987), 'A scale space and edge detection using anisotropic diffusion', *Proc. IEEE Computer Soc. Workshop on Computer Vision.*

L. Rudin, S. Osher and E. Fatemi (1992), 'Nonlinear total variation based noise removal algorithms', *Proc. Modélisations Matématiques pour le Traitement d'Images*, INRIA 149–179.

A. Rosenfeld and A. Kak (1982), *Digital Picture Processing* Vol. 1, Academic (New York).

G. Sapiro and A. Tannenbaum (1992a), 'On affine plane curve evolution', EE Pub 821, Department of Electrical Engineering, Technion Israel Institute of Technology, Haifa, Israel.

G. Sapiro and A. Tannenbaum (1992b), 'Affine shortening of non-convex plane curves', EE Pub 845, Department of Electrical Engineering, Technion Israel Institute of Technology, Haifa, Israel.

J. Serra (1982), *Image Analysis and Mathematical Morphology* Vol. 1, Academic (New York).

A. Treisman (1985), 'Preattentive processing in vision', *Comput. Vision, Graph. Image Process.* **31**, 156–177.

H. Voorhees and T. Poggio (1987), 'Detecting textons and texture boundaries in natural images', *Proc. Int. Conf. Computer Vision*, IEEE (New York), 250–258.

A. P. Witkin (1983), 'Scale-space filtering', *Proc. IJCAI (Karlsruhe)* 1019–1021.

A. Yuille (1988), 'The creation of structure in dynamic shape', *Proc. Second International Conference on Computer Vision (Tampa)* 685–689.

A. Yuille and T. Poggio (1986), 'Scaling theorems for zero crossings', *IEEE Trans. Patt. Anal. Machine Intell.* **8**.

S. W. Zucker (1976), 'Region growing: childhood and adolescence (survey)', *Comput. Graph. Image Process.* **5**, 382–399.

Acta Numerica (1994), *pp.* 61–143

Domain decomposition algorithms

Tony F. Chan
Department of Mathematics,
University of California at Los Angeles,
Los Angeles, CA 90024, USA
Email: chan@math.ucla.edu.

Tarek P. Mathew
Department of Mathematics,
University of Wyoming,
Laramie, WY 82071-3036, USA
Email: mathew@corral.uwyo.edu.

Domain decomposition refers to divide and conquer techniques for solving partial differential equations by iteratively solving subproblems defined on smaller subdomains. The principal advantages include enhancement of parallelism and localized treatment of complex and irregular geometries, singularities and anomalous regions. Additionally, domain decomposition can sometimes reduce the computational complexity of the underlying solution method.

In this article, we survey *iterative* domain decomposition techniques that have been developed in recent years for solving several kinds of partial differential equations, including elliptic, parabolic, and differential systems such as the Stokes problem and mixed formulations of elliptic problems. We focus on describing the salient features of the algorithms and describe them using easy to understand matrix notation. In the case of elliptic problems, we also provide an introduction to the convergence theory, which requires some knowledge of finite element spaces and elementary functional analysis.

* The authors were supported in part by the National Science Foundation under grant ASC 92-01266, by the Army Research Office under contract DAAL03-91-G-0150 and subcontract under DAAL03-91-C-0047, and by the Office for Naval Research under contract ONR N00014-92-J-1890.

CONTENTS

1	Introduction	62
2	Overlapping subdomain algorithms	70
3	Nonoverlapping subdomain algorithms	74
4	Introduction to the convergence theory	91
5	Some practical implementation issues	101
6	Multilevel algorithms	106
7	Algorithms for locally refined grids	110
8	Domain imbedding or fictitious domain methods	113
9	Convection–diffusion problems	117
10	Parabolic problems	121
11	Mixed finite elements and the Stokes problem	125
12	Other topics	128
	References	130

1. Introduction

Domain decomposition (DD) methods are techniques for solving partial differential equations based on a decomposition of the spatial domain of the problem into several subdomains. Such reformulations are usually motivated by the need to create solvers which are easily parallelized on coarse grain parallel computers, though sometimes they can also reduce the complexity of solvers on sequential computers. These techniques can often be applied directly to the partial differential equations, but they are of most interest when applied to discretizations of the differential equations (either by finite difference, finite element, spectral or spectral element methods). The primary technique consists of solving subproblems on various subdomains, while enforcing suitable continuity requirements between adjacent subproblems, till the local solutions converge (within a specified accuracy) to the true solution.

In this article, we focus on describing *iterative* domain decomposition algorithms, particularly on the formulation of preconditioners for solution by conjugate gradient type methods. Though many fast *direct* domain decomposition solvers have been developed in the engineering literature, see Kron (1953) and Przemieniecki (1963) (these are often called *substructuring or tearing* methods), the more recent developments have been based on the iterative approach, which is potentially more efficient in both time and storage. The earliest known iterative domain decomposition technique was proposed in the pioneering work of H. A. Schwarz in 1870 to prove the existence of harmonic functions on irregular regions which are the union of overlapping subregions. Variants of Schwarz's method were later studied by Sobolev (1936), Morgenstern (1956) and Babuška (1957). See also Courant and Hilbert (1962). The recent interest in domain decomposition was initiated in studies by Dinh, Glowinski and Périaux (1984), Dryja (1984), Golub and Mayers (1984), Bramble, Pasciak and Schatz (1986b), Bjørstad and Widlund (1986), Lions (1988), Agoshkov and Lebedev (1985) and Marchuk, Kuznetsov and Matsokin (1986), where the primary motivation was the inherent parallelism of these methods. There are not many general references that provide an overview of the field, but here are a few: discussions in Keyes and Gropp (1987), Canuto, Hussaini, Quarteroni and Zang (1988), Xu (1992a), Dryja and Widlund (1990), Hackbusch (1993), Le Tallec (1994) and the books of Lebedev (1986), Kang (1987) and Lu, Shih and Liem (1992) and the forthcoming book by Smith, Bjørstad and Gropp (1994). The best source of references remains the collection of conference proceedings: Glowinski, Golub, Meurant and Périaux (1988), Chan, Glowinski, Périaux and Widlund (1989, 1990), Glowinski, Kuznetsov, Meurant, Périaux and Widlund (1991), Chan, Keyes, Meurant, Scroggs and Voigt (1992a), Quarteroni (1993).

This article is conceptually organized in three parts. The first part (Sections 1 through 5) deals with second-order self-adjoint elliptic problems. The algorithms and theory are most mature for this class of problem and the topics here are treated in more depth than in the rest of the article. Most domain decomposition methods can be classified as either an overlapping or a nonoverlapping subdomain approach, which we shall discuss in Sections 2 and 3 respectively. A basic theoretical framework for studying the convergence rates will be summarized in Section 4. Some practical implementation issues will be discussed in Section 5. The second part (Sections 6–8) considers algorithms that are not, strictly speaking, domain decomposition methods, but that can be studied by the general framework set up in the first part. The key idea here is to extend the concept of the subdomains to that of subspaces. The topics include multilevel preconditioners (Section 6), locally refined grids (Section 7) and fictitious domain methods (Section 8). In the last part (Sections 9–12), we consider domain decomposition methods for more general problems, including convection–diffusion problems (Section 9), parabolic problems (Section 10), mixed finite element methods and the Stokes problems (Section 11). In Section 12, we provide references to algorithms for the biharmonic problem, spectral element methods, indefinite problems and nonconforming finite element methods. Due to space limitation, and the fact that both the theory and algorithms are generally less well developed for these problems, we do not treat Parts II and III in as much depth as in Part I. Our aim is instead to highlight some of the key ideas, using the framework and terminology developed in Part I, and to provide a guide to the vast developing literature.

We present the methods in algorithmic form, expressed in matrix notation, in the hope of making the article accessible to a broad spectrum of readers.

Given the space limitation, most of the theorems (especially those in Parts II and III) are stated without proofs, with pointers to the literature given instead. We also do not cover nonlinear problems or specific applications (e.g. CFD) of domain decomposition algorithms.

In the rest of this section, we introduce the main features of domain decomposition procedures by describing several algorithms based on the simpler case of *two subdomain* decomposition for solving the following general second-order self-adjoint, coercive elliptic problem:

$$Lu \equiv -\nabla \cdot (a(x,y)\nabla u) = f(x,y), \quad \text{in } \Omega, \quad u = 0 \quad \text{on } \partial\Omega. \tag{1.1}$$

We are particularly interested in the solution of its discretization (by either finite elements or finite differences) which yields a large sparse symmetric positive definite linear system:

$$Au = f. \tag{1.2}$$

1.1. Overlapping subdomain approach

Overlapping domain decomposition algorithms are based on a decomposition of the domain Ω into a number of overlapping subregions. Here, we consider the case of two overlapping subregions $\{\hat{\Omega}_1, \hat{\Omega}_2\}$ which form a *covering* of Ω; see Figure 1. We shall let $\Gamma_i, i = 1, 2$ denote the part of the boundary of Ω_i which is in the interior of Ω.

The basic Schwarz alternating algorithm to solve (1.1) starts with any suitable initial guess u^0 and constructs a sequence of improved approximations $u^1, u^2, \ldots$. Starting with the kth iterate u^k, we solve the following two subproblems on $\hat{\Omega}_1$ and $\hat{\Omega}_2$ successively with the most current values as boundary condition on the artificial interior boundaries:

$$\begin{cases} Lu_1^{k+1} &= f, & \text{on } \hat{\Omega}_1, \\ u_1^{k+1} &= u^k|_{\Gamma_1} & \text{on } \Gamma_1, \\ u_1^{k+1} &= 0, & \text{on } \partial\hat{\Omega}_1 \backslash \Gamma_1, \end{cases}$$

and

$$\begin{cases} Lu_2^{k+1} &= f, & \text{on } \hat{\Omega}_2, \\ u_2^{k+1} &= u_1^{k+1}|_{\Gamma_2} & \text{on } \Gamma_2, \\ u_2^{k+1} &= 0, & \text{on } \partial\hat{\Omega}_2 \backslash \Gamma_2. \end{cases}$$

The iterate u^{k+1} is then defined by

$$u^{k+1}(x,y) = \begin{cases} u_2^{k+1}(x,y) & \text{if } (x,y) \in \hat{\Omega}_2 \\ u_1^{k+1}(x,y) & \text{if } (x,y) \in \Omega \backslash \hat{\Omega}_2. \end{cases}$$

It can be shown that in the norm induced by the operator L, the iterates $\{u^k\}$ converge geometrically to the true solution u on Ω, i.e.

$$\|u - u^k\| \leq \rho^k \|u - u^0\|,$$

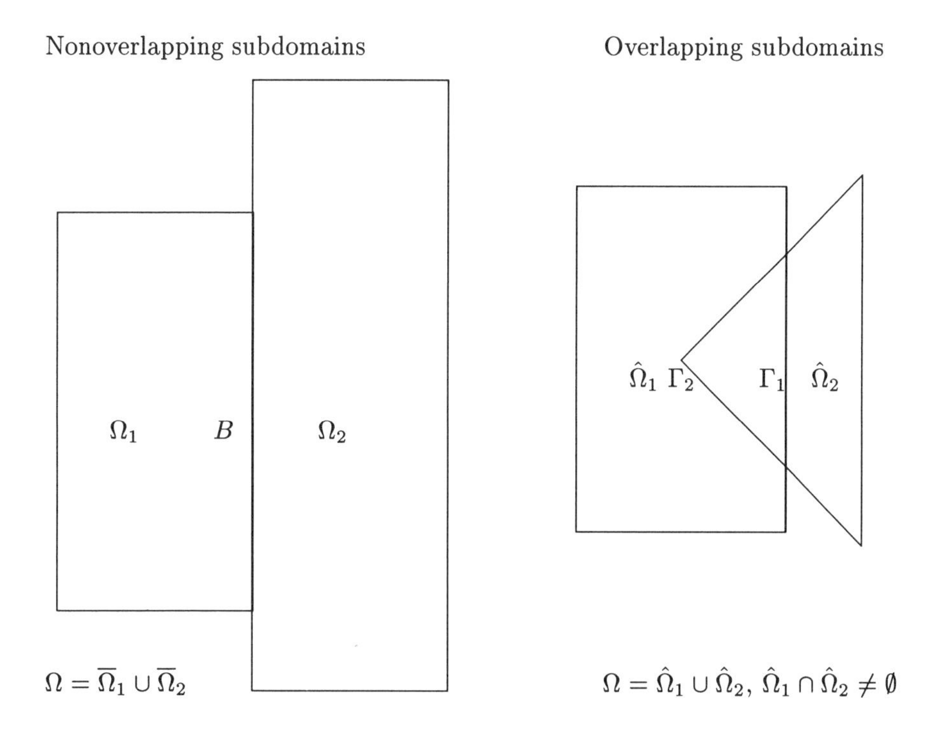

Fig. 1. Two subdomain decompositions.

where $\rho < 1$ depends on the choice of $\hat{\Omega}_1$ and $\hat{\Omega}_2$.

The above Schwarz procedure extends almost verbatim to discretizations of (1.1). We shall describe the discrete algorithm in matrix notation. Corresponding to the subregions $\{\hat{\Omega}_1, \hat{\Omega}_2\}$, let $\{\hat{I}_1, \hat{I}_2\}$ denote the indices of the nodes in the interior of domain $\hat{\Omega}_1$ and interior of $\hat{\Omega}_2$ respectively. Thus $\hat{I}_1$ and $\hat{I}_2$ form an overlapping set of indices for the unknown vector u. Let $\hat{n}_1$ be the number of indices in $\hat{I}_1$, and let $\hat{n}_2$ be the number of indices in $\hat{I}_2$. Due to overlap, $\hat{n}_1 + \hat{n}_2 > n$, where n is the number of unknowns in Ω.

Corresponding to each region $\hat{\Omega}_i$, we define a rectangular $n \times \hat{n}_i$ extension matrix R_i^T whose action extends by zero a vector of nodal values in $\hat{\Omega}_i$. Thus, given a subvector x_i of length $\hat{n}_i$ with nodal values at the interior nodes on $\hat{\Omega}_i$ we define:

$$(R_i^T x_i)_k = \begin{cases} (x_i)_k & \text{for } k \in \hat{I}_i \\ 0 & \text{for } k \in I - \hat{I}_i, \text{ where } I = \hat{I}_1 \cup \hat{I}_2. \end{cases}$$

The entries of the matrix R_i^T are ones or zeros. The transpose R_i of this extension map R_i^T is a restriction matrix whose action restricts a full vector x of length n to a vector of size $\hat{n}_i$ by choosing the entries with indices $\hat{I}_i$ corresponding to the interior nodes in $\hat{\Omega}_i$. Thus, $R_i x$ is the subvector of nodal values of x in the interior of $\hat{\Omega}_i$. The local subdomain matrices (corresponding to the discretization on $\hat{\Omega}_i$) are, therefore,

$$A_1 = R_1 A R_1^T, \quad A_2 = R_2 A R_2^T,$$

and these are principal submatrices of A.

The discrete version of the Schwarz alternating method, described earlier, to solve $Au = f$, starts with any suitable initial guess u^0 and generates a sequence of iterates $u^0, u^1, \ldots$ as follows

$$u^{k+1/2} = u^k + R_1^T A_1^{-1} R_1 (f - Au^k), \tag{1.3}$$
$$u^{k+1} = u^{k+1/2} + R_2^T A_2^{-1} R_2 (f - Au^{k+1/2}). \tag{1.4}$$

Note that this corresponds to a generalization of the block Gauss–Seidel iteration (with overlapping blocks) for solving (1.1). At each iteration, two subdomain solvers are required (A_1^{-1} and A_2^{-1}). Defining

$$P_i \equiv R_i^T A_i^{-1} R_i A, \quad i = 1, 2,$$

the convergence is governed by the iteration matrix $(I - P_2)(I - P_1)$, hence this is often called a *multiplicative* Schwarz iteration. With sufficient overlap, it can be proved that the above algorithm converges with a rate independent of the mesh size h (unlike the classical block Gauss–Seidel iteration).

We note that P_1 and P_2 are symmetric with respect to the A inner product (see Section 4), but not so for the iteration matrix $(I - P_2)(I - P_1)$. A *symmetrized* version can be constructed by iterating one more half-step with A_1^{-1} after equation (1.4). The resulting iteration matrix becomes $(I-P_1)(I-P_2)(I - P_1)$ which is symmetric with respect to the A inner product and therefore conjugate gradient acceleration can be applied.

An analogous block Jacobi version can also be defined:

$$u^{k+1/2} = u^k + R_1^T A_1^{-1} R_1 (f - Au^k), \tag{1.5}$$
$$u^{k+1} = u^{k+1/2} + R_2^T A_2^{-1} R_2 (f - Au^k). \tag{1.6}$$

This version is more parallelizable because the two subdomain solves can be carried out concurrently. Note that by eliminating $u^{k+1/2}$, we obtain

$$u^{k+1} = u^k + (R_1^T A_1^{-1} R_1 + R_2^T A_2^{-1} R_2)(f - Au^k).$$

This is simply a Richardson iteration on $Au = f$ with the following *additive Schwarz preconditioner* for A:

$$M_{\text{as}}^{-1} = R_1^T A_1^{-1} R_1 + R_2^T A_2^{-1} R_2.$$

The preconditioned system can be written as

$$M_{\text{as}}^{-1} A = P_1 + P_2,$$

which is symmetric with respect to the A inner product and can also be used with conjugate gradient acceleration. Again, for suitably chosen overlap (see Section 1), the condition number of the preconditioned system is bounded independently of h (unlike classical block Jacobi).

1.2. Nonoverlapping subdomain approach

Nonoverlapping domain decomposition algorithms are based on a partition of the domain Ω into various nonoverlapping subregions. Here, we consider a model partition of Ω into two nonoverlapping subregions Ω_1 and Ω_2, see Figure 1, with interface $B = \partial\Omega_1 \cap \partial\Omega$ (separating the two regions). Let $u = (u_1, u_2, u_B)$ denote the solution u restricted to Ω_1, Ω_2 and B respectively. Then, u_1, u_2 satisfy the following local problems:

$$\left\{\begin{array}{rcll} Lu_1 &=& f & \text{in } \Omega_1 \\ u_1 &=& 0 & \text{on } \partial\Omega_1 \backslash B \\ u_1 &=& u_B & \text{on } B \end{array}\right. \quad \text{and} \quad \left\{\begin{array}{rcll} Lu_2 &=& f & \text{in } \Omega_2 \\ u_2 &=& 0 & \text{on } \partial\Omega_2 \backslash B \\ u_2 &=& u_B & \text{on } B \end{array}\right. \qquad (1.7)$$

as well as the following *transmission boundary condition* on the continuity of the flux across B:

$$\boldsymbol{n}_1 \cdot (a \nabla u_1) = -\boldsymbol{n}_2 \cdot (a \nabla u_2) \ \text{ on } B,$$

where each $\boldsymbol{n}_i$ is the outward pointing normal vector to B from Ω_i. (We omit derivation of the above, but note that it can be obtained by applying integration by parts to the weak form of the problem.) Thus, if the value u_B of the solution u on B is known, the local solutions u_1 and u_2 can be obtained at the cost of solving two subproblems on Ω_1 and Ω_2 in parallel.

The main task in nonoverlapping domain decomposition is to determine the interface data u_B. To this end, an equation satisfied by u_B can be obtained by using the transmission boundary conditions. Let g denote arbitrary Dirichlet boundary data on B. Define $E_1 g$ and $E_2 g$ as solutions of the following local problems, on Ω_1 and Ω_2 respectively:

$$\left\{\begin{array}{rcll} L(E_1 g) &=& f & \text{in } \Omega_1 \\ E_1 g &=& 0 & \text{on } \partial\Omega_1 \backslash B \\ E_1 g &=& g & \text{on } B \end{array}\right. \quad \text{and} \quad \left\{\begin{array}{rcll} L(E_2 g) &=& f & \text{in } \Omega_2 \\ E_2 g &=& 0 & \text{on } \partial\Omega_2 \backslash B \\ E_2 g &=& g & \text{on } B. \end{array}\right. \qquad (1.8)$$

Then, by construction the boundary values of $E_1 g$ and $E_2 g$ match on B (and equal g). However, in general the *flux* of the two local solutions will not match on B, i.e.

$$\boldsymbol{n}_1 \cdot (a \nabla E_1 g) \neq -\boldsymbol{n}_2 \cdot (a \nabla E_2 g) \ \text{ on } B,$$

unless $g = u_B$. Define the following affine linear mapping T which maps the boundary data g on B to the jump in the flux across B:

$$T : g \longrightarrow \boldsymbol{n}_1 \cdot (a \nabla E_1 g) + \boldsymbol{n}_2 \cdot (a \nabla E_2 g).$$

Thus, the boundary value u_B of the true solution u, satisfies the equation

$$Tu_B = 0. \tag{1.9}$$

The map T is referred to as a Steklov–Poincaré operator, and is a pseudo-differential operator (Agoshkov, 1988; Quarteroni and Valli, 1990). A property of the map T (or a linear map derived from T since it is *affine* linear) is that it is symmetric, and positive definite with respect to the L^2 inner product on B. The discrete versions of system (1.9) can therefore be solved by preconditioned conjugate gradient methods.

We now consider the corresponding algorithm for solving the linear system $Au = f$. Based on the partition $\Omega = \Omega_1 \cup \Omega_2 \cup B$, let $I = I_1 \cup I_2 \cup I_3$ denote a partition of the indices in the linear system, where I_1 and I_2 consists of the indices of nodes in the interior of Ω_1 and Ω_2, respectively, while I_3 consists of the nodes on the interface B. Correspondingly, the unknowns u can be partitioned as $u = [u_1, u_2, u_3]^T$ and $f = [f_1, f_2, f_3]^T$, and the linear system (1.2) takes the following block form:

$$\begin{bmatrix} A_{11} & 0 & A_{13} \\ 0 & A_{22} & A_{23} \\ A_{13}^T & A_{23}^T & A_{33} \end{bmatrix} \begin{bmatrix} u_1 \\ u_2 \\ u_3 \end{bmatrix} = \begin{bmatrix} f_1 \\ f_2 \\ f_3 \end{bmatrix}. \tag{1.10}$$

Here, the blocks A_{12} and A_{21} are zero only under the assumption that the nodes in Ω_1 are not directly coupled to the nodes in Ω_2 (except through nodes on B), and this assumption holds true for finite element and low-order finite difference discretizations.

As in the continuous case, the problem $Au = f$ can be reduced to an equivalent system for the unknowns u_3 on the interface B. If u_3 is known, then u_1 and u_2 can be determined by using the first two block rows of (1.10):

$$u_1 = A_{11}^{-1}(f_1 - A_{13}u_3) \quad \text{and} \quad u_2 = A_{22}^{-1}(f_2 - A_{23}u_3).$$

Substituting for u_1 and u_2 in the third block row of (1.10), we obtain a reduced problem for the unknowns u_3:

$$Su_3 = \tilde{f}_3, \tag{1.11}$$

where $S \equiv \left(A_{33} - A_{13}^T A_{11}^{-1} A_{13} - A_{23}^T A_{22}^{-1} A_{23}\right)$ and $\tilde{f}_3 \equiv f_3 - A_{13}^T A_{11}^{-1} f_1 - A_{23}^T A_{22}^{-1} f_2$. The matrix S is referred to as the *Schur complement* of A_{33} in A, and the equation $Su_3 - \tilde{f}_3 = 0$ is a discrete approximation of the Steklov–Poincaré equation $Tu_B = 0$, enforcing the transmission boundary condition. The Schur complement S also plays a key role in the following block LU factorization of (1.10)

$$\begin{bmatrix} I & 0 & 0 \\ 0 & I & 0 \\ A_{13}^T A_{11}^{-1} & A_{23}^T A_{22}^{-1} & I \end{bmatrix} \begin{bmatrix} A_{11} & 0 & A_{13} \\ 0 & A_{22} & A_{23} \\ 0 & 0 & S \end{bmatrix} \begin{bmatrix} u_1 \\ u_2 \\ u_3 \end{bmatrix} = \begin{bmatrix} f_1 \\ f_2 \\ f_3 \end{bmatrix}, \tag{1.12}$$

from which (1.11) can also be derived.

Solving (1.11) by direct methods can be expensive since the Schur complement S is dense and, moreover, computing it requires as many solves of each A_{ii} system as there are nodes on B.

Therefore, it is common practice to solve the Schur complement system *iteratively* via preconditioned conjugate gradient methods. Each matrix–vector multiplication with S involves two subdomain solvers (A_{12}^{-1} and A_{22}^{-1}) which can be performed in parallel. It can be shown that the condition number of S is $\mathcal{O}(h^{-1})$ (which is better than that of A but can still be large) and therefore a good preconditioner is needed. Note that an advantage of the nonoverlapping approach over the overlapping approach is that the iterates are shorter vectors.

1.3. Main features of domain decomposition algorithms

The two preceding algorithms extend naturally to the case of many subdomains. However, a straightforward extension will not be *scalable*, i.e. the convergence rate will deteriorate as the number of subdomains increase. This is necessarily so because in the above algorithms, the only mechanism for sharing information is local, i.e. either through the interface or the overlapping regions. However, for elliptic problems the domain of dependence is global (i.e. the Green function is nonzero throughout the domain) and some way of transmitting *global* information is needed to make the algorithms scalable. One of the most commonly used mechanisms is to use *coarse spaces*, e.g. solving an appropriate problem on a coarser grid. This will be described in detail later.

In this sense, many of the domain decomposition algorithms can be viewed as a two-scale procedure, i.e. there is a fine grid with size h on which the solution is sought and on which the subdomain problems are solved, as well as a coarse grid with mesh size H which provides the global coupling between distant subdomains. The goal is to design the appropriate interaction of these two mechanisms so that the resulting algorithm has a convergence rate that is as insensitive to h and H as possible. In fact, in the literature on domain decomposition, a method is called *optimal* if its convergence rate is *independent* of h and H.

In practice, however, an *optimal* preconditioner does not necessarily provide the least execution time or minimal computational complexity. To achieve a computationally efficient algorithm requires paying attention to other factors, in addition to h and H. First of all, even though the number of iterations required by an optimal method can be bounded independent of h and H, one still has to ensure that it is not large. Second, each iteration step must not cost too much to implement. In addition, it would be desirable for the convergence rate to be insensitive to the variations in the coefficients of the elliptic problem, as well as the aspect ratios of the subdomains. We shall touch on some of these issues later.

We summarize here the key features of domain decomposition algorithms that we have introduced in this section, and which we shall study in some detail in the rest of this article:

1 domain decomposition as preconditioners with conjugate gradient acceleration;
2 overlapping versus nonoverlapping subdomain algorithms;
3 nonoverlapping algorithms involve solving a Schur complement system, using interface preconditioners;
4 additive versus multiplicative algorithms;
5 optimal preconditioners require solving a coarse problem;
6 the goal of achieving a convergence rate and efficiency independent of h, H, coefficients and geometry.

Notation We use the notation $\mathrm{cond}\,(M^{-1}A)$ to denote the condition number of the preconditioned system $M^{-1/2}AM^{-1/2}$, where M is symmetric and positive definite. We call a preconditioner M *spectrally equivalent* to A if $\mathrm{cond}\,(M^{-1}A)$ is bounded independently of the mesh sizes h and H, whichever is appropriate.

2. Overlapping subdomain algorithms

We now describe Schwarz algorithms based on *many overlapping subregions* to solve (1.1). We first discuss a commonly used technique for constructing an overlapping decomposition of Ω into p subregions $\hat{\Omega}_1, \ldots, \hat{\Omega}_p$. To this end, let $\Omega_1, \ldots, \Omega_p$ denote a nonoverlapping partition of Ω. For instance, each subregion Ω_i may be chosen as *elements* from a coarse finite element triangulation τ^H of Ω of mesh size H. Next, we extend each nonoverlapping region Ω_i to $\hat{\Omega}_i$, consisting of all points in Ω within a distance of βH from Ω_i where β ranges from 0 to $\mathcal{O}(1)$. See Figure 2 for an illustration of a two-dimensional rectangular region Ω partitioned into sixteen overlapping subregions.

Once the extended subdomains $\hat{\Omega}_i$ are defined, we define *restriction* maps R_i, *extension* maps R_i^T, and local matrices A_i corresponding to each subregion $\hat{\Omega}_i$ as follows. Let A be $n \times n$ and let $\hat{n}_i$ be the number of interior nodes in $\hat{\Omega}_i$. For each $i = 1, \ldots, p$, let $\hat{I}_i$ denote the indices of the nodes lying in the interior of $\hat{\Omega}_i$. Thus $\{\hat{I}_1, \ldots, \hat{I}_p\}$ form an overlapping collection of index sets. For each region $\hat{\Omega}_i$ let R_i denote the $n \times \hat{n}_i$ restriction matrix (whose entries consist of 1's and 0's) that restricts a vector x of length n to $R_i x$ of length $\hat{n}_i$, by choosing the subvector having indices in $\hat{I}_i$ (corresponding to the interior nodes in $\hat{\Omega}_i$). The transpose R_i^T of R_i is referred to as an *extension* or *interpolation* matrix, and it extends subvectors of length $\hat{n}_i$ on

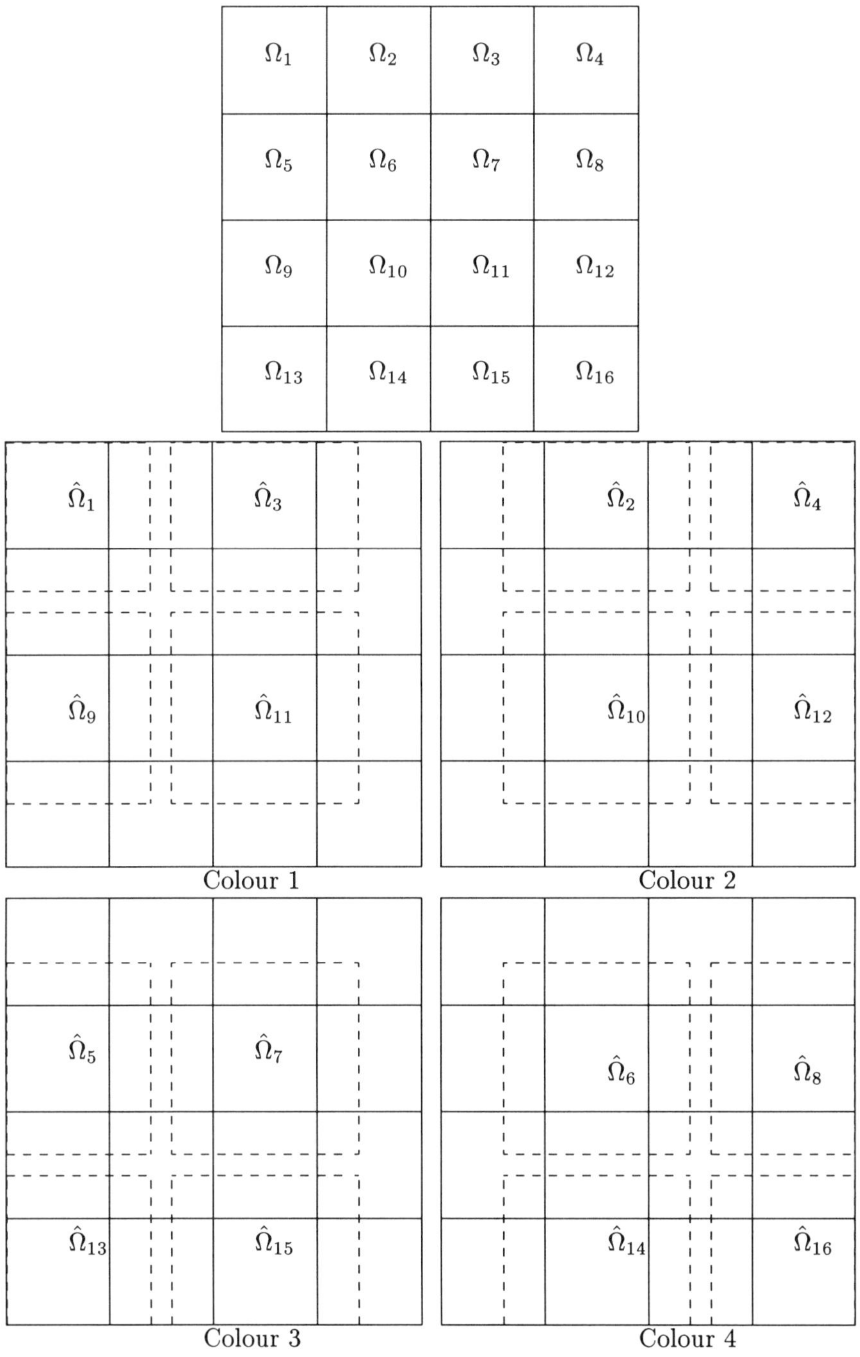

Fig. 2. Nonoverlapping subdomains Ω_i, overlapping subdomains $\hat{\Omega}_i$, 4 colours.

$\hat{\Omega}_i$ to vectors of length n using extension by zero to the rest of Ω. Finally, we let $A_i = R_i A R_i^T$, which is the local stiffness matrix corresponding to the subdomain $\hat{\Omega}_i$. Since R_i and R_i^T have entries of 1's and 0's, each A_i is a principal submatrix of A.

2.1. Additive Schwarz algorithms

The most straightforward generalization of the two subdomain additive Schwarz preconditioners described in Section 1 to the many subdomain case is the following:

$$M_{\mathrm{as},1}^{-1} = \sum_{i=1}^{p} R_i^T A_i^{-1} R_i.$$

Since the action of each term $R_i^T A_i^{-1} R_i z$ can be computed on separate processors, this immediately leads to coarse grain parallelism. The actions of R_i^T and R_i are *scatter–gather* operations, respectively, and it is not necessary to store the extension and restriction matrices.

The preconditioner $M_{\mathrm{as},1}$ is a straightforward generalization of the standard *block Jacobi* preconditioner to include overlapping blocks. However, the algorithm is not scalable because the convergence rate of this preconditioned iteration deteriorates as the number of subdomains p increases (i.e. as H decreases).

Theorem 1 There exists a positive constant C independent of H and h (but possibly dependent on the coefficients a) such that:

$$\mathrm{cond}\,(M_{\mathrm{as},1}^{-1} A) \leq C H^{-2} \left(1 + \beta^{-2}\right).$$

Proof. See Dryja and Widlund (1992a; 1989b). □

This deterioration in the convergence rate can be removed at a small cost by introducing a mechanism for global communication of information. There are several possible techniques for this, and here we will describe the most commonly used mechanism which is suitable only when the fine grid τ^h is a refinement of the coarse mesh τ^H. Accordingly, let R_H^T denote the standard interpolation map of coarse grid functions to fine grid functions (as in two-level *multigrid* methods). In the finite element context, R_H^T simply interpolates the nodal values from the coarse grid vertices to all the vertices on the fine grid, say by piecewise linear interpolation. Its transpose R_H is thus a weighted restriction map. If there are n_c coarse grid interior vertices, then R_H^T will be an $n \times n_c$ matrix. Indeed, if $\psi_1, \ldots, \psi_{n_c}$ are n_c column vectors representing the coarse grid nodal basis functions on the fine grid, then

$$R_H^T = [\psi_1, \ldots, \psi_{n_c}]\,.$$

Corresponding to the coarse grid triangulation τ^H, let A_H denote the coarse grid discretization of the elliptic problem, i.e. $A_H = R_H A R_H^T$. Then, the improved *additive Schwarz* preconditioner $M_{\text{as},2}$ is defined by

$$M_{\text{as},2}^{-1} = R_H^T A_H^{-1} R_H + \sum_{i=1}^{p} R_i^T A_i^{-1} R_i = \sum_{i=0}^{p} R_i^T A_i^{-1} R_i, \tag{2.1}$$

where we have let $R_0 = R_H$ and $A_0 = A_H$. The convergence rate using this preconditioner is independent of H (for sufficient overlap).

Theorem 2 There exists a positive constant C independent of H, h (but possibly dependent on the variation in the coefficients a) such that

$$\text{cond}\,(M_{\text{as},2}^{-1} A) \le C\left(1 + \beta^{-1}\right).$$

Proof. See Dryja and Widlund (1992a; 1989b), Dryja, Smith and Widlund (1993) and Theorems 14 and 16 in Section 4. □

2.2. Multiplicative Schwarz algorithms

The *multiplicative Schwarz* algorithm for many overlapping subregions can be analogously defined. Starting with an iterate u^k, we compute u^{k+1} as follows

$$u^{k+(i+1)/(p+1)} = u^{k+i/(p+1)} + R_i^T A_i^{-1} R_i (f - A u^{k+i/(p+1)}), \quad i = 0, 1, \ldots, p.$$

Theorem 3 The error $\|u - u^k\|$ in the kth iterate of the above multiplicative Schwarz algorithm satisfies

$$\|u - u^k\| \le \rho^k \|u - u^0\|,$$

where $\rho < 1$ is independent of h and H, and depends only on β and the coefficients a, and $\|\cdot\|$ is the A-norm.

Proof. See Bramble, Pasciak, Wang and Xu (1991) and Theorems 15 and 16. □

As for the additive Schwarz algorithm, if the coarse grid correction is dropped, then the convergence rate of the multiplicative algorithm will deteriorate as $\mathcal{O}(H^{-2})$ when $H \to 0$.

The multiplicative algorithm as stated above has less parallelism than the additive version. However, this can be improved through the technique of multicolouring, as follows. Each subdomain is identified with a colour such that subdomains of the same colour are disjoint. The multiplicative Schwarz algorithm then iterates sequentially through the different colours, but now all the subdomain systems of the same colour can be solved in parallel. Typically, only a small number of colours is needed, see Figure 2 for an example. We caution that the convergence rate of the multicoloured algorithm can depend on the ordering of the subdomains in the iteration and the increased parallelism may result in slower convergence (well known for the classical pointwise Gauss–Seidel method). However, this effect is less noticeable when a coarse grid solve is used.

The convergence bounds we have stated for both the additive and multiplicative Schwarz algorithms are valid in both two and three dimensions, but with possible *dependence* on the variation in the *coefficients* a. For large jumps in the coefficients, the convergence rate can deteriorate, but with maximum possible deterioration stated below.

Theorem 4 Assume that the coefficients a are constant (or mildly varying) within each coarse grid element. Then, for the additive Schwarz algorithm in two dimensions,

$$\mathrm{cond}\,(M_{\mathrm{as},2}^{-1}A) \leq C\,(1+\log(H/h))\,,$$

and in three dimensions,

$$\mathrm{cond}\,(M_{\mathrm{as},2}^{-1}A) \leq C\,(H/h)\,,$$

where C is independent of the jumps in the coefficients and the mesh parameters H and h, but dependent on the overlap parameter β.

Proof. See Dryja and Widlund (1987) and Dryja *et al.* (1993). □

Corresponding results exist for the multiplicative Schwarz algorithms and the deterioration in the convergence rate can be improved by the use of alternative *coarse spaces*, see preceding reference.

For a numerical study of Schwarz methods, see Gropp and Smith (1992).

3. Nonoverlapping subdomain algorithms

As we saw in Section 2, there are two kinds of coupling mechanisms present in an optimal Schwarz type algorithm based on many overlapping subregions: local coupling between adjacent subdomains provided by the overlapped regions, and global coupling between distant subdomains provided by the coarse grid problem. In the case of nonoverlapping approach, the Schur complement system represents the coupling between the nodes on the interface B and in order to obtain optimal convergence rates, a coarse grid solve is still needed. However, since there is no overlap between neighbouring subdomains, the local coupling must be provided by some other mechanism. The most often used method is to use *interface preconditioners*, i.e. an effective approximation to the part of the Schur complement matrix S that corresponds to the unknowns on the interface separating two neighbouring subdomains. (In two dimensions, the interface is an edge and in three dimensions it is a face.) We shall first describe such interface preconditioners in Section 3.1 in the context of two subdomain decomposition (where it is the only preconditioner needed). The case of many subregions is discussed in Section 3.2.

3.1. Two nonoverlapping subdomains: interface preconditioners

Consider the same setting as in Section 1, with Ω partitioned into two subdomains Ω_1 and Ω_2 separated by an interface B. We need a preconditioner M for the Schur complement $S \equiv A_{33} - A_{13}^T A_{11}^{-1} A_{13} - A_{23}^T A_{22}^{-1} A_{23}$.

(1) Exact eigen-decomposition of S: In some special cases, an exact eigen-decomposition of S can be derived from which the action of S^{-1} can be computed efficiently. For example, consider the five-point discretization of $-\Delta$ on a uniform grid of size h on the rectangular domain $\Omega = [0,1] \times [0, l_1 + l_2]$, which is partitioned into two subdomains $\Omega_1 = [0,1] \times [0, l_1]$ and $\Omega_2 = [0,1] \times [l_1, l_1 + l_2]$ with interface $B = \{(x,y) : y = l_1,\ 0 < x < 1\}$. We assume that the grid is $n \times (m_1 + 1 + m_2)$ with $l_i = (m_i + 1)h$, for $i = 1,2$ and $h = 1/(n+1)$. It was shown by Bjørstad and Widlund (1986) and Chan (1987) that

$$S = F\Lambda F,$$

where F is the orthogonal sine transform matrix:

$$(F)_{ij} = \sqrt{\frac{2}{n+1}} \sin\left(\frac{ij\pi}{n+1}\right),$$

Λ is a diagonal matrix with elements given by

$$(\Lambda)_i = \left(\frac{1+\gamma_i^{m_1+1}}{1-\gamma_i^{m_1+1}} + \frac{1+\gamma_i^{m_2+1}}{1-\gamma_i^{m_2+1}}\right) \sqrt{\sigma_i + \sigma_i^2/4},$$

where

$$\sigma_i = 4\sin^2\left(\frac{i\pi}{2(n+1)}\right) \quad \text{and} \quad \gamma_i = (1 + \sigma_i/2 - \sqrt{\sigma_i + \sigma_i^2/4})^2.$$

If m_1, m_2 are large enough, then two good approximations to S are:

$$M_{\rm GM} = F(\Sigma + \Sigma^2/4)^{1/2} F, \quad \text{and} \quad M_{\rm D} = F\Sigma^{1/2}F,$$

where $\Sigma = \text{diag}(\sigma_i)$. $M_{\rm D}$ was first used by Dryja (1982) in a more general setting. The improved preconditioner $M_{\rm GM}$ was later proposed by Golub and Mayers (1984).

Note that all the above preconditioners can be solved in $\mathcal{O}(n\log(n))$ operations using the Fast Sine Transform and it is easy to show that they are *spectrally equivalent* to S. In theory, this is true for *any* second-order elliptic operator. However, these preconditioners can be sensitive to the *aspect ratios* l_1 and l_2 and the coefficients (in the case of variable coefficients) on the subdomains. To apply this class of preconditioners to domains more general than a rectangle, and to provide some adaptivity to aspect ratios, Chan and Resasco (1985; 1987) suggested using the exact eigen-decomposition of a rectangle which approximates the given domain and shares the same interface. Exact eigen-decompositions have also been derived by Resasco (1990) for three-dimensional problems and unequal mesh sizes in each subdomain, and by Chan and Hou (1991) for five point stencils approximating general second-order constant coefficient elliptic problems (which provides some adaptivity to the coefficients).

(2) The Neumann–Dirichlet preconditioner (See Bjørstad and Widlund (1984), Bjørstad and Widlund (1986), Bramble *et al.* (1986b), Marini and Quarteroni (1989).) To describe this method, it is convenient to first write S in a form which reflects the contributions from Ω_1 and Ω_2 more explicitly. In either finite difference or finite element methods, the term A_{33} can be written as

$$A_{33} = A_{33}^{(1)} + A_{33}^{(2)},$$

where $A_{33}^{(i)}$ corresponds to the contribution to A_{33} from subdomain Ω_i (assuming the coefficients are zero on the adjacent subdomain). For instance, in the case of finite elements, $A_{33}^{(i)}$ is obtained by integrating the weak form on Ω_i. We can now write

$$S = S^{(1)} + S^{(2)},$$

where

$$S^{(i)} = A_{33}^{(i)} - A_{i3}^T A_{ii}^{-1} A_{i3}, \quad i = 1, 2.$$

Due to symmetry, $S^{(1)} = S^{(2)} = \frac{1}{2}S$ if the two subdomain problems are symmetric about the interface. This motivates the use of either $S^{(1)}$ or $S^{(2)}$ as a preconditioner for S even if the two subdomains are not equal. For example, a right-preconditioned system using $M_{\mathrm{ND}} = S^{(1)}$ has the form $(S^{(1)} + S^{(2)}){S^{(1)}}^{-1} = I + S^{(2)}{S^{(1)}}^{-1}$. It can be shown that the action of ${S^{(1)}}^{-1}$ on a vector v can be obtained by solving a problem on Ω_1 with v as Neumann boundary condition on the interface and extracting the solution values (Dirichlet values) on the interface:

$${S^{(1)}}^{-1} v = \begin{bmatrix} 0 & I \end{bmatrix} \begin{bmatrix} A_{11} & A_{13} \\ A_{13}^T & A_{33}^{(1)} \end{bmatrix}^{-1} \begin{bmatrix} 0 \\ v \end{bmatrix}.$$

It is proved in Bjørstad and Widlund (1986) that this preconditioner is spectrally equivalent to S.

(3) The Neumann–Neumann preconditioner One may notice a lack of symmetry in the Neumann–Dirichlet preconditioner in the choice of which subdomain to solve the Neumann problem on. The Neumann–Neumann preconditioner, first proposed by Bourgat, Glowinski, Le Tallec and Vidrascu (1989), is completely symmetric with respect to the two subdomains. Here the inverse of the preconditioner is given by

$$M_{\mathrm{NN}}^{-1} = \tfrac{1}{4} S^{(1)^{-1}} + \tfrac{1}{4} S^{(2)^{-1}}.$$

Obviously, the action of M_{NN}^{-1} requires solving a Neumann problem on each of the two subdomains. In addition to the added symmetry, this preconditioner is also more directly generalizable to the case of many subdomains and to three dimensions (see Sections 3.6 and 3.9).

(4) Probing preconditioner This purely algebraic technique, first proposed by Chan and Resasco (1985) and later refined in Keyes and Gropp (1987) and Chan and Mathew (1992), is motivated by the observation that the entries of the rows (and columns) of the matrix S often decay rapidly away from the main diagonal. This decay is faster than the decay of the Green function of the original elliptic operator. The idea in the probing preconditioner is to efficiently compute a banded approximation to S. Note that this would be easy if S was known explicitly because we could then simply take the central diagonals of S. However, recall that we want to avoid computing S explicitly. The technique used in probing is to find such an approximation by *probing* the action of S on a few carefully selected vectors. For example, *if* S were tridiagonal, then it can be exactly recovered by its action on the three vectors:

$$\begin{aligned} v_1 &= (1,0,0,1,0,0,\ldots)^T, \\ v_2 &= (0,1,0,0,1,0,\ldots)^T, \\ v_3 &= (0,0,1,0,0,1,\ldots)^T \end{aligned}$$

through a simple recursion. Since S is not exactly tridiagonal, the tridiagonal matrix M_{P} obtained by probing will not be equal to S, but it is often a very good preconditioner. Keyes and Gropp (1987) showed that if S were symmetric, then two probing vectors suffice to compute a symmetric tridiagonal approximation. For more details, see Chan and Mathew (1992), where it is proved that the conditioner number of $M_{\mathrm{P}}^{-1}S$ can be bounded by $\mathcal{O}(h^{-1/2})$ (hence M_{P} is not spectrally equivalent to S) but it adapts very well to the aspect ratios and the coefficient variations of the subdomains. It would seem ideal to combine the advantages of the probing technique with a spectrally equivalent technique but this has proved to be elusive.

(5) Multilevel preconditioners These techniques make use of the multilevel elliptic preconditioners to be discussed in Section 6 and adapt them to obtain preconditioners for the Schur complement interface system. We will not describe these methods in detail, but the main idea is simple to understand. If a change of basis from the standard nodal basis to a hierarchical nodal basis is used (assuming that the grid has a hierarchical structure), then a diagonal scaling often provides an effective preconditioner in the new basis. It can be shown rather easily that the Schur complement of the matrix A in the hierarchical basis is the same as that obtained by representing S with respect to the hierarchical basis on the interface B (i.e. by a multilevel change of basis restricted to the interface). Thus a good multilevel preconditioner for A automatically leads to a good multilevel preconditioner for S. The reader is referred to Smith and Widlund (1990) for using the hierarchical basis method of Yserentant (1986) and Tong, Chan and Kuo (1991) (see also Xu (1989)) for the multilevel nodal basis method of Bramble, Pasciak and Xu (1990). The resulting methods have optimal or almost optimal convergence rates.

3.2. Many nonoverlapping subdomains

Many of the preconditioners described in Section 3.1 for two nonoverlapping subdomains can be extended to the case of many nonoverlapping subregions. However, in the case of many subregions, these preconditioners need to be modified to take account of the more complex geometry of the interface, and to provide global coupling amongst the many subregions.

Let Ω be partitioned into p nonoverlapping regions of size $\mathcal{O}(H)$ with interface B separating them, see Figure 3:

$$\Omega = \Omega_1 \cup \cdots \cup \Omega_p \cup B, \quad \text{where } \Omega_i \cap \Omega_j = \emptyset \text{ for } i \neq j,$$

the interface B is given by: $B = \{\cup_{i=1}^{p} \partial\Omega_i\} \cap \Omega$. For $i = 1, \ldots, p$, let I_i denote the indices corresponding to the nodes in the interior of subdomain Ω_i, and let $I = \cup_{i=1}^{p} I_i$ denote the indices all nodes lying in the interior of subdomains. To minimize notation, we will use B to denote not only the interface, but also the indices of the nodes lying on B. Then, corresponding to the permuted indices $\{I, B\}$, the vector u can be partitioned as $u = [u_I, u_B]^T$, and $f = [f_I, f_B]^T$, and equation (1.2) can be written in block form as follows

$$\begin{bmatrix} A_{II} & A_{IB} \\ A_{IB}^T & A_{BB} \end{bmatrix} \begin{bmatrix} u_I \\ u_B \end{bmatrix} = \begin{bmatrix} f_I \\ f_B \end{bmatrix}. \tag{3.1}$$

For five-point stencils in two dimensions and seven-point stencils in three dimensions, A_{II} will be block diagonal, since the interior nodes in each subdomain will be decoupled from the interior nodes in other subdomains:

$$A_{II} = \text{blockdiag}\,(A_{ii}) = \begin{bmatrix} A_{11} & & 0 \\ & \ddots & \\ 0 & & A_{pp} \end{bmatrix}. \tag{3.2}$$

As in Section 1, the unknowns u_I can be eliminated resulting in a reduced system for u_B (the unknowns on B). We use the following block LU

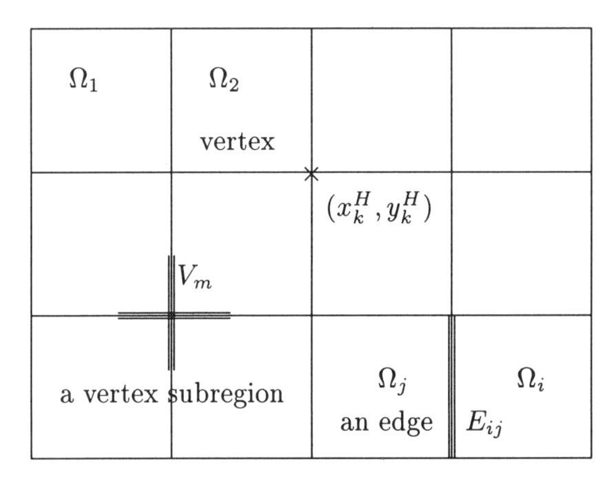

Fig. 3. A partition of Ω into 12 subdomains.

factorization of A:

$$A \equiv \begin{bmatrix} A_{II} & A_{IB} \\ A_{IB}^T & A_{BB} \end{bmatrix} = \begin{bmatrix} I & 0 \\ A_{IB}^T A_{II}^{-1} & I \end{bmatrix} \begin{bmatrix} A_{II} & 0 \\ 0 & S \end{bmatrix} \begin{bmatrix} I & A_{II}^{-1} A_{IB} \\ 0 & I \end{bmatrix}, \tag{3.3}$$

where the Schur complement matrix S is defined by

$$S = A_{BB} - A_{IB}^T A_{II}^{-1} A_{IB}.$$

Consequently, solving $Au = f$ based on the LU factorization above requires computing the action of A_{II}^{-1} twice, and S^{-1} once.

By eliminating u_I, we obtain

$$S u_B = \tilde{f}_S, \tag{3.4}$$

where $\tilde{f}_B \equiv f_B - A_{IB} A_{II}^{-1} f_I$. The Schur complement S in the case of many subdomains has similar properties to the two subdomain case. Here we only note that the condition number of S is approximately $\mathcal{O}(H^{-1}h^{-1})$ in the case of many subdomains, an improvement over the $\mathcal{O}(h^{-2})$ growth for A. The rest of this section will be devoted to the description of various preconditioners M for S in two and three dimensions.

3.3. Two-dimensional case: block Jacobi preconditioner M_1 For S

Here, we describe a block diagonal preconditioner M_1 which reduces the condition number of S from $\mathcal{O}(H^{-1}h^{-1})$ to $\mathcal{O}\left(H^{-2}\log^2(H/h)\right)$ (without involving global communication of information). A variant of this preconditioner was proposed by Bramble, Pasciak and Schatz (1986a), see also Widlund (1988), Dryja *et al.* (1993).

The preconditioner M_1 will correspond to an additive Schwarz preconditioner for S corresponding to a partition of the interface B into subregions. The interface B is partitioned as a union of *edges* E_i for $i = 1, \ldots, m$, and *vertices* V of the subdomains, see Figure 3:

$$B = \{E_1 \cup \cdots \cup E_m\} \cup V,$$

where the edges $E_i = \partial\Omega_j \cap \partial\Omega_l$ form the common boundary of two subdomains (excluding the endpoints). With duplicity of notation, we also denote by E_i the indices of the nodes lying on edge E_i, and use V to denote the indices of the vertices V. Corresponding to this ordering of indices, we partition $u_B = [u_{E_1}, \ldots, u_{E_m}, u_V]$, and obtain a block partition of S:

$$S = \begin{bmatrix} S_{E_1E_1} & S_{E_1E_2} & \cdots & S_{E_1E_m} & S_{E_1V} \\ S^T_{E_1E_2} & S_{E_2E_2} & \cdots & S_{E_2E_m} & S_{E_2V} \\ \vdots & \vdots & \ddots & \vdots & \vdots \\ S^T_{E_1\;E_m} & S^T_{E_2E_m} & \cdots & S_{E_mE_m} & S_{E_mV} \\ S^T_{E_1V} & S^T_{E_2V} & \cdots & S^T_{E_mV} & S_{VV} \end{bmatrix}.$$

Note that $S_{E_iE_j} = 0$ if E_i and E_j are not part of the same subdomain.

A block diagonal (Jacobi) preconditioner for S is:

$$M_1 = \begin{bmatrix} S_{E_1E_1} & 0 & \cdots & \cdots & 0 \\ 0 & S_{E_2E_2} & \ddots & & 0 \\ \vdots & \ddots & \ddots & \ddots & \vdots \\ \vdots & & \ddots & S_{E_mE_m} & 0 \\ 0 & \cdots & \cdots & 0 & S_{VV} \end{bmatrix}.$$

The preconditioner M_1 can also be described in terms of restriction and extension maps. For each edge E_i, let R_{E_i} denote the pointwise restriction map from B onto the nodes on E_i, and let $R^T_{E_i}$ denote the corresponding extension map. Similarly, let R_V denote the pointwise restriction map onto the vertices V, and let R^T_V denote extension by zero of nodal values on V to B. Then the block Jacobi preconditioner is defined by

$$M_1^{-1} \equiv \sum_{i=1}^{m} R^T_{E_i} S^{-1}_{E_iE_i} R_{E_i} + R^T_V S^{-1}_{VV} R_V.$$

Since this preconditioner does not involve global coupling between subdomains, its convergence rate deteriorates as $H \to 0$.

Theorem 5 There exists a constant C independent of H and h (but may depend on the coefficient a), such that

$$\mathrm{cond}\,(M_1^{-1}S) \le CH^{-2}\left(1 + \log^2(H/h)\right).$$

Proof. See Bramble *et al.* (1986a), Widlund (1988), Dryja *et al.* (1993). □

Since the $S_{E_iE_i}$s are not explicitly constructed, computing the action of $S^{-1}_{E_iE_i}$ poses a problem (similarly for S_{VV}). Fortunately, each $S_{E_iE_i}$ and S_{VV} can be replaced by efficient approximations. For example, the block entries $S_{E_iE_i}$ can be replaced by any suitable two subdomain interface preconditioner $M_{E_iE_i}$ discussed in Section 3.1, for instance:

$$M_{E_iE_i} \equiv \alpha_{E_i} F \Lambda^{1/2} F,$$

where α_{E_i} represents the average of the coefficient a in the two subdomains adjacent to E_i. Alternatively, the action of $S^{-1}_{E_iE_i}$ can be computed exactly, using

$$S^{-1}_{E_iE_i} z_{E_i} = \begin{bmatrix} 0 & 0 & I \end{bmatrix} A^{-1}_{\Omega_j \cup \Omega_k \cup E_i} \begin{bmatrix} 0 & 0 & z_{E_i} \end{bmatrix}^T, \tag{3.5}$$

where $E_i = \partial\Omega_j \cap \partial\Omega_k$, and $A_{\Omega_j\cup\Omega_k\cup E_i}$ is the 3×3 block partitioned stiffness matrix corresponding to the region $\Omega_j \cup \Omega_k \cup E_i$. Note that this involves solving a problem on $\Omega_j \cup \Omega_k \cup E_i$. The matrix S_{VV} may be approximated by the diagonal matrix A_{VV} (the principal submatrix of A corresponding to nodes on V).

3.4. Two-dimensional case: the Bramble–Pasciak–Schatz (BPS) preconditioner M_2 for S

The H^{-2} factor in the condition number of the block Jacobi preconditioner M_1 can be removed by incorporating some mechanism for global coupling, such as through a coarse grid problem based on the coarse triangulation $\{\Omega_i\}$. Accordingly, let R_H^T denote an interpolation map (say piecewise linear interpolation) from the nodal values on V (vertices of subdomains) onto all the nodes on B. Then, R_H can be viewed as the weighted restriction map from B onto V. Note that the range of R_H^T here is B instead of the whole domain.

A variant M_2 of the preconditioner proposed by Bramble *et al.* (1986a) is a simple modification of M_1:

$$M_2^{-1} = \sum_{i=1}^{m} R^T_{E_i} S^{-1}_{E_iE_i} R_{E_i} + R_H^T A_H^{-1} R_H, \tag{3.6}$$

where A_H is the coarse grid discretization as in Section 2. With the global communication of information, the rate of convergence of the algorithm becomes logarithmic in H/h.

Theorem 6 There exists a constant C independent of H, h such that

$$\text{cond}\,(M_2^{-1}S) \le C\left(1 + \log^2(H/h)\right).$$

In case the coefficients a are constant in each subdomain Ω_i, then C is also independent of a.

Proof. See Bramble *et al.* (1986a), Widlund (1988) and Dryja *et al.* (1993). □

As for the preconditioner M_1 to efficiently implement this algorithm, it is necessary to replace the subblocks $S_{E_iE_i}$ by suitable preconditioners, such as those described for the two subdomain case in Section 3.1, see also Chan, Mathew and Shao (1992b).

3.5. Two-dimensional case: vertex space preconditioner M_3 for S

The logarithmic growth $(1+\log(H/h))^2$ in the condition number of the preceding preconditioner M_2 can be eliminated at additional cost, by modifying the BPS algorithm to result in the *vertex space preconditioner* proposed by Smith (1990, 1992).

The basic idea is to include additional overlap between the subblocks used in the BPS preconditioner M_2. Recall that the Schur complement S is not block diagonal in the permutation $[E_1, \ldots, E_m, V]$, since adjacent edges are coupled, with $S_{E_iE_j} \neq 0$ whenever edges E_i and E_j are part of the boundary of the same subdomain Ω_i. This coupling was ignored in the preceding two preconditioners, and resulted in the logarithmic growth factor in the condition number. By introducing overlapping subblocks, one can provide sufficient approximation of this coupling, resulting in *optimal* convergence bounds.

Overlap in the decomposition of interface

$$B = \{E_1 \cup \cdots \cup E_m\} \cup V,$$

can be obtained by introducing *vertex regions* $\{VS_1, \ldots, VS_q\}$ centred about each vertex in V (assume there are q subdomain vertices):

$$B \subset \{E_1 \cup \cdots \cup E_m\} \cup V \cup \{VS_1 \cup \cdots VS_q\}.$$

The vertex regions VS_k are illustrated in Figure 3, and are defined as the cross shaped regions centred at each subdomain vertex (x_k^H, y_k^H) containing segments of length βH of all the edges E_i that emanate from it. Such vertex spaces were used earlier by Nepomnyaschikh (1984; 1986).

Corresponding to this overlapping cover of B, we denote the indices of the nodes that lie on E_i by E_i, the indices of the vertices by V, and the indices of the vertex region VS_i by VS_i. Thus

$$E_1 \cup \cdots \cup E_m \cup V \cup VS_1 \cdots \cup VS_q$$

form an overlapping collection of indices of all unknowns on B. As with the restriction and extension maps for the BPS, we let R_{VS_i} denote the restriction of full vectors to subvectors corresponding to the indices in VS_i. Its transpose $R_{VS_i}^T$ denotes the extension by zero of subvectors with indices VS_i to full vectors. The principal submatrix of S corresponding to the indices VS_i will be denoted $S_{VS_i} = R_{VS_i} S R_{VS_i}^T$. The *vertex space* preconditioner M_3 is an additive Schwarz preconditioner defined on this overlapping partition:

$$M_3^{-1} = \sum_{i=1}^{m} R_{E_i}^T S_{E_i E_i}^{-1} R_{E_i} + R_H^T A_H^{-1} R_H + \sum_{i=1}^{q} R_{VS_i}^T S_{VS_i}^{-1} R_{VS_i}. \qquad (3.7)$$

In general, the matrices S_{VS_i} are dense and expensive to compute. However, sparse approximations can be computed efficiently using the probing technique or modifications of Dryja's interface preconditioner by Chan *et al.* (1992b). Alternately, using the following approximation:

$$S_{VS_i}^{-1} z_{VS_i} \approx \left[\begin{array}{cc} 0 & I \end{array} \right] \left[\begin{array}{cc} A_{\Omega_{VS_i}} & A_{\Omega_{VS_i}, VS_i} \\ A_{\Omega_{VS_i}, VS_i}^T & A_{VS_i, VS_i} \end{array} \right]^{-1} \left[\begin{array}{c} 0 \\ z_{VS_i} \end{array} \right],$$

the action of $S_{VS_i}^{-1}$ can be approximated by solving a Dirichlet problem on a domain Ω_{VS_i} of diameter $2\beta H$ which contains VS_i and which is partitioned into a small number (four for rectangular regions) subregions by the interface VS_i.

The convergence rate of the vertex space preconditioned system is optimal in H and h (but may depend on variations in the coefficients).

Theorem 7 There exists a constant C_0 independent of H, h and β such that

$$\operatorname{cond}(M_3^{-1} S) \leq C_0 (1 + \beta^{-1}),$$

where C_0 may depend on the variations in a. There also exists a constant C_1 independent of H, h, and the jumps in a (provided a is constant on each subdomain Ω_i) but can depend on β such that

$$\operatorname{cond}(M_3^{-1} S) \leq C_1 (1 + \log(H/h)).$$

Proof. See Smith (1992), Dryja *et al.* (1993) and also Section 4. □

Thus, in the presence of large jumps in the coefficient a, the condition number bounds for the vertex space algorithm may deteriorate to $(1 + \log(H/h))$, which is the same growth as for the *BPS* preconditioner.

3.6. Two-dimensional case: Neumann–Neumann preconditioner M_4 for S

The Neumann–Neumann preconditioner for S in the case of many subdomains is a natural extension of the Neumann–Neumann algorithm for the case of two subregions, described in Section 3.1. This preconditioner was originally proposed by Bourgat *et al.* (1989), and extended by De Roeck (1989), De Roeck and Le Tallec (1991), Le Tallec, De Roeck and Vidrascu

(1991), Dryja and Widlund (1990; 1993a,b), Mandel (1992) and Mandel and Brezina (1992). There are several versions of the Neumann–Neumann algorithm, with the differences arising in the choice of a mechanism for global communication of information. We follow here a version due to Mandel and Brezina (1992), referred to as the *balancing domain decomposition* preconditioner.

Neumann–Neumann refers to the process of solving Neumann problems on each subdomain Ω_i during each preconditioning step. For each subdomain boundary $\partial\Omega_i$, let $R_{\partial\Omega_i}$ denote the pointwise restriction map (matrix) from nodes on B into nodes on $\partial\Omega_i \cap B$. Its transpose $R^T_{\partial\Omega_i}$ denotes an extension by zero of nodal values in $\partial\Omega_i \cap B$ to the rest of B. Corresponding to subdomain Ω_i, we denote the stiffness matrix of the Neumann problem by

$$A^{(i)} \equiv \begin{bmatrix} A^{(i)}_{II} & A^{(i)}_{IB} \\ A^{(i)^T}_{IB} & A^{(i)}_{BB} \end{bmatrix},$$

where $A^{(i)}_{II}$ is a principal submatrix of A corresponding to the nodes in the interior of Ω_i, $A^{(i)}_{IB}$ is a submatrix of A corresponding to the coupling between nodes in the interior of Ω_i and the nodes on the interface B restricted to $\partial\Omega_i$, and $A^{(i)}_{BB}$ corresponds to the coupling between the nodes on $\partial\Omega_i$ with contributions from Ω_i (in the finite element case, $A^{(i)}_{BB}$ is obtained by integrating the weak form on Ω_i for all the basis functions corresponding to the nodes on $\partial\Omega_i$).

For each subdomain Ω_i, we let $S^{(i)}$ denote the Schur complement with respect to the nodes on $\partial\Omega_i \cap B$ of the local stiffness matrix $A^{(i)}$:

$$S^{(i)} = A^{(i)}_{BB} - A^{(i)T}_{IB} A^{(i)^{-1}}_{II} A^{(i)}_{IB}. \tag{3.8}$$

The natural extension of the two subdomain Neumann–Neumann preconditioner is simply $\tilde{M}_4$:

$$\tilde{M}_4^{-1} = \sum_{i=1}^{p} R^T_{\partial\Omega_i} D_i \left(S^{(i)}\right)^{-1} D_i R_{\partial\Omega_i}, \tag{3.9}$$

where D_i is a diagonal weighting matrix. Note that $(S^{(i)})^{-1}v$ can be computed by a Neumann solve with v as Neumann data (see Section 3.1). This preconditioner is highly parallelizable, but it has two potential problems:

- The matrix $S^{(i)}$ is singular for interior subdomains since it corresponds to a Neumann problem on Ω_i. Accordingly, a compatibility condition must be satisfied, and additionally, the solution of the singular system will not be unique.
- There is no mechanism for global communication of information, and hence the condition number of the preconditioned system deteriorates at least as H^{-2}.

One way to rectify these two defects is the *balancing* procedure of Mandel and Brezina (1992). The residual is projected onto a subspace which automatically satisfies the compatibility conditions for each of the singular systems (as many as p constraints). Additionally, in a post processing step, a constant is added to the solution of each local singular system so that the residual remains in the appropriate subspace. This procedure also provides a mechanism for global communication of information. We omit the technical details, and refer the reader to Mandel and Brezina (1992). The singularity of the local Neumann problems also arises in a related method by Farhat and Roux (1992) where the interface compatibility conditions are enforced by a Lagrange multiplier approach.

The modified Neumann–Neumann preconditioner M_4 (with balancing) satisfies:

Theorem 8 There exists a constant C independent of H and h and the jumps in the coefficients a such that

$$\operatorname{cond}(M_4^{-1}S) \leq C\,(1+\log(H/h))^2 .$$

Proof. See De Roeck and Le Tallec (1991), Mandel and Brezina (1992), Dryja and Widlund (1993a). □

The Neumann–Neumann preconditioner has several attractive features:

- the subregions Ω_i need not be triangular or rectangular; they can have general shapes;
- no explicit computation of the entries of S;
- the rate of convergence is logarithmic in H/h and insensitive to large jumps in the coefficients a.

However, the Neumann–Neumann preconditioner requires twice as many subdomain solves per step as a multiplication with S.

3.7. Three-dimensional case: vertex space preconditioner M_1 for S

Constructing effective preconditioners for the Schur complement matrix S is more complicated in three dimensions. These difficulties arise in part from the increased dimension of the boundaries of three-dimensional regions, and is also, technically, from a weaker Sobolev inequality in three dimensions.

As in the two-dimensional case, we assume that Ω is partitioned into p nonoverlapping subregions with interface B:

$$\Omega = \Omega_1 \cup \cdots \cup \Omega_p \cup B, \quad \text{where } B = (\cup_{i=1}^{p} \partial\Omega_i) \cap \Omega.$$

For most of the three-dimensional algorithms we will describe, it will be assumed that the $\{\Omega_i\}$ consist of either tetrahedrons or cubes and form a coarse triangulation of Ω having mesh size H. The boundary $\partial\Omega_i$ of each tetrahedron or cube can be further partitioned into *faces, edges* and *vertices.* The *faces* F_{ij} = interior of $\partial\Omega_i \cap \partial\Omega_j$ are assumed to be open two-dimensional surfaces. The *edges* E_k are one-dimensional curves defined to be the intersection of the boundaries of two faces: $E_k = \partial F_{ij} \cap \partial F_{ln}$ excluding the endpoints. Finally, the *vertices* V are point sets which are the endpoints of edges.

As a prelude, we describe two preconditioners M_{1a} and M_{1b} related to the vertex space preconditioner M_1. Corresponding to the partition of B into faces, edges and subdomain vertices, we permute the unknowns on B as $x_B = [x_F, x_E, x_V]^T$, where F denote all the nodes on the faces, E corresponds to all the nodes on the edges E, while V denotes all the subdomain vertices. Thus, the matrix S has the following block form:

$$S = \begin{bmatrix} S_{FF} & S_{FE} & S_{FV} \\ S_{FE}^T & S_{EE} & S_{EV} \\ S_{FV}^T & S_{EV}^T & S_{VV} \end{bmatrix}.$$

The first preconditioner M_{1a} will be a block diagonal approximation of the above block partition of S, with the inclusion of a coarse grid model for global communication of information, see Dryja *et al.* (1993). Accordingly, for each of the subregions of B, let R_{F^i}, R_{E_k} and R_V denote the pointwise restriction map from B onto the nodes on face F_i, edge E_k and subdomain vertices V, respectively. Their transposes correspond to extensions by zero onto all other nodes on B. The principal submatrices of S corresponding to the nodes on F_i, E_k and V will be denoted by $S_{F_iF_i}$, $S_{E_kE_k}$ and S_{VV}, respectively. For the coarse grid problem, let R_H^T denote the interpolation map from the subdomain vertices V to all nodes on B. Then, its transpose R_H denotes a weighted restriction map onto the subdomain vertices V. The coarse grid matrix is then given by $A_H = R_H A R_H^T$.

In terms of the restriction and extension maps given above, M_{1a} is defined by

$$M_{1a}^{-1} = \sum_i R_{F_i}^T S_{F_iF_i}^{-1} R_{F_i} + \sum_k R_{E_k}^T S_{E_kE_k}^{-1} R_{E_k} + R_H^T A_H^{-1} R_H.$$

We note that the coupling terms $S_{F_iF_j}$ and $S_{E_iE_j}$ between adjacent faces and edges have been dropped. For finite element and finite difference discretizations, the blocks $S_{E_iE_i}$ can be shown to be well conditioned (indeed, for seven-point finite difference approximations on three-dimensional rectangular subdomains, $S_{E_iE_i} = A_{E_iE_i}$, since boundary data on the edges do not influence the solution in the interior of the region). Consequently, $S_{E_iE_i}$ may be effectively replaced by a suitably scaled multiple of the identity matrix $M_{E_iE_i}$:

$$S_{E_iE_i} \approx M_{E_iE_i} = h\sigma_{E_i} I_{E_i},$$

where σ_{E_i} represents the average of the coefficients a in the subdomains adjacent to edge E_i. The action of $S_{F_iF_i}^{-1}$ can be approximated by analogues of the two-dimensional interface preconditioners from Section 3.1 or by solving a Dirichlet problem using a principal submatrix of A corresponding to nodes on a region Ω_{F_i} partitioned by face F_i.

A related preconditioner M_{1b} can be obtained at a small additional cost. For this, we note that the principal submatrix S_{VV} of S (corresponding to the nodes on the subdomain vertices V) can be replaced by a suitably scaled diagonal matrix M_{VV}:

$$S_{VV} \approx M_{VV} \equiv h \,\mathrm{diag}\,(\sigma_{V^k}),$$

where σ_{V^k} is the average of the coefficients a in the subdomains adjacent to vertex V_k. The preconditioner M_{1b} is defined by

$$M_{1b}^{-1} = \sum_i R_{F_i}^T S_{F_iF_i}^{-1} R_{F_i} + \sum_k R_{E_k}^T S_{E_kE_k}^{-1} R_{E_k} + R_H^T A_H^{-1} R_H + R_V^T M_{VV}^{-1} R_V.$$

The following are condition number bounds for the two preconditioners given above.

Theorem 9 The preconditioner M_{1a} results in condition number of

$$\mathrm{cond}\left(M_{1a}^{-1} S\right) \le C_1 \frac{H}{h} \left(1 + \log(H/h)\right)^2,$$

where C_1 is independent of H, h and jumps in the coefficients a. The preconditioner M_{1b} results in improved condition number with respect to mesh parameters:

$$\mathrm{cond}\left(M_{1b}^{-1} S\right) \le C_2 \left(1 + \log(H/h)\right)^2,$$

where the coefficient C_2 may depend on the coefficients a.

Proof. See Dryja *et al.* (1993). □

We note that for smooth coefficients, M_{1b} is preferable to M_{1a} with improved condition number where the factor H/h has been eliminated.

The vertex space preconditioner of Smith (1992) in three dimensions corresponds to an *additive Schwarz* preconditioner for S, based on a suitable decomposition of the interface B into overlapping subregions and a coarse grid model. Accordingly, for each edge E_j, let $\hat{E}_j$ denote an extension consisting of all nodes on adjacent faces F_{ik} (but not adjacent edges or subdomain vertices) within a distance of βH from E^j. Similarly, corresponding to each subdomain vertex V_l, let $\hat{V}_l$ denote the vertex region consisting of all nodes in B within a distance of βH from vertex V_l. An overlapping partition of the interface B is then obtained:

$$B \subset (\cup_i F_i) \cup \left(\cup_k \hat{E}_k\right) \cup \left(\cup_l \hat{V}_l\right).$$

Corresponding to each overlapping subregion of the interface, define the pointwise restriction and extension maps as follows. Let $R_{\hat{E}_k}$, $R_{\hat{V}_l}$ and R_{F_i} denote the pointwise restriction map from B onto the nodes on $\hat{E}_k$, $\hat{V}_l$ and F_i, respectively. Their transposes correspond to an extension by zero onto the rest of the nodes on B. Accordingly, let $S_{F_iF_i}$, $S_{\hat{E}_k\hat{E}_k}$ and $S_{\hat{V}_l\hat{V}_l}$ denote the principal submatrices of S corresponding to the nodes on F_i, $\hat{E}_k$ and $\hat{V}_l$ respectively. As for the preconditioners M_{1a} and M_{1b}, R_H^T and R_H will denote the coarse grid interpolation map and weighted restriction map, respectively. The coarse grid discretization matrix is obtained by $A_H = R_H A R_H^T$.

The vertex space preconditioner M_1 is defined by

$$M_1^{-1} = \sum_i R_{F_i}^T S_{F_iF_i}^{-1} R_{F_i} + \sum_k R_{\hat{E}_k}^T S_{\hat{E}_k}^{-1} R_{\hat{E}_k} + \sum_l R_{\hat{V}_l}^T S_{\hat{V}_l\hat{V}_l}^{-1} R_{\hat{V}_l} + R_H^T A_H^{-1} R_H. \tag{3.10}$$

As in the two-dimensional case, the action of the inverses $S_{F_iF_i}^{-1}$, $S_{\hat{V}_l\hat{V}_l}^{-1}$ and $S_{\hat{E}_k\hat{E}_k}^{-1}$ can be approximated without explicit construction of S. These approximations can be obtained by solving linear systems with principal submatrices of A as coefficient matrices, corresponding to subregions Ω_{F_i}, $\Omega_{\hat{E}_k}$ and $\Omega_{\hat{V}^l}$ containing F_i, $\hat{E}_k$ and $\hat{V}^l$ respectively, see Dryja *et al.* (1993), or by extensions of techniques in Chan *et al.* (1992b).

The rate of convergence of the *vertex space* preconditioner is independent of H and h, provided β is uniformly bounded. However, it may depend on the variation in the coefficients a.

Theorem 10 There exists a constant C, independent of H and h, but depending on the coefficients a such that

$$\text{cond}\,(M_1^{-1}S) \le C(1 + \log^2(\beta^{-1})).$$

Proof. See Smith (1990) and Dryja and Widlund (1992b). □

3.8. Three-dimensional case: wirebasket preconditioners for S

Wirebasket algorithms were originally introduced in Bramble, Pasciak and Schatz (1989) (see also Dryja (1988)), and later modified and generalized by Smith (1991) and Dryja *et al.* (1993). These preconditioners for S involve computations on a *wirebasket* region W of B, and have almost optimal convergence rates with respect to mesh parameters and coefficients a (in case the coefficients are constant or mildly varying within each subdomain). The theoretical basis for the wirebasket method is an alternate coarse grid space based on a wirebasket region, which replaces the standard coarse grid problem. The interpolation map onto the wirebasket based coarse space has the favourable theoretical property that its bounds are independent of the variations in the coefficients and only mildly dependent on the mesh parameters (unlike the standard interpolation map onto the coarse grid).

We describe here a parallel wirebasket algorithm due to Smith (1991), see also Dryja *et al.* (1993).

The wirebasket preconditioners for S are based on a partition of the interface $B = F \cup W$ into *faces* F and a *wirebasket* W. As for the vertex space preconditioner described earlier, F will denote the collection of all the faces F_i. For each subdomain boundary $\partial\Omega_i$, define the ith wirebasket $W^{(i)}$ to consist of the union of all the edges and subdomain vertices lying on $\partial\Omega_i$:

$$W^{(i)} \equiv \bigcup_{E_k \subset \partial\Omega_i} E_k \bigcup_{V_j \subset \partial\Omega_i} V_j.$$

The *wirebasket* of B is defined to be the union of all the subdomain wirebaskets:

$$W \equiv \bigcup_{i=1}^{p} W^{(i)}.$$

Corresponding to the partition of the nodes $B = F \cup B$, the unknowns can be permuted: $x_B = [x_F, x_W]^T$, and the matrix S has the following block partition:

$$S = \begin{bmatrix} S_{FF} & S_{FW} \\ S_{FW}^T & S_{WW} \end{bmatrix}.$$

As for the vertex space algorithm, R_{F_i} will denote the pointwise restriction map onto nodes on F_i. Its transpose $R_{F_i}^T$ will denote extension by zero of nodal values on F_i to all the nodes on B. Next, corresponding to the wirebasket region W, there will be two kinds of restriction (and extension) maps, namely a pointwise restriction map R_W and a weighted restriction map $\hat{\mathcal{R}}_W$. For each i, the pointwise restriction map $R_{W^{(i)}}$ will restrict nodal values on B onto nodal values on the ith wirebasket $W^{(i)}$. Its transpose $R_{W^{(i)}}^T$ denotes the extension of nodal values on $W^{(i)}$ by zero to all nodes on B. Given a grid function u_W on W, the *wirebasket interpolation map* $\hat{\mathcal{R}}_W^T u_W$ extends the nodal values of u_W on W to the nodes on the faces as follows. On all the interior nodes on face F_i, the interpolant $\hat{\mathcal{R}}_W^T u_W$ is a constant equal to the average value of u_W on the boundary ∂F_i of face F_i:

$$\hat{\mathcal{R}}_W^T u_W = \begin{cases} u_W & \text{nodes } \in W \\ \text{average}(u_W)|_{\partial F_j} & \text{nodes } \in F_j. \end{cases}$$

Thus, its transpose $\hat{\mathcal{R}}_W$ will be a weighted restriction, mapping vectors u_B on B into vectors on W as follows:

$$\left(\hat{\mathcal{R}}_W u_B\right)_i = (u_B)_i + \sum_{k:i\in\partial F_k} \sum_{j\in F_k} \frac{(u_B)_j}{\dim(\partial F_k)}.$$

Next, let $z_{W^{(i)}}$ denote the vector whose entries are 1's for all indices on the ith wirebasket $W^{(i)}$. For $i = 1, \ldots, p$, define $\mathcal{B}^{(i)} = \rho_i\,(1 + \log(H/h))\, hI$

to be a diagonal matrix of the same size as the number of nodes on $W^{(i)}$, with $\rho_i = a|_{\Omega_i}$. Then, the matrix $\mathcal{B}$ is defined on the wirebasket W as a sum of the local matrices $\mathcal{B}^{(i)}$:

$$\mathcal{B} \equiv \sum_{i=1}^{p} R_{W^{(i)}}^T \mathcal{B}^{(i)} R_{W^{(i)}}.$$

Since $\mathcal{B}$ is the sum of several diagonal matrices, it will also be a diagonal matrix.

The wirebasket preconditioner M_2 of Smith (1991) has the following additive form:

$$M_2^{-1} = \sum_{i=1}^{m} R_{F_i}^T S_{F_iF_i}^{-1} R_{F_i} + \hat{\mathcal{R}}_W^T M_{WW}^{-1} \hat{\mathcal{R}}_W, \tag{3.11}$$

where the matrix M_{WW} is defined by its quadratic form:

$$u_W^T M_{WW} u_W = \sum_{i=1}^{p} \min_{\omega_i} \left(R_{W^{(i)}} u_W - \omega_i z_{W^{(i)}}\right)^T \mathcal{B}^{(i)} \left(R_{W^{(i)}} u_W - \omega_i z_{W^{(i)}}\right).$$

The terms $\omega_i z_{W^{(i)}}$ and the minimization are there to ensure that the local Schur complement $S^{(i)}$ and $M_2^{(i)}$ have the same null space spanned by $z_{W^{(i)}}$ (which in the case of scalar problems is $[1, \ldots, 1]^T$, but for systems such as elasticity, there may be several linearly independent null vectors).

The ease of inversion of M_{WW} is of course crucial to the efficiency of the preconditioner M_2. The linear system

$$M_{WW} x_W = f_W,$$

is equivalent, due to positive definiteness, to the following minimization problem:

$$\min_{x_W} \tfrac{1}{2} x_W^T M_{WW} x_W - x_W^T f_W,$$

and by substituting the quadratic form for M_{WW}, we obtain

$$\min_{x_W} \frac{1}{2} \sum_{i=1}^{p} \min_{\omega_i} \left(R_{W^{(i)}} x_W - \omega_i z_{W^{(i)}}\right)^T \mathcal{B}^{(i)} \left(R_{W^{(i)}} x_W - \omega_i z_{W^{(i)}}\right) - x_W^T f_W.$$

Differentiating with respect to all unknowns in x_W and with respect to $\omega_1, \ldots, \omega_p$, the following equivalent linear system is obtained:

$$\begin{cases} z_{W^{(i)}}^T \mathcal{B}^{(i)} \left(R_{W^{(i)}} x_W - \omega_i z_{W^{(i)}}\right) & = & 0 \text{ for } i = 1, \ldots, p, \\ \mathcal{B} x_W - \sum_{i=1}^{p} \omega_i R_{W^{(i)}}^T \mathcal{B}^{(i)} z_{W^{(i)}} & = & f_W. \end{cases}$$

If $\omega_1, \ldots, \omega_p$ are known, then x_W can be determined by solving the second block row (which is a diagonal system):

$$x_W = \mathcal{B}^{-1} \left(f_W + \sum_{i=1}^{p} \omega_i R_{W^{(i)}}^T \mathcal{B}^{(i)} z_{W^{(i)}} \right).$$

Substituting this into the first block row, we obtain

$$\left(z_{W^{(i)}}^T \mathcal{B}^{(i)} z_{W^{(i)}}\right) \omega_i - z_{W^{(i)}}^T \mathcal{B}^{(i)} R_{W^{(i)}} \mathcal{B}^{-1} \sum_{j=1}^{p} \omega_j R_W^T(j) \mathcal{B}^{(j)} z_{W^{(j)}}$$
$$= z_W^T(i) \mathcal{B}^{(i)} R_{W^{(i)}} \mathcal{B}^{-1} f_W.$$

Note that this $p \times p$ coefficient matrix for $\omega_1, \ldots, \omega_p$ can be computed, and it can be verified that it will be sparse. The resulting system for $\omega_1, \ldots, \omega_p$ can be solved using any sparse direct solver.

The convergence rate of this additive wirebasket algorithm of Smith (1991) is logarithmic in the number of unknowns per subdomain.

Theorem 11 If the coefficients a are mildly varying within each subdomain, there exists a constant C independent of H, h and a such that

$$\operatorname{cond}(M_2^{-1} S) \leq C(1 + \log(H/h))^2.$$

Proof. See Smith (1991), Dryja *et al.* (1993). □

For alternate wirebasket algorithms, we refer the reader to Bramble *et al.* (1989), Mandel (1989a), Dryja *et al.* (1993). The latter contains a wirebasket algorithm with condition number bounded by $1 + \log(H/h)$.

3.9. Three dimensions: Neumann–Neumann preconditioner M_3 for S

The Neumann–Neumann preconditioner for S in three dimensions is identical in form to the two-dimensional Neumann–Neumann preconditioner described earlier, and so the algorithm will not be repeated here. We mention here that an attractive feature of the Neumann–Neumann algorithm in three dimensions is that it does not require distinction between various subregions of the boundary $\partial\Omega_i$ of each subdomain (such as faces, edges, vertices and wirebaskets). Additionally, the almost optimal convergence rates are also valid for three-dimensional problems, see De Roeck and Le Tallec (1991), Dryja and Widlund (1990; 1993a), Mandel and Brezina (1992).

4. Introduction to the convergence theory

In this section, we provide a brief introduction to a theoretical framework for studying the convergence rates of the Schwarz (overlapping) and Schur complement (nonoverlapping) based domain decomposition methods discussed in this article (the Schwarz framework can also be used for analysing multilevel methods). Since the convergence rates of preconditioned conjugate gradient methods depend on the quotient of the extreme eigenvalues of the preconditioned matrix $M^{-1}A$ (which is assumed to be symmetric, positive definite in a suitable inner product), this theoretical framework involves techniques for estimating and bounding the extreme eigenvalues of the resulting preconditioned matrices. Additionally, in case of unaccelerated iterations based on matrix splittings, the framework provides a technique for estimating the spectral radius or norm of the error propagation matrix.

A prominent feature of the Schwarz algorithms that simplifies their convergence analysis is that the preconditioned matrices (or the error propagation matrices in case of unaccelerated iterations) can be expressed as sums (or products) of orthogonal projection matrices. The abstract framework described here, originated and evolved from convergence studies of the classical Schwarz alternating algorithm in a variational framework, see Lions (1988), Sobolev (1936), Babuška (1957) and Morgenstern (1956), with extensions and applications in the finite element context by Widlund (1988), Dryja and Widlund (1987; 1989b; 1990; 1993a), Matsokin and Nepomnyaschikh (1985), Nepomnyaschikh (1986), Bramble *et al.* (1991), Xu (1992a), and others. Nonvariational theories, in particular ones based on the maximum principle, have also been used to study domain decomposition methods, Miller (1965), Tang (1988), Lions (1989), Chan, Hou and Lions (1991a).

4.1. Abstract framework for additive and multiplicative Schwarz algorithms

Recall that the preconditioned system $M^{-1}A$ of the additive Schwarz preconditioner M is defined by

$$M^{-1}A = \sum_{i=0}^{p} R_i^T A_i^{-1} R_i A = \sum_{i=0}^{p} P_i,$$

where $P_i \equiv R_i^T A_i^{-1} R_i A$. (We have, for convenience, denoted the coarse grid problem $R_H^T A_H^{-1} R_H$ by $R_0^T A_0^{-1} R_0$.) When A is symmetric positive definite, the matrices P_i are orthogonal projection matrices in the A inner product, since

$$P_i P_i = R_i^T \left(A_i^{-1} R_i A R_i^T \right) A_i^{-1} R_i A = R_i^T A_i^{-1} R_i A = P_i,$$

and

$$A P_i = A R_i^T A_i^{-1} R_i A = P_i^T A.$$

Thus, the extreme eigenvalues of $M^{-1}A$ can be estimated by finding upper and lower bounds for the spectra of the sums of the orthogonal projections P_i. We describe the abstract framework for doing this in the following.

Let V be a Hilbert space with inner product $a(.,.)$ and let $V_0, \ldots, V_p$ be subspaces $V_i \subset V$. (In the matrix case, $a(u,v) \equiv u^T A v$.) For $i = 0, \ldots, p$, let P_i denote the orthogonal projection from V into V_i, i.e.

$$P_i u \in V_i \text{ satisfies } \; a(P_i u, v) = a(u,v) \;\; \forall v \in V_i.$$

Let N_c denote the minimum number of distinct colours so that the spaces $V_1, \ldots, V_p$ of the same colour are mutually orthogonal in the $a(.,.)$ inner product (note that the subspaces corresponding to disjoint subdomains will be mutually orthogonal, for domain decomposition algorithms). Then the following upper bound holds for the spectra of the additive operator $P_0 + \cdots + P_p$.

Theorem 12 $\lambda_{\max}(P_0 + \cdots + P_p) \leq N_c + 1$.

Proof. Recall that the spectral radius of any matrix A satisfies $\rho(A) \leq \|A\|$, and for orthogonal matrices the norm $\|P_i\| \leq 1$. Thus, an upper bound of $p+1$ is trivially obtained since the norm of each projection P_i is bounded by 1, and the sum of $p+1$ such projections gives a bound of $p+1$. The improved upper bound of $N_c + 1$ is obtained by noting that the sum of projections of the same colour, equals a projection onto the sum of the subspaces of the same colour. Consequently, there are only N_c projections for the colours, and projection P_0 onto the coarse grid. The result thus follows. □

A lower bound for a sum of the projections can be obtained, provided the spaces V_i satisfy the following property with constant C_0 that can be estimated.

Partition property of V_i *For any $u \in V$, there exists a constant $C_0 \geq 1$, such that the partition: $u = u_0 + \cdots + u_p$, where $u_i \in V_i$, satisfies*

$$\sum_{i=0}^{p} a(u_i, u_i) \leq C_0 a(u, u).$$

The lower bound for the sum of the projections can be estimated based on C_0, in a result described in Lions (1988), see also Dryja and Widlund (1987; 1989b). Similar ideas were developed earlier by Matsokin and Nepomnyaschikh (1985).

Theorem 13 Suppose the subspaces V_i for $i = 0, \ldots, p$, satisfy the partition property with constant $C_0 \geq 1$. Then,

$$\lambda_{\min}(P_0 + \cdots + P_p) \geq 1/C_0.$$

Proof. We shall use the Rayleigh quotient characterization:

$$\lambda_{\min}(P_0 + \cdots + P_p) = \min_{u \neq 0} \sum_{i=0}^{p} a(P_i u, u)/a(u, u).$$

For arbitrary $u \in V$, consider

$$a(u, u) = \sum_{i=0}^{p} a(u_i, u), \qquad \text{where } u = u_0 + \cdots + u_p.$$

Since P_i are projections, we obtain that $a(u_i, u) = a(u_i, P_i u)$. Now, applying the Schwarz inequality, we obtain

$$\sum_{i=0}^{p} a(u_i, u) = \sum_{i=0}^{p} a(u_i, P_i u) \leq \left(\sum_{i=0}^{p} a(u_i, u_i)\right)^{1/2} \left(\sum_{i=0}^{p} a(P_i u, P_i u)\right)^{1/2}.$$

By the partition property, we obtain that

$$a(u, u) \leq C_0^{1/2} a(u, u)^{1/2} \left(\sum_{i=0}^{p} a(P_i u, P_i u)\right)^{1/2}.$$

After cancellation this becomes

$$a(u, u)^{1/2} \leq C_0^{1/2} \left(\sum_{i=0}^{p} a(P_i u, P_i u)\right)^{1/2} = C_0^{1/2} \left(\sum_{i=0}^{p} a(P_i u, u)\right)^{1/2},$$

where the last equality follows since $a(P_i u, P_i u) = a(P_i u, u)$. Squaring both sides, the result gives a lower bound for the Rayleigh quotient. □

Combining the upper and lower bounds, we obtain:

Theorem 14 The condition number cond $(M^{-1}A)$ of the additive Schwarz preconditioned system is bounded by $(N_c + 1)C_0$.

Next, we estimate the convergence rate of the unaccelerated multiplicative Schwarz method. Analogous to the two subdomain case presented in Section 1, it can be easily derived that the error $e^n = u - u^n$ satisfies

$$e^{n+1} = (I - P_p) \cdots (I - P_0) e^n.$$

Thus:

$$\|e^n\| \leq \|(I - P_p) \cdots (I - P_0)\| \|e^n\|.$$

Clearly, $\|(I - P_p) \cdots (I - P_0)\| \leq 1$ in the norm generated by bilinear form $a(.,.)$, since the $(I - P_i)$ are also orthogonal projections with norms bounded by 1. Moreover, it is strictly less than 1 whenever $V = V_0 + \cdots + V_p$. More precisely, we have:

Theorem 15 Let V_i satisfy the partition property with constant C_0. Then the error propagation map of the multiplicative Schwarz iteration satisfies

$$\|(I - P_p) \cdots (I - P_0)\| \leq 1 - c/C_0 < 1,$$

where c is a constant that depends only on N_c but independent of p.

Proof. See Bramble *et al.* (1991). A precise expression for $0 < c < C_0$ is also given in Xu (1992a), Wang (1993), Cai and Widlund (1993). □

For the Schwarz algorithms based on the subdomains illustrated in Figure 2, the number of colours is $N_c = 4$. Analogous subdomain partitions in three dimensions yield $N_c = 8$. More generally, for most domain decomposition algorithms, N_c is a fixed number, independent of the number of subdomains. (However, for multilevel methods, N_c equals the number of levels, and then the colouring assumption must be replaced by a weaker assumption, see Bramble, Pasciak, Wang and Xu (1991), Xu (1992a), Yserentant (1986) and Griebel and Oswald (1993).) Thus, the rate of convergence depends critically on the *partition constant* C_0 and this will be estimated for finite element spaces in the next section.

4.2. A partition lemma for finite element spaces

In this section, following Dryja and Widlund (1987; 1992b) and Bramble *et al.* (1991), we describe a technique for estimating the partition constant C_0 for the basic overlapping Schwarz algorithms of Section 2.

Let $V^h(\Omega)$ denote the space of finite element functions defined on a quasi-uniform triangulation $\tau^h(\Omega)$, and let $V_i \equiv V^h(\Omega_i) \cap H_0^1(\hat{\Omega}_i)$ denote the finite element functions in $V^h(\Omega)$ which vanish outside $\hat{\Omega}_i$. Additionally, let $V_0 = V^H(\Omega)$ denote the space of finite element functions based on the coarse triangulation $\tau^H(\Omega)$ consisting of nonoverlapping elements $\Omega_1, \ldots, \Omega_p$.

We then have the following partition lemma.

Theorem 16 Let $a(.,.)$ denote the bilinear form associated with the elliptic problem in R^d for $d \leq 3$. The subspaces V_i defined above satisfy that for any $u \in V^h(\Omega)$, there exists $u_i \in V_i$ with

$$u = \sum_{i=0}^{p} \text{ and } \sum_{i=0}^{p} a(u_i, u_i) \leq C\left(1 + \beta^{-2}\right) a(u,u), \tag{4.1}$$

where C is a constant independent of H and h, but which depends on the coefficients.

Proof. We outline the proof only for the case of continuous piecewise linear finite element functions. Let $u_0 = Q_0 u_h$, where Q_0 is the L^2 orthogonal projection onto V_0. Then, by the H^1 stability of the L^2 projection, see Xu (1989) and Bramble and Xu (1991), we have

$$|u_0|_{H^1(\Omega)} \leq C|u_h|_{H^1(\Omega)}, \tag{4.2}$$

for some constant C independent of H and h. By using the equivalence between $a(.,.)$ and the H^1 norm, it follows from (4.2) that

$$a(u_0, u_0) \leq Ca(u_h, u_h). \tag{4.3}$$

Let I_H denote the finite element interpolation map onto the coarse space $V^H(\Omega)$. By using the best approximation property of Q_0 and applying the standard finite element interpolation error bound for $(u_h - I_H u_h)$ we obtain

$$\|u_h - u_0\|_{L^2(\Omega)} \leq \|u_h - I_H u_h\|_{L^2(\Omega)} \leq CH|u_h|_{H^1(\Omega)}. \tag{4.4}$$

Next, let $\chi_1, \ldots, \chi_p$ be a partition of unity, subordinate to the covering $\hat{\Omega}_1, \ldots, \hat{\Omega}_p$, satisfying:

$$0 \leq \chi_i \leq 1, \ \chi_i \in C_0^\infty\left(\hat{\Omega}_i\right), \text{ with } \sum_{i=1}^{p} \chi_i = 1, \text{ and } |\nabla \chi_i|_\infty \leq C\beta^{-1}H^{-1}.$$

Note that such a partition of unity exists due to the overlapping cover. We then define the following partition of $u_h - u_0$:

$$u_i = I_h\left(\chi_i\left(u_h - u_0\right)\right), \text{ for } i = 1, \ldots, p, \tag{4.5}$$

where I_h is the finite element interpolation onto $V^h(\Omega)$. We note that without the interpolation, the terms $\chi_i\left(u_h - u_0\right)$ will not be in the finite element space, since the product with χ_i is not piecewise polynomial. By linearity of the interpolant I_h, and the partition of unity, it follows that

$$u_1 + \cdots + u_p = u_h - u_0.$$

We now estimate the partition constant C_0 in several steps. To simplify the notation, C will denote a generic constant below. For each element $e \in \tau^h$, let $0 \leq \theta_e \leq 1$ be a constant such that $\|\chi_i - \theta_e\|_{L^\infty(e)} = \mathcal{O}(h/H)$ (e.g. $\theta_e = \chi_i(x_0)$ where x_0 is the centre of the element). Then, in element e we have

$$\begin{aligned} u_i &\equiv I_h\left(\chi_i(u_h - u_0)\right) \\ &= I_h\left((\chi_i - \theta_e)(u_h - u_0)\right) + I_h\left(\theta_e(u_h - u_0)\right) \\ &= I_h\left((\chi_i - \theta_e)(u_h - u_0)\right) + \theta_e\left(u_h - u_0\right), \end{aligned}$$

since θ_e is constant in element e.

By applying the triangle inequality to the gradient of the above expression, and using that $\theta_e \leq 1$, we obtain

$$|u_i|_{H^1(e)} \equiv \|\nabla u_i\|_{L^2(e)}^2 \leq 2\|\nabla I_h(\chi_i - \theta_e)(u_h - u_0)\|_{L^2(e)}^2 + 2\|\nabla(u_h - u_0)\|_{L^2(e)}^2.$$

By applying an inverse inequality (which states that $|v_h|_{H^1} \leq Ch^{-1}\|v_h\|_{L^2}$ for any finite element function v_h), and the fact that $\|I_h\left(fv_h\right)\|_{L^2(e)} \leq \|f\|_{L^\infty(e)}\|v_h\|_{L^2(e)}$ for any continuous function f, the first term on the right-hand side can be bounded by

$$Ch^{-2}\|I_h(\chi_i - \theta_e)(u_h - u_0)\|_{L^2(e)}^2 \leq Ch^{-2}\|\chi_i - \theta_e\|_{L^\infty(e)}^2\|I_h(u_h - u_0)\|_{L^2(e)}^2.$$

Since $\|\chi_i - \theta_e\|_{L^\infty(e)} = \mathcal{O}(h/H)$, this in turn can be bounded by

$$Ch^{-2}\left(\frac{h}{\beta H}\right)^2 \|u_h - u_0\|_{L^2(e)}^2.$$

Combining the above, we obtain

$$|u_i|_{H^1(e)} \leq \frac{C}{\beta^2 H^2}\|u_h - u_0\|_{L^2(e)}^2 + 2|u_h - u_0|_{H^1(e)}^2.$$

Summing over all i and noting that only a finite number of u_i (bounded by the minimum number of colors N_c) is nonzero on the element e, we obtain

$$\sum_{i=1}^{p} |u_i|_{H^1(e)} \leq \left(\frac{C}{\beta^2 H^2} \|u_h - u_0\|^2_{L^2(e)} + 2|u_h - u_0|^2_{H^1(e)} \right) N_c.$$

Summing over all elements e in Ω, we obtain

$$\sum_{i=1}^{p} |u_i|_{H^1(\Omega)} \leq \left(\frac{C}{\beta^2 H^2} \|u_h - u_0\|^2_{L^2(\Omega)} + C|u_h - u_0|^2_{H^1(\Omega)} \right) N_c.$$

Applying (4.4) to the first term and the triangle inequality to the second term on the right, we have

$$\sum_{i=1}^{p} |u_i|_{H^1(\Omega)} \leq C\beta^{-2}|u_h|^2_{H^1(\Omega)} + C|u_0|^2_{H^1(\Omega)}.$$

Using the H^1 stability of Q_0 in the second term on the right and the equivalence between the H^1 norm and the $a(.,.)$ norm, we obtain

$$\sum_{i=1}^{p} a(u_i, u_i) \leq C\left(1 + \beta^{-2}\right) a(u_h, u_h).$$

Adding (4.3), we obtain (4.1). □

Here, C is independent of h, H and β, but may depend on the coefficients, since we used the equivalence between the $a(.,.)$ norm and the H^1 norm. For an improved bound of $C\left(1 + \beta^{-1}\right)$ and for bounds which are valid independently of the jumps in the coefficients, we refer the reader to Dryja and Widlund (1992b).

4.3. Theory for Schur complement based methods

The convergence rate of Schur complement based methods depends on the spectrum of the preconditioned Schur matrix $M^{-1}S$. In this section, we will describe some techniques for estimating the extreme eigenvalues of some preconditioned Schur systems (mainly in two dimensions).

First, we prove the following equivalence between S and A. Given a vector x_B on the boundary B (see Section 3 for notation), define the discrete *harmonic extension* $Ex_B \equiv -A_{II}^{-1}A_{IB}x_B$. Then we have the following fundamental result:

Lemma 1

$$A[Ex_B, x_B]^T = [0, Sx_B]^T;$$

and

$$x_B^T S x_B = [Ex_B, x_B]A[Ex_B, x_B]^T.$$

Proof. Direct computation from the block factorization of A. □

Thus, the action of S on x_B can be obtained by first computing Ex_B, followed by a matrix product of A with $[Ex_B, x_B]^T$, and restricting the result to the nodes on the interface B.

This lemma provides a framework for constructing suitable preconditioners M for S: *if M is a matrix defined for vectors x_B, such that the M energy of x_B (i.e. $x_B^T M x_B$) approximates the A energy of the* discrete harmonic extension $[Ex_B, x_B]^T$, *then M can be used as a preconditioner for S, provided M can be easily inverted.*

Theorem 17 (**Trace Theorem**) There exists a continuous linear map $\gamma : H^1(\Omega) \longrightarrow L^2(\partial\Omega)$ such that $\gamma u = u|_{\partial\Omega}$ for smooth functions $u \in C^\infty(\overline{\Omega})$. Furthermore

$$\|\gamma u\|_{H^{1/2}(\partial\Omega)} \leq C\|u\|_{H^1(\Omega)},$$

for some positive constant C, where $H^{1/2}(\partial\Omega)$ is a fractional Sobolev norm.

Proof. See Nečas (1967) and Lions and Magenes (1972). □

The map γ is often referred to as the *trace* map. $H^{1/2}(\partial\Omega)$ is a fractional index Sobolev space which can be defined by interpolation between $H^1(\partial\Omega)$ and $H^0(\partial\Omega) = L^2(\partial\Omega)$ (we omit this description; see Lions and Magenes (1972)).

Using the trace theorem, we can prove the following fundamental property of harmonic functions.

Lemma 2 Let L be a second-order uniformly elliptic operator and u be a function defined on any region D, such that $Lu = 0$ in the interior of D. Then the $H^1(D)$ semi-norm of u on D is equivalent to the $H^{1/2}(\partial D)$ semi-norm of u on the boundary ∂D, i.e. there exist positive constants c and C such that

$$c|u|^2_{H^{1/2}(\partial D)} \leq |u|^2_{H^1(D)} \leq C|u|^2_{H^{1/2}(\partial D)}, \quad \text{for all } u \in H^1(D).$$

Proof. The left inequality follows from the trace theorem. The right inequality follows from elliptic regularity for harmonic functions, see Lions and Magenes (1972), Nečas (1967) for a proof. □

The corresponding result also holds for *discrete harmonic* functions, with constants c and C independent of mesh size h.

Theorem 18 If $u_h \in V^h(D)$ is a finite element function defined on a region D, such that u_h is discrete harmonic in D, then there exist constants c and C, independent of h such that

$$c|u_h|^2_{H^{1/2}(\partial D)} \leq |u_h|^2_{H^1(D)} \leq C|u_h|^2_{H^{1/2}(\partial D)}.$$

Proof. The left inequality follows from the trace theorem (as in the continuous case). The right inequality can be proved by using an *extension* theorem for finite element functions (which extends finite element functions defined on the boundary of a domain into the interior, such that the H^1 norm of the extension is bounded in terms of the $H^{1/2}$ norm of the boundary data), with a constant C independent of the mesh size h. Such an extension theorem was established by Widlund (1987), Bramble *et al.* (1986b), and Bjørstad and Widlund (1986). □

Thus, if a matrix M is the matrix representation of the bilinear form given by the $H^{1/2}(\partial D)$ inner product restricted to the finite element space $V^h(\partial D)$, then M is spectrally equivalent to S, the Schur complement obtained if $B = \partial\Omega$. The matrix M can be obtained by interpolation as follows.

4.4. Interface preconditioners for two-dimensional problems

Let K_B denote the discretization of $-\Delta$ on edge B, with zero boundary conditions on the vertices ∂B. Additionally, let M_B denote the mass matrix representing the $L^2(B)$ inner product on B. Then, the matrix representation J_B of the $H^{1/2}(B)$ bilinear form (or more precisely, the $H_{00}^{1/2}(B)$ bilinear form, see Lions and Magenes (1972)) is obtained by matrix interpolation between K_B and M_B as follows

$$J_B^{1/2} = [M_B, K_B]_{1/2} \equiv M_B^{1/2}\left(M_B^{-1/2} K_B M_B^{-1/2}\right)^{1/2} M_B^{1/2},$$

see Bjørstad and Widlund (1986) and Bramble *et al.* (1986b). Since M_B is spectrally equivalent to a scaled identity matrix, $J_B^{1/2}$ can be replaced by a scaled version of $K_B^{1/2}$, which is precisely Dryja's preconditioner M_{D} as presented in Section 3.1.

Theorem 19 For a two subdomain partition, the condition number of the preconditioned Schur matrix $J_B^{-1/2} S$ is bounded by a constant C independent of h.

Proof. By construction, $J_B^{1/2}$ is the matrix representation of the $H_{00}^{1/2}(B)$ inner product, therefore

$$u_B^T J_B^{1/2} u_B = \|u_B\|^2_{H_{00}^{1/2}(B)},$$

where we have used u_B to denote both a finite element function and its vector representation. By a variant of Theorem 18, $\|u_B\|^2_{H_{00}^{1/2}(B)}$ is spectrally equivalent to

$$[Eu_B, u_B] A [Eu_B, u_B]^T,$$

which in turn is spectrally equivalent to $u_B^T S u_B$ by Lemma 1. Therefore $J_B^{1/2}$ is spectrally equivalent to S. □

4.5. Many subdomain nonoverlapping algorithms

The theory for estimating the convergence rates of many subdomain preconditioners for S can often be reduced to estimates based on the Schwarz algorithms, see Dryja *et al.* (1993), Dryja and Widlund (1990; 1993a). Here, we sketch some of the basic ideas by considering the vertex space preconditioner M_{vs} of Smith (1992) in two dimensions:

$$M_{vs}^{-1} = \sum_k R_{E_k}^T S_{E_k E_k}^{-1} R_{E_k} + \sum_i R_{VS_i}^T S_{VS_i VS_i}^{-1} R_{VS_i} + R_H^T A_H^{-1} R_H.$$

In the following, we will assume that A_H is replaced by $S_H = R_H S R_H^T$, in which case the above preconditioner becomes an additive Schwarz preconditioner for S, based on an overlapping decomposition of the interface B:

$$B = \bigcup_k \{E_k\} \cup \left\{ \bigcup_l VS_l \right\},$$

and additionally the use of a coarse solver.

The preconditioned Schur matrix $M_{vs}^{-1} S$ can thus be written as a sum of projections, orthogonal in the S based inner product:

$$M_{vs}^{-1} S = \sum_k P_{E_k} + \sum_i P_{VS_i} + P_H,$$

where

$$P_{E_k} \equiv R_{E_k}^T S_{E_k E_k}^{-1} R_{E_k} S, \quad P_{VS_i} \equiv R_{VS_i}^T S_{VS_i VS_i}^{-1} R_{VS_i} S$$

and

$$P_H = R_H^T S_H^{-1} R_H S.$$

The condition number can be estimated in terms of a partition property with constant C_0 and the number of colours N_c.

We now sketch briefly, a technique for reducing this to using a corresponding partition for $V^h(\Omega)$ in the $a(\,.\,,\,.\,)$ based norm. First, corresponding to each subregion of the interface, we define a decomposition of Ω as follows. Let Ω_{E_k} be a subdomain of size $\mathcal{O}(H)$ containing E_k, and partitioned into two disjoint regions by E_ks (for instance, let Ω_{E_k} be the union of the two subdomains adjacent to E_k). Similarly, for each vertex region VS_i, let Ω_{VS_i} denote a subregion of Ω of size $\mathcal{O}(H)$ containing VS_i, and which is partitioned into a small number of disjoint subregions by VS_i (for instance, let Ω_{VS_i} be a rectangular or quadrilateral patch covering the vertex region VS_i). Then,

- Given u_B defined on B, extend it *discrete harmonically* into the subdomains: $[Eu_B, u_B]^T$.
- Next, partition $[Eu_B, u_B]^T$ (the extension) using the spaces $\{V^h(\Omega_{E_k})\}$, $\{V^h(\Omega_{VS_i})\}$ and coarse space V_0 with a partition constant C_0 that can be estimated by the same partition lemma (which was stated earlier). Thus,

$$[Eu_B, u_B]^T = \tilde{u}_0 + \sum_k \tilde{u}_{E_k} + \sum_i \tilde{u}_{VS_i},$$

with

$$\sum a(\tilde{u}_i, \tilde{u}_i) \leq C_0 a(Eu_B, Eu_B) = C_0 S(u_B, u_B),$$

where $\tilde{u}_i$ denotes the same partition, suitably re-indexed. The last equality follows from the equivalence between the S-energy and the A-energy of discrete harmonic extensions. The constant C_0 is bounded independent of H and h.
- Next, restrict each $\tilde{u}_i$ onto B to obtain a partition for u_B on B.
- Finally, use the equivalence between the S-energy and the A-energy of discrete harmonic extensions with the additional fact that the $a(.,.)$ energy of each $\tilde{u}_i$ is greater than the $a(.,.)$ energy of the discrete harmonic extension of its values on B.

By combining the results above, the partition constant for the Schur based algorithm can be estimated, see Dryja *et al.* (1993) for the details.

4.6. Summary of convergence bounds

In Table 1, we summarize the known condition number bounds for several of the preconditioners described in Sections 2 and 3. In the last two columns, we list condition number bounds that are most appropriate (tighter) when the coefficients are mildly varying and when the coefficients are discontinuous with possibly large jumps, respectively. $C(a)$ refers to a constant independent of H and h but dependent on the coefficients a, while C refers to a constant independent of H, h and a (provided a is mildly varying in each subdomain Ω_i). For the Schwarz and vertex space algorithms, β refers to the overlap parameter.

5. Some practical implementation issues

The focus of the previous sections were on the development of the basic components of domain decomposition algorithms (at a certain level of abstraction). In order to implement these algorithms efficiently, possibly on a parallel computer, there are other more practical matters to consider as well. In this section, we shall briefly touch on several of these issues.

Table 1. *Upper bounds for condition numbers of various algorithms.*

Algorithm	Eqn	Mild Coeff.	Disc. Coeff.
2D BPS	(3.6)	$C\left(1+\log^2(H/h)\right)$	$C\left(1+\log^2(H/h)\right)$
2D vertex space	(3.7)	$C(a)\left(1+\log^2(\beta^{-1})\right)$	$C(\beta)\left(1+\log(H/h)\right)$
3D vertex space	(3.10)	$C(a)\left(1+\log^2(\beta^{-1})\right)$	$C(\beta)(H/h)$
2D additive Schwarz	(2.1)	$C(a)\left(1+\beta^{-1}\right)$	$C(\beta)\left(1+\log(H/h)\right)$
3D additive Schwarz	(2.1)	$C(a)\left(1+\beta^{-1}\right)$	$C(\beta)(H/h)$
3D wirebasket	(3.11)	$C\left(1+\log^2(H/h)\right)$	$C\left(1+\log^2(H/h)\right)$
2D Neumann–Neumann	(3.9)	$C\left(1+\log^2(H/h)\right)$	$C\left(1+\log^2(H/h)\right)$
3D Neumann–Neumann	(3.9)	$C\left(1+\log^2(H/h)\right)$	$C\left(1+\log^2(H/h)\right)$

5.1. Inexact subdomain solvers

Every step of a domain decomposition iteration normally requires the exact solution of a subdomain problem, and perhaps also a coarse problem. Although this usually costs less than the solution of the original problem on the whole domain, it can still be quite expensive and it is natural to try to use a cheaper approximate solver instead. Also, when the iterates are still far from the true solution, it seems wasteful to solve these subdomain problems exactly. The issue here is how to incorporate these inexact solvers properly into the existing framework.

In most of the domain decomposition algorithms we have introduced so far, the exact solves involving A_i^{-1} and A_H^{-1} can be replaced by inexact solves $\tilde{A}_i^{-1}$ and $\tilde{A}_H^{-1}$, which can be standard elliptic preconditioners themselves (e.g. multigrid, ILU, SSOR, etc.). However, in order to rigorously prove that the conjugate gradient method converges, the inexact solvers $\tilde{A}_i^{-1}$ and $\tilde{A}_H^{-1}$ must be *fixed, linear* operators, e.g. they cannot be a few steps of an adaptive iterative method that depends on the vector being operated on (e.g. a few steps of the conjugate gradient method). In practice, however, solving the local problems approximately with a Krylov space method may work fine.

For the overlapping additive Schwarz methods the modification is straightforward. For example, the *Inexact Solve Additive Schwarz Preconditioner* is simply:

$$\tilde{M}_{\mathrm{as},2}^{-1} z = R_0^T \tilde{A}_H^{-1} R_H z + \sum_{i=1}^{p} R_i^T \tilde{A}_i^{-1} R_i z.$$

We caution, however, that replacing A_i by $\tilde{A}_i$ can potentially lead to divergence in multiplicative Schwarz iteration, unless the spectral radii

$$\rho(\tilde{A}_i^{-1} A_i) < 2,$$

see Bramble *et al.* (1991), Xu (1992a), Cai and Widlund (1993).

The Schur complement methods require more changes to accommodate inexact solves. For example, by replacing A_H^{-1} by $\tilde{A}_H^{-1}$ and $S_{E_iE_i}$ by $\tilde{S}_{E_iE_i}$ in the definitions of the Bramble–Pasciak–Schatz preconditioner M_2 (see (3.6)) and the vertex space preconditioner M_3 (see (3.7)), we can easily obtain relatively ill-conditioned inexact preconditioners $\tilde{M}_2$ and $\tilde{M}_3$ for S. The main difficulty is, however, that the evaluation of the product Sz_B still requires exact subdomain solves using A_{II}^{-1}. One way to get around this is to use an *inner* iteration using $\tilde{A}_i$ as a preconditioner for A_i in order to compute the action of A_{II}^{-1}. An alternative is to perform the iteration on the original system $Au = f$, and construct a preconditioner $\tilde{A}$ for A from the block factorization of A in equation (3.3) by replacing the terms A_{II} and S by $\tilde{A}_{II}$ and $\tilde{S}$, respectively, where $\tilde{S}$ can be either $\tilde{M}_2$ or $\tilde{M}_3$. However, care must be taken to scale $\tilde{A}_H$ and $\tilde{A}_i$ so that they are as close to A_H and A_i as possible respectively – it is not sufficient that the condition number of $\tilde{A}_H^{-1}A_H$ and $\tilde{A}_i^{-1}A_i$ be close to unity, because the scaling of the coupling matrix A_{IB} may be wrong. For more details, the reader is referred to Börgers (1989), Goovaerts (1989) and Goovaerts, Chan and Piessens (1991).

We note that, when set up properly, the use of inexact solvers does not compromise on the accuracy of the final converged solution – only the preconditioner is changed, see Gropp and Smith (1992).

5.2. The choice of the coarse grid size H

Another practical matter in implementing a domain decomposition algorithm is to decide how many subdomains to use, i.e. the coarse scale H. Since most of the domain decomposition algorithms we have described have convergence rates that are bounded independently (or only slightly dependent on) of H, the theory does not lead to a clear choice. If the fine grid is obtained as a refinement of a coarse grid, then H is naturally defined. Moreover, very often the choice of subdomains is dictated by geometric considerations, e.g. if the domain can be naturally decomposed into several subdomains with regular geometry on which fast solvers can be used. Finally, in a parallel setting, it is natural to match the number of subdomains to the number of processors available. The choice of H must take all these factors into account and there are no guidelines that will work in all situations.

However, from a purely computational complexity standpoint, it is possible to make a more rational decision based on minimizing the computational cost. Given h, it has been observed empirically (Keyes and Gropp, 1989; Smith, 1990; Gropp and Smith, 1992) that there often exists an optimal value of H which minimizes the total computational time for solving for the converged solution. A small H provides a better, but more expensive, coarse grid approximation, and requires solving more subdomain problems

Table 2. *Complexity of solvers on an n^3 grid with coarse grid size n_H. (MIC: modified incomplete Cholesky.)*

Basic solver	Complexity	Optimal n_H	Complexity of domain decomposition solver using optimal n_H
Multigrid	$\mathcal{O}(n^3)$	1	$\mathcal{O}(n^3)$
MIC	$\mathcal{O}(n^{3.5})$	$0.61n^{7/8}$	$\mathcal{O}(n^{3.06})$
Nested dissection	$\mathcal{O}(n^6)$	$0.93n^{2/3}$	$\mathcal{O}(n^4)$
Band-Cholesky	$\mathcal{O}(n^7)$	$0.95n^{7/11}$	$\mathcal{O}(n^{4.45})$
Solver n^α	$\mathcal{O}(n^\alpha), \alpha \to \infty$	$n^{1/2}$	$\mathcal{O}(n^{\alpha/2})$

of smaller size. A large H has the opposite effect. If we make the assumption that the *same solver* is used for the subdomain problems as well as for the coarse problems, and that the convergence rate is independent of H (which is true in practice for most optimal methods), then one can derive an asymptotically optimal value of H (Chan and Shao, 1993). For example, on a one-processor architecture, for a model problem on a uniform d-dimensional grid with mesh size h and a solver with complexity $\mathcal{O}(m^\alpha)$ on an m^d grid, the optimal choice is

$$H_{\text{opt}} = \left(\frac{\alpha}{\alpha - d}\right)^{1/(\alpha-d)} h^{\alpha/(2\alpha-d)},$$

and the complexity of the overall domain decomposition solver using H_{opt} is $\mathcal{O}(h^{-\alpha/(2\alpha-d)})$, which can be significantly smaller than $\mathcal{O}(h^{-\alpha})$, the complexity of using the same solver to solve the whole problem without using a domain decomposition method. For example, in three dimensions ($d = 3$), the complexities are summarized in Table 2, where $n \equiv 1/h$.

In a parallel environment, if we assume that each subdomain solve is performed in parallel on the individual processors, and that the coarse solve is performed on one of the processors, either sequentially after or in parallel with the subdomain solves, then it turns out, ignoring communication costs (whether this is valid depends on the problem size and the particular hardware), the optimal value of H is $H_{\text{opt}} = \sqrt{h}$, independent of α and d. The optimal number of processors is $n^{d/2}$, and the execution time using H_{opt} is $\mathcal{O}(n^{\alpha/2})$.

In practice, it may pay to empirically determine a near optimal value of H if the preconditioner is to be re-used many times. The above asymptotic results for the model problem can be used as a guide.

5.3. *Partition of the domain*

In addition to deciding *how many* subdomains to use, it is also necessary to *identify* them. Very often, the domain Ω is already discretized and the problem is to decompose the grid itself. This can be viewed as a graph partitioning problem. The geometry of the domain can usually provide some guidance, e.g. subdomains with regular geometry are preferable. In a parallel setting, it is also desirable to have connections (i.e. edges) between neighbouring subdomains to be minimized (which would in turn minimize the communication cost) and to have the load (e.g. the number of grid points) in each subdomain balanced. For a structured and quasi-uniformly refined grid, one can often do this decomposition at a coarse level either by inspection or by brute force. For unstructured grids, finding the optimal decomposition is an NP-complete problem. There have been several heuristic approaches proposed, including geometric approaches such as the recursive coordinate bisection method (Fox, 1988; Berger and Bokhari, 1987) and the inertia method (Farhat and Lesoinne, 1993); recursive graph based approaches such as the Kernighan and Lin (1970) exchange method, the minimum bandwidth method and the spectral partitioning method (Pothen, Simon and Liou, 1990); and global minimization techniques such as using simulated annealing (Williams, 1991). These techniques trade off efficiency with the ability to find good partitions, and it is not clear at this point which method is the best. Recent surveys can be found in Simon (1991) and Farhat and Lesoinne (1993).

5.4. *Solving the coarse problem in parallel*

The most natural way of mapping a domain decomposition algorithm onto a parallel architecture is to map the subdomains to individual processors. In this setting, the solution of the coarse problem often presents some difficulties because the data are scattered among all the processors. If not done carefully, the coarse solve can dominate the execution time of the domain decomposition method. There are several obvious alternatives:

1. keep the data in place and solve it using a parallel method with data exchanges at each step;
2. gather the data in *one* processor, solve there and broadcast the result;
3. gather the data to *all* processors and solve it on all of them in parallel.

According to Gropp (1992), the last two approaches are often better than the first and on typical architectures. For parallel implementations of domain decomposition methods, see Bjørstad and Skogen (1992) and Smith (1993).

5.5. *To overlap or not to overlap?*

There is no definitive answer to this question but here are some guidelines. First, the overlapping method is generally easier to describe, implement and understand. It is also easier to achieve an optimal convergence rate and often more robust. On the other hand, extra work is performed on the overlapped regions. Moreover, if the coefficients are discontinuous across the subdomains, the extended subdomains must necessarily have discontinuous coefficients, making their solution more problematic. Recently, Bjørstad and Widlund (1989) and Chan and Goovaerts (1992) have shown that there is a fundamental relationship between the two approaches: the overlapping method is equivalent to a nonoverlapping method with a specific interface preconditioner. One can think of the overlapping method implicitly computing the effect of this preconditioner by the extra operations performed on the overlapping region.

6. Multilevel algorithms

In recent years, much research and interest has been focused on the development of multilevel algorithms to solve elliptic problems, that provide alternative preconditioners to the standard multigrid method. These multilevel algorithms include, for instance, the hierarchical basis multigrid method of Yserentant (1986) and Bank, Dupont and Yserentant (1988), the BPX algorithm of Bramble *et al.* (1990), the multilevel algorithms of Axelsson and Vassilevski (1990), and the multilevel additive Schwarz algorithm of Zhang (1992b) (a similar idea was mentioned in the thesis of Xu (1989) and in Wang (1991)). Although strictly speaking these algorithms are not domain decomposition methods, they have similarities with Schwarz type domain decomposition methods (with inexact solves) where different grid levels and subspaces play the role of subregions, see for instance Xu (1992a). Additionally, a convergence theory has been developed that incorporates both multilevel and domain decomposition methods into a unified framework, see Xu (1992a) and Dryja and Widlund (1990).

6.1. *Background on multilevel discretizations*

Consider the Dirichlet boundary value problem for the elliptic problem (1.1) on Ω. In order to obtain a multilevel discretization of this problem, the domain Ω is first triangulated by a coarse grid $\tau^1(\Omega)$ consisting of elements of diameter h_1. By successive refinement of each element, (say by dividing each element into four pieces in two dimensions, etc) a refined triangulation $\tau^2(\Omega)$ is obtained with a mesh size of $h_2 = h_1/2$, and such that each element of $\tau^1(\Omega)$ is a union of elements of $\tau^2(\Omega)$. This procedure can be repeated a total of $J-1$ times, till the grid size $h_J = h_1/2^{J-1}$ on the finest level

J provides sufficient accuracy. We therefore have J nested triangulations $\tau^1(\Omega), \ldots, \tau^J(\Omega)$ of Ω.

On each grid level i, for $i = 1, \ldots, J$, we define the standard finite element space $V^{h_i}(\Omega) \subset H_0^1(\Omega)$ consisting of continuous piecewise linear functions based on a triangulation $\tau^i(\Omega)$, which vanish on the boundary $\partial\Omega$. Note that

$$V^{h_1}(\Omega) \subset V^{h_2}(\Omega) \subset \cdots \subset V^{h_J}(\Omega).$$

For $i = 1, \ldots, J$, we let A^{h_j} denote the stiffness matrix corresponding to the discretization of the elliptic problem on the jth level based on the finite element space $V^{h_j}(\Omega)$, and let M^{h_j} denote the mass matrix corresponding to the bilinear form generated by the L^2 inner product.

We now describe several multilevel preconditioners that correspond to additive Schwarz (additive subspace) preconditioners with suitably defined restriction maps R_j.

6.2. *The hierarchical basis multigrid method*

The hierarchical basis method of Yserentant (1986) and Bank *et al.* (1988) is based on a new multilevel *hierarchical* basis for the finite element space. Let I_j denote the standard finite element interpolation map:

$$I_j : V^{h_J}(\Omega) \to V^{h_j}(\Omega),$$

from the fine grid onto the nodal basis functions on grid level j. Then, by telescoping series, we obtain:

$$I_J = I_1 + (I_2 - I_1) + \cdots + (I_J - I_{J-1}).$$

Each of the terms $I_j - I_{j-1}$ represents grid functions on level j which are zero at the nodes corresponding to the coarser grid level $j-1$. The range of these interpolation maps $I_j - I_{j-1}$ (i.e. the new nodes on each level) will correspond to the 'subdomains' in a Schwarz (subspace) method.

The hierarchical basis multigrid preconditioner M for A is an additive subspace (Schwarz) preconditioner of the form:

$$M_{\text{hb}}^{-1} = \sum_{j=1}^{J} R_j^T D_j^{-1} R_j,$$

with restriction map $R_j \equiv I_j - I_{j-1}$, and where the local matrices $A_j = R_j A R_j^T$ are replaced by its diagonal D_j, resulting in an inexact solve. See Bank *et al.* (1988), Xu (1992a) for details. In two dimensions, cond $(M^{-1}A)$ is bounded by $\mathcal{O}(1+\log^2(h))$, but in three dimensions this bound deteriorates to $\mathcal{O}(h^{-1})$, see Yserentant (1986) and Ong (1989).

6.3. The BPX algorithm

The BPX preconditioner of Bramble *et al.* (1990) can also be viewed as an additive subspace (Schwarz) preconditioner:

$$M^{-1} \equiv \sum_{j=1}^{J} R_j^T A_j^{-1} R_j,$$

where R_j^T denotes the interpolation map from the jth grid level to the finest grid, and R_j corresponds to a weighted restriction. Additionally, the exact local matrices $A_j = R_j A R_j^T$ can be further approximated by $ch_j^{d-2} I$ for second-order uniformly elliptive problems without deterioration in the convergence rates. The resulting preconditioner is

$$M_{\text{BPX}}^{-1} \equiv \sum_{j=1}^{J} R_j^T h_j^{2-d} R_j.$$

The convergence rate of the BPX algorithm is optimal.

Theorem 20 There exists a constant C independent of h_i and J such that

$$\text{cond}\,(M_{\text{BPX}}^{-1} A_J) \leq C.$$

Proof. The original convergence bound due to Xu (1989) and Bramble *et al.* (1990) was J^2 (J with full elliptic regularity), i.e. deteriorated mildly with increasing number of levels. A different proof by Zhang (1992b) improved the bound to J. Bounds by Oswald (1991) are optimal, independent of J. For alternative proofs, see Griebel (1991) and Bornemann and Yserentant (1993). □

We note that when implementing the restriction and interpolation maps R_i and R_i^T respectively, it is easier and more efficient to obtain $R_i z$ from $R_{i+1} z$ as in a standard multigrid algorithm.

6.4. Multilevel additive Schwarz algorithm

We note that the above version of the BPX algorithm does not take into account the variation in the coefficients of the elliptic problems. In this section, we describe the multilevel additive Schwarz algorithm of Zhang (1992b; 1991) which generalizes the BPX algorithm by including overlapping subdomains on each grid level, and which takes coefficients into account in the preconditioning.

The multilevel Schwarz algorithm is based on the same J grid levels as the previous algorithms. However, the elements $\{e_{h_j}\}$ on grid level j are decomposed into a collection of N_j overlapping subdomains $\Omega_1^{h_j}, \ldots, \Omega_{N_j}^{h_j}$:

$$\Omega \subset \left(\Omega_1^{h_j} \cup \cdots \cup \Omega_{N_j}^{h_j}\right),$$

where the diameter of each jth level subdomain $\Omega_i^{h_j}$ is $\mathcal{O}(h_{j-1})$ (which is the size of the preceding coarser level). Additionally, it is assumed, that the size of the overlap between the adjacent subregions on grid level j, is βh_{j-1}.

For all the subdomains, on all the grid levels, the following interpolation maps are defined:

$$R^T_{\Omega_i^{h_j}} : V^{h_j}(\Omega_i^{h_j}) \cap H_0^1(\Omega_i^{h_j}) \longrightarrow V^{h_J}(\Omega),$$

where $R^T_{\Omega_i^{h_j}}$ is the extension map from the nodal values on the interior grid points in $\Omega_i^{h_j}$ on the jth grid level to the finest grid level J. Its transpose $R_{\Omega_i^{h_j}}$ is a weighted restriction map onto the interior nodes in subdomain $\Omega_i^{h_j}$ on the jth grid level. The local stiffness matrix corresponding to subregion $\Omega_i^{h_j}$ on the jth grid level is denoted $A_{\Omega_i^{h_j}}$, where

$$A_{\Omega_i^{h_j}} = R_{\Omega_i^{h_j}} A R^T_{\Omega_i^{h_j}},$$

is a principal submatrix of the jth level stiffness matrix A_{h_j}.

The multilevel additive Schwarz preconditioner M_{mlas} is defined by

$$M_{\mathrm{mlas}}^{-1} z = \sum_{j=1}^{J} \sum_{i=1}^{N_j} R^T_{\Omega_i^{h_j}} A^{-1}_{\Omega_i^{h_j}} R_{\Omega_i^{h_j}} z.$$

We note that this corresponds to a sum of additive Schwarz preconditioners on each grid level with suitably chosen subdomain sizes. The convergence rate of the multilevel additive Schwarz algorithm is described in the following theorem.

Theorem 21 Suppose that the mesh sizes satisfy: $h_i/h_{i-1} \leq cr$, where $r < 1$, and that the subregions on grid level j satisfy $\mathrm{Area}(\Omega_i^{h_j}) \approx h_{j-1}$. Then,

$$\mathrm{cond}\,(M_{\mathrm{mlas}}^{-1} A) \leq C(r, a),$$

where the constant $C(r, a)$ can depend on r and the coefficients a, but is independent of J and h_i.

Proof. See Zhang (1992b; 1991). □

Remarks

- The preconditioner M_{mlas} can be obtained as a special case of the BPX preconditioner by choosing the 'smoothing' operator in Xu (1989) to be the additive Schwarz preconditioner. Conversely, the BPX preconditioner for the discrete Laplacian can be obtained as a special case of the multilevel additive Schwarz algorithm by choosing each subdomain

$\Omega_i^{h_j}$ on the jth grid level to contain only one interior point from the jth grid level, i.e. with minimal overlap amongst subdomains on each grid level. In this case, the local matrices $A_{\Omega_i^{h_j}} = ch_j^{d-2}$ ($\Omega \subset R^d$) will be 1×1, and correspond to the diagonal entries of the stiffness matrix on the jth grid level A_{h_j}. Additionally: $\sum_{i=1}^{N_j} R_{\Omega_i^{h_j}} = R_j$, the weighted restriction map onto the jth grid level.

- We note that using the submatrices $A_{\Omega_i^{h_j}}$ on level j provides the scaling based on the coefficients and computing (or approximating) them involves some overhead cost.
- We may skip a few levels of refinement, and the convergence rate will depend only on the ratio of the relevant mesh sizes.
- A multiplicative version has been considered in Wang (1991).

7. Algorithms for locally refined grids

In this section we describe domain decomposition algorithms for solving the linear systems arising from discretizations of elliptic partial differential equations on composite grids obtained by local refinement on subregions of Ω. The discretizations we consider are based on the use of 'slave variables' on the interface separating the different refined regions, see Bramble, Ewing, Pasciak and Schatz (1988), McCormick (1989), Widlund (1989a). Our description will be brief, and our goal is to formulate the problem so that the same domain decomposition methodology of Schwarz methods can be applied. Indeed, a composite grid is the union of various 'subgrids' on different subregions, see Figure 4, and these 'subgrids' correspond to 'subdomains' in a Schwarz method.

7.1. Discretization of elliptic problems on locally refined grids

Consider the elliptic problem (1.1) on a domain Ω, which is triangulated by a quasi-uniform grid $\tau^h(\Omega)$ of mesh size h. The local refinement procedure is applied to a sequence of nested subregions: $\Omega_p \subset \cdots \subset \Omega_2 \subset \Omega_1 \equiv \Omega$. Starting with a quasi-uniform triangulation $\tau^{h_1}(\Omega_1)$ with mesh size h_1, all elements from this triangulation lying in Ω_2 are uniformly refined, for instance with mesh size $h_2 = h/2$ resulting in the local triangulation $\tau^{h_2}(\Omega_2)$. The process is repeated, with successive refinements on each nested subregion, with local triangulations $\tau^{h_i}(\Omega_i)$ for $i = 2, \ldots, p$, where $h_i = h_{i-1}/2$, see Figure 4.

Corresponding to each local grid $\tau^{h_i}(\Omega_i)$ let $V^{h_i}(\Omega_i) \subset H_0^1(\Omega_i)$ denote the space of continuous, piecewise linear finite element functions vanishing outside Ω_i. The composite finite space $V^{h_1,h_2,\ldots,h_p}$ is defined as the sum of the local spaces:

$$V^{h_1,h_2,\ldots,h_p} = V^{h_1}(\Omega_1) + V^{h_2}(\Omega_2) + \cdots + V^{h_p}(\Omega_p).$$

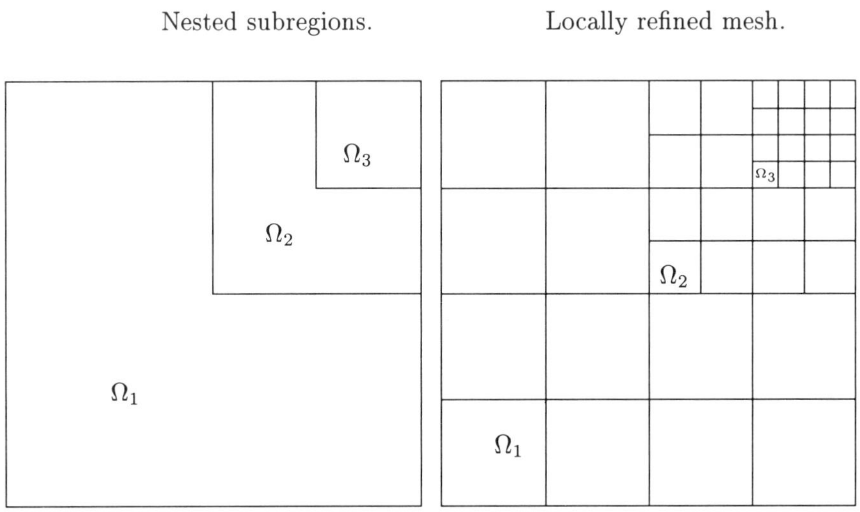

Nested subregions. Locally refined mesh.

Ω_3 Ω_2 Ω_1 Ω_3 Ω_2 Ω_1

Fig. 4. Nested subregions with repeated local refinement.

The elliptic problem is discretized using the standard Galerkin procedure based on the composite finite element space $V^{h_1,h_2,\ldots,h_p}$ resulting in a linear system

$$Au = f, \tag{7.1}$$

see McCormick (1984), Bramble *et al.* (1988) and Widlund (1989a) for the details.

Throughout the rest of this section, we will use $\Omega_i^{h_i}$ to denote $\tau^{h_i}(\Omega_i)$, the ith refined grid on Ω_i, where for $i = 1$ this corresponds to the initial triangulation of Ω. For $i = 1, \ldots, p$, we let $R^T_{\Omega_i^{h_i}}$ denote the interpolation (extension) map from $V^{h_i}(\Omega_i)$ to the composite grid $V^{h_1,h_2,\ldots,h_p}$ and let $R_{\Omega_i^{h_i}}$ denote the corresponding restriction map. The local stiffness matrices are given by $A_{\Omega_i^{h_i}} = R_{\Omega_i^{h_i}} A R^T_{\Omega_i^{h_i}}$.

7.2. The Bramble–Ewing–Pasciak–Schatz (BEPS) algorithm for solving two-level problems

For the case of just one level of refinement (i.e. $p = 2$), Bramble *et al.* (1988) proposed a preconditioner M_{BEPS} for system (7.1) that corresponds to a symmetrized multiplicative Schwarz preconditioner sweeping over the grids $\Omega_i^{h_i}$ for $i = 2, 1, 2$ respectively, with zero initial iterate. The BEPS preconditioner therefore involves inversion of $A_{\Omega_1^{h_1}}$ once and $A_{\Omega_2^{h_2}}$ twice.

We refer the reader to Bramble *et al.* (1988) for the algorithmic details and the proof of the following convergence theorem.

Theorem 22 There exists a constant C, independent of h_1 and h_2, such that

$$\operatorname{cond}(M_{\text{BEPS}}^{-1}A) \leq C.$$

For a more parallelizable variant of the BEPS preconditioner, see Bramble, Ewing, Parashkevov and Pasciak (1992).

7.3. The FAC and AFAC algorithms for composite grids

The FAC (*Fast Adaptive Composite Grid Method*) and AFAC (*Asynchronous Fast Adaptive Composite Grid Method*) algorithms (McCormick, 1984; Mandel and McCormick, 1989; Widlund, 1989b) for solving (7.1) can be viewed as multilevel generalizations of the BEPS algorithm. The FAC algorithm corresponds to a multiplicative Schwarz algorithm based on the 'subproblems' on the refined grids $\Omega_i^{h_i}$ with matrices $A_{\Omega_i^{h_i}}$, restriction and extension maps $R_{\Omega_i^{h_i}}$ and $R^T_{\Omega_i^{h_i}}$ respectively, for $i = 1, \ldots, p$, see McCormick (1989) and Widlund (1989b) for the algorithmic details and the proof of the following convergence theorem.

Theorem 23 The convergence factor ρ of the FAC iteration is independent of the mesh sizes h_i and the number of levels, p, and depends only on the ratio $\max\{h_i/h_{i-1}\}$ and on the ratio of volumes (or areas) $\max\{|\Omega_{i-1}|/|\Omega_i|\}$.

An additive preconditioner M_{FAC} corresponding to the FAC iteration is

$$M_{\text{FAC}}^{-1} f \equiv \sum_{i=1}^{p} R^T_{\Omega_i^{h_i}} A^{-1}_{\Omega_i^{h_i}} R_{\Omega_i^{h_i}} f.$$

The convergence is not as good as the multiplicative version.

Theorem 24 There exists a constant C, independent of the mesh sizes h_i and the number of levels p, such that

$$\operatorname{cond}(M_{\text{FAC}}^{-1}A) \leq Cp,$$

Proof. See Widlund (1989b) and McCormick (1989). □

Part of the reason why M_{FAC} is nonoptimal is that some of the grid points in the refined regions are redundantly accounted for by all coarser level terms in the preconditioner. In the AFAC preconditioner (see Mandel and McCormick (1989), Widlund (1989b)), this redundancy is removed explicitly and *optimal* convergence is restored.

We introduce the following additional notation.

- For $i = 2, \ldots, p$, we use $A_{\Omega_i^{h_{i-1}}}$ to denote the stiffness matrix obtained by discretizing the elliptic problem based on the triangulation $\tau^{h_{i-1}}(\Omega_i)$ on Ω_i, i.e. using the space $V^{h_{i-1}}(\Omega_i) \cap H_0^1(\Omega_i)$.
- For $i = 2, \ldots, p$, the following additional *extension* maps will be used:

$$R^T_{\Omega_i^{h_{i-1}}} : V^{h_{i-1}}(\Omega_i) \to V^{h_1,h_2,\ldots,h_p},$$

  which denotes extension of interior nodal values on the grid $\tau^{h_{i-1}}(\Omega_i)$ to the composite grid. Its transpose will be a weighted *restriction* map onto the nodes in $\tau^{h_{i-1}}(\Omega_i)$.

The AFAC preconditioner M_{AFAC} is defined by

$$M^{-1}_{\text{AFAC}} \equiv R^T_{\Omega^h} A^{-1}_{\Omega^h} R_{\Omega^h} + \sum_{i=2}^{p} \left(R^T_{\Omega_i^{h_i}} A^{-1}_{\Omega_i^{h_i}} R_{\Omega_i^{h_i}} - R^T_{\Omega_i^{h_{i-1}}} A^{-1}_{\Omega_i^{h_{i-1}}} R_{\Omega_i^{h_{i-1}}} \right).$$

Thus, the AFAC preconditioner requires solving two subproblems (with different grid sizes) on each refined subregion Ω_i.

Theorem 25 There exists a constant C, independent of the mesh sizes h_i and the number of levels p, and dependent only on the ratios of the mesh sizes h_{i-1}/h_i and the ratios of the areas (or volumes) of the refined regions, such that

$$\text{cond}\,(M^{-1}_{\text{AFAC}} A) \leq C.$$

Proof. See Widlund (1989b), Dryja and Widlund (1989a) and McCormick (1989). □

8. Domain imbedding or fictitious domain methods

A dual approach to domain decomposition is the *domain imbedding* or *fictitious domain* method (another name is *capacitance matrix* method), in which problems on irregular domains are imbedded into larger problems on regular domains (such as rectangles or cubes) on which fast solvers are available, and the solution to the original problem is obtained iteratively by solving a sequence of problems on the extended domain. We will follow here the approach of Buzbee, Dorr, George and Golub (1971), Proskurowski and Widlund (1976), O'Leary and Widlund (1979), Börgers and Widlund (1990), Proskurowski and Vassilevski (1994). A rich literature on fictitious domain methods is found in the Soviet literature, and we refer the reader to Astrakhantsev (1978), Lebedev (1986), Marchuk *et al.* (1986) and Finogenov and Kuznetsov (1988), for details and references. Recently, very interesting alternative approaches based on control theory and optimization have been proposed for fictitious domain methods, and we refer the reader to Atamian, Dinh, Glowinski, He and Périaux (1991).

In this section, we briefly describe two examples of domain imbedding methods for solving a coercive (positive definite) Helmholtz problem on a domain Ω_1:

$$-\Delta u + cu = f, \qquad \text{in } \Omega_1, \quad \text{where } c \geq 0$$

with either Dirichlet boundary conditions $u = g_{\mathrm{D}}$ or Neumann boundary conditions $\partial u / \partial n = g_{\mathrm{N}}$ on $\partial \Omega_1$. In case $c = 0$, then the Neumann boundary data g_{N} must satisfy the standard compatibility conditions with f.

We imbed Ω_1 in a regular domain (for instance a rectangle or cube) $\Omega \supset \Omega_1$ and define $\Omega_2 = \Omega - \Omega_1$. The interface separating the two subregions will be denoted by $B = \partial\Omega_1 \cap \partial\Omega_2$ (which may equal $\partial\Omega_1$, in case Ω_1 is completely imbedded in Ω). The extended elliptic problem on Ω, in the above case will be the same Helmholtz problem (assuming that c is constant). We assume that the extended problem on Ω is discretized (by either finite element or finite difference methods) resulting in the linear system $Au = f$. We partition the unknowns as $u = [u_1, u_2, u_3]^T$, where u_1 and u_2 corresponds to the interior nodes in Ω_1 and Ω_2, respectively, while u_3 corresponds to the nodes on the interface B separating the two regions. The extended linear system then has the following block form:

$$\begin{bmatrix} A_{11} & 0 & A_{13} \\ 0 & A_{22} & A_{23} \\ A_{13}^T & A_{23}^T & A_{33} \end{bmatrix} \begin{bmatrix} u_1 \\ u_2 \\ u_3 \end{bmatrix} = \begin{bmatrix} f_1 \\ f_2 \\ f_3 \end{bmatrix}, \tag{8.1}$$

where A_{ii} are the coefficient matrices corresponding to the Dirichlet problem on Ω_i, for $i = 1, 2$, etc.

In the following two subsections, we describe imbedding methods for solving Neumann and Dirichlet problems on Ω_1.

8.1. Preconditioner M_{N} for the Neumann problem on Ω_1

Here we describe a domain imbedding preconditioner for the Neumann problem on Ω_1, following the development in Börgers and Widlund (1990). Using the block ordering in (8.1), the linear system corresponding to the Neumann problem on Ω_1 is

$$A_{\mathrm{N}} \begin{bmatrix} u_1 \\ u_3 \end{bmatrix} \equiv \begin{bmatrix} A_{11} & A_{13} \\ A_{13}^T & A_{33}^{(1)} \end{bmatrix} \begin{bmatrix} u_1 \\ u_3 \end{bmatrix} = \begin{bmatrix} g_1 \\ g_3 \end{bmatrix},$$

where $A_{33}^{(1)}$ corresponds to the contribution to A_{33} from Ω_1. We note that this matrix may be singular, in case $c = 0$ for the Helmholtz problem, with $[1, \ldots, 1]^T$ in its null space. In such cases, care must be exercised to ensure that the conjugate gradient iterates remain orthogonal to the null space.

The action of the inverse M_{N}^{-1} of a domain imbedding preconditioner M_{N}

to the above problem is defined by

$$M_{\mathrm{N}}^{-1}\begin{bmatrix} g_1 \\ g_3 \end{bmatrix} \equiv \begin{bmatrix} I & 0 & 0 \\ 0 & 0 & I \end{bmatrix} A^{-1} \begin{bmatrix} I & 0 \\ 0 & 0 \\ 0 & I \end{bmatrix} \begin{bmatrix} g_1 \\ g_3 \end{bmatrix}.$$

This involves the solution of the extended problem with right-hand sides $g_1, 0$ and g_3 on Ω_1, Ω_2 and B respectively,

By using the block factorization of A, it can be easily verified that

$$M_{\mathrm{N}} = \begin{bmatrix} A_{11} & A_{13} \\ A_{13}^T & A_{33}^{(1)} + S^{(2)} \end{bmatrix},$$

where $S^{(2)} = A_{33}^{(2)} - A_{23}^T A_{22}^{-1} A_{23}$ is the Schur complement of the nodes on B with respect to the nodes in the domain Ω_2. Thus, the preconditioner M_{N} is a modification of the Neumann problem, by the addition of the Schur complement to a diagonal block. The convergence rate is optimal.

Theorem 26 The exists a constant C, independent of h, such that

$$\mathrm{cond}\,(M_{\mathrm{N}}^{-1} A_N) \leq C.$$

Proof. See Börgers and Widlund (1990). □

Finally, we note that the problem of choosing a grid on Ω that allows a fast solver, and whose restriction on Ω_1 allows for suitable discretization on Ω_1 is discussed at length in Börgers and Widlund (1990), where a triangulation algorithm is also described. Additionally, we note that exact solvers on Ω may be replaced by suitable inexact solvers, especially based on a topologically equivalent grid, without affecting the optimal convergence rate.

8.2. Capacitance matrix solution of the Dirichlet problem on Ω_1

Here, we consider the solution of the following linear system corresponding to the Dirichlet problem on Ω_1:

$$A_{11} u_1 = f_1.$$

Unfortunately, a straightforward modification of preconditioner M_{N} to the Dirichlet case, i.e.

$$\hat{M}_{\mathrm{D}}^{-1} f_1 \equiv \begin{bmatrix} I & 0 & 0 \end{bmatrix} A^{-1} \begin{bmatrix} I \\ 0 \\ 0 \end{bmatrix} f_1,$$

does not work very well. Indeed, $\mathrm{cond}\,(\hat{M}_{\mathrm{D}}^{-1} A_{11})$ grows as $\mathcal{O}(h^{-1})$, see Börgers and Widlund (1990). An alternative preconditioner based on the

Neumann problem for the exterior domain $\Omega_2 = \Omega - \Omega_1$, is described in the same article.

The solution procedure we describe for the Dirichlet problem will be based on a recently proposed capacitance matrix algorithm of Proskurowski and Vassilevski (1994). The solution of $A_{11}u_1 = f_1$ will be computed in a few stages, just as in Schur complement based domain decomposition methods, and it is based on the following two matrix identities relating the the solution u_1 on Ω_1 to the extended problem on Ω:

Lemma 3 Let the Schur complement of A be

$$S \equiv A_{33} - A_{13}^T A_{11}^{-1} A_{13} - A_{23}^T A_{22}^{-1} A_{23}.$$

Then the following identities hold:

$$(1) \quad A_{11}^{-1} = \begin{bmatrix} I & 0 & 0 \end{bmatrix} A^{-1} \left[\begin{bmatrix} I & 0 & 0 \\ 0 & I & 0 \\ 0 & 0 & I \end{bmatrix} - \begin{bmatrix} 0 & 0 & 0 \\ 0 & 0 & 0 \\ 0 & 0 & S \end{bmatrix} A^{-1} \right] \times \begin{bmatrix} I & 0 & 0 \end{bmatrix}^T,$$

$$(2) \quad C \equiv S^{-1} = \begin{bmatrix} 0 & 0 & I \end{bmatrix} A^{-1} \begin{bmatrix} 0 & 0 & I \end{bmatrix}^T.$$

Proof. This can be verified directly using the block factorization of A. □

The algorithm is a direct implementation of the first identity.

Capacitance matrix method for solving $A_{11}u_1 = f_1$

1 Solve $A \begin{bmatrix} y_1, & y_2, & y_3 \end{bmatrix}^T = \begin{bmatrix} f_1, & 0, & 0 \end{bmatrix}^T$.
2 Compute $w_3 = Sy_3$ by solving $Cw_3 = y_3$, using a preconditioned conjugate gradient method, with a matrix–vector product involving C computed by identity (2) in the lemma above (requiring solves with A). The inverse of any preconditioner for S (e.g. from Section 3) can be used as a preconditioner for C.
3 Solve $A \begin{bmatrix} v_1, & v_2, & v_3 \end{bmatrix}^T = \begin{bmatrix} 0, & 0, & w_3 \end{bmatrix}^T$.
4 Set $u_1 = y_1 - v_1$.

Theorem 27 For preconditioners M for C such that M^{-1} is a spectrally equivalent preconditioner for S, cond$\,(M^{-1}C)$ is bounded independent of the mesh size h.

Proof. See Proskurowski and Vassilevski (1994). □

We refer the reader to Atamian *et al.* (1991), and to Proskurowski and Vassilevski (1992) for domain imbedding algorithms for solving indefinite and nonsymmetric problems.

9. Convection–diffusion problems

In this section, we briefly describe some domain decomposition algorithms for solving the nonsymmetric linear systems arising from the discretization of convection–diffusion problems such as

$$-\epsilon\Delta u + \boldsymbol{b} \cdot \nabla u + c_0 u = f, \text{ in } \Omega, \ u = 0, \text{ on } \partial\Omega, \tag{9.1}$$

where $\epsilon > 0$ represents viscosity, $\boldsymbol{b}$ is a vector field and $c_0 \geq 0$. Though such problems are elliptic, they pose some difficulties for iterative solution. In case the diffusion term dominates the convection term, (such as when $\|\boldsymbol{b}\|h/\epsilon \ll 1$) most of the domain decomposition algorithms we have described, including the Schwarz and Schur methods, can be extended to solve the nonsymmetric problem, with suitable modifications such as replacing conjugate gradient methods by GMRES, BiCG, BiCGStab or QMR methods, see Freund, Golub and Nachtigal (1992). However, the convergence rates of the standard algorithms deteriorate as ϵ approaches zero, unless a coarse grid discretization of the original problem is solved exactly on a grid of size H, where $H < H_0$ (a constant), see Cai and Widlund (1992, 1993), Xu (1992b), Xu (1992c), Xu and Cai (1992). This coarse grid condition has been known in the multigrid literature. Additionally, for small diffusion, the solution is more strongly coupled along the characteristics of the convection problem, making the solution procedure sensitive to the ordering of nodes. Thus, the solution of these nonsymmetric problems by standard algorithms poses some difficulties when the convection term dominates.

In Sections 9.1 and 9.2, we describe the extension of several many subdomain overlapping and nonoverlapping algorithms to the nonsymmetric case. Following that, in Sections 9.3 and 9.4, we briefly describe alternative approaches that have been recently proposed by Gastaldi, Quarteroni and Sacchi-Landriani (1990), Glowinski, Périaux and Terrasson (1990b), and Ashby, Saylor and Scroggs (1992) based on two subdomain decompositions that couple elliptic and hyperbolic problems using an asymptotics approach.

Throughout this section, we will assume that problem (9.1) is discretized by a stable scheme (such as upwind finite differences, streamline diffusion finite elements or a scheme based on artificial viscosity), resulting in a linear system:

$$L(\epsilon)u = \epsilon A u + C u = f, \tag{9.2}$$

where $A = A^T > 0$ is the discretization of the Laplacian, and C corresponds to the discretization of the convection and the $c_0 u$ term.

9.1. Schwarz algorithms for convection–diffusion problems

As in Section 2, let $\hat{\Omega}_1, \ldots, \hat{\Omega}_p$ denote an overlapping covering of Ω, with corresponding restriction and extension maps R_i and R_i^T, respectively. The coarse grid restriction and extension maps will be denoted by R_H and R_H^T respectively.

A straightforward extension of the additive Schwarz preconditioner for $L(\epsilon)$ is defined by

$$M_{\text{as},1}^{-1} = R_H^T \left(\epsilon A_H + C_H\right)^{-1} R_H + \sum_{i=1}^{p} R_i^T \left(\epsilon A_i + C_i\right)^{-1} R_i,$$

where $\epsilon A_H + C_H = R_H L(\epsilon) R_H^T$ and $\epsilon A_i + C_i = R_i L(\epsilon) R_i^T$ are the coarse grid and local matrices, respectively. The corresponding linear system can be solved by any suitable nonsymmetric conjugate gradient like procedure. In the nonsymmetric case, we also have the following variant:

$$M_{\text{as},2}^{-1} = R_H^T \left(\epsilon A_H + C_H\right)^{-1} R_H + \sum_{i=1}^{p} R_i^T \left(\epsilon A_i\right)^{-1} R_i,$$

where the local convection–diffusion problems are replaced by more easily solvable (symmetric, positive definite) diffusion problems. The following convergence bounds have been established by Cai and Widlund (1993) and Xu and Cai (1992).

Theorem 28 There exists a maximum coarse grid size $H_0(\epsilon, h, \boldsymbol{b}, c_0)$ such that if $H < H_0(\epsilon, h, \boldsymbol{b}, c_0)$, then the rate of convergence of both the additive Schwarz preconditioned systems is independent of $H < H_0$ and h.

An explicit form for $H_0(\epsilon, h, \boldsymbol{b}, c_0)$ has not been derived in the literature (to the knowledge of the authors), but heuristically, it *may* depend on ϵ and h as

$$H_0 \approx \max\left\{\frac{\epsilon}{\|\boldsymbol{b}\|}, h\right\},$$

and this decreases as $\epsilon \to 0$. Consequently, the cost of solving the coarse grid problem can increase with smaller ϵ, and places some limitations on the convergence rate and efficiency of the algorithms, see Cai, Gropp and Keyes (1992).

The *multiplicative Schwarz* method can also be extended to the nonsymmetric case, analogously. However, to ensure convergence without acceleration, care must be exercised so that if approximation of the local problems are used, they must be spectrally close to the true local problems. We refer the reader to Xu (1992b), Cai and Widlund (1993), Xu and Cai (1992) and Wang (1993) for the details.

9.2. Schur complement based algorithms for convection–diffusion problems

As for the symmetric, positive definite case described in Section 3, we partition the domain Ω into p nonoverlapping subregions $\Omega_1, \ldots, \Omega_p$, with interface B. The block form of the system becomes:

$$\begin{bmatrix} L_{II} & L_{IB} \\ L_{IB}^T & L_{BB} \end{bmatrix} \begin{bmatrix} u_I \\ u_B \end{bmatrix} = \begin{bmatrix} f_I \\ f_B \end{bmatrix}, \tag{9.3}$$

where $L_{II} = \epsilon A_{II} + C_{II}$, etc. The Schur complement system is:

$$Su_B = \tilde{f}_B, \text{ where } S = L_{BB} - L_{IB}^T L_{II}^{-1} L_{IB}, \text{ and } \tilde{f}_B = f_B - L_{IB}^T L_{II}^{-1} f_I.$$

The solution procedure is analogous to the symmetric, positive definite case. Once u_B is determined, u_I is obtained as $u_I = L_{II}^{-1}(f_I - L_{IB}u_B)$.

The nonsymmetric Schur complement system can be solved by a preconditioned iterative method (in conjunction with GMRES or suitable algorithms), with any of the preconditioners of Section 3. However, as previously noted, care must be exercised so that the size of the coarse grid problem is sufficiently small with $H < H_0$. For instance, the nonsymmetric BPS preconditioner has the form:

$$M_{\text{BPS}}^{-1} = R_H^T L_H^{-1} R_H + \sum_{i=1}^{n} R_{E_i}^T S_{E_i E_i}^{-1} R_{E_i},$$

where the edge problems $S_{E_i E_i}$ can be replaced by preconditioners applicable in the symmetric, positive definite case, or preferably preconditioners that adapt to the convection term. We refer the reader to Cai and Widlund (1993), D'Hennezel (1992) and Chan and Keyes (1990) for the details.

For a numerical comparison of both Schwarz and Schur complement algorithms, see Cai *et al.* (1992).

9.3. Elliptic–hyperbolic approximation of convection–diffusion problems

Classical asymptotics based studies of singular perturbation problems have much in common with domain decomposition. Typically, the domain is decomposed into two regions, one corresponding to a boundary or interior layer region and referred to as the *inner region*, where the full viscous problem is solved, and an *outer region*, where the inviscid or hyperbolic problem is solved. The inner and outer solutions are required to satisfy certain compatibility conditions on the interface or region of overlap between the two subregions. In problems where asymptotic expansions may not be tractable, an alternative is to use numerical approximations in each of the subregions, and to couple the solutions together using matching conditions. Several detailed and interesting studies have been conducted in the domain decomposition framework, and we provide references to some of the literature.

For second-order scalar elliptic convection diffusion problems, Gastaldi *et al.* (1990) proposed a mixed elliptic–hyperbolic approximation of the convection diffusion problem. The domain is partitioned into two nonoverlapping subregions: Ω_{E}, where the full elliptic problem is solved, and Ω_{H} where the hyperbolic problem obtained by dropping the viscous term is solved. They proposed new *transmission boundary conditions* coupling the two subproblems, obtained by using a vanishing viscosity procedure. Additionally, a Dirichlet–Neumann type iterative procedure was proposed that solves the resulting mixed, elliptic–hyperbolic approximation of the convection diffusion problem. Theoretical and numerical estimates of the approximation error and convergence rates are provided in Gastaldi *et al.* (1990) and the references contained therein. A detailed theory has now been developed by Quarteroni and Valli (1990) for various heterogeneous approximations, and studies are being conducted for the compressible Navier–Stokes equations.

An alternative approach based on overlapping subregions was used by Glowinski *et al.* (1990b) for coupling the viscous and inviscid compressible Navier–Stokes equations. The domain is decomposed into two overlapping subregions corresponding to viscous and inviscid regions, and a least-squares minimization is applied to a functional of the two solutions on the region of overlap. The resulting least-squares problem is then solved via a nonlinear GMRES procedure.

For alternative studies, more closely aligned with classical boundary layer expansions, we refer the reader to Hedstrom and Howes (1990), Chin, Hedstrom, McGraw and Howes (1986), Gropp and Keyes (1993), and to Garbey (1992) and Scroggs (1989), for studies on conservation laws. An interesting domain decomposition method based on an approximate factorization of the convection–diffusion operator was recently proposed by Nataf and Rogier (1993).

9.4. Block preconditioners for convection–diffusion problems

In this section, we briefly describe an alternative block matrix preconditioner (without coarse grid solves) for the nonsymmetric linear system (9.2). This preconditioner was recently proposed by Ashby *et al.* (1992), motivated by matched asymptotic expansions, and is referred to as the *physically motivated domain decomposition preconditioner.*

We consider a decomposition of Ω into two regions, a hyperbolic region Ω_{H} and an elliptic region Ω_{E}, with an overlap of width equal to one grid size. Corresponding to this partition, the unknowns can be ordered $u = [u_1, u_2]^T$, where u_1 corresponds to the interior unknowns in the hyperbolic region Ω_{H} and u_2 corresponds to the interior unknowns in the elliptic region Ω_{E}. Note that due to one grid overlap, there are no 'boundary unknowns'. The linear system (9.2) then takes on the block form:

$$\begin{bmatrix} \epsilon A_{11} + C_{11} & \epsilon A_{12} + C_{12} \\ \epsilon A_{12}^T + C_{21} & \epsilon A_{22} + C_{22} \end{bmatrix} \begin{bmatrix} u_{\mathrm{H}} \\ u_{\mathrm{E}} \end{bmatrix} = \begin{bmatrix} f_{\mathrm{H}} \\ f_{\mathrm{E}} \end{bmatrix}.$$

Based on the above block partition, the physically motivated domain decomposition preconditioner M_{pmdd} of Ashby *et al.* (1992) is defined by

$$M_{\text{pmdd}} = \begin{bmatrix} C_{11} & 0 \\ \epsilon A_{12}^T + C_{21} & \epsilon A_{22} + C_{22} \end{bmatrix}. \tag{9.4}$$

It is block lower triangular and inverting it involves inverting the two diagonal blocks. The motivation for setting the diffusion term to zero in the $(1,1)$ block is that it then corresponds to a hyperbolic problem on region Ω_{H} (analogous to asymptotic expansions for singular perturbation problems). For most direction fields $\boldsymbol{b}$, and for upwind finite difference discretizations, C_{11} can be inverted by 'marching along characteristics'. That is, if the subregion Ω_{H} is suitably chosen, the indices of the nodes in Ω_{H} may be reordered to produce a *lower triangular matrix* C_{11}, which can be easily solved since it is sparse. The block $\epsilon A_{22} + C_{22}$ may be more difficult to invert, since in the elliptic region the grid may be refined, and the diffusion term may dominate the convection term. In such cases, it may be suitable to replace $\epsilon A_{22} + C_{22}$ by the symmetric, positive definite matrix ϵA_{22} (or suitable parallelizable preconditioners).

Numerical tests conducted in Chan and Mathew (1993) indicate that on uniform grids, with suitably chosen elliptic and hyperbolic regions, the convergence rate of the M_{pmdd} preconditioned system improves as $\epsilon \to 0$, for fixed mesh size h. However, for fixed ϵ, as $h \to 0$, the convergence rate deteriorates mildly. It is speculated in Chan and Mathew (1993) that this deterioration may be due to the approximation of the elliptic term by a hyperbolic term in Ω_{H}. However, since in general, the mesh size does not need refinement on the hyperbolic region Ω_{H}, but only in the boundary layer region Ω_{E}, the above algorithm may be more robust with respect to local refinement in Ω_{E}.

A variant of this method was studied in Chan and Mathew (1993), and corresponds to a matrix version of the Dirichlet–Neumann preconditioner for the elliptic–hyperbolic approximation of Gastaldi *et al.* (1990), and is based on the use of a Neumann problem on Ω_{E}. In matrix terms, both preconditioners correspond to variants of the classical block Gauss–Seidel preconditioner, i.e. a block lower triangular matrix, whose diagonal blocks are modified to permit ease of solvability.

10. Parabolic problems

In this section, we briefly describe domain decomposition algorithms for solving the linear systems obtained by implicit discretization of parabolic problems. We consider the following model parabolic problem for $(x,t) \in$

$\Omega \times [0,T]$:

$$\begin{cases} u_t = \nabla \cdot (a\nabla u) + f, & \text{on } \Omega \times [0,T], \\ u(x,0) = u_0(x), & \text{on } \Omega, \\ u(x,t) = 0, & \text{on } \partial\Omega \times [0,T]. \end{cases} \tag{10.1}$$

To be specific, we consider a discretization by finite differences in space and backward Euler in time, resulting in

$$\begin{cases} (u^{n+1} - u^n)/\tau = -Au^{n+1} + f^{n+1}, \\ u^0 = u_0^h, \end{cases}$$

where A is a symmetric positive definite matrix corresponding to the discretization of $-\nabla \cdot (a\nabla u)$ and τ is the time step. At each time step, the following linear system must be solved:

$$(I + \tau A)\, u^{n+1} = u^n + \tau f^{n+1}. \tag{10.2}$$

Similar equations are obtained for Crank–Nicholson in time, and finite elements in space. The implicit system (10.2) corresponds to a discretization of the elliptic operator $L(\tau)u = \tau u - \nabla \cdot (a\nabla u)$ and, consequently, most of the domain decomposition algorithms of Sections 2 and 3 are applicable. However, there are some crucial differences that make this system easier to solve: the condition number of $I + \tau A$ is bounded by $\mathcal{O}(\tau h^{-2})$ which can be relatively smaller than cond(A) if τ is small (say $\tau = \mathcal{O}(h)$ or $\tau = \mathcal{O}(h^2)$). Consequently:

- The entries of the Green function $(I + \tau A)^{-1}$ can be shown to decay more rapidly away from the diagonal than the entries of A^{-1}, and so depending on τ, a *coarse grid problem may not be required* for global communication of information.
- It is possible to use just one iteration of the domain decomposition method and still maintain a stable approximation preserving the local truncation error.

In Section 10.1, Schwarz algorithms are described for (10.2), with modifications in the coarse problem. In Sections 10.2 and 10.3, algorithms that require only *one* iteration are described.

10.1. Schwarz preconditioners for parabolic problems

We follow here the development due to Lions (1988) and Cai (1991; 1993). As in Section 2, we decompose Ω into an overlapping covering $\hat{\Omega}_1, \ldots, \hat{\Omega}_p$, with corresponding restriction and extension maps R_i and R_i^T, respectively. Similarly, R_H and R_H^T will denote the restriction and interpolation maps corresponding to the coarse grid. The local submatrices will be denoted $L_i(\tau) \equiv I_i + \tau A_i = R_i\,(I + \tau A)\, R_I^T$, and the coarse grid problem by $L_H(\tau) \equiv$

$R_H\,(I+\tau A)\,R_H^T$. We define two additive Schwarz preconditioners for $L(\tau) \equiv I+\tau A$:

$$M_{\mathrm{as},1}^{-1} = \sum_{i=1}^{p} R_i^T L_i(\tau)^{-1} R_i,$$

and

$$M_{\mathrm{as},2}^{-1} = \sum_{i=1}^{p} R_i^T L_i(\tau)^{-1} R_i + R_H^T L_H(\tau)^{-1} R_H.$$

The following convergence results are proved in Cai (1991).

Theorem 29 If $\tau \le CH^2$, then cond $(M_{\mathrm{as},1}^{-1} L(\tau))$ is bounded by a constant C_1 independent of τ, H and h. For larger τ, cond $(M_{\mathrm{as},2}^{-1} L(\tau))$ is bounded by a constant C_2 independent of τ, H and h.

Thus, if $\tau \le CH^2$, then a preconditioner without a coarse model may be used effectively. However, if τ is large, a coarse grid correction term must be used in order to maintain a constant rate of convergence. Similar results hold for multiplicative Schwarz methods and for Schur complement based methods. We refer the reader to Cai (1991) for the details.

10.2. One iteration based approximations: overlapping subdomains

As mentioned before, it is possible to obtain approximate solutions w^{n+1} of system (10.2) that are accurate to within the local truncation error of the true numerical solution u^{n+1}:

$$\|w^{n+1} - u^{n+1}\| \le \mathcal{O}(\epsilon),$$

where $\mathcal{O}(\epsilon)$ is the local truncation error, and which can be constructed by solving only one problem on suitably chosen subdomains. Here, we briefly describe one such algorithm proposed by Kuznetsov (1991; 1988) and Meurant (1991).

Kuznetsov's method is based on the observation that the entries in the ith row of the discrete Green function $G(\tau)$ (where $G(\tau) = (I+\tau A)^{-1}$) decays rapidly as the distance between the nodes $\{x_i\}$ increases, specifically

$$|G_{ij}(\tau)| \le \epsilon, \text{ when } |x_i - x_j| \ge c\sqrt{\tau}\log(\epsilon^{-1}). \tag{10.3}$$

Thus, if the right-hand side of equation (10.2) has support in a subregion Ω_i, then the solution will decay rapidly with distance with a rate of decay given by (10.3).

Accordingly, let $\Omega_1, \ldots, \Omega_p$ denote a partition of Ω into p *nonoverlapping subregions*, and let $\hat{\Omega}_i \supset \Omega_i$ denote an extension of Ω_i containing all points in Ω within a distance of $c\sqrt{\tau}\log(\epsilon^{-1})$. Thus, $\hat{\Omega}_1, \ldots, \hat{\Omega}_p$ form an overlapping covering of Ω, as in Schwarz algorithms. To approximately solve

$$(I + \tau A)\, u^{n+1} = g,$$

the right-hand side is first partitioned as

$$g = g_1 + \cdots + g_p, \quad \text{where support}(g_i) \subset \overline{\Omega}_i.$$

(Such a partition can be obtained, for instance, analogously to the construction in the proof of the partition lemma in Theorem 16.) Next, solve the following problem on each extended subdomain $\hat{\Omega}_i$:

$$L_{\hat{\Omega}_i} u_i = g_i, \quad \text{for } i = 1, \ldots, p,$$

where $L_{\hat{\Omega}_i} \equiv R_{\hat{\Omega}_i}(I + \tau A)R^T_{\hat{\Omega}_i}$ denotes the principal submatrix of $I + \tau A$ corresponding to the interior nodes on $\hat{\Omega}_i$. The approximate solution w^{n+1} is defined as

$$w^{n+1} \equiv u_1 + \cdots + u_p.$$

The following error bound is proved in Kuznetsov (1988).

Theorem 30 If the extended subdomains have overlap of size

$$\mathcal{O}(\sqrt{\tau} \log(\epsilon^{-1})),$$

the error satisfies

$$\|w^{n+1} - u^{n+1}\| \le \mathcal{O}(\epsilon).$$

Thus, for instance, when the time step $\tau = h$ and $\epsilon = h^2$, the overlap should be approximately $\mathcal{O}(\sqrt{h} \log(h))$. Consequently, the extended subdomains must have a minimum overlap of the size prescribed above in order for the truncation error to be acceptable. This provides a constraint on the choice of subdomains. The case of convection diffusion problems is discussed in Kuznetsov (1990).

10.3. Alternative one iteration based approximations

An alternative algorithm that provides an approximate solution of (10.2) was proposed by Dryja (1991) and corresponds to a domain decomposed matrix splitting (fractional step method) involving two nonoverlapping subregions. The resulting scheme can be shown to be unconditionally stable. Unfortunately, the discretization error of the splitting scheme becomes the square root of the discretization error of the original scheme, see Dryja (1991) for the details. It is possible to recover the original discretization error by using an alternative splitting, see Laevsky (1992; 1993).

Kuznetsov (1988) proposed an explicit–implicit scheme to solve parabolic problems based on a partition of Ω into nonoverlapping regions. The boundary value of u^{n+1} on the interface B is first computed using an explicit method (or even an implicit scheme) in a small neighbourhood of B. Using these boundary values, Dirichlet problems can be solved on each subdomain to provide the solution u^{n+1} on the whole domain Ω. This idea is particularly appealing on grids containing regions of refinement.

Another alternative approach was proposed by Dawson and Du (1991), Dawson, Du and Dupont (1991), in which the domain is partitioned into many nonoverlapping subdomains with interface B. Special basis function are constructed having support in a small 'tube' of width $\mathcal{O}(H)$ containing the interface B. In the first step approximate boundary values are computed on B using these special basis functions (involving some overhead cost). Finally, using these boundary values, the solution u^{n+1} is determined at the interior of the subdomains.

11. Mixed finite elements and the Stokes problem

In this section, we briefly describe some domain decomposition methods for solving the linear systems arising from mixed finite element discretizations of elliptic problems and discretizations of the steady Stokes equations (see Girault and Raviart (1986), Brezzi and Fortin (1991) for details on mixed finite element discretizations). Studies of domain decomposition methods for mixed finite element discretizations of elliptic problems were initiated by Glowinski and Wheeler (1988), while studies of domain decomposition for the Stokes problem were initiated by Lions (1988), Fortin and Aboulaich (1988), Bramble and Pasciak (1988) and Quarteroni (1989).

The mixed formulation of an elliptic problem: $-\nabla \cdot (a\nabla p) = f$ on Ω, with Neumann boundary conditions $\boldsymbol{n} \cdot a\nabla p = g$ on $\partial\Omega$ is given by

$$\begin{cases} a^{-1}\boldsymbol{u} + \nabla p &= 0, & \text{in } \Omega, & \text{Darcy's law} \\ \nabla \cdot \boldsymbol{u} &= f, & \text{in } \Omega, & \text{Conservation of mass} \\ \boldsymbol{n} \cdot \boldsymbol{u} &= -g, & \text{in } \partial\Omega, & \text{Flux boundary condition} \end{cases}$$

where the compatibility condition

$$\int_\Omega f \,\mathrm{d}x + \int_{\partial\Omega} g \,\mathrm{d}s = 0$$

is assumed. The Stokes problem with Dirichlet boundary conditions for the velocity u is

$$\begin{cases} -\nu\Delta\boldsymbol{u} + \nabla p &= f, & \text{in } \Omega, \\ \nabla \cdot \boldsymbol{u} &= 0, & \text{in } \Omega, \\ \boldsymbol{u} &= 0, & \text{on } \partial\Omega. \end{cases}$$

In both problems $\boldsymbol{u}$ refers to the velocity and p to the pressure.

After discretization, both these problems result in linear systems of the following form:

$$\begin{bmatrix} A & B^T \\ B & 0 \end{bmatrix} \begin{bmatrix} u \\ p \end{bmatrix} = \begin{bmatrix} f \\ g \end{bmatrix}, \tag{11.1}$$

where u is the discrete velocity unknowns and p is the discrete pressure unknowns. Note that (11.1) is symmetric but indefinite and cannot be solved directly by the conjugate gradient method. Such systems are usually solved by block matrix and optimization based solution procedures. The square matrix A is symmetric and positive definite for both the Stokes and mixed case. In particular, A is block diagonal in the Stokes case, with diagonal blocks corresponding to discretization of the Laplacian. In the mixed elliptic case, A corresponds to a discretization of a^{-1}, the inverse of the coefficients a of the elliptic problem. The matrix B^T is rectangular and represents a discretization of the gradient, while its transpose B represents a discretization of the divergence operator. In many applications B^T has a null space spanned by $[1, \ldots, 1]^T$.

11.1. Methods based on elimination of the velocity

A simple procedure to solve (11.1) is to eliminate u and solve the reduced system for p:

$$Sp \equiv -BA^{-1}B^T p = g - BA^{-1}f,$$

after which u can be obtained by $u = A^{-1}(f - B^T p)$. Note that the Schur complement S is negative definite and hence a conjugate gradient type method can be used. Each matrix–vector product with S can be computed at the cost of solving a linear system with coefficient matrix A.

For the Stokes problem, it can be shown that S is well conditioned and requires no preconditioning. However, the matrix A is block diagonal with diagonal blocks corresponding to the Laplacian, and domain decomposition preconditioners can be applied to A. We refer the reader to Bramble and Pasciak (1988) for the details, where the Stokes problem is reformulated as a positive definite linear system and, additionally, a nonoverlapping domain decomposition algorithm is described. See Rusten and Winther (1992) and Rusten (1991) for an interesting algorithm for preconditioning the entire system without eliminating either u or p.

For the mixed elliptic case, the operator S is not well conditioned, and the above elimination method is not as attractive, see Wheeler and Gonzalez (1984). However, if a *dual* formulation of the mixed problem is used, see Arnold and Brezzi (1985), then the resulting Schur complement for the pressure becomes a nonconforming finite element discretization of the corresponding elliptic problem for the pressure. Efficient domain decomposition preconditioners have been proposed for such nonconforming discretizations (corresponding to the Schur complement S in the *dual* formulation), see

Cowsar, Mandel and Wheeler (1993), Cowsar (1993), Sarkis (1993) and Meddahi (1993).

11.2. Methods based on divergence free velocities

An alternative to algorithms based on elimination of the velocity are those in which the pressure is implicitly eliminated. These methods are based on the observation that the pressure corresponds to a Lagrange multiplier in the following constrained minimization problem:

$$\min \tfrac{1}{2}u^T Au - u^T f, \quad \text{subject to} \quad Bu = g,$$

see Girault and Raviart (1986), Lions (1988), Glowinski and Wheeler (1988) and Quarteroni (1989). In particular, if the divergence constraint $Bu = g$ can be reduced to $Bu = 0$, (i.e. the feasible set of velocities corresponds to a linear subspace of divergence free velocities), then in this subspace the problem becomes positive definite because

$$\begin{bmatrix} u \\ p \end{bmatrix}^T \begin{bmatrix} A & B^T \\ B & 0 \end{bmatrix} \begin{bmatrix} u \\ p \end{bmatrix} = u^T Au + 2p^T Bu = u^T Au > 0. \tag{11.2}$$

This positive definiteness provides the basis for applying standard conjugate gradient methods to determine the minimum velocity solution within the feasible set of velocities satisfying the divergence constraint.

Based on the space of divergence free velocities, Glowinski and Wheeler (1988) proposed a nonoverlapping domain decomposition method for mixed finite element discretizations of elliptic problems. In subsequent articles, Glowinski, Kinton and Wheeler (1990a) and Cowsar and Wheeler (1991) proposed improved preconditioners for the corresponding Schur complement system. Nonoverlapping algorithms for the Stokes problem were proposed by Bramble and Pasciak (1988) and Quarteroni (1989).

Schwarz alternating algorithms for the Stokes problem were proposed in Lions (1988) and Fortin and Aboulaich (1988), see also Pahl (1993), and are based on implicit elimination of the pressure. They were extended to the case of mixed finite element discretizations of elliptic problems in Mathew (1989; 1993a) and Ewing and Wang (1991). In the following, we briefly describe the basic linear algebraic issues for formulating Schwarz algorithms in the mixed case.

Two issues need to be addressed in order to define a Schwarz method involving subproblems of the form:

$$\begin{bmatrix} * & * & * & * \\ * & A_i & B_i^T & * \\ * & B_i & 0 & * \\ * & * & * & * \end{bmatrix} \begin{bmatrix} * \\ u_i \\ p_i \\ * \end{bmatrix} = \begin{bmatrix} * \\ W_i \\ F_i \\ * \end{bmatrix}, \tag{11.3}$$

after some suitable reordering of (11.1). They are:

1 The submatrix B_i may be singular, depending on the boundary conditions, due to the nonuniqueness of the pressure. In such a case, F_i must have mean value zero if $[1, \ldots, 1]^T$ spans the null space of B_i (as is often the case). This corresponds to the compatibility condition for solvability of the subproblem.

2 When B_i is singular, due to the nonuniqueness of the local pressure solution p_i, its mean value on the subregion is arbitrary and should be suitably prescribed in order to compute a globally defined pressure p_h.

The first difficulty can be handled by reducing the problem to one involving divergence free velocities. The second difficulty can be treated by sequentially modifying the local pressure solutions so that they have the same mean value with adjacent pressures on the regions of overlap. The algorithm can now be outlined:

1 Determine a velocity u^* satisfying $Bu^* = g$. Then, the correction $\tilde{u} = u - u^*$ to the velocity satisfies: $B\tilde{u} = 0$, and all subsequent local subproblems will be compatible with the zero flux data.

2 Next, apply the Schwarz methods to compute the divergence free velocity $\tilde{u}$ by solving local problems which have the same form as (11.1) and which involves local velocities and pressures.

3 Finally, determine a global pressure using the local pressures determined in Step 2.

We refer the reader to Ewing and Wang (1991) and Mathew (1993a) for the details.

For suitable choices of overlapping subregions and a coarse mesh, as in Section 2, the convergence rates of the additive and multiplicative Schwarz algorithms in the mixed elliptic case has been shown to be independent of H and h, see Ewing and Wang (1991) and Mathew (1993b).

12. Other topics

In this section, we provide some references to several domain decomposition procedures that we do not have space to discuss in any details. A good source of references is the set of conference proceedings mentioned in the introduction.

12.1. Biharmonic problem

For conforming finite element discretizations of the biharmonic problem based on *Hermite* finite elements, the additive and multiplicative Schwarz algorithms as well as the multilevel Schwarz algorithms have been developed with optimal convergence rates, see Zhang (1991; 1992a,c). See also Scapini (1990). Nonoverlapping domain decomposition algorithms based on

Hermite elements are discussed in Sun and Zou (1991), Hoffmann and Zou (1992), and Scapini (1991).

Algorithms for finite difference discretizations of the biharmonic equation (such as the 13-point stencil) pose additional difficulties. Indeed, if a nonoverlapping decomposition is used for such discretizations, the *interface* must consist of *two lines* in order to decouple the local subproblems. This requires modifications in the usual construction of interface preconditioners for the Schur complement system. We refer the reader to Tsui (1991) and Chan, Weinan and Sun (1991b). Thus far, to the knowledge of the authors, the Schwarz algorithms have not been studied in the case of finite difference discretizations of the biharmonic problem.

12.2. Spectral, spectral element and p version finite elements

For a general discussion on domain decomposition for spectral methods, we refer the reader to Canuto *et al.* (1988), and for a discussion on spectral element methods to Bernardi, Maday and Patera (1989), Maday and Patera (1989), Bernardi, Debit and Maday (1989), Bernardi and Maday (1992) and Fischer and Rønquist (1993).

The Schwarz algorithm for spectral methods was proposed in Morchoisne (1984), Canuto and Funaro (1988). More recently, Dirichlet–Neumann-type domain decomposition algorithms were proposed by Funaro, Quarteroni and Zanolli (1988). For boundary layer and elliptic–hyperbolic problems, spectral methods are described in Gastaldi *et al.* (1990). Applications and techniques of pseudospectral domain decomposition methods in fluid dynamics are described by Phillips (1992).

The earliest domain decomposition algorithm for p version finite elements was proposed by Babuška, Craig, Mandel and Pitkäranta (1991), in two dimensions. Since then, algorithms similar to the Neumann–Neumann, wirebasket and Schwarz methods have been developed for p version finite elements having almost optimal convergence rates (polylogarithmic growth in p). We refer the reader to Mandel (1989; 1990), for Neumann–Neumann and wirebasket-type algorithms, and to Pavarino (1992, 1993a,b) and Pavarino and Widlund (1993) for Schwarz, local refinement and wirebasket type algorithms for p version finite elements.

12.3. Indefinite Helmholtz problems

The solution of the indefinite Helmholtz problem

$$-\Delta u - k^2 u = f$$

is a difficult problem for large k (by domain decomposition or other methods). For a discussion of Schwarz algorithms for solving indefinite problems, we refer the reader to Cai and Widlund (1992); the convergence rate depends on the size of the coarse grid used. Nonoverlapping domain decomposition algorithms were recently proposed by Despres (1991), and by Ernst and Golub (1992) (for the complex Helmholtz equation). Alternative approaches based on fictitious domains are described in Proskurowski and Vassilevski (1992) and Atamian *et al.* (1991).

12.4. Nonconforming finite elements

Domain decomposition algorithms (cf Neumann–Neumann and Schwarz) have been developed for solving nonconforming finite element discretizations of elliptic problems, such as the Crouzeix–Raviart elements and dual mixed finite element discretizations, see Arnold and Brezzi (1985). Sarkis (1993) proposed several extensions of the Neumann–Neumann algorithm to the nonconforming case. In the context of dual formulations, related algorithms were independently proposed by Cowsar *et al.* (1993). Versions of the Schwarz algorithm were proposed by Cowsar (1993) and Meddahi (1993).

Acknowledgements The authors wish to thank Olof Widlund for many helpful discussions on various topics presented in this article. We have greatly benefitted from comments and helpful suggestions on an earlier draft of this article by Patrick Ciarlet, Max Dryja, Patrick Le Tallec, Barry Smith, Olof Widlund, Jinchao Xu and Jun Zou. Our sincere thanks to all of them.

REFERENCES

V.I. Agoshkov (1988), 'Poincaré–Steklov operators and domain decomposition methods in finite dimensional spaces', in *First Int. Symp. Domain Decomposition Methods for Partial Differential Equations* (R. Glowinski, G.H. Golub, G.A. Meurant and J. Périaux, eds), SIAM (Philadelphia, PA).

V.I. Agoshkov and V.I. Lebedev (1985), 'The Poincaré–Steklov operators and the domain decomposition methods in variational problems', in *Computational Processes and Systems* Nauka (Moscow) 173–227. In Russian.

D.N. Arnold and F. Brezzi (1985), 'Mixed and nonconforming finite element methods: Implementation, post processing and error estimates', *Math. Model. Numer. Anal.* **19**, 7–32.

S.F. Ashby, P.E. Saylor and J.S. Scroggs (1992), 'Physically motivated domain decomposition preconditioners', in *Proc. Second Copper Mountain Conf. on Iterative Methods*, Vol. 1, Comput. Math. Group, University of Colorado at Denver.

G.P. Astrakhantsev (1978), 'Method of fictitious domains for a second-order elliptic equation with natural boundary conditions', *USSR Comput. Math. Math. Phys.* **18**, 114–121.

C. Atamian, Q.V. Dinh, R. Glowinski, J. He and J. Périaux (1991), 'Control approach to fictitious-domain methods application to fluid dynamics and electromagnetics', in *Fourth Int. Symp. Domain Decomposition Methods for Partial Differential Equations* (R. Glowinski, Y.A. Kuznetsov, G.A. Meurant, J. Périaux and O. Widlund, eds), SIAM (Philadelphia, PA).

O. Axelsson and P. Vassilevski (1990), 'Algebraic multilevel preconditioning methods, II', *SIAM J. Numer. Anal.* **27**, 1569–1590.

I. Babuška (1957), 'Über Schwarzsche Algorithmen in partielle Differentialgleichungen der mathematischen Physik', *ZAMM* **37**(7/8), 243–245.

I. Babuška, A. Craig, J. Mandel and J. Pitkäranta (1991), 'Efficient preconditioning for the *p*-version finite element method in two dimensions', *SIAM J. Numer. Anal.* **28**(3), 624–661.

R.E. Bank, T.F. Dupont and H. Yserentant (1988), 'The hierarchical basis multigrid method', *Numer. Math.* **52**, 427–458.

M. Berger and S. Bokhari (1987), 'A partitioning strategy for nonuniform problems on multiprocessors', *IEEE Trans. Comput.* **36**, 570–580.

C. Bernardi and Y. Maday (1992), 'Approximations spectral de problèmes aux limites elliptiques', in *Mathematiques et Applications*, Vol. 10, Springer (Paris).

C. Bernardi, Y. Maday and A. Patera (1989), 'A new nonconforming approach to domain decomposition: the mortar element method', *Nonlinear Partial Differential Equations and their Applications* (H. Brezis and J.L. Lions, eds), Pitman (London).

C. Bernardi, N. Debit and Y. Maday (1990), 'Coupling finite element and spectral methods', *Math. Comput.* **54**, 21–39.

P.E. Bjørstad and M. Skogen (1992), 'Domain decomposition algorithms of Schwarz type, designed for massively parallel computers', *Fifth Int. Symp. Domain Decomposition Methods for Partial Differential Equations* (T.F. Chan, D.E. Keyes, G.A. Meurant, J.S. Scroggs and R.G. Voigt, eds), SIAM (Philadelphia, PA).

P.E. Bjørstad and O.B. Widlund (1984), 'Solving elliptic problems on regions partitioned into substructures', *Elliptic Problem Solvers II* (G. Birkhoff and A. Schoenstadt, eds), Academic (London), 245–256.

P.E. Bjørstad and O.B. Widlund (1986), 'Iterative methods for the solution of elliptic problems on regions partitioned into substructures', *SIAM J. Numer. Anal.* **23**(6), 1093–1120.

P.E. Bjørstad and O.B. Widlund (1989), 'To overlap or not to overlap: A note on a domain decomposition method for elliptic problems', *SIAM J. Sci. Stat. Comput.* **10**(5), 1053–1061.

C. Börgers (1989), 'The Neumann–Dirichlet domain decomposition method with inexact solvers on the subdomains', *Numer. Math.* **55**, 123–136.

C. Börgers and O.B. Widlund (1990), 'On finite element domain imbedding methods', *SIAM J. Numer. Anal.* **27**(4), 963–978.

F. Bornemann and H. Yserentant (1993), 'A basic norm equivalence for the theory of multilevel methods', *Num. Math.* **64**, 455–476.

J.-F. Bourgat, R. Glowinski, P. Le Tallec and M. Vidrascu (1989), 'Variational formulation and algorithm for trace operator in domain decomposition calculations', in *Second Int. Conf. on Domain Decomposition Methods* (T. Chan, R. Glowinski, J. Périaux and O. Widlund, eds), SIAM (Philadelphia, PA).

J.H. Bramble and J.E. Pasciak (1988), 'A preconditioning technique for indefinite systems resulting from mixed approximations of elliptic problems', *Math. Comput.* **50**, 1–18.

J.H. Bramble and J. Xu (1991), 'Some estimates for a weighted L^2 projection', *Math. Comput.* **56**, 463–476.

J.H. Bramble, R.E. Ewing, J.E. Pasciak and A.H. Schatz (1988), 'A preconditioning technique for the efficient solution of problems with local grid refinement', *Comput. Meth. Appl. Mech. Engrg* **67**, 149–159.

J.H. Bramble, R.E. Ewing, R.R. Parashkevov and J.E. Pasciak (1992), 'Domain decomposition methods for problems with partial refinement', *SIAM J. Sci. Comput.* **13(1)**, 397–410.

J.H. Bramble, J.E. Pasciak and A.H. Schatz (1986a), 'The construction of preconditioners for elliptic problems by substructuring, I', *Math. Comput.* **47**, 103–134.

J.H. Bramble, J.E. Pasciak and A.H. Schatz (1986b), 'An iterative method for elliptic problems on regions partitioned into substructures', *Math. Comput.* **46**(173), 361–369.

J.H. Bramble, J.E. Pasciak and A.H. Schatz (1989), 'The construction of preconditioners for elliptic problems by substructuring, IV', *Math. Comput.* **53**, 1–24.

J.H. Bramble, J.E. Pasciak and J. Xu (1990), 'Parallel multilevel preconditioners', *Math. Comput.* **55**, 1–22.

J.H. Bramble, J.E. Pasciak, J. Wang and J. Xu (1991), 'Convergence estimates for product iterative methods with applications to domain decomposition', *Math. Comput.* **57**(195), 1–21.

S.C. Brenner (1993), 'Two-level additive Schwarz preconditioners for nonconforming finite element methods', in *Proc. Seventh Int. Symp. on Domain Decomposition Methods for Partial Differential Equations*, to appear.

F. Brezzi and M. Fortin (1991), *Mixed and Hybrid Finite Element Methods*, Springer (Berlin).

B.L. Buzbee, F. Dorr, J. George and G. Golub (1971), 'The direct solution of the discrete poisson equation on irregular regions', *SIAM J. Numer. Anal.* **11**, 722–736.

X.-C. Cai (1991), 'Additive Schwarz algorithms for parabolic convection–diffusion equations', *Numer. Math.* **60**(1), 41–61.

X.-C. Cai (1993), 'Multiplicative Schwarz methods for parabolic problems', *SIAM J. Sci. Comput.*, to appear.

X.-C. Cai and O. Widlund (1992), 'Domain decomposition algorithms for indefinite elliptic problems', *SIAM J. Sci. Comput.* **13**(1), 243–258.

X.-C. Cai and O. Widlund (1993), 'Multiplicative Schwarz algorithms for some nonsymmetric and indefinite problems', *SIAM J. Numer. Anal.* **30**(4), 936–952.

X.-C. Cai, W.D. Gropp and D.E. Keyes (1992), 'A comparison of some domain decomposition algorithms for nonsymmetric elliptic problems', in *Fifth Int. Symp. on Domain Decomposition Methods for Partial Differential Equations* (T.F. Chan, D.E. Keyes, G.A. Meurant, J.S. Scroggs and R.G. Voigt, eds), SIAM (Philadelphia, PA).

C. Canuto and D. Funaro (1988), 'The Schwarz algorithm for spectral methods', *SIAM J. Numer. Anal.* **25**(1), 24–40.

C. Canuto, M.Y. Hussaini, A. Quarteroni and T.A. Zang (1988), *Spectral Methods in Fluid Dynamics*, Springer (Berlin).

T.F. Chan (1987), 'Analysis of preconditioners for domain decomposition', *SIAM J. Numer. Anal.* **24**(2), 382–390.

T.F. Chan and J. Shao (1993), 'Optimal coarse grid size in domain decomposition', Technical Report 93-24, UCLA CAM Report, Los Angeles, CA 90024-1555.

T.F. Chan and D. Goovaerts (1992), 'On the relationship between overlapping and nonoverlapping domain decomposition methods', *SIAM J. Matrix Anal. Appl.* **13**(2), 663.

T.F. Chan and T.Y. Hou (1991), 'Eigendecompositions of domain decomposition interface operators for constant coefficient elliptic problems', *SIAM J. Sci. Comput.* **12**(6), 1471–1479.

T.F. Chan and D.F. Keyes (1990), 'Interface preconditioning for domain decomposed convection diffusion operators', in *Third Int. Symp. on Domain Decomposition Methods for Partial Differential Equations* (T. Chan, R. Glowinski, J. Périaux and O. Widlund, eds), SIAM (Philadelphia, PA).

T.F. Chan and T.P. Mathew (1992), 'The interface probing technique in domain decomposition', *SIAM J. Matrix Anal. Appl.* **13**(1), 212–238.

T.F. Chan and T. Mathew (1993), 'Domain decomposition preconditioners for convection diffusion problems', in *Domain Decomposition Methods for Partial Differential Equations* (A. Quarteroni, ed.), American Mathematical Society (Providence, RI), to appear.

T.F. Chan and D.C. Resasco (1985), 'A survey of preconditioners for domain decomposition', Technical Report /DCS/RR-414, Yale University.

T.F. Chan and D.C. Resasco (1987), 'Analysis of domain decomposition preconditioners on irregular regions', in *Advances in Computer Methods for Partial Differential Equations - VI* (R. Vichnevetsky and R. Stepleman, eds), IMACS, 317–322.

T.F. Chan, T.Y. Hou and P.L. Lions (1991a), 'Geometry related convergence results for domain decomposition algorithms', *SIAM J. Numer. Anal.* **28**(2), 378.

T.F. Chan, E. Weinan and J. Sun (1991b), 'Domain decomposition interface preconditioners for fourth order elliptic problems', *Appl. Numer. Math.* **8**, 317–331.

T.F. Chan, D.E. Keyes, G.A. Meurant, J.S. Scroggs and R.G. Voigt, eds (1992a), *Fifth Conf. on Domain Decomposition Methods for Partial Differential Equations*, SIAM (Philadelphia, PA).

T.F. Chan, T.P. Mathew and J.-P. Shao (1992b), 'Efficient variants of the vertex space domain decomposition algorithm', Technical Report CAM 92-07, Department of Mathematics, UCLA, to appear in *SIAM J. Sci. Comput.*

T.F. Chan, R. Glowinski, J. Périaux and O. Widlund, eds (1989), *Second Int. Conf. on Domain Decomposition Methods*, SIAM (Philadelphia, PA).

T. Chan, R. Glowinski, J. Périaux and O. Widlund, eds (1990), *Third Int. Symp. on Domain Decomposition Methods for Partial Differential Equations*, SIAM (Philadelphia, PA).

R.C.Y. Chin, G.W. Hedstrom, J.R. McGraw and F.A. Howes (1986), 'Parallel computation of multiple scale problems', in *New Computing Environments: Parallel, Vector and Systolic* (A. Wouk, ed.) SIAM (Philadelphia, PA).

R. Courant and D. Hilbert (1962), *Methods of Mathematical Physics*, Vol. 2, Interscience (New York).

L.C. Cowsar (1993), 'Dual variable Schwarz methods for mixed finite elements', Technical Report TR93-09, Department of Mathematical Sciences, Rice University.

L.C. Cowsar and M.F. Wheeler (1991), 'Parallel domain decomposition method for mixed finite elements for elliptic partial differential equations', in *Fourth Int. Symp. on Domain Decomposition Methods for Partial Differential Equations* (R. Glowinski, Y.A. Kuznetsov, G.A. Meurant, J. Périaux and O. Widlund, eds), SIAM (Philadelphia, PA).

L.C. Cowsar, J. Mandel and M.F. Wheeler (1993), 'Balancing domain decomposition for mixed finite elements', Technical Report TR93-08, Department of Mathematical Sciences, Rice University.

C.N. Dawson and Q. Du (1991), 'A domain decomposition method for parabolic equations based on finite elements', in *Fourth Int. Symp. on Domain Decomposition Methods for Partial Differential Equations* (R. Glowinski, Y.A. Kuznetsov G.A. Meurant, J. Périaux and O. Widlund, eds), SIAM (Philadelphia, PA).

C. Dawson, Q. Du and T.F. Dupont, (1991), 'A finite difference domain decomposition algorithm for numerical solution of the heat equation', *Math. Comput.* **57**, 195.

Y.-H. De Roeck (1989), 'A local preconditioner in a domain-decomposed method', Technical Report TR89/10, Centre Européen de Recherche et de Formation Avancée en Calcul Scientifique, Toulouse, France.

Y.-H. De Roeck and P. Le Tallec (1991), 'Analysis and test of a local domain decomposition preconditioner', in *Fourth Int. Symp. on Domain Decomposition Methods for Partial Differential Equations* (R. Glowinski, Y. Kuznetsov, G. Meurant, J. Périaux and O. Widlund, eds), SIAM (Philadelphia, PA).

B. Despres (1991), 'Methodes de decomposition de domaines pour les problemes de propagation d'ondes en regime harmoniques', PhD thesis, University of Paris IX, Dauphine.

Q. Dinh, R. Glowinski and J. Périaux (1984), 'Solving elliptic problems by domain decomposition methods with applications', in *Elliptic Problem Solvers II* (G. Birkhoff and A. Schoenstadt, eds), Academic (New York), 395–426.

M. Dryja (1982), 'A capacitance matrix method for Dirichlet problem on polygon region', *Numer. Math.* **39**, 51–64.

M. Dryja (1984), 'A finite element-capacitance method for elliptic problems on regions partitioned into subregions', *Numer. Math.* **44**, 153–168.

M. Dryja (1988), 'A method of domain decomposition for 3-D finite element problems', in *First Int. Symp. on Domain Decomposition Methods for Partial Differential Equations* (R. Glowinski, G.H. Golub, G.A. Meurant and J. Périaux, eds) SIAM (Philadelphia, PA).

M. Dryja (1989), 'An additive Schwarz algorithm for two- and three-dimensional finite element elliptic problems', in *Second Int. Conf. on Domain Decomposition Methods* (T. Chan, R. Glowinski, J. Périaux and O. Widlund, eds), SIAM (Philadelphia, PA).

M. Dryja (1991), 'Substructuring methods for parabolic problems', in *Fourth Int. Symp. on Domain Decomposition Methods for Partial Differential Equations*

(R. Glowinski, Y.A. Kuznetsov, G.A. Meurant, J. Périaux and O. Widlund, eds), SIAM (Philadelphia, PA).

M. Dryja and O.B. Widlund (1987), 'An additive variant of the Schwarz alternating method for the case of many subregions', Technical Report 339, also Ultracomputer Note 131, Department of Computer Science, Courant Institute.

M. Dryja and O.B. Widlund (1989a), 'On the optimality of an additive iterative refinement method', in *Proc. Fourth Copper Mountain Conf. on Multigrid Methods*, SIAM (Philadelphia, PA) 161–170.

M. Dryja and O.B. Widlund (1989b), 'Some domain decomposition algorithms for elliptic problems', in *Iterative Methods for Large Linear Systems* (L. Hayes and D. Kincaid, eds), Academic (San Diego, CA), 273–291.

M. Dryja and O.B. Widlund (1990), 'Towards a unified theory of domain decomposition algorithms for elliptic problems', in *Third Int. Symp. on Domain Decomposition Methods for Partial Differential Equations* (T. Chan, R. Glowinski, J. Périaux and O. Widlund, eds), SIAM (Philadelphia, PA).

M. Dryja and O.B. Widlund (1992a), 'Additive Schwarz methods for elliptic finite element problems in three dimensions', in *Fifth Conf. on Domain Decomposition Methods for Partial Differential Equations* (T.F. Chan, D.E. Keyes, G.A. Meurant, J.S. Scroggs and R.G. Voigt, eds), SIAM (Philadelphia, PA).

M. Dryja and O.B. Widlund (1992b), 'Domain decomposition algorithms with small overlap', Technical Report 606, Department of Computer Science, Courant Institute, to appear in *SIAM J. Sci. Comput.*

M. Dryja and O.B. Widlund (1993a), 'Schwarz methods of Neumann-Neumann type for three-dimensional elliptic finite element problems', Technical Report 626, Department of Computer Science, Courant Institute.

M. Dryja and O.B. Widlund (1993b), 'Some recent results on Schwarz type domain decomposition algorithms', in *Sixth Conf. on Domain Decomposition Methods for Partial Differential Equations* (A. Quarteroni, ed.), American Mathematical Society (Providence, RI), to appear. Technical Report 615, Department of Computer Science, Courant Institute.

M. Dryja, B.F. Smith and O.B. Widlund (1993), 'Schwarz analysis of iterative substructuring algorithms for elliptic problems in three dimensions', Technical Report 638, Department of Computer Science, Courant Institute. *SIAM J. Numer. Anal.*, submitted.

O. Ernst and G. Golub (1992), 'A domain decomposition approach to solving the helmholtz equation with a radiation boundary condition', Technical Report 92-08, Stanford University, Computer Science Department, Numerical Analysis Project, Stanford, CA 94305.

R.E. Ewing and J. Wang (1991), 'Analysis of the Schwarz algorithm for mixed finite element methods', *RAIRO, Math. Modell. Numer. Anal.* **26**(6), 739–756.

C. Farhat and M. Lesoinne (1993), 'Automatic partitioning of unstructured meshes for the parallel solution of problems in computational mechanics', *Int. J. Numer. Meth. Engrg* **36**, 745–764.

C. Farhat and F.X. Roux (1992), 'An unconventional domain decomposition method for an efficient parallel solution of large scale finite element systems', *SIAM J. Sci. Comput.* **13**, 379–396.

S.A. Finogenov and Y.A. Kuznetsov (1988), 'Two-stage fictitious components method for solving the Dirichlet boundary value problem', *Sov. J. Numer. Anal. Math. Modell.* **3**(4), 301–323.

P.F. Fischer and E.M. Rønquist (1993), 'Spectral element methods for large scale parallel Navier–Stokes calculations', in *Second Int. Conf. on Spectral and High Order Methods for PDE's.* Proc. ICOSAHOM 92, a conference held in Montpellier, France, June 1992. To appear in *Comput. Meth. Appl. Mech. Engrg.*

M. Fortin and R. Aboulaich (1988), 'Schwarz's decomposition method for incompressible flow problems', in *First Int. Symp. on Domain Decomposition Methods for Partial Differential Equations* (R. Glowinski, G.H. Golub, G.A. Meurant and J. Périaux, eds), SIAM (Philadelphia, PA).

G. Fox (1988), 'A review of automatic load balancing and decomposition methods for the hypercube', in *Numerical Algorithms for Modern Parallel Computers* (M. Schultz, ed.), Springer (Berlin), 63–76.

R.W. Freund, G.H. Golub and N. Nachtigal (1992), 'Iterative solution of linear systems', *Acta Numerica*, Cambridge University Press (Cambridge), 57–100.

D. Funaro, A. Quarteroni and P. Zanolli (1988), 'An iterative procedure with interface relaxation for domain decomposition methods', *SIAM J. Numer. Anal.* **25**, 1213–1236.

M. Garbey (1992), 'Domain decomposition to solve layers and singular perturbation problems', in *Fifth Conf. on Domain Decomposition Methods for Partial Differential Equations* (T.F. Chan, D.E. Keyes, G.A. Meurant, J.S. Scroggs and R.G. Voigt, eds), SIAM (Philadelphia, PA).

F. Gastaldi, A. Quarteroni and G. Sacchi-Landriani (1990), 'On the coupling of two-dimensional hyperbolic and elliptic equations: analytical and numerical approach', in *Domain Decomposition Methods for Partial Differential Equations* (T. Chan, R. Glowinski, J. Périaux and O. Widlund, eds), SIAM (Philadelphia, PA) 23–63.

V. Girault and P.-A. Raviart (1986), *Finite Element Approximation of the Navier–Stokes Equations: Theory and Algorithms*, Springer (Berlin).

R. Glowinski and M.F. Wheeler (1988), 'Domain decomposition and mixed finite element methods for elliptic problems', in *First Int. Symp. on Domain Decomposition Methods for Partial Differential Equations* (R. Glowinski, G.H. Golub, G.A. Meurant and J. Périaux, eds), SIAM (Philadelphia, PA).

R. Glowinski, G.H. Golub, G.A. Meurant and J. Périaux, eds (1988), *Proc. First Int. Symp. on Domain Decomposition Methods for Partial Differential Equations*, SIAM (Philadelphia, PA).

R. Glowinski, W. Kinton and M.F. Wheeler (1990a), 'Acceleration of domain decomposition algorithms for mixed finite elements by multi-level methods', in *Third Int. Symp. on Domain Decomposition Methods for Partial Differential Equations* (T. Chan, R. Glowinski, J. Périaux and O. Widlund, eds), SIAM (Philadelphia, PA).

R. Glowinski, Y.A. Kuznetsov, G.A. Meurant, J. Périaux and O. Widlund, eds (1991), *Fourth Int. Symp. on Domain Decomposition Methods for Partial Differential Equations*, SIAM (Philadelphia, PA).

R. Glowinski, J. Périaux and G. Terrasson (1990b), 'On the coupling of viscous and inviscid models for compressible fluid flows via domain decomposition', in *Domain Decomposition Methods for Partial Differential Equations* (T. Chan, R. Glowinski, J. Périaux and O. Widlund, eds), SIAM (Philadelphia, PA).

G. Golub and D. Mayers (1984), 'The use of preconditioning over irregular regions', in *Computing Methods in Applied Sciences and Engineering, VI* (R. Glowinski and J.L. Lions, eds), North-Holland (Amsterdam, New York, Oxford) 3–14.

D. Goovaerts (1989), 'Domain decomposition methods for elliptic partial differential equations', PhD thesis, Department of Computer Science, Catholic University of Leuven.

D. Goovaerts, T. Chan and R. Piessens (1991), 'The eigenvalue spectrum of domain decomposed preconditioners', *Appl. Numer. Math.* **8**, 389–410.

M. Griebel (1991), 'Multilevel algorithms considered as iterative methods on indefinite systems', Inst. für Informatik, Tech. Univ. Munchen, SFB 342/29/91A.

M. Griebel and P. Oswald (1993), 'Remarks on the abstract theory of additive and multiplicative Schwarz algorithms', Inst. für Informatik, Tech. Univ. Munchen, SFB 342/6/93A.

W.D. Gropp (1992), 'Parallel computing and domain decomposition', in *Fifth Conf. on Domain Decomposition Methods for Partial Differential Equations* (T.F. Chan, D.E. Keyes, G.A. Meurant, J.S. Scroggs and R.G. Voigt, eds), SIAM (Philadelphia, PA).

W. Gropp and D. Keyes (1993), 'Domain decomposition as a mechanism for using asymptotic methods', in *Asymptotic and Numerical Methods for Partial Differential Equations with Critical Parameters* (H. Kaper and M. Garbey, eds), Vol. 384, NATO ASI Series C, 93–106.

W. Gropp and B. F. Smith (1992), 'Experiences with domain decomposition in three dimensions: overlapping Schwarz methods', Mathematics and Computer Science Division, Argonne National Laboratory, to appear in *Proc. Sixth Int. Symp. on Domain Decomposition Methods.*

W. Hackbusch (1993), *Iterative Methods for Large Sparse Linear Systems*, Springer (Heidelberg).

G.W. Hedstrom and F.A. Howes (1990), 'Domain decomposition for a boundary value problem with a shock layer', in *Third Int. Symp. on Domain Decomposition Methods for Partial Differential Equations* (T.F. Chan, R.Glowinski, J. Périaux and O. Widlund, eds), SIAM (Philadelphia, PA).

F. D'Hennezel (1992), 'Domain decomposition method with non-symmetric interface operator', *Fifth Conf. on Domain Decomposition Methods for Partial Differential Equations* (Tony F. Chan, David E. Keyes, Gérard A. Meurant, Jeffrey S. Scroggs and Robert G. Voigt, eds), SIAM (Philadelphia, PA).

K.-H. Hoffmann and J. Zou (1992), 'Solution of biharmonic problems by the domain decomposition method', Technical Report No. 387, DFG-SPP, Technical University of Munich.

Kang, L. S. (1987), *Parallel Algorithms and Domain Decomposition*, Wuhan University Press (China). In Chinese.

B. Kernighan and S. Lin (1970), 'An efficient heuristic procedure for partitioning graphs', *Bell Systems Tech. J.* **29**, 291–307.

D.E. Keyes and W.D. Gropp (1987), 'A comparison of domain decomposition techniques for elliptic partial differential equations and their parallel implementation', *SIAM J. Sci. Comput.* **8**(2), s166–s202.

D.E. Keyes and W.D. Gropp (1989), 'Domain decomposition with local mesh refinement', Technical Report YALEU/DCS/RR-726, Yale University.

G. Kron (1953), 'A set of principles to interconnect the solutions of physical systems', *J. Appl. Phys.* **24**(8), 965.

Y.A. Kuznetsov (1988), 'New algorithms for approximate realization of implicit difference schemes', *Sov. J. Numer. Anal. Math. Modell.* **3**, 99–114.

Y.A. Kuznetsov (1990), 'Domain decomposition methods for unsteady convection–diffusion problems', in *IXth Int. Conf. on Computing Methods in Applied Science and Engineering* (R. Glowinski and J.L. Lions, eds), INRIA (Paris) 327–344.

Y.A. Kuznetsov (1991), 'Overlapping domain decomposition methods for fe-problems with elliptic singular perturbed operators', in *Fourth Int. Symp. on Domain Decomposition Methods for Partial Differential Equations* (R. Glowinski, Y.A. Kuznetsov, G.A. Meurant, J. Périaux and O. Widlund, eds), SIAM (Philadelphia, PA).

Yu. M. Laevsky (1992), 'Direct domain decomposition method for solving parabolic equations', Preprint no. 940, Novosibirsk, Computing Center Siberian Branch Academy of Sciences. In Russian.

Yu. M. Laevsky (1993), 'On the domain decomposition method for parabolic problems', *Bull. Novosibirsk Comput. Center* **1**, 41–62.

P. Le Tallec (1994), 'Domain decomposition methods in computational mechanics', *J. Comput. Mech. Adv.*, to appear.

P. Le Tallec, Y.-H. De Roeck and M. Vidrascu (1991), 'Domain-decomposition methods for large linearly elliptic three-dimensional problems', *J. Comput. Appl. Math.* **34**.

V.I. Lebedev (1986), *Composition Methods*, USSR Academy of Sciences, Moscow. In Russian.

P.L. Lions (1988), 'On the Schwarz alternating method. I.', in *First Int. Symp. on Domain Decomposition Methods for Partial Differential Equations* (R. Glowinski, G.H. Golub, G.A. Meurant and J. Périaux, eds), SIAM (Philadelphia, PA).

P.L. Lions (1989), 'On the Schwarz alternating method. II', in *Second Int. Conf. on Domain Decomposition Methods* (T. Chan, R. Glowinski, J. Périaux and O. Widlund, eds), SIAM (Philadelphia, PA).

J.-L. Lions and E. Magenes (1972), *Nonhomogeneous Boundary Value Problems and Applications*, Vol. I, Springer (New York, Heidelberg, Berlin).

Lu Tao, T. Shih and C. Liem (1992), *Domain Decomposition Methods: New Numerical Techniques for Solving PDE*, Science Publishers (Beijing, China).

Y. Maday and A.T. Patera (1989), 'Spectral element methods for the Navier–Stokes equations', in *State of the Art Surveys in Computational Mechanics* (A.K. Noor and J.T. Oden, eds), ASME (New York).

J. Mandel (1989a), 'Efficient domain decomposition preconditioning for the *p*-version finite element method in three dimensions', Technical Report, Computational Mathematics Group, University of Colorado at Denver.

J. Mandel (1989b), 'Two-level domain decomposition preconditioning for the p-version finite element version in three dimensions', *Int. J. Numer. Meth. Engrg*, **29**, 1095–1108.

J. Mandel (1990), 'Hierarchical preconditioning and partial orthogonalization for the p-version finite element method', in *Third Int. Symp. on Domain Decomposition Methods for Partial Differential Equations* (T.F. Chan, R. Glowinski, J. Périaux and O. Widlund, eds), SIAM (Philadelphia, PA).

J. Mandel (1992), 'Balancing domain decomposition', *Commun. Numer. Meth. Engrg* **9**, 233–241.

J. Mandel and M. Brezina (1992), 'Balancing domain decomposition: Theory and computations in two and three dimensions', Technical Report, Computational Mathematics Group, University of Colorado at Denver.

J. Mandel and S. McCormick (1989), 'Iterative solution of elliptic equations with refinement: The model multi-level case', in *Second Int. Conf. on Domain Decomposition Methods* (T. Chan, R. Glowinski, J. Périaux and O. Widlund, eds), SIAM (Philadelphia, PA).

G.I. Marchuk, Y.A. Kuznetsov and A.M. Matsokin (1986), 'Fictitious domain and domain decomposition methods', *Sov. J. Numer. Anal. Math. Modell.* **1**, 3–61.

L.D. Marini and A. Quarteroni (1989), 'A relaxation procedure for domain decomposition methods using finite elements', *Numer. Math.* **56**, 575–598.

T.P. Mathew (1989), 'Domain decomposition and iterative refinement methods for mixed finite element discretisations of elliptic problems', PhD thesis, Courant Institute of Mathematical Sciences. Technical Report 463, Department of Computer Science, Courant Institute.

T.P. Mathew (1993a), 'Schwarz alternating and iterative refinement methods for mixed formulations of elliptic problems, part I: Algorithms and numerical results', *Num. Math.* **65**(4), 445–468.

T.P. Mathew (1993b), 'Schwarz alternating and iterative refinement methods for mixed formulations of elliptic problems, part II: Theory', *Num. Math.* **65**(4), 469–492.

A.M. Matsokin and S.V. Nepomnyaschikh (1985), 'A Schwarz alternating method in a subspace', *Sov. Math.* **29(10)**, 78–84.

S. McCormick (1984), 'Fast adaptive composite grid (FAC) methods', in *Defect Correction Methods: Theory and Applications* (K. Böhmer and H.J. Stetter, eds), Computing Supplement 5, Springer (Wien), 115–121.

S.F. McCormick (1989), *Multilevel Adaptive Methods for Partial Differential Equations*, SIAM (Philadelphia, PA).

S. Meddahi (1993), 'The Schwarz algorithm for a Raviart–Thomas mixed method', preprint, to appear.

G.A. Meurant (1991), 'Numerical experiments with a domain decomposition method for parabolic problems on parallel computers', in *Fourth Int. Symp. on Domain Decomposition Methods for Partial Differential Equations* (R. Glowinski, Y.A. Kuznetsov, G.A. Meurant, J. Périaux and O. Widlund, eds), SIAM (Philadelphia, PA).

K. Miller (1965), 'Numerical analogs of the Schwarz alternating procedure', *Numer. Math.* **7**, 91–103.

Y. Morchoisne (1984), 'Inhomogeneous flow calculations by spectral methods: Mono-domain and multi-domain techniques', in *Spectral Methods for Partial Differential Equations* (R.G. Voigt, D. Gottlieb and M.Y. Hussaini, eds), SIAM-CBMS, 181–208.

D. Morgenstern (1956), 'Begründung des alternierenden Verfahrens durch Orthogonalprojektion', *ZAMM* **36**, 7–8.

F. Nataf and F. Rogier (1993), 'Factorization of the advection-diffusion operator and domain decomposition method', in *Asymptotic and Numerical Methods for Partial Differential Equations with Critical Parameters* (H. Kaper and M. Garbey, eds), Vol. 384, NATO ASI Series C, 123–133.

J. Nečas (1967), *Les Méthodes Directes en Théorie des Équations Elliptiques*, Academia (Prague).

S.V. Nepomnyaschikh (1984), 'On the application of the method of bordering for elliptic mixed boundary value problems and on the difference norms of $W_2^{1/2}(S)$'. In Russian.

S.V. Nepomnyaschikh (1986), 'Domain decomposition and Schwarz methods in a subspace for the approximate solution of elliptic boundary value problems', Computing Center of the Siberian Branch of the USSR Academy of Sciences, Novosibirsk, USSR.

D.P. O'Leary and O.B. Widlund (1979), 'Capacitance matrix methods for the helmholtz equation on general three-dimensional regions', *Math. Comput.* **33**, 849–879.

M. Ong (1989), 'Hierarchical basis preconditioners for second-order elliptic problems in three dimensions', Technical Report 89–3, Department of Applied Maths, University of Washington, Seattle.

P. Oswald (1991), 'On discrete norm estimates related to multilevel preconditioners in the finite element method', in *Proc. Int. Conf. Theory of Functions, Varna 91*, to appear.

S. Pahl (1993), 'Domain decomposition for the Stokes problem', Master's thesis, University of Witwatersrand, Johannesburg.

L.F. Pavarino (1992), Domain decomposition algorithms for the *p*-version finite element method for elliptic problems, PhD thesis, Courant Institute of Mathematical Sciences, Department of Mathematics.

L.F. Pavarino (1993a), 'Schwarz methods with local refinement for the *p*-version finite element method', Technical Report TR93-01, Rice University, Department of Computational and Applied Mathematics, submitted to *Numer. Math.*

L.F. Pavarino (1993b), 'Some Schwarz algorithms for the spectral element method', in *Sixth Conf. on Domain Decomposition Methods for Partial Differential Equations* (A. Quarteroni, ed.), American Mathematical Society (Providence, RI), to appear. Technical Report 614, Department of Computer Science, Courant Institute.

L.F. Pavarino and O.B. Widlund (1993), 'Iterative substructuring methods for *p*-version finite elements in three dimensions', Technical Report, Courant Institute of Mathematical Sciences, Department of Computer Science, to appear.

T.N. Phillips (1992), 'Pseudospectral domain decomposition techniques for the Navier–Stokes equations', in *Fifth Int. Symp. on Domain Decomposition*

Methods for Partial Differential Equations (T.F. Chan, D.E. Keyes, G.A. Meurant, J.S. Scroggs and R.G. Voigt, eds), SIAM (Philadelphia, PA).

A. Pothen, H. Simon and K. Liou (1990), 'Partitioning sparse matrices with eigenvector of graphs', *SIAM J. Math. Anal. Appl.* **11**(3), 430–452.

W. Proskurowski and P. Vassilevski (1992), 'Preconditioning nonsymmetric and indefinite capacitance matrix problems in domain imbedding', Technical Report, UCLA, CAM Report 92-48, University of California, Los Angeles. *SIAM J. Sci. Comput.*, to appear.

W. Proskurowski and P. Vassilevski (1994), 'Preconditioning capacitance matrix problems in domain imbedding', *SIAM J. Sci. Comput.*, to appear.

W. Proskurowski and O.B. Widlund (1976), 'On the numerical solution of helmholtz's equation by the capacitance matrix method', *Math. Comput.* **30**, 433–468.

J.S. Przemieniecki (1963), 'Matrix structural analysis of substructures', *Amer. Inst. Aero. Astro. J.* **1**(1), 138–147.

A. Quarteroni (1989), 'Domain decomposition algorithms for the Stokes equations', in *Second Int. Conf. on Domain Decomposition Methods* (T. Chan, R. Glowinski, J. Périaux and O. Widlund, eds), SIAM (Philadelphia, PA).

A. Quarteroni, ed. (1993), *Domain Decomposition Methods in Science and Engineering*, American Mathematical Society (Providence, RI).

A. Quarteroni and A. Valli (1990), 'Theory and applications of Steklov–Poincaré operators for boundary-value problems: the heterogeneous operator case', in *Proc. 4th Int. Conf. on Domain Decomposition Methods, Moscow* (T. Chan, R. Glowinski, J. Périaux and O. Widlund, eds), SIAM (Philadelphia, PA).

D.C. Resasco (1990), 'Domain decomposition algorithms for elliptic partial differential equations', PhD thesis, Department of Computer Science, Yale University.

T. Rusten (1991), 'Iterative methods for mixed finite element systems', University of Oslo.

T. Rusten and R. Winther (1992), 'A preconditioned iterative method for saddle point problems', *SIAM J. Matrix Anal.* **13**(3), 887.

M. Sarkis (1993), 'Two-level Schwarz methods for nonconforming finite elements and discontinuous coefficients', Technical Report 629, Department of Computer Science, Courant Institute of Mathematical Sciences, New York University.

F. Scapini (1990), 'The alternating Schwarz method applied to some biharmonic variational inequalities', *Calcolo* **27**, 57–72.

F. Scapini (1991), 'A decomposition method for some biharmonic problems', *J. Comput. Math.* **9**, 291–300.

H.A. Schwarz (1890), *Gesammelte Mathematische Abhandlungen*, Vol. 2, Springer (Berlin) 133–143. First published in *Vierteljahrsschrift der Naturforschenden Gesellschaft* in Zürich, Vol. 15, 1870, 272–286.

J.S. Scroggs (1989), 'A parallel algorithm for nonlinear convection diffusion equations', in *Third Int. Symp. on Domain Decomposition Methods for Partial Differential Equations*, SIAM (Philadelphia, PA).

H. Simon (1991), 'Partitioning of unstructured problems for parallel processing', *Comput. Sys. Engrg* **2**(2/3), 135–148.

B.F. Smith (1990), 'Domain decomposition algorithms for the partial differential equations of linear elasticity', PhD thesis, Courant Institute of Mathematical Sciences. Technical Report 517, Department of Computer Science, Courant Institute.

B.F. Smith (1991), 'A domain decomposition algorithm for elliptic problems in three dimensions', *Numer. Math.* **60**(2), 219–234.

B.F. Smith (1992), 'An optimal domain decomposition preconditioner for the finite element solution of linear elasticity problems', *SIAM J. Sci. Comput.* **13**(1), 364–378.

B.F. Smith (1993), 'A parallel implementation of an iterative substructuring algorithm for problems in three dimensions', *SIAM J. Sci. Comput.* **14**(2), 406–423.

B.F. Smith and O.B. Widlund (1990), 'A domain decomposition algorithm using a hierarchical basis', *SIAM J. Sci. Comput.* **11**(6), 1212–1220.

B.F. Smith, P. Bjørstad and W.D. Gropp (1994), 'Domain decomposition: algorithms, implementations and a little theory', to appear.

S.L. Sobolev (1936), 'L'algorithme de Schwarz dans la théorie de l'elasticité', *C. R. (Dokl.) Acad. Sci. URSS* **IV**((XIII) 6), 243–246.

J. Sun and J. Zou (1991), 'Ddm preconditioner for 4th order problems by using B-spline finite element method', in *Fourth Int. Symp. on Domain Decomposition Methods for Partial Differential Equations* (R. Glowinski, Y.A. Kuznetsov, G.A. Meurant, J. Périaux and O. Widlund, eds), SIAM (Philadelphia, PA).

W. P. Tang (1988), 'Schwarz splitting and template operators', Department of Computer Science, Stanford University.

C.H. Tong, T.F. Chan and C.C.J. Kuo (1991), 'A domain decomposition preconditioner based on a change to a multilevel nodal basis', *SIAM J. Sci. Comput.* **12**, 1486–1495.

W. Tsui (1991), 'Domain decomposition of biharmonic and Navier–Stokes equations', PhD thesis, Department of Mathematics, University of California, Los Angeles.

J. Wang (1991), 'Convergence analysis of Schwarz algorithms and multilevel decomposition iterative methods: Part I', *Proc. IMACS Int. Symp. on Iterative Methods in Linear Algebra* (R. Beauwens and P. De Groen, eds), (Belgium).

J. Wang (1993), 'Convergence analysis of schwarz algorithms and multilevel decomposition iterative methods: Part II', *SIAM J. Numer. Anal.* **30**(4), 953–970.

M.F. Wheeler and R. Gonzalez (1984), 'Mixed finite element methods for petroleum reservoir engineering problems', in *Computing Methods in Applied Sciences and Engineering, VI* (R. Glowinski and J.L. Lions, eds), North-Holland (New York) 639–658.

O.B. Widlund (1987), 'An extension theorem for finite element spaces with three applications', in *Numerical Techniques in Continuum Mechanics: Notes on Numerical Fluid Mechanics*, Vol. 16, (W. Hackbusch and K. Witsch, eds), Friedr. Vieweg und Sohn (Braunschweig/Wiesbaden) 110–122. Proc. Second GAMM Seminar, Kiel, January, 1986.

O.B. Widlund (1988), 'Iterative substructuring methods: Algorithms and theory for elliptic problems in the plane', in *First Int. Symp. on Domain Decomposition*

Methods for Partial Differential Equations (R. Glowinski, G.H. Golub, G.A. Meurant and J. Périaux, eds), SIAM (Philadelphia, PA).

O.B. Widlund (1989a), 'Iterative solution of elliptic finite element problems on locally refined meshes', in *Finite Element Analysis in Fluids* (T.J. Chung and G.R. Karr, eds), University of Alabama in Huntsville Press (Huntsville, Alabama), 462–467.

O.B. Widlund (1989b), 'Optimal iterative refinement methods', in *Second Int. Conf. on Domain Decomposition Methods* (T. Chan, R. Glowinski, J. Périaux and O. Widlund, eds), SIAM (Philadelphia, PA).

R. Williams (1991), 'Performance of dynamic load balancing algorithms for unstructured mesh calculations', *Concurrency* **3**, 457–481.

J. Xu (1989), 'Theory of multilevel methods', PhD thesis, Cornell University.

J. Xu (1992a), 'Iterative methods by space decomposition and subspace correction', *SIAM Rev.* **34**, 581–613.

J. Xu (1992b), 'A new class of iterative methods for nonselfadjoint or indefinite problems', *SIAM J. Numer. Anal.* **29**(2), 303–319.

J. Xu (1992c), 'Iterative methods by SPD and small subspace solvers for nonsymmetric or indefinite problems', *Fifth Conf. on Domain Decomposition Methods for Partial Differential Equations* (T.F. Chan, D.E. Keyes, G.A. Meurant, J.S. Scroggs and R.G. Voigt, eds) SIAM (Philadelphia, PA).

J. Xu and X.-C. Cai (1992), 'A preconditioned GMRES method for nonsymmetric or indefinite problems', *Math. Comput.* **59**, 311–319.

H. Yserentant (1986), 'On the multi-level splitting of finite element spaces', *Numer. Math.* **49**, 379–412.

X. Zhang (1991), 'Studies in domain decomposition: Multilevel methods and the biharmonic Dirichlet problem', PhD thesis, Courant Institute, New York University.

X. Zhang (1992a), 'Domain decomposition algorithms for the biharmonic Dirichlet problem', in *Fifth Conf. on Domain Decomposition Methods for Partial Differential Equations* (T.F. Chan, D.E. Keyes, G.A. Meurant, J.S. Scroggs and R.G. Voigt, eds), SIAM (Philadelphia, PA).

X. Zhang (1992b), 'Multilevel Schwarz methods', *Num. Math.* **63**(4), 521–539.

X. Zhang (1992c), 'Multilevel Schwarz methods for the biharmonic dirichlet problem', Technical Report CS-TR2907 (UMIACS-TR-92-60), University of Maryland, Department of Computer Science, submitted to *SIAM J. Sci. Comput.*

Acta Numerica (1994), *pp.* 145–202

Aspects of the numerical analysis of neural networks

S.W. Ellacott
Department of Mathematical Sciences
University of Brighton
Moulsecoomb, Brighton BN2 4GJ
England
E-mail: swe@unix.bton.ac.uk

This article starts with a brief introduction to neural networks for those unfamiliar with the basic concepts, together with a very brief overview of mathematical approaches to the subject. This is followed by a more detailed look at three areas of research which are of particular interest to numerical analysts.

The first area is approximation theory. If K is a compact set in $\mathbb{R}^n$, for some n, then it is proved that a semilinear feedforward network with one hidden layer can uniformly approximate any continuous function in $C(K)$ to any required accuracy. A discussion of known results and open questions on the degree of approximation is included. We also consider the relevance of radial basis functions to neural networks.

The second area considered is that of learning algorithms. A detailed analysis of one popular algorithm (the delta rule) will be given, indicating why one implementation leads to a stable numerical process, whereas an initially attractive variant (essentially a form of steepest descent) does not. Similar considerations apply to the backpropagation algorithm. The effect of filtering and other preprocessing of the input data will also be discussed systematically.

Finally some applications of neural networks to numerical computation are considered.

CONTENTS

1	An introduction to neural networks	146
2	Density and approximation by neural networks	150
3	Numerical analysis of learning algorithms	174
4	Some numerical applications of neural networks	191
5	Concluding remarks	199
	References	200

1. An introduction to neural networks

1.1. A network to compute 'XOR'

A neural network is a model of computation based loosely on the mammalian brain. Rather than give a formal definition we illustrate by a simple example. Figure 1 shows a network designed to compute the 'exclusive-or' (XOR) function. Each input unit takes a single scalar input. In general these may take any real value, but for this particular example the inputs are restricted to the values 0 or 1. Thus the set of possible input vectors is

$$\{(0,0)^T, (0,1)^T, (1,0)^T, (1,1)^T\}.$$

The network is required to compute the output '0' if the two inputs are the same or '1' if they are different. It does this in the following way. The vertices of the graph shown as circles are called *units* or *neurons*. The input units shown with no inscribed numbers simply pass their inputs to their output edges which in this context are called *links* or *synapses*. They are multiplied by the *weights* shown on these links and summed at the input to the next unit. Suppose for instance the input vector is $(1,0)^T$. The input to the unit inscribed '1.5' is thus $1 \times 1 + 1 \times 0 = 1$. In this type of network, the '1.5' itself is a threshold value. Since the total input $1 < 1.5$, the output of the unit is 0. If the original input vector were $(1,1)^T$, the input to this unit would be $2 > 1.5$, so this unit would output 1. The output of a unit as a function of the given input is called the *activation function* of the unit. With the original input vector $(1,0)^T$, we see that the input to the unit

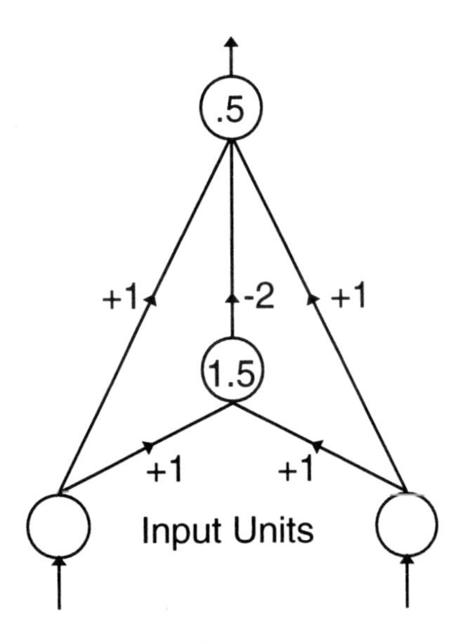

Fig. 1.

inscribed '.5' is thus $1 \times 1 - 2 \times 0 + 1 \times 0 = 1$. Since $1 > .5$, the output of the whole network is 1 as required. The reader might like to verify in a similar manner that the network correctly computes the exclusive-or of the other three possible input vectors.

In this case the network is being used as a pattern classifier: the input patterns are separated into classes. For this example there are just two classes according to whether the XOR of the inputs is 0 or 1. Thus a single binary output is sufficient. In other cases we may require more than two classes. This can be achieved by allowing the output to take more values or, alternatively by using more than one output. Originally neural nets were thought of as logical devices like conventional computers. Pattern classification is thus the classical application. However, it is now becoming clear that much of that phenomenon that we think of as intelligence is not really binary in this way. A tennis player attempting to hit a ball is required to produce a complex muscular response to equally complex visual, tactile and kinesthetic input data. The problem is one of control theory at a complicated mutlivariate level. The human nervous system and brain in this context is best thought of as a nonlinear analogue controller with learning capability. Control applications of neural nets constitute a major application area (see e.g. Warwick *et al.*, 1992; Werbos, 1992), as do pattern recognition problems in speech and vision where it is difficult to get good models to analyse conventionally (e.g. Linggard and Nightingale, 1992). The ability of neural systems to 'learn' (see Section 3) means that they can produce usable models on the basis of examples *without* the need of a formal axiomatic analysis.

Humans can also recognize patterns and structure in data when the required output is not known. In statistical language, the problem then is one of clustering rather than classification. This is rather more difficult than simply learning known patterns, but it can also be tackled by neural networks. The most popular network for this problem is that due to Kohonen (see, for example, Wasserman (1990) Ch. 4). We will not discuss the Kohonen net here, but note that the mathematical problems are somewhat similar to the networks we do consider.

There is a vast range of variations on these general ideas: the interested reader should consult one of the many textbooks available on this subject. For an introductory treatment, see Aleksander and Morton (1990), Simpson (1990) or Wasserman (1989). A deeper work, although now a little dated, is Rumelhart and McClelland (1986).

1.2. Perceptrons and multilayer perceptrons

In this article we shall concentrate on the simplest of all neural networks, the *perceptron*, and an extension of this called the *multilayer perceptron*

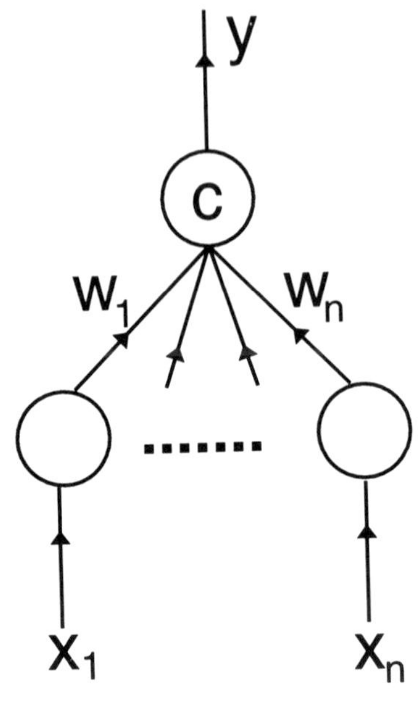

Fig. 2.

or *semilinear feedforward network*. The density results of Section 2 refer to the latter. It is the most popular of all neural network architectures, probably because it is relatively easy to understand and use. (However, many of the ideas are much more widely applicable and the formulation of the backpropagation algorithm given in Section 3 is certainly much more general than is required just for the multilayer perceptron.) We will look briefly at some other architectures including the Hopfield net in Section 4.

Figure 2 shows a simple perceptron with a single output. The units are interpreted as in Figure 1 with the input units having identity activation function and the output unit having a simple threshold. Thus denoting the input vector by $\mathbf{x}$ and the weight vector by $\mathbf{w}$ (both in $\mathbb{R}^n$) it is easy to see that the output y is 1 if $\mathbf{w}^T\mathbf{x} > c$ and 0 if $\mathbf{w}^T\mathbf{x} \leq c$. So for a fixed weight vector $\mathbf{w}$ and threshold c, the network divides $\mathbb{R}^n$ into two half spaces separated by a hyperplane. For obvious reasons, the perceptron is described as a linear network. Considering the case $n = 2$, observe that the pairs of inputs required to produce outputs 1 or 0 for the exclusive-or function are at diagonally opposite corners of a square *so they cannot be separated by a perceptron.* This simple observation delayed the development of neural networks for many years until tools for handling nonlinear networks such as Figure 1 became available.

In spite of this restriction, it is worth studying the perceptron as it constitutes the simplest case of many other network architectures and can give very useful insight into their behaviour. The multilayer perceptron shown in

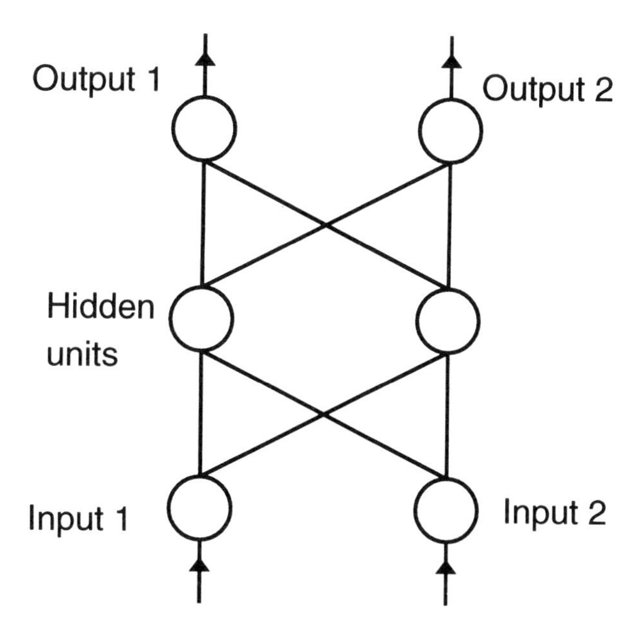

Fig. 3.

Figure 3 is the most obvious and straightforward generalization. It consists simply of layers of perceptrons connected together in cascade. Observe that the AND and OR logical functions *can* be separated linearly. If we wish to classify sets of points into two sets, we can certainly do it with three perceptrons in cascade: the first layer divides the plane into half-planes, the second can AND these to produce polygons and the third can OR these polygons to assign them to a required class. We can therefore construct any desired partition of $\mathbb{R}^n$ into polygonal regions, and classify these regions. This process is illustrated for the two-dimensional case in Figure 4. Originally it was believed that this was the minimum number of layers needed to solve the classification problem, but it is now known that, in principle, two perceptrons in cascade (i.e. only one hidden layer) is sufficient. A proof of this will be given in Section 2. However, it does seem plausible to suppose that nets with two or more hidden layers might be more efficient. This question is still open.

In Figure 3 we have shown just two neurons in each layer. In practice the input layer must match the dimension of the input vectors and the output layer provides the number of desired outputs (usually small). However, the number of units in the hidden layers may be chosen by the designer. For the present we may still regard the units as having thresholds although in fact this is not usually the way they are implemented, as will be described in Section 2.

1.3. Mathematical approaches to neural networks

The theory of conventional computing devices is mostly a matter of discrete mathematics. Indeed, computation has inspired considerable advances in this branch of mathematics (Taylor, 1993). However as we have already seen, neural nets can be considered as analogue devices so the required mathematics is much more classical. The problems of classification and clustering traditionally belong to the statisticians. The structure of the classification space can be analysed using statistical decision theory (Amari, 1990). If we attempt to understand the actual behaviour of biological neurons we are in the realm of mathematical modelling. For example, some authors have considered architectures involving coupled oscillators and/or chaos theory (Jones, 1992). Neurons can be thought of as simple, statistically identical 'particles' linked by the synapses. Thus the large-scale behaviour of assemblies of neurons has much in common with the statistical physics of gasses (Venkataraman and Athithan, 1991). Dynamic behaviour can be introduced into networks in many different ways. Some network architectures are recursive (consider the output of a multilayer perceptron being fed back in as part of the input, or the Hopfield net introduced in Section 4 below). Others are defined by or approximated by differential equations. In fact the whole panoply of dynamical systems and control theory underlies the study or neural networks in a manner too pervasive to be adequately surveyed here. Differential topology has its adherents (Wang *et al.*, 1992). Indeed the theory of neural networks would appear to be almost as chaotic as their dynamic behaviour. It seems certain many of the results must be duplicated in different papers under different names and using different languages. There is a real need for the subject to develop its own coherent structure, rather than borrowing from a host of other mathematical disciplines. Notwithstanding this remark, we will now proceed to investigate neural networks from the viewpoint of numerical analysis!

2. Density and approximation by neural networks

A natural question arising from the ideas developed in Section 1 is to consider what sets of points a given network can classify. We have already seen that a multilayer perceptron can separate any finite sets of points in $\mathbb{R}^n$. Let us describe this a little more carefully. Let A and B be two finite sets in $\mathbb{R}^n$. In Figure 4, A might consist of the points labelled o and B the points labelled χ. Suppose we wish the network to produce output 1 for points in A and 0 for points in B. Clearly it is possible to construct a finite set of polygons $P_1 \cdots P_k$ such that

$$A \subset Q := \cup P_j \quad \text{and} \quad B \cap P_j = \phi \quad \text{for } j = 1, \ldots, k. \tag{2.1}$$

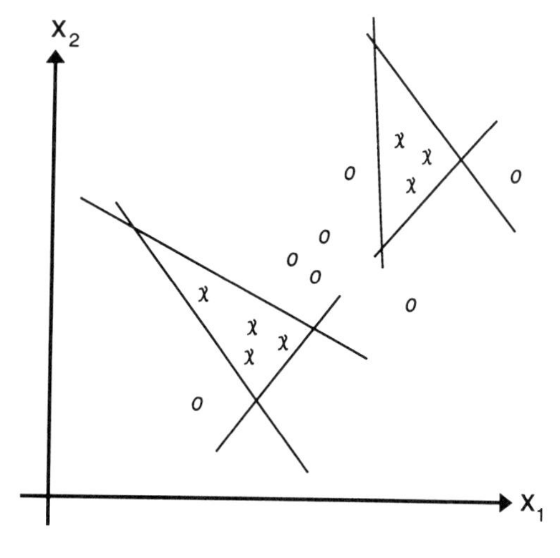

Fig. 4.

Each P_j consists of a finite intersection of half spaces. It can thus be obtained by a network computing the logical AND function which is linearly separable. The union to include A can then be obtained by a network computing the OR function: OR is also linearly separable. This approach is natural and simple, but it is difficult to take it very far. Moreover it only applies to discrete logical functions. We would like our networks to be able to cope with continuous problems such as the control problems involved in balancing a rocket or catching a ball. A different viewpoint proves more fruitful.

As before, we regard our inputs as vectors in $\mathbb{R}^n$. The output $\mathbf{y}$ of the network is a vector in $\mathbb{R}^m$ where usually $m \ll n$. (In many cases $m = 1$.) The network thus computes a function $\mathbf{g} : \mathbb{R}^n \to \mathbb{R}^m$ which we regard as an approximation to some other function $\mathbf{f} : \mathbb{R}^n \to \mathbb{R}^m$. The point classification problem discussed above can be put into this context by choosing $m = 1$ and $\mathbf{f}$ to be the characteristic function of the set Q in (2.1) (i.e. $\mathbf{f}(x) = 1$ if $x \in Q$ and 0 otherwise). This viewpoint means that neural networks can be discussed using methods derived from approximation theory. The point sets A and B are conveniently regarded as interpolation or sample points for approximation of the function $\mathbf{f}$. (Sometimes networks are actually constructed this way: the radial basis function networks discussed in Section 2.3 are of this type (Broomhead and Lowe, 1988; Mason and Parks, 1992).) As constructed here the function $\mathbf{f}$ is not continuous; however since the point sets A and B are finite, it is clearly possible to overcome this with some smoothing process.

The question of what a neural net can compute may thus be restated in approximation theoretic terms. Specifically we wish to know if our set

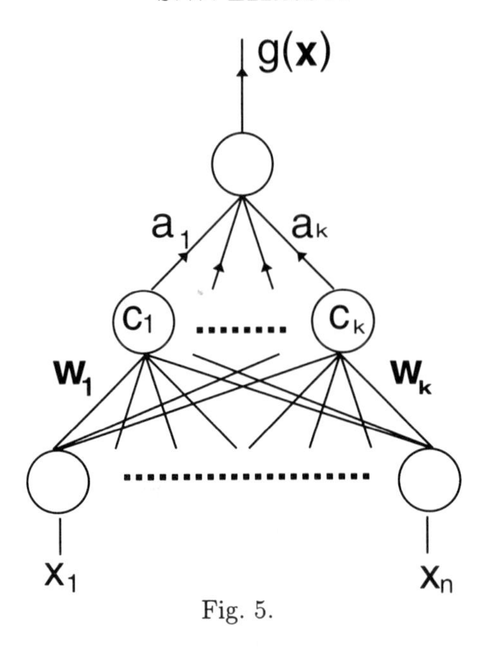

Fig. 5.

of possible functions **g**, corresponding to our particular class of network, is dense in some suitable function space which includes our target function **f**. This question is the subject of this section. In fact nearly all the results just consider the case of a single output network $m = 1$, so we will also make this simplification. Our functions are thus (scalar) real valued, so we drop the vector bold type and just refer to them as g and f.

In view of the difficulty of dealing with nondifferentiable and discontinuous functions, it is usual to use a smooth activation function, instead of a threshold, for the units. (The activation function was explained in section 1.1. It is the function that relates the sum of the inputs to a given unit to the output.) Note that the activation function $\sigma : \mathbb{R} \to \mathbb{R}$ Various restrictions need to be put on σ to make a practical network, but we will introduce these as required.

The results to be discussed next refer to the case of an multilayer perceptron with a *single* hidden layer of k units. The form of the network is shown in Figure 5. The weight vector relating the inputs to the jth hidden neuron is denoted $\mathbf{w}_j$. Thus for a given input $\mathbf{x}$, the input to this unit is $\mathbf{w}_j^T\mathbf{x}$. We assume that each of the hidden units has identical activation function σ, but that a 'threshold like' shift of the argument by a real scalar c_j is permitted. So the output from the jth hidden unit is

$$\sigma(\mathbf{w}_j^T\mathbf{x} + c_j).$$

Functions of this form are called *ridge functions*. The name derives from the fact that they are obviously constant on the hyperplane $\mathbf{w}_j^T\mathbf{x} = \text{constant}$. In two dimensions, this means that the contours of the function form straight ridges. An immediate consequence of this observation is the fact that no nontrivial continuous ridge function can be in $L_1(\mathbb{R}^n)$ if $n > 1$. To see this, choose $\mathbf{x}_0$ for which $\sigma(\mathbf{w}_j^T\mathbf{x}_0 + c_j) \neq 0$. Then integrate $|\sigma|$ over the infinite strip $|\mathbf{w}_j^T\mathbf{x} - \mathbf{w}_j^T\mathbf{x}_0| < \delta$, where δ is chosen sufficiently small that σ does not vanish in the region.

Now we denote the weight connecting the jth hidden unit to the output by a_j. The output function g of the network is therefore

$$g(\mathbf{x}) = \sum_{j}^{k} a_j\sigma(\mathbf{w}_j^T\mathbf{x} + c_j). \tag{2.2}$$

Activation functions σ used in practice have the property of being monotonic increasing, bounded and *sigmoidal*, which means that the limits at $\pm\infty$ are 1 and 0 respectively. Except for the threshold function, they are also continuous and smooth. The most popular choice is

$$\sigma(x) = 1/(1 + \exp(-x)). \tag{2.3}$$

However, the density proofs do not use all these conditions. For the basic results only continuity or uniform continuity is required, plus in some cases the condition that σ be sigmoidal.

We are interested, then, in approximation by linear combinations of ridge functions. The first papers to establish that one hidden layer is sufficient, i.e. that functions of the form (2.2) are dense, were Cybenko (1989), Hornik *et al.* (1989) and Funahashi (1989). However, simpler and sharper proofs have since superseded this work. An excellent survey of this topic is already available: Light (1992), to which the current author is considerably indebted. Rather than merely repeat the contents of this survey, we will adopt a more synthetic and comparative approach, taking aspects from the methods of the various authors and considering also some work by Mhaskar which postdates the Light survey. Another survey, which deals with some approximation theory issues in neural nets in a more practical way, is Mason and Parks (1992), but this contains little analytical detail. We will also concentrate on the particular practical issues involved for neural networks, but we shall attempt to explain and simplify some of the analytical details which tend to be rather technical in this field. In Sections 2.1 and 2.2 we will give two complete and quite different proofs of the fundamental density result, together with some other interesting sidelights. In the rest of Section 2 we discuss more briefly two other relevant aspects of the interaction of neural computing and approximation theory, namely radial basis networks, and networks using finite length arithmetic.

2.1. Direct approaches to density

Several proofs of the density result start by considering the one-dimensional case and we will also adopt this strategy in this subsection. A direct constructive approximation operator has been devised for this case by Chen, Chen and Liu (1991) (their result is most easily found in Light (1992).) It is of the type known as a *quasi interpolant* which means that it is based on linear combinations of function values. (It does not actually interpolate of course.) The operator is, in fact, similar to the well known Bernstein operator for polynomial approximation. The basic idea is to approximate the function f by piecewise constants: a fairly natural approach in a neural net context where the activation functions are generally thought of as smoothed thresholds. We start by recalling a basic measure of continuity. Let K be a compact set in $\mathbb{R}$ Recall that if $f \in C(K)$, the *modulus of continuity* of f is defined as

$$\omega(f,\delta) = \sup_{\substack{x,y \in K \\ |x-y|<\delta}} |f(x) - f(y)|.$$

Since f is continuous, ω is finite and tends monotonically to zero as $\delta \to 0$. The modulus of continuity encapsulates various ideas of smoothness of f that might be introduced. For example, suppose that f is Lipschitz, i.e. $|f(x) - f(y)| < L|x-y|$ for some real number L and for all $x, y \in K$. Then we have immediately $\omega(f,\delta) \leq L\delta$. Similar estimates apply for α-Lipschitz and for differentiable functions. The modulus of continuity gives a simple estimate of how well f can be approximated by piecewise constants. For simplicity (but without much loss of generality as we can always rescale) we choose K to be the interval [0,1], and let $n \in \mathbb{N}$ We consider the step function $h_n(x)$ which takes the value $f(\nu/n)$ in the interval $\nu/n \leq x =< (\nu+1)/n$. Obviously this gives $\|f - h_n\|_\infty \leq \omega(f, 1/n)$. (Actually we could reduce the constant from 1 to $\frac{1}{2}$ by evaluating at mid-points but this choice simplifies the notation.) It is convenient to write

$$h_n(x) = f(0) + \sum_{\nu=1}^{\mu} \{f(\nu/n) - f((\nu-1)/n)\} \tag{2.4}$$

where μ is the largest integer which does not exceed nx.

Now consider a continuous sigmoidal function such as (2.3). If we replace x by ax for some $a > 1$, we steepen σ in the transitional region around $x = 0$. In fact as $a \to \infty$, $\sigma(ax) \to 0$ or 1, according as $x < 0$ or $x > 0$. In other words σ converges pointwise to a simple threshold function with the value $\sigma(0)$ at 0. Thus $\sigma(ax) - \sigma(a(x-1))$ will approach the unit step function which takes the value 1 on (0,1) and 0 elsewhere. In view of the discontinuities at 0 and 1, this convergence cannot be uniform. But Chen *et al.* (1991) noticed that if one combines the constructions of this and the previous paragraph to approximate f, one *does* get uniform approximation, since the size of the discontinuity approaches 0 as $n \to \infty$. More specifically, suppose σ is continuous (on $\mathbb{R}$) and sigmoidal. They define A_n to be the smallest positive integer such that

$$|\sigma(x)| \leq n^{-1} \text{ for } x \leq -A_n \quad \text{and} \quad (1-n^{-1}) \leq \sigma(x) \leq (1+n^{-1}) \text{ for } x \geq A_n.$$

Then they define the quasi interpolant g_n as

$$g_n(x) = f(0) + \sum_{\nu=1}^{n} \{f(\nu/n) - f((\nu-1)/n)\}\sigma(A_n(nx-\nu)) \tag{2.5}$$

for $x \in [0,1]$. Observe that this is precisely the approximation obtained by the construction described above: for those values of ν in the summation with $x < \nu$, $\sigma(A_n(nx-\nu))$ is approximately zero. A careful estimate yields the following theorem.

Theorem 2.1 There exists a constant c such that for $f \in C[0,1]$,

$$\|f - g_n\|_\infty \leq c\omega(f, 1/n).$$

(Here the uniform norm is taken on the interval [0,1].) Note that c is independent of f and in fact we may choose $c = 4 + 2S$ where $S = \sup|\sigma(x)|$ for $x \in \mathbb{R}$, i.e. the norm of σ taken on the whole of $\mathbb{R}$.

Proof. We have $\|f - g_n\|_\infty = \|f - h_n + h_n + g_n\|_\infty \leq \|f - h_n\|_\infty + \|h_n - g_n\|_\infty$. We already know $\|f - h_n\|_\infty \leq \omega(f, 1/n)$ so only the second term need be considered. Now for any $x \in [0,1]$ with μ defined as in (2.5) we have

$$\begin{aligned} h_n(x) - g_n(x) &= f(0) + \sum_{\nu=1}^{\mu} \{f(\nu/n) - f((\nu-1)/n\}\{1 - \sigma(A_n(nx-\nu))\} \\ &\quad + \sum_{\nu=\mu+1}^{n} \{f(\nu/n) - f((\nu-1)/n)\}\sigma(A_n(nx-\nu)). \end{aligned}$$

Now $\nu \leq \mu - 1$ implies $nx - \nu \geq 1$ so $|1 - \sigma(A_n(A_n - \nu))| \leq 1/n$ by the definition of A_n. Similarly $\nu \geq \mu + 2$ implies $|\sigma(A_n(nx-\nu))| \leq 1/n$. Thus

$$\begin{aligned} |h_n(x) - g_n(x)| &\leq \omega(f, 1/n) + |\{f(\mu/n) - f((\mu-1)/n\} \\ &\quad + \{f((\mu+1)/n)\}\sigma(A_n(nx-\mu-1))|. \end{aligned}$$

The second term on the right-hand side is bounded by $2(1+S)\omega(f, 1/n)$, which completes the proof. □

We remark in passing that if σ is monotonic then $S = 1$ and c in Theorem 2.1 may be chosen as 6.

Now we need to pass to the n-dimensional case. There are two well known methods of passing from one-dimensional to higher-dimensional approximations: the blending operator and the tensor product. The former method has not to this author's knowledge, been applied to neural nets at all: the 'infinite interpolation' properties of blending operators seem likely to cause severe problems. However, the tensor product approach offers more hope.

To illustrate both the idea and the problems we will consider briefly the two-dimensional case. Suppose we have two sets of basis functions $\{\phi_1, \dots, \phi_\mu\}$ and $\{\psi_1, \dots, \psi_n\}$ where $\phi_i, \psi_j : \mathbb{R} \to \mathbb{R}$. The *tensor product basis* is the set of $\mu \times \nu$ functions

$$\zeta_{i,j}(x, y) = \phi_i(x)\psi_j(y).$$

Sometimes one can construct a two-dimensional approximation using the tensor product basis by applying a one-dimensional approximation operator in each dimension: for example two-dimensional orthogonal expansions can be constructed in this way. In practice the two sets are usually the same type of function (e.g. both polynomials or both trigonometric functions) although μ and ν may of course be different. Now let us consider what happens if we apply this construction to ridge functions. For simplicity we assume that the same function σ is to be used for x and y. So typical one-dimensional ridge functions will be $\sigma(a_i x + c_i)$ and $\sigma(b_j y + d_j)$. The tensor product basis thus consists of functions of the form

$$\sigma(a_i x + c_i)\sigma(b_j y + d_j).$$

In general this does *not* give a two-dimensional ridge function so we will not land up with a neural net approximation of the form (2.2). However, there is one particular choice of σ for which the construction *does* work, namely $\sigma(x) = \exp(x)$. For then we get

$$\begin{aligned} \sigma(a_i x + c_i)\sigma(b_j y + d_j) &= \exp(a_i x + c_i)\exp(b_j y + d_j) \\ &= \exp(a_i x + b_j y + c_i d_j) \\ &= \sigma(a_i x + b_j y + c_i d_j). \end{aligned}$$

This observation has been used by several authors to produce n-dimensional ridge function approximations. The basic idea is to prove the density of the ridge functions for the special case of $\sigma(x) = \exp(x)$ and then to use a one-dimensional result such as Theorem 2.1 to approximate the exponential function by linear combinations of the desired σ. If we are not interested in constructive methods then a simple application of the Stone–Weierstrass theorem (Cheney, 1966, p. 190) will do for the first stage (Diaconis and Shashahani, 1984; see also Hornik *et al.*, 1989). To avoid writing down explicit lincar combinations of the form (2.2) all the time, we introduce the following definition: a set of functions is said to be *fundamental* in a space if linear combinations of them are dense in that space.

Theorem 2.2 Let K be a compact set in $\mathbb{R}^n$. Then the set E of functions of the form $\mu(\mathbf{x}) = \exp(\mathbf{a}^T\mathbf{x})$, where $a \in \mathbb{R}^n$, is fundamental in $C(K)$.

Proof. By the Stone–Weierstrass Theorem we need only show that the set forms an algebra and separates points. Suppose $\mathbf{x} \in K$. We first have

$$\exp(\mathbf{a}^T\mathbf{x})\exp(\mathbf{b}^T\mathbf{x}) = \exp(\mathbf{a}^T\mathbf{x} + \mathbf{b}^T\mathbf{x}) = \exp((\mathbf{a}^T + \mathbf{b}^T)\mathbf{x}).$$

The set also contains the function '1': simply choose $\mathbf{a} = 0$. This establishes that E is an algebra. It remains to show that E separates the points of K. So let $\mathbf{x}, \mathbf{y} \in K$ with $\mathbf{x} \neq \mathbf{y}$. Set $\mathbf{a} = (\mathbf{x} - \mathbf{y})$. Then $\mathbf{a}^T(\mathbf{x} - \mathbf{y}) \neq 0$ so $\mathbf{a}^T\mathbf{x} \neq \mathbf{a}^T y$. Thus $\exp(\mathbf{a}^T\mathbf{x}) \neq \exp(\mathbf{a}^T\mathbf{y})$. The proof is complete. □

Before considering more constructive versions of this result let us complete the density proof.

Theorem 2.3 Let K be a compact set in $\mathbb{R}^n$. Then the set F of functions of the form $g(x)$, defined by (2.2) with σ a continuous sigmoidal function, is dense in $C(K)$.

Proof. Let $f \in C(K)$. For any $\epsilon > 0$, there exists (by Theorem 2.2) a finite number m of vectors $\mathbf{a}_i$ such that

$$\left\| f - \sum_{i=1}^{m} \exp(\mathbf{a}_i^T\mathbf{x}) \right\|_\infty < \epsilon/2.$$

Since there are only m scalars $\mathbf{a}_i^T\mathbf{x}$, we may find a finite interval including all of them. Thus there exists a number Γ such that $\exp(\mathbf{a}_i^T\mathbf{x}) = \exp(\Gamma y)$ where $y = (\mathbf{a}_i^T\mathbf{x}/\Gamma) \in [0,1]$. Then Theorem 2.1 tells us that the function $\exp(\Gamma y)$ can be approximated by linear combinations functions of the form $\sigma(\mathbf{w}_j^T\mathbf{x} + c_j)$ with a uniform error less than $\epsilon/(2m)$, from which the desired result easily follows. □

Sun and Cheney (see e.g. Light (1992) p. 4) have a more sophisticated version of this argument which shows that the elements of the vectors $\mathbf{w}_j$ and the constants c_j may be chosen to be rational numbers (i.e. there is a countable fundamental set). We will not give the details of their result as it complicates the proof significantly. However it is possible to draw the same conclusion by adopting a more constructive approach to Theorem 2.2. Observe first that Γ in the proof of Theorem 2.3 can be chosen to be an integer, and the numbers A_n, n and ν in (2.5) are also integers. The only problem therefore is to show that the vectors $\mathbf{a}$ in Theorem 2.2 can be chosen with rational elements. Moreover, this is also the only part of the argument above that is not constructive. Chen *et al.* (1991) and Mhaskar and Micchelli (1992) get round this by handling the approximation problem of Theorem 2.2 more explicitly. We will not give full analytical details, but indicate the method of attack. Once again, this is based on a tensor product basis. For simplicity, suppose that K is the Cartesian product of closed intervals. Then if a function $f \in C(K)$ is piecewise smooth, it is well known that f may be expanded in a multivariate Fourier series which will converge uniformly. (If f or K is more complicated, simple expansion is not sufficient, but methods of classical approximation theory may still be used to obtain uniform approximations.) Multiplying out terms one obtains an approximation to f as a linear combination of functions of the form

$$\begin{aligned}\exp(im_1x_1)\exp(im_2x_2)\exp(im_3x_3)\ldots\exp(im_nx_n)\\ = \exp\{\mathrm{i}(m_1x_1 + m_2x_2 + \ldots + m_nx_n)\},\end{aligned}$$

where $\mathrm{i}^2 = -1$ and the $m_j \in \mathbb{N}$. This is precisely of the form required for Theorem 2.2 except for the introduction of complex numbers. Actually the terms occur in complex conjugate pairs so the simplest way to proceed is to use trigonometric functions instead of the exponential function in Theorem 2.2. This will make no difference as far as proof of Theorem 2.3 is concerned.

The method of the previous paragraph also allows the classical Jackson theorems for trigonometric approximation to be used (Cheney, 1966, pp. 139–49) to obtain an estimate of the degree of approximation like that of Theorem 2.1. This question of rate of convergence of approximations is obviously of considerable importance. If f is smooth and we use smooth approximating functions such as (2.3) we might hope to get better convergence than the simple $\mathcal{O}(1/n)$ implied by Theorem 2.1. Apart from a paper by Mhaskar which we consider shortly, very little attention has been given to this issue. Although natural in a neural net context, approximating by piecewise constants as in Theorem 2.1 is a rather odd approach from the viewpoint of classical approximation theory since obviously it means that we cannot possibly do better than $\mathcal{O}(1/n)$.

To get a better degree of approximation a different starting point is required: one such approach will be discussed in Section 2.2. But it is also important to consider which functions we are trying to approximate. To solve the classification problem described at the start of Section 2, approximation by discontinuous functions may seem more natural. Even in analogue applications, it is well known that time optimal controllers may be discontinuous. (See Sagan (1969, pp. 295–97) for a simple example.) Thus, while it is certainly interesting to discuss degree of approximation of smooth functions by smooth networks, it would also be of value to consider degree of approximation to discontinuous functions by nonsmooth or discontinuous networks. Clearly this will require abandonment of the uniform norm and yet the classical L_p norms are not obviously appropriate either. It will be necessary to examine the applications to derive appropriate measures of smoothness and error: these may well be Sobolev or measure theoretic based. In addition we observe that Figure 4 might suggest that two hidden layers rather than one would be preferable. The argument discussing this figure makes use of geometric ideas of closeness: points which are close together in space can be handled by groups of neurons which deal only with these points. The success of spline approximations suggests that this is a desirable property, and it also makes sense from the point of view of understanding knowledge organization in networks.

Mhaskar (1993) does not solve these problems but he does have some interesting new insights. Already in Mhaskar and Micchelli (1992) the idea of a kth degree sigmoidal function is introduced. A function σ is said to be kth *degree sigmoidal* if

$$\sigma(x)/x^k \to 1 \quad \text{as } x \to \infty$$

and

$$\sigma(x)/x^k \to 0 \quad \text{as } x \to -\infty.$$

The case $k = 0$ recovers the ordinary sigmoidal functions. Functions of this type can be used to approximate a spline of degree k. The idea of Mhaskar and Micchelli was to replace the piecewise constant function (2.4) by a spline of higher degree, thus obtaining a better degree of approximation to smooth f. They then approximate the spline terms by kth degree sigmoidal functions to obtain an approximation of the form (2.2) with a better degree of approximation to f if $k > 0$. (The analytical details are conveniently found in Light (1992, pp. 28–30).) Of course σ is no longer the conventional sigmoidal function normally used for neural nets. In the later article Mhaskar uses these ideas in new way. First he deals with the multivariate case directly, approximating the multivariate f by a tensor product spline. He then considers the problem of approximating f by a neural network with a fixed number of neurons but arranged in more than one layer. Both the cases $k = 0$ and $k > 0$ are considered; the details are slightly different for the two cases although the basic idea is the same. Multilayer networks involve compositions of linear combinations of sigmoidal functions. Mhaskar employs the fact that the polynomial terms in the splines can themselves be decomposed as compositions of linear functions which can be approximated by the sigmoidal functions. Several interesting results are obtained on the degree of approximation obtainable by this method when $k > 0$. Unfortunately the technical details of Mhaskar's arguments are quite complicated, but there is no doubt that the article will repay close study. But of course, k sigmoidal functions are not the functions actually used in neural nets.

2.2. Dual space and convolution methods of approximation

Readers familiar with approximation theory will be aware that as well as the direct constructive approach to density, it is also possible to achieve such results by methods based on integrals, generally based either on dual space arguments or convolutions. These methods can be used to address the question of density of networks, the most recent work here being the convolution technique of Xu *et al.* (1991) who obtain constructive approximations this way. Although their work requires rather more stringent conditions on σ than the methods of the previous section, it does have certain advantages. In particular, being based on the use of quadrature formulae, it would seem to present the possibility of a line of attack for estimating the degree of approximation to smooth f.

Most of this section will therefore be devoted to convolutions, but first in the interests of completeness we consider dual space methods. This topic is of at least historical interest, since Cybenko's original proof was of this type (Cybenko, 1989). Since this approach may be rather mystifying to approximation-theory nonspecialists, we will first describe the fundamental idea of the method. Consider a normed vector space X over $\mathbb{R}$ The (bounded) *dual space* of X, denoted by X', is the space of all bounded linear functionals on X. (A linear functional is a linear mapping from X to $\mathbb{R}$) It has a natural norm defined by

$$\|l\| = \sup_{\substack{\mathbf{x}\in X \\ \|\mathbf{x}\|=1}} |l(\mathbf{x})|. \tag{2.6}$$

X' is always a Banach space, even if X is not. (Readers not familiar with this construction at all are advised to consult a suitable textbook such as Kreyszig (1978, pp. 119–25).)

Now let V be a subspace of X. We wish to know whether V is dense in X. The relevance of the dual space is shown by the following theorem.

Theorem 2.4 V is dense in X if and only if the only linear functional $l \in X'$ for which $l(\mathbf{v}) = 0$ for all $v \in V$ is the trivial one $l(\mathbf{x}) \equiv 0$.

Proof. Suppose first that V is dense in X. Suppose also that l is a linear functional l such that $l(v) = 0$ for $\mathbf{v} \in V$. Let $\mathbf{x} \in X$. For any $\epsilon > 0$ we have $\mathbf{v} \in V$ with $\|\mathbf{x} - \mathbf{v}\| < \epsilon$. Then $|l(\mathbf{x})| = |l(\mathbf{x}) - l(\mathbf{v})| = |l(\mathbf{x} - \mathbf{v})| \leq \|l\|\|\mathbf{x} - \mathbf{v}\| < \|l\|\epsilon$. Since this is true for any $\epsilon > 0$, we must have $l(\mathbf{x}) = 0$. This establishes the 'only if' part of the theorem.

Now suppose that V is *not* dense in X. Then there is a $\mathbf{x} \in X$ and a number $\delta > 0$ such that $\|\mathbf{x} - \mathbf{v}\| > \delta$ for all $\mathbf{v}$ in V. Let W be the space spanned by $\mathbf{x}$ and the space V, i.e. the set of all linear combinations $\alpha\mathbf{x} + \mathbf{v}$, where $\alpha \in \mathbb{R}$ and $\mathbf{v} \in V$. Note that the number α here is unique, for if $\alpha_1 x + \mathbf{v}_1 = \alpha_2\mathbf{x} + \mathbf{v}_2$ we have $(\alpha_1 - \alpha_2)\mathbf{x} = \mathbf{v}_2 - \mathbf{v}_2$, whence we must have $\alpha_1 = \alpha_2$ since $\mathbf{x} \notin V$. Thus we can define the following linear functional on $W : l(\alpha\mathbf{x} + \mathbf{v}) = \alpha$. Note that $l(\mathbf{x}) = 1$ and $l(\mathbf{v}) = 0$ for all $\mathbf{v} \in V$. Now if $\mathbf{w} = \alpha\mathbf{x} + \mathbf{v}$ with $\alpha \neq 0$,

$$\|\alpha\mathbf{x} + \mathbf{v}\| = |\alpha|\|\mathbf{x} + \alpha^{-1}\mathbf{v}\| \geq |l(w)|\delta$$

so

$$|l(w)| \leq \|\mathbf{w}\|/\delta.$$

On the other hand if $\alpha = 0$, $|l(w)| = 0$ so the inequality above holds trivially. This shows that l is a nontrivial bounded linear functional on W. By the Hahn–Banach Theorem (Kreyszig, 1978 p. 214) l may be extended to a bounded linear functional on the whole of X. This completes the proof. □

Thus if we want to establish density of V in X we need only show that any linear functional which annihilates V is, in fact, the zero functional. This may at first sight seem a harder task than the original one. However for most function spaces used in practice it is possible to get a concrete representation of X' and its norm. This makes the task tractable. In fact it may be shown (Kreyszig, 1978, p. 227) that for the case of $X = C[a,b]$, any linear functional may be written

$$l(f) = \int_a^b f(x)\,\mathrm{d}w(x), \tag{2.7}$$

where w is a function of bounded variation. (Readers unfamiliar with the interpretation of this integral are again referred to Kreyszig (1978, p. 226), but it is convenient to use the Lebesgue integral here rather than the Riemann integral employed by Kreyszig.) So to establish the density of one-dimensional sigmoidal functions (compare Theorem 2.1) we need only show that if the integral (2.7) vanishes whenever f is a sigmoidal function, then necessarily w is constant. This is fairly straightforward. We have

$$0 = \int_a^b \sigma(kx + l)\,\mathrm{d}w(x) \tag{2.8}$$

for all $k, l \in \mathbb{Z}$. Once again we adopt the basic idea of making k large enough so that the integrand looks like a step function. For any $p, q \in \mathbb{Z}$ with $p/q \in [a,b]$, define

$$r(x) = \begin{cases} 0 & a < p/q \\ \sigma(l) & x = p/q \\ 1 & p/q < x \leq b. \end{cases}$$

Now consider the expression $\sigma(nq(t - p/q))$. As $n \to \infty$, this expression converges pointwise to r on $[a,b]$. By the Lebesgue Dominated Convergence Theorem we conclude that

$$0 = \int_a^b r(x)\,\mathrm{d}w(x) = \int_{(p/q)^+}^b \mathrm{d}w(x) + \sigma(l)(w(p/q^+) - w(p/q^-)).$$

The final term denotes the jump in w at the point p/q. Notice that the integral term does not depend on l so we first let $l \to -\infty$. By the definition of a sigmoidal function, $\sigma(l) \to 0$. We conclude that the integral term is zero. By subtraction we may deduce that

$$0 = \int_t^s \mathrm{d}w(x) \tag{2.9}$$

whenever s, t are rational and (by the Lebesgue Dominated Convergence Theorem using sequences of rational numbers to converge monotonically to the required end points) we deduce that in fact (2.9) holds for any $t, s \in [a, b]$. But this integral is precisely $w(s) - w(t)$, showing that w is, in fact, constant as required.

Although this result captures the main significance of Theorem 2.1, it is not as powerful since it does not yield the $\omega(f, 1/n)$ estimate, and it is not clear how such estimates could be obtained by this approach. So we will look instead at the use of convolutions.

The idea of the convolution method is to construct a kernel based on the sigmoidal functions which can be used to approximate the reproducing property of the Dirac generalized function. The convolution itself and the kernel are, in turn, approximated by quadrature formulae thus yielding the required approximation. The material discussed here is essentially that of Xu *et al.* (1991) as expounded in that article and in Light (1992). However, our development will differ in one or two respects.

We first require the basic reproducing property. A problem presents itself immediately in that the standard Theorem 2.5 below, as quoted by Xu *et al.* (Stein and Weiss, 1971, p. 10) requires the function f to be uniformly continuous on the whole of $\mathbb{R}^n$. In the approximation-theory context, it is more natural to consider a compact subset K. We therefore need to extend f from K to $\mathbb{R}^n$ in a suitable way. Xu *et al.* take K to be $[-\nu, \nu]$ and do not give the extension explicitly, but clearly such an extension is possible. We use $\mathrm{d}V(\mathbf{x})$ to denote the volume element $\mathrm{d}x_1\, \mathrm{d}x_2 \ldots \mathrm{d}x_n$. (Both Stein and Weiss, and Xu *et al.* use just $\mathrm{d}x$, but we prefer to emphasize that a volume integral is involved.) To avoid confusion here we also use $\|\cdot\|$ to denote the ordinary Euclidean norm on $\mathbb{R}^n$. Norms on the function spaces will be subscripted 1 or ∞ as appropriate. We also need to extend the notion of modulus of continuity to $\mathbb{R}^n$: this is simply defined to be

$$\omega(f, \delta) = \sup_{\substack{\mathbf{x},\mathbf{y} \in K \\ \|\mathbf{x}-\mathbf{y}\| < \delta}} |f(\mathbf{x}) - f(\mathbf{y})|.$$

Here $K \subset \mathbb{R}^n$ does not need to be compact, but we do need f to be uniformly continuous on K. We will give a proof of the required convolution result here, and add a couple of bounds that in some cases might enable results to be sharpened. Although we will not discuss these in detail in the rest of the article, it seems worthwhile to give them in order to stimulate further research on sharp estimates.

Theorem 2.5 Let f be bounded and uniformly continuous on $\mathbb{R}^n$ and let

$g \in L_1(\mathbb{R}^n)$ with

$$\int_{\mathbb{R}^n} g(\mathbf{x})\,\mathrm{d}V(\mathbf{x}) = 1.$$

Define $g_m(\mathbf{x}) = m^n g(m\mathbf{x})$. Then

(a) $f * g_m(x)$ converges uniformly to f as $m \to \infty$.

(b) For any $R > 0$,

$$\|f * g_m - f\|_\infty \le \omega(f, 2R/m)\|g\|_1 + 2\|f\|_\infty \int_{\|s\|>R} |g(\mathbf{s})|\,\mathrm{d}V(\mathbf{s}),$$

where $\|.\|_\infty$ is taken over the whole of $\mathbb{R}^n$.

(c) As an alternative to condition (b), suppose that f is Lipschitz with constant Λ and that

$$M = \int_{\mathbb{R}^n} \|\mathbf{x}\||g(\mathbf{x})|\,\mathrm{d}V(\mathbf{x}) < \infty.$$

Then $\|f * g_m - f\|_\infty \le M\Lambda/m$.

Proof. First observe that

$$\begin{aligned}\int_{\mathbb{R}^n} g_m(\mathbf{x})\,\mathrm{d}V(\mathbf{x}) &= \int_{\mathbb{R}^n} m^n g(m\mathbf{x})\,\mathrm{d}V(\mathbf{x}) = \int_{\mathbb{R}^n} g(m\mathbf{x})\,\mathrm{d}V(m\mathbf{x}) \\ &= 1 \quad (\text{setting } \mathbf{y} = m\mathbf{x}).\end{aligned}$$

Hence

$$(f * g_m)(\mathbf{x}) - f(\mathbf{x}) = \int_{\mathbb{R}^n} (f(\mathbf{x}-\mathbf{t}) - f(\mathbf{x}))g_m(\mathbf{t})\,\mathrm{d}V(\mathbf{t})$$

so

$$\begin{aligned}|(f * g_m)(\mathbf{x}) - f(\mathbf{x})| &\le \int_{\mathbb{R}^n} |f(\mathbf{x}-\mathbf{t}) - f(\mathbf{x})||g_m(\mathbf{t})|\,\mathrm{d}V(\mathbf{t}) \\ &= \int_{\mathbb{R}^n} |f(\mathbf{x}-\mathbf{t}) - f(\mathbf{x})||g(m\mathbf{t})|m^n\,\mathrm{d}V(\mathbf{t}) \\ &= \int_{\mathbb{R}^n} |f(\mathbf{x}-\mathbf{s}/m) - f(\mathbf{x})||g(\mathbf{s})|\,\mathrm{d}V(\mathbf{s}) \quad (2.10) \\ &\quad \text{where } \mathbf{s} = m\mathbf{t} \\ &\le \int_{\mathbb{R}^n} \omega(f, \|\mathbf{s}\|/m)|g(\mathbf{s})|\,\mathrm{d}V(\mathbf{s}).\end{aligned}$$

Clearly $\omega(f, \|\mathbf{s}\|/m)$ is an integrable function of f and converges monotonically pointwise to zero as $m \to \infty$. Hence the integral on the right goes to zero by the Monotone Convergence Theorem. This establishes (a).

To get (b) we simply split the integration in (2.10) into the two regions $\|\mathbf{s}\| \le R$ and $\|\mathbf{s}\| > R$ and bound each term.

Part (c) is also straightforward: simply write $|f(\mathbf{x}-\mathbf{s}/m) - f(\mathbf{x})| < \Lambda\|\mathbf{s}\|/m$ in (2.10). □

The bound given in part (b) of the theorem does not tend to zero as $m \to \infty$ unless g has compact support, in which case we can choose R so that the second term is zero. Since we are going to construct g as an integral of our sigmoidal functions, compact support will be hard to achieve. On the other hand, we shall often find that $g(\mathbf{x})$ goes to zero very rapidly as $\|\mathbf{x}\| \to \infty$, so the bound might well be useful for finite m. Moreover the conditions of part (c) are likely to hold in many cases of practical interest. But this will still not yield any better estimates than those we already have from Section 2.1. Nevertheless, this approach does seem likely to reward further work with sharper estimates.

Equipped with this basic tool on uniform convergence of convolutions, let us now consider the work Xu *et al.* They construct a convolution kernel g as

$$g(\mathbf{x}) = \alpha_{n-1}^{-1} \int_{S^{n-1}} \phi(\mathbf{x}^T \mathbf{u})\, \mathrm{d}S^{n-1}(\mathbf{u}), \tag{2.11}$$

where S^{n-1} is the unit sphere in $\mathbb{R}^n$ (i.e. the set $\{\mathbf{u} \in \mathbb{R}^n \mid \|\mathbf{u}\| = 1\}$ and α_{n-1} is the 'surface area' obtained as the value of the integral with $\phi \equiv 1$. They work on a suitable compact set K in $\mathbb{R}^n$. Their choice is actually the n-dimensional interval $K = [-a, a]^n$ for some real a: the actual choice makes little difference in the following proof. We assume that $f \in C(K)$ and, so that we can apply Theorem 2.5, extended f continuously to $\mathbb{R}^n$ in such a way that $f(\mathbf{x}) = 0$ for $\mathbf{x} \in 2K$ where for any $t > 0$,

$$tK = \{\mathbf{x} \in \mathbb{R}^n \mid \mathbf{x}/t \in K\}.$$

Since f is then continuous on the bounded set $2K$ we conclude that f is uniformly continuous and bounded on $\mathbb{R}^n$. The fundamental result is:

Theorem 2.6 Let $K = [-a, a]^n$ for some real a, and let $\phi \in C(\mathbb{R})$ be uniformly continuous. Suppose g is defined by (2.11). If (i) $g \in L_1(\mathbb{R}^n)$ and (ii)

$$\int_{\mathbb{R}^n} g(\mathbf{x})\, \mathrm{d}V(\mathbf{x}) \neq 0,$$

then the set of functions of the form $\phi(\mathbf{x}^T\mathbf{a} + c)$, $\mathbf{a} \in \mathbb{R}^n$, $c \in \mathbb{R}$, is fundamental in $C(K)$.

Proof. Let $f \in C(K)$. We extend f to a bounded uniformly continuous function on $\mathbb{R}^n$ as described in the paragraph before the theorem. Rescaling ϕ if necessary, we can assume that the integral of g is 1. Now suppose we are given $\epsilon > 0$. In view of Theorem 2.5 we may choose m such that

$$\|f * g - f\|_\infty \le \epsilon/3,$$

where here we restrict $\|\cdot\|_\infty$ to K. (Note that there would appear to be a minor error in the the proof of this theorem in both Xu *et al.* (1991, p. 12) and Light (1992, p. 18). The bound in terms of $\omega(f, m^{-1})$ is not, in fact, correct; compare Theorem 2.5(b). Nevertheless the bound as given above is valid.)

The next step is to approximate the convolution by a quadrature formula. We have

$$(f * g_m)(\mathbf{x}) = \int_{2K} g_m(\mathbf{x}-\mathbf{y}) f(\mathbf{y})\,\mathrm{d}V(\mathbf{y}) = \int_{2mK} g(m\mathbf{x}-\mathbf{z}) f(\mathbf{z}/m)\,\mathrm{d}V(\mathbf{z}). \tag{2.12}$$

It is a simple consequence of (2.11) that g is continuous: indeed $\omega(g, \delta) \leq \omega(\phi, \delta)$. Since the integrand is continuous and $2mK$ is compact we can approximate it by a quadrature formula for any particular value of $\mathbf{x}$. But we need a bound uniform over $\mathbf{x}$. For the present we will just use a simple Riemann sum approximation as do Xu *et al.* (However to get sharper estimates it will prove necessary to give closer attention to the approximation of (2.12).) For any $\delta > 0$, let P be a partition of $2mK$ into a finite disjoint family of Borel sets, each of diameter at most δ. For each $A \in P$, choose $\mathbf{z}_A \in A$ and define

$$b_A = \int_A f(\mathbf{z}/m)\,\mathrm{d}V(\mathbf{z}).$$

So

$$\begin{aligned}
&\left| \int_{2mK} g(m\mathbf{x}-\mathbf{z}) f(\mathbf{z}/m)\,\mathrm{d}V(\mathbf{z}) - \sum_{A\in P} b_A g(m\mathbf{x} - A) \right| \\
&\quad \leq \sum_{A\in P} \int_A |g(m\mathbf{x}-\mathbf{z}) - g(m\mathbf{x}-\mathbf{z}_A)| |f(\mathbf{z}/m)|\,\mathrm{d}V(\mathbf{z}) \\
&\quad \leq \omega(g, \delta) \sum_{A\in P} \int_A f(\mathbf{z}/m)|\,\mathrm{d}V(\mathbf{z}) \\
&\quad = m^n \omega(g, \delta) \int_{2K} |f(\mathbf{y})|\,\mathrm{d}V(\mathbf{y}).
\end{aligned}$$

Hence we can choose δ and P so that this error is less than $\epsilon/3$.

Now we apply a similar argument to (2.11). For any $\theta > 0$, let Q be a partition of S^{n-1} into a finite disjoint collection of Borel sets of diameter at most θ. For any $B \in Q$, Set

$$c_B = \alpha_{n-1}^{-1} \int_B \mathrm{d}S^{n-1}(\mathbf{u})$$

and choose $\mathbf{u}_B \in B$. We get in a similar fashion to the argument above

$$|g(m\mathbf{x}-\mathbf{z}_A) - \sum_{B\in Q} c_B \phi((m\mathbf{x}-\mathbf{z}_A)^T \mathbf{u}_B)| \leq \omega(\phi, \|m\mathbf{x}-\mathbf{z}_A\|\theta).$$

But $\mathbf{z}_A \in 2mK$ so $\mathbf{z}_A/m \in 2K$. Hence for any $\mathbf{x} \in K, \|m\mathbf{x} - \mathbf{z}_A\| = m\|\mathbf{x} - \mathbf{z}_A/m\| \leq 3R$, where R is the diameter of K. Thus we can choose

θ and B so that the right-hand side of the inequality above is less than $\|f\|_1\epsilon/3$, whence

$$\left|\sum_{A\in P} b_A\Big(g(m\mathbf{x}-\mathbf{z}_A)-\sum_{B\in Q} c_B\phi((m\mathbf{x}-\mathbf{z}_A)^T\mathbf{u}_B)\Big)\right| \le \epsilon/3.$$

Finally, putting these three approximations together, we find that, for any $\mathbf{x}\in K$,

$$\left|f(\mathbf{x})-\sum_{A\in P}\sum_{B\in Q} b_A c_B \phi(m\mathbf{x}^T\mathbf{u}_B-\mathbf{z}_A^T\mathbf{u}_B\right| \le \epsilon. \quad \square$$

Thus, to establish that the functions $\phi(\mathbf{x}^T\mathbf{a}+c)$, $\mathbf{a}\in\mathbb{R}^n$, $c\in\mathbb{R}$ are fundamental, we have two points to check: first that g defined by (2.11) is in $L_1(\mathbb{R}^n)$; and second that it has a nonzero integral. We shall find that we cannot simply take $\phi=\sigma$ where σ is our required sigmoidal function; we will need to take a linear combination of σ terms.

The first question we need to consider is how fast g must go to zero to be in $L_1(\mathbb{R}^n)$. We need to do nothing more sophisticated than to bound $g(\mathbf{x})$ by a suitable power of $r=\|x\|$ when r is large. The next lemma tells us what power is required.

Lemma 2.7 With $r=\|\mathbf{x}\|, \mathbf{x}\in\mathbb{R}^n$, and $q,R\in\mathbb{R}$ with $R>0$, we have

$$\int_{\|\mathbf{x}\|\ge R} r^{-q}\,\mathrm{d}V(\mathbf{x})<\infty$$

if and only if $q>n$.

Proof. Denoting the sphere of radius r by S_r^{n-1}, we have for $\rho>R$

$$\int_{\rho\ge\|\mathbf{x}\|\ge R} r^{-q}\,\mathrm{d}V(\mathbf{x})=\int_R^\rho r^{-q}\int_{S_r^{n-1}}\mathrm{d}S_r^{n-1}\,\mathrm{d}r.$$

But since S_r^{n-1} is an $(n-1)$-dimensional manifold,

$$\int_{S_r^{n-1}}\mathrm{d}S_r^{n-1}=r^{n-1}\alpha_{n-1}.$$

(Recall α_{n-1} is the area of the sphere of radius 1: see (2.11).) Thus

$$\int_{\|\mathbf{x}\|\ge R}^{\rho} r^{-q}\,\mathrm{d}V(\mathbf{x})=\alpha_{n-1}\int_R^\rho r^{-q+n-1}\,D \quad =\alpha_{n-1}\left[\frac{r^{n-q}}{n-q}\right]_R^\rho \quad \text{for } q\ne n.$$

Thus the limit as $\rho\to\infty$ exists only if $q>n$. The case $q=n$ yields a logarithmic integral which also goes to infinity. $\square$

Thus to show that $g\in L_1(\mathbb{R}^n)$, it is sufficient to show that $g(\mathbf{x})=o(\|\mathbf{x}\|^{-n})$ as $\|\mathbf{x}\|\to\infty$. To do this Xu *et al.* find an alternative form of (2.11).

Lemma 2.8 Let g be defined by (2.11). Then $g(\mathbf{x}) = g_0(\phi, r)$ where $r = \|\mathbf{x}\|$ and

$$g_0(\phi, r) = \frac{\alpha_{n-2}}{\alpha_{n-1}} \int_{-1}^{1} \phi(rs)(1-s^2)^{(n-3)/2}\,\mathrm{d}s, \tag{2.13}$$

$$= \frac{\alpha_{n-2}}{\alpha_{n-1}} \int_{-r}^{r} r^{2-n}\phi(t)(r^2-t^2)^{(n-3)/2}\,\mathrm{d}t, \quad r \neq 0. \tag{2.14}$$

(A function such as g which depends only on r is said to be *radial.*)

Proof. We may assume $\|\mathbf{x}\| \neq 0$: this point can be 'filled in' as a limiting case since both (2.11) and (2.13) depend continuously on $\mathbf{x}$. So we may choose a coordinate system with its pole in the direction of $\mathbf{x}$. Let $\mathbf{w}$ be a unit vector in the direction of $\mathbf{x}$, whence $\mathbf{w} = \mathbf{x}/\mathbf{r}$. Then any point $\mathbf{u} \in S^{n-1}$ can be expressed as $\mathbf{u} = \mathbf{w}\cos\theta + \mathbf{v}$, where $\cos\theta = \mathbf{u}^T\mathbf{x}/r$, and $\mathbf{v}$ is a unit vector perpendicular to $\mathbf{x}$ with $\|\mathbf{v}\| = \sin\theta$. Moreover we cover the whole of S^{n-1} as θ varies from 0 to π and v takes all directions orthogonal to $\mathbf{x}$. So

$$g(\mathbf{x}) = \alpha_{n-1}^{-1} \int_{S^{n-1}} \phi(\mathbf{x}^T\mathbf{u})\,\mathrm{d}S^{n-1}(\mathbf{u})$$

$$= \alpha_{n-1}^{-1} \int_0^{\pi} \phi(r\cos\theta) \int_{S^{n-2}_{\sin\theta}} \mathrm{d}S^{n-2}_{\sin\theta}(\mathbf{v})\,\mathrm{d}\theta$$

where, as in the proof of Lemma 2.7, $S^{n-2}_{\sin\theta}$ denotes the $(n-2)$-dimensional sphere of radius $\sin\theta$. The inner integral is thus $\alpha_{n-2}\sin^{n-2}\theta$. This establishes that g is indeed a radial function and we can define g_0. Moreover, we have

$$g_0(\phi, r) = \frac{\alpha_{n-2}}{\alpha_{n-1}} \int_0^{\pi} \phi(r\cos\theta)\sin^{n-3}\theta\sin\theta\,\mathrm{d}\theta.$$

Now put $s = \cos\theta$, so $\mathrm{d}s = -\sin\theta\,\mathrm{d}\theta$ and $\sin^{n-3}\theta = (1-s^2)^{(n-3)/2}$ as required. (2.14) is obtained by substituting $t = rs$. □

Now, how do we choose ϕ to give g in $L_1(\mathbb{R}^n)$? The essential requirement is that $\phi(t)$ goes to zero quickly enough at $\pm\infty$. Let us now consider sigmoidal functions specifically. We note from (2.13) that g vanishes if ϕ is odd, so we might as well choose choose ϕ even, although this is not essential. Sigmoidal functions σ tend to 1 at $+\infty$, so let us first define

$$\psi(t) = \sigma(1+t) + \sigma(1-t) - 1. \tag{2.15}$$

Note that ψ is even and goes to zero at $\pm\infty$. Since we expect our functions σ to approximate step functions, it is reasonable to suppose that ψ goes to zero quickly. More specifically, let us suppose that σ is continuous and is such that

$$|\psi(t)| \leq K|t|^{-q} \quad q > n-2 \tag{2.16}$$

for some real K. Note that if σ is the usual choice (2.3), this condition actually holds for *any* $q > 0$ as ψ goes to zero exponentially. The next step in the argument is to expand the kernel in (2.14). Write $\lambda = (n-3)/2$. From now on we will assume $n > 2$: $n = 2$ requires special treatment (Xu *et al.*, 1991, p. 10), but we will not bother with this here. Let us first consider the case n odd, so that λ is a nonnegative integer. Hence $(r^2 - t^2)^\lambda$ is simply a polynomial in r and t, and the integral (2.14) may be taken termwise. We get

$$g_0(\phi, r) = r^{2-n} \sum_{j=0}^{\lambda} \beta_j(r) r^{2\lambda - 2j} \tag{2.17}$$

where

$$\beta_j(r) = \alpha_{n-2}\alpha_{n-1}^{-1}\,{}^\lambda C_j \int_{-r}^{r} \phi(t) t^{2j}\,\mathrm{d}t. \tag{2.18}$$

(Here ${}^\lambda C_j$ is the usual binomial coefficient.) The condition (2.16) means that all the β_j converge as $r \to \infty$. Hence:

Lemma 2.9 Let n be odd and ψ satisfy (2.16) and g be defined by (2.11) with $\psi = \phi$. A necessary and sufficient condition for g to be in $L_1(\mathbb{R}^n)$ is that

$$\int_0^\infty \psi(t) t^{2j}\,\mathrm{d}t = 0, \quad j = 0, \ldots, (n-3)/2. \tag{2.19}$$

Proof. If (2.19) fails for some j, we see from (2.17) that as $r \to \infty$, $g_0(\phi, r)$ would behave like r^{-p}, where

$$p = n - 2 - 2\lambda + 2j \le n - 2 < n$$

(compare Lemma 2.7). To get the sufficiency we note that (2.17) and (2.19) together imply that

$$g_0(\phi, r) = -r^{2-n} \sum_{j=0}^{\lambda} \gamma_j(r) r^{2\lambda - 2j} \tag{2.20}$$

with

$$\begin{aligned}
\gamma_j(r) &= \alpha_{n-2}\alpha_{n-1}^{-1}\,{}^\lambda C_j \int_{|t|>r} \psi(t) t^{2j}\,\mathrm{d}t \\
&\le \alpha_{n-2}\alpha_{n-1}^{-1}\,{}^\lambda Cj \int_{|t|>r} K|t|^{-q} t^{2j}\,\mathrm{d}t \\
&= 2K\alpha_{n-2}\alpha_{n-1}^{-1}\,{}^\lambda C_j \int_r^\infty t^{2j-q}\,\mathrm{d}t \\
&= -2K\alpha_{n-2}\alpha_{n-1}^{-1}\,{}^\lambda C_j r^{2j-q+1}
\end{aligned}$$

since by hypothesis $q > n - 2 > 2j$. Substituting this bound into (2.20) we find that g_0 goes to zero at least as fast as r to the power

$$(2-n+2\lambda-2j+2j-q+1)=(2-n+n-3-q+1)=-q, \quad \text{and } q>n.$$

(At at least the same rate as ψ, in fact.) Thus $g(\|x\|)$ goes to zero sufficiently fast. □

At first sight it appears from from (2.17), (2.18) and (2.19) that there is little hope of finding a nonzero g. But Xu *et al.* observed that this is not the case. First we will see that the moment condition (2.19) poses no serious difficulty. If we replace $\phi(t)$ by $\phi(pt)$ where $p>1$, we have

$$\int_0^\infty \phi(pt)t^{2j}\,\mathrm{d}t = p^{-(2j+1)}\int_0^\infty \phi(\mathrm{s})s^{2j}\,\mathrm{d}s \quad \text{by the substitution } s=pt. \tag{2.21}$$

If we have ψ (not identically zero) satisfying (2.16) we may choose (say) $p=2$ and define $\psi_0(t)=\psi(t)$, and $\psi_j(t)=\psi_{j-1}(t)-2^{2j+1}\psi_{j-1}(2t)$, $j=0,\ldots,\lambda$. Observe that ψ_j will still satisfy (2.16) with the same order q but with the previous K replaced at each stage by $(1+2^{2j+1-q})K$. Also (2.21) means that the moment condition of (2.19) is satisfied for powers of t up to j: thus $\phi(t)=\psi_\lambda(t)$ will satisfy (2.19) for all the required values of j. (The reader might be concerned that ψ_λ could vanish identically. However this will turn out to be impossible in the context we are going to use the result.)

It remains to show that the resulting g cannot have zero integral. Xu *et al.* prove the following elegant result.

Lemma 2.10 Let n be odd and ψ satisfy the conditions of Lemma 2.10 including (2.19). Suppose also ψ is even. Then

$$\int_{\mathbb{R}^n} g(\mathbf{x})\,\mathrm{d}V(\mathbf{x}) = -2\alpha_{n-2}\tau_n\int_0^\infty \psi(t)t^{n-1}\,\mathrm{d}t$$

where

$$\tau_n = \int_0^1 r(1-r^2)^{(n-3)/2}\,\mathrm{d}r > 0.$$

Proof.

$$\begin{aligned}
\int_{\mathbb{R}^n} g(\mathbf{x})\,\mathrm{d}V(\mathbf{x}) &= \alpha_{n-1}\int_0^\infty r^{n-1}g_0(r)\,\mathrm{d}r, \\
&\qquad \text{as in the proof of Lemma 2.7} \\
&\qquad \text{with } r^{-q} \text{ replaced by } g_0(r). \\
&= 2\alpha_{n-2}\int_0^\infty r\int_0^r \psi(t)(r^2-t^2)^{(n-3)/2}\,\mathrm{d}t\,\mathrm{d}r, \\
&\qquad \text{by (2.14) since } \psi \text{ is even} \\
&= -2\alpha_{n-2}\int_0^\infty r\int_r^\infty \psi(t)(r^2-t^2)^{(n-3)/2}\,\mathrm{d}t\,\mathrm{d}r \\
&\qquad \text{by (2.19)}
\end{aligned}$$

$$= -2\alpha_{n-2} \int_0^\infty \psi(t) \int_0^t r(r^2 - t^2)^{(n-3)/2}\, \mathrm{d}r\, \mathrm{d}t,$$

by Fubini's theorem.

But the inner integral is a constant multiplied by t^{n-1}, and by putting $t = 1$ we see that the value of the constant is τ_n. τ_n is certainly strictly positive, as its integrand is positive except at 0 and 1. □

Note that it is no real restriction that ψ be even; we can make it so as in (2.15). Putting all this together we arrive at the following theorem.

Theorem 2.11 Let $\sigma \in C(\mathbb{R})$ and ψ defined by (2.15). Suppose that n is odd and K is defined as for Theorem 2.5. Suppose also that ψ satisfies (2.16), and that

$$\int_0^\infty \psi(t)t^{n-1}\, \mathrm{d}t \neq 0.$$

Then the set of functions $\sigma(\mathbf{x}^T\mathbf{a} + c)$, $\mathbf{a} \in \mathbb{R}^n$, $c \in \mathbb{R}$, is fundamental in $C(K)$. In particular, this is true if σ is defined by (2.3).

Proof. This result is basically just an application of Theorems 2.6 and 2.11, but we do have to worry about the moment condition (2.19). Instead of ψ in (2.19) we must use ψ_λ as defined in the paragraph after (2.21). By a similar argument to (2.21) we find that for each j in this definition,

$$\int_0^\infty \psi_j(t)t^{n-1}\, \mathrm{d}t = (1 - 2^{2j+1-n}) \int_0^\infty \psi_{j-1}(t)\, \mathrm{d}t.$$

Since $j \leq (n-3)/2$, so the power cannot be zero, we find that the integral vanishes for $j = \lambda$ if and only if it vanishes for $j = 0$. But by definition $\psi_0 = \psi$.

If σ is as defined in (2.3), then a routine calculation shows $\psi(t) = (1 - \mathrm{e}^{-2})/(1 + \mathrm{e}^{-2} + 2\mathrm{e}^{-1}\cosh(t)) > 0$ for all t. So the integral in the statement of the theorem does not vanish. □

We remark also that the restriction $q > n$ in condition (2.16) can also be relaxed if ψ is sufficiently well behaved at ∞. More specifically, suppose

$$\psi(t) \simeq Ht^{-q} \quad \text{when } q \text{ is large.}$$

Then a binomial expansion shows that

$$\psi(t+1) - \psi(t-1) \simeq -Hqt^{-(q+1)}.$$

(In fact (2.15) gives one higher order for ψ than σ.) We may repeat this process until the exponent is greater than n. However, we might then have difficulty in showing that the integral in Theorem 2.11 does not vanish!

The case n even is rather less satisfactory. Lemma 2.9 still holds with in this case $j = 0, \ldots, (n-2)/2$ in (2.19). (Note the increased upper limit.)

The argument is essentially the same, although a little more care is needed since the expansion of the kernel is no longer finite in (2.17): we have to replace the upper limit of the sum by ∞. Then we need to justify the termwise integration: details are given in Light (1992, p. 16). However, the increased power of t that must be annihilated causes a problem. The first nonvanishing power of t is n. Thus the first nonvanishing power of r is $(2-n)+(-n-3+n) = -(n+1)$. To obtain the integral of g itself we must multiply by r^{n-1} and integrate (compare the proof of Lemma 2.10). The leading power to be integrated is therefore r^{-2} which integrates to a $1/r$ term. This suggests that the integral of g will in fact vanish. Xu *et al.* devote several pages of analysis to justifying this formally (a process which unfortunately tends to hide the fact that it is essentially a power counting argument). Since the integral of g vanishes, we cannot apply Theorem 2.5 directly to the case n even. However this is overcome by averaging in one higher dimension. Specifically, define

$$h(\mathbf{x}) = \alpha_n^{-1} \int_{S^n} \phi(\mathbf{x}'^T \mathbf{u})\, \mathrm{d}S^n(\mathbf{u}),$$

where for $\mathbf{x} = (x_1, x_2, \ldots, x_n) \in \mathbb{R}^n$, $\mathbf{x}' = (x_1, x_2, \ldots, x_n, 0) \in \mathbb{R}^{n+1}$. Using similar arguments to those above, we can show that h is a suitable kernel.

Now we have not actually used the fact that σ is sigmoidal, only that ψ defined by (2.15) be uniformly continuous and satisfy (2.16). Xu *et al.* also use a slightly weaker condition than (2.16), although it amounts to the same thing for practical σs. But apart from these minor considerations Theorem 2.11 is actually weaker than Theorem 2.3 since it requires stronger conditions on σ. The reader may therefore wonder why we have expended so much effort on it. However, in our opinion, the approach of Xu *et al.* offers at least two attractive features that make it worthy of serious study. First, it is a *direct* mutlivariate approach, avoiding the 'tensor product problem' discussed in Section 2.1. It thus gives some insight as to how the linear functionals corresponding to the weights in the first layer of the network might distribute information to the hidden nodes. Second the approach offers at least hope of providing sharper estimates for smooth f. This is because it is based on well understood principles of convolution and quadrature.

2.3. Radial basis networks

In Section 2.1 we met some interesting connections between neural networks and radial functions. Radial functions can also be employed more directly in neural computation. Look again at the function (2.2) which represents the function computed by a multilayer perceptron with one hidden layer. As we have already considered, the c_j quantities can be considered roughly as thresholds; they raise or lower the value at which the sigmoidal function σ switches from its asymptotic 0 value at negative arguments to its 1 value at positive arguments. The more important part of the argument is the inner product $\mathbf{w}_j^T\mathbf{x}$. Assuming that the input vectors $\mathbf{x}$ all have a similar normalization, we see that the input to the network node corresponding to the term $\sigma(\mathbf{w}_j^T\mathbf{x} + c_j)$ depends on the projection of the input vector $\mathbf{x}$ onto the weight vector $\mathbf{w}_j$. In other words the weight vectors represent 'test vectors' or in artificial intelligence language 'features' against which each input vector $\mathbf{x}$ is tested for a match. The closer the alignment, the higher the response. Perceptron type networks are essentially row projection networks, a fact which we will investigate in much more detail in Section 3. For the moment, however, let us consider a different approach.

Suppose that instead of identifying features as row vectors, we identify them as points. In a classification problem each point $\mathbf{w}_j$ might be selected as a typical representative of a known class, or if we do not know suitable *a priori* classes they might simply be distributed in some sensible fashion about the input space. Instead of measuring the projection of $\mathbf{x}$ onto each $\mathbf{w}_j$, we consider the distance. We still have a network with the same topology as Figure 5, but now the input to a given unit is $\|\mathbf{x} - \mathbf{w}_j\|$. The function computed by the network becomes

$$g(\mathbf{x}) = \sum_j^k a_j \sigma(\|\mathbf{x} - \mathbf{w}_j\|). \tag{2.22}$$

Note that the meaning of the second layer weight a_j remains unchanged. We therefore have a linear combination of *radial basis functions*. The corresponding network architecture is known as a *radial basis network*. There is an extensive literature of approximation by radial basis functions: see Powell (1992). Indeed this theory is much more advanced and better understood than that of ridge function approximation so it is superfluous to go into details here. We will just explain how radial basis networks are normally applied. The architecture was first introduced by Broomhead and Lowe (1988) and this article remains a good explanation of the basic method. However, certain *caveats* need to be made in referring to this article now. First, it was believed at the time of writing that multilayer perceptrons required *two* hidden layers to solve the classification problem. The radial basis architecture was proposed partly as a solution to this problem. However, as we have already seen, it is now known that only one hidden layer is needed, at least to obtain density. (The question of whether more layers give a better degree of approximation is still open, although many authors believe that the answer will turn out to be affirmative.) In the model of Broomhead and Lowe we need to choose the centres $\mathbf{w}_j$: they cannot be learned (at least not without resorting to a nonlinear algorithm which is precisely what the authors wished to avoid). Only the a_j are adapted in the fitting process. One could, in principle, choose the weight vectors in (2.2) and therefore solve a linear problem for those a_j in exactly the same way. Furthermore some of the comments of the authors on the backpropagation method could be construed as slightly misleading: we will come back to this in the next section. But notwithstanding these cautions, the idea remains a good one in view of the well developed theory and good numerical properties of radial basis functions. The method has retained its adherents. Mason and Parks (1992) survey this topic in a little more detail, and consider some more recent work.

The actual application is straightforward. First we need to choose a suitable activation function σ. Broomhead and Lowe recommend either the Gaussian $\sigma(r) = \exp(-r^2)$ or a multiquadric $\sigma(r) = (c_2 + r^2)^{1/2}$. We then assume that our function f that we wish to approximate is given at a set of points $\{\mathbf{x}_1, \ldots, \mathbf{x}_t\}$: this is the usual situation in classification problems (see the introductory remarks at the start of Section 2). The a_j are chosen simply to minimize

$$\sum_{j=1}^{t}(f(\mathbf{x}_j) - g(\mathbf{x}_j))^2.$$

This is a standard least-squares problem, which Broomhead and Lowe suggest is solved by computing the Moore–Penrose pseudoinverse (see, e.g., Ben-Israel and Greville (1974)). For simple networks the problem can be solved explicitly: the authors discuss the XOR problem (see Section 1.1) in some detail.

2.4. The effect of rounding on the approximation

Our work so far assumes that the weights can be evaluated to arbitrary precision. In a practical network, especially if implemented in hardware, one may only be able to store them to eight or 16 bits. There may be no point in using a very accurate network if its realization introduces large errors.

This problem has been encountered by various authors: the most systematic treatment would appear to be that of Brause (1992). Finite precision machine arithmetic is a classical issue in numerical analysis, but little deep theoretical work has been done in this context. The approach of Brause and others is largely experimental, backed up with some heuristic considerations and simple analysis. Brause's paper is a little off putting to numerical analysts at first sight, since it is expressed in the language of information theory. However, it is actually not difficult to come to grips with. His idea is to measure the error for a network with a given fixed system information, by which he means that a fixed *total* number of bits may be used to express the weights. So if we attempt to improve the approximation by introducing more neurons (and hence more weights) we must pay for that by storing the weights to a lower precision. He then takes two test problems and computes approximations (using an approximate minimax criterion) for fixed information, trading off the number of weights against the precision. He discovers that for each test problem there is a well defined optimum precision giving the best achievable error (a result which will not surprise anyone who has attempted to use finite differences to compute derivatives!).

3. Numerical analysis of learning algorithms

We now turn our attention to some algorithmic aspects of neural networks. We are going to consider the so-called *supervised* learning problem which we will pose as follows (see also Section 2.3).

We are given at a set of points or *patterns* $\{\mathbf{x}_1, \ldots \mathbf{x}_t\}$ in $\mathbb{R}^n$. Associated with each pattern $\mathbf{x}_j$ there is a desired response y_j. For simplicity we will assume here that the network is to have a single output so y_j is a scalar. (For the single layer perceptron to be considered first, we shall show that this involves no loss of generality: the outputs can be treated separately as described in Section 3.1. For multilayer networks this is not the case, but nevertheless the single output case is sufficient to illustrate the situation.) We think of the y_j as being values of some function f which we want the network to 'learn': it is supposed to identify some generic features of the mapping so that when presented with an unknown input $\mathbf{x}$ it will estimate the corresponding y. In mathematical terms we would think of this process as interpolation (or extrapolation), but in learning theory it is called *generalization.* To start with, we need some measure of error. In Section 2 we used $\|\cdot\|_\infty$, but this is unlikely to lead to easy algorithms. Moreover the fact that we are working in $\mathbb{R}^n$ and using inner products suggest using a Hilbert space formulation. Of course, this is not the only choice: entropic measures also have their advocates particularly among the information theory fraternity. (See, e.g., Bichsel and Seitz (1989).) However, the basic simplicity of the least-squares approach means that it remains the most popular. So we formulate our learning problem as one of least-squares optimization. Suppose the actual output of our network for a given set of parameters is $g(\mathbf{x})$. For example, for a one-hidden-layer perceptron we have (2.2) with the parameters a_j, $\mathbf{w}_j$ and c_j, $j = 1, \ldots, k$ to be chosen. Then our task is simply to minimize $\sum(y_j - g(\mathbf{x}_j))^2$. Here the sum is over the t patterns $\mathbf{x}_j$, $j = 1, \ldots, t$. (For the case of nonscalar output we need to sum over the outputs as well.) As is usual in such problems we do not necessarily need a true minimum: a 'good' solution will do us. (Before getting down to work on this problem it is worth mentioning the *unsupervised learning problem.* In this case we do not know the desired outputs but wish to cluster the patterns into subsets apparently sharing common features. The most successful approach to this problem seems to be that of Kohonen: see Wasserman (1989, Ch. 4).)

Now having defined our problem a few general words on learning algorithms are in order. First, as expressed in the previous paragraph, the unsupervised learning problem is the classical one of nonlinear least squares. If n is not too large, it can be (and often has been) treated by standard optimization techniques. On the other hand, for very difficult problems simulated annealing or genetic algorithms can be applied. But we propose to look at the classic learning algorithms of the delta rule/backpropagation family, which remain the most popular approaches at least at present. An essential feature of these is that we only permit the patterns $\mathbf{x}_j$ to be presented to the system sequentially: we do not have them there all at once. Apart from the pragmatic consideration of popularity, there are two reasons for this restriction, one philosophical and one practical.

- First, mammals do not generally learn by considering a whole set of data at once. They learn from examples presented sequentially. Humans are not very good at considering lots of cases at once, and as far as we can tell, animals cannot do it at all. Yet they *do* successfully learn. To some extent, our least-squares error criterion is justified as a model of learning by the fact that it can tolerate this sequential restriction.
- Second, for very large n, second-order methods may simply be beyond the capabilities of the available hardware. Serial hardware may lack sufficient memory or computing power, and massively parallel machines present severe implementation problems. Sequential learning algorithms are memory efficient and naturally parallel: as such they deserve wider consideration even for problems which are not naturally formulated as 'learning': row projection methods are perhaps due for a renaissance!

Now it may be that in the long term, stochastic algorithms will prove more biologically plausible. But all of these are basically gradient descent methods with 'tricks' to avoid local minima. (Sometimes, particularly with genetic algorithms, the tricks are very sophisticated but the generalization remains valid.) So analysis of the simpler deterministic algorithms will not be wasted. Basically such analysis amounts to study of the underlying search geometry of the least-squares problem, which applies equally to the stochastic versions.

Finally, a remark to sceptical numerical analysts! It is often stated that backpropagation is 'merely steepest descent' and therefore unworthy of consideration by serious mathematicians. We shall show here that while it is certainly a gradient descent method, it is *not* steepest descent. In fact it has much better stability properties than steepest descent; one reason indeed for considering its use more widely!

The work described here first appeared in Ellacott (1990, 1992, 1993b,c). We mostly consider versions of the linear delta rule algorithm. However a justification of this linearization as an approximation to the nonlinear backpropagation method will be included.

3.1. The delta rule

We begin by considering the simplest of all neural models, the basic perceptron. Figure 2 shows a perceptron with a single output. We will briefly consider the case of a multiple output perceptron, so the output is a vector and the weights form a matrix. To avoid a superfluity of subscripts, denote the training vectors (generically) by $\mathbf{x}$ and *desired* output vectors by $\mathbf{y}$. We will ignore the threshold c and instead treat the problem as one of approximation. (It is possible to make c learnable as well by including an extra input fixed at 1, but we need not consider this here.) If we can approximate $\mathbf{y}$ sufficiently well by the network, then obviously a suitable choice of c will solve the classification problem if this is what we are interested in. Let W be the weight matrix. In summary, then, we wish to find W such that $\mathbf{y} \approx W\mathbf{x}$ for all pairs ($\mathbf{x}$,$\mathbf{y}$) of patterns and corresponding outputs. In general it is impossible to satisfy this exactly, so we seek a W for which the result holds approximately. The idea of a *learning algorithm* is as follows: we supply a set $\{\mathbf{x}_1, \mathbf{x}_2, \ldots, \mathbf{x}_t\}$ of input patterns in $\mathbb{R}^n$ and for each $\mathbf{x}_i$ we supply the corresponding output $\mathbf{y}_i$. The system uses these pattern pairs to update its estimate of the desired weight matrix W. As we have already remarked, a learning algorithm is at heart simply an optimization process, but it has the special feature that the patterns are supplied serially rather than simultaneously as in standard least-squares approximation and optimization. The process of applying a learning algorithm is called *training* the network. Once training is complete, the system will be presented with previously unknown patterns $\mathbf{x}$ and used to predict the corresponding $\mathbf{y}$.

Although not the original perceptron algorithm, the following method known as the *delta rule* is generally accepted as the best way to train such a network. We assume initially that W is updated after each training pattern. The change in W when the pattern $\mathbf{x}$ is presented is given (Rumelhart and McClelland, 1986, p. 322) by

$$(\delta W)_{ji} = \eta(y_j - (W\mathbf{x})_j)\mathbf{x}_i,$$

where η is a parameter to be chosen called the *learning rate*, and $(W\mathbf{x})_j$ denotes the jth element of $W\mathbf{x}$. Thus if the error term in brackets is (say) positive, we will add a component of $\mathbf{x}$ to each row of W, increasing the output of the network for this pattern. Conversely if the error is negative, a component of $\mathbf{x}$ is subtracted, reducing the output for this pattern. In fact, we can simplify matters here by observing that there is no coupling between the rows of W in this formula: the new jth row of W depends *only* on the old jth row. This enables us to drop the subscript j, denoting y_j just by y, and the jth row of W by the vector $\mathbf{w}^T$. Hence without loss of generality we return to the single output perceptron (Figure 2). We get

$$\delta w_i = \eta(y - \mathbf{w}^T \mathbf{x})x_i,$$

so

$$\delta \mathbf{w} = \eta(y - \mathbf{w}^T \mathbf{x})\mathbf{x}.$$

Thus given a current iterate weight vector $\mathbf{w_k}$,

$$\begin{aligned} \mathbf{w_{k+1}} &= \mathbf{w_k} + \delta \mathbf{w_k} \\ &= \mathbf{w_k} + \eta(y - \mathbf{w_k}^T \mathbf{x})\mathbf{x} \\ &= (I - \eta \mathbf{x}\mathbf{x}^T)\mathbf{w_k} + \eta y \mathbf{x}. \end{aligned} \tag{3.1}$$

The final equation is obtained by transposing the (scalar) quantity in brackets. Note the bold subscript $\mathbf{k}$ here, denoting the kth iterate, not the kth element. Observe also that the second equation makes clear what the delta rule actually does: it adds a suitable multiple of the current pattern $\mathbf{x}$ to the current weight vector (compare the discussion in Section 2.3). It is possible to analyse this iteration in the asymptotic case as $\eta \to 0$, but it is not used this way in practice. It is more relevant to consider a fixed η (Ellacott, 1990). We now prove some results about this iteration: the first lemma is a special case of a well known result (see, e.g., Oja (1983, p. 18)). The proof is a direct verification.

Lemma 3.1 Let $B = (I - \eta \mathbf{x}\mathbf{x}^T)$. Then B has only two distinct eigenvalues: $1 - \eta\|\mathbf{x}\|^2$ corresponding to the eigenvector $\mathbf{x}$ and 1 corresponding to the subspace of vectors orthogonal to $\mathbf{x}$. (Here $\|\cdot\|$ denotes the usual Euclidean norm.)

As an immediate consequence (see Isaacson and Keller (1966, p. 10, equation (11))) we obtain

Lemma 3.2 Provided $0 \leq \eta \leq 2/\|\mathbf{x}\|^2$, we have $\|B\| = \rho(B) = 1$, where $\rho(B)$ is the spectral radius of B.

Now suppose we actually have t pattern vectors $\mathbf{x}_1, \ldots, \mathbf{x}_t$. We will assume temporarily that these span the space of input vectors, i.e. that the set of pattern vectors contains n linearly independent ones. (This restriction will be removed later.)

Now for each pattern vector $\mathbf{x_p}$, we will have a different matrix B, say $B_p = (I - \eta \mathbf{x_p}\mathbf{x_p}^T)$. Let $\Lambda = B_t B_{t-1} \ldots B_1$.

Lemma 3.3 If $0 < \eta < 2/\|\mathbf{x_p}\|^2$ holds for each training pattern $\mathbf{x_p}$, and if the $\mathbf{x_p}$ span, then $\|\Lambda\| < 1$.

Proof. By definition, there exists $\mathbf{v}$ such that $\|\Lambda\| = \|\Lambda \mathbf{v}\|$ and $\|\mathbf{v}\| = 1$.

Thus $\|\Lambda\| = \|B_t B_{t-1} \dots B_1 \mathbf{v}\| \le \|B_t B_{t-1} \dots B_2\| \|B_1 \mathbf{v}\|$ (from the definition of the norm). We identify two cases:

1 If $\mathbf{v}^T \mathbf{x}_1 \neq 0$, $\|B_1 \mathbf{v}\| < 1$, since the component of $\mathbf{v}$ in the direction of $\mathbf{x}$ is reduced (see Lemma 3.1: if this is not clear write $\mathbf{v}$ in terms of $\mathbf{x}$ and the perpendicular component, and apply B_1 to it.) On the other hand $\|B_t B_{t-1} \dots B_2\| \le \|B_t\| \|B_{t-1}\| \dots \|B_2\| = 1$.

2 If $\mathbf{v}^T \mathbf{x}_1 = 0$, then $B_1 \mathbf{v} = \mathbf{v}$ (Lemma 3.1). Hence

$$\|\Lambda\| = \|B_t B_{t-1} \dots B_2 \mathbf{v}\|$$

and we may carry on removing Bs until Case 1 applies. Note that $\mathbf{v}$ cannot be orthogonal to all the $\mathbf{x_p}$ since by hypothesis they span. □

A common way to apply the delta rule is to apply patterns $\mathbf{x_1}, \mathbf{x_2}, \dots \mathbf{x_t}$ in order, and then to start again cyclically with $\mathbf{x}_1$. The presentation of one complete set of patterns is called an *epoch*. Assuming this is the strategy employed, iteration (3.1) yields

$$\mathbf{w_{k+t}} = \Lambda \mathbf{w_k} + \eta \mathbf{h} \tag{3.2a}$$

where Λ is as defined above and

$$\mathbf{h} = y_1 (B_t B_{t-1} \dots B_2) \mathbf{x_1} + \dots y_{t-1} B_t \mathbf{x_{t-1}} + y_t \mathbf{x_t}. \tag{3.2b}$$

Here, of course, y_p denotes the target y value for the pth pattern, not the pth element of a vector. Note that the Bs and hence $\mathbf{h}$ depend on η and the $\mathbf{x}$s, but *not* on the current $\mathbf{w}$.

Since δW in the delta rule is proportional to the error in the outputs, we get a fixed point of (3.1) only if all these errors can be made zero, which obviously is not true in general. Hence the iteration (3.1) does not in fact converge in the usual sense. On the other hand, we have shown (Lemma 3.2) that provided the $\mathbf{x_p}$ span the space of input vectors, then for sufficiently small η, $\|\Lambda\| < 1$. Hence the mapping $\mathbf{F}(\mathbf{w}) = \Lambda \mathbf{w} + \eta \mathbf{h}$ satisfies

$$\|\mathbf{F}(\mathbf{w}) - \mathbf{F}(\mathbf{v})\| = \|\Lambda(\mathbf{w} - \mathbf{v})\| \le \|\Lambda\| \|\mathbf{w} - \mathbf{v}\|,$$

i.e. it is contractive with contraction parameter $\|\Lambda\|$. It follows from the Contraction Mapping Theorem that the iteration (3.2a) does have a unique fixed point. Now if there exists a $\mathbf{w}$ that makes all the errors zero, then it is easy to verify that this $\mathbf{w}$ is a fixed point of (3.1) and hence also of (3.2a). Otherwise, (3.1) has no fixed points, and the fixed point of (3.2a) depends on η: we denote it by $\mathbf{w}(\eta)$. In the limit, as the iteration (3.1) runs through the patterns, it will generate a limit cycle of vectors $\mathbf{w_k}$ returning to $\mathbf{w}(\eta)$ after the cycle of t patterns has been completed.

Since $\mathbf{w}(\eta)$ is a fixed point of (3.2a) we have (writing $\mathbf{h} = \mathbf{h}(\eta)$ and $\Lambda = \Lambda(\eta)$ to emphasize the dependence)

$$\mathbf{w}(\eta) = \Lambda(\eta) \mathbf{w}(\eta) + \eta \mathbf{h}(\eta). \tag{3.3}$$

Now what can we conclude about $\mathbf{w}(\eta)$? Let us denote by $\mathbf{w}^*$ the weight vector $\mathbf{w}$ (unique since the $\mathbf{x_p}$ span) that minimizes

$$\epsilon^2 = \sum_{p=1}^{t} (y_p - \mathbf{w}^T \mathbf{x_p})^2. \tag{3.4}$$

Denote by X the matrix whose columns are $\mathbf{x}_1, \mathbf{x}_2, \ldots, \mathbf{x}_t$, and let

$$L = XX^T = \sum_{p=1}^{t} \mathbf{x_p} \mathbf{x_p}^T. \tag{3.5}$$

Then $\mathbf{w}^*$ satisfies the normal equations

$$L\mathbf{w}^* = \sum_{p=1}^{t} y_p \mathbf{x_p} = \mathbf{h}(0).$$

The second equality follows from (3.2*b*), observing that all the B matrices tend to the identity as $\eta \to 0$. On the other hand from (3.3) we get

$$H(\eta)\mathbf{w}(\eta) = \mathbf{h}(\eta) \quad \text{where} \quad H(\eta) = (I - \Lambda)/\eta.$$

Since by hypothesis the patterns span, L^{-1} exists. We define the *condition number* $\kappa(L) := \|L^{-1}\|\|L\|$. Moreover L is symmetric and positive definite, so $\kappa(L)$ is equal to the ratio of the largest and smallest eigenvalues of L (compare Isaacson and Keller (1966, p. 10, equation (11))).

A standard result on the solutions of linear equations (Isaacson and Keller, 1966, p. 37) gives, provided $\|L - H(\eta)\| < 1/\|L^{-1}\|$,

$$\frac{\|\mathbf{w}(\eta) - \mathbf{w}^*\|}{\|\mathbf{w}^*\|} \leq \frac{\kappa(L)}{1 - \|L^{-1}\|\|L - H(\eta)\|_2} \left(\frac{\|\mathbf{h}(\eta) - \mathbf{h}(0)\|}{\|\mathbf{h}(0)\|} + \frac{\|L - H(\eta)\|}{\|L\|} \right)$$

but

$$\Lambda(\eta) = \prod_{p=1}^{t} (I - \eta \mathbf{x_p} \mathbf{x_p}^T)$$

and considering powers of η in this product we obtain

$$\begin{aligned} \Lambda(\eta) &= I - \eta \sum_{p=1}^{t} \mathbf{x_p} \mathbf{x_p}^T + \mathcal{O}(\eta^2) \\ &= I - \eta L + \mathcal{O}(\eta^2). \end{aligned}$$

Thus $H(\eta) = L + \mathcal{O}(\eta)$. Also an examination of the products in (3.2*b*) reveals $\mathbf{h}(\eta) = \mathbf{h}(0) + \mathcal{O}(\eta)$. Putting all this together gives most of the following theorem.

Theorem 3.4 Suppose that the pattern vectors $\mathbf{x_p}$ span $\mathbb{R}^n$, and that $\mathbf{w}^*$ is (as above) the weight vector which minimizes the least-squares error of the outputs over all patterns. If the delta rule is applied with fixed η satisfying the condition of Lemma 3.3, then the weights will converge to a limit cycle. Let $\mathbf{w}(\eta)$ be any member of the limit cycle, then as $\eta \to 0$,

(a) $\|\mathbf{w}(\eta) - \mathbf{w}^*\| = \mathcal{O}(\eta)$.
(b) If $\epsilon(\eta)$ is the root-mean-square error corresponding to $\mathbf{w}(\eta)$ (see (3.4)), and ϵ^* is the corresponding error for $\mathbf{w}^*$, then $\epsilon(\eta) - \epsilon^* = \mathcal{O}(\eta^2)$.

Proof. Convergence to a limit cycle has already been established. (a) follows from the remarks immediately preceding the theorem. (b) is simply the observation that for a least-squares approximation problem, the vector of errors for the best vector $\mathbf{w}^*$ is orthogonal to the space of possible $\mathbf{w}$s, so an $\mathcal{O}(\eta)$ error in $\mathbf{w}^*$ yields only an $\mathcal{O}(\eta^2)$ increase in the root mean square error. □

Unfortunately the rate of convergence is proportional to $\kappa(L)$, and as we shall see in the next subsection this can be large.

Finally we need to consider what happens when the $\mathbf{x_p}$ do not span the input pattern space. In this case it follows from Lemma 3.1 that the iteration (3.1) leaves the orthogonal complement of the span invariant. By decomposing the input space into the span and its orthogonal complement, a straightforward modification of the argument above shows that (3.2) is contractive on the span of the input patterns, so we still get convergence to a limit cycle. However, then L fails to be invertible, and discussion of the behaviour as $\eta \to 0$ requires examination of the singular vectors of L.

An interesting sidelight on the argument above is to consider what would happen if we presented the patterns during each epoch in *random* order, while still insisting that the whole set of patterns is presented precisely once during the epoch. With this approach, each epoch would still give a contractive mapping, but this mapping would be different for each choice of order. Since there is a large but finite number of such permutations of order, we have an iterated function scheme (Falconer, 1990, Ch. 9). We may expect a fractal attractor, even though this iteration is linear.

3.2. The 'Epoch Method'

Since we are assuming that we have a fixed and finite set of patterns $\mathbf{x_p}, \mathbf{p} = 1, \ldots t$, an alternative strategy is not to update the weight vector until the whole epoch of patterns has been presented. This idea is initially attractive since we shall see that this actually generates the steepest-descent direction for the least-squares error. We will call this the *epoch method* to distinguish it from the usual delta rule. This leads to the iteration

$$\mathbf{w_{k+1}} = \mathbf{w_k} - \eta \sum_{p=1}^{t} (\mathbf{x_p} \mathbf{x_p}^T) \mathbf{w_k} + \eta \sum_{p=1}^{t} (y_p \mathbf{x_p})$$

$$= \Omega \mathbf{w_k} - \eta \sum_{p=1}^{t} (y_p \mathbf{x_p}), \tag{3.6}$$

where

$$\Omega = (I - \eta X X^T) = (I - \eta L).$$

(3.6) is, of course, the equivalent of (3.2a), not (3.1), since it corresponds to a complete epoch of patterns. There is no question of limit cycling, and, indeed, a fixed point will be a true least-squares minimum $\mathbf{w}^*$. To see this, put $\mathbf{w_{k+1}} = \mathbf{w_k} = \mathbf{w}^*$ and observe that (3.6) reduces to the normal equations for the least-squares problem. Moreover, the iteration (3.6) is simply steepest descent for the least-squares problem, applied with a fixed step length. Unfortunately, however, there is a catch! To see what this is, we need to examine the eigenvalues of Ω.

Clearly $L = XX^T$ is symmetric and positive semidefinite. Thus it has real nonnegative eigenvalues. In fact, provided the $\mathbf{x_p}$ span, it is (as is well known) strictly positive definite. The eigenvalues of Ω are $(1 - \eta) \times$ (the corresponding eigenvalues of L), and for a strictly positive definite matrix all the eigenvalues must be strictly positive. Thus we have for η sufficiently small, $\rho(\Omega) = \|\Omega\| < 1$.

Hence the iteration (3.6) will converge, provided the patterns span and η is sufficiently small. But how small does η have to be? (Recall that for the usual delta rule we need only the condition of Lemma 3.3.) To answer this question we need more precise estimates for the spectrum of L and the norm of Ω. From these we will be able to see why the epoch algorithm does not always work well in practice.

Suppose $L = XX^T$ has eigenvalues λ_j, $j = 1 \ldots n$, with

$$0 < \lambda_n \leq \lambda_{n-1} \leq \ldots \leq \lambda_1 = \rho(XX^T) = \|XX^T\| = \|X^T\|^2.$$

The eigenvalues of Ω are $(1 - \eta\lambda_1) \leq (1 - \eta\lambda_2) \leq \ldots \leq (1 - \eta\lambda_n)$, and $\rho(\Omega) = \max\{|1 - \eta\lambda 1|, |1 - \eta\lambda_n|\}$. (Observe that Ω is positive definite for small η, but ceases to be so when η becomes large.) Now

$$\lambda_1 = \|X^T\|^2 = \max_{\|\mathbf{v}\|=1} \|X^T\mathbf{v}\|^2 = \max_{\|\mathbf{v}\|=1} \mathbf{v}^T XX^T \mathbf{v} \leq \sum_{p=1}^{t} \|\mathbf{x_p}\|^2. \tag{3.7}$$

On the other hand, we can get a lower bound by substituting a particular $\mathbf{v}$ into the expression on the right-hand side of (3.7). For instance, we have for any k, $k = 1, \ldots, t$,

$$\lambda_1 \geq \frac{1}{\|\mathbf{x_k}\|} \left(\sum_{p=1}^{t} \mathbf{x_k^T x_p} \right)^2 \geq \|\mathbf{x_k}\|. \tag{3.8}$$

Now consider a particular case.

Suppose the $\mathbf{x_p}$ cluster around two vectors $\mathbf{u}$ and $\mathbf{v}$ which are mutually ortho*normal*. If these represent two classes which are to be separated, we are in an ideal situation for machine learning: the pattern classes are in two widely separated convex sets. However, even in this case the behaviour of the epoch method is not good. If the clusters are of equal size, we have from the first inequality in (3.8)

$$\lim_{\epsilon \to 0} \lambda_1 \geq /2 \quad \text{and since the rank of } L = XX^T \text{ collapses to 2,} \quad \lim_{\epsilon \to 0} \lambda_n = 0.$$

Thus, unlike the ordinary delta rule for which the convergence condition depends only on the norm of the individual patterns, for the epoch method (i.e. steepest descent) we may require an arbitrary small μ to get convergence. As promised we have shown that the delta rule is much more stable.

3.3. Generalization to nonlinear systems

As we saw in Section 1, the usefulness of linear neural systems is limited, since many pattern recognition problems are not linearly separable. We need to generalize to nonlinear systems such as the backpropagation algorithm for the multilayer perceptron. Clearly we can only expect this type of analysis to provide a local result: global behaviour is likely to be more amenable to dynamical systems or control theory approaches. Nevertheless, a local analysis can be useful in discussing the asymptotic behaviour near a local minimum.

The obvious approach to this generalization is to attempt the 'next simplest' case, i.e. the backpropagation algorithm. However, this method looks complicated when written down explicitly: in fact much more complicated than it actually is! A more abstract line of attack turns out to be both simpler and more general. We will define a general nonlinear delta rule, of which backpropagation is a special case. For the linear network the dimension of the input space and the number of weights are the same: n in our previous notation. Now we will let M denote the *total number* of weights and n the input dimension.

So the input patterns $\mathbf{x}$ to our network are in $\mathbb{R}^n$, and we have a vector $\mathbf{w}$ of parameters in $\mathbb{R}^M$ describing the particular instance of our network: i.e. the vector of synaptic weights. For a single layer perceptron with m outputs, the 'vector' $\mathbf{w}$ is the the $m \times n$ weight matrix, and thus $M = mn$. For a multilayer perceptron, $\mathbf{w}$ is the Cartesian product of the weight matrices in each layer. For a general system with m outputs, the network computes a function $g : \mathbb{R}^M \times \mathbb{R}^n \to \mathbb{R}^m$. Say

$$\mathbf{v} = \mathbf{g}(\mathbf{w}, \mathbf{x}),$$

where $\mathbf{v} \in \mathbb{R}^m$. We equip $\mathbb{R}^M$, $\mathbb{R}^m$ and $\mathbb{R}^n$ with the Euclidean norm. For pattern $\mathbf{x_p}$, denote the corresponding output by $\mathbf{v_p}$, i.e.

$$\mathbf{v_p} = \mathbf{g}(\mathbf{w}, \mathbf{x_p}).$$

We assume that $\mathbf{g}$ is Frechêt differentiable with respect to $\mathbf{w}$, and denote by $D = D(\mathbf{w}, \mathbf{x})$ the $m \times M$ matrix representation of the derivative with respect to the standard basis. Readers unfamiliar with Frechêt derivatives may prefer to think of this as the gradient vector: for $m = 1$ it is precisely the row vector representing the gradient when $\mathbf{g}$ is differentiated with respect to the elements of $\mathbf{w}$. Thus, for a small change $\delta\mathbf{w}$ and fixed $\mathbf{x}$, we have (by the definition of the derivative)

$$\mathbf{g}(\mathbf{w} + \delta\mathbf{w}, \mathbf{x}) = \mathbf{g}(\mathbf{w}, \mathbf{x}) + D(\mathbf{w}, \mathbf{x})\delta\mathbf{w} + o(\|\delta\mathbf{w}\|). \tag{3.9}$$

On the other hand for given $\mathbf{w}$, corresponding to a particular pattern $\mathbf{x_p}$, we have a desired output $\mathbf{y_p}$ and thus an error ϵ_p given by, say,

$$\epsilon_p^2 = (\mathbf{y_p} - \mathbf{v_p})^T(\mathbf{y_p} - \mathbf{v_p}) = \mathbf{q_p}^T\mathbf{q_p}. \tag{3.10}$$

The total error is obtained by summing the ϵ_p^2s over the t available patterns, thus

$$\epsilon^2 = \sum_{p=1}^{t} \epsilon_p^2.$$

An ordinary descent algorithm will seek to minimize ϵ^2. However, the class or methods we are considering generate, not a descent direction for ϵ^2, but rather successive steepest descent directions for ϵ_p^2. Now for a change $\delta\mathbf{q_p}$ in $\mathbf{q_p}$ we have from (3.10)

$$\begin{aligned}\delta\epsilon_p^2 &= (\mathbf{q_p} + \delta\mathbf{q_p})^T(\mathbf{q_p} + \delta\mathbf{q_p}) - \mathbf{q_p}^T\mathbf{q_p} \\ &= 2\delta\mathbf{q_p}^T\mathbf{q_p} + \delta\mathbf{q_p}^T\delta\mathbf{q_p}.\end{aligned}$$

Since $\mathbf{y_p}$ is fixed,

$$\delta\mathbf{q_p} = -\delta\mathbf{v_p} = -D(\mathbf{w}, \mathbf{x_p})\delta\mathbf{w} + o(\|\delta\mathbf{w}\|) \quad \text{by (3.9)}.$$

Thus

$$\begin{aligned}\delta\epsilon_p^2 &= -2(D(\mathbf{w}, \mathbf{x_p})\delta\mathbf{w})^T(\mathbf{y_p} - \mathbf{g}(\mathbf{w}, \mathbf{x_p})) + o(\|\delta\mathbf{w}\|) \\ &= -2\delta\mathbf{w}^T(D(\mathbf{w}, \mathbf{x_p}))^T(\mathbf{y_p} - \mathbf{g}(\mathbf{w}, \mathbf{x_p})) + o(\|\delta\mathbf{w}\|).\end{aligned}$$

Hence, ignoring the $o(\|\delta\mathbf{w}\|)$ term, and for a fixed size of small change $\delta\mathbf{w}$, the largest decrease in ϵ_p^2 is obtained by setting

$$\delta\mathbf{w} = \eta(D(\mathbf{w}, \mathbf{x_p}))^T(\mathbf{y_p} - \mathbf{g}(\mathbf{w}, \mathbf{x_p})).$$

This is the generalized delta rule. Compare this with the single output linear perceptron, for which the second term in this expression is scalar with

$$\mathbf{g}(\mathbf{w}, \mathbf{x_p}) = \mathbf{w}^T\mathbf{x}_p,$$

and the derivative is the gradient vector (considered as a row vector) obtained by differentiating this with respect to $\mathbf{w}$, i.e. $\mathbf{x_p}^T$. Thus we indeed have a generalization of (3.1). Given a kth weight vector $\mathbf{w_k}$, we have

$$\begin{aligned}\mathbf{w_{k+1}} &= \mathbf{w_k} + \delta\mathbf{w_k} \\ &= \mathbf{w_k} + \eta(D(\mathbf{w_k}, \mathbf{x_p}))^T(\mathbf{y_p} - \mathbf{g}(\mathbf{w_k}, \mathbf{x_p})). \qquad (3.11)\end{aligned}$$

To proceed further, we need to make evident the connection between (3.11) and the analysis of Section 3.1. However, there is a problem in that, guided by the linear case considered above, we actually expect a limit cycle rather than convergence to a minimum. Nevertheless it is necessary to fix attention to some neighbourhood of a local minimum, say $\mathbf{w}^*$, of the least-squares error ϵ: clearly we cannot expect any global contractivity result as in general ϵ may have many local minima, as is well known in the backpropagation case. Now from (3.10) and (3.11) we obtain (assuming continuity and uniform boundedness of D in a neighbourhood of $\mathbf{w}^*$),

$$\begin{aligned}\mathbf{w_{k+1}} &= \mathbf{w_k} + \eta(D(\mathbf{w_k}, \mathbf{x_p}))^T(\mathbf{y_p} - \mathbf{g}(\mathbf{w}^*, \mathbf{x_p}) - D(\mathbf{w}^*, \mathbf{x_p})(\mathbf{w_k} - \mathbf{w}^*)) \\ &\quad + o(\|\mathbf{w_k} - \mathbf{w}^*\|) \\ &= (I - \eta D(\mathbf{w_k}, \mathbf{x_p})^T D(\mathbf{w}^*, \mathbf{x_p}))\mathbf{w_k} + \eta(D(\mathbf{w_k}, \mathbf{x_p}))^T \\ &\quad \times(\mathbf{y_p} - \mathbf{g}(\mathbf{w}^*, \mathbf{x_p}) + D(\mathbf{w}^*, \mathbf{x_p})\mathbf{w}^*) + o(\|\mathbf{w_k} - \mathbf{w}^*\|). \qquad (3.12)\end{aligned}$$

The connection between (3.11) and (3.1) is now clear. Observe that the iteration matrix $(I - \eta D(\mathbf{w_k}, \mathbf{x_p})^T D(\mathbf{w}^*, \mathbf{x_p}))$ is not exactly symmetric in this case, although it will be nearly so if $\mathbf{w_k}$ is close to $\mathbf{w}^*$. More precisely, let us assume that $D(\mathbf{w}, \mathbf{x})$ is Lipschitz continuous at $\mathbf{w}^*$, uniformly over the space of pattern vectors $\mathbf{x}$. Then we have

$$\begin{aligned}\mathbf{w_{k+1}} &= (I - \eta D(\mathbf{w}^*, \mathbf{x_p})^T D(\mathbf{w}^*, \mathbf{x_p}))\mathbf{w_k} + \eta(D(\mathbf{w}^*, \mathbf{x_p}))^T \\ &\quad \times(\mathbf{y_p} - \mathbf{g}(\mathbf{w}^*, \mathbf{x_p}) + D(\mathbf{w}^*, \mathbf{x_p})\mathbf{w}^*) + \mathcal{O}(\|\mathbf{w_k} - \mathbf{w}^*\|). \qquad (3.13)\end{aligned}$$

Suppose we apply the patterns $\mathbf{x_1}, \ldots \mathbf{x_t}$ cyclically, as for the linear case. If we can prove that the linearized part (i.e. what we would get if we applied (3.13) without $\mathcal{O}$ term) of the mapping $\mathbf{w_k} \to \mathbf{w_{k+t}}$ is contractive, it will follow by continuity that there is a neighbourhood of $\mathbf{w}^*$ within which the whole mapping is contractive. This is because, by hypothesis, we have only a finite number of patterns. To establish contractivity of the linear part, we may proceed as follows.

First observe that

$$D(\mathbf{w}^*, \mathbf{x_p})^T D(\mathbf{w}^*, \mathbf{x_p})$$

is positive semidefinite. Thus for η sufficiently small,

$$\|I - \eta D(\mathbf{w}^*, \mathbf{x_p})^T D(\mathbf{w}^*, \mathbf{x_p})\| \leq 1.$$

We may decompose the space of weight vectors into the span of the eigenvectors corresponding to zero and nonzero eigenvalues respectively. These spaces are orthogonal complements of each other, as the matrix is symmetric. On the former space, the iteration matrix does nothing. On the latter space it is contractive provided

$$\eta < 1/\rho(D(\mathbf{w}^*, \mathbf{x_p})^T D(\mathbf{w}^*, \mathbf{x_p})).$$

We may then proceed in a similar manner to Lemma 3.3, provided the contractive subspaces for each pattern between them span the whole weight space. (If this condition fails then a difficulty arises, since the linearized product mapping will have norm 1, so the nonlinear map could actually be expansive on some subspace. We will not pursue this detail here.) For the single output case, $D(\mathbf{w}^*, \mathbf{x_p})$ is just a row vector, and we can identify the eigenvectors explicitly as in Lemma 3.1.

The *backpropagation rule* (Rumelhart and McClelland, 1986, pp. 322–328) used in many neural net applications is a special case of this. The name backpropagation derives from the fact that for an multilayer perceptron, the necessary terms in this expression can be calculated recursively back from the top layer. However, this is not relevant to our analysis here. We should make it clear, however, that backpropagation is rarely used in this 'pure' form. Rumelhart and McClelland themselves advocate the use of a 'momentum term' which is somewhat analogous to the Levenberg Marquardt method used in classical optimization (Moré, 1978). Moreover, the literature abounds with acceleration techniques. Fombellida and Destiné (1992) discuss two of the most popular: the *delta-bar-delta* and *quickprop* methods. They do some numerical comparisons and actually suggest a hybrid of the two methods as the most effective. However, none of the methods appears to have been subjected to any serious numerical analysis! A novel approach to accelerating backpropagation has been suggested by Almeida and Silva (1992). This is somewhat related to the work to be discussed in Section 3.4, so we defer consideration of this paper until Section 4.

3.4. The singular value decomposition and principal components

Since we now know that the backpropagation rule can be realistically considered as behaving locally like the delta rule, it makes sense to return to a closer study of the linear algorithm. Several interesting results can be obtained from Singular Value Decomposition (SVD). This is unsurprising in view of the well known connections between neural nets and statistical decision theory. Unfortunately they are easily obtained only for the algorithm in its 'epoch' form (3.6). This is a pity in view of the previous analysis, but since the algorithms are at least asymptotically the same for small η, they seem nevertheless worth having. Not all of the results in this section are really new, but it is difficult to find a formal and coherent exposition of them in the literature. This attempt at a systematic description is thus worthwhile.

Firstly, we can provide a simple explanation for the well known phenomenon of *overgeneralization* reported in many practical studies with neural networks. This is the observation that better results may well be obtained if the iteration is *not* continued to convergence. These problems are closely related to the issue of nonspanning patterns which we have already encountered. In many network applications such as vision, we may have a very large number of free weights. For example, even a medium resolution 64×64 image will have 4096 pixels. If we feed this into the network without any compression we will have at least this many weights. If we are training the network to recognize (say) a certain object in a set of images, it is most unlikely that we will have enough data to prevent the problem being severely underdetermined. But in fact, the delta rule (even in epoch form) can cope with this if the number of iterations is restricted: it includes a kind of built in compression. Recall (3.6):

$$\mathbf{w_{k+1}} = \Omega \mathbf{w_k} - \eta \sum_{p=1}^{t} (y_p \mathbf{x_p}),$$

where $\Omega = (I - \eta X X^T)$. We decompose X in singular value form. (See, e.g., the chapters by Wilkinson and Dennis in Jacobs (1977, pp. 3–53 and 269–312) respectively. Also Chapter 6 of Ben-Israel and Greville (1974).) Specifically we may write

$$X = PSQ^T \tag{3.14}$$

where P and Q are orthogonal and S is diagonal (but not necessarily square). Recall that in this context $\mathbf{y}$ is not a single output vector but the vector of single outputs over all the patterns. We find

$$\begin{aligned} \mathbf{w_{k+1}} &= (I - \eta PSS^T P^T)\mathbf{w_k} - \eta X\mathbf{y}, \\ &= P(I - \eta SS^T)P^T \mathbf{w_k} - \eta PSQ^T \mathbf{y} \end{aligned}$$

or, with $\mathbf{z_k} = P^T \mathbf{w_k}$ and $\mathbf{u} = P^T \mathbf{y}$,

$$\mathbf{z_{k+1}} = (I - \eta SS^T)\mathbf{z_k} - \eta S\mathbf{u}. \tag{3.15}$$

At this point the notation becomes a little messy: let us denote by $(\mathbf{z_k})_i$ the ith element of $\mathbf{z_k}$. These elements are decoupled by SVD. More specifically, suppose X has r nonzero singular values (the diagonal elements of S) $\nu_1 \geq \nu_2 \geq \ldots \geq \nu_n$, (3.15) when written elementwise gives

$$(\mathbf{z_{k+1}})_i = (1 - \eta \nu_i^2)(\mathbf{z_k})_i - \eta \nu_i u_i, \quad \text{for } i = 1, \ldots, r$$

and

$$(\mathbf{z_{k+1}})_i = (\mathbf{z_k})_i \quad \text{for } i = r+1, \ldots, n.$$

Assuming that η is sufficiently small to guarantee convergence (i.e. all terms $(1 - \eta\nu_i^2) < 1$), it is easy to see that convergence will be very much faster for the $(\mathbf{z_k})_i$ corresponding to the larger singular values. This is exactly what we would like. Since P and Q are orthogonal matrices their rows and columns have norm 1. Thus we see from (3.15) that the large singular values correspond to the actual information in the pattern data X. (This approach is called *principal component analysis.*) The delta rule (in epoch form at least) has the nice property of converging on the principal components of the data *first*. Unfortunately it is very hard to tell from the iteration when this has occurred since small singular values can make a large contribution to the least-squares error. This explains the phenomenon of *overgeneralization.* Initially the iteration picks out significant features in the variability of the data. Continued iteration makes it try to separate insignificant features or noise.

In view of the problems of slow convergence and underdetermination, many authors have commented on the advisability of performing some preprocessing of the input patterns before feeding them to the network. Often (not always, of course) the preprocessing suggested is linear. At first sight this seems to be a pointless exercise, for if the raw input data vector is $\mathbf{x}$ with dimension n', say, the preprocessing operation is represented by the $n \times n'$ matrix T, W is the input matrix of the net and we denote by the vector $\mathbf{h}$ the input to the next layer of the net, then

$$\mathbf{h} = WT\mathbf{x}. \tag{3.16}$$

Obviously, *the theoretical representational power of the network is the same as one with unprocessed input and input matrix* WT. However, this does *not* mean that these preprocessing operations are useless. We can identify at least the following three uses of preprocessing.

1 to reduce work by reducing dimension and possibly using fast algorithms (e.g. the FFT or wavelet transform) (so we do not want to increase the contraction parameter in the delta rule iteration);
2 to improve the search geometry by removing principal components of the data and corresponding singular values that are irrelevant to the classification problem;
3 to improve the stability of the iteration by removing near zero singular values (which correspond to noise) and clustering the other singular values near to 1: i.e. in the language of numerical analysis to *precondition* the iteration.

We will not address all these three points explicitly here. Instead we will derive some theoretical principles with the aid of which the issues may be attacked. The first point to consider is the effect of the filter on the stability of the learning process. For simplicity, we again consider only the linear epoch algorithm here.

We hope, of course, that a suitable choice of filter will make the learning properties better, but the results here show that whatever choice we make, the dynamics will not be made much worse unless the filter has very bad singular values. In particular, we show that if the filter is an orthogonal projection, then the gradient descent mapping with filtering will be at least as contractive as the unfiltered case.

We see from (3.6) that the crucial issue is the relationship between the unfiltered update matrix

$$\Omega = (I - \eta X X^T) \tag{3.17}$$

and its filtered equivalent

$$(I - \eta T X X^T T^T) = \Omega' \tag{3.18}$$

say.

Note that these operators may be defined on spaces of different dimension: indeed for a sensible filtering process we would expect the filter T to involve a significant dimension reduction. Recall that Ω in (3.17) is $n \times n$ and let us take Ω' to be $n' \times n'$. Note also that for purposes of comparison we have assumed the learning rates η are the same.

A natural question is to try to relate the norms of these two operators, and hence the rate of convergence of the corresponding iterations. As before, we suppose $L = XX^T$ has eigenvalues λ_j, $j = 1 \ldots n$, with

$$0 < \lambda_n \leq \lambda_{n-1} \leq \ldots \leq \lambda_1 = \rho(XX^T) = \|XX^T\| = \|X^T\|^2.$$

(Note here we assume the $\mathbf{x}$s span so $\lambda_n \neq 0$. In terms of the singular values ν_i of X, $\nu_i^2 = \lambda_i$.)

We need to relate the eigenvalues of XX^T with those of $TXX^TT^T = L'$, say. Let L' have eigenvalues $\mu_1 \geq \mu_2 \geq \ldots \geq \mu_{n'} > 0$ and T have singular values $\sigma_1 \geq \sigma_2 \geq \ldots \geq \sigma_n > 0$. Note that we are assuming T has full rank n' and so has no nonzero singular values. This is a reasonable assumption since there is no point in using a filter which has a nontrivial kernel. We should reduce the codomain dimension of the operator instead. For example, an orthogonal projection is formally defined as a mapping from (say) $\mathbb{R}^n$ to itself. However, in practice, if we use an orthogonal expansion as a filter, we will reduce the dimension by choosing an orthogonal basis for the image and ignoring the rest of the basis required to span $\mathbb{R}^n$.

Proposition 3.5 With the notation above, $\mu_1 \leq \sigma_1^2 \lambda_1$ and $\mu_{n'} \geq \sigma_{n'}^2 \lambda_n$.

Proof. The first inequality is straightforward. Since L and L' are symmetric

$$\mu_1 = \|TXX^TT^T\| \leq \|T\|\|XX^T\|\|T^T\| = \sigma_1^2\lambda_1.$$

The second inequality is slightly more difficult. Let $\mathbf{u_n}$ be the normalized eigenvector of L' corresponding to μ_n. Then

$$\mu_n = \mathbf{u_n}^T\mu_n\mathbf{u_n} = \mathbf{u_n}^TTXX^TT^T\mathbf{u_n} = \|X^TT^T\mathbf{u_n}\|^2.$$

But $\|X^TT^T\mathbf{u_n}\| \geq \lambda_n^{1/2}\sigma_{n'}$ as may be found by writing both matrices in terms of their SVDs. □

This result means that $\|\Omega'\|$ cannot be much larger than $\|\Omega\|$ if T has singular values close to 1.

Corollary 3.6 Let T be a truncated orthogonal expansion, or any other filter that is the restriction of an orthogonal projection to the orthogonal complement of its kernel (e.g. unweighted local averaging: see Ellacott (1993a)). Then with filtering applied the epoch method will converge at least as fast (as expressed by its contraction parameter) as the unfiltered version.

Proof. All the singular values of an orthogonal projection are either 0 (corresponding to the kernel) or 1 (corresponding to the image). It follows from Proposition 3.5 that the norm of Ω' in (3.18) cannot be greater than that of Ω in (3.17). □

The result above gives us some insight into the uses of filters for data compression, although its extension to the nonlinear case is not obvious: filters are applied to the input of a multilayer perceptron, whereas to employ this result directly we would need to apply them to the tangent space: compare (3.13). Let us turn now to the issue of preconditioning. An ideal choice of filter to act as a preconditioner would not require knowledge of the particular data set under consideration, but this would seem to be an almost impossible requirement since the matrix Ω is defined in terms of this data. The best one might hope for is something that would work for large classes of data sets in a particular context such as vision or speech recognition. In other words we might try to derive information from the problem domain, and use this to construct the filter. As an illustration of the difficulties, we show that the theoretically optimal preconditioner for the delta rule in epoch form is both easily described and completely useless! Suppose, as above, X has SVD PSQ^T. We set the filter matrix T to be the Moore–Penrose pseudoinverse of X (see, e.g., Ben-Israel and Greville (1974)) which we denote by $X^{\#}$. So

$$T = X^{\#} = QS^{\#}P^T.$$

Then

$$TX = QS^{\#}P^TPSQ^T = QS^{\#}SQ^T.$$

Thus (with the same notation as before Proposition 3.5)

$$L' = TXX^TT^T = QS^{\#}SS^TS^{\#T}Q^T,$$

and $S^{\#}SS^TS^{\#T}$ is a diagonal matrix with diagonal elements either 0 or 1. Thus all the eigenvalues of L' are either 0 or 1 and, indeed, if the $\mathbf{x}$s span so that XX^T has no zero eigenvalues, then all the eigenvalues of L' are 1. With $\eta = 1$, the iteration will converge in a single iteration. This is not surprising, since once we know $X^{\#}$, the least-squares solution for $\mathbf{w}$ may be given explicitly! (For the nonlinear case we would need to compute the local pseudoinverses for the relevant tangent vectors.)

A modification of the approach which might be slightly more practicable is just to remove the large eigenvalues of XX^T based on computation of the dominant singular values, and corresponding singular vectors, of X. We present an algorithm for removing the principal components one at a time. Whether an approach based on removal of individual singular values is going to be very useful for the interesting case of very large n is debatable: it may help if the data matrix X is dominated by a few principal components with large singular values but otherwise it it likely to be too inefficient. However, the method does suggest ways forward. (A recent paper (Oja, 1992) also has some relevance to this problem, as does the method of Almeida and Silva (1992) which we consider in Section 4.1.)

The first stage is to compute the largest eigenvalue and corresponding eigenvector of XX^T. This may be carried out by the power method (Isaacson and Keller, 1966, p. 147) at the same time as the ordinary delta rule iteration: the computation can be performed by running through the patterns one at a time, just as for the delta rule itself. We get a normalized eigenvector $\mathbf{p_1}$ of XX^T corresponding to the largest eigenvalue λ_1 of XX^T. Set

$$T = I + (\lambda_1^{-1/2} - 1)\mathbf{p_1}\mathbf{p_1}^T.$$

A routine calculation shows that TXX^TT^T has the same eigenvectors as XX^T, and the same eigenvalues but with λ_1 replaced by 1. Each pattern $\mathbf{x_p}$ should then be multiplied by T, and, since we are now iterating with different data, the current weight estimate $\mathbf{w}$ should be multiplied by T^{-1}. It is easy to check that

$$T^{-1} = I + (\lambda_1^{1/2} - 1)\mathbf{p_1}\mathbf{p_1}^T.$$

Basically the same idea can be used for the iteration with the weights updated after each pattern. However there is a problem in that the update matrix Λ is not exactly symmetric, although it is nearly so for small η. This could be overcome by computing the right as well as left eigenvectors of

$(\Lambda - I)/\eta$, but unfortunately this would require presenting the patterns in *reverse* order: somewhat inconvenient for a neural system. Another possibility is to perform two cycles of the patterns, with the patterns in reverse order on the second cycle. The composite iteration matrix $\Lambda^T\Lambda$ will then be symmetric. Although space and the requirements of simplicity do not permit a full discussion here, there is no reason in principle why this algorithm should not be applied to the nonlinear case.

4. Some numerical applications of neural networks

The most successful applications of neural networks have been in pattern recognition areas such as speech, vision and nonlinear control, where satisfactory existing models do not exist. The power of the approach is the ability of the network to construct its own model. However it is also possible to design networks to solve some standard mathematical problems. We will conclude our survey by looking at some of these. Of course, it is not suggested that these methods will out-perform standard algorithms when run on conventional machines. So why should we study these methods? First, as we saw in the previous section, there are close connections between linear algebra and the methods of filtering and data compression used in neural network applications. To design suitable filters we need networks to perform linear algebra calculations. Second, neural network algorithms are naturally parallel. They lend themselves easily to implementation on array processors. Moreover we do not in fact even need the power of current parallel machines. The whole point of neural networks is that they use large arrays of very simple nonprogrammable processors. Neural network chips are already starting to appear. When these become large enough and cheap enough it will become possible to design hard-wired circuitry to perform a range of standard tasks. Of particular interest are problems such as the travelling salesman problem which involve optimization on graphs. This problem is of course NP complete so we cannot guarantee to find an optimum solution. But it turns out that we can find good solutions quickly *if* we can build a large enough neural network. Thus in this last section we focus on the two issues of linear algebra and optimization.

4.1. Linear algebra applications

We first observe that the delta rule itself may be regarded as a row-projection method for solving linear equations in the least-squares sense. If we have a single output linear perceptron with pattern vectors $\mathbf{x_1}, \ldots, \mathbf{x_t}$ forming the columns of a matrix X, and corresponding required outputs $y_1, \ldots, y_t$, formed into a vector $\mathbf{y}$, then the delta rule will approximately minimize $\|\mathbf{y} - \mathbf{w}^T X\|$. Similarly if we use a t output net and make the output for $\mathbf{x_p}$ the pth column of I, the delta rule will approximately minimize $\|I - WX\|_S$,

The norm here is the Schur matrix norm: this is simply the square root of the sum of squares of all the elements. Thus the delta rule will attempt to invert X and will do so exactly if X is nonsingular, since then a true fixed point will exist (compare the remarks following (3.2)). Not only can we avoid storing the whole matrix X at once, but we can compute the rows of W individually as well.

Other networks and learning schemes have been proposed to perform various calculations concerned with least-squares approximation and principal components. There are several of these around, with variations on the basic ideas. We will look first at an example based on Baldi and Hornik (1989) but we modify it somewhat. Consider a multilayer perceptron with n inputs and n outputs. We have one hidden layer with m units. This network is illustrated in Figure 3. However for this network the activation of the hidden units (as well as the input and output units) is simply the identity function $\sigma(x) = x$. Thus if W is the matrix of weights in the first layer, and V is the matrix of weights in the second layer, the output for input vector $\mathbf{x}$ is simply $VW\mathbf{x}$. Now for any given $\mathbf{x}$ we specify our target output $\mathbf{y}$ as $\mathbf{y} = \mathbf{x}$. Obviously if we have $m \geq n$, then any V and W with $VW = I$ would achieve this exactly. But what happens for $m < n$? More specifically, let our input patterns $\mathbf{x}_j$, $j = 1, \ldots, t$, be the columns of an $n \times t$ matrix X. We train the network, perhaps by backpropagation, to minimize $\sum \|\mathbf{x_j} - \mathbf{g}(\mathbf{x_j})\|_2^2$, where the sum is over j and we recall that for any particular choice of V and W, $\mathbf{g}(\mathbf{x}) = VW\mathbf{x}$. Our minimization problem can be restated as follows: find V and W so that $\|X - VWX\|_S$ is minimized. As we have seen, training with a finite learning rate η will not in fact solve the least-squares problem exactly, but let us suppose that η is sufficiently small that for practical purposes we have the true minimum. (Since this is actually a linear problem there are no local minima in this case.) For convenience, we will also assume that the weight matrices are constrained so that $V = W^T$ (this restriction is not essential but it makes it easier to see what is going on). To understand what the matrix W does here, we write X in terms of its SVD $X = PSQ^T$ where P and Q are orthogonal and S is diagonal (but not necessarily square). The diagonal elements of S are the singular values $\nu_1 \geq \nu_2 \geq \ldots \geq \nu_n$. Suppose $\operatorname{rank}(X) = r$ so that in fact $\nu_{r+1} \cdots \nu_n = 0$. The crucial stage is to find a matrix H satisfying $\operatorname{rank}(H) \leq m$ and which minimizes $\|X - PHP^TX\|_S$. Once we have H it is not difficult to factorize PHP^T to get W and V. But for any H, (since the Schur norm of a matrix is unchanged if we multiply by an orthogonal matrix)

$$
\begin{aligned}
\|X - PHP^TX\|_S^2 &= \|(I - PHP^T)X\|_S^2 \\
&= \|(I - PHP^T)PSQ^T\|_S^2 \\
&= \|(P - PH)S\|_S^2 = \|(I - H)S\|_S^2
\end{aligned}
$$

$$= \sum_{i=1}^{r} \nu_i^2 \Big\{ (1 - h_{ii})^2 + \sum_{j \neq i} h_{ji}^2 \Big\}.$$

Obviously, at the minimum H is diagonal. But we require $\text{rank}(H) \le m$. Thus the minimizing H is obtained by setting $h_{ii} = 1$, $i = 1, \ldots, \min(r, m)$ and all the other elements to zero. If $r \le m$, there is no loss of information in this process, and the patterns $\mathbf{x_p}$ are reconstructed exactly. If $r > m$, then the total error over all patterns is given by the square root of

$$\sum_{i=m+1}^{r} \nu_i^2. \tag{4.1}$$

It remains to perform the factorization $VW = PHP^T$. While the choice of H is unique, this is not so for the factorization. However, since PHP^T is symmetric, it makes sense to set $V = W^T$ as suggested above. In fact we have for the minimizing H,

$$HH^T = H$$

whence

$$PHP^T = PHH^TP^T = PH(PH)^T.$$

PH has (at most) m nonzero columns: we may take these as V and make $W = V^T =$ the first m rows of H^TP^T. Then $VW = PHP^T$ as required. The rows of W are those eigenvectors of XX^T corresponding to the largest singular values. The effect of W is to project the input patterns $\mathbf{x_p}$ onto the span of these vectors. If $r \le m$ (which is certainly the case, for instance, if the number of patterns $t \le m$) then the t n-vectors $\mathbf{x_p}$ are compressed by W onto m-vectors $\mathbf{y_p}$ with no loss of information, since we can recover X as W^TY. Here, of course, the columns of Y are the $\mathbf{y_p}$s. More usefully, even if $r > m$ there will be little loss in this compression provided the quantity (4.1) is small.

Moreover, by definition, VWX is a best rank m approximation to X in terms of the Schur norm, and can be constructed using the trained network by feeding the columns of X through the network one at a time to get the columns of VWX. It is well known and not hard to show that this matrix is also a best rank m approximation with respect to the matrix 2 norm: the proof is left for the reader!

A rather similar idea to that proposed in the last few paragraphs of Section 3 is used by Almeida and Silva (1992). However they propose using a separate perceptron to decorrelate the data. The weight matrix of this perceptron is the filter T so that the output is simply $T\mathbf{x}$. The aim is to find T such that $TXX^TT^T = I$ (compare (3.18)). Starting with $T = I$, they update T according to the rule

$$T_{n+1} = (1 + \alpha)T_n - \alpha(T_n L T_n^T)T_n \tag{4.2}$$

where, as in Section 3, $L = XX^T$. Now if L has full rank, certainly there exists a nonsingular T^* such that

$$T^* L T^{*T} = I. \tag{4.3}$$

To see this write $L = PDP^T$, where P is orthogonal, and let

$$T^* = D^{-1/2} P^T.$$

To see that it is a fixed point, simply set T^* satisfying (4.3) as T_n in (4.2). Almeida and Silva give as a sufficient condition for convergence

$$\alpha < \min\{\tfrac{1}{2}, (3\rho(L) - 1)^{-1}\}.$$

However, it is perhaps worth pointing out that convergence is sublinear, for let

$$F(T) = (1+\alpha)T - \alpha(TLT^T)T \tag{4.4}$$

and suppose T^* satisfies (4.3). Direct calculation yields for $h \in \mathbb{R}$ and any $n \times n$ matrix S

$$\begin{aligned} F(T^* + hS) &= T^* + (1+\alpha)hS - \alpha T^* L T^{*T} hS - \alpha(hSLT^{*T}T^* \\ &\quad + hT^* L S^T T^*) + \mathcal{O}(h^2) \\ &= T^* + (1+\alpha)hS - \alpha hS - \alpha(hS + hT^* L^S T T^*) + \mathcal{O}(h^2) \end{aligned}$$

since T^* satisfies (4.3) and this condition also implies $LT^{*T}T^* = I$. Thus

$$\begin{aligned} F(T^* + hS) &= T^* + (1-\alpha)hS - \alpha h(T^{*T})^{-1} S^T T^* + \mathcal{O}(h^2) \\ &= (I + (1-\alpha)hS(T^*)^{-1} - \alpha h(T^{*T})^{-1} S^T)T^* + \mathcal{O}(h^2). \end{aligned}$$

Now if we choose S so that $(T^*)^{-1}S$ is antisymmetric we obtain

$$\begin{aligned} F(T^* + hS) &= (I + hS(T^*)^{-1})T^* + \mathcal{O}(h^2) \\ &= T^* + hS + \mathcal{O}(h^2). \end{aligned}$$

Thus we will not get linear convergence near the fixed point. This is not a true learning network anyway since all the patterns are required at once. Hence use of an optimization routine might be preferable. Or simply compute the SVD instead! Almeida and Silva do give a version which uses the patterns one at a time, but this requires a sequence of αs tending to zero. Thus it seems that a better update rule than (4.2) would be worth seeking. In spite of these problems the particular interest of this method is that Almeida and Silva have performed numerical experiments with networks interleaving ordinary backpropagation layers with layers trained by (4.2). Improved convergence of the backpropagation was found on two test problems.

We remark finally that another way to satisfy (4.3) is actually to compute the principal components of X. Oja *et al.* (1992) describe a network to do this.

4.2. Hopfield nets and graph optimization

The Hopfield net (see, e.g., Wasserman (1989, Ch. 6)) is a dynamic net which is of historical importance as the first usable nonlinear neural network. As such it was a significant factor in the revival of interest in connectionist models in the 1980s. The most important application of Hopfield nets today is in the field of graph theoretic optimization. We will first describe the Hopfield architecture, and then use it to address the travelling salesman problem.

Unlike the multilayer perceptron, the Hopfield net is not layered: all neuronal units are treated equally. Each is connected to every other unit with a *bidirectional* link. Thus, if the weight of the connection from unit i to unit j is w_{ij}, we have $w_{ij} = w_{ji}$. No connections are permitted from a unit to itself, i.e. $w_{ii} = 0$ (actually for neurons that can output only values 0 or 1, this is not a real restriction: for example see the discussion of the travelling salesman problem below). The topology of an n unit Hopfield net is thus the complete digraph on n vertices. In addition to taking input from the other nodes, each unit can also take a (constant) input. There are no output units as such: the output from each unit is formed into a state vector $\mathbf{x}$ (so the ith element of $\mathbf{x}$ is the output of the ith unit). Usually (and certainly for our application) $\mathbf{x}$ is a binary vector, i.e. each element is either 0 or 1. Since the connections of the network are recursive, it will evolve in time. In the basic Hopfield model, single neurons are 'fired' (i.e. updated) at a time, in random order (biologically plausible) or cyclically. (Sometimes later workers have chosen, on grounds of simplicity and speed, to update all the units at once in discrete time steps. As we shall see this does have disadvantages.) Let us form the weights into a (symmetric) matrix W and let $\mathbf{q}$ be the vector of inputs. At the kth time step we have a state vector $\mathbf{x_k}$. So at the $k+1$st time step, the input to the jth unit is the jth element of the vector $W\mathbf{x_k}+\mathbf{q}$. The activation of the unit is by thresholding: let p_j be the threshold on the jth unit and form these into a vector $\mathbf{p}$. Thus, denoting the jth element of a vector $\mathbf{y}$ by $(\mathbf{y})_j$, we fire the jth neuron by updating $(\mathbf{x_{k+1}})_j$ as

$$(\mathbf{x_{k+1}})_j = \begin{cases} 1 & \text{if } (W\mathbf{x_k} + \mathbf{q} - \mathbf{p})_j > 0, \\ 0 & \text{if } (W\mathbf{x_k} + \mathbf{q} - \mathbf{p})_j < 0, \\ (\mathbf{x_k})_j & \text{if } (W\mathbf{x_k} + \mathbf{q} - \mathbf{p})_j = 0. \end{cases} \tag{4.5}$$

(Sometimes other choices are made in the equality case, but this one is sensible: see below.) For single unit updating, we simply apply this rule for one particular j and leave the other elements unchanged. If we are updating all neurons at once, we apply it for all j.

Associated with the network state vector $\mathbf{x}$ is an energy functional

$$E(\mathbf{x}) = -\tfrac{1}{2}\mathbf{x}^T W\mathbf{x} + \mathbf{q}^T\mathbf{x} - \mathbf{p}^T\mathbf{x}, \tag{4.6}$$

which serves as a Liapunov function. Observe that although this is usually referred to as the energy, W is *not* positive definite. The condition $w_{ii} = 0$ guarantees that trace$(W) = 0$, so W must have negative eigenvalues. This applies equally to subspaces obtained by deleting rows and corresponding columns of W. Thus, the good news is that the minimum of E must occur not at a stationary point but at a vertex of the hypercube defined by the condition $0 \leq (\mathbf{x})_j \leq 1$, $j = 1, \ldots, n$, i.e. at a valid state vector $\mathbf{x}$. (So we can avoid the thresholding and use a continuous model instead, if we wish, but we will stick to the threshold model here.) This result is a special case of the Cohen–Grossberg Theorem, see e.g., Simpson (1990). The bad news is that we can get local minima.

A simple calculation shows that the change $\delta E = E(\mathbf{x_{k+1}}) - E(\mathbf{x_k})$ in the energy is given by

$$\delta E = -(W\mathbf{x_k} + \mathbf{q} - \mathbf{p})^T(\mathbf{x_{k+1}} - \mathbf{x_k}) - \tfrac{1}{2}(\mathbf{x_{k+1}} - \mathbf{x_k})^T W(\mathbf{x_{k+1}} - \mathbf{x_k}). \quad (4.7)$$

Now consider the inner product $(W\mathbf{x_k} + \mathbf{q} - \mathbf{p})^T(\mathbf{x_{k+1}} - \mathbf{x_k})$. Perhaps $(\mathbf{x_{k+1}} - \mathbf{x_k})_j = 0$ for some j, in which case no contribution to the inner product is made by the jth term. However, unless we have reached a stationary point of the iteration, there must be one or more js for which $(\mathbf{x_{k+1}} - \mathbf{x_k})_j \neq 0$. We identify two cases. Possibly, $(W\mathbf{x_k} + \mathbf{q} - \mathbf{p})_j > 0$. From (4.5) we must have $(\mathbf{x_{k+1}})_j = 1$ whence necessarily $(\mathbf{x}k)_j = 0$ and $(\mathbf{x_{k+1}} - \mathbf{x_k})_j = 1$. The jth term thus contributes a positive value to the inner product in this case. Conversely if $(W\mathbf{x_k} + \mathbf{q} - \mathbf{p})_j < 0$, $(\mathbf{x_{k+1}})_j = 0$, $(\mathbf{x_k})_j = 1$ and $(\mathbf{x_{k+1}} - \mathbf{x_k})_j = -1$. Again a positive contribution is made. Thus the first term of (4.7) constitutes a decrease in energy unless $(\mathbf{x_{k+1}} - \mathbf{x_k})_j = 0$. In this case, of course, we make no change in energy but moving on to another neuron will stop us getting stuck. Since W is not positive definite, we cannot guarantee that the second term in (4.4) is negative. One reason for firing single neurons, rather than updating them all at once, is that at most one element of $(\mathbf{x_{k+1}} - \mathbf{x_k})$, say the jth, can be nonzero. Then $(\mathbf{x_{k+1}} - \mathbf{x_k})^T W(\mathbf{x_{k+1}} - \mathbf{x_k}) = w_{jj} = 0$. Moreover by updating singly we will ultimately search in all possible directions from a given point until a reduction in energy is found, thus ensuring that a fixed point is actually a local minimum of the energy. If we use global updating we just have to hope that $\|\mathbf{x_{k+1}} - \mathbf{x_k}\|$ is small and that the term is therefore negligible. (Of course we could switch from global to single updating if the former fails to give a decrease in energy.)

4.3. The travelling salesman problem

In optimization applications, Hopfield nets are not trained. Instead we set up an energy functional corresponding to the function to be optimized, together with penalty functions for any constraints, and 'hard-wire' the weight matrix

W so that this energy is actually (4.6). We then simply start the net off and wait for it to reach a minimum. We illustrate this process by applying it to the travelling salesman problem, an application first discussed by Hopfield and Tank (see Wasserman (1989, Ch. 6) for this and other applications of Hopfield nets.) Suppose we have m cities. We wish to visit each city precisely once, returning to the starting point. Any such ordering of the cities is known as a *tour*. The aim is to find the tour of minimum length. Since the problem is NP complete, we will not expect to find the optimum solution every time: we will be satisfied with a method that produces reasonably good tours reliably. Let d_{ij} denote the distance between the ith and jth cities, so $d_{ij} = d_{ji}$. To set this problem up as a Hopfield net, we allocate m neurons to each city. Thus the net will have $n = m^2$ neurons. (This may seem excessive, but a large travelling salesman problem of say 100 cities will only require 10 000 units: potentially well within the capabilities of VLSI circuits when we remember that each unit is very simple.) Think of the neurons being arranged in a table of m rows of m units, each row corresponding to a city. Now a 1 in the jth neuron of any row means that that city will be visited jth. Thus for a valid tour, there must be exactly one 1 in each row. Since the cities are visited sequentially (we cannot be in two places at once), there must also be exactly one 1 in each column. So in total there will be m 1s in a valid tour table. To construct our energy functional we introduce penalty functions for each of these constraints. Let the state vector of the net be $\mathbf{x}$, allocated by rows of the table so that the first m elements refer to city 1, the second m elements refer to city 2 etc. Recall that $\mathbf{x}$ may contain only 0s or 1s. Denote by $\mathbf{e}$ the vector containing all 1s. We first introduce a term

$$(\mathbf{e}^T\mathbf{x} - m)^2 = \mathbf{x}^T(\mathbf{e}\mathbf{e}^T)\mathbf{x} - 2m\mathbf{x} + m^2.$$

This term will vanish if and only if $\mathbf{x}$ has exactly m 1s. The m^2 term can be ignored since this is constant and we are constructing an energy functional to minimize. The matrix $\mathbf{e}\mathbf{e}^T$, all of whose entries are 1, presents a difficulty, as it does not have zero diagonal. However, let $A = \mathbf{e}\mathbf{e}^T - I$. Then $\mathbf{x}^T(\mathbf{e}\mathbf{e}^T)\mathbf{x} = \mathbf{x}^T A\mathbf{x} + \mathbf{x}^T\mathbf{x}$. Moreover, since the entries of $\mathbf{x}$ can be only 0 or 1, we have $\mathbf{x}^T\mathbf{x} = \mathbf{e}^T\mathbf{x}$. So we can use

$$\mathbf{x}^T A\mathbf{x} + (1 - 2m)\mathbf{e}^T\mathbf{x}. \tag{4.8}$$

The other constraints are a little more complicated as we have to pick out the sections of the state vector corresponding to individual cities. This is conveniently achieved by introducing the following matrices. Let B be defined as follows: $b_{ii} = 0$, $i = 1, \ldots, n$. Also $b_{ij} = 0$ if $(\mathbf{x})_i$ and $(\mathbf{x})_j$ represent different cities or, in other words, if $[i/m] \neq [j/m]$, where $[r]$ denote the greatest integer *strictly* less than r. Finally $b_{ij} = 1$ if $(\mathbf{x})_i$ and $(\mathbf{x})_j$ represent the same city or, in other words, if $[i/m] = [j/m]$ but $i \neq j$.

The matrix B is thus symmetric and block diagonal. Each block on the diagonal corresponds to a particular city. The quadratic form

$$\mathbf{x}^T B \mathbf{x} \tag{4.9}$$

therefore has the following effect. Each block of B multiplies together and sums distinct entries of $\mathbf{x}$ corresponding to the *same* city, which will be greater than zero if two elements corresponding to the same city are nonzero. Finally these sums are themselves summed over the cities. (Note: if this is not clear, it helps to write out (4.9) for, say, $m = 3$.) Remembering that $\mathbf{x}$ can contain only 1s and 0s. we see that (4.9) will achieve its minimum value of 0 if each row of our unit table has at most one 1, i.e. if city is visited at most once. Similarly we need to ensure that we do not try and visit two cities at once, i.e. that each column of our table of units has at most one 1. This is encoded by

$$\mathbf{x}^T C \mathbf{x}, \tag{4.10}$$

where $c_{ii} = 0$, $c_{ij} = 1$ if i modulo $m = j$ modulo m but $i \neq j$, $c_{ij} = 0$ otherwise. Finally we need a term which actually reduces the length of tours. Basically, this simply requires us to read off the tour order from our table and add up the distances, but of course we need to write this process as a symmetric quadratic form. For each city, we have a 1 in the position of the table telling us when we visit it. The 1 in the previous column tells us where we got there from. Since there can be at most one 1 in this previous column, we may as well multiply the corresponding distance by the element of $\mathbf{x}$ corresponding to that entry, multiply that by the element of $\mathbf{x}$ corresponding to the current city, and sum. In order to preserve symmetry, it is best to look in both the previous and next columns. This results in each leg of the tour being included twice, but remember that we are not actually going to calculate it. We are merely using it to get the weights for our network. Of course, we also need to include the return to the starting city. The quadratic form is thus

$$\mathbf{x}^T F \mathbf{x}, \tag{4.11}$$

where we think of F as being made up of m^2 $m \times m$ blocks. The p, qth block corresponds to a trip from the pth city to the qth. Let F_{pq} be this $m \times m$ submatrix. Then $F_{pp} = 0$. For $p \neq q$, each entry in F_{pq} will be either d_{pq} or 0. The nonzero elements will occur only if the trip from city p to city q is actually made. Thus we need to put d_{pq} on the sub and super diagonals of F_{pq} . But we must also allow for the possibility that p is the first city and q the last, or *vice versa*. Therefore we must also set $(F_{pq})_{1m} = (F_{pq})_{m1} = d_{pq}$. The other elements of F_{pq} are 0.

Now choose parameters $\alpha \gg \beta, \gamma \gg \theta$. Comparing (4.6) with (4.8)–(4.11) we may set $W = \alpha A + \beta B + \gamma C + \theta F$ and $q = (1 - 2m)\mathbf{e}^T$. The network will first try to find a tour with n 1s, causing the (4.8) term to vanish. It will then try to find a 1 in each row and column, eliminating the (4.9) and (4.10) terms. Finally it will try to reduce the length of the tour. We need to set the thresholds to (say) $\frac{1}{2}$. In fact Hopfield and Tank used not this step function activation, but $\frac{1}{2}(1+\tanh(\mathbf{x}/\mathbf{u}_0)$ (Wasserman, 1989, p. 109) where $\mathbf{x}$ is the total input to the unit and $\mathbf{u}_0$ is a further parameter. The behaviour is reportedly very dependent on the choice of parameters. Van den Bout and Miller (1988) discuss this in detail and suggest improvements. A detailed analysis of when the method will give a valid tour, and how bounds on the optimal solution may be obtained, has been provided by Aiyer *et al.* (1989) and in two papers, one by the same authors and the other by de Carvalho and Barbosa in INNC 90 (1990, pp. 245–248 and pp. 249–253, respectively). This proceedings also contains several other related papers in its section on optimization (pp. 245–297).

5. Concluding remarks

Neural networks represent an important new tool for computation. Numerical analysts can and have made significant contributions to the field. As we have seen, the topic can stimulate new research in approximation theory and optimization algorithms. Equally, we saw in Section 4 that methods of neural computation can provide new tools for numerical computation. In particular, it seems plausible that the Hopfield net could be applied to processor assignment problems in parallel computation, particularly since, by definition, suitable parallel hardware would be available to implement it! There are other issues in neural net research which have hardly been treated formally at all. For example, the claim is often made that networks are fault tolerant in that deleting the connection to a unit, or even a whole unit, degrades the performance only marginally rather than catastrophically as it would with a conventional computing system. The argument for this is that networks store information in a distributed way, so only a little information will be lost. This is certainly the case if the network is constructed as in Section 2.2, as deleting one of the terms from the quadrature formulae will not destroy convergence. But networks are not constructed this way in practice. Little serious analysis of fault tolerance has been attempted. Applying numerical analysis to neural networks will not only be useful in applications, but should provide a new stimulus to numerical analysis itself.

Acknowledgement

I would like to thank Professor W. Light, of the University of Leicester, for some useful discussions on the material in Section 2.2.

REFERENCES

I. Aleksander and H. Morton (1990), *An Introduction to Neural Computing*, Chapman and Hall (London).

L. B. Almeida and F. M. Silva (1992), 'Adaptive decorrelation', in *Artificial Neural Networks 2* (I. Aleksander and J. Taylor, eds), Vol. 2, North-Holland (Amsterdam) 149–156.

S. I. Amari (1990), 'Mathematical foundations of neurocomputing', *Proc. IEEE* **78**, 1143–1463.

S. V. B. Aiyer, M. Niranjan and F. Fallside (1989), 'A theoretical investigation into the performance of the Hopfield model', *Tech. Report, CUED/F-INFENG/TR 36*, Cambridge University Engineering Department, Cambridge, CB2 1PZ, England.

P. Baldi and K. Hornik (1989), 'Neural networks and principal component analysis: learning from examples without local minima', *Neural Networks* **2**, 53–58.

A. Ben-Israel and T. N. E. Greville (1974), *Generalised Inverses, Theory and Applications*, Wiley (Chichester).

M. Bichsel and P. Seitz (1989), 'Minimum class entropy: a maximum information approach to layered networks', *Neural Networks* **2**, 133–141.

R. W. Brause (1992), 'The error bounded descriptional complexity of approximation networks', *Fachberiech Informatik*, J W Goethe University, Frankfurt am Main, Germany.

D. S. Broomhead and D. Lowe (1988), 'Multivariable function interpolation and adaptive networks', *Complex Systems* **2**, 321–355.

T. Chen, H. Chen and R. Liu (1991), 'A constructive proof and extension of Cybenko's approximation theorem', in *Computing Science and Statistics: Proc. 22nd Symp. on the Interface*, Springer (Berlin) 163–168.

E. W. Cheney (1966), *Introduction to Approximation Theory*, McGraw-Hill (New York).

G. Cybenko (1989), '∞ approximation by superpositions of a sigmoidal function', *Math. Control-Signals Systems* **2**, 303–314.

P. Diaconis and M. Shashahani (1984), 'On nonlinear functions of linear combinations', *SIAM J. Sci. Statist. Comput.* **5**, 175–191.

S. W. Ellacott (1990), 'An analysis of the delta rule', *Proc. Int. Neural Net Conf., Paris* Kluwer (Deventer) 956–959.

S. W. Ellacott (1993a), 'The numerical analysis approach', in *Mathematical Approaches to Neural Networks* (J.G. Taylor, ed.), North Holland (Amsterdam) 103–138.

S. W. Ellacott (1993b), 'Techniques for the mathematical analysis of neural networks', *J. Appl. Comput. Math.* to appear.

S. W. Ellacott (1993c), 'Singular values and neural network algorithms', in *Proc. British Neural Network Society Meeting, February 1993*.

K. Falconer (1990), *Fractal Geometry*, Wiley (New York).

M. Fombellida and J. Destiné (1992), 'The extended quickprop', in *Artificial Neural Networks 2* (I. Aleksander and J. Taylor), Vol. 2, North-Holland (Amsterdam) 973–977.

K.-I. Funahashi (1989), 'On the approximate realization of continuous mappings by neural networks', *Neural Networks* **2**, 183–192.

K. Hornik K, M. Stinchcombe and H. White (1989), 'Multilayer feedforward networks are universal approximators', *Neural Networks* **2**, 359–366.

INNC 90 (1990), *Proc. Int. Neural Network Conf. (9–13 July 1990, Palais de Congres, Paris, France)* Kluwer (Deventer).

E. Isaacson and H. B. Keller (1966), *Analysis of Numerical Methods*, Wiley (New York).

D. Jacobs (ed.) (1977), *The State of the Art in Numerical Analysis*, Academic Press (New York).

A. J. Jones (1992), 'Neural computing applications to prediction and control', Department of Computing, Imperial College, London, United Kingdom.

E. Kreyszig (1978), *Introductory Functional Analysis with Applications*, Wiley (New York).

B. Linggard and C. Nightingale (eds) (1992), *Neural Networks for Images, Speech and Natural Language*, Chapman and Hall (London).

W. Light (1992), 'Ridge function, sigmoidal functions and neural networks', in *Approximation Theory VII* (E. W. Cheney, C. K. Chui and L. L. Schumaker, eds) Academic (Boston) 1–44.

J. C. Mason and P. C. Parks (1992), 'Selection of neural network structures – some approximation theory guidelines', Ch. 8, in *Neural Networks for Control and Systems*, (K. Warwick, G. W. Irwin and K. J. Hunt, eds) *IEE Control Engineering Series no. 46*, Peter Peregrinus (Letchworth).

H. N. Mhaskar and C. Micchelli (1992), 'Approximation by superposition of sigmoidal functions', *Adv. Appl. Math.* **13**, 350–373.

H. N. Mhaskar (1993), 'Approximation properties of a multilayered feedforward artificial neural network', *Adv. Comput. Math.* **1**, 61–80.

J. J. Moré (1978), 'The Levenberg–Marquardt algorithm, implementation and theory', in *Proc. Dundee Biennial Conf. on Numerical Analysis 1977*, (G. A. Watson, ed.), *Springer Lecture Notes in Mathematics no. 630*, Springer (Berlin) 105–116.

E. Oja (1983), *Subspace Methods of Pattern Recognition*, Research Studies Press (Letchworth, UK).

E. Oja (1992), 'Principal components, minor components and linear neural networks', *Neural Networks* **5**, 927–935.

E. Oja, H. Ogawa and J. Wangviwattana (1992), 'PCA in fully parallel neural networks', in *Artificial Neural Networks 2* (I. Aleksander and J. Taylor, eds), Vol. 2, North-Holland (Amsterdam) 199–202.

M. J. D. Powell (1992), 'The theory of radial basis functions approximation in 1990', in *Advances in Numerical Analysis* (W. Light, ed.), Vol. II, Oxford University Press (Oxford), 105–210.

D. E. Rumelhart and J. L. McClelland (1986), *Parallel and Distributed Processing: Explorations in the Microstructure of Cognition*, Vols 1 and 2, MIT (Cambridge, MA).

H. Sagan (1969), *Introduction to the Calculus of Variations*, McGraw-Hill (New York).

P. K. Simpson (1990), *Artificial Neural Systems: Foundations, Paradigms, Applications and Implementations*, Pergamon Press (New York).

E. Stein and G. Weiss (1971), *Introduction to Fourier Analysis on Euclidean Spaces*, Princeton University Press (Princeton, USA).

J.G. Taylor (ed.) (1993), *Mathematical Approaches to Neural Networks*, North Holland (Amsterdam).

D. E. Van den Bout and T. K. Miller (1988), 'A travelling salesman objective function that works', *Proc. IEEE Conf. on Neural Networks*, Vol. 2, SOS Printing (San Diego, CA) 299–304.

G. Venkataraman and G. Athithan G (1991), 'Spin glass, the travelling salesman problem, neural networks and all that', *Prãmana J. Phys.* **36**, 1–77.

Z. Wang, M. T. Tham and A. J. Morris (1992), 'Multilayer feedforward neural networks: a cannonical form approximation of nonlinearity', Department of Chemical and Process Engineering, University of Newcastle upon Tyne, Newcastle upon Tyne, NE1 7RU, United Kingdom.

K. Warwick, G. W. Irwin and K. J. Hunt (1992), 'Neural networks for control and systems', *IEE Control Engineering Series No. 46*, Peter Peregrinus (Letchworth).

P. D. Wasserman (1989), *Neural Computing: Theory and Practice*, Van Nostrand Reinhold (New York).

P. J. Werbos (1992), 'Neurocontrol: where it is going and why it is crucial', in *Artificial Neural Networks 2* (I. Aleksander and J. Taylor, eds), Vol. 1, North-Holland (Amsterdam) 61–68.

Y. Xu, W. A. Light and E. W. Cheney (1991), 'Constructive methods of approximation by ridge functions and radial functions'. Address of first author: Department of Mathematics, University of Arkansas at Little Rock, Little Rock, AR 72204, USA.

Acta Numerica (1994), *pp.* 203–267

A review of pseudospectral methods for solving partial differential equations

Bengt Fornberg
Corporate Research
Exxon Research and Engineering Company
Annandale, NJ 08801, USA
E-mail: bfornbe@erenj.com

David M. Sloan
Department of Mathematics
University of Strathclyde
Glasgow G1 1XH, Scotland
E-mail: caas10@computer-centre-sun.strathclyde.ac.uk

CONTENTS

1 Introduction 1
2 Introduction to spectral methods via orthogonal functions 3
3 Introduction to PS methods via finite differences 8
4 Key properties of PS approximations 21
5 PS variations/enhancements 36
6 Comparisons of computational cost – FD versus PS methods 47
Appendix A. Implementation of Tau, Galerkin and Collocation (PS) for a 'toy' problem 51
Appendix B. Fortran code and test driver for algorithm to find weights in FD formulae 56
References 57

1. Introduction

Finite Difference (FD) methods approximate derivatives of a function by *local* arguments (such as $\mathrm{d}u(x)/\,\mathrm{d}x \approx (u(x+h) - u(x-h))/2h$, where h is a small grid spacing) – these methods are typically designed to be exact for polynomials of low orders. This approach is very reasonable: since the derivative is a local property of a function, it makes little sense (and is costly) to invoke many function values far away from the point of interest.

In contrast, spectral methods are *global*. The traditional way to introduce them starts by approximating the function as a sum of very smooth basis functions:

$$u(x) = \sum_{k=0}^{N} a_k \Phi_k(x),$$

where the $\Phi_k(x)$ are, for example, Chebyshev polynomials or trigonometric functions – and then differentiate these exactly. In the context of solving time-dependent Partial Differential Equations (PDEs), this approach has notable strengths:

1. for analytic functions, errors typically decay (as N increases) at an *exponential* rather than at a (much slower) *polynomial* rate;
2. the method is virtually dissipation-free (in the context of solving high Reynolds number fluid flows, the low physical dissipation will not be overwhelmed by large numerical dissipation);
3. the approach is (surprisingly) powerful for many cases with nonsmooth or even discontinuous functions;
4. especially in several space dimensions, the relatively coarse grids which suffice for most accuracy requirements allow very time- and memory-effective calculations.

However, there are also factors which might cause difficulties or inefficiencies: certain boundary conditions; irregular domains; strong shocks; variable resolution requirements in different parts of a large domain; partly incomplete theoretical understanding.

In many applications where these disadvantages are not present (or they can somehow be overcome), FD or FE (Finite Element) methods do not even come close in efficiency. However, the situation in most major applications turns out less clear-cut than this. At present, spectral methods are highly successful in several areas such as turbulence modelling, weather prediction, nonlinear waves, seismic modelling etc. and the list is growing (see, for example, Boyd (1989) for examples and references).

Spectral methods have been a tool for analytic studies of differential equations since the days of Fourier (1822). The idea of using them for numerical solutions of ordinary differential equations (ODEs) goes back at least to Lanczos (1938). Their current popularity for PDEs dates back to the early 1970s and the works of Kreiss and Oliger (1972) and Orszag (1972) – facilitated by the Fast Fourier Transform (FFT) algorithm presented by Cooley and Tukey (1965).

Although spectral methods are normally introduced in the way we have indicated (through expansions using smooth global functions – the topic of Section 2), there is a useful alternative: pseudospectral (PS) methods can be seen as a special case of high-order FD methods (Fornberg 1975, 1987, 1990a,b). The introduction to Section 3 lists some of the advantages this latter approach offers. In the remaining sections, key properties/variations of PS methods are discussed (using whichever viewpoint is most illuminating for the issue being discussed).

To obtain a uniform style, this review has been written and illustrated by BF from jointly prepared material. As a consequence, readers looking for functional analysis or proofs of technicalities are hereby warned not to waste their time proceeding beyond this point.

For one of us (BF), the interest in PS methods goes back to his PhD in 1972 under the supervision of H.-O. Kreiss. For later interactions, we wish, in particular, to acknowledge discussions with L.N. Trefethen.

2. Introduction to spectral methods via orthogonal functions

Spectral methods are usually described in the way we first indicated – as expansions based on global functions. Given a differential equation with boundary conditions, the idea is to approximate a solution $u(x)$ by a finite sum $v(x) = \sum_{k=0}^{N} a_k \Phi_k(x)$ (in the case of a time-dependent problem, $u(x,t)$ approximated by $v(x,t)$ and $a_k(t)$). Two main questions arise:

1 from which function class to choose $\Phi_k(x)$, $k = 0, 1, \ldots$ and
2 how to determine the expansion coefficients a_k.

These are addressed in Sections 2.1 and 2.2. Section 2.3 introduces cardinal functions and differentiation matrices – important tools for both understanding and computation (to be discussed in greater generality in Section 3.4 following the FD-based introduction to PS methods).

Reviews of this 'classical' approach to spectral methods can be found in Gottlieb and Orszag (1977), Voigt *et al.* (1984), Boyd (1989), Mercier (1989) and Funaro (1992).

2.1. Function classes

Three requirements need to be met:

1 The approximations $\sum_{k=0}^{N} a_k \Phi_k(x)$ to $v(x)$ must converge rapidly (at least for reasonably smooth functions).
2 Given coefficients a_k, the determination of b_k such that

$$\frac{\mathrm{d}}{\mathrm{d}x}\left(\sum_{k=0}^{N} a_k \Phi_k(x)\right) = \sum_{k=0}^{N} b_k \Phi_k(x) \tag{2.1}$$

should be efficient.
3 It should be possible to convert rapidly between coefficients a_k, $k = 0, \ldots, N$ and the values for the sum $v(x_i)$ at some set of nodes x_i, $i = 0, \ldots, N$.

Periodic problems The choice here is easy – *trigonometric expansions* satisfy all the requirements. The first two are immediate; the third was satisfied in 1965 through the FFT algorithm.

Non-periodic problems In this case, *trigonometric expansions* fail on requirement 1 – an irregularity will arise where the periodicity is artificially imposed. In the case of a discontinuity, a 'Gibbs' phenomenon' will occur (see Section 2.3). The coefficients a_n then decrease only like $\mathcal{O}(1/N)$ as $N \to \infty$.

Truncated *Taylor expansions* $v(x) = \sum_{k=0}^{N} a_k x^k$ will also fail on requirement 1: convergence over $[-1, 1]$ requires extreme smoothness of $v(x)$, i.e. analyticity throughout the unit circle.

The function class that has proven, by far, the most successful is *orthogonal polynomials* of Jacobi type with Chebyshev and Legendre polynomials as the most important special cases (cf. Table 1). These polynomials arise in many contexts:

- Gaussian integration formulae achieve high accuracy by using zeros of orthogonal polynomials as nodes.
- Singular Sturm–Liouville eigensystems are well known to offer superb bases for approximation – the Jacobi polynomials are the only polynomials arising in this way.
- Truncated expansions in *Legendre polynomials* are optimal in the L^2-norm (for max-norm approximations of smooth functions, truncated *Chebyshev expansions* are particularly accurate).
- Interpolation at the Chebyshev nodes

  $$x_k = -\cos(\pi k/N), \; k = 0, 1, \ldots, N$$

  give polynomials P_N^{CH} which are always within a very small factor of the optimal in max-norm approximation of any function $f(x)$:

  $$\|f - P_N^{\mathrm{CH}}\| \leq (1 + \Lambda_N^{\mathrm{CH}})\|f - P_N^{\mathrm{OPT}}\|.$$

  Here Λ_N^{CH} is known as the Lebesgue constant of order N for Chebyshev interpolation. It depends only on N; properties of f affect $\|f - P_N^{\mathrm{OPT}}\|$ as described by Jackson's theorems (Cheney, 1966; Powell, 1981). Λ_N^{CH} is smaller than the constant for interpolation using Legendre expansions and far superior to the disastrous one for equi-spaced interpolation (more on this 'Runge phenomenon' in Section 3.3).

  $$\Lambda_N^{\mathrm{CH}} = \mathcal{O}(\log N), \quad \Lambda_N^{\mathrm{LEG}} = \mathcal{O}(\sqrt{N}), \quad \Lambda_N^{\mathrm{EQI}} = \mathcal{O}(2^N/N \log N).$$

  For references on the three Lebesgue constants, see Rivlin (1969, p. 90), Szegő (1959, p. 336) and Trefethen and Weideman (1991) respectively.

These points confirm that Jacobi polynomials satisfy requirement 1. Because of the first derivative recursions (and the lack of explicit x-dependence

Table 1. *Jacobi polynomials' fact sheet*

	LEGENDRE	CHEBYSHEV	JACOBI
Weight function $W(x)$	1	$\frac{1}{\sqrt{1-x^2}}$	$(1-x)^\alpha(1+x)^\beta$ $\quad(\alpha=\beta=0$ Legendre) $\alpha>-1,\ \beta>-1$ $\quad(\alpha=\beta=-\frac{1}{2}$ Chebyshev)
First few polynomials	1 x $\frac{3}{2}x^2-\frac{1}{2}$ $\frac{5}{2}x^3-\frac{3}{2}x$ $\frac{35}{8}x^4-\frac{15}{4}x^2+\frac{3}{8}$ $\frac{63}{8}x^5-\frac{35}{4}x^3+\frac{15}{8}x$	1 x $2x^2-1$ $4x^3-3x$ $8x^4-8x^2+1$ $16x^5-20x^3+5x$	1 $\frac{1}{2}(2+\alpha+\beta)x+\frac{1}{2}(\alpha-\beta)$ $\frac{1}{8}(2+\alpha+\beta)(4+\alpha+\beta)x^2+\frac{1}{4}(\alpha-\beta)(3+\alpha+\beta)x$ $\quad+\frac{1}{8}[(\alpha-\beta)^2-(4+\alpha+\beta)]$ General n : $2^{-n}\sum_{k=0}^{n}\binom{n+\alpha}{k}\binom{n+\beta}{n-k}(x-1)^{n-k}(x+1)^k$
Orthogonality $\int_{-1}^{1}\Phi_m\Phi_n W\,\mathrm{d}x$	$0 \quad : m\neq n$ $\frac{2}{2n+1} \quad : m=n$	$0 \quad : m\neq n$ $\pi \quad : m=n=0$ $\frac{\pi}{2} \quad : m=n>0$	$0 \quad : m\neq n$ $\frac{2^{\alpha+\beta+1}\Gamma(n+\alpha+1)\Gamma(n+\beta+1)}{(2n+\alpha+\beta+1)n!\Gamma(n+\alpha+\beta+1)} \quad : m=n$
Three term recursion	$(n+1)L_{n+1}$ $-(2n+1)xL_n$ $+nL_{n-1}=0$	$T_{n+1}-2xT_n$ $+T_{n-1}=0$	$2(n+1)(n+\alpha+\beta+1)(2n+\alpha+\beta)P_{n+1}$ $-[(2n+\alpha+\beta+1)(\alpha^2-\beta^2)$ $\quad+(2n+\alpha+\beta)(2n+\alpha+\beta+1)(2n+\alpha+\beta+2)x]P_n$ $+2(n+\alpha)(n+\beta)(2n+\alpha+\beta+2)P_{n-1}=0$
Differential equation	$(1-x^2)L_n''-2xL_n'$ $+n(n+1)L_n=0$	$(1-x^2)T_n''-xT_n'$ $+n^2T_n=0$	$(1-x^2)P_n''+[(\beta-\alpha)-(\alpha+\beta+2)x]P_n'$ $+n(n+\alpha+\beta+1)P_n=0$
First derivative recursion	L_{n+1}' $=(2n+1)L_n+L_{n-1}'$	$\frac{1}{n+1}T_n'$ $=2T_n+\frac{1}{n-1}T_{n-1}'$	$2(2n+\alpha+\beta)(n+\alpha+\beta)(n+\alpha+\beta+1)P_{n+1}'$ $+2(\alpha-\beta)(n+\alpha+\beta)(2n+\alpha+\beta+1)P_n'$ $-2(n+\alpha)(n+\beta)(2n+\alpha+\beta+2)P_{n-1}'$ $=(n+\alpha+\beta)(2n+\alpha+\beta)(2n+\alpha+\beta+1)(2n+\alpha+\beta+2)P_n$

in these, see Table 1), requirement 2 is met. In the case of Chebyshev polynomials, the relation sought in (2.1) becomes

$$\begin{bmatrix} 1 & 0 & -\frac{1}{2} & & & & \\ & \frac{1}{4} & 0 & -\frac{1}{4} & & & \\ & & \frac{1}{6} & 0 & -\frac{1}{6} & & \\ & & & \frac{1}{8} & 0 & -\frac{1}{8} & \\ & & & & \ddots & \ddots & \ddots \\ & & & & & \frac{1}{2N-4} & 0 \\ & & & & & & \frac{1}{2N-2} \end{bmatrix} \times \begin{bmatrix} b_0 \\ b_1 \\ b_2 \\ b_3 \\ \vdots \\ b_{N-2} \\ b_{N-1} \end{bmatrix} = \begin{bmatrix} a_1 \\ a_2 \\ a_3 \\ a_4 \\ \vdots \\ a_{N-1} \\ a_N \end{bmatrix}. \quad (2.2)$$

and similarly for Legendre and Jacobi expansions. Requirement 3 is clearly satisfied for the Chebyshev case if we choose $x_i = -\cos(\pi i/N)$, $i = 0, \ldots, N$ (cosine FFT) – as we shall see later, the additional cost in the other cases need not be prohibitive. For these reasons, Chebyshev (and to a lesser extent, Legendre) polynomials have become the almost universally preferred choice for nonperiodic spectral approximations.

The special case of Jacobi polynomials $P_n^{(\alpha,\beta)}(x)$ with $\alpha = \beta$ goes under its own name – Gegenbauer or Ultraspherical polynomials (again including Legendre and Chebyshev as special cases). For an overview of orthogonal polynomials, see, for example, Sansone (1959) and Szegő (1959).

Formulas similar to (2.1) are also available when (2.1) is generalized, for example, to

$$x^q \frac{\mathrm{d}^p}{\mathrm{d}x^p}\left(\sum_{k=0}^{N} a_k \phi_k(x)\right) = \sum_{k=0}^{N} b_k \phi_k(x), \quad p, q = 1, 2, \ldots$$

(see the Appendix in Gottlieb and Orszag (1977)). These are essential for Tau and Galerkin – but not for collocation (PS) approximations; see the next section and Appendix 1.

2.2. *Techniques to determine the expansion coefficients*

The three main techniques to determine expansion coefficients a_k are Tau, Galerkin and Collocation (PS). In all cases, we consider the residual $R_n(x)$ (or $R_n(x,t)$) when the expansion is substituted into the governing equation. We need to keep this residual small across the domain and also to satisfy the boundary conditions.

- In the Tau technique the a_k have to be selected so that the boundary conditions are satisfied and the residual has to be made orthogonal to as many of the basis functions as possible.
- For the Galerkin technique we combine the original basis functions into a new set in which all the functions satisfy the boundary conditions.

Then the residual is required to be orthogonal to as many of these new basis functions as possible.

- Finally the Collocation (PS) technique is similar to the Tau one: the a_k have to be selected so that the boundary conditions are satisfied but the residual is made zero at as many (suitably chosen) spatial points as possible.

Implementation details for a model problem are given in Appendix 1.

The Tau method was first used by Lanczos (1938). The Galerkin idea is central to FE methods. Spectral (global) versions of it have been in use since the mid-1950s. The FFT algorithm as well as contributions by Orszag (1969, 1970; on ways to deal with nonlinearities) have contributed to its current usage. The collocation approach was first used for PDEs with periodic solutions by Kreiss and Oliger (1972). It was referred to as the 'pseudospectral method' in Orszag (1972).

The collocation (PS) method can be viewed as a method for finding numerical approximations to derivatives at grid points. Then, in a FD-like manner, the governing equations are satisfied pointwise in physical space. This makes the PS method particularly easy to apply to equations with variable coefficients and nonlinearities, as these only give rise to products of numbers rather than to problems of determining the expansion coefficients for products of expansions.

The rest of this review article will focus entirely on the PS method.

2.3. Cardinal functions – example of a differentiation matrix

The concepts of Cardinal Functions (CFs) and of Differentiation Matrices (DMs) are both theoretically and numerically useful well beyond the realm of methods derived from orthogonal functions. Therefore, we postpone the main discussion of these until Section 4.3 (when our background is more general) and consider them here only in the case of the Fourier PS method.

The trigonometric polynomial which interpolates periodic data can be thought of as a weighted sum of CFs, each with the property of having unit value at one of the data points and zero at the rest. This is very similar to how Lagrange's interpolation formula works – the main difference being that, in this Fourier case, all the CFs are just translates of each other. For references to CFs, see E.T. Whittaker (1915), J.M. Whittaker (1927), Stenger (1981).

Assume for simplicity that we have an odd number $n = 2m + 1$ of grid points, at locations $x_i = i/(m + \frac{1}{2})$, $i = -m, \ldots, -1, 0, 1, \ldots, m$ in $[-1, 1]$. By inspection,

$$\Phi_m(x) = \frac{1}{m+\frac{1}{2}}\{\tfrac{1}{2} + \cos \pi x + \cos 2\pi x + \ldots + \cos m\pi x\} = \frac{\sin(m+\frac{1}{2})\pi x}{(2m+1)\sin \frac{1}{2}\pi x} \tag{2.3}$$

is of the right form and satisfies

$$\Phi_m(x_i) = \begin{cases} 1, i = 0 & [\pm(2m+1), \pm(4m+2), \\ & \text{etc., if periodically extended}] \\ 0 & \text{otherwise.} \end{cases}$$

Figure 1 displays the CF $\Phi_8(x)$ and illustrates how its translates add up to give the trigonometric interpolant to a step function, and the Gibbs' phenomenon.

From (2.3) it follows that

$$\frac{\mathrm{d}}{\mathrm{d}x}\Phi_m(x_i) = \begin{cases} 0, i = 0 & [\pm(2m+1), \pm(4m+2), \text{ etc.,} \\ & \text{if periodically extended}] \\ \dfrac{(-1)^i \pi}{2\sin[i\pi/(2m+1)]} & \text{otherwise.} \end{cases}$$

The PS derivative $v'(x_k)$ of a vector of data values $v(x_k), k = -m, \ldots, m$ can therefore be obtained as a matrix $\times$ vector product. The element at position (i, j) of this matrix is equal to $\mathrm{d}\Phi_m(x_{i-j})/\mathrm{d}x$. This matrix is the DM for the periodic PS method (first derivative, odd number of points).

3. Introduction to PS methods via finite differences

FD formulae of increasing orders of accuracy provide not only an alternative introduction to PS methods (for both the periodic and nonperiodic cases); they suggest generalizations and offer additional insights.

- Orthogonal polynomials/functions lead only to a small class of possible spectral methods – the FD viewpoint allows many generalizations. For example, all classical orthogonal polynomials cluster the nodes quadratically towards the ends of the interval – this is often, but not always best.
- An FD viewpoint offers a chance to explore 'intermediate' methods between low-order FD and PS. One might consider methods of not quite as high order as the Chebyshev spectral method and with nodes not clustered quite as densely – possibly trading some accuracy for stability and simplicity.
- Two separate ways to view any method always give more opportunities to understand/improve/analyse it.
- Many special enhancements have been designed for FD methods. Viewing PS methods as a special case of FD methods often makes it easier to carry over such ideas. Examples include staggered grids, upwind techniques, enhancements at boundaries, polar and spherical coordinates, etc. (discussed in Section 5).
- Comparisons between PS and FD methods can be made more consistent. PS methods represent limits of increasingly accurate FD methods

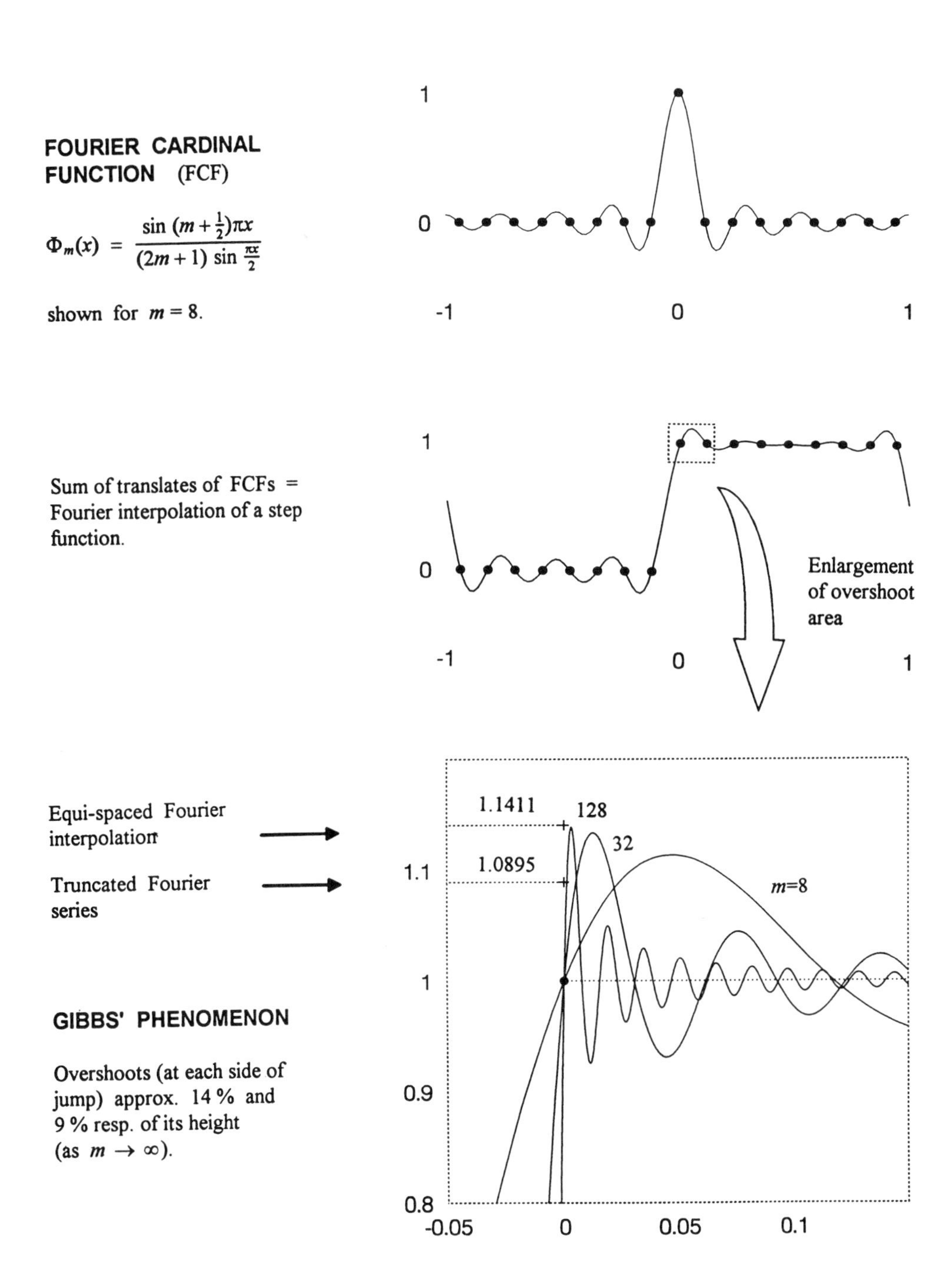

Fig. 1. Fourier CFs and Gibbs' phenomenon.

– the FD viewpoint provides a unifying framework in which to understand and interpret all these methods.

Sections 3.1 and 3.2 provide some general material on FD approximations, allowing us in Section 3.3 to discuss different types of node distribution. In Section 3.4, we derive the DM for the case that was considered in Section 2.3 (periodic problem, equi-spaced grid) and obtain an identical result – hence the methods are equivalent. In Section 3.5, this equivalence is seen to be very general.

3.1. Algorithm to find FD weights on arbitrarily spaced grids

Centred FD formulae for equi-spaced grids are readily available from tables and can be derived by symbolic manipulation of difference operators. For example, the centred approximations (at a grid point) to the first derivative are:

$$\begin{aligned} f'(x) &= [\quad -\tfrac{1}{2}f(x-h)+0f(x)+\tfrac{1}{2}f(x+h) \quad]/h+\mathcal{O}(h^2) \\ &= [\tfrac{1}{12}f(x-2h)-\tfrac{2}{3}f(x-h)+0f(x)+\tfrac{2}{3}f(x+h)-\tfrac{1}{12}f(x+2h)]/h+\mathcal{O}(h^4) \\ &\quad \vdots \\ &\quad \text{etc.,} \end{aligned}$$

exact for all polynomials of degrees $2, 4, \ldots$ resp. It is convenient to collect weights like these as is done in Table 2 (for the kth derivative, divide with h^k).

Another equi-spaced case of interest is one-sided stencils which are often necessary use at boundaries. Table 3 shows some weights in this case. The weights for the first derivative are the ones that arise in backward differentiation formulae for ODEs, see Lambert (1991).

To explore the general properties of FD schemes (and, more importantly, to use such schemes), it is desirable to have a simple algorithm for the more general problem:

Given: $x_0, x_1, \ldots, x_n$: grid points (nonrepeated, otherwise arbitrary)

ξ: point $x = \xi$ at which the approximations are wanted (may, but need not be at a grid point)

m: highest order of derivative of interest

Find: weights $c_{i,j}^k$ such that the approximations

$$\left.\frac{\mathrm{d}^k f}{\mathrm{d}x^k}\right|_{x=\xi} \approx \sum_{j=0}^{i} c_{i,j}^k f(x_j), \quad k = 0, 1, \ldots, m, \ i = k, k+1, \ldots, n$$

are all optimal.

A short and fast algorithm for this was discovered only recently (Fornberg, 1988b, in more detail 1992):

Table 1. *Weights for some centred FD formulae on an equi-spaced grid: D, order of derivative; A, order of accuracy.*

		Approximations at $x = 0$				(x = coordinates at nodes)				
D	A	-4	-3	-2	-1	0	1	2	3	4
0	∞					1				
1	2				$-\frac{1}{2}$	0	$\frac{1}{2}$			
	4			$\frac{1}{12}$	$-\frac{2}{3}$	0	$\frac{1}{2}$	$-\frac{1}{12}$		
	6		$-\frac{1}{60}$	$\frac{3}{20}$	$-\frac{3}{4}$	0	$\frac{3}{4}$	$-\frac{3}{20}$	$\frac{1}{60}$	
	8	$\frac{1}{280}$	$-\frac{4}{105}$	$\frac{1}{5}$	$-\frac{4}{5}$	0	$\frac{4}{5}$	$-\frac{1}{5}$	$\frac{4}{105}$	$-\frac{1}{280}$
2	2				1	-2	1			
	4			$-\frac{1}{12}$	$\frac{4}{3}$	$-\frac{5}{2}$	$\frac{4}{3}$	$-\frac{1}{12}$		
	6		$\frac{1}{90}$	$-\frac{3}{20}$	$\frac{3}{2}$	$-\frac{49}{18}$	$\frac{3}{2}$	$-\frac{3}{20}$	$\frac{1}{90}$	
	8	$-\frac{1}{560}$	$\frac{8}{315}$	$-\frac{1}{5}$	$\frac{8}{5}$	$-\frac{205}{72}$	$\frac{8}{5}$	$-\frac{1}{5}$	$\frac{8}{315}$	$-\frac{1}{560}$
3	2			$-\frac{1}{2}$	1	0	-1	$\frac{1}{2}$		
	4		$\frac{1}{8}$	-1	$\frac{13}{8}$	0	$-\frac{13}{8}$	1	$-\frac{1}{8}$	
	6	$-\frac{7}{240}$	$\frac{3}{10}$	$-\frac{169}{120}$	$\frac{61}{30}$	0	$-\frac{61}{30}$	$\frac{169}{120}$	$-\frac{3}{10}$	$\frac{7}{240}$
4	2			1	-4	6	-4	1		
	4		$-\frac{1}{6}$	2	$-\frac{13}{2}$	$\frac{28}{3}$	$-\frac{13}{2}$	2	$-\frac{1}{6}$	
	6	$\frac{7}{240}$	$-\frac{2}{5}$	$-\frac{122}{15}$	$\frac{91}{8}$	$-\frac{122}{15}$	$\frac{169}{60}$	$-\frac{2}{5}$	$\frac{7}{240}$	

$c^0_{0,0} := 1, \ \alpha := 1$
for $i := 1$ to n
    $\beta := 1$
    for $j := 0$ to $i - 1$
        $\beta := \beta(x_i - x_j)$
        for $k := 0$ to $\min(i, m)$
            $c^k_{i,j} := ((x_i - \xi)c^k_{i-1,j} - kc^{k-1}_{i-1,j})/(x_i - x_j)$
    for $k := 0$ to $\min(i, m)$
        $c^k_{i,j} := \alpha(kc^{k-1}_{i-1,i-1} - (x_{i-1} - \xi)c^k_{i-1,i-1})/\beta$
    $\alpha := \beta$

Notes.

1 Any noninitialized quantity referred to is assumed to be equal to zero.
2 Only four operations are needed for each weight (to leading order; note that the subtractions $x_i - \xi$ and $x_i - x_j$ can be moved out of the innermost loop).

Table 2. *Weights for some one-sided FD formulae on an equi-spaced grid: D, order of derivative; A, order of accuracy.*

D	A	Approximations at $x = 0$ (x = coordinates at nodes)								
		0	1	2	3	4	5	6	7	8
0	∞	1								
1	1	-1	1							
	2	$-\frac{3}{2}$	2	$-\frac{1}{2}$						
	3	$-\frac{11}{6}$	3	$-\frac{3}{2}$	$\frac{1}{3}$					
	4	$-\frac{25}{12}$	4	-3	$\frac{4}{3}$	$-\frac{1}{4}$				
	5	$-\frac{137}{60}$	5	-5	$\frac{10}{3}$	$-\frac{5}{4}$	$\frac{1}{5}$			
	6	$-\frac{49}{20}$	6	$-\frac{15}{2}$	$\frac{20}{3}$	$-\frac{15}{4}$	$\frac{6}{5}$	$-\frac{1}{6}$		
	7	$-\frac{363}{140}$	7	$-\frac{21}{2}$	$\frac{35}{3}$	$-\frac{35}{4}$	$\frac{21}{5}$	$-\frac{7}{6}$	$\frac{1}{7}$	
	8	$-\frac{761}{280}$	8	-14	$\frac{56}{3}$	$-\frac{35}{2}$	$\frac{56}{5}$	$-\frac{14}{3}$	$\frac{8}{7}$	$-\frac{1}{8}$
2	1	1	-2	1						
	2	2	-5	4	-1					
	3	$\frac{35}{12}$	$-\frac{26}{3}$	$\frac{19}{2}$	$-\frac{14}{3}$	$\frac{11}{12}$				
	4	$\frac{15}{2}$	$-\frac{77}{6}$	$\frac{107}{6}$	-13	$\frac{61}{12}$	$-\frac{5}{6}$			
	5	$\frac{203}{45}$	$-\frac{87}{5}$	$\frac{117}{4}$	$-\frac{254}{9}$	$\frac{33}{2}$	$-\frac{27}{5}$	$\frac{137}{180}$		
	6	$\frac{469}{90}$	$-\frac{223}{10}$	$\frac{879}{20}$	$-\frac{949}{18}$	41	$-\frac{201}{10}$	$\frac{1019}{180}$	$-\frac{7}{10}$	
	7	$\frac{29531}{5040}$	$-\frac{962}{35}$	$\frac{621}{10}$	$-\frac{4006}{45}$	$\frac{691}{8}$	$-\frac{282}{5}$	$\frac{2143}{90}$	$-\frac{206}{35}$	$\frac{363}{560}$
3	1	-1	3	-3	1					
	2	$-\frac{5}{2}$	9	-12	7	$-\frac{3}{2}$				
	3	$-\frac{17}{4}$	$\frac{71}{4}$	$-\frac{59}{2}$	$\frac{49}{2}$	$-\frac{41}{4}$	$\frac{7}{4}$			
	4	$-\frac{49}{8}$	29	$-\frac{461}{8}$	62	$-\frac{307}{8}$	13	$-\frac{15}{8}$		
	5	$-\frac{967}{120}$	$\frac{638}{15}$	$-\frac{3929}{40}$	$\frac{389}{3}$	$-\frac{2545}{24}$	$\frac{268}{5}$	$-\frac{1849}{120}$	$\frac{29}{15}$	
	6	$-\frac{801}{80}$	$\frac{349}{6}$	$-\frac{18353}{120}$	$\frac{2391}{10}$	$-\frac{1457}{6}$	$\frac{4891}{30}$	$-\frac{561}{8}$	$\frac{527}{30}$	$-\frac{469}{240}$
4	1	1	-4	6	-4	1				
	2	3	-14	26	-24	11	-2			
	3	$\frac{35}{6}$	-31	$\frac{137}{2}$	$-\frac{242}{3}$	$\frac{107}{2}$	-19	$\frac{17}{6}$		
	4	$\frac{28}{3}$	$-\frac{111}{2}$	142	$-\frac{1219}{6}$	176	$-\frac{185}{2}$	$\frac{82}{3}$	$-\frac{7}{2}$	
	5	$\frac{1069}{80}$	$-\frac{1316}{15}$	$\frac{15289}{60}$	$-\frac{2144}{5}$	$\frac{10993}{24}$	$-\frac{4772}{15}$	$\frac{2803}{20}$	$-\frac{536}{15}$	$\frac{967}{240}$

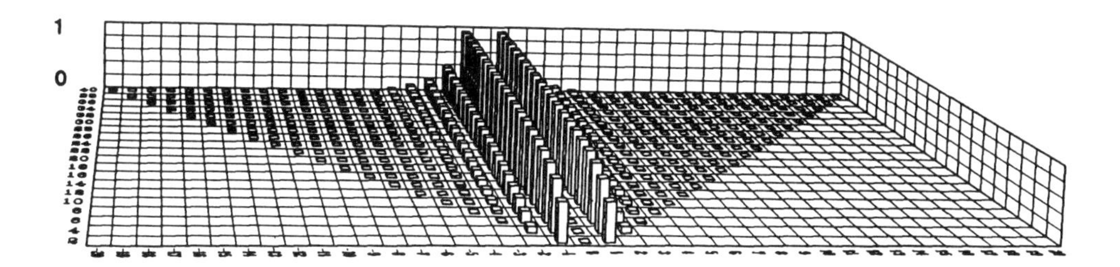

Fig. 2. Magnitude of weights for centred approximations to the first derivative on an equi-spaced grid (cf. Table 2).

3 The calculation of the weights is numerically stable (however, especially in the case of high derivatives, *applying* FD weights to a function may lead to severe cancellations and loss of significant digits).

4 The special case $m = 0$ offers the fastest way known to perform polynomial interpolation at a single point (in particular, significantly faster than the classical algorithms by Aitken and Neville).

5 If we are only interested in the weights for the stencils based on all the grid points $x_j, j = 0, 1, \ldots, n$ (and not in the lower-order stencils based on fewer points), we can omit the first of the two subscripts for c (i.e. it suffices to declare a two-dimensional array to hold $c_{0 \text{ to } n}^{0 \text{ to } m}$ – the 'overwriting' that will occur internally in the algorithm will be safe).

A Fortran code (with test driver) for this algorithm is given in Appendix 2.

3.2. Growth rates of FD weights on equi-spaced grids

Figures 2 and 3 illustrate how the magnitudes of the weights for the first derivative grow with increasing orders of accuracy (cf. Tables 2 and 3).

In the centred case, approximations of increasing orders of accuracy converge to a limit method of formally infinite-order of accuracy. For the first derivative ($m = 1$), this can be seen directly from the closed form expression for the weights $c_{p,j}^1$ (p (even) = order of accuracy, j = x-position of weight):

$$c_{p,j}^1 = \begin{cases} \dfrac{(-1)^{j+1}(\frac{1}{2}p)!^2}{j(\frac{1}{2}p+j)!(\frac{1}{2}p-j)!} & j = \pm 1, \pm 2, \ldots, \pm \frac{1}{2}p \\ 0 & j = 0. \end{cases}$$

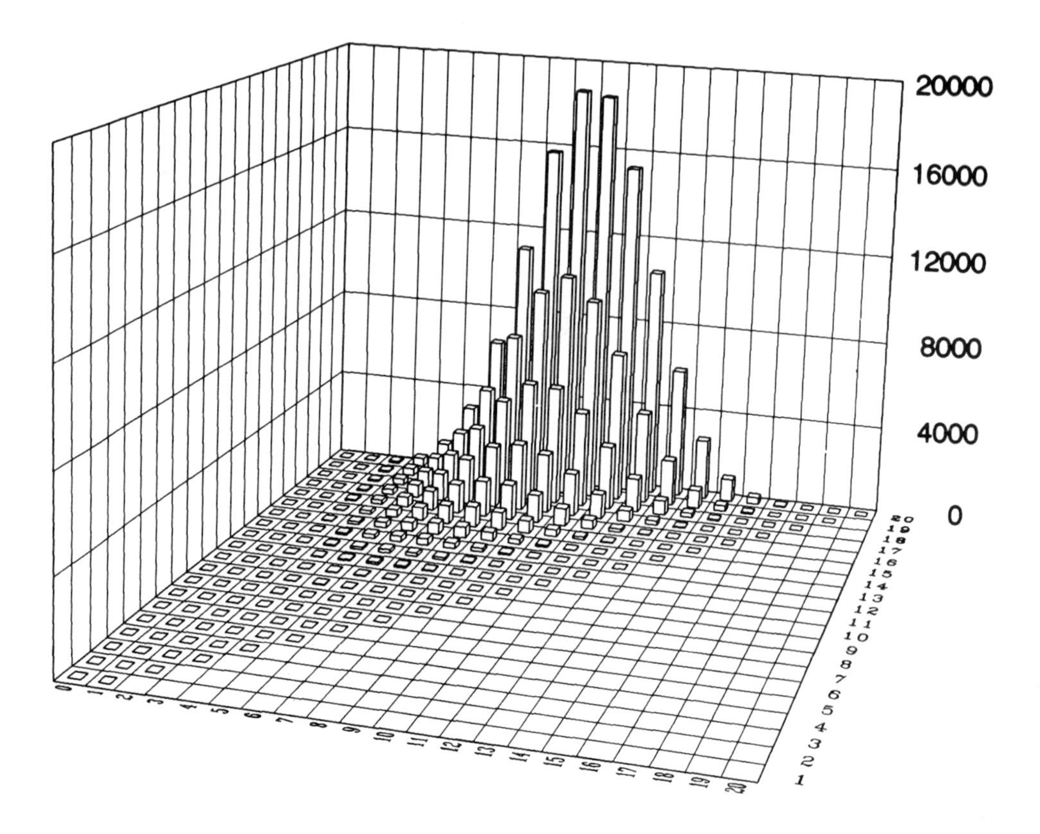

Fig. 3. Magnitude of weights for one-sided approximations to the first derivative on an equi-spaced grid (cf. Table 3).

Clearly, the limit for $p \to \infty$ exists and

$$c^1_{\infty,j} = \begin{cases} \dfrac{(-1)^{j+1}}{j} & j = \pm 1, \pm 2, \ldots \\ 0 & j = 0. \end{cases} \tag{3.1}$$

Beyond the second derivative, for which

$$c^2_{p,j} = 2c^1_{p,j}/j, \quad j = \pm 1, \pm 2, \ldots, \pm \tfrac{1}{2}p, \quad c^2_{p,0} = -2\sum_{i=1}^{p/2} 1/i^2,$$

closed form expressions for weights become very complicated. However, that does not affect the ease with which they can be calculated (using the algorithm in Section 3.1) or the existence of simple limits. For the second derivative, the limit becomes

$$c^2_{\infty,j} = \begin{cases} \dfrac{2(-1)^{j+1}}{j^2} & j = \pm 1, \pm 2, \ldots \\ -\tfrac{1}{3}\pi^2 & j = 0. \end{cases}$$

For higher derivatives the decay rates alternate between $\mathcal{O}(1/j)$ and $\mathcal{O}(1/j^2)$ for odd and even derivatives respectively (exact formulae for

$$c^m_{\infty,j}, \quad m = 1, 2, \ldots$$

are given in Fornberg (1990a)).

The situation is very different for one-sided approximations. The closed form expression for the first derivative is

$$c^1_{p,j} = \begin{cases} \dfrac{(-1)^{j+1}}{j}\begin{pmatrix} p \\ j \end{pmatrix} & j = 1, 2, \ldots, p \\ -\sum_{i=1}^{p} 1/i & j = 0. \end{cases}$$

The magnitudes of the weights form (nearly) a Gaussian distribution, which becomes increasingly peaked at the centre of the stencil while growing in height exponentially with p ($\sim \pi^{-1/2}p^{-3/2}2^{p+3/2}$). For higher derivatives, the general character and growth rates remain similar. Partly one-sided approximations initially grow more slowly but will also ultimately diverge exponentially. In the case of the first derivative (just described), the asymptotic rate is multiplied by a factor of $s!/p^s$ if the derivative is evaluated s steps in from the boundary.

3.3. Generalized node distributions

The previous discussion about the size of the weights in centred versus one-sided FD formulae suggests, for a nonperiodic problem, that the nodes need to be concentrated towards the ends of the interval (to offset the loss of accuracy which would otherwise arise because of the very large weights).

Let $\mu(x)$ denote how the *density* of a node distribution varies over $[-1, 1]$ (i.e. the distance between adjacent nodes decreases like $1/n\mu(x)$ as n increases. To prevent changes in $\mu(x)$ affecting the total number of nodes that fit into $[-1, 1]$, we require $\int_{-1}^{1} \mu(x)\,\mathrm{d}x = 1$. A one-parameter family

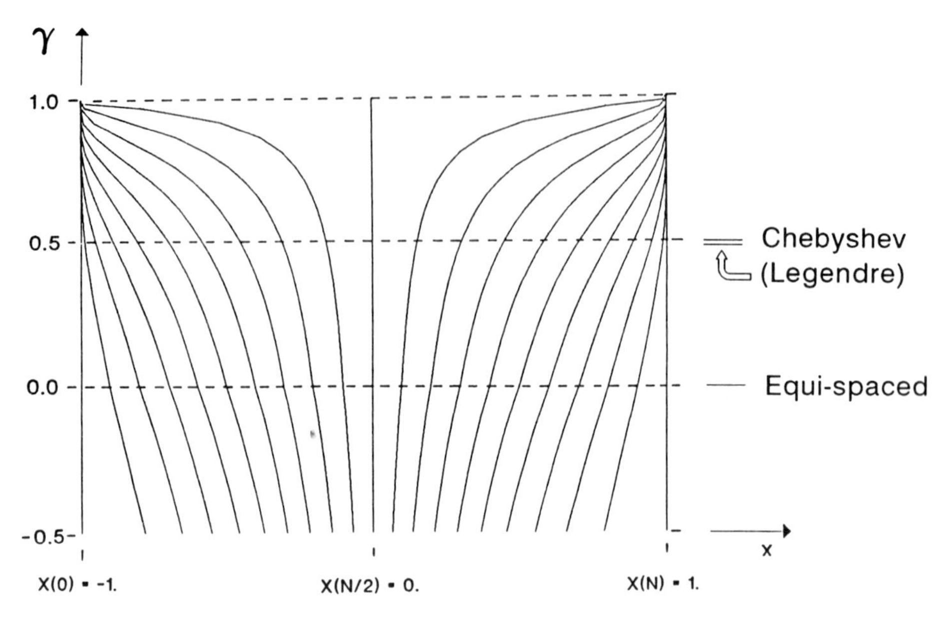

Fig. 4. Distribution of nodes corresponding to density function $\mu_\gamma(x)$, shown for $\gamma \in [-0.5, 1]$ ($n = 20$). The Legendre distribution (of extrema) is not obtained exactly for any γ, but $\gamma \approx 0.4785$ gives the closest (least-squares) fit in this case of $n = 20$. This difference (to 0.5 – it vanishes as $n \to \infty$) is illustrated at the right edge of the figure.

of density functions is outlined below, followed by two special cases that it incorporates:

Density function	Node locations $x_j,\ j = 0, 1, \ldots, n$	Comments
$\mu_\gamma(x) = c_\gamma/(1 - x^2)^\gamma$	$j/n = \int_{-1}^{x_j} \mu_\gamma(x)\,dx$	$\gamma < 1$; $c_\gamma = \dfrac{\Gamma(\frac{3}{2} - \gamma)}{\pi^{1/2}\Gamma(1 - \gamma)}$
$\mu_0(x) \equiv \frac{1}{2}$	$x_j = -1 + 2j/n$	$\gamma = 0$; equi-spaced
$\mu_{1/2}(x) = 1/\pi\sqrt{1 - x^2}$	$x_j = -\cos(\pi j/n)$	$\gamma = \frac{1}{2}$; Chebyshev

Figure 4 shows how the nodes move as a function of γ. One key question is whether quadratic clustering (the case $\gamma = 0.5$) is necessary. Several general arguments suggest this in the limit of $n \to \infty$ (with the FD approach to PS methods, the effects of other clusterings can be explored in special cases – see Section 5):

All classical orthogonal polynomials feature quadratic node clustering at the ends. Changing α and β in the weight function $(1 - x)^\alpha(1 - x)^\beta$ for Jacobi polynomials will still leave the nodes quadratically clustered

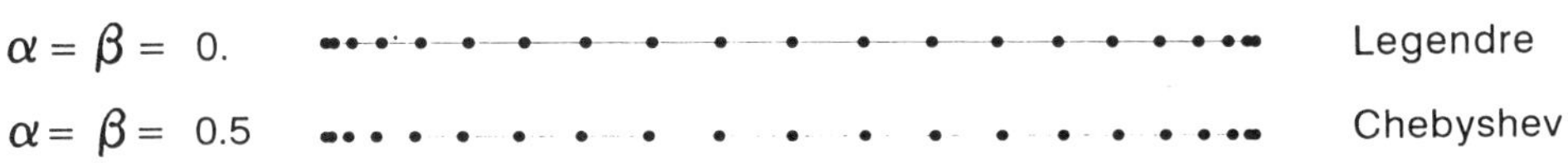

$\alpha = \beta =$ 0. Legendre

$\alpha = \beta =$ 0.5 Chebyshev

Fig. 5. Difference between the location of extrema for Legendre and Chebyshev polynomials ($n = 20$).

(this follows, for example, from their differential equation, see Table 1). Figure 5 compares the nodes (extrema) for Legendre and Chebyshev polynomials of order 20 (corresponding to $\alpha = \beta = 0$ and $\alpha = \beta = \frac{1}{2}$) – there is hardly any noticeable difference. Figure 4 also illustrates this, i.e. how small an effect changing $\alpha = \beta$ has compared with changes in γ.

To get the least possible interpolation error, the nodes must cluster as in the Chebyshev case. A heuristic argument for this goes as follows: let $p_n(x)$ be the interpolation polynomial of degree n to $f(x)$ on $[-1, 1]$. The remainder term is

$$f(x) - p_n(x) = \frac{1}{(n+1)!} f^{(n+1)}(\xi) \prod_{j=0}^{n} (x - x_j)$$

for some $\xi \in [-1, 1]$. The only part that can be controlled by re-positioning the nodes x_j is the product. Since the highest order term is $1 \cdot x^{n+1}$, the question becomes: which polynomial of that form stays smallest over $[-1, 1]$? This is a well known property of Chebyshev polynomials.

With any other type of clustering, convergence will require f to be analytic in some domain away from the interval $[-1, 1]$. In a complex $z = x + iy$ plane, a Taylor series converges in the *largest circle* around the expansion point that is free of singularities. This result generalizes to interpolating polynomials (when the nodes are distributed over an interval rather than all lumped at one point) as follows (Krylov (1962), Ch. 12, Markushevich (1967); general results on polynomial interpolation can further be found in Walsh (1960), Davis (1975), Gaier (1987) etc.).

Given a node density $\mu(x)$ (on $[-1, 1]$), form the potential function

$$\phi(z) = -\int_{-1}^{1} \mu(x) \log |z - x| \, dx \quad + \text{constant}. \tag{3.2}$$

Then $p_n(z)$ converges to $f(z)$ inside the *largest equi-potential curve* that does not enclose any singularity of $f(z)$ (and diverges outside it).

Figure 6(a) shows (as a thick straight line above the x-axis, at centre of the figure) the graph of $\mu_0(x) \equiv \frac{1}{2}$ and a matching equi-spaced set of nodes on the x-interval $[-1, 1]$ (barely visible). The heavy contour line on the

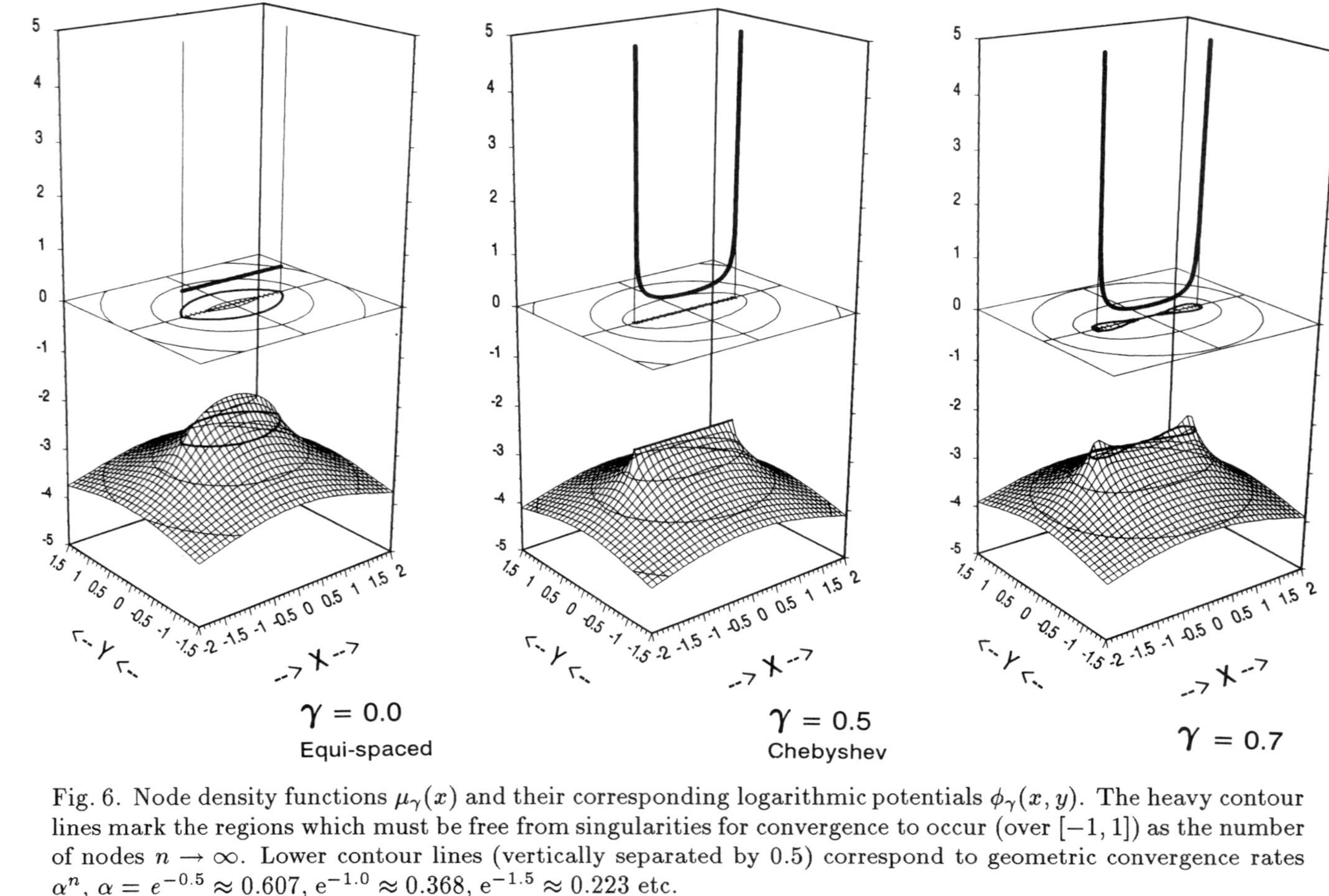

Fig. 6. Node density functions $\mu_\gamma(x)$ and their corresponding logarithmic potentials $\phi_\gamma(x, y)$. The heavy contour lines mark the regions which must be free from singularities for convergence to occur (over $[-1, 1]$) as the number of nodes $n \to \infty$. Lower contour lines (vertically separated by 0.5) correspond to geometric convergence rates α^n, $\alpha = e^{-0.5} \approx 0.607$, $\mathrm{e}^{-1.0} \approx 0.368$, $\mathrm{e}^{-1.5} \approx 0.223$ etc.

potential surface

$$\phi(z) = \tfrac{1}{2}\mathrm{Re}[(1-z)\log(1-z) - (-1-z)\log(-1-z)] + C$$

surrounds the smallest domain that includes $[-1,1]$ and is bounded by an equipotential curve. The function $f(z)$ must therefore be analytic everywhere within this domain for convergence to occur on $[-1,1]$. Any singularity within this domain restricts convergence to a still smaller equi-potential region, leading to the 'Runge phenomenon' – divergence near the ends of the interval.

In the Jacobi polynomial case

$$\mu(x) = 1/(\pi\sqrt{1-x^2}).$$

Equation (3.2) can then again be evaluated in closed form:

$$\phi(z) = -|\log|z + \sqrt{z^2-1}|| + C$$

(like the previous formula, correct for all complex values of z when selecting the conventional branches for the logarithm and square root functions).

Figure 6(b) shows how the potential surface $\phi(z)$ forms a perfectly flat ridge of the potential surface along $[-1,1]$. This clearly looks optimal – the only possibility that convergence to $f(z)$ on $[-1,1]$ does not require $f(z)$ to be analytic anywhere off the interval $[-1,1]$.

Finally, Figure 6(c) shows what happens when the nodes cluster still more densely towards the ends ($\gamma = 0.7$) – convergence again requires analyticity outside the interval.

3.4. Example of a differentiation matrix

We consider the same situation as in Section 2.3 – the first derivative approximated on a periodic, equi-spaced grid. Instead of using trigonometric interpolation, we employ the limiting FD formula (3.1) – of infinite width, but possible to apply since periodic data can be repeated indefinitely. Assuming $N = 2M+1$ and adjusting the weights for a mesh spacing of $h = 2/N$ (rather than $h = 1$ as in (3.1)), we get

$$c^1_{\infty,j} = \begin{cases} \dfrac{N}{2}\dfrac{(-1)^{j+1}}{j} & j \neq 0 \\ 0 & j = 0. \end{cases}$$

As Figure 7 illustrates, period-wide sections of the stencil can be added together to create an equivalent stencil covering only one period of the data. Its weights become

$$d^1_{\infty,j} = \begin{cases} \dfrac{N}{2}(-1)^{j+1}\displaystyle\sum_{k=-\infty}^{\infty}\frac{(-1)^k}{j+Nk} = \frac{1}{2}\sum_{k=-\infty}^{\infty}\frac{(-1)^k}{k+(j/N)} & \begin{matrix} j=\pm1,\pm2,\dots,\pm M \\ (\pm(M+1),\dots,\pm(N-1)) \end{matrix} \\ 0 & j = 0. \end{cases}$$

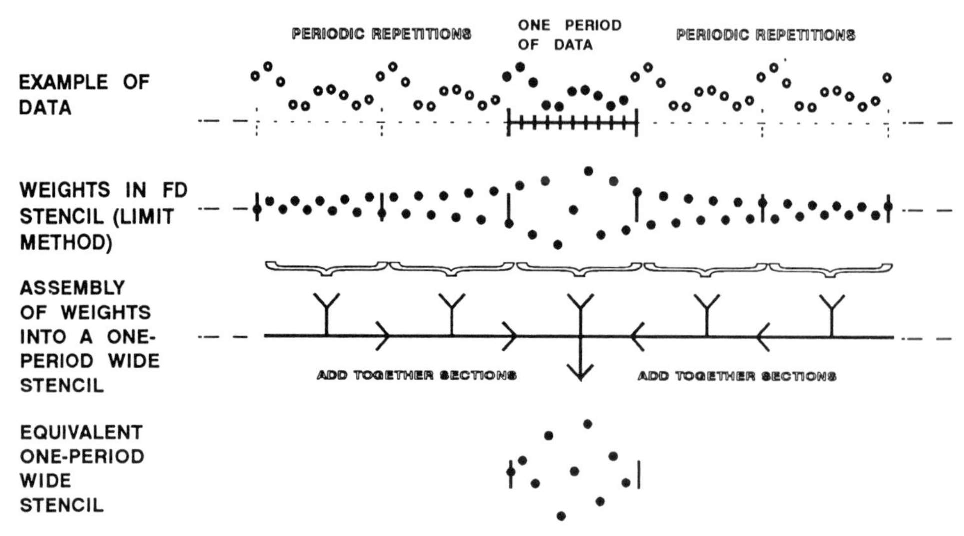

Fig. 6. Application of the limiting FD method to periodic data.

The DM is cyclic: its i, jth element is $d^1_{\infty,i-j}$. Noting the identity

$$\sum_{k=-\infty}^{\infty} \frac{(-1)^k}{k+x} = \frac{\pi}{\sin \pi x},$$

we get

$$D_{i,j} = \begin{cases} \dfrac{\pi(-1)^{i-j}}{2\sin(\pi(i-j)/N)} & i \neq j \\ 0 & i = j. \end{cases} \tag{3.3}$$

3.5. Equivalence of PS methods and limits of FD methods

Periodic case The DMs derived in Sections 2.3 and 3.4 are identical – hence the two methods are equivalent. With only little additional effort this can be shown to generalize to derivatives of any order, to even numbers of points, to 'staggered grids' (a topic discussed in Section 5.3) etc.; for details, see Fornberg (1990a).

Nonperiodic case No periodic data extensions are now available. The order of accuracy for the approximations corresponds to the number of grid points (rather than being formally infinite). The PS method now turns out to be equivalent to using the FD approximations whose stencils extend over all the grid points. This can be seen as follows.

PS approach Consider data given at $n+1$ points on $[-1, 1]$ (distributed, for example, according to the zeros or extrema of some orthogonal polynomial – as is customary in PS collocation; however, their distribution is irrelevant for our present argument.) By means of expansion in these polynomials, the PS method provides the exact derivative of the interpolation polynomial going through the data at these points.

FD approach With no periodic data extensions, we can, at best, consider FD stencils which are as wide as the grid is wide. To approximate derivatives at the grid points, the FD weights have to be calculated separately for each point. Every one of these approximations will be exact for any nth-degree polynomial. In particular, they are all exact for the interpolating polynomial.

For any given data and distribution of (distinct) nodes, the interpolating polynomial of minimal degree is unique. Since both approaches give the exact results for this polynomial, they will always give the same results – hence, the approaches are equivalent.

4. Key properties of PS approximations

In the previous sections, we have repeatedly referred to the exponential convergence rate of spectral methods for analytic functions. This is discussed in more detail in Section 4.1. When functions are not smooth, PS theory is much less clear. An approximation can appear very good in one error norm and, at the same time, very bad in another. As illustrated in Section 4.2, PS performance can still be very impressive – this is exploited in the major PS applications. The concluding Sections 4.3–4.5 deal with implementation issues; primarily differentiation matrices and their influence on time stepping procedures.

4.1. Convergence of PS methods for smooth functions

Nonperiodic case Polynomial interpolation of smooth functions based on the Chebyshev nodes (as well as expansions in Chebyshev polynomials) are well known to provide approximations with nearly uniform accuracy over $[-1, 1]$ whereas interpolation based on equi-spaced points can diverge near the ends (the 'Runge phenomenon'). The potential functions described in Section 3.3 provide a general tool for addressing issues like these – when and with what rates convergence will occur.

In the Chebyshev case ($\gamma = 0.5$), the relationship between the potential contours and the convergence rates becomes particularly simple. For a convergence rate α^N, $\alpha \in (0,1)$, on $x \in [-1,1]$, the nearest singularity is located on the ellipse

$$\frac{x^2}{(\frac{1}{2}(\alpha + 1/\alpha))^2} + \frac{y^2}{(\frac{1}{2}(\alpha - 1/\alpha))^2} = 1, \tag{4.1}$$

(with foci at ± 1). Conversely, given the location of the nearest singularity, this equation gives the convergence factor α. The derivation of (4.1) requires no potential considerations – it follows from the form of Lagrange's interpolation formula and requires no potential considerations – it follows from the form of Lagrange's interpolation formula and noting that

$$T_n(x) = \tfrac{1}{2}(z^n + 1/z^n),$$

where

$$\tfrac{1}{2}(z + 1/z) = x.$$

If we here consider x and z as complex variables, the ellipses (4.1) in the x-plane correspond to circles in the z-plane, centred at the origin and with radii $1/\alpha$.

The fundamental result about exponential convergence rates given in the caption to Figure 6 is valid for any node distribution functions – not just those defined through the parameter γ.

To use these analytic results to illustrate how convergence depends on the smoothness of a function $f(x)$ (f and x now real), let us consider the two-parameter class of functions

$$f_{\xi,\eta}(x) = \frac{1}{1 + ((x-\xi)/\eta)^2}.$$

Graphically, these functions have a 'hump' of unit height, centred at $x = \xi$, with widths (and radius of curvature at the tips) proportional to η. The equi-spaced PS method will be just borderline converging/diverging when the closest singularity of $f_{\xi,\eta}(x)$ (it has only two singularities – they are located at $x = \xi + i\eta$) falls on the equipotential curve passing through $x = \pm 1$ (drawn bold in Figure 6(a)). Figure 8(a) shows these most peaked functions$f_{\xi,\eta}(x)$ for different values of $\xi \in [-1,1]$. The equi-spaced PS method is clearly much better able to resolve high curvatures near the ends of $[-1,1]$ than in the interior.

This issue was also discussed by Solomonoff and Turkel (1989). They quote the closed form expression for $\phi_0(x,y)$, but select an inappropriate branch for an arctan which arises, leading to some flawed results.

However, it is also equally clear that (in this form) the equi-spaced non-periodic PS method is quite useless – no matter how many points are employed, it will not converge over $[-1,1]$ for functions any less smooth than those shown in Figure 8(a).

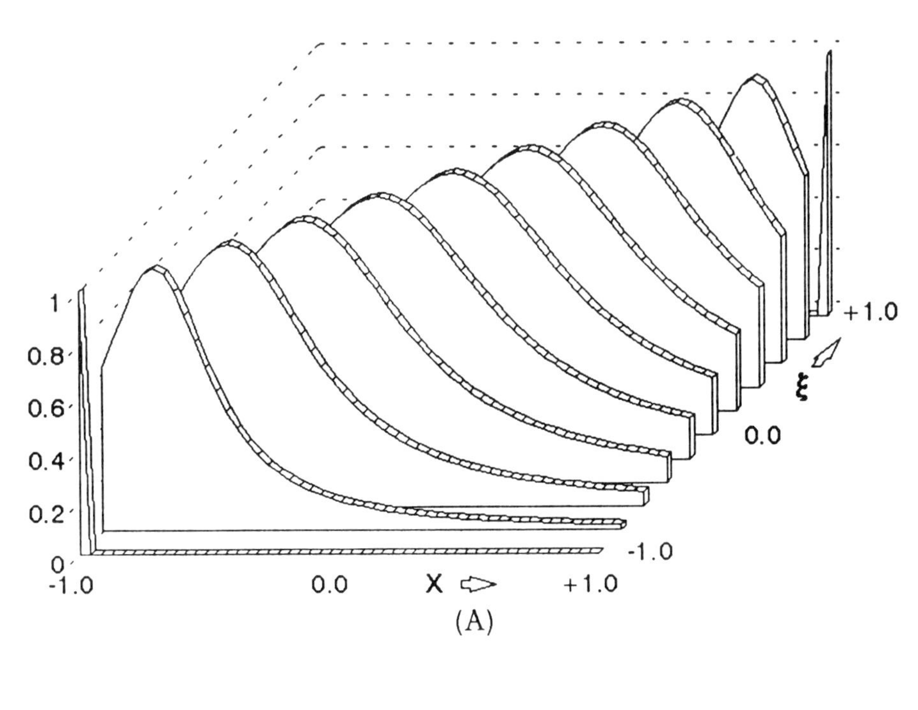

(A)

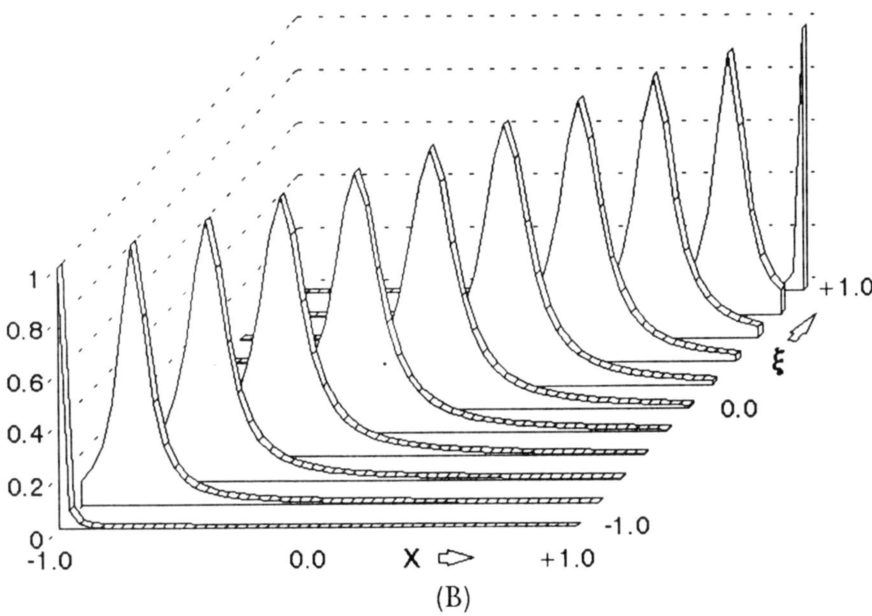

(B)

Fig. 8. Functions $f_{\xi,\eta}(x) = 1/\{1 + [(x - \xi)/\eta]^2\}$ with minimal η (i.e. maximal curvature at the tip) such that for the equi-spaced nonperiodic PS method, there is borderline convergence/divergence as $n \to \infty$ (a), and for the Chebyshev PS method there is convergence as α^n for $\alpha = 0.9$ (b).

Interpolation at the Chebyshev nodes will work for any ξ and any $\eta > 0$. However, if we require a 'reasonable' convergence rate, say $\alpha = 0.9$ (i.e. approximately a factor of 10^{-6} for every 130 node points), the situation is again somewhat similar, see Figure 8(b). Once more, the highest resolution is obtained near the boundaries. This has been exploited frequently (for example to resolve boundary layers in fluid mechanics). Note, however, that this is *not* an immediate consequence of the grid being finer there (this effect was no less prominent in the case of equi-spaced grids).

Periodic case For the periodic PS method (on $[-1, 1]$), the formula corresponding to (4.1) becomes

$$y = \pm \frac{2}{\pi} \ln \alpha \tag{4.2}$$

(related to the fact that a Fourier series converges in the widest horizontal strip around the x-axis that is free from singularities).

General discussion The relative resolution ability of different methods is sometimes expressed in the number of points needed per wavelength. For a Fourier expansion, this number is 2 (Kreiss and Oliger, 1972). For nonperiodic PS methods, it is π in the Chebyshev case (Gottlieb and Orszag, 1977) and 6 in the equi-spaced case (Weideman and Trefethen, 1988).

Figure 9 compares the curves given by (4.1) and (4.2) for $\alpha = 0.5$ and $\alpha = 0.9$. The ratio $2/\pi$ between points per wavelength for the periodic and Chebyshev PS methods follows from the ratio of the y-axis intercepts (values at $x = 0$) of these curves as $\alpha \to 1$. (Again, this is *not* a direct consequence of the fact that the Chebyshev grid happens to be $2/\pi$ as dense as the equi-spaced one at this location.)

In every instance of PS methods applied to functions analytic in some neighbourhood of $[-1, 1]$, the convergence takes the form $\mathcal{O}(\alpha^n)$ (discussed so far only for interpolation but clearly true as well – with the same α – for approximations to any derivative). This rate distinguishes spectral methods from FD and FE methods (where the rate for a pth-order method would be $\mathcal{O}(1/n^p)$; polynomial rather than exponential convergence). Whether or not there happens to be any classical family of orthogonal polynomials associated with the PS method is quite irrelevant.

We can only indicate one starting point here for analysis comparing FD against PS methods. Assume periodicity and $N + 1$ grid points spaced $h = 2/N$ apart within the period $[-1, 1]$. The range of Fourier modes $e^{i\omega x}$ that can be represented on this grid is $-\omega_{\max} \le \omega \le \omega_{\max}$, where $\omega_{\max} = \pi/h$.

Any mode ω outside $[-\omega_{\max}, \omega_{\max}]$ will, on the grid, appear equivalent to a mode within this range – an 'aliasing' error. How much is present of the different modes depends on the regularity of the function we approximate.

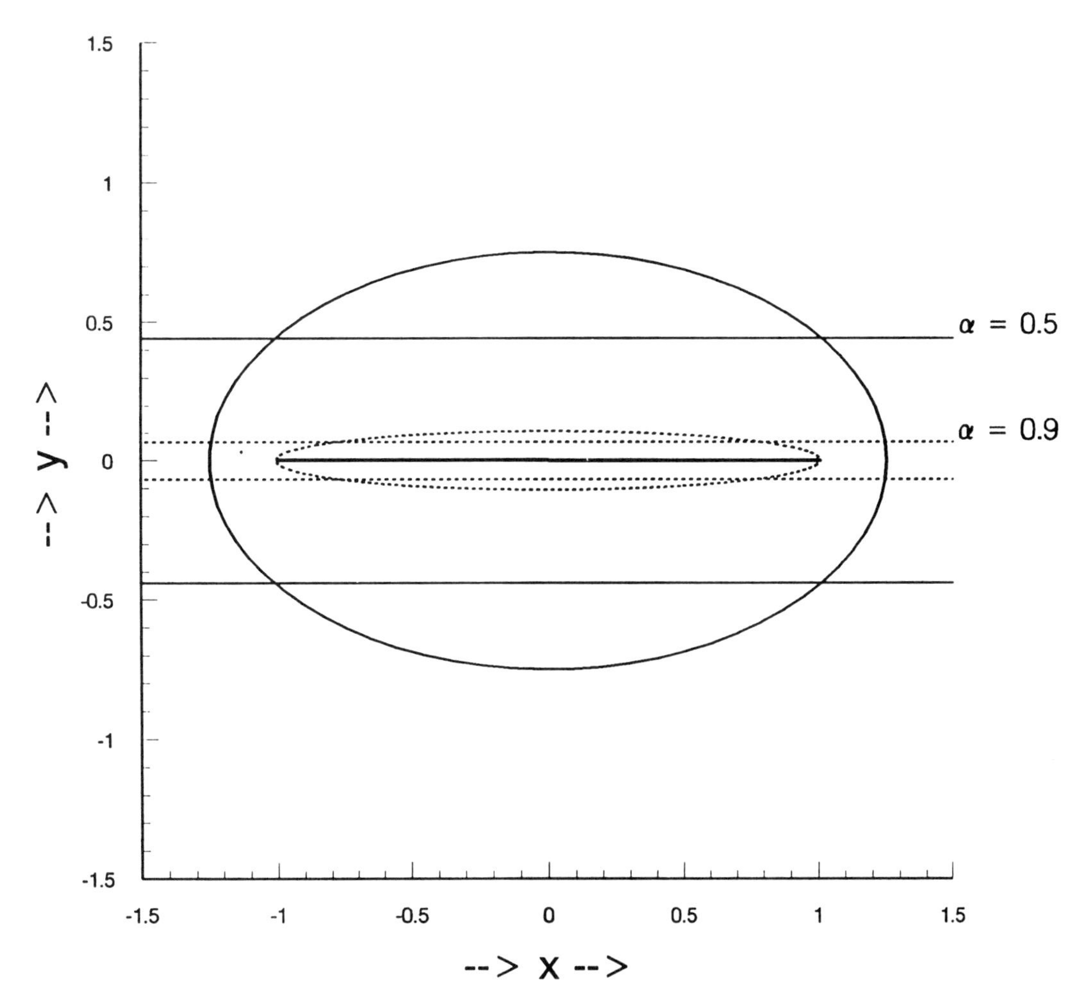

Fig. 9. Comparisons of domains in complex x-plane which need to be free of singularities to obtain convergence rates α^n, $\alpha = 0.5$ and $\alpha = 0.9$. Chebyshev method: ellipses with foci 1; equation (4.1). Periodic PS method: horizontal strips around the x-axis; equation (4.2).

Suppose we want to approximate $\mathrm{d}/\mathrm{d}x$. For a mode $\mathrm{e}^{\mathrm{i}\omega x}$, the exact answer should be

$$\frac{\mathrm{d}}{\mathrm{d}x}\mathrm{e}^{\mathrm{i}\omega x} = \mathrm{i}\omega\mathrm{e}^{\mathrm{i}\omega x}.$$

With FD2, we get

$$D_2\mathrm{e}^{\mathrm{i}\omega x} = \frac{\mathrm{e}^{\mathrm{i}\omega(x+h)} - \mathrm{e}^{\mathrm{i}\omega(x-h)}}{2h} = \mathrm{i}\frac{\sin \omega h}{h}\mathrm{e}^{\mathrm{i}\omega x} = \mathrm{i}f(2,\omega,h)\mathrm{e}^{\mathrm{i}\omega x}. \tag{4.3}$$

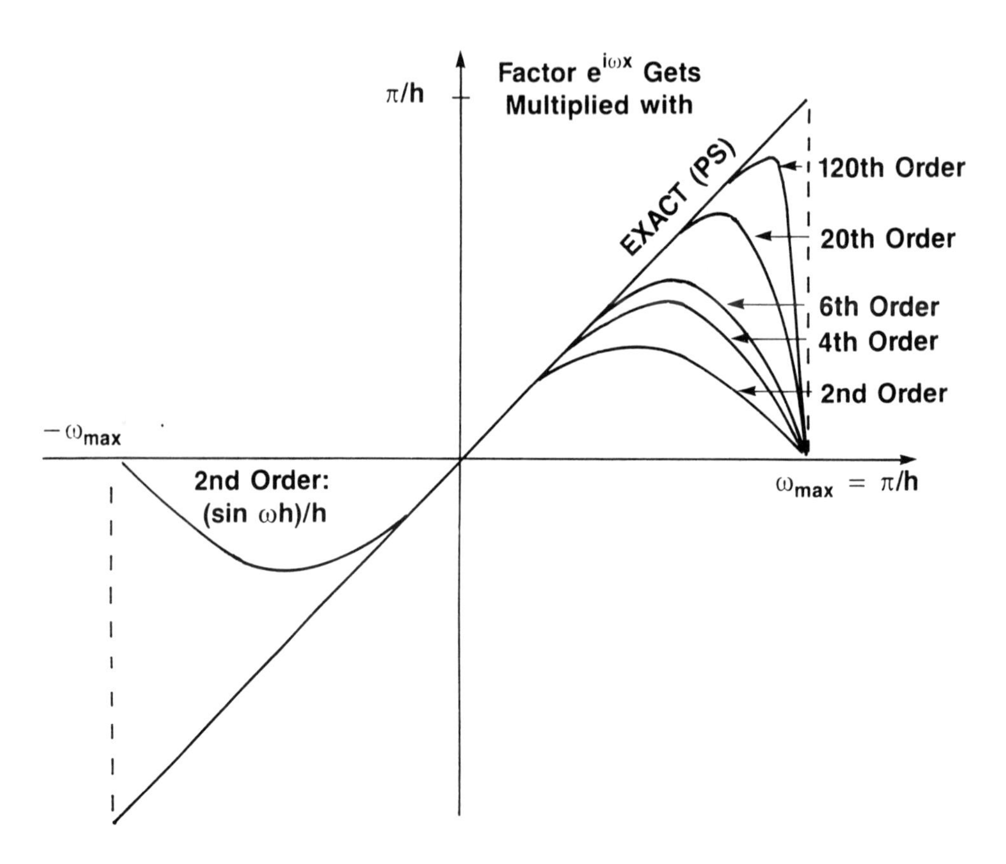

Fig. 10. Multiplicative factors $f(p, \omega, h)$ arising when the pth-order FD approximation for $\mathrm{d}/\mathrm{d}x$ is applied to $\mathrm{e}^{\mathrm{i}\omega x}$.

For the centred pth-order FD scheme ($p = 2, 4, 6, \ldots$), we get similarly

$$f(p, \omega, h) = \left\{ \frac{\sin \omega h}{h} \sum_{k=0}^{\frac{1}{2}p-1} \frac{(k!)^2}{(2k+1)!} 2^{2k} (\sin \tfrac{1}{2}\omega h)^{2k} \right\}. \tag{4.4}$$

Figure 10 compares $f(p, \omega, h)$ to the exact result ω. For $p = 2$, only a fraction of the Fourier modes present are treated even nearly correctly.

As $p \to \infty$, convergence is seen to occur as in a Taylor expansion – the number of correct derivatives at the origin is the same as the order of the FD scheme. It can make sense to give up some of the (unnecessarily high) accuracy for low ω (i.e. for long waves) in exchange for keeping the accuracy within some uniform tolerance over a wider ω range (or over a specific narrow frequency band relevant to a particular application). Such types of compact FD schemes can be very effective, for example in three-dimensional seismic modelling, see Holberg (1987), Mittet *et al.* (1988) and Kindelan *et al.* (1990).

Solomonoff (1994) presents still another approach to generate FD schemes that are optimized in application-specific ways (i.e. rather than being designed to be exact for polynomials of as high orders as possible). He notes that such schemes can be made less vulnerable to the Runge phenomenon.

Lele (1992) displays figures similar to Figure 10, also including various compact schemes that attain high orders by means of including additional unknowns at the grid points (for example the values of derivatives as well as function values).

In the $p = \infty$ limit (the periodic PS method), the only errors are 'aliasing' errors. Due to variable coefficients and/or nonlinearities, high Fourier modes outside the range $[-\omega_{\max}, \omega_{\max}]$ are generated and then possibly mistreated. One approach to controlling such errors is to apply weak damping (dissipation). However, as we will see, anybody who views 'aliasing' only as a source of errors is missing out on one of the most important (and intriguing) strengths of PS methods.

4.2. Convergence of PS methods for nonsmooth functions

The PS method sometimes performs very well even in the cases of nonsmooth functions. Several of the major PS applications depend on this (turbulence modelling, weather forecasting, seismic modelling etc.).

As an illustration, let us consider the one-dimensional acoustic wave equation

$$\left\{ \begin{array}{l} u_t = v_x \\ v_t = c^2(x) u_x \end{array} \right. \qquad \text{where } c(x) = \left\{ \begin{array}{ll} 1 & -1 < x < 0, \\ \frac{1}{2} & 0 < x < 1, \end{array} \right. \tag{4.5}$$

periodic outside $[-1, 1]$.

We will be using grids such that 0 and ± 1 fall half-way between grid points, thus saving us from having to decide on the values of $c(x)$ at these locations.

In each of the intervals $[-1, 0]$ and $[0, 1]$, equation (4.5) supports solutions travelling to the right ($\Rightarrow$) and to the left ($\Leftarrow$) of the following forms:

$$\text{In } [-1,0] \left\{ \begin{array}{ll} u(x,t) = -v(x,t) & \Rightarrow \\ u(x,t) = +v(x,t) & \Leftarrow \end{array} \right. \quad \text{in } [0,1] \left\{ \begin{array}{ll} u(x,t) = -2v(x,t) & \Rightarrow \\ u(x,t) = +2v(x,t) & \Leftarrow . \end{array} \right.$$

We choose as initial condition

$$u(x, 0) = 2v(x, 0) = \exp(-1600(x - \tfrac{1}{4})^2).$$

Figure 11 shows the time evolution for $u(x, t)$ – the one for $v(x, t)$ is qualitatively similar. After the pulses have hit the interfaces at $x = 0$ and $x = 1$ numerous times – on each occasion generating two outgoing pulses (trans-

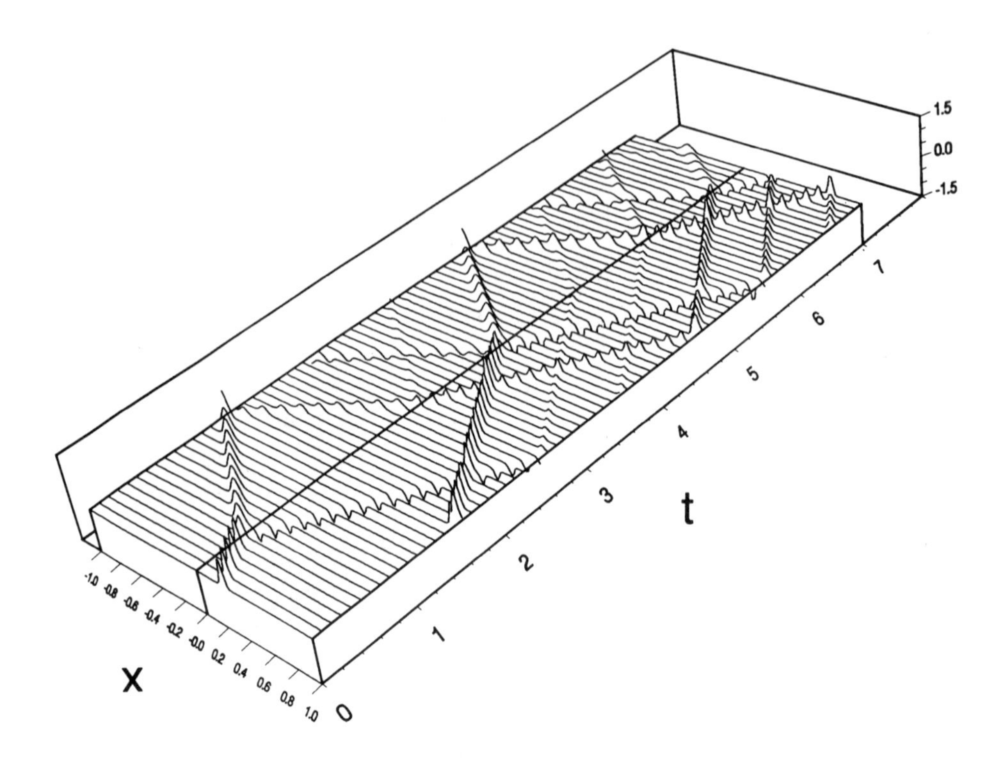

Fig. 11. Time evolution of $u(x,t)$ solving equation (4.5).

mitted and reflected) – the analytic solution at $t = 7$ consists of just three pulses.

The periodic second- and fourth-order FD and the PS methods give at $t = 7$ the results shown in Figure 12.

No numerical smoothing has been applied in any of these cases. The time integration was performed with leap-frog (centred second-order FD in time) with a sufficiently small time step that the errors which are seen are all due to the spatial discretizations. Many other ODE solvers could have been used equally well (such as Runge–Kutta, Adams Bashforth etc.).

We can note the following points.

- Already with $N = 64$, the PS method has retained considerable accuracy (in spite of the initial pulse being only about two grid points wide). For the higher values of N, the performance of the PS method is nearly flawless, and far superior to that of the FD methods.
- One might have expected the PS method to develop Gibbs-type oscillations. It is instead the FD methods which develop problems with the high modes. The generally used remedy against spurious high-frequency oscillations is to apply some viscous damping – preferably as little as possible to avoid smearing out the pulses themselves. This

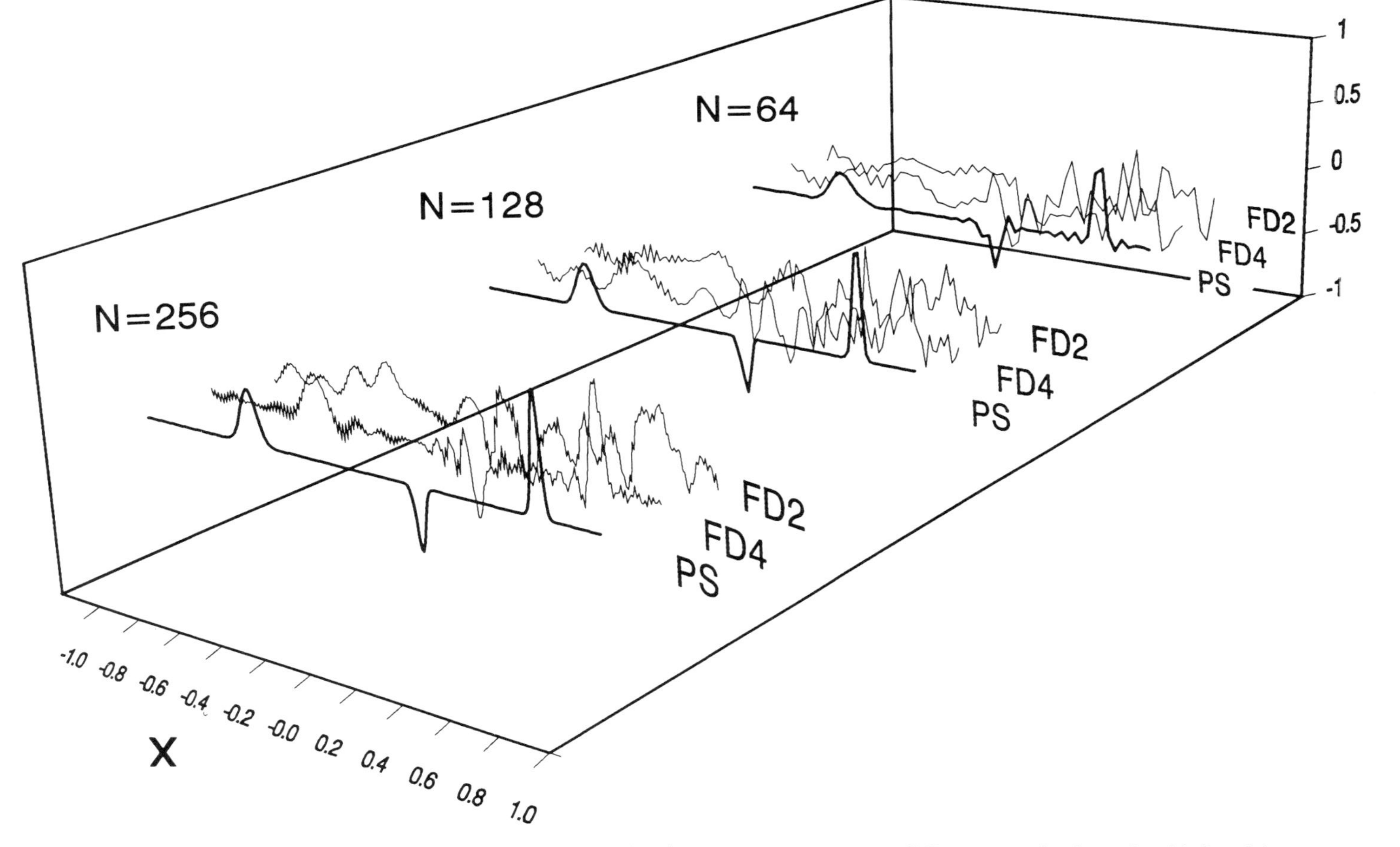

Fig. 12. Comparison of numerical solutions for $u(x, t)$ at time $t = 7$ using different methods and grid densities.

example shows that, for the PS method, often very little suffices (to use none – as in this example – is unnecessarily risky).

The difference between the methods lies not so much in the size of the local errors, as in how these accumulate or cancel over time. In the case of the PS method, relatively large errors cancel systematically.

In Fornberg (1987, 1988a), many similar tests were carried out for the two-dimensional elastic wave equation (a system of five first-order equations supporting both pressure and shear waves; see Figures 17 and 24) – with very similar results.

If Gibbs oscillations do arise in PS calculations, spectral pointwise accuracy can still sometimes be recovered in smooth regions by postprocessing. The paper 'Don't suppress the wiggles – they're trying to tell you something' (Greshko and Lee, 1981) discusses this and points out that viscous damping during a calculation (even when applied with good intent) can lead to an irretrievable loss of information. Although the convergence for step solutions is bad in most common norms (Majda *et al.*, 1978), Abarbanel *et al.* (1986) note this need not be the case for certain 'negative Sobolev norms'. Spectral accuracy of 'moments' provide information that can be used to restore Gibbs-affected solutions.

For many nonlinear equations, the discontinuties that arise are not of 'shock'-type but rather like contact discontinuities which are quite passively transported around in a linear fashion (for example the case in direct simulations of turbulence and – to a lesser extent – in weather forecasting). In such cases, it may suffice to add very little viscosity and rely on the method's ability to handle linear situations.

The problem with more severe nonlinearities is primarily that they can introduce couplings, disrupting the delicate cancellation process on which the PS method for nonsmooth functions is dependent. One idea is to add some more viscosity (just enough to be able to exploit the PS method's power in the case of smooth solutions but not so much that the solution itself gets too severely affected). Spectrally accurate solutions can sometimes be obtained in this way even for shock problems. Different versions of this idea have been proposed. One is the 'Spectral Viscosity Method' (Tadmor, 1989). Further discussions on this and similar methods can be found in Cai *et al.* (1989, 1992), Tadmor (1990, 1993) and Maday *et al.* (1993).

4.3. Differentiation matrices

For both computational and theoretical purposes, it is often convenient to collect all the weights for the approximations at the grid points in a 'differentiation matrix' (DM; cf. Sections 2.3 and 3.4). Finding the derivatives of a vector of data becomes a matrix × vector multiplication.

The relative efficiencies of straightforward matrix × vector multiplications

($\mathcal{O}(n^2)$ operations) versus FFT-based Chebyshev recursions ($\mathcal{O}(n \log n)$ operations) have been compared many times. The estimates for the point of break-even ranges at least from $n = 16$ (Canuto *et al.*, 1988) to n around 100 (Taylor *et al.*, 1984) – the point can be higher still for vector and parallel machines. Furthermore, Solomonoff (1992) shows how a restructuring of the matrix $\times$ vector multiplication for DMs can nearly double the speed of this approach.

The 'Fast Multipole Method' achieves an $\mathcal{O}(n \log n)$ operation count for arbitrary node distributions (Boyd, 1992). However, the proportionality constant is *much* higher than for FFTs – the approach appears not to be competitive in the present context.

Beylkin *et al.* (1991) describe a wavelet approach for converting between finite Legendre and Chebyshev expansions – but again, with a large constant in the $\mathcal{O}(n \log n)$.

Figures 13(a)–(d) illustrate what the DMs look like for the second derivative in the case of $n = 20$ (i.e. 21 grid points if both ends are included, 20 within the period for periodic problems). Figure 13(a) shows the (periodic) stencil $[1\ -2\ 1]/h^2$ and (b) the periodic PS matrix; (c) and (d) show the nonperiodic equi-spaced and Chebyshev matrices ($\gamma = 0$ and 0.5 respectively). Large elements are seen in the top and bottom rows (corresponding to approximations near the boundaries).

For nonperiodic problems, the algorithm in Section 3.1 (code in Appendix B) can be used to generate DMs very conveniently. In the PS case:

```
                                                      Specify size of grid
      PARAMETER (N=...  , M=...  )                    and highest derivative
      DIMENSION X(0:N),C(0:N,0:N,0:M),DM(0:N,0:N,M)
      DO 10 I=0,N
10       X(I) = ....                                  Specify the grid points
         DO 20 I=0,N
            CALL WEIGHTS (X(I),X(0),N,M,C)
            DO 20 L=1,M
               DO 20 J=0,N
20                DM(I,J,L) = C(J,N,L)                DM(*,*,L)
                                                      contains now the DMs
                                                      for the Lth deriv.,
.....                                                 L=1,2,...,M.
```

The computer time taken generating DMs is seldom critical. However, if this has to be done many times, the code above should not be used (since it fails to exploit the fact that all the separate calls to `WEIGHTS` are based on the same grid – some intermediate quantities need not be recalculated repeatedly).

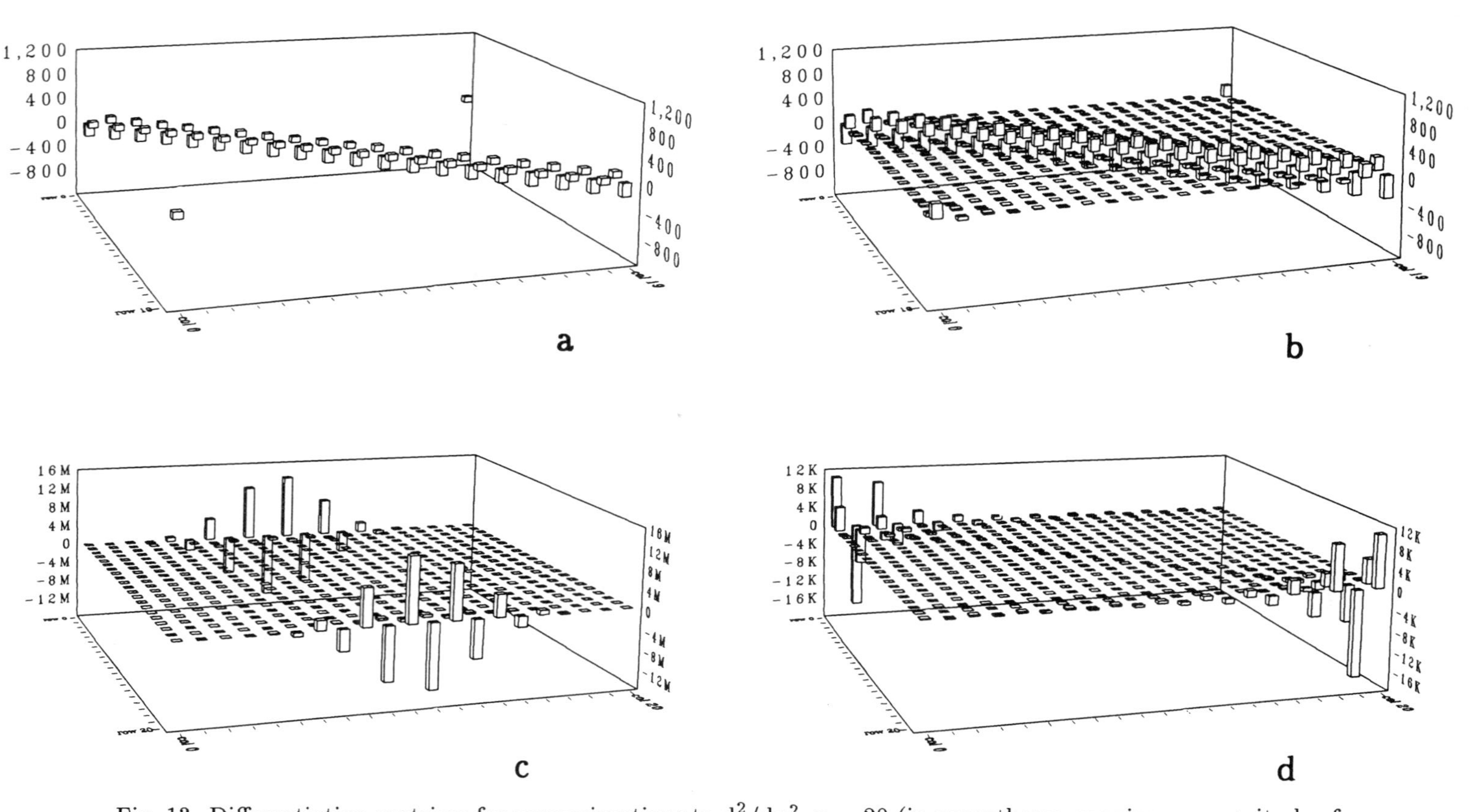

Fig. 13. Differentiation matrices for approximations to $\mathrm{d}^2/\mathrm{d}x^2$, $n = 20$ (in parentheses, maximum magnitude of DM element): (a) second-order FD, periodic (200.); (b) PS, periodic (331.); (c) PS, non-periodic, equi-spaced grid (13×10^6); (d) PS, non-periodic, Chebyshev (17×10^3).

With

$$a_k = \prod_{\substack{i=0 \\ i \neq k}}^{n} (x_k - x_i), \quad F_k(x) = \frac{1}{a^k} \prod_{\substack{i=0 \\ i \neq k}}^{n} (x - x_i),$$

Lagrange's interpolation formula becomes

$$p_n(x) = \sum_{k=0}^{n} f(x_k) F_k(x).$$

Relatively straightforward manipulation of this (Nielsen, 1956, pp. 150–154) allows the elements of D^1 to be computed in, to leading order, only four operations per element:

$$D^1_{jk} = \begin{cases} \dfrac{a_j}{a_k(x_j - x_k)} & j \neq k \\ \displaystyle\sum_{\substack{i=0 \\ i \neq k}}^{n} \frac{1}{x_k - x_i} & j = k. \end{cases}$$

DMs for higher derivatives can, in this case, be obtained as matrix powers of D^1. A much less costly recursion is also available – D^p can be obtained from D^{p-1} in only five operations per element ($p = 2, 3, \ldots$) (Huang and Sloan, 1993, Welfert, 1993). Welfert also notes:

- The PS literature contains many instances of authors assuming the relation $D^p = (D^1)^p$ when it does not hold (for example it fails for the periodic PS method if the number of points is even). A sufficient condition for this relation is presented.
- Closed form expressions for D^1_{jk} and D^2_{jk} become particularly simple for many cases of orthogonal polynomials. A comprehensive list has been collected.

Rounding error propagations within different methods for calculating Chebyshev DMs are discussed by Breuer and Everson (1992).

4.4. Eigenvalues of differentiation matrices

A major difficulty with nonperiodic PS methods is that their DMs tend to have very large spurious eigenvalues (in addition to their physically relevant ones). This adversely affects time stepping techniques (to be discussed in Section 4.5). Many of the special techniques in Section 5 are designed to (partially) overcome this. In order to provide a background for these discussions, we will describe some typical eigenvalue (EV) distributions.

Example 1 Periodic PS; advection equation $u_t = u_x$.

The DM is derived explicitly both in Sections 2.3 and 3.4 (cf. equation (3.3)). The DM is anti-symmetric; its eigenvalues are equi-spaced on the

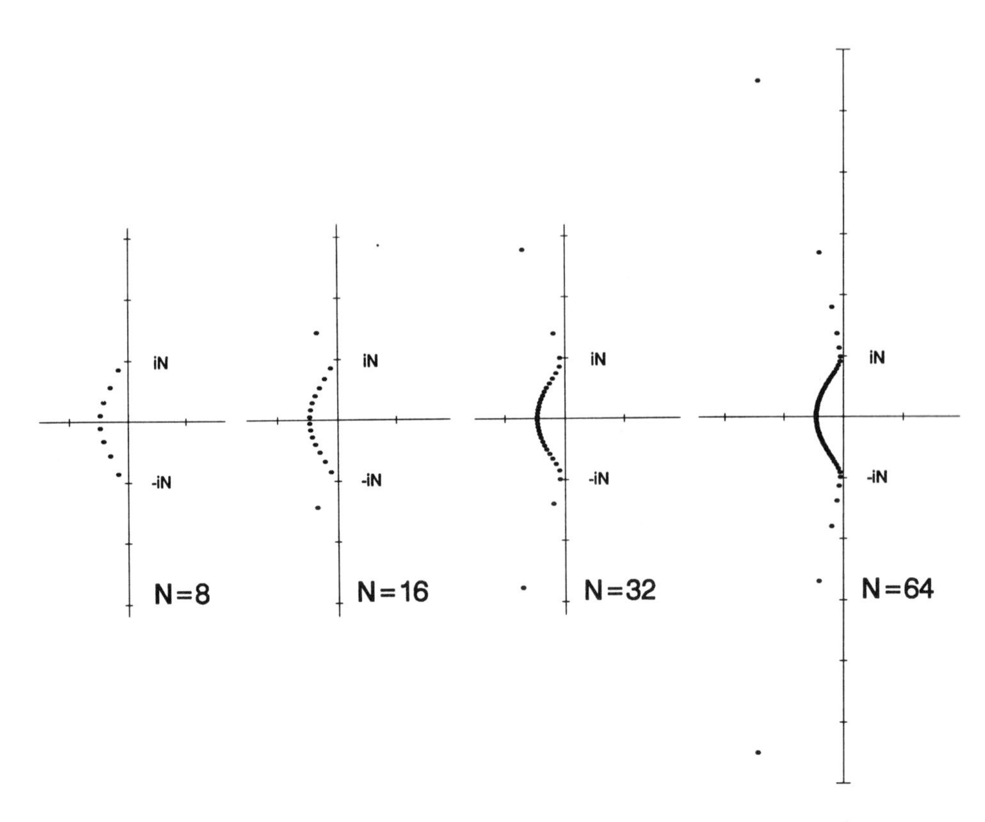

Fig. 14. Eigenvalues of Chebyshev DMs for an advection equation.

imaginary axis between $-N(\pi/2)\,\mathrm{i}$ and $+N(\pi/2)\,\mathrm{i}$ (when N is odd – very minor differences for N even).

Example 2 Chebyshev PS; advection equation $u_t = u_x$, $u(1) = 0$.

Figure 14 shows the EVs for $N = 8$, 16, 32 and 64. Although most of the EVs converge to a curve in the left half-plane between $-\,\mathrm{i}N$ and $+\,\mathrm{i}N$, a few spurious 'outliers' diverge at rates proportional to N.

Trefethen and Trummer (1987) note that the small (physical) eigenvalues for large values of N exhibit a very large sensitivity to rounding errors (however, still leaving them distributed along very distinct paths).

When the grid points are instead distributed as the zeros of Legendre polynomials $L_N(x)$, Dubiner (1987) noted that the spurious outliers were absent in this model problem. However, the EV sensitivity remains large (and EVs alone fail to fully describe stability issues as the DMs are highly nonnormal matrices). It is questionable whether use of Legendre polynomials offers any practical advantage (Trefethen, 1988).

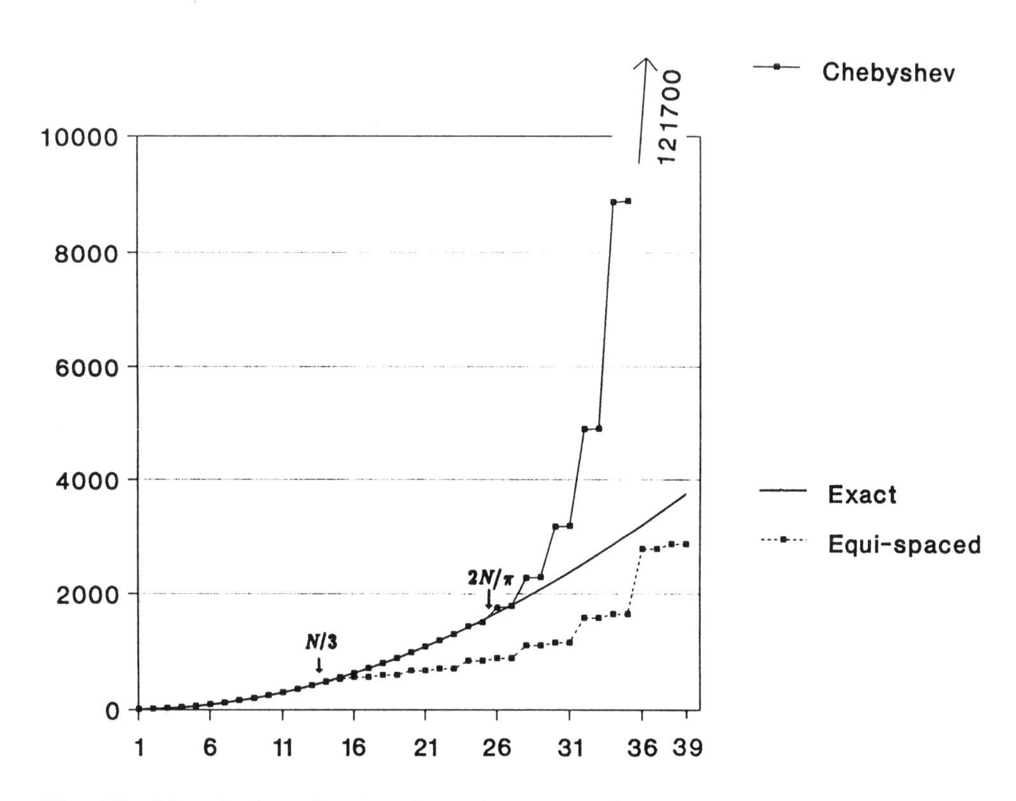

Fig. 15. Magnitudes of eigenvalues of PS DMs (for equi-spaced and Chebyshev grids) compared to analytic eigenvalues; $N = 40$.

Example 3 Chebyshev and equi-spaced nonperiodic PS; heat equation $u_t = u_{xx}$, $u(\pm 1) = 0$.

The continuous problem $u_{xx} = \lambda x$, $u(\pm 1) = 0$ has the EV $\lambda_k = -(k\pi/2)^2$, $k = 1, 2, \ldots$. The EVs of the Chebyshev DM for u_{xx} are all real and negative (Gottlieb and Lustman, 1983). In the case of $N = 20$, this DM is the matrix in Figure 13(d) with the first and last rows and columns removed. Figure 15 compares the magnitude of the EVs with those for the equi-spaced nonperiodic PS method (cf. Figure 13(c); in this case, many higher EVs are complex) and the exact ones. The portions $2/\pi$ and $1/3$ of the EVs are spectrally accurate in the two cases (cf. the numbers of points per wavelength 2, 6 and π mentioned in Section 4.1).

PS methods can also be devised for infinite domains. For eigenvalues of 'Hermite' and 'rational spectral' PS DMs, see Weideman (1992). Other such cases include Laguerre eigenvalues (Funaro, 1992) and sinc eigenvalues (Stenger, 1981).

4.5. Time-stepping methods and stability conditions

Stability (meaning that the numerical solution remains bounded up to a fixed time T as time and space steps Δt and $\Delta x \to 0$) is essential because of the Lax Equivalence Theorem which can be stated as follows.

Lax Equivalence Theorem For a well-posed linear problem, a consistent approximation converges if and only if it is stable.

Stability analysis for spectral methods is simple in only one case – the Fourier PS method for a periodic constant coefficient problem. The DMs are cyclic with known eigenvalues. Assuming again for simplicity $n = 2m+1$, the eigenfunctions (in one dimension) are $\mathrm{e}^{\mathrm{i}\omega x}$, $\omega = -m, \ldots, m$ with eigenvalues $\partial/\partial x \leftrightarrow \mathrm{i}\pi\omega$, $\partial^2/\partial x^2 \leftrightarrow -\pi^2\omega^2$ etc., $(\omega = -m, \ldots, m)$. Had we instead used second-order FD in space, we would have obtained $\partial/\partial x \leftrightarrow \mathrm{i}\sin(\pi\omega\Delta x)/\Delta x$ (cf. (4.3)), $\partial^2/\partial x^2 \leftrightarrow -4(\sin(\pi\omega\Delta x))^2/\Delta x^2$ etc. The stability restrictions when time stepping (of the forms $\Delta t/\Delta x <$ constant and $\Delta t/\Delta x^2 <$ constant respectively) therefore have constants $1/\pi$ and $4/\pi^2$ times those that arise with second-order FD methods (i.e. they are hardly any more severe).

For higher-order FD methods, similar ratios can be read from the maximum values of the curves in Figure 10 and their generalization to higher derivatives (the equivalents of (4.4) are given in Fornberg, 1990a).

Stability conditions like $\Delta t/\Delta x <$ constant and $\Delta t/\Delta x^2 <$ constant are normally not restrictive in connection with spectral methods. With, say, a fourth-order Runge–Kutta method in time and a better-than-eighth-order PS method in space, $\Delta t/\Delta x^2 <$ constant is needed anyway to make the temporal accuracy match the spatial one. However, conditions on $\Delta t/\Delta x^p$, $p > 2$ will arise in many nonperiodic PS cases. One of the main issues in designing (nonperiodic) spectral methods is to circumvent these.

For more realistic problems, several complications arise (both for FD and for PS methods):

- For variable coefficients or nonlinearities, stability for all problems with 'frozen' coefficients is neither necessary nor sufficient for stability (Kreiss, 1962, Richtmyer and Morton, 1967).
- With boundaries present, local mode analysis for the interior needs to be complemented by 'GKS' analysis at the boundaries (Kreiss, 1968, 1970, Gustafson, Kreiss and Sundström, 1972); for simplified versions of this, see Trefethen (1983), Goldberg and Tadmor (1985)).

Additional problems are more specific to PS methods:

- One needs to distinguish between 'Lax stability' (fixed T and $\Delta t \to 0$) and 'eigenvalue stability' (fixed Δx and Δt as $T \to \infty$). For the highly nonnormal DMs that arise from nonperiodic PS methods, large growths can initially arise if the norms are large even if all eigenvalues fall within (or on) the unit circle (cf. the Kreiss matrix theorem, in Richtmyer and Morton (1967)).

- PS methods can be unstable even when the corresponding FD methods of increasing orders are *all* stable. Tadmor (1987) addresses this phenomenon in connection with a linear model equation $u_t = c(x)u_x$. Reddy and Trefethen (1990) use 'pseudospectra' to provide further insight into this and similar phenomena.

'Energy methods' provide a powerful general tool for PS analysis, e.g. Gottlieb *et al.* (1980, 1987, 1991). However, due to their technical complexity, we restrict the discussion here to eigenvalue stability. Although limited, it can still provide useful guidelines for selecting time integrators.

The most common time-stepping approach is the 'Method of Lines' (MOL) which amounts to discretizing in space only and then applying a 'packaged' ODE solver (based, for example, on Runge–Kutta or backwards differentiation) to the resulting system of ODEs.

This approach allows the ODE solver to be developed and analysed separately from the spatial discretization method. The user need not be concerned with many tedious issues like starting techniques for multi-step methods, time step and order adjustments etc.

For constant coefficients, the MOL gives rise to a system of ODEs

$$\begin{bmatrix} u \end{bmatrix}_t = \begin{bmatrix} \quad A \quad \end{bmatrix} \begin{bmatrix} u \end{bmatrix} + \begin{bmatrix} f \end{bmatrix},$$

where A is the differentiation matrix for the approximation (which we now assume to be diagonalizable). For a first cut at assessing what kind of ODE package to select, the stability regions of the time integrator have to be compared with the eigenvalues of A.

Stability regions An ODE solver is *stable* for Δt and (complex) λ if the numerical solution to $u_t = \lambda u$ does not grow with t. It is called *A-stable* if it is stable for all λ in the negative half-plane (the ideal situation – this matches the same property of the analytic solution $u(t) = \mathrm{e}^{\lambda t}$). A-stability can seldom be achieved for methods of high accuracies. (cf. the 'Dahlquist barriers' (Dahlquist, 1956, 1985)). Instead, for most methods we find that $\lambda \Delta t$ must lie within some smaller domain than the full left half-plane. If any eigenvalues λ of A happen to be large in magnitude, this restricts Δt.

Example Forward Euler:

$$\begin{aligned} v(t + \Delta t) &= v(t) + \Delta t \lambda v(t) \Rightarrow \\ v(t + k\Delta t) &= (1 + \lambda \Delta t)^k v(t) \Rightarrow \end{aligned}$$

Stable if $|1 + \lambda\Delta t| \le 1 \Rightarrow$ stability region (values of $\lambda\Delta t$ giving no growth) is a circle with radius 1 centred at -1.

However, very low accuracy and no stability coverage along the imaginary axis makes this method very unattractive.

Commonly used ODE solvers represent compromises between low operation counts, high accuracies and large stability domains. They include many Runge–Kutta (RK) schemes, Adams-type methods and, for 'stiff' problems (with some eigenvalues far away in the left half-plane) BDF methods. For discussions on ODE solvers, see, for example, Gear (1971), Shampine and Gordon (1975), Lambert (1991), Hairer *et al.* (1987), Hairer and Wanner (1991). Many stability domains are illustrated in Sand and Østerby (1979).

5. PS variations/enhancements

Up to this point, we have described 'basic' PS implementations. However, many variations are possible, offering advantages in different respects. In this section, we discuss a few of these.

5.1. Use of additional information from the governing equations

This idea (like most others) is best described through the use of examples. To use it the problem must first be manipulated analytically (for example by repeated differentiation) to provide more information than is immediately available from its original formulation.

Example 1 Exploiting additional derivative information at the boundaries for the eigenvalue problem $u_{xx} = \lambda u$, $u(\pm 1) = 0$.

Since $u(\pm 1) = 0$, clearly also $u''(\pm 1) = 0$ and $u''''(\pm 1) = 0$ (for this example, we ignore that this pattern continues indefinitely and that u becomes periodic). The information on u'' and u'''' can be exploited in different ways:

A: Reduce the largest spurious EVs (cf. Figure 15).
Each 'extra' boundary condition has a corresponding (one-sided) difference stencil. From each row of the DM (as shown in Figure 13(c) and (d)), we can subtract any multiple of these stencils. These multiples can be choosen to minimize the sum of the squares of the elements of the resulting DM. As shown in Fornberg (1990b), this procedure much improves the conditioning of the DM without any degradation in its spectral accuracy:

$\Rightarrow$ Using only $u''(\pm 1) = 0$ and $u''''(\pm) = 0$ reduces the largest matrix elements of the DMs shown in Figures 13(c) and (d) (edge elements not included) to about 500 – down from about 500 000 and 7000 respectively!).

⇒ Each time an 'extra' boundary condition is applied to a Chebyshev-type approximation, the largest (remaining) spurious eigenvalue gets changed to *exactly* zero.

B: Increase the accuracy of the computed EVs.
Given k extra relations (for example information at the boundaries), we can proceed as follows:

1. Introduce k new fictitious grid points (anywhere – inside or outside the domain).
2. Find the $N+1$ stencils for u_{xx} which are accurate at the original grid point locations $x_0, x_1, \ldots, x_N$ respectively, but which extend also over the fictitious points.
3. Find the k stencils which express the k extra pieces of information (again extending over all grid points – original and fictitious).
4. Add/subtract multiples of these k stencils from the ones calculated in Step 2 – so that the weights at all the fictitious points are eliminated.

The resulting stencils for u_{xx} become exact for all polynomials of degree $N+k$ which take prescribed values at $x_0, x_1, \ldots, x_N$ *and* satisfy the k extra relations. They are therefore more accurate, but still no more costly to apply than straightforward PS approximations also extending over $x_0, x_1, \ldots, x_N$ (these would be exact only for polynomials up to degree N).

The idea of introducing points outside a boundary and then eliminating them again using boundary conditions is often used with FD methods. The PS case is remarkable in that the location of these (temporary) points turns out to have no influence at all on the final result (apart from rounding errors). In this PS case, they offer a very convenient way of generating stencils satisfying 'side conditions' without the need for any additional analytical devices.

Another boundary situation for which large improvements can be achieved is the treatment of the origin in polar coordinates. This is discussed further in Section 5.6. The example below (from Fornberg (1994)) demonstrates this idea in a case of axial symmetry.

Example 2 Exploiting symmetry at a boundary.
Bessel's equation arises from Poisson's equation in the case of axial symmetry. Consider its eigenvalue problem

$$u'' + \frac{1}{r}u' - \frac{n^2}{r^2}u = -\lambda u, \quad n = 0, 1, \ldots, \ u(0) \text{ bounded}, \ u(1) = 0.$$

The exact eigenvalues $\lambda_{n,k}$, $k = 1, 2, \ldots$ satisfy $J_n(\sqrt{\lambda_{n,k}}) = 0$.
We compare two approximation methods:

1 Note that

$$\begin{cases} n = 0 & u'(0) = 0 \\ n \neq 0 & u(0) = 0 \end{cases}$$

and use straightforward Chebyshev approximation on [0,1]. Grid points are located at $r_k = (1 - \cos k\pi/N)/2$, $k = 0, 1, \ldots, N$.

2 Note that

$$\begin{cases} n \text{ even} & u(r) \text{ even} \\ n \text{ odd} & u(r) \text{ odd.} \end{cases}$$

Consider FD stencils extending over $[-1, 1]$, but use symmetry to reduce actual calculations to within [0,1]. Grid points are located at $r_k = \sin k\pi/2N$, $k = 0, 1, \ldots, N$ (i.e. no clustering at $r = 0$).

Figure 16 compares the accuracies in the numerical EVs obtained through these two methods for $n = 7$, $k = 1$ ($\lambda_{7,1} \approx 122.9$; cf. Gottlieb and Orszag (1977, pp. 152–153)). The values for method 1 are taken from Huang and Sloan (1993a).

Further variations of boundary implementations are discussed by Canuto and Quarteroni (1987) and Funaro and Gottlieb (1988).

Morals When using PS methods, always consider whether the FD viewpoint can offer any advantages (in accuracy, simplicity, flexibility, etc.)

The fundamental reason for clustering grid points at the ends of an interval is to compensate for the large error terms in one-sided approximations. The more information we can exploit at boundaries, the less we should cluster.

5.2. Use of different PS approximations for different terms in an equation

When a single variable appears more than once in an equation, it is normally approximated in a similar way at each instance. However, Huang and Sloan (1993b,c) note two situations when it is better to use different types of PS approximation.

Example 1 Solve the singular perturbation problem

$$\epsilon u'' + u' = 1, \quad x \in [-1, 1], \quad 1 \gg \epsilon > 0.$$

Straightforward centred second-order FD approximations for both u' and u'' give an oscillatory solution with $\mathcal{O}(1)$ errors across $[-1, 1]$ for any N when $\epsilon < 1/N$. Approximating u_x by the one-sided FD stencil $[u(x+h) - u(x)]/h$ reduces the errors to $\mathcal{O}(1/N)$. When using Chebyshev PS approximations, we can similarly approximate u' with stencils based on all grid points *but* the one at $x = -1$. In the limit of $\epsilon \to 0$, this gives spectral accuracy across $[-1, 1]$ (rather than $\mathcal{O}(1)$ errors).

For another idea to solve this problem with a PS method, see Eisen and Heinrichs (1992).

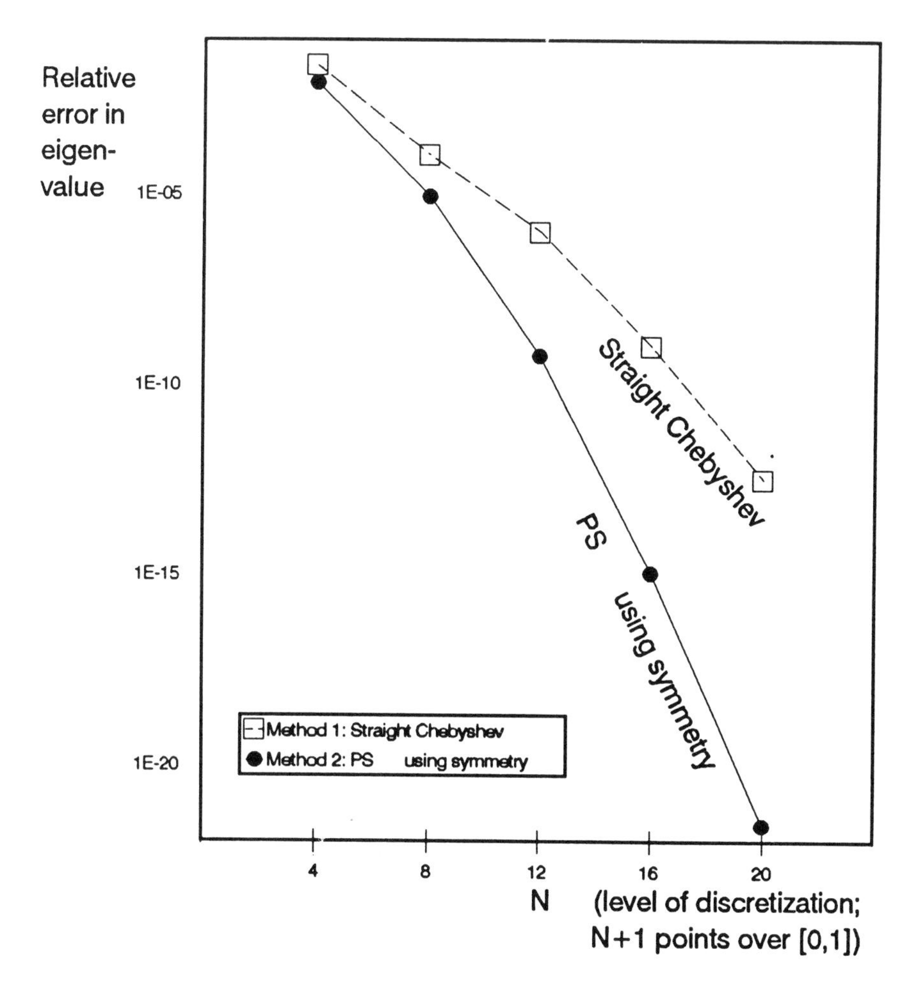

Fig. 16. Errors in eigenvalue $\lambda_{7,1}$ for Bessel's equation when approximated using two different strategies.

Example 2 Solve the eigenvalue problem $u'''' + 4u''' = \lambda u''$, $x \in [-1, 1]$, $u(\pm 1) = u'(\pm 1) = 0$.

Problems similar to this arise, for example, in linearized stability analysis in fluid mechanics. Spurious EVs denote in this case EVs appearing incorrectly in the right half-plane, suggesting physical instabilities that do not exist.

If we approximate all derivatives of u on a Chebyshev grid, incorporating $u'(\pm 1) = 0$ as in Example 1, Part B, Section 5.1, we will get spurious EVs. However, ignoring $u'(\pm 1) = 0$ when approximating u'' overcomes this.

In both these examples, variable coefficients would have added no complications (as is usually the case for PS methods – in sharp contrast to spectral Galerkin or Tau methods).

5.3. Staggered grids

When using an FD method, it is customary to compute values for each unknown at each grid point. Figure 17 illustrates an alternative, which can be employed quite frequently. Even for equations with only one unknown variable, a similar staggering can sometimes be used effectively from time level to time level. The idea is to gain accuracy – derivative approximations at 'half-way' points are often much more accurate than at grid points.

For the first derivative:

		Second-order accuracy	Leading error terms	Ratio of error
Reg.	$f'(x) =$	$\{\frac{1}{2}f(x-h)$		
		$+\frac{1}{2}f(x+h)\}/h$	$+\frac{1}{6}h^2 f'''(x)\ldots$	
Stag.	$f'(x) =$	$\{f(x-\frac{1}{2}h)$		
		$+\frac{1}{2}f(x+\frac{1}{2}h)\}/h$	$+\frac{1}{24}h^2 f'''(x)\ldots$	$\frac{1}{4} = 0.25$
		Fourth-order accuracy		
Reg.	$f'(x) =$	$\{\frac{1}{12}f(x-2h)$		
		$-\frac{2}{3}f(x-h)$		
		$+\frac{2}{3}f(x+h)$		
		$-\frac{1}{12}f(x+2h)\}/h$	$-\frac{1}{30}h^4 f^{v}(x)+\ldots$	
Stag.	$f'(x) =$	$\{\frac{1}{24}f(x-\frac{3}{2}h)$		
		$-\frac{9}{8}f(x-\frac{1}{2}h)$		
		$+\frac{9}{8}f(x+\frac{1}{2}h)$		
		$-\frac{1}{24}f(x+\frac{3}{2}h)\}/h$	$-\frac{3}{640}h^4 f^{v}(x)+\ldots$	$\frac{9}{64} \approx 0.141$

For approximations of order p, the ratio of error terms turns out to be

$$\left\{\frac{p!}{2^p\{(\frac{1}{2}p)!\}^2}\right\}^2 \approx \frac{2}{\pi p}.$$

Since the periodic PS method can be viewed as the limit of $p \to \infty$, this suggests that the idea of staggering would also be advantageous in that case.

Another suggestive argument follows from comparing the weights in the stencils. Figure 18(a) shows the magnitudes of the weights for increasingly accurate approximations to the first derivative as is also displayed in the right half of Figure 2. In the limit, they become $(-1)^\nu/\nu$, $\nu = 1, 2, \ldots$. For the staggered approximations, the limit is much more local in nature: $(-1)^{\nu+1/2}/\pi\nu^2$, $\nu = \frac{1}{2}, \frac{3}{2}, \frac{5}{3}, \ldots$. Since the derivative is a local property of a function, a more compact approximation makes more sense than one relying on extensive cancellation of distant contributions.

For nonperiodic problems, staggering can be achieved, for example, by using grids based on Chebyshev extrema (as usual) and Chebyshev zeros.

Staggering turns out to be advantageous for odd derivatives (first, third,

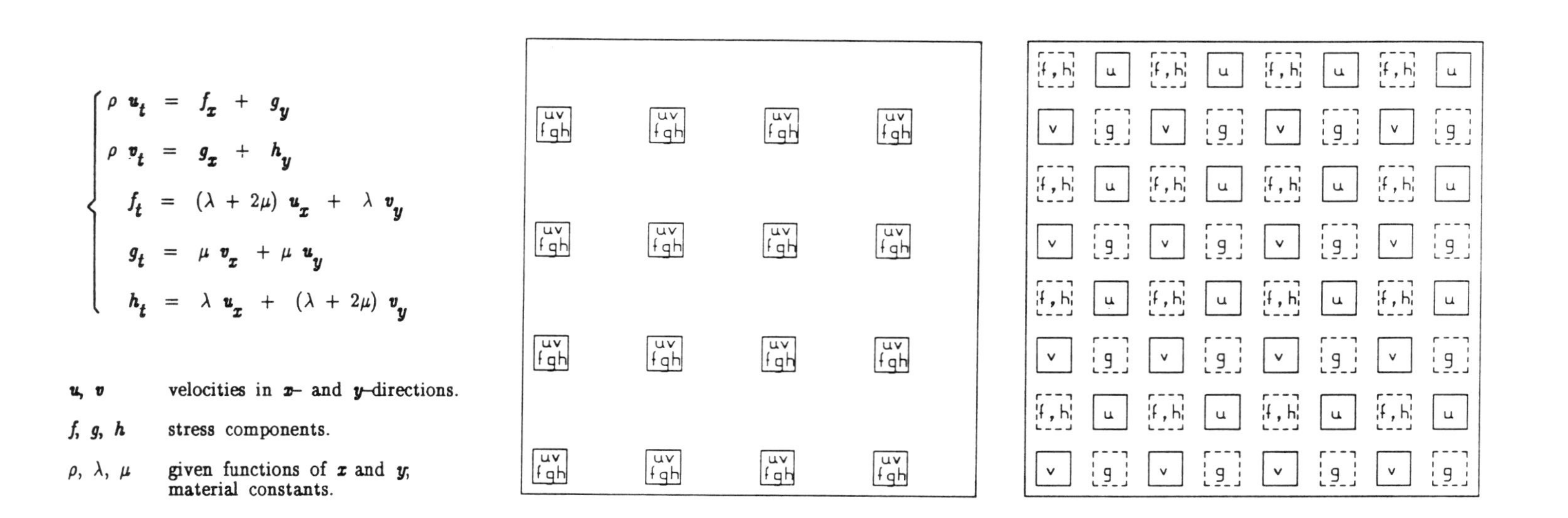

Fig. 17. Example of staggered grid arrangement.

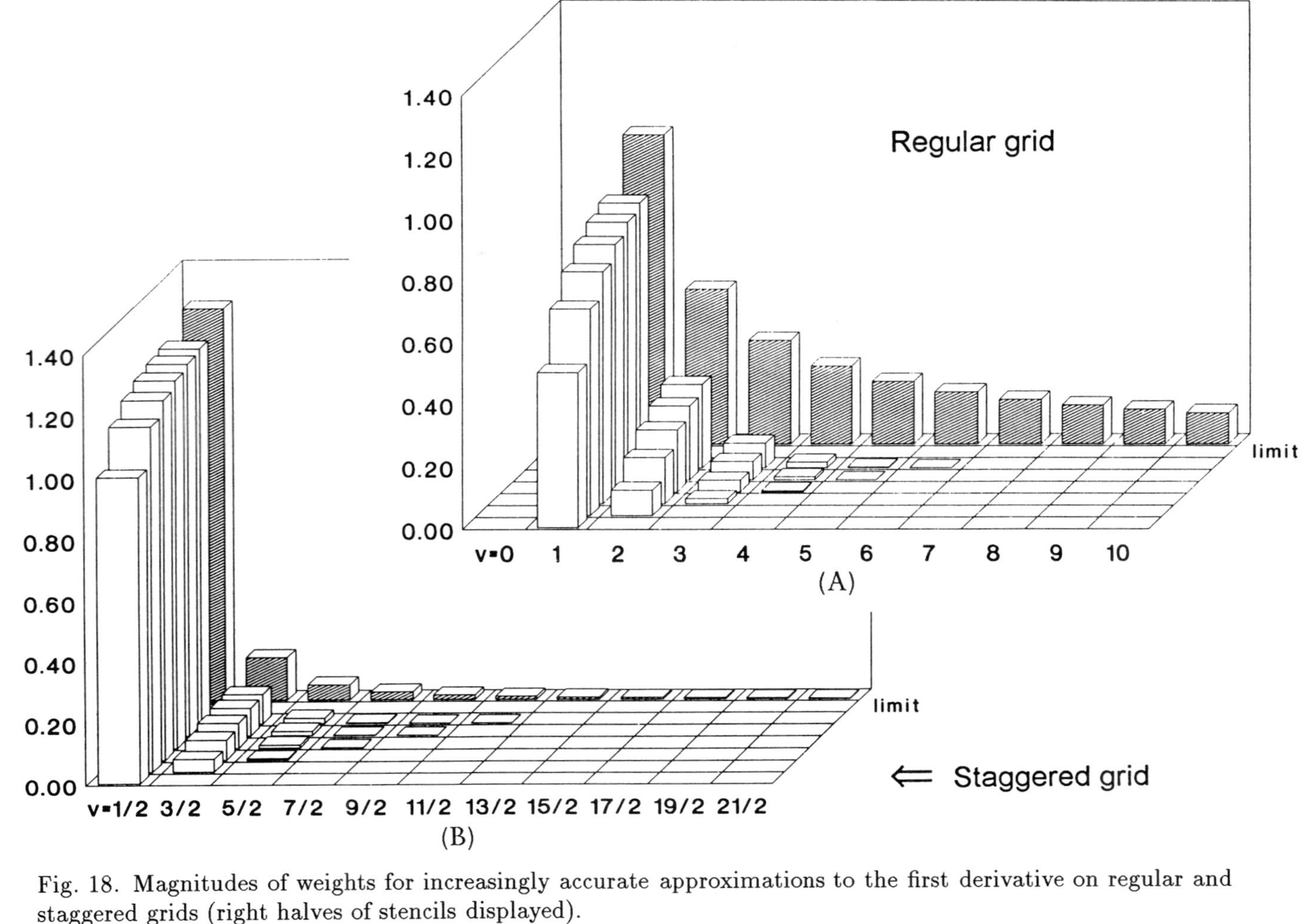

Fig. 18. Magnitudes of weights for increasingly accurate approximations to the first derivative on regular and staggered grids (right halves of stencils displayed).

etc.) whereas regular grids are better for even derivatives. For analysis and references on grid staggering in connection with high-order methods, see Fornberg (1990a).

5.4. Preconditioning

The large spurious eigenvalues in many PS DMs can make explicit time stepping methods very costly (forcing the use of extremely small time steps). Implicit methods often have unbounded stability domains, but they require the solution of a full linear system every time step.

The idea of preconditioning works in any number of dimensions, but is easiest to illustrate in one dimension.

Example Preconditioning for a Chebyshev PS solution of a two-point boundary value problem $u''(x) = f$, where $u(\pm 1)$ and f are given.

Chebyshev PS discretization (viewed as a FD method) gives rise to a linear system $Cu = f$, where C is the Chebyshev DM for $\mathrm{d}^2/\mathrm{d}x^2$ (as shown in Figure 13(d), but with the edge rows and columns removed). C is neither symmetric nor diagonally dominant – standard iterative techniques will not converge.

Let F be the second-order FD DM based on the same Chebyshev grid (with elements obtained using the algorithm in Section 3.1). Since F is tri-diagonal, it is easily inverted (in higher dimensions, one can, for example, use alternating direction arrangements for the tri-diagonal matrices). The system to be solved can be written as $[F^{-1}C]u = g$, where $g = F^{-1}f$. The matrix $F^{-1}C$ is illustrated in Figure 19. Any standard iterative technique can be used to rapidly obtain the (spectrally accurate) vector u.

For matrices of this form (near-symmetric, diagonally dominant), convergence of some methods improve if the ratio of largest-to-smallest eigenvalues is lowered. For $F^{-1}C$, $\lambda_{\max}/\lambda_{\min} \to \frac{1}{4}\pi^2$ as $N \to \infty$ (Haldenwang *et al.*, 1984). Using higher-order FD methods, this ratio can be lower still (Phillips *et al.*, 1986) but savings may be off-set by a higher cost of applying F^{-1}. In contrast, we can note that for the (nonsymmetric, nondiagonally dominant) matrix C, $\lambda_{\max}/\lambda_{\min}$ grows like $\mathcal{O}(N^4)$.

For odd derivatives, FD preconditioning normally requires the use of staggered grids. This is discussed for Chebyshev methods by Hussaini and Zang (1984) and Funaro (1987). Mulholland and Sloan (1992) considers FD preconditioners for staggered approximations to $\partial^3/\partial x^3$ (applicable, for example, to the Korteweg–de Vries equation).

FE-based preconditioners have been discussed by Canuto and Quarteroni (1985), Deville and Mund (1985) and Canuto and Pietra (1987). General references on preconditioning include Canuto *et al.* (1988) and Boyd (1989).

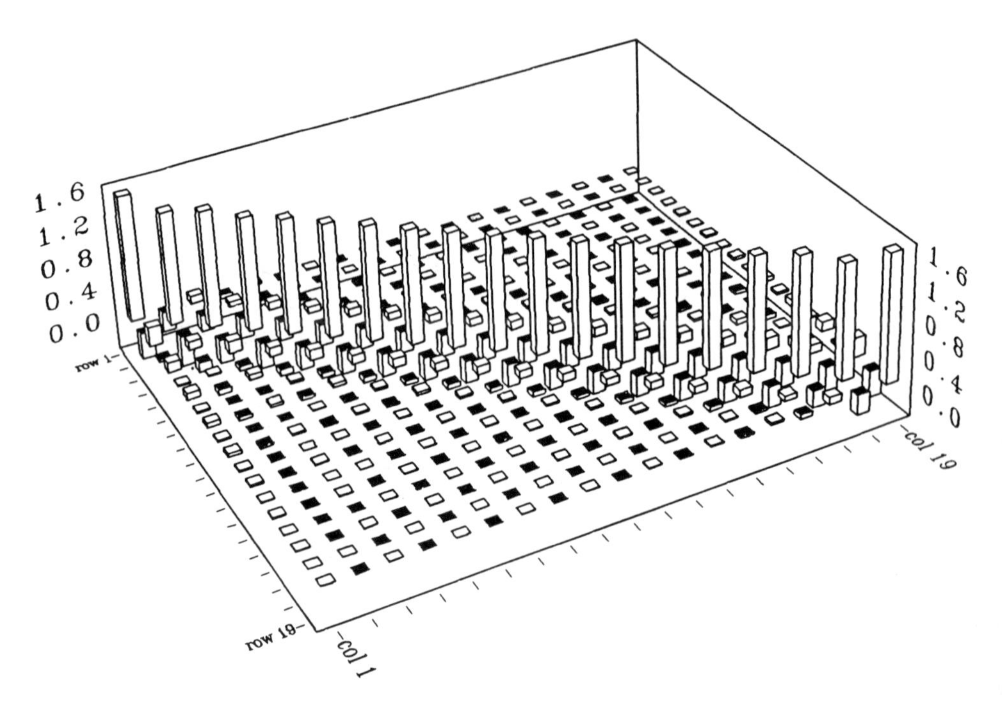

Fig. 19. Display of the matrix $F^{-1}C$ resulting from second-order FD preconditioning of the Chebyshev DM C for $\mathrm{d}^2/\mathrm{d}x^2$ (with $n = 20$).

5.5. *Improved conditioning through change of variable*

All the approximations to the derivatives that we have considered so far (for nonperiodic problems) have been based on differentiating interpolating polynomials. The difficulties at boundaries have been linked to large weights in one-sided stencils – in turn a consequence of the rapid growth of high-degree polynomials at increasing distances from the origin.

Kosloff and Tal-Ezer (1993) proposed to change first the independent variable x into y through $x = \arcsin(\alpha y)/\arcsin(\alpha)$ (both x and $y \in [-1, 1]$, the parameter $\alpha \in [0, 1]$). In the governing equations, $\partial/\partial x$ needs then to be replaced by

$$\frac{\arcsin(\alpha)}{\alpha}\sqrt{1-(\alpha y)^2}\,\frac{\partial}{\partial y}$$

(and similarly for higher derivatives). Applying a standard Chebyshev PS method in the y variable corresponds, in the x variable, to working with nonpolynomial basis functions.

In the limit of $\alpha \to 0$, y is equal to x, and we have the regular Chebyshev PS method. As $\alpha \to 1$, the x grid approaches uniform spacing. Close to this limit, the Chebyshev polynomials (in the y variable) have, in the x variable, become stretched to resemble trigonometric functions. This reduces the spurious EVs. In Example 2 of Section 4.4, they decrease from $\mathcal{O}(N^2)$ to $\mathcal{O}(N)$. Figure 20(a) shows how they move in this case when α increases from 0 (as in Figure 14) to 0.9. Figure 20(b) shows the effect this has on the accuracy of different Fourier modes.

Compared to the standard Chebyshev PS method, this procedure offers reduced spurious EVs, better conditioned DMs (much closer to being normal matrices) and a wider range of accurately treated Fourier modes in exchange for less accuracy for the lowest Fourier modes.

If we had just moved the grid points towards equi-spaced locations without the accompanying change of variable, we would suffer all the problems of the Runge phenomenon (cf. Figures 3 and 8(a)) – disastrous growth of condition number and an inability to approximate anything but extremely smooth functions.

The properties of this method are similar to those of medium-to-high order FD schemes – no conclusive efficiency comparisons have yet been carried out.

The idea of changing variable to improve the Chebyshev PS method was proposed earlier by Bayliss *et al.* (1989) for quite a different purpose – to achieve additional grid clustering at interior locations where extra resolution might be needed (see also Bayliss and Turkel (1992)).

5.6. PS methods in polar and spherical coordinates

Separation of variables for the Laplacian operator in three-dimensional polar coordinates leads to a class of functions called 'spherical harmonics' – combinations of trigonometric and Gegenbauer polynomials. These offer a complicated (but workable) base for spectral methods. For an overview of this approach, see, e.g., Boyd (1989, Ch. 15).

An alternative is outlined below, for simplicity first for polar coordinates in the plane. Instead of generating PS methods from some set of basis functions, we start from the basic FD premise that derivatives in different spatial directions can be approximated entirely separately from each other. In each direction, we thus use the most appropriate FD scheme of maximal order (typically one-dimensional Fourier or Chebyshev PS approximations).

Polar coordinates A polar coordinate system on the unit circle can be obtained through

$$\begin{cases} x = r\cos\theta \\ y = r\sin\theta \end{cases} \qquad 0 \le r \le 1, \ -\pi \le \theta \le \pi.$$

At $r = 0$, all θ positions collapse into one physical grid point – therefore requiring only one governing equation. At this location, one can use a Cartesian x–y-based FD stencil (free of the singularities that might have been introduced by the polar coordinate formulation).

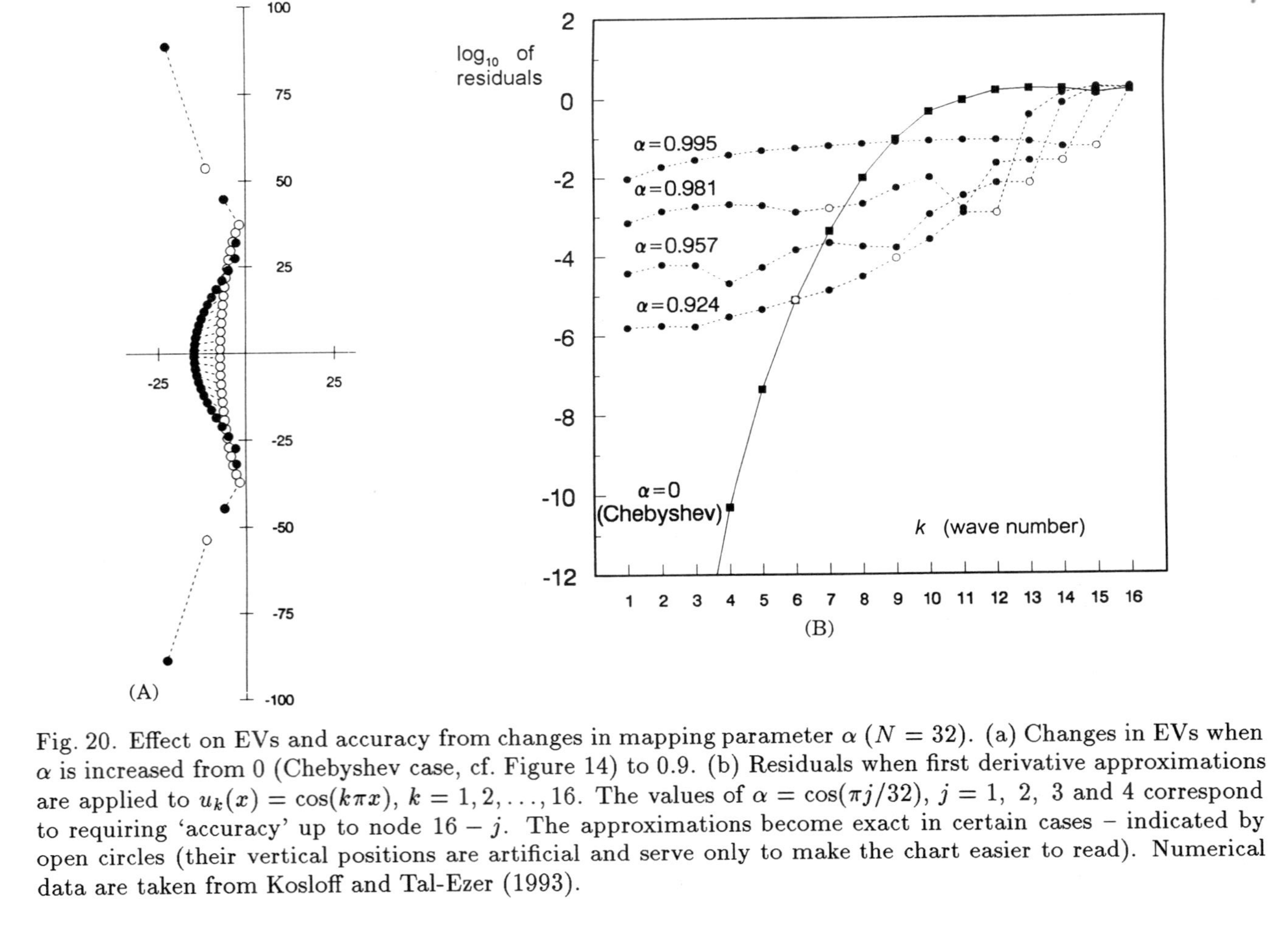

Fig. 20. Effect on EVs and accuracy from changes in mapping parameter α ($N = 32$). (a) Changes in EVs when α is increased from 0 (Chebyshev case, cf. Figure 14) to 0.9. (b) Residuals when first derivative approximations are applied to $u_k(x) = \cos(k\pi x)$, $k = 1, 2, \ldots, 16$. The values of $\alpha = \cos(\pi j/32)$, $j = 1,\ 2,\ 3$ and 4 correspond to requiring 'accuracy' up to node $16 - j$. The approximations become exact in certain cases – indicated by open circles (their vertical positions are artificial and serve only to make the chart easier to read). Numerical data are taken from Kosloff and Tal-Ezer (1993).

One might be tempted to proceed by using Fourier PS in θ and Chebyshev PS in r. A much better alternative is to consider $-1 \leq r \leq 1$, $0 \leq \theta \leq \pi$ (instead of $0 \leq r \leq 1$, $-\pi \leq \theta \leq \pi$). There is then no longer any reason to refine the grid in the r direction near $r = 0$:

- saves grid points;
- higher-order accuracy of PS stencils in r directions (since they extend over twice as many grid points – cf. Example 2 in Section 5.1);
- less severe two-dimensional point clustering near the origin;
- high degree of smoothing in the θ direction for small r values provides favourable CFL (Courant–Friedrichs–Levy) stability conditions without damaging overall accuracy;
- Fourier PS available as before in the θ direction.

Spherical coordinates We consider a surface φ, θ – grid as shown in Figure 21. The dotted arrows indicate how periodicity can be implemented in both φ and θ. The observation for two-dimensional coordinates carry over;

- for θ near $\pm\pi$, polar stability can be enhanced by smoothing in the φ direction;
- an r direction can again be added as in two-dimensions (with $-1 \leq r \leq 1$ and halving the angular domain in case $r = 0$ is in the region of interest).

This PS method has been tested for convective flow in different directions over the surface of a sphere (Fornberg, 1994). Its performance turns out to be entirely unharmed by the presence of polar singularities – the accuracy is as high as is typical for one-dimensional periodic problems.

6. Comparisons of computational cost – FD versus PS methods

High-order FD and the PS methods are particularly advantageous in cases of high smoothness of solution (but note again the discussion in Section 4.2), stringent error requirement, long time integrations and more than one space dimension.

Since the PS methods for periodic and nonperiodic problems are quite different, they will be discussed separately.

6.1. Periodic problems

To obtain more precise insights into how the formal order of a method affects the accuracy, we consider the model problem $\partial u/\partial t + \partial u/\partial x = 0$ on $[-1, 1]$, integrated in time from 0 to 2 (the time it takes the analytical solution $u(x,t)$ to move once across the period). The data in Figure 10 can be recast

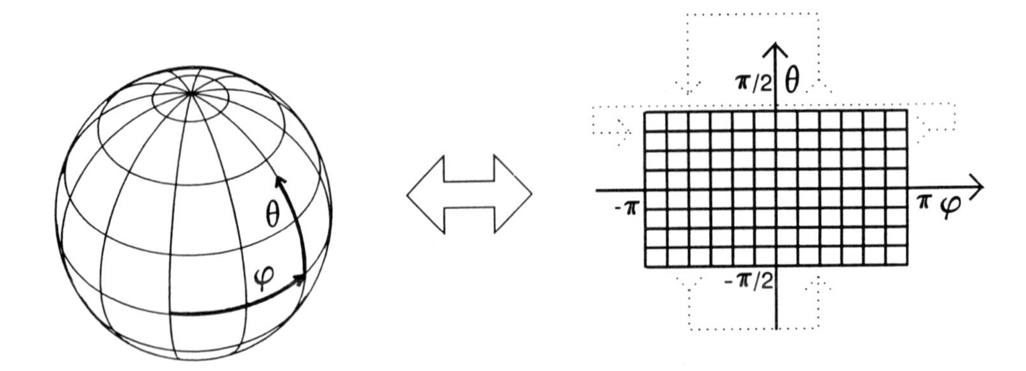

Fig. 20. Grid arrangement and periodicities for spherical coordinates.

into Figure 22 (for details, see Fornberg, 1987). The following explains how to interpret this figure.

Using an accurate time integrator, Fourier modes in the numerical solution of this equation will develop phase but not amplitude errors. If a phase error is π, that mode will have the wrong sign and will not add any accuracy to a Fourier expansion. Here we consider (somewhat arbitrarily) a mode to be 'accurate' if its phase error is less than $\pi/4$.

A second-order FD method with $N_{\mathrm{GP}} = 500$ (i.e. 500 grid points in the spatial direction) is seen to give the same accuracy (have the same horizontal position in the figure) as a fourth-order FD method with $N_{\mathrm{G}} \approx 160$ and a PS method with $N_{\mathrm{G}} \approx 32$. The numbers on the axes indicate: horizontally (approximately location '16' in the example above), modes up to $\sin(16\pi x)$, $\cos(16\pi x)$ are 'accurate' at the end of the integration; vertically, in the different cases the number of grid points needed per wavelength.

The governing equations for the test case shown in Figures 23(a) and (b) are those in Figure 17 (here using a 'regular' grid). A sharp 'pressure wave' pulse is sent down through an elastic medium which carries both pressure and shear waves with lower velocities near the centre of the domain. Following focusing and the subsequent development of a cusp-shaped wave front, Figure 23(b) shows second- and fourth-order FD and PS results. In the three cases, comparable accuracies are obtained on grids of densities 512×512, 128×128 and 32×32 respectively (in quite good agreement with the discussion just above). With the PS method implemented by FFTs, the relative computer times scale as 20:2:1. In three dimensions, these numbers would become 300:8:1. For memory requirements, the differences become even larger: 256:16:1 in two dimensions and 4096:64:1 in three dimensions.

The largest cost–benefits from high-order methods arise in two and three

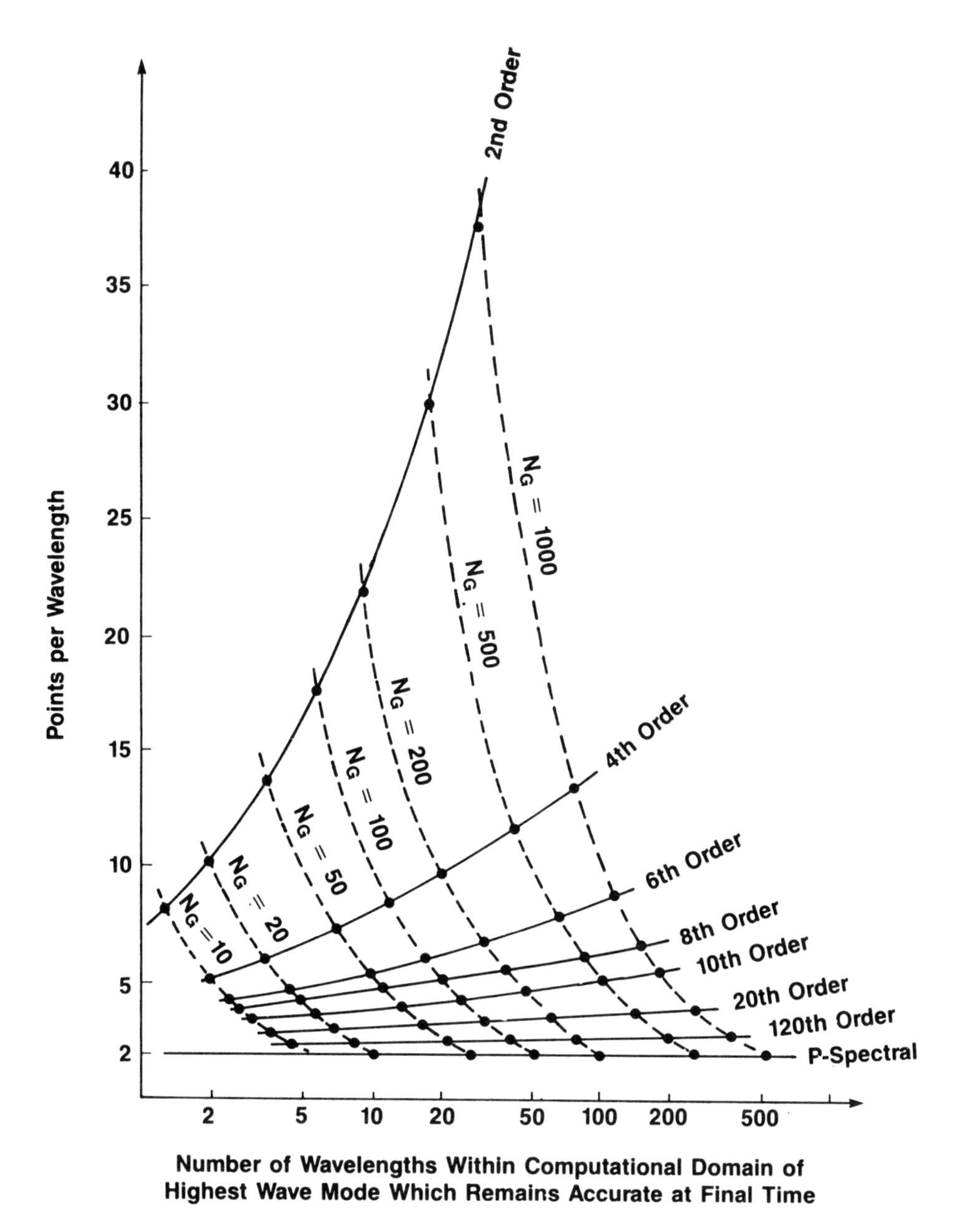

Fig. 22. Relations between grid densities and obtained accuracies when applying different methods to a model problem.

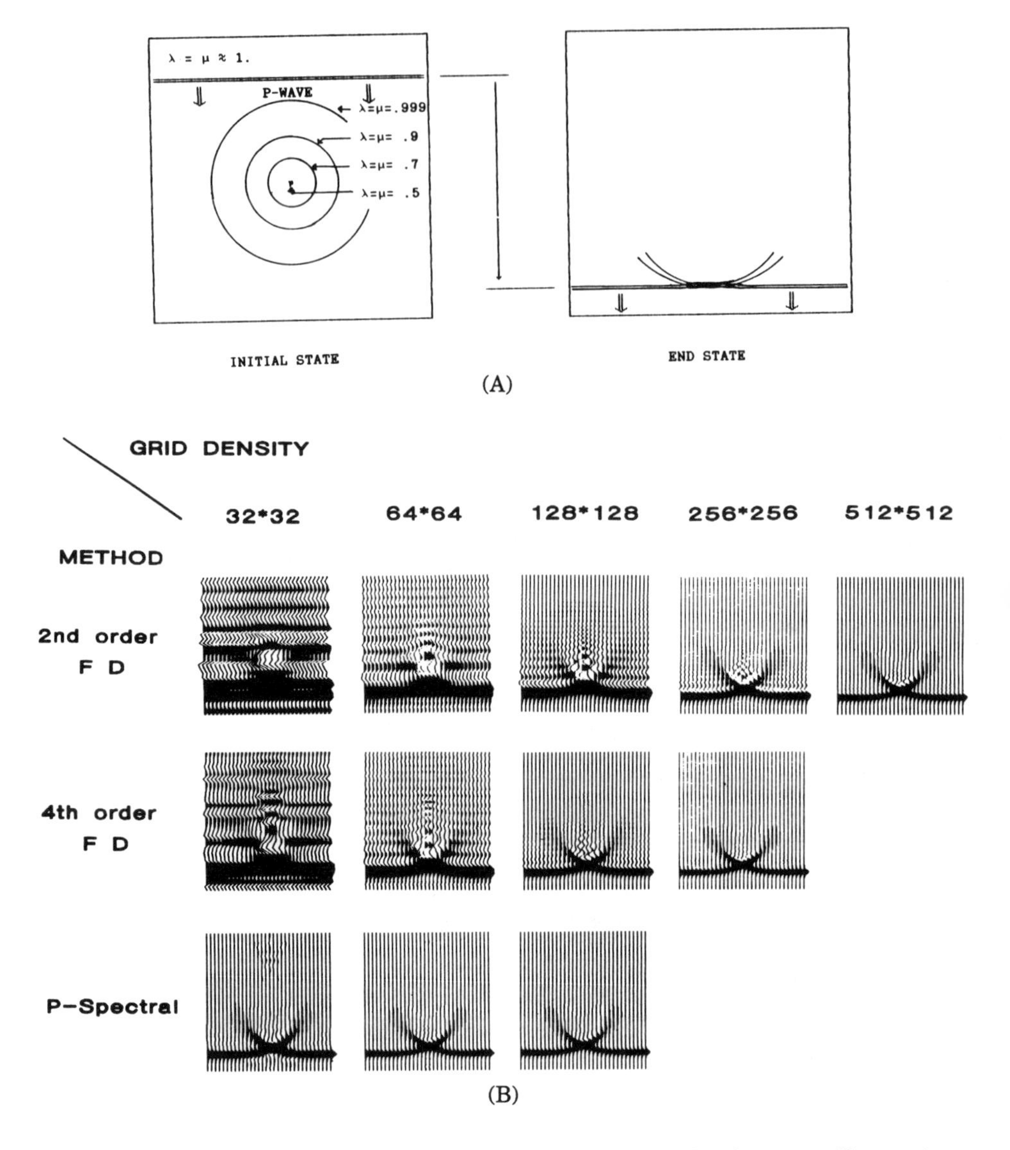

Fig. 23. (a) Contour curves for the variable medium and schematic illustrations of the initial and end states of the test runs. (b) Numerical results for the test problem (variable f displayed). Comparison between different methods and grid densities.

dimensions. Hou and Kreiss (1993) note that in one dimension and with near-singular solutions (for example with thin internal layers to resolve), fourth- and sixth-order FD methods sometimes match (or even exceed) the PS method in efficiency.

6.2. Nonperiodic problems

In this case it is more difficult to provide any single (and still simple) test example that is general enough to be meaningful. Passing from periodic to nonperiodic problems, PS methods encounter many more problems than (low-order) FD methods:

FD: Some more care is needed in stability analysis.

PS: Grid clustering is necessary near edges. This leads to

- conditioning and stability problems (especially notable when time stepping),
- need for preconditioners,
- the prevalence of spurious EVs, especially for high derivatives (cf. Merryfield and Shizgal (1993), on the KdV equation – in sharp contrast to a very favourable situation for periodic PS methods (Fornberg and Whitham, 1978),
- reduced ability to resolve Fourier modes (need π versus 2 points per wavelength).

Formal order of accuracy is the same as the number of grid points – not infinite as in the periodic case (but the significance of this 'philosophical' difference is unclear).

Performance in nonsmooth cases is less well known.

The many successes of Chebyshev-type PS methods in a wide range of applications prove that the added complications are often outweighed by the advantage of exponential accuracy.

One convenient way to keep many options open when developing application codes is to write a FD code of variable order of accuracy on a grid with variable density (using the algorithm in Section 3.1 and Appendix 2). By simply changing parameter values, one can then explore (and exploit) the full range of methods from low-order FD on a uniform grid to Chebyshev (Legendre etc.) and other PS methods. (Obviously, it is also desirable to structure codes so that time stepping methods (if present) are easily interchangeable.) The optimal selections may well turn out to depend not only on the problem type but also on the solution regimes that are studied, the accuracy that is desired etc.

Appendix A. Implementation of Tau, Galerkin and Collocation (PS) for a 'toy' problem

We consider the following model problem

$$u_{xx} + u_x - 2u + 2 = 0, \quad -1 \le x \le 1,$$

$$u(-1) = u(1) = 0,$$

and approximate the exact solution

$$u(x) = 1 - \frac{\sinh(2)e^x + \sinh(1)e^{-2x}}{\sinh(3)}$$

by

$$v(x) = \sum_{k=0}^{4} a_k T_k(x).$$

From (2.1) and (2.2) it follows that the residual

$$R(x) = v_{xx} + v_x - 2v + 2 = \sum_{k=0}^{4} A_k T_k(x) \tag{A.1}$$

satisfies

$$\begin{bmatrix} A_0 \\ A_1 \\ A_2 \\ A_3 \\ A_4 \end{bmatrix} = \begin{bmatrix} -2 & 1 & 4 & 3 & 32 \\ 0 & -2 & 4 & 24 & 8 \\ 0 & 0 & -2 & 6 & 48 \\ 0 & 0 & 0 & -2 & 8 \\ 0 & 0 & 0 & 0 & -2 \end{bmatrix} \begin{bmatrix} a_0 \\ a_1 \\ a_2 \\ a_3 \\ a_4 \end{bmatrix} + \begin{bmatrix} 2 \\ 0 \\ 0 \\ 0 \\ 0 \end{bmatrix}. \tag{A.2}$$

The matrix is obtained as

$$\begin{bmatrix} 0 & 0 & 4 & 0 & 32 \\ & 0 & 0 & 24 & 0 \\ & & 0 & 0 & 48 \\ & & & 0 & 0 \\ & & & & 0 \end{bmatrix} + \begin{bmatrix} 0 & 1 & 0 & 3 & 0 \\ & 0 & 4 & 0 & 8 \\ & & 0 & 6 & 0 \\ & & & 0 & 8 \\ & & & & 0 \end{bmatrix} - 2 \begin{bmatrix} 1 & & & & \\ & 1 & & & \\ & & 1 & & \\ & & & 1 & \\ & & & & 1 \end{bmatrix}$$

corresponding to v_{xx}, v_x and $-2v$ respectively. In general, $\partial^p v/\partial x^p$, the matrix becomes A^p where A is the inverse of the matrix in (2.1) with all zero first column and last row added (a consequence of the shift in the indices between the two column vectors in (2.2)). This procedure to find the elements of A generalizes immediately to Jacobi ploynomials. Closed form expressions for $A^p_{i,j}$ (the element at row i, column j, $0 \le i, j \le n$) turn out to be very simple for both Legendre and Chebyshev expansions. In the Chebyshev case:

$$A_{i,j} = \begin{cases} 1/c_i \times 2j & j > i, i+j \text{ odd}, \\ 0 & \text{otherwise}, \end{cases}$$

$$A^2_{i,j} = \begin{cases} 1/c_i \times (j-i)j(j+i), & j > i, i+j \text{ even}, \\ 0, & \text{otherwise}, \end{cases}$$

where

$$c_i = \begin{cases} 2, & i = 0 \text{ odd}, \\ 1, & i > 0, \end{cases}$$

and

$$A^{p+2}_{i,j} = A^p_{i,j} \times (j-i-p)(j+i+p)(j-i+p)(j+i-p)/[16p(p+1)], \quad p \geq 1$$

(can be shown using the theorem in Karageorghis (1988); for Legendre polynomials, see Phillips (1988)).

Enforcing the boundary conditions $v(-1) = v(1) = 0$ leads to

$$\begin{bmatrix} 1 & 1 & 1 & 1 & 1 \\ 1 & -1 & 1 & -1 & 1 \end{bmatrix} \begin{bmatrix} a_0 \\ a_1 \\ a_2 \\ a_3 \\ a_4 \end{bmatrix} = \begin{bmatrix} 0 \\ 0 \end{bmatrix}. \tag{A.3}$$

Ideally, we would like to get $A_i = 0$, $i = 0, 1, \ldots, 4$ while still satisfying (A.3). However, this would mean satisfying seven relations with only five free parameters a_i, $i = 0, 1, \ldots, 4$. The three spectral methods differ in how they approximate this overdetermined system.

Tau Require $R(x)$ (A.1) to be orthogonal to $T_k(x)$, $k = 0, 1, 2$:

$$\int_{-1}^{1} \frac{R(x)T_k(x)}{\sqrt{1-x^2}}\,\mathrm{d}x = 0 \quad \Rightarrow \quad \begin{array}{l} A_0 = 0, \\ A_1 = 0, \\ A_2 = 0. \end{array}$$

The top three lines of (A.2) together with (A.3) give

$$[a_0, \ldots, a_4] = [0.2724, -0.0444, -0.2562, 0.0444, -0.0162].$$

Galerkin Create from $T_0, \ldots, T_4$ three basis functions Φ_2, Φ_3, Φ_4 which satsfy both boundary conditions:

$$\begin{aligned} \Phi_2(x) &= T_2(x) - T_0(x), \\ \Phi_3(x) &= T_3(x) - T_1(x), \\ \Phi_4(x) &= T_4(x) - T_0(x). \end{aligned}$$

Then $v(x) = \sum_{k=2}^{4} c_k \Phi_k(x)$ which is equal to $\sum_{k=0}^{4} a_k T_k(x)$ constrained by (A.3). Require $R(x)$ to be orthogonal to $\Phi_k(x)$, $k = 2, 3, 4$:

$$\int_{-1}^{1} \frac{R(x)\Phi_k(x)}{\sqrt{1-x^2}}\,\mathrm{d}x = 0 \quad \Rightarrow \begin{bmatrix} 2 & 0 & -1 & 0 & 0 \\ 0 & 1 & 0 & -1 & 0 \\ 2 & 0 & 0 & 0 & -1 \end{bmatrix} \begin{bmatrix} A_0 \\ A_1 \\ A_2 \\ A_3 \\ A_4 \end{bmatrix} = \begin{bmatrix} 0 \\ 0 \\ 0 \end{bmatrix}.$$

Together with (A.2) and (A.3):

$$[a_0, \ldots, a_4] = [0.2741, -0.0370, -0.2593, 0.0370, -0.0148].$$

Collocation (PS) Force $R(x_i) = 0$ at $x_i = \cos(i\pi/4)$, $i = 1, 2, 3$:

$$\begin{bmatrix} 1 & 1/\sqrt{2} & 0 & -1/\sqrt{2} & -1 \\ 1 & 0 & -1 & 0 & 1 \\ 1 & -1/\sqrt{2} & 0 & 1/\sqrt{2} & -1 \end{bmatrix} \begin{bmatrix} A_0 \\ A_1 \\ A_2 \\ A_3 \\ A_4 \end{bmatrix} = \begin{bmatrix} 0 \\ 0 \\ 0 \end{bmatrix}.$$

The section of the discrete cosine transform matrix has the entries $T_k(x_i) = \cos ki\pi/4$, $k = 0, \ldots, 4$, $i = 1, 2, 3$. Together with (A.2) and (A.3):

$$[a_0, \ldots, a_4] = [0.2473, -0.0371, -0.2600, 0.0143].$$

In exact arithmetic,

$$[a_0, \ldots, a_4] = \left[\tfrac{48}{175}, -\tfrac{13}{350}, -\tfrac{13}{50}, \tfrac{13}{350}, -\tfrac{1}{70}\right]$$

and the values at the node locations $x_i, i = 0, \ldots, 4$ become

$$0, \tfrac{101}{350}, +\tfrac{13}{350}\sqrt{2}, \tfrac{13}{25}, \tfrac{101}{350}, -\tfrac{13}{350}\sqrt{2}, 0].$$

This description of the PS approach followed the style of those for Tau and Galerkin, but gave no indication why the PS approach is more flexible than the other two in cases of variable coefficients and nonlinearities. We therefore describe the PS method again, this time in terms of nodal values rather than expansion coefficients. If ν_i denote the approximations at the nodes $x_i = \cos i\pi/4$, $i = 0, 1, \ldots, 4$, the first and second derivatives of the interpolating polynomial become

$$\begin{bmatrix} \nu_{x\,0} \\ \nu_{x\,1} \\ \nu_{x\,2} \\ \nu_{x\,3} \\ \nu_{x\,4} \end{bmatrix} = \begin{bmatrix} -\frac{11}{2} & 4+2\sqrt{2} & -2 & 4-2\sqrt{2} & -\frac{1}{2} \\ -1-\frac{1}{2}\sqrt{2} & \frac{1}{2}\sqrt{2} & \sqrt{2} & -\frac{1}{2}\sqrt{2} & 1-\frac{1}{2}\sqrt{2} \\ \frac{1}{2} & -\sqrt{2} & 0 & \sqrt{2} & -\frac{1}{2} \\ -1+\frac{1}{2}\sqrt{2} & \frac{1}{2}\sqrt{2} & -\sqrt{2} & -\frac{1}{2}\sqrt{2} & 1+\frac{1}{2}\sqrt{2} \\ \frac{1}{2} & -4+2\sqrt{2} & 2 & -4-2\sqrt{2} & \frac{11}{2} \end{bmatrix} \begin{bmatrix} \nu_0 \\ \nu_1 \\ \nu_2 \\ \nu_3 \\ \nu_4 \end{bmatrix} \tag{A.4}$$

and

$$\begin{bmatrix} \nu_{xx\,0} \\ \nu_{xx\,1} \\ \nu_{xx\,2} \\ \nu_{xx\,3} \\ \nu_{xx\,4} \end{bmatrix} = \begin{bmatrix} 17 & -20-6\sqrt{2} & 18 & -20+6\sqrt{2} & 5 \\ 5+3\sqrt{2} & -14 & 6 & -2 & 5-3\sqrt{2} \\ -1 & 4 & -6 & 4 & -1 \\ 5-3\sqrt{2} & -2 & 6 & -14 & 5+3\sqrt{2} \\ 5 & -20+6\sqrt{2} & 18 & -20-6\sqrt{2} & 17 \end{bmatrix} \begin{bmatrix} \nu_0 \\ \nu_1 \\ \nu_2 \\ \nu_3 \\ \nu_4 \end{bmatrix} \tag{A.5}$$

These matrices are examples of 'differentiation matrices' – discussed in numerous places in this review. Section 4.3 describes how their elements can be obtained very conveniently.

Enforcing

$$R(x) = \nu_{xx} + \nu_x - 2\nu + 2 = 0$$

at the node points x_k, $k = 1, 2, 3$ and the boundary conditions $\nu_0 = \nu_4 = 0$ lead to

$$\begin{bmatrix} -16+\frac{1}{2}\sqrt{2} & 6+\sqrt{2} & -2-\frac{1}{2}\sqrt{2} \\ 4-\sqrt{2} & -8 & 4+\sqrt{2} \\ -2+\frac{1}{2}\sqrt{2} & 6-\sqrt{2} & -16-\frac{1}{2}\sqrt{2} \end{bmatrix} \begin{bmatrix} \nu_1 \\ \nu_2 \\ \nu_3 \end{bmatrix} = \begin{bmatrix} -2 \\ -2 \\ -2 \end{bmatrix} \tag{A.6}$$

with the same solution as before:

$$\begin{bmatrix} \nu_1 \\ \nu_2 \\ \nu_3 \end{bmatrix} = \begin{bmatrix} \frac{101}{350} + \frac{13}{350}\sqrt{2} \\ \frac{13}{25} \\ \frac{101}{350} - \frac{13}{350}\sqrt{2} \end{bmatrix} \tag{A.7}$$

Had variable coefficients been present, their values at the nodes would have been used to multiply the rows of (A.4), (A.5) etc., when assembling (A.6).

Figure A.1 shows how the accuracy increases with n – in all three cases featuring an exponential rate of convergence. For comparison, curves for second- and fourth-order FD approximations (on equi-spaced grids) are also included.

Appendix B. Fortran code and test driver for algorithm to find weights in FD formulae

```
      SUBROUTINE WEIGHTS (XI,X,N,M,C)
C     INPUT PARAMETERS:
C       XI  POINT AT WHICH THE APPROXIMATIONS ARE TO BE ACCURATE
C       X   X-COORDINATES FOR THE GRID POINTS, ARRAY DIMENSIONED X(0:N)
C       N   THE GRID POINTS ARE AT X(0),X(1), ...  X(N) (I.E. N+1 IN ALL)
C       M   HIGHEST ORDER OF DERIVATIVE TO BE APPROXIMATED
C
```

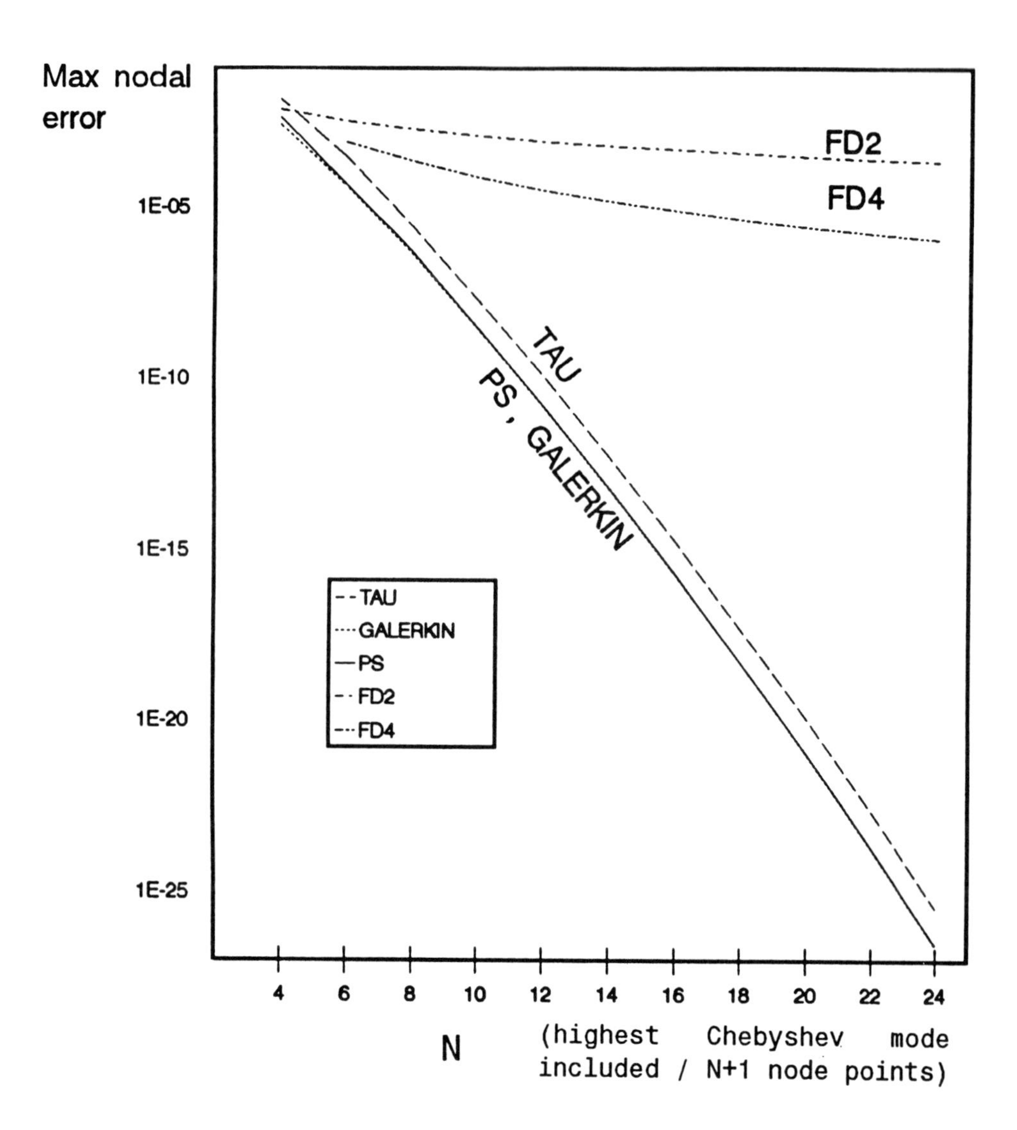

Fig. A.1. Maximum nodal errors for different methods when applied to 'toy problem' in Appendix 1 – comparison between three spectral implementations and equispaced FD methods of second and fourth order.

```
C     OUTPUT PARAMETERS:
C       C  WEIGHTS, ARRAY DIMENSIONED C(0:N,0:N,0:M).
C          ON RETURN, THE ELEMENT C(K,J,I) CONTAINS THE WEIGHT TO BE
C          APPLIED AT X(K) WHEN THE I:TH DERIVATIVE IS APPROXIMATED
C          BY A STENCIL EXTENDING OVER X(0),X(1), ...  , X(J).
C
C
      DIMENSION X(0:N),C(0:N,0:N,0:M)
      C(0,0,0) = 1.
      C1       = 1.
      C4       = X(0)-XI
      DO 40 J=1,N
        MN = MIN(J,M)
        C2 = 1.
        C5 = C4
        C4 = X(J)-XI
        DO 20 K=0,J-1
          C3 = X(J)-X(K)
          C2 = C2*C3
          IF (J.LE.M) C(K,J-1,J)=0.
          C(K,J,0) = C4*C(K,J-1,0)/C3
          DO 10 I=1,MN
10          C(K,J,I) = (C4*C(K,J-1,I)-I*C(K,J-1,I-1))/C3
20        CONTINUE
        C(J,J,0) = -C1*C5*C(J-1,J-1,0)/C2
        DO 30 I=1,MN
30        C(J,J,I) = C1*(I*C(J-1,J-1,I-1)-C5*C(J-1,J-1,I))/C2
40      C1 = C2
      RETURN
      END
```

Note If N is very large, the calculation of the variable C2 might cause overflow (or underflow). For example, in generating extensions of Tables 2 and 3, this problem arises when `N!` exceeds the largest possible number (i.e. `N > 34` in typical 32 bit precision with 3×10^{38} as the largest number; `N > 965` in CRAY single precision (64 bit word length, 15 bit exponent, largest number approximately 10^{2465})). In such cases, scaling of `C1` and `C2` (only used in forming the ratio `C1/C2`) should be added to the code.

The following test program will print out all the entries in Table 3 (including a table of coefficients for the zeroth derivative – interpolation weights – trivial here since approximations are requested at a grid point).

```
      PROGRAM TEST
      PARAMETER (M=4,N=8)
      DIMENSION X(0:N),C(0:N,0:N,0:M)
```

```
      DO 10 I=0,N
10      X(I) = I
      CALL WEIGHTS (0.,X,N,M,C)
      DO 30 I=0,M
        DO 20 J=I,N
20        WRITE (6,40) (C(K,J,I),K=0,J)
30      WRITE (6,*)
40    FORMAT (1X,9F8.3)
      STOP
      END
```

All the data in Table 2 can similarly be obtained from a single call to `SUBROUTINE WEIGHTS` by initializing

`X(0:8)` to `/0, −1,1,−2,2,−3,3,−4,4/`

(and ignore every second line of the output).

REFERENCES

S. Abarbanel, D. Gottlieb and E. Tadmor (1986), 'Spectral methods for discontinuous problems', in *Numerical Methods for Fluid Dynamics II* (K.W. Morton and M.J. Baines, eds), Clarendon Press (Oxford), 129–153.

A. Bayliss, D. Gottlieb, B.J. Matkowsky and M. Minkoff (1989), 'An adaptive pseudospectral method for reaction diffusion problems', *J. Comput. Phys.* **81**, 421–443.

A. Bayliss and E. Turkel (1992), 'Mappings and accuracy for Chebyshev pseudospectral approximations', *J. Comput. Phys.* **101**, 349–359.

G. Beylkin, R. Coifman and V. Rokhlin (1991), 'Fast wavelet transforms and numerical algorithms', *Comm. Pure Appl. Math.* **44**, 141–183.

J.P. Boyd (1989), *Chebyshev and Fourier Spectral Methods*, Springer (New York)

J.P. Boyd (1992), 'Multipole expansions and pseudospectral cardinal functions: a new generalization of the fast Fourier transform', *J. Comput. Phys.* **103**, 184–186.

K.S. Breuer and R.M. Everson (1992), 'On the errors incurred calculating derivatives using Chebyshev polynomials', *J. Comput. Phys.* **99**, 56–67.

W. Cai, D. Gottlieb and C.-W. Shu (1989), 'Essentially nonoscillatory spectral Fourier methods for shock wave calculation', *Math. Comput.* **52**, 389–410.

W. Cai, D. Gottlieb and C.W. Shu (1992), 'One-sided filters for spectral Fourier approximation of discontinuous functions', *SIAM J. Numer. Anal.* **29**, 905–916.

C. Canuto, M.Y. Hussaini, A. Quarteroni and T. Zang (1988), *Spectral Methods in Fluid Dynamics*, Springer (New York).

C. Canuto and P. Pietra (1987), 'Boundary and interface conditions with a FE preconditioner for spectral methods'. Report No. 553, I.A.N., Pavia University, Italy.

C. Canuto and A. Quarteroni (1985), 'Preconditioned minimal residual methods for Chebyshev spectral calculations', *J. Comput. Phys.* **60**, 315–337.

C. Canuto and A. Quarteroni (1987), 'On the boundary treatment in spectral method for hyperbolic systems', *J. Comput. Phys.* **71**, 100–110.

E.W. Cheney (1966), *Introduction to Approximation Theory*, McGraw-Hill (New York).

J.W. Cooley and J.W. Tukey (1965), 'An algorithm for the machine calculation of complex Fourier series', *Math. Comput.* **19**, 297–301.

G. Dahlquist (1956), 'Convergence and stability in the numerical integration of ordinary differential equations', *Math. Scand.* **4**, 33–53.

G. Dahlquist (1985), '33 years of numerical instability, part I', *BIT* **25**, 188–204.

P.J. Davis (1975), *Interpolation and Approximation*, Dover (New York).

P.J. Davis and P. Rabinowitz (1984), *Methods of Numerical Integration*, 2nd Edn, Academic Press (London, New York).

M. Deville and E. Mund (1985), 'Chebyshev PS solution of second-order elliptic equations with finite element preconditioning', *J. Comput. Phys.* **60**, 517–553.

M. Dubiner (1987), 'Asymptotic analysis of spectral methods', *J. Sci. Comput.* **2**, 3–31.

H. Eisen and W. Heinrichs (1992), 'A new method of stabilization for singular perturbation problems with spectral methods', *SIAM J. Numer. Anal.* **29**, 107–122.

B. Fornberg (1975), 'On a Fourier method for the integration of hyperbolic equations', *SIAM J. Numer. Anal.* **12**, 509–528.

B. Fornberg (1987), 'The pseudospectral method: comparisons with finite differences for the elastic wave equation', *Geophysics* **52**, 483–501.

B. Fornberg (1988a), 'The pseudospectral method: accurate representation of interfaces in elastic wave calculations', *Geophysics* **53**, 625–637.

B. Fornberg (1988b), 'Generation of finite difference formulae on arbitrarily spaced grids', *Math. Comput.* **51**, 699–706.

B. Fornberg (1990a), 'High order finite differences and the pseudospectral method on staggered grids', *SIAM J. Num. Anal.* **27**,904–918.

B. Fornberg (1990b), 'An improved pseudospectral method for initial-boundary value problems', *J. Comput. Phys.* **91**, 381–397.

B. Fornberg (1992), 'Fast generation of weights in finite difference formulas', in *Recent Developments in Numerical Methods and Software for ODEs/DAEs/PDEs* (G.D. Byrne and W.E. Schiesser, eds), World Scientific (Singapore), 97–123.

B. Fornberg (1994), 'A pseudospectral approach for polar and spherical geometries', *SIAM J. Sci. Comput.*, submitted.

B. Fornberg and G.B. Whitham (1978), 'A numerical and theoretical study of certain nonlinear wave phenomena', *Phil. Trans. Roy. Soc. London* A **289**, 373–404.

J.B.J. Fourier (1822), *Théorie analytique de la chaleur*, Paris.

D. Funaro (1987), 'A preconditioned matrix for the Chebyshev differencing operator', *SIAM J. Numer. Anal.* **24**, 1024–1031.

D. Funaro (1992), *Polynomial Approximation of Differential Equations*, Lecture Notes in Physics 8, Springer (Berlin).

D. Funaro and D. Gottlieb (1988), 'A new method of imposing boundary conditions in pseudospectral approximations of hyperbolic equations', *Math. Comput.* **51**, 599–613.

D. Gaier (1987), *Lectures on Complex Approximation*, Birkhäuser (Boston).

C.W. Gear (1971), *Numerical Solution of Ordinary and Partial Differential Equations*, Prentice Hall (Englewood Cliffs, NJ).

M. Goldberg and E. Tadmor (1985), 'Convenient stability criteria for difference approximations of hyperbolic initial-boundary value problems', *Math. Comput.* **44**, 361–377.

D. Gottlieb and L. Lustman (1983), 'The spectrum of the Chebyshev collocation operator for the heat equation', *SIAM J. Numer. Anal.* **20**, 909–921.

D. Gottlieb, L. Lustman and E. Tadmor (1987), 'Stability analysis of spectral methods for hyperbolic initial-boundary value problems', *SIAM J. Num. Anal.* **24**, 241–258.

D. Gottlieb and S.A. Orszag (1977), *Numerical Analysis of Spectral Methods*, SIAM (Philadelphia).

D. Gottlieb and E. Tadmor (1991), 'The CFL condition for spectral approximation to hyperbolic BVPs', *Math. Comput.* **56**, 565–588.

D. Gottlieb and E. Turkel (1980), 'On time discretization for spectral methods', *Stud. Appl. Math.* **63**, 67–86.

P.M. Greshko and R.L. Lee (1981), 'Don't suppress the wiggles – they're trying to tell you something', *Computers and Fluids* **9**, 223–253.

B. Gustafsson, H.-O. Kreiss and A. Sundstrom (1972), 'Stability theory of difference approximations for mixed initial-boundary value problems II', *Math. Comput.* **26**, 649–685.

E. Hairer, S.P. Nørsett and G. Wanner (1987), *Solving Ordinary Differential Equations I – Nonstiff Problems*, Springer (Berlin).

E. Hairer and G. Wanner (1991), *Solving Ordinary Differential Equations II – Stiff and Differential-Algebraic Problems*, Springer (Berlin).

P. Haldenwang, G. Labrosse, S. Abboudi and M. Deville (1984), 'Chebyshev 3-D spectral and 2-D pseudospectral solvers for the Helmholtz equation', *J. Comput. Phys.* **55**, 115–128.

O. Holberg (1987), 'Computational aspects of the choice of operator and sampling interval for numerical differentiation in large-scale simulation of wave phenomena', *Geophys. Prospecting* **35**, 629–655.

Y.-C. C. Hou and H.-O. Kreiss (1993), 'Comparison of finite difference and the pseudo-spectral approximations for hyperbolic equations', to be published.

W. Huang and D.M. Sloan (1993a), 'Pole condition for singular problems: the pseudospectral approximation', *J. Comput. Phys.* **107**, 254–261.

W. Huang and D.M. Sloan (1993b), 'A new pseudospectral method with upwind features', *IMA. J. Numer. Anal.* **13**, 413–430.

W. Huang and D.M. Sloan (1993c), 'The pseudospectral method for solving differential eigenvalue problems', *J. Comput. Phys.*, to appear.

M.Y. Hussaini and T.A. Zang (1984), 'Iterative spectral methods and spectral solution to compressible flows', in *Spectral Methods for PDEs* (R. Voigt, D. Gottlieb and M.Y. Hussaini, eds), SIAM (Philadelphia).

A. Karageorghis (1988), 'A note on the Chebyshev coefficients of the general order derivative of an infinitely differentiable function', *J. Comput. Appl. Math.* **21**, 129–132.

M. Kindelan, A. Kamel and P. Sguazzero (1990), 'On the construction and efficiency of staggered numerical differentiators for the wave equation', *Geophysics* **55**, 107–110.

D. Kosloff and H. Tal-Ezer (1993), 'A modified Chebyshev pseudospectral method with an $\mathcal{O}(N^{-1})$ time step restriction', *J. Comput. Phys.* **104**, 457-469.

H.-O. Kreiss (1962), 'Über die Stabilitätsdefinition für Differenzengleichungen die partielle Differentialgleichungen approximieren', *Nordisk Tidskr. Informationsbehandling* **2**, 153–181.

H.-O. Kreiss (1968), 'Stability theory for difference approximations of mixed initial boundary value problems I', *Math. Comput.* **22**, 703–714.

H.-O. Kreiss (1970), 'Initial boundary value problems for hyperbolic systems', *Comm. Pure Appl. Math.* **23**, 277–288.

H.-O. Kreiss and J. Oliger (1972), 'Comparison of accurate methods for the integration of hyperbolic equations', *Tellus* **XXIV**, 199–215.

V.I. Krylov (1962), *Approximate Calculation of Integrals*, Macmillan (New York).

J.D. Lambert (1991), *Numerical Methods for Ordinary Differential Systems: The Initial Value Problem*, Wiley (Chichester, UK).

C. Lanczos (1938), 'Trigonometric interpolation of empirical and analytical functions', *J. Math. Phys.* **17**, 123–199.

S.K. Lele (1992), 'Compact finite difference schemes with spectral-like resolution', *J. Comput. Phys.* **103**, 16–42.

Y. Maday, S.M.O. Kaber and E. Tadmor (1993), 'Legendre PS viscosity methods for nonlinear conservation laws', *SIAM J. Numer. Anal.* **30**, 321–342.

A. Majda, J. McDonough and S. Osher (1978), 'The Fourier method for non-smooth initial data', *Math. Comput.* **32**, 1041–1081.

A.I. Markushevich (1967), *Theory of Functions of a Complex Variable*, Vol. III (Transl. by R.A. Silverman), Prentice Hall (New York).

B. Mercier (1989), *An Introduction to the Numerical Analysis of Spectral Methods*, Springer (Berlin).

W.J. Merryfield and B. Shizgal (1993), 'Properties of collocation third-derivative operators', *J. Comput. Phys.* **105**, 182–185.

R. Mittet, O. Holberg, B. Arntsen and L. Amundsen (1988), 'Fast finite difference modeling of 3-D elastic wave equation', *Society of Exploration Geophysics Expanded Abstracts* **I**, 1308–1311.

L.S. Mulholland and D.M. Sloan (1992), 'The role of preconditioning in the solution of evolutionary PDEs by implicit Fourier PS methods', *J. Comput. Appl. Math.* **42**, 157–174.

K.L. Nielsen (1956), *Methods in Numerical Analysis*, Macmillan (New York).

S.A. Orszag (1969), 'Numerical methods for the simulation of turbulence', *Phys. Fluids Suppl. II*, **12**, 250–257.

S.A. Orszag (1970), 'Transform method for calculation of vector coupled sums: Application to the spectral form of the vorticity equation', *J. Atmos. Sci.* **27**, 890–895.

S.A. Orszag (1972), 'Comparison of pseudospectral and spectral approximations', *Stud. Appl. Math.* **51**, 253–259.

T.N. Phillips (1988), 'On the Legendre coefficients of a general-order derivative of an infinitely differentiable function', *IMA J. Numer. Anal.* **8**, 455–459.

T.N. Phillips, T.A. Zang and M.Y. Hussaini (1986), 'Preconditioners for the spectral multigrid method', *IMA J. Numer. Anal.* **6**, 273–292.

M.J.D. Powell (1981), *Approximation Theory and Methods*, Cambridge University Press (Cambridge).

S.C. Reddy and L.N. Trefethen (1990), 'Lax-stability of fully discrete spectral methods via stability regions and pseudo-eigenvalues', *Comput. Meth. Appl. Mech. Engrg* **80**, 147–164.

R.D. Richtmyer and K.W. Morton (1967), *Difference Methods for Initial-value Problems*, 2nd Edn, Wiley (New York).

T.J. Rivlin (1969), *An Introduction to the Approximation of Functions*, Dover (New York).

J. Sand and O. Østerby (1979), 'Regions of absolute stability', Report DAIMI PP-102, Computer Science Department, Aarhus University, Denmark.

G. Sansone (1959), *Orthogonal Functions*, Interscience (New York).

L.F. Shampine and M.K. Gordon (1975), *Computer Solution of Ordinary Differential Equations*, W.H. Freeman (San Francisco).

A. Solomonoff (1992), 'A fast algorithm for spectral differentiation', *J. Comput. Phys.* **98**, 174–177.

A. Solomonoff (1994), 'Bayes finite difference schemes', *SIAM J. Num. Anal.*, submitted.

A. Solomonoff and E. Turkel (1989), 'Global properties of pseudospectral methods', *J. Comput. Phys.* **81**, 239–276.

F. Stenger (1981), 'Numerical methods based on Whittaker cardinal, or sinc functions', *SIAM Review* **23**, 165–224.

G. Szegő (1959), *Orthogonal Polynomials*, American Mathematical Society (Washington, D.C.).

E. Tadmor (1987), 'Stability analysis of finite-difference, pseudospectral and Fourier–Galerkin approximations for time-dependent problems', *SIAM Review* **29**, 525–555.

E. Tadmor (1989), 'Convergence of spectral methods for nonlinear conservation laws', *SIAM J. Num. Anal.* **26**, 30–44.

E. Tadmor (1990), 'Shock capturing by the spectral viscosity method', *Comput. Meth. Appl. Mech. Engrg* **80**, 197–208.

E. Tadmor (1993),'Superviscosity and spectral approximations of nonlinear conservation laws', in *Numerical Methods for Fluid Dynamics IV* (M.J. Baines and K.W. Morton, eds), Oxford University Press (Oxford), 69–82.

T.D. Taylor, R.S. Hirsh and N.M. Nadworny (1984), 'Comparison of FFT, direct inversion and conjugate gradient methods for use in pseudospectral methods', *Comput. Fluids* **12**, 1–9.

L.N. Trefethen (1983), 'Group velocity interpretation of the stability theory of Gustafsson, Kreiss and Sundström', *J. Comput. Phys.* **49**, 199–217.

L.N. Trefethen (1988), 'Lax-stability vs. eigenvalue stability of spectral methods', in *Numerical Methods for Fluid Dynamics III* (K.W. Morton and M.J. Baines, eds), Clarendon Press (Oxford), 237–253.

L.N. Trefethen and M.R. Trummer (1987), 'An instability phenomenon in spectral methods', *SIAM J. Numer. Anal.* **24**, 1008–1023.

L.N. Trefethen and J.A.C. Weideman (1991), 'Two results on polynomial interpolation in equally spaced points', *J. Approx. Theory* **65**, 247–260.

R.G. Voigt, D. Gottlieb and M.Y. Hussani (eds) (1984), *Spectral Methods for Partial Differential Equations*, SIAM (Philadelphia).

J.L. Walsh (1960), *Interpolation and Approximation by Rational Functions in the Complex Domain*, Colloquium Publications of the Amer. Math. Soc., 20, 3rd Edn.

J.A.C. Weideman (1992), 'The eigenvalues of Hermite and rational spectral DMs', *Numer. Math.* **61**, 409–432.

J.A.C. Weideman and L.N. Trefethen (1988), 'The eigenvalues of second-order spectral differentiation matrices', *SIAM J. Numer. Anal.* **25**, 1279–1298.

B.D. Welfert (1993), 'A remark on pseudospectral differentiation matrices', submitted to *SIAM J. Num. Anal.*

E.T. Whittaker (1915), 'On the functions which are represented by the expansions of the interpolation theory', *Proc. Roy. Soc. Edinburgh* **35**, 181–194.

J.M. Whittaker (1927), 'On the cardinal function of interpolation theory', *Proc. Edinburgh Math. Soc. Ser. 1*, **2**, 41–46.

Acta Numerica (1994), *pp.* 269–378

Exact and approximate controllability for distributed parameter systems

R. Glowinski
University of Houston, Houston, Texas, USA
Université Pierre et Marie Curie, Paris, France
C.E.R.F.A.C.S., Toulouse, France

J.L. Lions
Collège de France
Rue d'Ulm, 75005 Paris, France

This is the first part of an article which will be continued in the next volume of *Acta Numerica*. References in the text to Sections 2 to 7 refer to the continuation.

CONTENTS

0 INTRODUCTION 271
0.1 What is it all about 271
0.2 Motivation 272
0.3 Topologies and numerical methods 273
0.4 Choice of control 274
0.5 Relaxation of the controllability notion 275
0.6 Various remarks 275
1 DISTRIBUTED AND POINTWISE CONTROL FOR LINEAR DIFFUSION EQUATIONS 276
1.1 First example 276
1.2 Approximate controllability 279
1.3 Formulation of the approximate controllability problem 280
1.4 Dual problem 281
1.5 Direct solution to the dual problem 282
1.6 Penalty arguments 285
1.7 L^∞ cost functions and bang–bang controls 288
1.8 Numerical methods 294
1.9 Relaxation of controllability 320
1.10 Pointwise control 324
References 376

CONTENTS 1995

The contents of the rest of this article (to appear in *Acta Numerica 1995*) are as follows:

2. BOUNDARY CONTROL

2.1. Dirichlet control (I): formulation of the control problems
2.2. Dirichlet control (II): optimality conditions and dual formulations
2.3. Dirichlet control (III): iterative solution of the control problems
2.4. Dirichlet control (IV): approximation of the control problems
2.5. Dirichlet control (V): iterative solution of the fully discrete dual problem (2.128)
2.6. Dirichlet control (VI): numerical experiments
2.7. Neuman control (I): formulation of the control problems
2.8. Neuman control (II): optimality conditions and dual formulations
2.9. Neuman control (III): conjugate gradient solution of the dual problem (2.196)
2.10. Neuman control (IV): iterative solution of the dual problem (2.212), (2.213)

3. CONTROL OF THE STOKES SYSTEM

3.1. Generalities. Synopsis
3.2. Formulation of the Stokes system. A fundamental controllability result
3.3. Two approximate controllability problems
3.4. Optimality conditions and dual problems
3.5. Iterative solution of the control problems
3.6. Time discretization of the control problem (3.19)
3.7. Numerical experiments

4. CONTROL OF NONLINEAR DIFFUSION SYSTEMS

4.1. Generalities. Synopsis
4.2. An example of a noncontrollable nonlinear system
4.3. Pointwise control of the viscous Burgers equation
4.4. Controllability and the Navier–Stokes equations

5. DYNAMIC PROGRAMMING FOR LINEAR DIFFUSION EQUATIONS

5.1. Introduction. Synopsis
5.2. Derivation of the Hamilton–Jacobi–Bellman equation
5.3. Some remarks

6. WAVE EQUATIONS

6.1. Wave equations: Dirichlet boundary control
6.2. Approximate controllability

6.3. Formulation of the approximate controllability problem
6.4. Dual problems
6.5. Direct solution of the dual problem
6.6. Exact controllability and new functional spaces
6.7. On the structure of space E
6.8. Numerical methods for the Dirichlet boundary controllability of the wave equation
6.9. Experimental validation of the filtering procedure of Section 6.8.7 via the solution of the test problem of Section 6.8.5
6.10. Other boundary controls
6.11. Distributed controls for wave equations
6.12. Dynamic programming
6.13. On the application of controllability methods to the solution of the Helmholtz equation at large wave numbers
6.14. Further problems

7. COUPLED SYSTEMS

7.1. A problem from thermoelasticity
7.2. The limit cases $\alpha \to 0$ and $\alpha \to +\infty$
7.3. Approximate partial controllability
7.4. Approximate controllability via penalty

0. INTRODUCTION

0.1. What is it all about

We consider a system whose *state* is given by the solution y to a Partial Differential Equation (PDE) of evolution, and which contains *control functions*, denoted by v.

Let us write in a formal fashion for the time being. The state equation is written

$$\frac{\partial y}{\partial t} + \mathcal{A}(y) = \mathcal{B}v, \tag{0.1}$$

where y is a scalar or a vector valued function.

In (0.1), $\mathcal{A}$ is a set of Partial Differential Operators (PDO), linear or nonlinear (at least for the time being). In (0.1), v denotes the control and $\mathcal{B}$ maps the 'space of controls' into the state space. It goes without saying that all this has to be made precise. This will be the task of the following sections.

The PDE (0.1) should include *boundary conditions*. We do not make them explicit here. They are supposed to be contained in the abstract formulation (0.1), where v can be applied inside the domain $\Omega \subset \mathbb{R}^d$ where (0.1) is considered (one says that v is a *distributed* control) or on the boundary Γ of Ω (or a part of it). One says then that v is *a boundary control.*

If v is applied at points of Ω, v is said to be a *pointwise* control.

One has also to add *initial conditions* to (0.1): if we assume that $t = 0$ is the initial time, then these initial conditions are given by

$$y|_{t=0} = y_0, \tag{0.2}$$

y_0 being a given element of the state space.

It will be assumed that, given v (in a suitable space), problem (0.1), (0.2) (and the boundary conditions included in formulation (0.1)) *uniquely defines a solution.*

It is a function (scalar or vector valued) of $x \in \Omega$, $t > 0$, and of y_0 and v. *We shall denote by* $y(v)(= \{x,t\} \to y(x,t;v))$ *this solution.*

We shall denote by $y(t;v)$ the function $x \to y(x,t;v)$.

Then (0.2) can be written as

$$y(0;v) = y_0. \tag*{(0.2)$_*$}$$

Remark 0.1 The notions introduced below can be extended to situations where the *uniqueness* of the solution to (0.1), (0.2) is *not known.* We are thinking here of the Navier–Stokes equations in $\Omega \subset \mathbb{R}^d$, $d = 3$, and the equations related to it.

We can now introduce the notion of *controllability*, either exact or approximate.

Let $T > 0$ *be given* and let y_T be a given element of the state space. We want to 'drive the system' from y_0 at $t = 0$ to y_T at $t = T$, i.e. we want to find v such that

$$y(T;v) = y_T. \tag{0.3}$$

If this is possible for every y_T in the state space, one says that the system is *controllable* (or *exactly controllable*).

If – as we shall see in most of the examples – condition (0.3) is too strict, it is natural to replace it by

$$y(T;v) \text{ belongs to a 'small' neighbourhood of } y_T. \tag{0.4}$$

If this is possible, one says that the system is *approximately controllable.* Otherwise the system is *not* controllable.

Before starting with precise examples, we want to say a few words concerning the motivation for studying these problems.

0.2. Motivation

There are several aspects which make controllability problems important in practice.

Aspect 1 At a *given time horizon* we want the system under study to behave *exactly* as we wish (or in a manner arbitrarily close to it).

Problems of this type are common in science and engineering: we would like, for example, to have the temperature (or pressure) of a system equal or very close to a given value – globally or locally – at a given time. *Chemical engineering* is an important source of such problems, a typical example in that direction being the design of car catalytic converters; in this example chemical reactions have to take place leading to the 'destruction' at a given time horizon (very small in practice) of the polluting chemicals contained in the exhaust gases (the modelling and numerical simulation of catalytic converter systems are discussed in Engquist, Gustafsson and Vreeburg (1978), Friedman (1988, Ch. 7); see also Friend (1993)).

Aspect 2 For *linear* systems, it is known (cf. Russell (1978)) that exact controllability is equivalent to the possibility of *stabilizing* the system.

Stabilization problems abound, in particular in (large) composite structures, the so called 'multi-body systems' made of many different parts, which can be considered as three-, two- or one-dimensional and which are 'tied' together by *junctions* and *joints*. The modelling and analysis of such systems is the subject of many interesting studies. We wish to mention here Ph. Ciarlet and his collaborators (see, for example, Ciarlet (1990a,b) and Ciarlet, Le Dret and Nzengwa (1989)), Hubert and Palencia (1989), Lagnese and Leugering (1994), Simo and his collaborators (see for example, Laursen and Simo (1994)), Park and his collaborators (see for example Park, Chiou and Downer (1990) and Downer, Park and Chiou (1992)).

Studying controllability is *one* approach to stabilization (Lions, 1988a).

Aspect 3 Controllability and reversibility. Suppose we have a system which *was* in a state z_1 at time $-t_0$, $t_0 > 0$, and which is *now* in the state y_0.

We would like to have the system *returning* to a state as close as possible to z_1, i.e. $y_T = z_1$. If this is possible, it means some kind of 'reversibility'. What we have in mind here are *environment systems*; should they be 'local' or 'global' in the space variables?

Noncontrollable subsystems can suffer 'irreversible' changes (cf. Lions (1990) and Diaz (1991)).

We return now to the general questions of Section 0.1, making them more precise before giving examples.

0.3. Topologies and numerical methods

The topology of the state space appears explicitly in condition (0.4). It is obvious that approximate controllability *depends* on the choice of the topology on the state space, i.e. of the state space, itself. Actually *exact* controllability depends on the choice of the state space as well.

The choice of the state space is therefore an obviously fundamental issue for the *theory*.

We want to emphasize that it is *also* a fundamental issue from the *numerical point of view*.

If one has (as we shall see in several situations) exact or approximate controllability in a very general space (which can include elements which are not distributions but 'ultra distributions') and *not* in a classical space of smooth (or sufficiently smooth) functions, then the numerical approximation will *necessarily* develop singularities and 'remedies' should be based on knowledge of the topology where theoretical convergence takes place. We shall return to these issues in the following sections; some of them have been addressed in, e.g., Dean, Glowinski and Li (1989), Glowinski, Li and Lions (1990), Glowinski and Li (1990), Glowinski 1992a), where various filtering techniques are discussed in order to eliminate the numerical singularities mentioned above.

In the following section we shall address the question, also very general in nature, namely: *how to choose the control*?

0.4. Choice of control

Let us return to the general formulation (0.1), (0.2), (0.3) (or (0.4)). If there exists *one* control v achieving these conditions, then there exist in general *infinitely many other* vs also achieving these conditions. Which one should we choose and how?

A most important question is: how to choose the *norm* (we are always working in Banach or Hilbert spaces) for the vs? This is related to the *topology* of the state space. It is indeed clear that the regularity (or irregularity!) properties of v and of y in (0.1) are related. Let us assume that a norm $|||v|||$ is chosen.

Once this choice is made, a natural formulation of the problem is then to find

$$\inf |||v|||, \tag{0.5}$$

among all vs such that (0.1), (0.2), (0.3) (or (0.4)) take place.

Remark 0.2 There is still some flexibility here, since problem (0.5) makes sense if one replaces $|||\cdot|||$ by a *stronger* norm. This remark may be of *practical* interest, as we shall see later on.

Remark 0.3 One can meet questions of controllability for systems depending on 'small' parameters. Two classical (by now) examples are:

(i) *singular perturbations*,

(ii) *homogenization* which is important for the controllability of structures made of *composite materials*.

In these situations one has to introduce either *families* of norms in (0.5) or norms *equivalent* to $|||\cdot|||$ but which depend on the homogenization parameter.

0.5. Relaxation of the controllability notion

Let us return again to (0.1), (0.2).

Condition (0.3) concerns the state y itself. In a 'complex system' this condition can be (and will be in general) unnecessarily strong.

We may want some *subsystem* to behave according to our wishes. We may also want *average* values to behave accordingly, etc.

A general formulation is as follows.

We consider an operator

$$C \in \mathcal{L}(Y, \mathcal{H}), \tag{0.6}$$

where Y is the state space (chosen!) and where $\mathcal{H}$ is another Banach or Hilbert space (the *observation* space). Think, for instance, of C as being an *averaging* operator.

Then we 'relax' (0.3) (respectively (0.4)) as follows

$$Cy(T;v) = h_T, \quad h_T \text{ given in } \mathcal{H} \tag{0.7}$$

(respectively

$$Cy(T;v) \text{ belongs to some neighbourhood of } h_T \text{ in } \mathcal{H}). \tag{0.8}$$

Then we consider (0.5) where v is subject to (0.7) (respectively (0.8)).

0.6. Various remarks

Remark 0.4 For most examples considered in this article, the control function is either distributed (or pointwise) or of a boundary nature. It can also be a *geometrical* one. Namely we can consider the domain Ω as variable or, to be more precise, at least a part of the boundary of Ω is variable, and we want to 'move this part of the boundary' in order to drive the system from a given state to another one. In summary we look for *controllability by a suitable variable geometry.* Problems of this type are discussed in Bushnell and Hefner (1990); they mostly concern *drag reduction* for viscous flow (see also Sellin and Moses (1989)). We shall return to this on other occasions.

Remark 0.5 Some recent events have shown the importance of *stealth technologies.* The related problems are very complicated from the modelling, mathematical, numerical and engineering points of view; several approaches can be envisaged (they do not exclude one another) such as active control, passive control through well chosen coating materials and/or well chosen shape, use of decoy strategies, etc. Indeed these methods can be applied for planes and submarines as well. These problems justify a book in themselves and will not be specifically addressed here. We think, however, that various notions related to controllability including the recently introduced concept of *sentinels* can be most helpful in the formulation and solution of stealth problems. It is also worth mentioning that the *exact controllability* based solution methods for the *Helmholtz equation at large wave numbers*, described in Section 6.13, have been motivated by stealth issues.

1. DISTRIBUTED AND POINTWISE CONTROL FOR LINEAR DIFFUSION EQUATIONS

1.1. First example

Let Ω be a bounded open set in $\mathbb{R}^d$ ($d \leq 3$ in the applications).

Remark 1.1 The 'boundedness' hypothesis is by no means a strict necessity.
We shall also assume that $\Gamma = \partial\Omega$ is 'sufficiently smooth', which is also not mandatory.

Let $\mathcal{O} \subset \Omega$ be an open subset of Ω.

Remark 1.2 We emphasize here at the very beginning that $\mathcal{O}$ can be *arbitrarily 'small'*.

The control function v will be with support in $\bar{\mathcal{O}}$. It is a *distributed control.*
The state equation is given by

$$\frac{\partial y}{\partial t} + Ay = v\chi_{\mathcal{O}} \text{ in } \Omega \times (0,T), \tag{1.1}$$

where $\chi_{\mathcal{O}}$ is the characteristic function of $\mathcal{O}$ and where A is a *second-order elliptic operator*, with variable coefficients. *The coefficients of A can also depend on t.*

Example 1.1 A typical elliptic operator A is the one defined by

$$Ay = -\sum_{i=1}^{d} \frac{\partial}{\partial x_i} \sum_{j=1}^{d} a_{ij} \frac{\partial y}{\partial x_j} + \mathbf{V}_0 \cdot \boldsymbol{\nabla} y, \tag{1.2}$$

where, in (1.2), $\boldsymbol{\nabla} = \{\partial/\partial x_i\}_{i=1}^d$ and where

(i) The coefficients a_{ij} belong to $L^\infty(\Omega)$ $\forall i,j, 1 \leq i,j \leq d$, and the matrix function $(a_{ij})_{1\leq i,j\leq d}$ satisfies

$$\sum_{i=1}^{d}\sum_{j=1}^{d} a_{ij}(x)\xi_i\xi_j \geq \alpha\|\boldsymbol{\xi}\|^2 \;\forall \boldsymbol{\xi} = \{\xi_i\}_{i=1}^d \in \mathbb{R}^d, \text{ a.e. in } \Omega, \tag{1.3}$$

with $\alpha > 0$ and $\|\cdot\|$ the canonical Euclidean norm of $\mathbb{R}^d$.

(ii) The vector $\mathbf{V}_0$ is *divergence-free* (i.e. $\boldsymbol{\nabla} \cdot \mathbf{V}_0 = 0$) and belongs to $(L^\infty(\Omega))^d$.

(iii) We have used the *dot product* notation for the canonical Euclidean scalar product of $\mathbb{R}^d$, i.e.

$$\boldsymbol{\eta} \cdot \boldsymbol{\xi} = \sum_{i=1}^{d} \eta_i \xi_i \ \forall \boldsymbol{\eta} = \{\eta_i\}_{i=1}^d, \ \boldsymbol{\xi} = \{\xi_i\}_{i=1}^d \in \mathbb{R}^d.$$

If the above hypothesis on the a_{ij}s and $\mathbf{V}_0$ are satisfied, then the *bilinear form* $a(\cdot\,,\cdot)$ defined by

$$a(y,z) = \sum_{i=1}^{d} \sum_{j=1}^{d} \int_\Omega a_{ij} \frac{\partial y}{\partial x_j} \frac{\partial z}{\partial x_i} \,\mathrm{d}x + \int_\Omega \mathbf{V}_0 \cdot \boldsymbol{\nabla} y z \,\mathrm{d}x \tag{1.4}$$

is *continuous* over $H^1(\Omega) \times H^1(\Omega)$; it is also *strongly elliptic* over $H_0^1(\Omega) \times H_0^1(\Omega)$ since we have, from (1.3) and from $\boldsymbol{\nabla} \cdot \mathbf{V}_0 = 0$, the following relation

$$a(y,y) \geq \alpha \int_\Omega |\boldsymbol{\nabla} y|^2 \,\mathrm{d}x \ \forall y \in H_0^1(\Omega). \tag{1.5}$$

If $\mathbf{V}_0 = \mathbf{0}$ and if $a_{ij} = a_{ji} \ \forall i,j,\ 1 \leq i,\ j \leq d$, then the bilinear form $a(\cdot\,,\cdot)$ is *symmetric*.

Above, $H^1(\Omega)$ and $H_0^1(\Omega)$ are the *functional spaces* defined as follows

$$H^1(\Omega) = \{\varphi \mid \varphi \in L^2(\Omega),\ \partial\varphi/\partial x_i \in L^2(\Omega) \ \forall i = 1, \ldots, d\}, \tag{1.6}$$

and

$$H_0^1(\Omega) = \{\varphi \mid \varphi \in H^1(\Omega),\ \varphi = 0 \text{ on } \Gamma\}, \tag{1.7}$$

respectively. Equipped with the classical *Sobolev norm*

$$\|\varphi\|_{H^1(\Omega)} = \left(\int_\Omega (\varphi^2 + |\boldsymbol{\nabla}\varphi|^2) \,\mathrm{d}x \right)^{1/2},$$

and with the corresponding *scalar product*

$$(\varphi, \psi)_{H^1(\Omega)} = \int_\Omega (\varphi\psi + \boldsymbol{\nabla}\varphi \cdot \boldsymbol{\nabla}\psi) \,\mathrm{d}x,$$

$H^1(\Omega)$ and $H_0^1(\Omega)$ are *Hilbert spaces*.

Since Ω is bounded,

$$\varphi \to \left(\int_\Omega |\boldsymbol{\nabla}\varphi|^2 \,\mathrm{d}x \right)^{1/2}$$

defines a norm over $H_0^1(\Omega)$ which is *equivalent* to the above $H^1(\Omega)$ norm, the corresponding scalar product being

$$\{\varphi, \psi\} \to \int_\Omega \boldsymbol{\nabla}\varphi \cdot \boldsymbol{\nabla}\psi \,\mathrm{d}x.$$

If we denote by $H^{-1}(\Omega)$ the *dual space* of $H_0^1(\Omega)$, then the above operator A is *linear* and *continuous* from $H^1(\Omega)$ into $H^{-1}(\Omega)$ and is an *ismorphism* from $H_0^1(\Omega)$ *onto* $H^{-1}(\Omega)$. □

Back to (1.1), and motivated by the class of elliptic operators discussed in the above example, we shall suppose from now on that operator A is *linear and continuous* from $H^1(\Omega)$ into $H^{-1}(\Omega)$ and that it satisfies the following (*ellipticity*) property

$$\langle A\varphi, \varphi\rangle \geq \alpha\|\varphi\|^2_{H^1(\Omega)} \; \forall\varphi \in H_0^1(\Omega),$$

where, in the above relation, α is a *strictly positive* constant and where $\langle\cdot,\cdot\rangle$ denotes the *duality pairing* between $H^{-1}(\Omega)$ and $H_0^1(\Omega)$. Operator A is *symmetric* over $H_0^1(\Omega)$ if

$$\langle A\varphi, \psi\rangle = \langle A\psi, \varphi\rangle \; \forall\varphi, \psi \in H_0^1(\Omega).$$

The *bilinear* form

$$\{\varphi, \psi\} \to \langle A\varphi, \psi\rangle : H_0^1(\Omega) \times H_0^1(\Omega) \to \mathbb{R}$$

will be denoted by $a(\cdot,\cdot)$ and is symmetric if and only if A is self-adjoint.

In order to fix ideas and to make things as simple as possible, we add to (1.1) the following boundary condition, of *Dirichlet type*,

$$y = 0 \text{ on } \Sigma = \Gamma \times (0, T). \tag{1.8}$$

The initial condition is

$$y(0) = y_0, \tag{1.9}$$

where y_0 is given in $L^2(\Omega)$.

We shall assume that

$$v \in L^2(\mathcal{O} \times (0, T)). \tag{1.10}$$

We emphasize that this is a *choice* which is by no means compulsory. We shall return to this. We begin with (1.10) since it is the *simplest* possible choice, at least from a theoretical point of view.

It is a well known fact (cf. for instance Lions (1961), Lions and Magenes (1968)) that (1.1), (1.8), (1.9) admits a unique solution (denoted sometimes as $t \to y(t; v)$, with $y(t; v) = x \to y(x, t; v)$) which has the following properties

$$y \in L^2(0, T; H_0^1(\Omega)), \partial y/\partial t \in L^2(0, T; H^{-1}(\Omega)), \tag{1.11}$$

$$y \text{ is continuous from } [0, T] \to L^2(\Omega). \tag{1.12}$$

We are going to study the *(approximate) controllability* of problem (1.1), (1.8), (1.9).

1.2. Approximate controllability

As a preliminary remark, we note that *exact* controllability is going to be difficult. Indeed, if we assume that the coefficients of A are smooth (respectively real analytic) then the solution y is, at time $T > 0$, *smooth outside* $\mathcal{O}$ (respectively *real analytic outside* $\mathcal{O}$).

Therefore if y_T is given in $L^2(\Omega)$ – which is a natural choice if we take (1.12) into account – the condition of exact controllability

$$y(T) = y_T$$

will be, in general, *impossible*.

This will become more precise below. For the time being, we start with the *approximate controllability*. In that direction a key result is given by the following

Proposition 1.1 When v spans $L^2(\mathcal{O} \times (0,T)), y(T;v)$ spans an affine subspace which is dense in $L^2(\Omega)$.

Proof.

(i) Let Y_0 be the solution to (1.1), (1.8), (1.9) for $v = 0$. Then $y(T;v) - Y_0(T)$ describes a subspace of $L^2(\Omega)$ and we have to show the *density* of this subspace. It amounts to proving the above density result assuming $y_0 = 0$.

(ii) We then apply the *Hahn–Banach theorem*, as in Lions (1968) (so that the present proof is *not constructive*).

Let us consider indeed an element $f \in L^2(\Omega)$ such that

$$(y(T;v), f)_{L^2(\Omega)} = 0 \ \forall v \in L^2(\mathcal{O} \times (0,T)). \tag{1.13}$$

We introduce ψ as the solution to

$$-\frac{\partial \psi}{\partial t} + A^* \psi = 0 \text{ in } \Omega \times (0,T), \tag{1.14}$$

where A^* is the adjoint operator of A and where ψ also satisfies

$$\psi = 0 \text{ on } \Sigma, \tag{1.15}$$

$$\psi(x,T) = f(x). \tag{1.16}$$

Then multiplying (1.14) by $y(v)$ and applying *Green's formula*, we obtain

$$(y(T;v), f)_{L^2(\Omega)} = \iint_{\mathcal{O}\times(0,T)} \psi v \, \mathrm{d}x \, \mathrm{d}t. \tag{1.17}$$

Therefore (1.13) is equivalent to

$$\psi = 0 \text{ in } \mathcal{O} \times (0,T). \tag{1.18}$$

It then follows from the Mizohata (1958) uniqueness theorem, that

$$\psi \equiv 0 \text{ in } \Omega \times (0,T) \tag{1.19}$$

so that $f = 0$, which proves the proposition. □

Remark 1.3 Mizohata's theorem supposes that the coefficients of A are sufficiently smooth (cf. also Saut and Scheurer (1987)).

Remark 1.4 A similar density property holds true if v spans, say, the space of those functions which are C^∞ and with compact support in $\mathcal{O} \times (0,T)$. This fact gives a lot of flexibility to the formulation which follows.

Remark 1.5 Suppose we would like to drive the system at time T 'close' to a state y_T *containing some singularities.* To fix ideas (but there is also much flexibility here) suppose that

$$y_T \in H^{-1}(\Omega). \tag{1.20}$$

Then it *may* be sensible to admit fairly *general controls*, such as

$$v \in L^2(0,T; H^{-1}(\mathcal{O})) \tag{1.21}$$

or even more general ones. We shall not pursue these lines here.

1.3. Formulation of the approximate controllability problem

As we have seen in Section 1.2, we do not restrict the generality by assuming that $y_0 = 0$ (it amounts to replacing y_T by $y_T - Y_0(T)$).

Let B be the unit ball of $L^2(\Omega)$. We want

$$y(T;v) \text{ to belong to } y_T + \beta B, \beta > 0 \text{ (arbitrarily small).} \tag{1.22}$$

According to Proposition 1.1 there are controls vs (actually infinitely many such vs) such that (1.22) holds true. Among all these vs, we want to find those which are solutions to the following minimization problem:

$$\inf_v \frac{1}{2} \iint_{\mathcal{O}\times(0,T)} v^2 \, dx \, dt, \quad v \in L^2(\mathcal{O} \times (0,T)), \ y(T;v) \in y_T + \beta B. \tag{1.23}$$

In fact problem (1.23) admits a unique solution. We want to construct numerical approximation schemes to find it.

Before we proceed, a few remarks are now in order.

Remark 1.6 All that is stated above is true with

$$\begin{aligned} T > 0 &\quad \text{arbitrarily small,} \\ \mathcal{O} \subset \Omega &\quad \text{arbitrarily 'small',} \\ \beta > 0 &\quad \text{also arbitrarily small.} \end{aligned}$$

Letting $\beta \to 0$ *will be, in general, impossible.* This will be made explicit below.

Remark 1.7 Choices other than (1.23) are possible. The 'obvious' candidates are

$$\inf_v \|v\|_{L^1(\mathcal{O}\times(0,T))}, \quad v \in L^1(\mathcal{O} \times (0,T)), \quad y(T;v) \in y_T + \beta B, \tag{1.24}$$

or

$$\inf_v \|v\|_{L^\infty(\mathcal{O}\times(0,T))}, \quad v \in L^\infty(\mathcal{O} \times (0,T)), \quad y(T;v) \in y_T + \beta B. \tag{1.25}$$

Other – more subtle – choices may be of interest. We shall return to this below.

1.4. Dual problem

We are going to apply the *Duality Theory of Convex Analysis* to problem (1.23).

We define the following functionals and operator

$$F_1(v) = \tfrac{1}{2} \iint_{\mathcal{O}\times(0,T)} v^2 \, dx \, dt, \tag{1.26}$$

$$F_2(f) = \begin{cases} 0 & \text{for } f \text{ in } L^2(\Omega), \quad f \in y_T + \beta B \\ +\infty & \text{otherwise,} \end{cases} \tag{1.27}$$

(F_2 is a 'proper' convex functional)

$$Lv = y(T;v), \tag{1.28}$$

so that

$$L \in \mathcal{L}(L^2(\mathcal{O} \times (0,T)); L^2(\Omega)). \tag{1.29}$$

Then problem (1.23) where the infimum is taken over all vs satisfying (1.22) is *equivalent* to the following minimization problem

$$\inf_{v\in L^2(\mathcal{O}\times(0,T))} [F_1(v) + F_2(Lv)]. \tag{1.30}$$

We can now apply the duality theorem of W. Fenchel and T.R. Rockafellar (cf. Ekeland and Temam (1974)). It gives

$$\inf_{v\in L^2(\mathcal{O}\times(0,T))} [F_1(v) + F_2(Lv)] = - \inf_{f\in L^2(\Omega)} [F_1^*(L^*f) + F_2^*(-f)] \tag{1.31}$$

where F_i^* is the conjugate function of F_i and L^* is the adjoint operator of L.

We have

$$\begin{aligned} F_1^*(v) &= \sup_{\hat{v}\in L^2(\mathcal{O}\times(0,T))} [(v,\hat{v}) - F_1(\hat{v})] = F_1(v), \\ F_2^*(f) &= \sup_{\hat{f}\in y_T+\beta B} (f,\hat{f}) = (f, y_T) + \beta\|f\|, \end{aligned}$$

where $\|f\|$ = norm of f in $L^2(\Omega)$ and where (f, y_T) = scalar product of f

and y_T in $L^2(\Omega)$. We now compute L^*. Given f in $L^2(\Omega)$, we define ψ as the solution to (1.14)–(1.16).

Then, one verifies easily (actually one uses (1.17)) that

$$L^* f = \psi\chi_{\mathcal{O}}, \qquad \chi_{\mathcal{O}} = \text{characteristic function of } \mathcal{O}. \tag{1.32}$$

Therefore (1.31) *gives*

$$\inf_{v \in L^2(\mathcal{O}\times(0,T))} [F_1(v) + F_2(Lv)]$$
$$= - \inf_{\hat{f} \in L^2(\Omega)} \left[\frac{1}{2} \iint_{\mathcal{O}\times(0,T)} \hat{\psi}^2 \, dx \, dt - (\hat{f}, y_T) + \beta \|\hat{f}\| \right], \tag{1.33}$$

where $\hat{\psi}$ *is the solution to*

$$-\frac{\partial \hat{\psi}}{\partial t} + A^* \hat{\psi} = 0 \text{ in } \Omega \times (0,T), \quad \hat{\psi}(T) = \hat{f}, \quad \hat{\psi} = 0 \text{ on } \Sigma. \tag{1.34}$$

Minimizing the functional on the right-hand side of (1.33), *where the state function is now given by* (1.34), *is the dual problem.*

Remark 1.8 Problem (1.33), (1.34) admits a unique solution. Let f denote this solution. Then *the solution* u *to problem* (1.23) *is given by*

$$u = \psi\chi_{\mathcal{O}}, \tag{1.35}$$

where ψ is the solution to (1.34) corresponding to $\hat{f}$.

Remark 1.9 We now want to give constructive algorithms for finding the solution to the dual problem, hence for the solution to the primal problem (using (1.35)).

Remark 1.10 As is classical in questions of this sort, relation (1.33) leads to lower and upper bounds, hence to some error estimates.

1.5. Direct solution to the dual problem

Given f in $L^2(\Omega)$, let us set

$$[f] = \|\psi\|_{L^2(\mathcal{O}\times(0,T))}. \tag{1.36}$$

We observe that $[f]$ is a norm on $L^2(\Omega)$. Indeed, if $[f] = 0$ then $\psi = 0$ in $\mathcal{O} \times (0,T)$, hence (according to the proof of Proposition 1.1) $f = 0$ follows.

Let us now introduce a *variational inequality* expressing that f realizes the minimum on the right-hand side of (1.33).

It is given by

$$\iint_{\mathcal{O}\times(0,T)} \psi(\hat{\psi} - \psi) \, dx \, dt - (y_T, \hat{f} - f) + \beta\|\hat{f}\| - \beta\|f\| \geq 0 \; \forall \hat{f} \in L^2(\Omega), \tag{1.37}$$

where $\hat{\psi}$ is the solution to (1.34) corresponding to $\hat{f}$.

Using (1.36), this is equivalent to

$$[f, \hat{f} - f] - (y_T, \hat{f} - f) + \beta\|\hat{f}\| - \beta\|f\| \geq 0 \;\forall \hat{f} \in L^2(\Omega). \tag{1.38}$$

Let us introduce the 'adjoint' state function y defined by

$$\frac{\partial y}{\partial t} + Ay = \psi\chi_{\mathcal{O}} \text{ in } \Omega \times (0,T), \quad y(0) = 0, \quad y = 0 \text{ on } \Sigma. \tag{1.39}$$

Multiplying the first equation in (1.39) by $\hat{\psi} - \psi$ gives

$$\iint_{\mathcal{O}\times(0,T)} \psi(\hat{\psi} - \psi)\, dx\, dt = (y(T), \hat{f} - f). \tag{1.40}$$

Let us set

$$y(T) = y(T; f) = \Lambda f, \tag{1.41}$$

where, given f, one computes ψ by (1.34) and then y by (1.39).

Then (1.37) (or (1.38)) can be written

$$(\Lambda f, \hat{f} - f) - (y_T, \hat{f} - f) + \beta\|\hat{f}\| - \beta\|f\| \geq 0 \;\forall \hat{f} \in L^2(\Omega). \tag{1.42}$$

Remark 1.11 The equivalence between problems (1.38) and (1.42) relies on the following relation

$$[f, \hat{f}] = (\Lambda f, \hat{f}) \forall f, \hat{f} \in L^2(\Omega). \tag{1.43}$$

Remark 1.12 Operator Λ satisfies

$$\Lambda \in \mathcal{L}(L^2(\Omega); L^2(\Omega)), \quad \Lambda = \Lambda^*, \;\Lambda \geq 0. \tag{1.44}$$

It follows from (1.44) that the (unique) solution to problem (1.42) is also the solution to

$$\inf_{\hat{f}\in L^2(\Omega)} [\tfrac{1}{2}(\Lambda \hat{f}, \hat{f}) - (y_T, \hat{f}) + \beta\|\hat{f}\|]. \tag{1.45}$$

We can summarize by the following

Theorem 1.1 (i) *We have the identity*

$$\inf_{\substack{v \\ y(T;v)\in y_T + \beta B}} \tfrac{1}{2} \iint_{\mathcal{O}\times(0,T)} v^2\, dx\, dt$$
$$= - \inf_{\hat{f}\in L^2(\Omega)} \left[\tfrac{1}{2} \iint_{\mathcal{O}\times(0,T)} \hat{\psi}^2\, dx\, dt - (y_T, \hat{f}) + \beta\|\hat{f}\|\right], \tag{1.46}$$

where $\hat{\psi}$ *is given by* (1.34).

(ii) *The unique solution* f *of the dual problem is the solution of* (1.45) *where* Λ *is defined by* (1.41), *i.e.* $\Lambda f = y(T)$ *where*

$$-\frac{\partial \psi}{\partial t} + A^*\psi = 0 \text{ in } \Omega \times (0,T), \quad \psi(T) = f, \;\psi = 0 \text{ on } \Sigma \tag{1.47$_1$}$$

$$\frac{\partial y}{\partial t} + Ay = \psi \chi_{\mathcal{O}} \text{ in } \Omega \times (0,T), \quad y(0) = 0, y = 0 \text{ on } \Sigma. \tag{1.47$_2$}$$

(iii) *The unique solution u of* (1.46) *is given by*

$$u = \psi \chi_{\mathcal{O}}. \tag{1.48}$$

Application As a corollary – *which we have to make precise!* – one obtains the general principle of a solution method, namely

(i) Guess the solution f of problem (1.46).
(ii) Compute the corresponding value of ψ.
(iii) Use an iterative method to compute the inf in f, using the right-hand side of (1.46) or using (1.45).

This will be the task of Section 1.8. Before that several remarks have to be made.

Remark 1.13 The optimal control v – with respect to the *choice* of

$$\tfrac{1}{2} \iint_{\mathcal{O}\times(0,T)} v^2 \, \mathrm{d}x \, \mathrm{d}t$$

as the quantity to minimize – is given by (1.48), where ψ is the solution of the parabolic equation $(1.47)_1$. Therefore ψ is *smooth* (the smoother the coefficients of A, the smoother ψ will be). *In other words, u is smooth.*

This remark *excludes* the possibility of finding an optimal control of the 'bang–bang' type.

Of course trying to find an optimal control satisfying some kind of bang–bang principle is by no means compulsory! But knowing in advance that such a property holds true may be of some help.

The first idea which comes to mind is to replace

$$\|v\|_{L^2(\mathcal{O}\times(0,T))} \text{ by } \|v\|_{L^\infty(\mathcal{O}\times(0,T))};$$

this possibility will be discussed in Section 1.7.

Remark 1.14 (Further comments on exact controllability.) Exact controllability corresponds to $\beta = 0$ in (1.45), or, equivalently, to

$$\inf_{\hat{f}} [\tfrac{1}{2}[\hat{f}]^2 - (y_T, \hat{f})], \hat{f} \in L^2(\Omega). \tag{1.49}$$

Let us denote by $\widehat{L^2(\Omega)}$ the completion of $L^2(\Omega)$ for the norm $[\hat{f}]$. Due to the smoothness properties of parabolic equations, $\widehat{L^2(\Omega)}$ will contain (except for the case, without practical interest, where $\mathcal{O} = \Omega$) very singular distributions and even distributions of infinite order (outside $\bar{\mathcal{O}}$), i.e. ultra-distributions.

Then (1.49) admits a unique solution f_0 iff

$$y_T \in (\widehat{L^2(\Omega)})' \tag{1.50}$$

the dual of $\widehat{L^2(\Omega)}$, when $L^2(\Omega)$ is identified with its dual. It means that exact controllability is possible iff y_T belongs to a '*very small*' space, namely $(\widehat{L^2(\Omega)})'$.

We also have the following convergence result: let f_β be the unique solution to (1.46), *then* $f_\beta \to f_0$ *in* $L^2(\Omega)$ *as* $\beta \to 0$ *iff* $y_T \in (\widehat{L^2(\Omega)})'$.

Remark 1.15 Another way of expressing this is to observe that Λ is an *isomorphism* from $\widehat{L^2(\Omega)}$ onto its dual. This is closely related to the *Hilbert Uniqueness Method* (HUM) as introduced in Lions (1988a,b).

1.6. Penalty arguments

In problems where there are many constraints of a different nature, penalty arguments can be used in a very large number of ways.

In the present section we are going to 'penalize' the constraint

$$y(T;v) \text{ belongs to } y_T + \beta B. \tag{1.51}$$

This can also be done in many ways!

One possibility is to introduce a *smooth* functional over $L^2(\Omega)$ which is zero on the ball $y_T + \beta B$, and > 0 outside the ball; let $\mu(\cdot)$ be such a functional. Then one can consider

$$\begin{gathered}\inf_v \left[\tfrac{1}{2} \iint_{\mathcal{O}\times(0,T)} v^2 \, \mathrm{d}x \, \mathrm{d}t + k\mu(y(T;v))\right], \\ v \in L^2(\mathcal{O} \times (0,T)), \quad k > 0 \text{ 'large'}.\end{gathered} \tag{1.52}$$

Another possibility is the following. One introduces

$$J_k(v) = \tfrac{1}{2} \iint_{\mathcal{O}\times(0,T)} v^2 \, \mathrm{d}x \, \mathrm{d}t + \frac{k}{2}\|y(T;v) - y_T\|^2, \tag{1.53}$$

where $k > 0$ is 'large' and where $\|\cdot\|$ denotes the $L^2(\Omega)$ norm.

Then one considers the problem

$$\inf_v J_k(v), \quad v \in L^2(\mathcal{O} \times (0,T)). \tag{1.54}$$

This problem *admits a unique solution*, denoted by u_k. Let us verify the following result:

$$\left\{\begin{array}{l}\text{There exists } k \text{ large enough such that the solution } u_k \text{ of (1.54)} \\ \text{satisfies } \|y(T;u_k) - y_T\| \le \beta.\end{array}\right. \tag{1.55}$$

Before proving (1.55) let us make the following remark.

Remark 1.16 It follows from (1.55) that u_k is, for k large enough, *one* control such that $y(T; u_k) \in y_T + \beta B$. Of course it has no reasons to coincide with the solution u_β of

$$\inf \tfrac{1}{2} \iint_{\mathcal{O}\times(0,T)} v^2 \,\mathrm{d}x\,\mathrm{d}t, \quad v \in L^2(\mathcal{O} \times (0,T)), \quad y(T; v) \in y_T + \beta B.$$

Remark 1.17 The proof to follow is *not* constructive, therefore it does not give a 'constructive choice' for k which is a difficulty since β 'disappears' in problem (1.54). *We make below a constructive proposal for the choice of* k.

Remark 1.18 Of course, given k, the *optimality system* for problem (1.54) is quite classical. One obtains

$$\begin{cases} \dfrac{\partial y}{\partial t} + Ay = \psi\chi_{\mathcal{O}} \text{ in } \Omega \times (0,T), \; y(0) = 0, \; y = 0 \text{ on } \Sigma, \\ -\dfrac{\partial \psi}{\partial t} + A^*\psi = 0 \text{ in } \Omega \times (0,T), \; \psi(T) = k(y_T - y(T)), \; \psi = 0 \text{ on } \Sigma. \end{cases} \tag{1.56}$$

The optimal control u_k is given by $\psi\chi_{\mathcal{O}}$ where ψ is the solution obtained by solving (1.56).

It is worth noticing that if one denotes the function $\psi(T)$ by f, then f satisfies the functional equation

$$(k^{-1}\mathbf{I} + \Lambda)f = y_T, \tag{1.57}$$

where operator Λ is still defined by (1.41) (see Section 1.5).

Proof of (1.55). Given $\varepsilon > 0$, there exists a control w such that

$$\|y(T; w) - y_T\| \leq \varepsilon. \tag{1.58}$$

This follows from the approximate controllability result and it is not constructive.

Then

$$J_k(u_k) \leq \tfrac{1}{2} \iint_{\mathcal{O}\times(0,T)} w^2 \,\mathrm{d}x\,\mathrm{d}t + \frac{k\varepsilon^2}{2}, \tag{1.59}$$

so that

$$\|y(T; u_k) - y_T\|^2 \leq \frac{1}{k} \iint_{\mathcal{O}\times(0,T)} w^2 \,\mathrm{d}x\,\mathrm{d}t + \varepsilon^2. \tag{1.60}$$

We choose $\varepsilon = \beta/\sqrt{2}$, then w is chosen so that (1.58) holds and we choose k such that

$$\frac{1}{k} \iint_{\mathcal{O}\times(0,T)} w^2 \,\mathrm{d}x\,\mathrm{d}t \leq \tfrac{1}{2}\beta^2.$$

Then (1.60) implies (1.55). □

Remark 1.19 In general (i.e. for y_T generically given in $L^2(\Omega)$) the above process does *not* converge as $k \to +\infty$ (otherwise it would give exact controllability at the limit!).

Remark 1.20 It will remain to solve (1.56) *if we have a way to choose* k. This is what we propose now.

Duality on $J_k(v)$:
We introduce

$$F_1(v) = \tfrac{1}{2} \iint_{\mathcal{O}\times(0,T)} v^2 \, dx \, dt, \quad F_2(f) = \tfrac{1}{2} k \|f - y_T\|^2, \; Lv = y(T; v). \tag{1.61}$$

We have

$$\inf_v J_k(v) = \inf_v (F_1(v) + F_2(Lv))$$

and using *duality* as in previous sections (and with similar notation), we obtain

$$\inf_v J_k(v) = - \inf_{\hat{f} \in L^2(\Omega)} (F_1^*(L^* \hat{f}) + F_2^*(-\hat{f})). \tag{1.62}$$

This leads to the following dual problem:
Let $\hat{\psi}$ be defined by

$$-\frac{\partial \hat{\psi}}{\partial t} + A^* \hat{\psi} = 0 \text{ in } \Omega \times (0,T), \quad \hat{\psi}(T) = \hat{f}, \; \hat{\psi} = 0 \text{ on } \Sigma. \tag{1.63}$$

Then the dual problem is to find

$$\inf_{\hat{f} \in L^2(\Omega)} \left[\tfrac{1}{2} \iint_{\mathcal{O}\times(0,T)} \hat{\psi}^2 \, dx \, dt - (\hat{f}, y_T) + \frac{1}{2k} \|\hat{f}\|^2 \right], \tag{1.64}$$

or, equivalently,

$$\inf_{\hat{f} \in L^2(\Omega)} \left[\tfrac{1}{2} (\Lambda \hat{f}, \hat{f}) - (\hat{f}, y_T) + \frac{1}{2k} \|\hat{f}\|^2 \right], \tag{1.65}$$

which is in turn equivalent to the *linear* problem (1.57). Problem (1.57), (1.65) has the following variational formulation

$$\begin{cases} f \in L^2(\Omega) \; \forall \hat{f} \in L^2(\Omega), \text{ we have} \\ \int_\Omega (\Lambda f) \hat{f} \, dx + \frac{1}{k} \int_\Omega f \hat{f} \, dx = \int_\Omega y_T \hat{f} \, dx. \end{cases} \tag{1.66}$$

Taking $\hat{f} = f$ in (1.66), we obtain

$$\int_\Omega (\Lambda f) f \, dx + \frac{1}{k} \|f\|^2 = \int_\Omega y_T f \, dx. \tag{1.67}$$

We now compare problem (1.64), (1.65) to the problem (1.42), (1.45); it follows from Sections 1.4 and 1.5 that the solution f^* of (1.42), (1.45) satisfies the following *variational inequality*

$$\begin{cases} f^* \in L^2(\Omega)\ \forall \hat{f} \in L^2(\Omega), \text{ we have} \\ \int_\Omega (\Lambda f^*)(\hat{f} - f^*)\,dx + \beta\|\hat{f}\| - \beta\|f^*\| \geq \int_\Omega y_T(\hat{f} - f^*)\,dx, \end{cases} \tag{1.68}$$

which implies in turn (take $\hat{f} = 0$ and $\hat{f} = 2f^*$ in (1.68)) that

$$\int_\Omega (\Lambda f^*) f^*\,dx + \beta\|f^*\| = (y_T, f^*). \tag{1.69}$$

Suppose now that $f = f^*$; it follows then from (1.67), (1.69) that

$$\frac{1}{k}\|f\|^2 = \beta\|f\|,$$

i.e., if $f \neq 0$,

$$k = \|f\|/\beta. \tag{1.70}$$

We propose consequently the following rule:

After a few iterations, where k is given *a priori*, we take k variable with n and defined by

$$k_n = \frac{1}{\beta}\|f^n\|. \tag{1.71}$$

Remark 1.21 It follows from Remark 1.18 and from (1.65) that problem (1.64) is equivalent to (1.57), namely

$$(k^{-1}\mathbf{I} + \Lambda)f = y_T. \tag{1.72}$$

On the other hand, it follows from Section 1.5 that the minimization problem on the right-hand side of (1.33) is equivalent to the 'equation' (it is indeed an inclusion).

$$y_T \in \beta\partial j(f) + \Lambda f, \tag{1.73}$$

where $\partial j(\cdot)$ denotes the *subgradient* (see e.g. Ekeland and Temam (1974) for this notion) of the *convex* functional $j(\cdot)$ defined by

$$j(\hat{f}) = \|\hat{f}\|_{L^2(\Omega)}\ \forall \hat{f} \in L^2(\Omega). \tag{1.74}$$

Intuitively, problem (1.72) being *linear* is easier to solve than (1.73) which is nonlinear, nondifferentiable, etc. In fact, we shall see in Section 1.8 that if one has a method for solving problem (1.72), it can be used in a very simple way to solve problem (1.73).

1.7. L^∞ cost functions and bang–bang controls

We consider the same 'model' problem as before, namely

$$\frac{\partial y}{\partial t} + Ay = v\chi_{\mathcal{O}} \text{ in } \Omega \times (0,T) = Q,\ y(0) = 0,\ y = 0 \text{ on } \Sigma. \tag{1.75}$$

Given $T > 0$ and given $y_T \in L^2(\Omega)$, we consider those control vs such that

$$y(T) \in y_T + \beta B, \tag{1.76}$$

where, in (1.76), β is a positive number and B is the unit ball of $L^2(\Omega)$. Next, we consider the following *control problem*

$$\inf \|v\|_{L^\infty(\mathcal{O}\times(0,T))}, \tag{1.77}$$

where v is subjected to (1.75), (1.76).

A few remarks are in order.

Remark 1.22 This remark is purely technical. The space described by $y(T;v)$ is *dense* in $L^2(\Omega)$ when v spans the space of the C^∞ functions with compact support in $\mathcal{O} \times (0,T)$, so that the infimum in (1.77) is *always* a finite number, *no matter how small* $\beta(> 0)$ *is.*

Remark 1.23 The choice of the L^∞ norm in (1.77) is less convenient than the choice of the L^2 norm, but is not an unreasonable choice. It leads to new difficulties, essentially due to the *nondifferentiability* of the L^∞ norm (and of any power of it). We explain below what to do in order to proceed with this type of cost function, which leads to *bang–bang* type results (see below).

Remark 1.24 Of course, one can more generally consider

$$\inf \|v\|_{L^s(\mathcal{O}\times(0,T))}, \tag{1.78}$$

where s is chosen arbitrarily in $[1, +\infty]$, i.e.

$$1 \leq s \leq +\infty. \tag{1.79}$$

Indeed, if $s \in (1, +\infty)$ it is more convenient to use $v \to s^{-1}\|v\|^s_{L^s(\mathcal{O}\times(0,T))}$ as the cost function, since it has better *differentiability properties* and does not change the solution of problem (1.78).

Let us consider the case $s = 1$; then for *any* v in $L^1(\mathcal{O}\times(0,T))$ the function $y(T;v)$ belongs to $L^2(\Omega)$ if and only if $d \leq 2$ (see, e.g., Ladyzenskaya, Solonnikov and Ural'ceva (1968) for this result). Actually, this does not modify the statement of problem (1.78) (with $s = 1$), since if $d > 2$, we can always restrict ourselves to those controls v in $L^1(\mathcal{O} \times (0,T))$, such that $y(T;v) \in L^2(\Omega)$.

Remark 1.25 There is still another variant that we shall not consider in this article, namely to replace in (1.76) the unit ball B of $L^2(\Omega)$ by the unit ball of $L^r(\Omega)$. We refer to Fabre, Puel and Zuazua (1993) for a discussion of this case.

Remark 1.26 For technical reasons (the explanation for which will appear later on) we are going to consider the problem in the following form

$$\inf \tfrac{1}{2}\|v\|^2_{L^\infty(\mathcal{O}\times(0,T))}, \tag{1.80}$$

or

$$\inf \tfrac{1}{2}\|v\|^2_{L^s(\mathcal{O}\times(0,T))}, \tag{1.81}$$

with v subjected to (1.75), (1.76)

Synopsis In the following, we propose an approximation method for problem (1.80), which leads to: (i) numerical methods; and (ii) connections with one result from Fabre *et al.* (1993).

The results in the above reference have been found by a duality approach, which leads – among other things – to some very interesting formulae; we will present these formulae.

Approximation by penalty and regularization I We begin by considering the following problem

$$\inf J^s_k(v) \tag{1.82}$$

where, in (1.82), the cost function $J^s_k(\cdot)$ is defined by

$$J^s_k(v) = \tfrac{1}{2}\|v\|^2_{L^s} + \tfrac{1}{2}k\|y(T;v) - y_T\|^2_{L^2(\Omega)}, \tag{1.83}$$

and where in (1.83), L^s stands for $L^s(\mathcal{O}\times(0,T))$ and $y(\cdot\,, v)$ is the solution of (1.75). The idea here is to have $k(>0)$ *large* to 'force' (*penalty*) the final condition $y(T;v) = y_T$, and to have s large, as an approximation of $s = +\infty$ (*regularization*). Problem (1.82) has a *unique* solution and we are going to write the corresponding *optimality conditions*. We can easily verify that

$$\frac{\mathrm{d}}{\mathrm{d}\lambda}(\tfrac{1}{2}\|v+\lambda\hat{v}\|^2_{L^s})|_{\lambda=0} = \|v\|^{2-s}_{L^s}\iint_{\mathcal{O}\times(0,T)} v|v|^{s-2}\hat{v}\,\mathrm{d}x\,\mathrm{d}t\ \forall v,\hat{v}\in L^s. \tag{1.84}$$

The *quadratic* part of $J^s_k(\cdot)$ gives no problem and we verify easily that if we denote by $\nabla J^s_k(\cdot)$ the *derivative* of $J^s_k(\cdot)$ we have

$$\nabla J^s_k(v)\in L^{s'}\ \forall v\in L^s,\ \text{with } s' = s/(s-1), \tag{1.85}$$

and, from (1.84),

$$\left\{\begin{aligned}\iint_{\mathcal{O}\times(0,T)} \nabla J^s_k(v)\hat{v}\,\mathrm{d}x\,\mathrm{d}t = {} & \|v\|^{2-s}_{L^s}\iint_{\mathcal{O}\times(0,T)} v|v|^{s-2}\hat{v}\,\mathrm{d}x\,\mathrm{d}t\\ & -\iint_{\mathcal{O}\times(0,T)} p\hat{v}\,\mathrm{d}x\,\mathrm{d}t\\ \forall v,\hat{v}\in L^s, &\end{aligned}\right. \tag{1.86}$$

where, in (1.86), p is the solution of the *adjoint state equation*

$$-\frac{\partial p}{\partial t} + A^*p = 0 \text{ in } Q,\quad p(T) = k(y_T - y(T;v)),\ p = 0 \text{ on } \Sigma, \tag{1.87}$$

with, in (1.87), $y(T; v)$ obtained from v by (1.75); above the exponent s' is the *conjugate* of s, since $1/s + 1/s' = 1$.

Let us denote by u *the* solution of problem (1.82); it satisfies $\nabla J_k^s(u) = 0$, which implies (from (1.86)) that

$$\|u\|_{L^s}^{2-s} u|u|^{s-2} = p\chi_{\mathcal{O}}, \tag{1.88}$$

where, in (1.88), we still denote by p the particular solution of the adjoint system (1.87) corresponding to $v = u$. Relation (1.88) is equivalent to

$$u = \|p\|_{L^s}^{2-s'} p|p|^{s'-2}\chi_{\mathcal{O}}. \tag{1.89}$$

We have therefore obtained the following *optimality system* for problem (1.82):

$$\begin{cases} \dfrac{\partial y}{\partial t} + Ay = \|p\|_{L^s}^{2-s'} p|p|^{s'-2}\chi_{\mathcal{O}} \text{ in } Q,\ y(0) = 0,\ y = 0 \text{ on } \Sigma, \\ -\dfrac{\partial p}{\partial t} + A^*p = 0 \text{ in } Q,\ p(T) = k(y_T - y(T)),\ p = 0 \text{ on } \Sigma. \end{cases} \tag{1.90}$$

The above result holds for any fixed s arbitrarily large and the same observation applies to k.

The optimality system (1.90) *has a unique solution and the optimal control* u *is given by relation* (1.89).

Approximation by penalty and regularization II Suppose now that $s \to +\infty$, i.e. $s' \to 1$ in (1.90), the parameter k being fixed. We make the assumption (actually it is a *conjecture*; see Fabre *et al.* (1993) for a discussion of this issue) that

$$p \neq 0 \text{ a.e. in } \Omega \times (0, T) \tag{1.91}$$

(except if $p \equiv 0$). Then the limit of (1.90) is given by

$$\begin{cases} \dfrac{\partial y}{\partial t} + Ay = \|p\|_{L^1} \operatorname{sign} p\chi_{\mathcal{O}} \text{ in } Q,\ y(0) = 0,\ y = 0 \text{ on } \Sigma, \\ -\dfrac{\partial p}{\partial t} + A^*p = 0 \text{ in } Q,\ p(T) = k(y_T - y(T)),\ p = 0 \text{ on } \Sigma. \end{cases} \tag{1.92}$$

Remark 1.27 We observe that (1.92) has been obtained by taking the limit in (1.90) as $s \to +\infty$. This convergence result is not difficult to prove if we suppose that (1.91) holds; see Fabre *et al.* (1993), for further details and results.

Remark 1.28 It follows from (1.92) (or (1.89)) that the *optimal control* u is given by

$$u = \|p\|_{L^1} \operatorname{sign} p\chi_{\mathcal{O}}, \tag{1.93}$$

which is a *bang–bang* result.

Remark 1.29 What has been discussed above is simple thanks to the choice of (1.80) as control problem, which leads in turn to the *regularized* and *regularized–penalized* problems (1.81) and (1.82). This approach and the corresponding results are closely related to those in Fabre *et al.* (1993); in fact, these authors start from the *dual formulation* which is discussed below.

Dual formulation I We can use *duality* as in the L^2 case. We obtain therefore the following duality relation

$$\inf_{v \in L^s} [\tfrac{1}{2}\|v\|^2_{L^s} + \tfrac{1}{2}k\|y(T;v) - y_T\|^2_{L^2(\Omega)}]$$
$$= -\inf_{\hat{f}} \left[\tfrac{1}{2}\|\hat{\psi}\|^2_{L^{s'}} + \frac{1}{2k}\|\hat{f}\|^2_{L^2(\Omega)} - (y_T, \hat{f})_{L^2(\Omega)}\right], \tag{1.94}$$

where, in (1.94), $\hat{\psi}$ is obtained from $\hat{f}$ via the solution of

$$-\frac{\partial \hat{\psi}}{\partial t} + A^* \hat{\psi} = 0 \text{ in } Q, \quad \hat{\psi}(T) = \hat{f}, \ \hat{\psi} = 0 \text{ on } \Sigma. \tag{1.95}$$

As already mentioned in Remark 1.29, Fabre *et al.* (1993), start from the formulation (1.94), (1.95) *directly* with $s' = 1$; this has to be understood in the following manner: one considers as the *primal problem*

$$\inf_{\hat{f} \in L^2(\Omega)} \left[\tfrac{1}{2}\|\hat{\psi}\|^2_{L^1} + \frac{1}{2k}\|\hat{f}\|^2_{L^2(\Omega)} - (y_T, \hat{f})_{L^2(\Omega)}\right], \tag{1.96}$$

with $\hat{\psi}$ still defined by (1.95); then the *dual* problem is

$$\inf_{v \in L^\infty} [\tfrac{1}{2}\|v\|^2_{L^\infty} + \tfrac{1}{2}k\|y(T;v) - y_T\|^2_{L^2(\Omega)}]. \tag{1.97}$$

Dual formulation II What we want to achieve is (1.76), namely

$$y(T;v) \in y_T + \beta B.$$

Using the *penalized* formulation one obtains $y(T;v)$ 'close' to y_T. In order to have $y(T;v)$ satisfying (1.76) one has to choose k in a suitable fashion. This can be done as follows.

Observe first that, from (1.94), the *dual* problem of problem (1.82) is given by

$$\inf_{\hat{f} \in L^2(\Omega)} \left[\tfrac{1}{2}\|\hat{\psi}\|^2_{L^{s'}} + \frac{1}{2k}\|\hat{f}\|^2_{L^2(\Omega)} - (y_T, \hat{f})_{L^2(\Omega)}\right]. \tag{1.98}$$

Let us denote by f the solution of problem (1.98); it satisfies (with obvious notation) the following *variational equation* in $L^2(\Omega)$:

$$\begin{cases} \hat{f} \in L^2(\Omega) \ \forall \hat{f} \in L^2(\Omega) \text{ we have} \\ \|\psi\|^{2-s'}_{L^{s'}} \displaystyle\iint_{\mathcal{O}\times(0,T)} |\psi|^{s'-2}\psi\hat{\psi}\, dx\, dt + \frac{1}{k}\int_\Omega f\hat{f}\, dx = \int_\Omega y_T \hat{f}\, dx, \end{cases} \tag{1.99}$$

which implies in turn that

$$\|\psi\|_{L^{s'}}^2 + \frac{1}{k}\|f\|_{L^2(\Omega)}^2 = (y_T, f)_{L^2(\Omega)}. \tag{1.100}$$

Consider now the control problem

$$\inf_v \tfrac{1}{2}\|v\|_{L^s}^2, \ v \text{ satisfies } (1.75), (1.76). \tag{1.101}$$

Its dual problem is given by

$$\inf_{\hat{f}\in L^2(\Omega)} [\tfrac{1}{2}\|\hat{\psi}\|_{L^{s'}}^2 + \beta\|\hat{f}\|_{L^2(\Omega)} - (y_T, \hat{f})_{L^2(\Omega)}]. \tag{1.102}$$

Denote by f^* the solution of problem (1.102); it satisfies the following *variational inequality* in $L^2(\Omega)$

$$\begin{cases} f^* \in L^2(\Omega)\ \forall \hat{f} \in L^2(\Omega) \text{ we have} \\ \|\psi^*\|_{L^{s'}}^{2-s'} \displaystyle\iint_{\mathcal{O}\times(0,T)} |\psi^*|^{s'-2}\psi^*(\hat{\psi} - \psi^*)\,\mathrm{d}x\,\mathrm{d}t \\ \quad +\beta\|\hat{f}\|_{L^2(\Omega)} - \beta\|f^*\|_{L^2(\Omega)} \geq (y_T, \hat{f} - f^*)_{L^2(\Omega)}. \end{cases} \tag{1.103}$$

Taking successively $\hat{f} = 0$ and $\hat{f} = 2f^*$ in (1.103), we obtain

$$\|\psi^*\|_{L^{s'}}^2 + \beta\|f^*\|_{L^2(\Omega)} = (y_T, f^*)_{L^2(\Omega)}. \tag{1.104}$$

The positive member β being given, we look for k such that $f = f^*$, which implies in turn that $\psi = \psi^*$ and therefore that the primal problems (1.82) and (1.101) have the same solution $u(= \|\psi\|_{L^{s'}}^{2-s'}\psi|\psi|^{s'-2}\chi_{\mathcal{O}})$. Suppose that $f = f^*$, it follows then from (1.100) and (1.104) that

$$\frac{1}{k}\|f\|_{L^2(\Omega)}^2 = \beta\|f\|_{L^2(\Omega)}.$$

If $\|f\|_{L^2(\Omega)} \neq 0$ we thus have

$$k = \|f\|_{L^2(\Omega)}/\beta. \tag{1.105}$$

From (1.105) we have the following approach to solving problem (1.102) using the solution methods for problem (1.98):

Suppose that we have an *iterative* procedure producing $f^1, f^2, \ldots f^n, \ldots$; we shall use a constant parameter k for several iterations and then a variable one defined by

$$k = \|f^n\|_{L^2(\Omega)}/\beta. \tag{1.106}$$

We shall conclude this section with the following remark.

Remark 1.30 A control problem closely related to those discussed above is the one defined by

$$\inf_{v\in\mathcal{C}_f} \tfrac{1}{2}\|y(T;v) - y_T\|_{L^2(\Omega)}^2, \tag{1.107}$$

where, in (1.107), $\mathcal{C}_f$ is the *closed convex* subset of $L^\infty(\mathcal{O} \times (0,T))$ defined by

$$\mathcal{C}_f = \{v \mid v \in L^\infty(\mathcal{O} \times (0,T)), |v(x,t)| \leq C \text{ a.e. in } \mathcal{O} \times (0,T)\}. \tag{1.108}$$

In fact, problem (1.107) is fairly easy to solve if we have solution methods for problem (1.82) with $s = 2$; such methods will be discussed in the following Section 1.8, together with applications to the solution of problems such as (1.107).

1.8. Numerical methods

1.8.1. Generalities. Synopsis.

In this section, we shall address the *numerical solution of the approximate controllability* problems discussed in the preceding sections (the notation of which is kept); we shall start our discussion with the solution of the following two fundamental control problems:

First control problem This is defined by

$$\inf_{v \in \mathcal{U}_f} \frac{1}{2} \iint_{\mathcal{O}\times(0,T)} v^2 \, dx \, dt \tag{1.109}$$

with $\mathcal{U}_f$ defined by

$$\mathcal{U}_f = \{v \mid v \in L^2(\mathcal{O} \times (0,T)),\ y(T) \in y_T + \beta B\}, \tag{1.110}$$

where, in (1.110), *the target function* y_T is given in $L^2(\Omega)$, B is the unit ball of $L^2(\Omega)$, β is a positive parameter and where the state function y is the solution of the following *parabolic problem*

$$\frac{\partial y}{\partial t} + Ay = v\chi_{\mathcal{O}} \text{ in } Q = \Omega \times (0,T), \tag{1.111}$$

$$y(0) = y_0 (\in L^2(\Omega)), \tag{1.112}$$

$$y = 0 \text{ on } \Sigma = \Gamma \times (0,T). \tag{1.113}$$

Control problem (1.109) has a unique solution.

Second control problem This is defined by

$$\inf_{v \in L^2(\mathcal{O}\times(0,T))} \left[\frac{1}{2} \iint_{\mathcal{O}\times(0,T)} v^2 \, dx \, dt + \frac{k}{2} \|y(T) - y_T\|^2_{L^2(\Omega)}\right], \tag{1.114}$$

where, in (1.114), k is *a positive* parameter and y is still defined by (1.111)–(1.113).

Control problem (1.114) has a unique solution.

The solution of the control problems (1.109) and (1.114) can be achieved by methods acting *directly* on the control v; these methods have the advantage of being easy to generalize (in principle) to control problems for *nonlinear* state equations as shown in future sections. In the particular case of problems (1.109) and (1.114) where the state equation (namely (1.111)–(1.113)) is *linear* and the cost functions *quadratic*, instead of solving (1.109) and (1.114) directly, we can solve equivalent problems obtained by applying *Convex Duality Theory*, as already shown in Sections 1.5 and 1.6. In fact, these dual problems can be viewed as *identification* problems for the *final data* of a *backward* (in time) *adjoint equation*, in the spirit of the *Reverse Hilbert Uniqueness Method* (RHUM) introduced in Lions (1988b); from our point of view, these dual problems are better suited to numerical calculations than the original ones (for a discussion concerning the exact and approximate *boundary* controllability of the *heat equation*, which includes numerical methods, see Glowinski (1992b) and Carthel, Glowinski and Lions (1994)).

It follows from Section 1.5 (respectively Section 1.6) that the dual problem to (1.109) (respectively (1.114)) is defined by the following *variational inequality*

$$\begin{cases} f \in L^2(\Omega)\ \forall \hat{f} \in L^2(\Omega) \text{ we have} \\ (\Lambda f, \hat{f} - f)_{L^2(\Omega)} + \beta\|\hat{f}\|_{L^2(\Omega)} - \beta\|f\|_{L^2(\Omega)} \\ \quad \geq (y_T - Y_0(T), \hat{f} - f)_{L^2(\Omega)} \end{cases} \tag{1.115}$$

(respectively by the following *linear* equation

$$(k^{-1}\mathbf{I} + \Lambda)f = y_T - Y_0(T)), \tag{1.116}$$

where in (1.115), (1.116), operator Λ is the one defined in Section 1.5, and where the function Y_0 is defined by

$$\frac{\partial Y_0}{\partial t} + AY_0 = 0 \text{ in } Q, \tag{1.117}$$

$$Y(0) = y_0, \tag{1.118}$$

$$Y_0 = 0 \text{ on } \Sigma. \tag{1.119}$$

In the following subsections we shall discuss the numerical solution of problem (1.116) by methods combining *conjugate gradient algorithms to finite difference and finite element discretizations.* We shall then apply the resulting methodology to the solution of *nonlinear* problem (1.115).

1.8.2. Conjugate gradient solution of problem (1.116).

From now on we shall denote by $(\cdot\,,\cdot)$ and $\|\cdot\|$ the canonical $L^2(\Omega)$-scalar product and $L^2(\Omega)$-norm, respectively. The various approximations of problem (1.116) can be solved by iterative methods closely related to the algorithm discussed in this section.

Writing (1.116) in *variational form* we obtain

$$\begin{cases} f \in L^2(\Omega), \\ k^{-1}(f, \hat{f}) + (\Lambda f, \hat{f}) = (y_T - Y_0(T), \hat{f}) \ \forall \hat{f} \in L^2(\Omega). \end{cases} \tag{1.120}$$

From the *symmetry*, *continuity* and *positive-definiteness* of the bilinear form $\{f, \hat{f}\} \to (\Lambda f, \hat{f})$, the variational problem (1.120) is a particular case of the following general problem

$$\begin{cases} u \in V, \\ a(u, v) = L(v) \ \forall v \in V, \end{cases} \tag{1.121}$$

where:

(i) V is a real *Hilbert space* for the scalar product $(\cdot, \cdot)$ and the corresponding norm $\| \cdot \|$.

(ii) $a : V \times V \to \mathbb{R}$ is bilinear, continuous, symmetric and V-elliptic (i.e. $\exists \alpha > 0$ such that

$$a(v, v) \geq \alpha \|v\|^2 \ \forall v \in V).$$

(iii) $L : V \to \mathbb{R}$ is *linear and continuous.*

If properties (i) to (iii) hold, then problem (1.121) has a *unique* solution (for this result which goes back to Hilbert, see e.g. Lions (1968), Ekeland and Temam (1974), Glowinski (1984)).

Problem (1.121) can be solved by the following *conjugate gradient algorithm:*

$$u^0 \in V \text{ is given;} \tag{1.122}$$

solve

$$g^0 \in V, \quad (g^0, v) = a(u^0, v) - L(v) \ \forall v \in V, \tag{1.123}$$

and set

$$w^0 = g^0. \quad \square \tag{1.124}$$

For $n \geq 0$, u^n, g^n, w^n *being known, compute* u^{n+1}, g^{n+1}, w^{n+1} *as follows.*

$$\varrho_n = \|g^n\|^2 / a(w^n, w^n) \tag{1.125}$$

and take

$$u^{n+1} = u^n - \varrho_n w^n. \tag{1.126}$$

Solve

$$g^{n+1} \in V, \quad (g^{n+1}, v) = (g^n, v) - \varrho_n a(w^n, v) \ \forall v \in V, \tag{1.127}$$

and compute

$$\gamma_n = \|g^{n+1}\|^2 / \|g^n\|^2, \tag{1.128}$$

$$w^{n+1} = g^{n+1} + \gamma_n w^n. \tag{1.129}$$

Do $n = n+1$ *and go to* (1.125).

Concerning the convergence of algorithm (1.122)–(1.129) it can be shown (cf. Daniel (1970)) that

$$\|u^n - u\| \leq c\|u^0 - u\| \left(\frac{\sqrt{\nu_a}-1}{\sqrt{\nu_a}+1}\right)^n, \tag{1.130}$$

where u is the solution of (1.121), and where the *condition member* ν_a of $a(\cdot,\cdot)$ is defined by $\nu_a = \|A\|\|A^{-1}\|$, where A is the unique operator in $\mathcal{L}(V,V)$ defined by

$$a(v,w) = (Av, w)\ \forall v, w \in V.$$

Application to the solution of problem (1.116) Before applying algorithm (1.122)–(1.129) to the solution of problem (1.116), let us recall the definition of operator Λ; it follows from Section 1.5, relation (1.41), that operator Λ is defined by

$$\Lambda f = \varphi(T), \tag{1.131}$$

where the function φ is obtained from f as follows.

Solve the *backward* equation

$$-\frac{\partial \psi}{\partial t} + A^*\psi = 0 \text{ in } Q, \quad \psi = 0 \text{ on } \Sigma,\ \psi(T) = f, \tag{1.132}$$

and then the *forward* equation

$$\frac{\partial \varphi}{\partial t} + A\varphi = \psi\chi_{\mathcal{O}} \text{ in } Q, \quad \varphi = 0 \text{ on } \Sigma,\ \varphi(0) = 0. \tag{1.133}$$

Applying now algorithm (1.122)–(1.129) to problem (1.116), we obtain the following iterative method (of *conjugated gradient* type);

$$f^0 \textit{ is given in } L^2(\Omega); \tag{1.134}$$

solve first

$$-\frac{\partial p^0}{\partial t} + A^*p^0 = 0 \textit{ in } Q, \quad p^0 = 0 \textit{ on } \Sigma,\ p^0(T) = f^0, \tag{1.135}$$

and set

$$u^0 = p^0\chi_{\mathcal{O}}. \tag{1.136}$$

Solve now

$$\frac{\partial y^0}{\partial t} + Ay^0 = u^0 \textit{ in } Q, \quad y^0 = 0 \textit{ on } \Sigma,\ y^0(0) = y_0, \tag{1.137}$$

compute

$$g^0 = k^{-1}f^0 + y^0(T) - y_T, \tag{1.138}$$

and set

$$w^0 = g^0. \quad \square \tag{1.139}$$

Then, for $n \geq 0$, assuming that f^n, g^n, w^n are known compute f^{n+1}, g^{n+1}, w^{n+1} as follows.

Solve

$$-\frac{\partial \bar{p}^n}{\partial t} + A^* \bar{p}^n = 0 \text{ in } Q, \quad \bar{p}^n = 0 \text{ on } \Sigma, \; \bar{p}^n(T) = w^n \tag{1.140}$$

and set

$$\bar{u}^n = \bar{p}^n \chi_{\mathcal{O}}. \tag{1.141}$$

Solve

$$\frac{\partial \bar{y}^n}{\partial t} + A\bar{y}^n = \bar{u}^n \text{ in } Q, \quad \bar{y}^n = 0 \text{ on } \Sigma, \; \bar{y}^n(0) = 0 \tag{1.142}$$

and compute

$$\bar{g}^n = k^{-1} w^n + \bar{y}^n(T), \tag{1.143}$$

$$\varrho_n = \|g^n\|^2 / (\bar{g}^n, w^n), \tag{1.144}$$

and then

$$f^{n+1} = f^n - \varrho_n w^n, \tag{1.145}$$

$$g^{n+1} = g^n - \varrho_n \bar{g}^n. \tag{1.146}$$

If $\|g^{n+1}\| / \|g^0\| \leq \varepsilon$, take $f = f^{n+1}$ and solve (1.132) to obtain $u = \psi \chi_{\mathcal{O}}$, the solution of problem (1.114); *if the above stopping test is not satisfied, compute*

$$\gamma_n = \|g^{n+1}\|^2 / \|g^n\|^2, \tag{1.147}$$

and then

$$w^{n+1} = g^{n+1} + \gamma_n w^n. \quad \square \tag{1.148}$$

Do $n = n+1$ and go to (1.140).

Remark 1.31 It is fairly easy to show that

$$\|k^{-1}\mathbf{I} + \Lambda\| = k^{-1} + \|\Lambda\|, \quad \|(k^{-1}\mathbf{I} + \Lambda)^{-1}\| = k,$$

implying that the *condition number* of the bilinear from in the left-hand side of (1.120) is equal to $\|\Lambda\|k + 1$. It follows from this result, and from (1.130), that the number of iterations of algorithm (1.134)–(1.148) necessary to obtain convergence varies like $\sqrt{k}\ln\varepsilon^{-1}$ for *large* values of k.

1.8.3. Time discretization of problem (1.116).

The crucial point here is to approximate properly the operator Λ defined by (1.131)–(1.133) in Section 1.8.2. Assuming that T is *bounded* and that operator A is independent of t, we introduce a *time discretization step*, defined by $\Delta t = T/N$, where N is a *positive integer*. Using an *implicit Euler* time discretization scheme, we approximate (1.132) by

$$\psi^{N+1} = f, \quad f \in L^2(\Omega); \tag*{$(1.149)_1$}$$

then, assuming that ψ^{n+1} *is known, we solve the following Dirichlet problem for* $n = N, N-1, \ldots, 1$,

$$-\frac{\psi^{n+1} - \psi^n}{\Delta t} + A^*\psi^n = 0 \ in \ \Omega, \quad \psi^n = 0 \ on \ \Gamma, \tag{1.149$_2$}$$

where $\psi^n \sim \psi(n\Delta t)(\psi(n\Delta t) : x \to \psi(x, n\Delta t))$. *Next, using similar notation, we approximate* (1.133) *by*

$$\varphi^0 = 0, \tag{1.150$_1$}$$

then assuming that φ^{n-1} *is known, we solve the following Dirichlet problem for* $n = 1, \ldots, N$,

$$\frac{\varphi^n - \varphi^{n-1}}{\Delta t} + A\varphi^n = \psi^n \chi_{\mathcal{O}} \ in \ \Omega, \quad \varphi^n = 0 \ on \ \Gamma. \tag{1.150$_2$}$$

Finally, we approximate Λ *by* $\Lambda^{\Delta t}$ *defined by*

$$\Lambda^{\Delta t} f = \varphi^N. \quad \square \tag{1.151}$$

From the *ellipticity properties* of A *and* A^* (see Section 1.1), *the Dirichlet* problems $(1.149)_2$ and $(1.150)_2$ have *a unique* solution; we furthermore have the following

Theorem 1.2 Operator $\Lambda^{\Delta t}$ is symmetric and positive semi-definite from $L^2(\Omega)$ into $L^2(\Omega)$.

Proof. Consider a pair $\{f, \hat{f}\} \in L^2(\Omega) \times L^2(\Omega)$. We have then

$$(\Lambda^{\Delta t} f, \hat{f}) = \int_\Omega \varphi^N \hat{\psi}^{N+1} \, \mathrm{d}x. \tag{1.152}$$

We also have, since $\varphi^0 = 0$,

$$\Delta t \sum_{n=1}^{N} \left[\varphi^n \left(\frac{\hat{\psi}^{n+1} - \hat{\psi}^n}{\Delta t} \right) + \hat{\psi}^n \left(\frac{\varphi^n - \varphi^{n-1}}{\Delta t} \right) \right] = \varphi^N \hat{\psi}^{N+1}. \tag{1.153}$$

Integrating (1.153) over Ω and taking $(1.149)_2$ into account we obtain

$$\begin{aligned} (\Lambda^{\Delta t} f, \hat{f}) &= \int_\Omega \varphi^N \hat{\psi}^{N+1} \, \mathrm{d}x \\ &= \Delta t \sum_{n=1}^{N} \int_\Omega (\varphi^n A^* \hat{\psi}^n - \hat{\psi}^n A\varphi^n) \, \mathrm{d}x + \Delta t \sum_{n=1}^{N} \int_{\mathcal{O}} \psi^n \hat{\psi}^n \, \mathrm{d}x \\ &= \Delta t \sum_{n=1}^{N} \int_{\mathcal{O}} \psi^n \hat{\psi}^n \, \mathrm{d}x, \end{aligned} \tag{1.154}$$

which completes the proof of the theorem. $\square$

Next, we compute the discrete analogue of Y_0 via

$$Y_0^0 = y_0, \tag{1.155$_1$}$$

and for $n = 1, \ldots, N$, *assuming that* Y_0^{n-1} *is known, solve the following (well-posed) elliptic problem*

$$\frac{Y_0^n - Y_0^{n-1}}{\Delta t} + AY_0^n = 0 \text{ in } \Omega, \quad Y_0^n = 0 \text{ on } \Gamma. \qquad (1.155)_2$$

Finally, we approximate problem (1.116) by

$$\begin{cases} f^{\Delta t} \in L^2(\Omega), \\ (k^{-1} f^{\Delta t} + \Lambda^{\Delta t} f^{\Delta t}, \hat{f}) = (y_T - Y_0^N, \hat{f}) \ \forall \hat{f} \in L^2(\Omega). \end{cases} \qquad (1.156)$$

Problem (1.156) can be solved by a time discrete analogue of algorithm (1.134)–(1.148).

Remark 1.32 The *Euler schemes* which have been used to time discretize problem (1.116) are only *first-order accurate*; for some applications this may require very small time steps Δt to obtain an acceptable level of accuracy. A simple way to improve this situation is to use second-order schemes like those described in Section 1.8.5 (variants of these schemes have been successfully used to solve boundary controllability problems for the *heat equation* in Carthel *et al.* (1994)).

1.8.4. Full discretization of problem (1.116).

We suppose from now on – and for simplicity – that Ω and $\mathcal{O}$ are *polygonal* domains of $\mathbb{R}^2$ (for nonpolygonal domains Ω and/or $\mathcal{O}$ we shall approximate them by polygonal domains). We introduce then a first *finite element triangulation* $\mathcal{T}_h$ of Ω (h: largest length of the edges of the triangles of $\mathcal{T}_h$) as in Dean, Glowinski and Li (1989), Glowinski, Li and Lions (1990) and Glowinski (1992a); we suppose that both $\bar{\Omega}$ and $\bar{\mathcal{O}}$ are unions of triangles of $\mathcal{T}_h$. Next, we approximate $H^1(\Omega)$, $L^2(\Omega)$ and $H_0^1(\Omega)$ by the following *finite-dimensional* spaces (with P_1 the space of the polynomials in two variables of degree ≤ 1)

$$V_h = \{v_h \mid v_h \in C^0(\bar{\Omega}), v_h|_T \in P_1 \ \forall T \in \mathcal{T}_h\} \qquad (1.157)$$

and

$$V_{0h} = \{v_h \mid v_h \in V_h, v_h = 0 \text{ on } \Gamma\} \quad (= V_h \cap H_0^1(\Omega)), \qquad (1.158)$$

respectively. We introduce now a second finite element triangulation $\mathcal{T}_H$ of Ω (we may take $\mathcal{T}_h = \mathcal{T}_H$, but the idea here is to have $\mathcal{T}_H$ *coarser* than $\mathcal{T}_h$) and we associate with $\mathcal{T}_H$ the following two finite-dimensional spaces

$$E_H = \{\hat{f}_H \mid \hat{f}_H \in C^0(\bar{\Omega}), \hat{f}_H|_T \in P_1 \ \forall T \in \mathcal{T}_H\}, \qquad (1.159)$$

$$E_{0H} = \{\hat{f}_H \mid \hat{f}_H \in E_H, \ \hat{f}_H = 0 \text{ on } \Gamma\} \quad (= E_H \cap H_0^1(\Omega)). \qquad (1.160)$$

Since *closure of* $H_0^1(\Omega)$ *in* $L^2(\Omega) = L^2(\Omega)$ we can use either V_h or V_{0h} (respectively E_H or E_{0H}) to approximate $L^2(\Omega)$.

At this stage, it is convenient to (re)introduce $a : H_0^1(\Omega) \times H_0^1(\Omega) \to \mathbb{R}$, the *bilinear* form associated with the elliptic operator A; it is defined by

$$a(y, z) = \langle Ay, z\rangle \ \forall y, z \in H_0^1(\Omega), \tag{1.161}$$

where, in (1.161), $\langle \cdot , \cdot \rangle$ denotes the duality pairing between $H^{-1}(\Omega)$ and $H_0^1(\Omega)$. Similarly we have

$$a(z, y) = \langle A^*y, z\rangle \ \forall y, z \in H_0^1(\Omega). \tag{1.162}$$

From the properties of operator A (see Section 1.1), the above bilinear form is *continuous* over $H_0^1(\Omega) \times H_0^1(\Omega)$ and $H_0^1(\Omega)$-*elliptic*.

We approximate problem (1.116) by

$$\begin{cases} f_{hH}^{\Delta t} \quad \in E_{0H} \ \forall \hat{f}_H \in E_{0H} \text{ we have} \\ \displaystyle\int_\Omega (k^{-1} f_{hH}^{\Delta t} + \Lambda_{hH}^{\Delta t} f_{hH}^{\Delta t})\hat{f}_H \, \mathrm{d}x = \int_\Omega (y_T - Y_{0h}^N)\hat{f}_H \, \mathrm{d}x, \end{cases} \tag{1.163}$$

where, in (1.163), Y_{0h}^N is obtained from the *full discretization* of problem (1.117)–(1.119), namely

$$Y_{0h}^0 = y_{0h} \text{ with } y_{0h} (\in V_h) \quad \text{an approximation of } y_0; \tag{1.164$_1$}$$

for $n = 1, \ldots, N$, *assuming that* Y_{0h}^{n-1} *is known, compute* Y_{0h}^n *via the solution of the following (approximate and well-posed) elliptic problem.*

$$\begin{cases} Y_{0h}^n \in V_{0h}, \\ \displaystyle\int_\Omega \frac{Y_{0h}^n - Y_{0h}^{n-1}}{\Delta t} v_h \, \mathrm{d}x + a(Y_{0h}^n, v_h) = 0 \ \forall v_h \in V_{0h}. \end{cases} \tag{1.164$_2$}$$

The operator $\Lambda_{hH}^{\Delta t}$ is defined by

$$\Lambda_{hH}^{\Delta t} f_H = \varphi_h^N \ \forall f_H \in E_{0H}, \tag{1.165}$$

where in order to compute φ_h^N we solve sequentially the following two discrete parabolic problems:

First problem

$$\psi_h^{N+1} = f_H; \tag{1.166$_1$}$$

then for $n = N, N-1, \ldots, 1$, *we compute* ψ_h^n *from* ψ_h^{n+1} *via the solution of the following discrete Dirichlet problem*

$$\int_\Omega \frac{\psi_h^n - \psi_h^{n+1}}{\Delta t} v_h \, \mathrm{d}x + a(v_h, \psi_h^n) = 0 \ \forall v_h \in V_{0h}; \ \psi_h^n \in V_{0h}. \tag{1.166$_2$}$$

Second problem

$$\varphi_h^0 = 0; \tag{1.167$_1$}$$

for $n = 1, \ldots, N$, *we compute* φ_h^n *from* φ_h^{n-1} *via the solution of the following discrete Dirichlet problem*

$$\int_\Omega \frac{\varphi_h^n - \varphi_h^{n-1}}{\Delta t} v_h \, dx + a(\varphi_h^n, v_h) = \int_{\mathcal{O}} \psi_h^n v_h \, dx \ \forall v_h \in V_{0h}; \ \varphi_h^n \in V_{0h}. \tag{1.167$_2$}$$

The discrete elliptic problems $(1.166)_2$ and $(1.167)_2$ have a *unique* solution (this follows from the properties of the bilinear form $a(\cdot\,,\cdot)$).

Concerning the properties of $\Lambda_{hH}^{\Delta t}$ we can prove the following *fully discrete* analogue of relation (1.154):

$$\int_\Omega (\Lambda_{hH}^{\Delta t} f_H) \hat{f}_H \, dx = \Delta t \sum_{n=1}^{N} \int_{\mathcal{O}} \psi_h^n \hat{\psi}_h^n \, dx \ \forall f_H, \hat{f}_H \in E_{0H}, \tag{1.168}$$

which shows that operator $\Lambda_{hH}^{\Delta t}$ is *symmetric* and *positive semi-definite*, implying in turn that problem (1.163) has a *unique solution* and can be solved by a *conjugate gradient algorithm* (described in the following Section 1.8.5).

Remark 1.33 We can apply the *trapezoidal rule* to evaluate the various $L^2(\Omega)$-scalar products taking place in (1.163), (1.164), (1.166), (1.167) (see Glowinski *et al.* (1990) and Glowinski (1992a) for more details about the use of *numerical integration* in the context of control problems).

Remark 1.34 Instead of E_{0H} we can take the space E_H to approximate problem (1.116); the corresponding approximate problem is still well posed and can be solved by a conjugate gradient algorithm.

1.8.5. Iterative solution of problem (1.163).

From the properties of $\Lambda_{hH}^{\Delta t}$ shown in the previous section, the bilinear form in (1.163) is *symmetric* and *positive definite* (in fact *uniformly* with respect to h, H and Δt). Thus, problem (1.163) can be solved by a *conjugate gradient algorithm* which is a discrete analogue of algorithm (1.134)–(1.148).

Description of the algorithm For simplicity, we shall drop the subscripts h, H and superscript Δt from $f_{hH}^{\Delta t}$.

Initialization

$$f_0 \text{ is given in } E_{0H}; \tag{1.169}$$

assuming that p_0^{n+1} *is known, solve the following discrete Dirichlet problem for* $n = N, \ldots, 1$

$$\int_\Omega \frac{p_0^n - p_0^{n+1}}{\Delta t} v \, dx + a(v, p_0^n) = 0 \ \forall v \in V_{0h}; \ p_0^n \in V_{0h}, \tag{1.170$_1$}$$

with

$$p_0^{N+1} = f_0, \tag{1.170$_2$}$$

and set

$$u_0^n = p_0^n \mid_{\mathcal{O}} . \tag{1.171}$$

Assuming that y_0^{n-1} *is known, solve for* $n = 1, \ldots, N$, *the following (well-posed) discrete problem*

$$\int_\Omega \frac{y_0^n - y_0^{n-1}}{\Delta t} v \, \mathrm{d}x + a(y_0^n, v) = \int_{\mathcal{O}} u_0^n v \, \mathrm{d}x \; \forall v \in V_{0h}; \; y_0^n \in V_{0h}, \qquad (1.172)_1$$

with

$$y_0^0 = y_{0h}. \qquad (1.172)_2$$

Finally, solve the following variational problem

$$\begin{cases} g_0 \in E_{0H}, \\ \displaystyle\int_\Omega g_0 \hat{f} \, \mathrm{d}x = \int_\Omega (k^{-1} f_0 + y_0^N - y_T) \hat{f} \, \mathrm{d}x \; \forall \hat{f} \in E_{0H}, \end{cases} \tag{1.173}$$

and set

$$w_0 = g_0. \quad \square \tag{1.174}$$

Then for $m \geq 0$, *assuming that* f_m, g_m, w_m *are known, compute* f_{m+1}, g_{m+1}, w_{m+1} *as follows.*

Assuming that $\bar{p}_m^{n+1}$ *is known, solve for* $n = N, \ldots, 1$, *the following (well-posed) problem*

$$\int_\Omega \frac{\bar{p}_m^n - \bar{p}_m^{n+1}}{\Delta t} v \, \mathrm{d}x + a(v, \bar{p}_m^n) = 0 \; \forall v \in V_{0h}; \; \bar{p}_m^n \in V_{0h}, \qquad (1.175)_1$$

with

$$\bar{p}_m^{N+1} = w_m, \qquad (1.175)_2$$

and set

$$\bar{u}_m^n = \bar{p}_m^n \mid_{\mathcal{O}} . \tag{1.176}$$

Assuming that $\bar{y}_m^{n-1}$ *is known, solve for* $n = 1, \ldots, N$, *the following (well-posed) problem*

$$\int_\Omega \frac{\bar{y}_m^n - \bar{y}_m^{n-1}}{\Delta t} v \, \mathrm{d}x + a(\bar{y}_m^n, v) = \int_{\mathcal{O}} \bar{u}_m^n v \, \mathrm{d}x \; \forall v \in V_{0h}; \; \bar{y}_m^n \in V_{0h}, \qquad (1.177)_1$$

with

$$\bar{y}_m^0 = 0. \qquad (1.177)_2$$

Next, solve

$$\begin{cases} \bar{g}_m \in E_{0H}, \\ \displaystyle\int_\Omega \bar{g}_m \hat{f} \, \mathrm{d}x = \int_\Omega (k^{-1} w_m + \bar{y}_m^N) \hat{f} \, \mathrm{d}x \; \forall \hat{f} \in E_{0H}, \end{cases} \tag{1.178}$$

and compute

$$\rho_m = \frac{\int_\Omega |g_m|^2 \, dx}{\int_\Omega \bar{g}_m w_m \, dx}, \tag{1.179}$$

and then

$$f_{m+1} = f_m - \rho_m w_m, \tag{1.180}$$

$$g_{m+1} = g_m - \rho_m \bar{g}_m. \tag{1.181}$$

If $\|g_{m+1}\|_{L^2(\Omega)}/\|g_0\|_{L^2(\Omega)} \leq \varepsilon$, *take* $f = f_{m+1}$ *and solve (1.166) (with* $f_H = f$*) to obtain* $u^n = \psi^n |_{\mathcal{O}}$, *for* $n = 1, \ldots, N$; *if the above stopping test is not satisfied, compute*

$$\gamma_m = \frac{\|g_{m+1}\|^2_{L^2(\Omega)}}{\|g_m\|^2_{L^2(\Omega)}}, \tag{1.182}$$

and then

$$w_{m+1} = g_{m+1} + \gamma_m w_m. \quad \Box \tag{1.183}$$

Do $m = m + 1$ *and go to (1.175).*

Remark 1.35 The computer implementation of algorithm (1.169)–(1.183) requires the solution of the *discrete Dirichlet problems* $(1.170)_1$, $(1.172)_1$ and $(1.175)_1$ and $(1.177)_1$; to solve these (linear) problems we can use either *direct methods* (such as *Cholesky's* if the blinear form $a(.,.)$ is *symmetric*) or *iterative methods* (such as *conjugate gradient, relaxation, multigrid,* etc.). To initialize the iterative methods we shall use the solution of the corresponding problem at the previous time step.

A variant of algorithm (1.169)–(1.183) has been employed in Carthel *et al.* (1994), to solve exact and approximate boundary controllability problems for the heat equation; see also Section 2.5 (*Acta Numerica 1995*).

1.8.6. On the use of second-order accurate time discretization schemes for the solution of problem (1.114).

We now complete Remark 1.32 and closely follow Carthel *et al.* (1994, Section 4.6).

1.8.6.1. Generalities. The numerical methods described in Sections 1.8.3 to 1.8.5 rely on a *first-order accurate* time discretization scheme (namely the *backward Euler* scheme). In order to decrease the computational cost for a *given accuracy* (or increase the accuracy for the same computational cost), it makes sense to use higher order time discretization schemes. A natural choice in that direction seems to be the *Crank–Nicolson* scheme (see, e.g. Raviart and Thomas (1988, Ch. 7)) since it is a *one-step, second-order accurate* time discretization scheme, which is, in addition, no more complicated to implement in practice than the backward Euler scheme. Unfortunately, it is well known that the Crank–Nicolson scheme is not well suited (unless one takes Δt of the order of h^2) to simulate *fast transient phenomena* and/or to carry out numerical integration on *long time intervals* $[0, T]$. From these drawbacks a more natural choice is the *two-step implicit* scheme described next which is *second-order accurate*, has much better properties than Crank–Nicolson concerning fast transients and long time intervals, and which is no more complicated to implement in practice than the backward Euler scheme (for a discussion of multistep schemes applied to the time discretization of parabolic problems, see, e.g. Thomee (1990, Section 6)).

1.8.6.2. A second-order accurate time approximation of problem (1.114). In order to solve the control problem (1.114) via the solution of the functional equation (1.116), the crucial point is – again – to properly approximate the operator Λ and the function Y_0 defined in Section 1.8.1.

Approximation of operator Λ Focusing on time discretization, we approximate Λ by $\Lambda^{\Delta t}$ defined as follows (we use the notation of Section 1.8.3).

Let us consider $f \in L^2(\Omega)$, then

$$\Lambda^{\Delta t} f = 2\varphi^{N-1} - \varphi^{N-2}, \tag{1.184}$$

where to obtain $\varphi^{N-2}, \varphi^{N-1}$ we solve first for $n = N-1, \dots, 1$ the following (well-posed) Dirichlet problem

$$\frac{\frac{3}{2}\psi^n - 2\psi^{n+1} + \frac{1}{2}\psi^{n+2}}{\Delta t} + A^*\psi^n = 0 \text{ in } \Omega, \quad \psi^n = 0 \text{ on } \Gamma, \tag{1.185}$$

with

$$\psi^N = 2f, \quad \psi^{N+1} = 4f, \tag{1.186}$$

then, with $\varphi^0 = 0$,

$$\frac{\varphi^1 - \varphi^0}{\Delta t} + (\tfrac{2}{3}A\varphi^1 + \tfrac{1}{3}A\varphi^0) = \tfrac{2}{3}\psi^1\chi_{\mathcal{O}} \text{ in } \Omega, \quad \varphi^1 = 0 \text{ on } \Gamma, \tag{1.187}$$

and, finally, for $n = 2, \dots, N-1$,

$$\frac{\frac{3}{2}\varphi^n - 2\varphi^{n-1} + \frac{1}{2}\varphi^{n-2}}{\Delta t} + A\varphi^n = \psi^n\chi_{\mathcal{O}} \text{ in } \Omega, \quad \varphi^n = 0 \text{ on } \Gamma. \quad \square \tag{1.188}$$

It can be shown that

$$\int_\Omega (\Lambda^{\Delta t} f)\hat{f}\,dx = \Delta t \sum_{n=1}^{N-1} \int_{\mathcal{O}} \psi^n \hat{\psi}^n\,dx \;\forall f,\ \hat{f} \in L^2(\Omega),$$

i.e. Theorem 1.2 still holds for this new operator $\Lambda^{\Delta t}$ (in fact, $\Lambda^{\Delta t}$ has been defined so that the above relation holds; see also Remark 1.36).

Approximation of Y_0 To compute the discrete analogue of Y_0, we take

$Y_0^0 = y_0$ and we solve the Dirichlet problem

$$\frac{Y_0^1 - Y_0^0}{\Delta t} + (\tfrac{2}{3}AY_0^1 + \tfrac{1}{3}AY_0^0) = 0 \text{ in } \Omega, \quad Y_0^1 = 0 \text{ on } \Gamma, \tag{1.189$_1$}$$

and then for $n = 2, \ldots, N-1$,

$$\frac{\tfrac{3}{2}Y_0^n - 2Y_0^{n-1} + \tfrac{1}{2}Y_0^{n-2}}{\Delta t} + AY_0^n = 0 \text{ in } \Omega, \quad Y_0^n = 0 \text{ on } \Gamma. \tag{1.189$_2$}$$

Approximation of problem (1.116) We approximate problem (1.116) by

$$\begin{cases} f^{\Delta t} \in L^2(\Omega);\ \forall \hat{f} \in L^2(\Omega) \ \text{ we have} \\ (k^{-1}f^{\Delta t} + \Lambda^{\Delta t} f^{\Delta t}, \hat{f})_{L^2(\Omega)} = (y_T - 2Y_0^{N-1} + Y_0^{N-2}, \hat{f})_{L^2(\Omega)}. \end{cases} \tag{1.190}$$

Problem (1.190) can be solved by a discrete analogue of algorithm (1.134)–(1.148). Also, the finite element discretization discussed in Section 1.8.4 can be applied easily to problem (1.190) and the resulting fully discrete problem can be solved by a variant of the conjugate gradient algorithm (1.169)–(1.183).

Remark 1.36 The definition of $\Lambda^{\Delta t}$ via relations (1.184) to (1.188), may look somewhat artificial; in fact, it can be shown that the control obtained via the solution of (1.190) is the *unique* solution of the following (time discrete) control problem:

$$\min_{\{v^n\}_{n=1}^{N-1} \in (L^2(\mathcal{O}))^{N-1}} J^{\Delta t}(v^1, \ldots, v^{N-1}), \tag{1.191}$$

where, in (1.191), we have

$$J^{\Delta t}(v_1, \ldots, v^{N-1}) = \frac{\Delta t}{2} \sum_{n=1}^{N-1} \int_{\mathcal{O}} |v^n|^2 \, dx + \frac{k}{2} \|2y^{N-1} - y^{N-2} - y_T\|^2_{L^2(\Omega)} \tag{1.192}$$

and where y^{N-2}, y^{N-1} are obtained from $\{v^n\}_{n=1}^{N-1}$ via the solution of the following discrete parabolic problem:

$$y^0 = y_0, \tag{1.193}$$

$$\frac{y^1 - y^0}{\Delta t} + A\left(\tfrac{2}{3}y^1 + \tfrac{1}{3}y^0\right) = \tfrac{2}{3}v^1 \chi_{\mathcal{O}} \text{ in } \Omega, \quad y^1 = 0 \text{ on } \Gamma, \tag{1.194}$$

and for $n = 2, \ldots, N-1$,

$$\frac{\tfrac{3}{2}y^n - 2y^{n-1} + \tfrac{1}{2}y^{n-2}}{\Delta t} + Ay^n = v^n \chi_{\mathcal{O}} \text{ in } \Omega, \quad y^n = 0 \text{ on } \Gamma. \tag{1.195}$$

In principle, $2y^{N-1} - y^{N-2}$ is an $\mathcal{O}(|\Delta t|^2)$ accurate approximate value of $y(T)$ obtained by *extrapolation*.

Remark 1.37 A variant of the previously mentioned second-order time discretization scheme has been successfully applied in Carthel *et al.* (1994, Section 7) to the solution of exact and approximate boundary controllability for the heat equation; see also Section 2.5 (*Acta Numerica 1995*).

1.8.7. Convergence of the approximate solutions of problems (1.114).

In this section, we shall discuss the *convergence* of the solution of the *fully discrete* problem (1.163) – and of the corresponding approximate solution of problem (1.114) – as $\{\Delta t, h, H\} \to 0$. Problem (1.163) has been defined in Section 1.8.4 (whose notation is kept) by

$$\begin{cases} f_{hH}^{\Delta t} \in E_{0H} \ \forall \hat{f}_H \in E_{0H} \quad \text{we have} \\ \displaystyle\int_\Omega (k^{-1} f_{hH}^{\Delta t} + \Lambda_{hH}^{\Delta t} f_{hH}^{\Delta t}) \hat{f}_H \, \mathrm{d}x = \int_\Omega (y_T - Y_{0h}^N) \hat{f}_H \, \mathrm{d}x. \end{cases} \tag{1.196}$$

Concerning the convergence of $\{f_{hH}^{\Delta t}\}_{\{\Delta t, h, H\}}$ as $\{\Delta t, h, H\} \to 0$, we have the following

Theorem 1.3 We suppose that

$$\lim_{h \to 0} \|y_{0h} - y_0\|_{L^2(\Omega)} = 0, \tag{1.197}$$

and

$$\begin{cases} \text{the angles of } \mathcal{T}_h \text{ are uniformly bounded away} \\ \text{from 0 (i.e. } \exists \theta_0 > 0, \text{ such that } \theta \geq \theta_0 \ \forall \theta \text{ angle of } \mathcal{T}_h \ \forall h). \end{cases} \tag{1.198}$$

Then

$$\lim_{\{\Delta t, h, H\} \to 0} \|f_{hH}^{\Delta t} - f\|_{L^2(\Omega)} = 0, \tag{1.199}$$

$$\lim_{\{\Delta t, h, H\} \to 0} \|\psi_{hH}^{\Delta t} \chi_{\mathcal{O}} - u\|_{L^2(\mathcal{O} \times (0,T))} = 0, \tag{1.200}$$

where, in (1.199), f and $f_{hH}^{\Delta t}$ are the solutions of problems (1.116) and (1.163), (1.196), respectively and where, in (1.200), u is the solution of the control problem (1.114) and $\psi_{hH}^{\Delta t} \chi_{\mathcal{O}}$ the discrete control corresponding to $f_{hH}^{\Delta t}$ via (1.166), with $\psi_h^{N+1} = f_{hH}^{\Delta t}$ in $(1.166)_1$.

Proof. To simplify the presentation, we split the proof into several steps.

(i) *Estimates.* Taking $\hat{f}_H = f_{hH}^{\Delta t}$ in (1.196) we obtain, since operator $\Lambda_{hH}^{\Delta t}$ is *positive semi-definite* (see Section 1.8.4, relation (1.168)), that

$$\|f_{hH}^{\Delta t}\|_{L^2(\Omega)} \leq k \|y_T - Y_{0h}^N\|_{L^2(\Omega)} \ \forall \{\Delta t, h, H\}. \tag{1.201}$$

It follows then from standard results on the *finite element approximation of parabolic problems* (see, e.g. Raviart and Thomas (1988, Ch. 7, Section 7.5) and Fujita and Suzuki (1991, Ch. 2, Section 8)) that Properties (1.197),

(1.198) imply that

$$\lim_{\{\Delta t,h\}\to 0} \|Y_{0h}^N - Y_0(T)\|_{L^2(\Omega)} = 0, \tag{1.202}$$

where Y_0 is *the* solution of the parabolic problem (1.117)–(1.119). It follows from (1.201) that the family $\{Y_{0h}^N\}_{\{\Delta t,h\}}$ is *bounded* in $L^2(\Omega)$ which implies, in turn, that the right-hand side of (1.201) and therefore

$$\{\|f_{hH}^{\Delta t}\|_{L^2(\Omega)}\}_{\{\Delta t,h,H\}}$$

are *bounded*. Since the family $\{f_{hH}^{\Delta t}\}_{\{\Delta t,h,H\}}$ is bounded in $L^2(\Omega)$ we can extract a subsequence – still denoted by $\{f_{hH}^{\Delta t}\}_{\{\Delta t,h,H\}}$ – such that

$$\lim_{\{\Delta t,h,H\}\to 0} f_{hH}^{\Delta t} = f^* \text{ weakly in } L^2(\Omega). \tag{1.203}$$

(ii) *Weak convergence.* To show that $f^* = f$, it is convenient to introduce Π_H, the $L^2(\Omega)$-projection operator from $L^2(\Omega)$ into E_{0H}; we have

$$\lim_{H\to 0} \|\Pi_H \hat{f} - \hat{f}\|_{L^2(\Omega)} = 0 \ \forall \hat{f} \in L^2(\Omega). \tag{1.204}$$

In Lemma 1.1 which follows later, we shall prove that

$$\lim_{\{\Delta t,h,H\}\to 0} \|\Lambda_{hH}^{\Delta t}\Pi_H \hat{f} - \Lambda\hat{f}\|_{L^2(\Omega)} = 0 \ \forall \hat{f} \in L^2(\Omega). \tag{1.205}$$

It follows then from (1.196), (1.201)–(1.205) and from the *symmetry* of operators Λ and $\Lambda_{hH}^{\Delta t}$ that, $\forall \hat{f} \in L^2(\Omega)$,

$$\begin{aligned}
&\lim_{\{\Delta t,h,H\}\to 0} \int_\Omega (k^{-1} f_{hH}^{\Delta t} + \Lambda_{hH}^{\Delta t} f_{hH}^{\Delta t})\Pi_h \hat{f}\, dx \\
&= \lim_{\{\Delta t,h,H\}\to 0} \left[k^{-1}\int_\Omega f_{hH}^{\Delta t}(\Pi_h \hat{f})\, dx + \int_\Omega (\Lambda_{hH}^{\Delta t})\Pi_H \hat{f} f_{hH}^{\Delta t}\, dx\right] \\
&= \int_\Omega (k^{-1} f^* + \Lambda f^*)\hat{f}\, dx \\
&= \lim_{\{\Delta t,h,H\}\to 0} \int_\Omega (y_T - Y_{0h}^N)\Pi_H \hat{f}\, dx = \int_\Omega (y_T - Y_0(T))\hat{f}\, dx.
\end{aligned}$$

Thus, we have proved (if (1.205) holds) that f^* is a solution of problem (1.116); since (1.116) has a *unique* solution we have $f^* = f$ and also the fact that the *whole* family $\{f_{hH}^{\Delta t}\}_{\{\Delta t,h,H\}}$ converges to f as $\{\Delta t, h, H\} \to 0$.

(iii) *Strong convergence.* Let us introduce $\bar{f}_{hH}^{\Delta t} = f_{hH}^{\Delta t} - \Pi_H f$; we clearly have

$$\lim_{\{\Delta t,h,H\}\to 0} \bar{f}_{hH}^{\Delta t} = 0 \text{ weakly in } L^2(\Omega). \tag{1.206}$$

We also have, $\forall\{\Delta t, h, H\}$,

$$k^{-1}\|\bar{f}_{hH}^{\Delta t}\|_{L^2(\Omega)}^2 \leq \int_\Omega (k^{-1}\bar{f}_{hH}^{\Delta t} + \Lambda_{hH}^{\Delta t}\bar{f}_{hH}^{\Delta t})\bar{f}_{hH}^{\Delta t}\, dx. \tag{1.207}$$

Concerning the right-hand side of (1.207) we have, from (1.196),

$$
\begin{aligned}
&(k^{-1}\bar{f}_{hH}^{\Delta t} + \Lambda_{hH}^{\Delta t}\bar{f}_{hH}^{\Delta t}, \bar{f}_{hH}^{\Delta t})_{L^2(\Omega)} \\
&\quad = (k^{-1}f_{hH}^{\Delta t} + \Lambda_{hH}^{\Delta t}f_{hH}^{\Delta t}, \bar{f}_{hH}^{\Delta t})_{L^2(\Omega)} - (k^{-1}\Pi_H f + \Lambda_{hH}^{\Delta t}\Pi_H f, \bar{f}_{hH}^{\Delta t}) \\
&\quad = (y_T - Y_{0h}^N, \bar{f}_{hH}^{\Delta t})_{L^2(\Omega)} - (k^{-1}\Pi_H f + \Lambda_{hH}^{\Delta t}\Pi_H f, \bar{f}_{hH}^{\Delta t}.
\end{aligned}
$$

Taking the limit in the above relations and in (1.207) as $\{\Delta t, h, H\} \to 0$, we obtain from (1.201)–(1.206) that

$$
0 \le \underline{\lim}_{\{\Delta t,h,H\}\to 0}\|\bar{f}_{hH}^{\Delta t}\|_{L^2(\Omega)} \le \overline{\lim}_{\{\Delta t,h,H\}\to 0}\bar{f}_{hH}^{\Delta t}\|_{L^2(\Omega)} \le 0; \tag{1.208}
$$

we have thus proved that

$$
\lim_{\{\Delta t,h,H\}\to 0}\|\bar{f}_{hH}^{\Delta t}\|_{L^2(\Omega)} = 0,
$$

which combined with (1.204) (with $\hat{f} = f$) implies in turn the convergence property (1.199).

(iv) *Convergence of the discrete control.* The solution u of the control problem (1.114) satisfies $u = \psi\chi_{\mathcal{O}}$, where ψ is the solution of the parabolic problem (1.132) when $\psi(T) = f$, f being the solution of problem (1.116). Similarly, we associate the solution $f_{hH}^{\Delta t}$ of problem (1.163), (1.196) with the solution $\{\psi_{hH}^n\}_{n=1}^N$ of problem (1.166) when $\psi_h^{N+1} = f_{hH}^{\Delta t}$ in $(1.166)_1$ or, equivalently, the piecewise constant function $\psi_{hH}^{\Delta t}$ of t, defined by

$$
\psi_{hH}^{\Delta t} = \sum_{n=1}^{N}\psi_{hH}^n I_n, \tag{1.209}
$$

where I_n is the *characteristic function of* $(0,T)\cap((n-1/2)\Delta t, (n+1/2)\Delta t)$.

Since $\lim_{\{\Delta t,h,H\}\to 0}\|f_{hH}^{\Delta t} - f\|_{L^2(\Omega)} = 0$, it follows from Raviart and Thomas (1988), and from Lemma 1.1, that

$$
\lim_{\{\Delta t,h,H\}\to 0}\|\psi_{hH}^{\Delta t} - \psi\|_{L^2(Q)} = 0
$$

which implies in turn that

$$
\lim_{\{\Delta t,h,H\}\to 0}\|\psi_{hH}^{\Delta t}\chi_{\mathcal{O}} - u\|_{L^2(\mathcal{O}\times(0,T))} = 0,
$$

i.e. relation (1.200) holds. □

The proof of Theorem 1.3 will be complete once we have proved the following

Lemma 1.1 Suppose that the angle condition (1.198) holds and consider a family $\{\hat{f}_H\}_H$ of E_{0H} such that

$$
\lim_{H\to 0}\|\hat{f}_H - \hat{f}\|_{L^2(\Omega)} = 0. \tag{1.210}
$$

If with $\hat{f}_H$ we associate $\Lambda_{hH}^{\Delta t}\hat{f}_H, \{\hat{\psi}_h\}_{n=1}^N, \{\hat{\varphi}_h\}_{n=1}^N$ via (1.165)–(1.167), respectively, we then have

$$\lim_{\{\Delta t,h,H\}\to 0} \|\hat{\psi}_{hH}^{\Delta t} - \hat{\psi}\|_{L^2(Q)} = 0, \tag{1.211}$$

$$\lim_{\{\Delta t,h,H\}\to 0} \|\Lambda_{hH}^{\Delta t}\hat{f}_H - \Lambda\hat{f}\|_{L^2(\Omega)} = 0, \tag{1.212}$$

where, in (1.211) $\hat{\psi}_{hH}^{\Delta t}$ is defined from $\{\hat{\psi}_h\}_{n=1}^N$ by (1.209) and where $\hat{\psi}$ is the solution of

$$-\frac{\partial\hat{\psi}}{\partial t} + A^*\hat{\psi} = 0 \text{ in } Q, \quad \hat{\psi} = 0 \text{ on } \Sigma, \quad \hat{\psi}(T) = \hat{f}. \tag{1.213}$$

Proof. (i) Proof of (1.211). For convenience, extend $\{\hat{\psi}_h\}_{n=1}^N$ to $n = 0$ by solving $(1.166)_2$ for $n = 0$, and still denote by $\hat{\psi}_{hH}^{\Delta t}$ the function $\sum_{n=0}^N \hat{\psi}_h^n I_n$; it follows from Raviart and Thomas (1988), that

$$\lim_{\{\Delta t,h,H\}\to 0} \|\hat{\psi}_{hH}^{\Delta t} - \hat{\psi}\|_{L^\infty(0,T;L^2(\Omega))} = 0 \tag{1.214}$$

if we can show that

$$\lim_{\{\Delta t,h,H\}\to 0} \|\hat{\psi}_h^N - \hat{f}\|_{L^2(\Omega)} = 0. \tag{1.215}$$

To show (1.215), observe first that $\hat{\psi}_h^N$ is the unique solution of the discrete elliptic problem

$$\begin{cases} \hat{\psi}_h^N \in V_{0h}, \\ \int_\Omega \hat{\psi}_h^N v_h \, dx + \Delta t a(v_h, \hat{\psi}_h^N) = \int_\Omega \hat{f}_H v_h \, dx \;\forall v_h \in V_{0h}. \end{cases} \tag{1.216}$$

Taking $v_h = \hat{\psi}_h^N$ in (1.216), we obtain, from the $H_0^1(\Omega)$ ellipticity of $a(\cdot,\cdot)$ (see Section 1.1), from the Schwarz inequality in $L^2(\Omega)$, and from (1.210), that

$$\|\hat{\psi}_h^N\|_{L^2(\Omega)} \le C \;\forall\{\Delta t, h, H\}, \tag{1.217}$$

$$(\Delta t)^{1/2}\|\hat{\psi}_h^N\|_{H_0^1(\Omega)} \le C \;\forall\{\Delta t, h, H\}, \tag{1.218}$$

where, in (1.217), (1.218) (and in the following), C denotes various quantities independent of $\Delta t, h, H$.

Since from (1.217)), $\{\hat{\psi}_h^N\}_{\{\Delta t,h,H\}}$ is bounded in $L^2(\Omega)$, we can extract a subsequence – still denoted by $\{\hat{\psi}_h^N\}_{\{\Delta t,h,H\}}$ – such that

$$\lim_{\{\Delta t,h,H\}\to 0} \hat{\psi}_h^N = \hat{f}^* \text{ weakly in } L^2(\Omega). \tag{1.219}$$

Consider, next, $v \in \mathcal{D}(\Omega)$ and denote by $r_h v$ the *linear interpolate* of v on $\mathcal{T}_h$; since the *angle condition* (1.198) holds, it follows from, e.g., Ciarlet (1978;

1991), Raviart and Thomas (1988), Glowinski (1984, Appendix 1), that

$$\lim_{h \to 0} \|r_h v - v\|_{H_0^1(\Omega)} = 0 \ \forall v \in \mathcal{D}(\Omega). \tag{1.220}$$

Take now $v_h = r_h v$ in (1.216); it follows then from (1.218), (1.220), and from the continuity of $a(\,.\,,\,.\,)$ over $H_0^1(\Omega) \times H_0^1(\Omega)$ that

$$\left| \int_\Omega \hat{\psi}_h^N r_h v \, dx - \int_\Omega \hat{f}_H r_h v \, dx \right| \leq C \|v\|_{H_0^1(\Omega)} |\Delta t|^{1/2} \ \forall v \in \mathcal{D}(\Omega). \tag{1.221}$$

Taking the limit in (1.221), as $\{\Delta t, h, H\} \to 0$, it follows then from (1.210), (1.219), (1.220) that

$$\int_\Omega \hat{f}^* v \, dx = \int_\Omega \hat{f} v \, dx \ \forall v \in \mathcal{D}(\Omega). \tag{1.222}$$

Since $\mathcal{D}(\Omega)$ is *dense* in $L^2(\Omega)$, it follows from (1.222) that $\hat{f}^* = \hat{f}$ and also that the whole family $\{\hat{\psi}_h^N\}_{\{\Delta t, h, H\}}$ converges weakly to $\hat{f}$. To prove the strong convergence, observe that

$$\begin{aligned}
\int_\Omega |\hat{\psi}_h^N - \hat{f}|^2 \, dx &= \int_\Omega |\hat{f}|^2 \, dx - 2 \int_\Omega \hat{\psi}_h^N \hat{f} \, dx + \int_\Omega |\hat{\psi}_h^N|^2 \, dx \\
&\leq \int_\Omega |\hat{f}|^2 \, dx - 2 \int_\Omega \hat{\psi}_h^N \hat{f} \, dx + \int_\Omega |\hat{\psi}_h^N|^2 \, dx \\
&\quad + \Delta t a(\hat{\psi}_h^N, \hat{\psi}_h^N) \\
&= \int_\Omega |\hat{f}|^2 \, dx - 2 \int_\Omega \hat{\psi}_h^N \hat{f} \, dx + \int_\Omega \hat{f}_H \hat{\psi}_h^N \, dx.
\end{aligned} \tag{1.223}$$

It follows then from (1.210), (1.223) and from the weak convergence of $\{\hat{\psi}_h^N\}_{\{\Delta t, h, H\}}$ to $\hat{f}$ in $L^2(\Omega)$ that the convergence property (1.215) holds; it implies (1.214) and therefore (1.211).

(ii) *Proof of (1.212).* We associate the solution $\hat{\psi}$ of (1.213) with the solution $\hat{\varphi}$ of

$$\frac{\partial \hat{\varphi}}{\partial t} + A\hat{\varphi} = \hat{\psi}|_{\mathcal{O}} \text{ in } Q, \quad \hat{\varphi} = 0 \text{ on } \Sigma, \quad \hat{\varphi}(0) = 0. \tag{1.224}$$

We then have

$$\Lambda \hat{f} = \hat{\varphi}(T). \tag{1.225}$$

Similarly, we associate $\{\hat{\psi}_h^n\}_{n=0}^N$ with $\{\hat{\varphi}_h^n\}_{n=0}^N$ defined by

$$\hat{\varphi}_h^0 = 0, \tag*{$(1.226)_1$}$$

and, for $n = 1, \ldots, N$, by the solution of the following discrete elliptic problems

$$\begin{cases} \hat{\varphi}_h^n \in V_{0h}, \\ \displaystyle\int_\Omega \frac{\hat{\varphi}_h^n - \hat{\varphi}_h^{n-1}}{\Delta t} v_h \, dx + a(\hat{\varphi}_h^n, v_h) = \int_{\mathcal{O}} \hat{\psi}_h^n v_h \, dx \ \forall v_h \in V_{0h}. \end{cases} \tag*{$(1.226)_2$}$$

We have

$$\Lambda_{hH}^{\Delta t}\hat{f}_H = \hat{\varphi}_h^N. \tag{1.227}$$

In order to prove (1.212) it is quite convenient to associate with $\hat{\psi}$ *the family* $\{\hat{\theta}_h^n\}_{n=0}^N$ defined by

$$\hat{\theta}_h^0 = 0, \tag{1.228$_1$}$$

and, for $n = 1, \ldots, N$, by the following discrete elliptic problems

$$\begin{cases} \hat{\theta}_h^n \in V_{0h}, \\ \displaystyle\int_\Omega \frac{\hat{\theta}_h^n - \hat{\theta}_h^{n-1}}{\Delta t} v_h \, dx + a(\hat{\theta}_h^n, v_h) = \int_{\mathcal{O}} \hat{\psi}(n\Delta t) v_h \, dx \ \forall v_h \in V_{0h}. \end{cases} \tag{1.228$_2$}$$

Let us define $\hat{\varphi}_{hH}^{\Delta t}$ and $\hat{\theta}_{hH}^{\Delta t}$ by

$$\hat{\varphi}_{hH}^{\Delta t} = \sum_{n=1}^N \hat{\varphi}_h^n I_n, \tag{1.229}$$

$$\hat{\theta}_{hH}^{\Delta t} = \sum_{n=1}^N \hat{\theta}_h^n I_n, \tag{1.230}$$

respectively. Since

$$\hat{\psi} \in C^0([0,T]; L^2(\Omega)),$$

it follows from Raviart and Thomas (1988), that

$$\lim_{\{\Delta t, h, H\} \to 0} \hat{\theta}_{hH}^{\Delta t} = \hat{\varphi} \text{ strongly in } L^2(0,T; H_0^1(\Omega)), \tag{1.231}$$

$$\lim_{\{\Delta t, h, H\} \to 0} \max_{1 \le n \le N} \|\hat{\theta}_h^n - \hat{\varphi}(n\Delta t)\|_{L^2(\Omega)} = 0. \tag{1.232}$$

Actually, similar convergence results hold for $\hat{\varphi}_{hH}^{\Delta t}$. To show them, denote by $\hat{\bar{\varphi}}_{hH}^{\Delta t}$ the difference $\hat{\varphi}_{hH}^{\Delta t} - \hat{\theta}_{hH}^{\Delta t}$; we clearly have

$$\hat{\bar{\varphi}}_h^0 = 0, \tag{1.233$_1$}$$

and for $n = 1, \ldots, N$

$$\begin{cases} \hat{\bar{\varphi}}_h^n \in V_{0h} \\ \displaystyle\int_\Omega \frac{\hat{\bar{\varphi}}_h^n - \hat{\bar{\varphi}}_h^{n-1}}{\Delta t} v_h \, dx + a(\hat{\bar{\varphi}}_h^n, v_h) = \int_{\mathcal{O}} (\hat{\psi}_h^n - \hat{\psi}(n\Delta t)) v_h \, dx \ \forall v_h \in V_{0h}. \end{cases} \tag{1.233$_2$}$$

Take $v_h = \hat{\bar{\varphi}}_h^n$ in $(1.233)_2$ and remember that $a(v,v) \ge \alpha\|v\|^2 \ \forall v \in H_0^1(\Omega)$, with $\alpha > 0$ (see Section 1.1); we then have from the Schwarz inequality in

$L^2(\Omega)$ and from the relation

$$2\alpha\beta \leq c\alpha^2 + c^{-1}\beta^2 \ \forall \alpha, \beta \in \mathbb{R} \ \forall c > 0,$$

that

$$\tfrac{1}{2\Delta t}(\|\hat{\bar{\varphi}}_h^n\|^2_{L^2(\Omega)} - \|\hat{\bar{\varphi}}_h^{n-1}\|^2_{L^2(\Omega)}) + \alpha\|\hat{\bar{\varphi}}_h^n\|^2_{H_0^1(\Omega)} \leq \|\hat{\psi}_h^n - \hat{\psi}(n\Delta t)\|_{L^2(\Omega)}\|\hat{\bar{\varphi}}_h^n\|_{L^2(\Omega)},$$

which implies, in turn, since the injection from $H_0^1(\Omega)$ into $L^2(\Omega)$ is continuous, that $\forall n \geq 1, \forall c > 0$, we have

$$\begin{aligned}\tfrac{1}{2\Delta t}(\|\hat{\bar{\varphi}}_h^n\|^2_{L^2(\Omega)} - \|\hat{\bar{\varphi}}_h^{n-1}\|^2_{L^2(\Omega)}) + \gamma\|\hat{\bar{\varphi}}_h^n\|^2_{L^2(\Omega)} \\ \leq \tfrac{1}{2}\left(\frac{1}{c}\|\hat{\psi}_h^n - \hat{\psi}(n\Delta t)\|^2_{L^2(\Omega)} + c\|\hat{\bar{\varphi}}_h^n\|^2_{L^2(\Omega)}\right),\end{aligned} \tag{1.234}$$

where, in (1.234), γ is a *positive constant.*

Taking $c = 2\gamma$ in (1.234), we obtain

$$\|\hat{\bar{\varphi}}_h^n\|^2_{L^2(\Omega)} - \|\hat{\bar{\varphi}}_h^{n-1}\|^2_{L^2(\Omega)} \leq \frac{\Delta t}{2\gamma}\|\hat{\psi}_h^n - \hat{\psi}(n\Delta t)\|^2_{L^2(\Omega)} \ \forall n = 1, \ldots, N,$$

which implies, by summation from $n = 1$ to $n = N$, that

$$\|\hat{\bar{\varphi}}_h^N\|^2_{L^2(\Omega)} \leq \frac{\Delta t}{2\gamma}\sum_{n=1}^{N}\|\hat{\psi}_h^n - \hat{\psi}(n\Delta t)\|^2_{L^2(\Omega)}. \tag{1.235}$$

It follows then from (1.211) and (1.235) that $\lim_{\{\Delta t,h,H\}\to 0}\|\hat{\varphi}_h^N - \hat{\theta}_h^N\|_{L^2(\Omega)} = 0$, which combined with (1.232) implies that

$$\lim_{\{\Delta t,h,H\}\to 0}\|\hat{\varphi}_h^N - \hat{\varphi}(T)\|_{L^2(\Omega)} = 0. \tag{1.236}$$

Finally, relations (1.225), (1.227) and (1.236) imply the convergence result (1.212).

1.8.8. Solution methods for problem (1.115).

In this section, we discuss the solution of the *variational inequality* (1.115), which is *equivalent* to the control problem (1.109) (via a *duality* argument). We observe that (1.115) can also be written as the following nonlinear (multivalued) equation in $L^2(\Omega)$

$$y_T - Y_0(T) \in \Lambda f + \beta\partial j(f), \tag{1.237}$$

where, in (1.237), $\partial j(f)$ denotes the *subgradient* (see, e.g., Ekeland and Temam (1974) for this concept) at f of the *convex functional* $j(\cdot)$ defined by

$$j(\hat{f}) = \|\hat{f}\|_{L^2(\Omega)} \ \forall \hat{f} \in L^2(\Omega).$$

Equation (1.237) strongly suggests the use of *operator splitting methods* like those discussed in, for example, P.L. Lions and Mercier (1979) and Glowinski and Le Tallec (1989). A simple way to derive such methods is to associate with (1.237) a *time-dependent* equation (for a *pseudo-time* τ) such as

$$\begin{cases} \frac{\partial f}{\partial \tau} + \Lambda f + \beta \partial j(f) = y_T - Y_0(T), \\ f(0) = f_0 (\in L^2(\Omega)). \end{cases} \tag{1.238}$$

Next, we use time discretization by operator splitting to integrate (1.238) from $\tau = 0$ to $\tau = +\infty$ in order to capture the steady-state solution of (1.238), namely the solution of (1.237).

A natural choice to integrate (1.238) is the *Peaceman–Rachford* scheme (cf. Peaceman and Rachford (1955)), which for the present problem provides

$$f^0 = f_0; \tag{1.239}$$

then, for $k \geq 0$, compute $f^{k+1/2}$ and f^{k+1}, from f^k, by solving

$$\frac{f^{k+1/2} - f^k}{\Delta\tau/2} + \beta \partial j(f^{k+1/2}) + \Lambda f^k = y_T - Y_0(T), \tag{1.240}$$

and

$$\frac{f^{k+1} - f^{k+1/2}}{\Delta\tau/2} + \beta \partial j(f^{k+1/2}) + \Lambda f^{k+1} = y_T - Y_0(T), \tag{1.241}$$

where $\Delta\tau (> 0)$ is a (pseudo) time discretization step. The *convergence* of $\{f^k\}_{k \geq 0}$ to the solution f of (1.115), (1.237) is a direct consequence of Lions and Mercier (1979), Gabay (1982; 1983) and Glowinski and Le Tallec (1989); the convergence results shown in the above references apply to the present problem since operator Λ (respectively function $j(\cdot)$) is *linear, continuous* and *positive definite* (respectively *convex* and *continuous*) over $L^2(\Omega)$.

A variant of this algorithm is given by the following θ-scheme (where $0 < \theta \leq 1/3$; see, e.g., Glowinski and Le Tallec (1989)):

$$f^0 = f_0; \tag{1.242}$$

then, for $k \geq 0$, compute $f^{k+\theta}, f^{k+1-\theta}, f^{k+1}$ from f^k by solving

$$\frac{f^{k+\theta} - f^k}{\theta \Delta\tau} + \beta \partial j(f^{k+\theta}) + \Lambda f^k = y_T - Y_0(T), \tag{1.243}$$

$$\frac{f^{k+1-\theta} - f^{k+\theta}}{(1-2\theta)\Delta\tau} + \beta \partial j(f^{k+\theta}) + \Lambda f^{k+1-\theta} = y_T - Y_0(T), \tag{1.244}$$

$$\frac{f^{k+1} - f^{k+1-\theta}}{\theta \Delta\tau} + \beta \partial j(f^{k+1}) + \Lambda f^{k+1-\theta} = y_T - Y_0(T). \tag{1.245}$$

In practice, it may pay to use a *variable* $\Delta\tau$. Concerning now the solution of the various subproblems in the above two algorithms, we can make the following observations:

(i) Assuming that we know how to solve problems (1.240) and (1.243), (1.245), the functions f^{k+1} and $f^{k+1-\theta}$ are obtained via the solution of linear problems similar to the one the solution of which has been discussed in Sections 1.8.2 to 1.8.7; in particular, we can use the conjugate gradient algorithm (1.169)–(1.183) to solve finite element approximations of problems (1.239) and (1.244).

(ii) Problems (1.240) and (1.243), (1.245) are fairly easy to solve. Consider, for example, problem (1.240); it is clearly equivalent to the following minimization problem

$$\begin{cases} f^{k+1/2} \in L^2(\Omega), \\ J_k(f^{k+1/2}) \leq J_k(v) \ \forall v \in L^2(\Omega), \end{cases} \tag{1.246}$$

with

$$J_k(v) = \tfrac{1}{2}\int_\Omega |v|^2 \, \mathrm{d}x + \beta\tfrac{1}{2}\Delta\tau\|v\|_{L^2(\Omega)} - \int_\Omega f^k v \, \mathrm{d}x$$
$$-\frac{\Delta\tau}{2}\int_\Omega (y_T - Y_0(T) - \Lambda f^k) v \, \mathrm{d}x \ \forall v \in L^2(\Omega). \tag{1.247}$$

To solve problem (1.246), we define $f_*^{k+1/2}$ as

$$f_*^{k+1/2} = f^k + \tfrac{1}{2}\Delta\tau(y_T - Y_0(T) - \Lambda f^k) \tag{1.248}$$

and observe that the solution of problem (1.246) is clearly of the form

$$f^{k+1/2} = \lambda^{k+1/2} f_*^{k+1/2} \text{ with } \lambda^{k+1/2} \geq 0. \tag{1.249}$$

To obtain $\lambda^{k+1/2}$ we minimize with respect to λ, the polynomial

$$\|f_*^{k+1/2}\|^2_{L^2(\Omega)}(\tfrac{1}{2}\lambda^2 - \lambda) + \tfrac{1}{2}\beta\Delta\tau\|f_*^{k+1/2}\|_{L^2(\Omega)}\lambda (= J_k(\lambda f_*^{k+1/2})).$$

We obtain then (since $\lambda^{k+1/2} \geq 0$)

$$\begin{cases} \lambda^{k+1/2} = 1 - \tfrac{1}{2}\beta\Delta\tau/\|f_*^{k+1/2}\|_{L^2(\Omega)} & \text{if } \|f_*^{k+1/2}\|_{L^2(\Omega)} \geq \beta\Delta\tau/2, \\ \lambda^{kl+1/2} = 0 \quad \text{if } \|f_*^{k+1/2}\|_{L^2(\Omega)} < \beta\Delta\tau/2. \end{cases} \tag{1.250}$$

The same method applies to the solution of problems (1.243) and (1.244).

Remark 1.38 Concerning the calculation of f^{k+1} we shall use equation (1.240) to rewrite (1.241) as

$$\frac{f^{k+1} - 2f^{k+1/2} + f^k}{\Delta\tau/2} + \Lambda f^{k+1} = \Lambda f^k, \tag{1.251}$$

which is better suited for practical computations. A similar observation holds for the calculation of $f^{k+1-\theta}$ in (1.244).

1.8.9. Splitting methods for nonquadratic cost functions and control constrained problems.

1.8.9.1. Generalities. In Section 1.7, we have considered control problems such as (or closely related to)

$$\min_{v \in L^s(\mathcal{O}\times(0,T))} \left[\frac{1}{s}\int_{\mathcal{O}\times(0,T)} |v|^s \, dx \, dt + \tfrac{1}{2}k\|y(T) - y_T\|^2_{L^2(\Omega)}\right], \tag{1.252}$$

where, in (1.252), $s \in [1, +\infty), k > 0$ and where y is defined by (1.111)–(1.113); the case $s = 2$ has been treated in Sections 1.8.2 to 1.8.7. Solving (1.252) for large values of s provide solutions close to those obtained with cost functions containing terms such as $\|v\|_{L^\infty(\mathcal{O}\times(0,T))}$.

Another control problem of interest is defined by

$$\min_{v \in \mathcal{C}_f} \tfrac{1}{2}\|y(T) - y_T\|^2_{L^2(\Omega)}, \tag{1.253}$$

with

$$\mathcal{C}_f = \{v \mid v \in L^\infty(\mathcal{O}\times(0,T)), |v(x,t)| \le C \ \text{ a.e. in } \mathcal{O}\times(0,T)\}$$

and y still defined by (1.111)–(1.113).

The *convex* set $\mathcal{C}_f$ is clearly *closed* in $L^2(\mathcal{O}\times(0,T))$; we shall denote by $I_{\mathcal{C}_f}$ its *characteristic function* in $L^2(\mathcal{O}\times(0,T))$.

Problem (1.253) is clearly *equivalent* to

$$\min_{v \in L^2(\mathcal{O}\times(0,T))} [I_{\mathcal{C}_f}(v) + \tfrac{1}{2}\|y(T) - y_T\|^2_{L^2(\Omega)}]. \tag{1.254}$$

In the following subsections we shall show that problems (1.252) and (1.253), (1.254) are fairly easy to solve if one has a solver for problem (1.114) (i.e. for problem (1.252) when $s = 2$).

1.8.9.2. Solution of problem (1.252). Suppose for the time being that $s > 1$ and let us denote by $J(\cdot)$ the *strictly convex* functional defined by

$$J(v) = \frac{1}{s}\iint_{\mathcal{O}\times(0,T)} |v|^s \, dx \, dt + \tfrac{1}{2}k\|y(T) - y_T\|^2_{L^2(\Omega)}, \tag{1.255}$$

where, in (1.255), y is obtained from v via (1.111)–(1.113). Define next $J_1(\cdot)$ and $J_2(\cdot)$ by

$$J_1(v) = \frac{1}{s}\iint_{\mathcal{O}\times(0,T)} |v|^s \, dx \, dt \tag{1.256}$$

and

$$J_2(v) = \tfrac{1}{2}k\|y(T) - y_T\|^2_{L^2(\Omega)}, \tag{1.257}$$

where y is obtained from v via (1.111)–(1.113), respectively. Both functions are clearly differentiable in $L^s(\mathcal{O} \times (0,T))$ and we have

$$\langle J_1'(v), w \rangle = \iint_{\mathcal{O}\times(0,T)} |v|^{s-2} v w \, dx \, dt \ \forall v, \ w \in L^s(\mathcal{O} \times (0,T)), \quad (1.258)$$

$$\langle J_2'(v), w \rangle = - \iint_{\mathcal{O}\times(0,T)} p w \, dx \, dt \ \forall v, \ w \in L^s(\mathcal{O} \times (0,T)), \quad (1.259)$$

where, in (1.258), (1.259), $\langle .,. \rangle$ denotes the *duality pairing* between $L^{s'}(\mathcal{O} \times (0,T))$ and $L^s(\mathcal{O} \times (0,T))(s' = s/(s-1))$ and where p is the solution of the *adjoint state equation*

$$-\frac{\partial p}{\partial t} + A^* p = 0 \text{ in } Q, \quad p = 0 \text{ on } \Sigma, \ p(T) = k(y_T - y(T)). \quad (1.260)$$

If u is *the* solution of the control problem (1.252), it is characterized by $J'(u) = 0$, which here takes the following form:

$$J_1'(u) + J_2'(u) = 0. \quad (1.261)$$

In order to solve (1.252), via (1.261), we follow the approach taken in Section 1.8.8 and we associate with (1.261) the following (pseudo) time-dependent problem in $L^s(\mathcal{O} \times (0,T))$:

$$\begin{cases} \dfrac{\partial u}{\partial \tau} + J_1'(u) + J_2'(u) = 0, \\ u(0) = u_0. \end{cases} \quad (1.262)$$

To obtain the *steady-state solution* of (1.262) (i.e. the solution of (1.252), (1.261)) we integrate (1.262) from 0 to $+\infty$ by *operator splitting*; if one uses the *Peaceman–Rachford scheme* (see Section 1.8.8), we obtain

$$u^0 = u_0, \quad (1.263)$$

and for $n \geq 0$, assuming that u^n is known

$$\frac{u^{n+1/2} - u^n}{\frac{1}{2}\Delta\tau} + J_1'(u^{n+1/2}) + J_2'(u^n) = 0, \quad (1.264)$$

$$\frac{u^{n+1} - u^{n+1/2}}{\frac{1}{2}\Delta\tau} + J_1'(u^{n+1/2}) + J_2'(u^{n+1}) = 0. \quad (1.265)$$

Equation (1.264) can also be written

$$\frac{u^{n+1/2} - u^n}{\frac{1}{2}\Delta\tau} + |u^{n+1/2}|^{s-2} u^{n+1/2} - p^n \chi_{\mathcal{O}} = 0, \quad (1.266)$$

where p^n is obtained from u^n via

$$\frac{\partial y^n}{\partial t} + Ay^n = u^n \chi_{\mathcal{O}} \text{ in } Q, \quad y^n = 0 \text{ on } \Sigma,\ y^n(0) = y_0, \tag{1.267}$$

$$-\frac{\partial p^n}{\partial t} + A^* p^n = 0 \text{ in } Q, \quad p^n = 0 \text{ on } \Sigma,\ p^n(T) = k(y_T - y^n(T)). \tag{1.268}$$

We thus obtain $u^{n+1/2}$ from u^n by solving the *nonlinear problem*

$$u^{n+1/2} + \tfrac{1}{2}\Delta\tau |u^{n+1/2}|^{s-2} u^{n+1/2} = u^n + \tfrac{1}{2}\Delta\tau p^n \chi_{\mathcal{O}}. \tag{1.269}$$

Problem (1.269) can be solved *pointwise* in $\mathcal{O} \times (0,T)$; at almost every point of $\mathcal{O} \times (0,T)$ (in practice at the nodes of a *finite difference* or *finite element grid*) we shall have to solve a *one variable* equation of the form

$$\xi + \tfrac{1}{2}\Delta\tau |\xi|^{s-2}\xi = b, \tag{1.270}$$

which has, $\forall b \in \mathbb{R}$, a *unique* solution.

Problem (1.265) is equivalent to the following *minimization problem*

$$\min_{v \in L^2(\mathcal{O}\times(0,T))} j_n(v), \tag{1.271}$$

where $j_n(.)$ is defined by

$$\begin{aligned} j_n(v) = \tfrac{1}{2} &\iint_{\mathcal{O}\times(0,T)} v^2 \, dx\, dt \\ - &\iint_{\mathcal{O}\times(0,T)} (u^{n+1/2} + \tfrac{1}{2}\Delta\tau |u^{n+1/2}|^{s-2} u^{n+1/2}) v \, dx\, dt \\ + &\tfrac{1}{4} k \Delta\tau \|y(T) - y_T\|^2_{L^2(\Omega)}, \end{aligned} \tag{1.272}$$

with y obtained from v via (1.111)–(1.113). Problem (1.271), (1.272) is a simple variant of problem (1.114); it can therefore be solved by the numerical methods described in Sections 1.8.2 to 1.8.7.

Remark 2.39 From a formal point of view the above method still applies if $s = 1$. In such a case we shall replace (1.269) by the minimization problem

$$\begin{aligned} \min_{v \in L^2(\mathcal{O}\times(0,T))} \Big[&\tfrac{1}{2} \iint_{\mathcal{O}\times(0,T)} v^2 \, dx\, dt \\ + &\tfrac{1}{2}\Delta\tau \iint_{\mathcal{O}\times(0,T)} |v| \, dx\, dt - \iint_{\mathcal{O}\times(0,T)} (u^n + \tfrac{1}{2}\Delta\tau p^n) v \, dx\, dt \Big] \end{aligned} \tag{1.273}$$

whose solution $u^{n+1/2}$ is given (in *closed form*) by

$$\left\{ \begin{aligned} &u^{n+1/2}(x,t) = 0 \text{ if } |(u^n + \tfrac{1}{2}\Delta\tau p^n)(x,t)| \le \tfrac{1}{2}\Delta\tau, \{x,t\} \in \mathcal{O} \times (0,T), \\ &u^{n+1/2}(x,t) = (u^n + \tfrac{1}{2}\Delta\tau p^n)(x,t) - \tfrac{1}{2}\Delta\tau \text{ sgn } (u^n + \tfrac{1}{2}\Delta\tau p^n)(x,t) \\ &\quad \text{if } |(u^n + \tfrac{1}{2}\Delta\tau p^n)(x,t)| > \tfrac{1}{2}\Delta\tau, \{x,t\} \in \mathcal{O} \times (0,T). \end{aligned} \right. \tag{1.274}$$

Concerning now the calculation of u^{n+1}, we observe that this function is the solution of

$$\frac{u^{n+1} - 2u^{n+1/2} + u^n}{\frac{1}{2}\Delta\tau} + J_2'(u^{n+1}) = J_2'(u^n),$$

which is equivalent to the minimization problem

$$\min_{v \in L^2(\mathcal{O}\times(0,T))} \left[\frac{1}{2} \iint_{\mathcal{O}\times(0,T)} v^2 \,\mathrm{d}x\,\mathrm{d}t + \frac{1}{4}\Delta\tau k \|y(T) - y_T\|^2_{L^2(\Omega)} - \iint_{\mathcal{O}\times(0,T)} (2u^{n+1/2} - u^n - \tfrac{1}{2}\Delta\tau p^n) v \,\mathrm{d}x\,\mathrm{d}t \right], \tag{1.275}$$

where y is a function of v via the solution of (1.111)–(1.113). Problem (1.275) is also a variant of problem (1.114).

Remark 1.40 Equation (1.262) and algorithm (1.263)–(1.265) are largely formal if $1 \leq s < 2$; however they make full sense for the discrete analogues of problem (1.252) obtained by finite difference and finite element approximations close to those discussed in Sections 1.8.2 to 1.8.7.

1.8.9.3. Solution of problem (1.253), (1.254). We follow the approach taken in Section 1.8.9.2; we introduce therefore J_1 and J_2 defined by

$$J_1(v) = I_{C_f}(v), \tag{1.276}$$

and

$$J_2(v) = \frac{1}{2}\|y(T) - y_T\|^2_{L^2(\Omega)}, \tag{1.277}$$

y obtained from v via (1.111)–(1.113), respectively. The solution u of problem (1.253), (1.254) is *characterized* therefore by

$$0 \in \partial J_1(u) + J_2'(u) \tag{1.278}$$

where, in (1.278), $\partial J_1(.)$ is the subgradient of $J_1(.)$, and where $J_2'(.)$ is defined by (1.259), (1.260) with $k = 1$.

We associate with (1.278) the following (pseudo) time-dependent problem in $L^2(\mathcal{O} \times (0,T))$:

$$\begin{cases} \dfrac{\partial u}{\partial \tau} + \partial J_1(u) + J_2'(u) = 0, \\ u(0) = u_0 (\in C_f). \end{cases} \tag{1.279}$$

Applying as in Section 1.8.9.2 the *Peaceman–Rachford scheme*, we obtain

$$u^0 = u_0, \tag{1.280}$$

and for $n \geq 0$, assuming that u^n is known

$$\frac{u^{n+1/2} - u^n}{\frac{1}{2}\Delta\tau} + \partial J_1(u^{n+1/2}) + J_2'(u^n) = 0, \tag{1.281}$$

$$\frac{u^{n+1} - u^{n+1/2}}{\frac{1}{2}\Delta\tau} + \partial J_1(u^{n+1/2}) + J_2'(u^{n+1}) = 0. \tag{1.282}$$

Equation (1.281) is equivalent to the minimization problem

$$\min_{v \in C_f} \left[\frac{1}{2} \iint_{\mathcal{O}\times(0,T)} v^2 \, dx \, dt - \iint_{\mathcal{O}\times(0,T)} p^n v \, dx \, dt \right], \tag{1.283}$$

where, in (1.283), p^n is obtained from u^n via (1.267), (1.268) with $k = 1$; we have then

$$u^{n+1/2}(x,t) = \min(C, \max(-C, p^n(x,t))), \quad \text{a.e. on } \mathcal{O} \times (0,T). \tag{1.284}$$

Summing (1.281) and (1.282) implies that

$$\frac{u^{n+1} - 2u^{n+1/2} + u^n}{\frac{1}{2}\Delta\tau} + J_2'(u^{n+1}) = J_2'(u^n),$$

which is equivalent to the minimization problem

$$\min_{v \in L^2(\mathcal{O}\times(0,T))} \left[\frac{1}{2} \iint_{\mathcal{O}\times(0,T)} v^2 \, dx \, dt + \frac{1}{4}\Delta\tau \|y(T) - y_T\|^2_{L^2(\Omega)} \right.$$
$$\left. - \iint_{\mathcal{O}\times(0,T)} (2u^{n+1/2} - u^n - \tfrac{1}{2}\Delta\tau p^n) v \, dx \, dt \right], \tag{1.285}$$

where y is a function of v via (1.111)–(1.113). Problem (1.285) is a variant of problem (1.114).

1.9. Relaxation of controllability

1.9.1. Generalities.

Let $\mathcal{H}$ be a *Hilbert space* and let C be a *linear operator* such that

$$C \in \mathcal{L}(L^2(\Omega); \mathcal{H}), \tag{1.286}$$

and

$$\text{the range of } C \text{ is dense in } \mathcal{H}. \tag{1.287}$$

We consider again the *state equation*

$$\frac{\partial y}{\partial t} + Ay = v\chi_{\mathcal{O}} \text{ in } Q, \quad y(0) = 0, \quad y = 0 \text{ on } \Sigma, \tag{1.288}$$

and we look now for the solution of

$$\inf \frac{1}{2} \iint_{\mathcal{O}\times(0,T)} v^2 \, dx \, dt, \tag{1.289}$$

for all vs such that

$$Cy(T; v) \in h_T + \beta B_{\mathcal{H}}, \tag{1.290}$$

where h_T is given in $\mathcal{H}$ and where $B_{\mathcal{H}}$ denotes the *unit ball* of $\mathcal{H}$.

Remark 1.41 If $\mathcal{H} = L^2(\Omega)$, $C =$ identity, and if $h_T = y_T$, then problem (1.289) is exactly the control problem discussed before.

1.9.2. Examples of operators C.

Example 1.2 Let ω be an *open* set in Ω and χ_ω be its *characteristic function.* Then

$$Cy = y\chi_\omega \tag{1.291}$$

corresponds to

$$\mathcal{H} = L^2(\omega). \tag{1.292}$$

Here, we want to reach (or to get close to) a given state *on the subset* ω.

Example 1.3 Let $g_1, \dots, g_N$ be N given elements of $L^2(\Omega)$, *linearly independent.* Then

$$Cy = \{(y, g_i)_{L^2(\Omega)}\}_{i=1}^N, \tag{1.293}$$

corresponds to $\mathcal{H} = \mathbb{R}^N$.

The same considerations as in previous sections apply. Let us write down explicitly the dual formulation of (1.289), (1.290), in the particular cases of Examples 1.2 and 1.3.

1.9.3. Dual formulation in the case of Example 1.2.

Let $\hat{f}$ be given in $L^2(\omega)$. We introduce $\hat{\psi}$ defined by

$$-\frac{\partial\hat{\psi}}{\partial t} + A^*\hat{\psi} = 0 \text{ in } Q, \quad \hat{\psi}(T) = \hat{f}\chi_\omega, \quad \hat{\psi} = 0 \text{ on } \Sigma. \tag{1.294}$$

The *dual problem* is then

$$\inf_{\hat{f}\in L^2(\omega)} \left[\frac{1}{2}\iint_{\mathcal{O}\times(0,T)} \hat{\psi}^2 \,dx\,dt - (\hat{f}, h_T)_{L^2(\omega)} + \beta\|\hat{f}\|_{L^2(\omega)}\right]. \tag{1.295}$$

If f is the (unique) solution of problem (1.295) the solution u of the corresponding control problem (1.289) is given by $u = \psi\chi_{\mathcal{O}\times(0,T)}$, where ψ is the solution of (1.294) corresponding to $\hat{f} = f$.

1.9.4. Dual formulation in the case of Example 1.3.

Let $\hat{\mathbf{f}} = \{\hat{f}_i\}_{i=1}^N$ be given in $\mathbb{R}^N$. We define $\hat{\psi}$ by

$$-\frac{\partial\hat{\psi}}{\partial t} + A^*\hat{\psi} = 0 \text{ in } Q, \quad \hat{\psi}(T) = \sum_{i=1}^N \hat{f}_i g_i, \quad \hat{\psi} = 0 \text{ on } \Sigma. \tag{1.296}$$

The dual problem is then (analogous to (1.295) but with $L^2(\omega)$ replaced by $\mathbb{R}^N$):

$$\inf_{\hat{\mathbf{f}}\in\mathbb{R}^N} \left[\frac{1}{2}\iint_{\mathcal{O}\times(0,T)} \hat{\psi}^2 \,dx\,dt - (\hat{f}, h_T)_{\mathbb{R}^N} + \beta\|\hat{f}\|_{\mathbb{R}^N}\right]. \tag{1.297}$$

If f is the (unique) solution of problem (1.297) the solution u of the corresponding control problem (1.289) is given by $u = \psi\chi_{\mathcal{O}\times(0,T)}$, where ψ is the solution of (1.296) corresponding to $\hat{f} = f$.

1.9.5. Further comments.

Remark 1.42 We can also consider *time averages* as shown in Lions (1993).

Concerning now the numerical solution of problem (1.289), it can be achieved by numerical methods directly inspired by those discussed in Section 1.8. In particular, it is quite convenient to introduce an operator $\Lambda \in \mathcal{L}(\mathcal{H}, \mathcal{H}')$ which will play for problem (1.289) the role played for problems (1.109) and (1.114) by the operator Λ defined in Section 1.5 (see also Section 1.8.2).

Considering, first, Example 1.2, the dual problem (1.295) can also be written as

$$\begin{cases} f \in L^2(\omega), \\ (\Lambda f, \hat{f} - f)_{L^2(\omega)} + \beta\|\hat{f}\|_{L^2(\omega)} - \beta\|f\|_{L^2(\omega)} \\ \quad \geq (h_T, \hat{f} - f)_{L^2(\omega)} \ \forall \hat{f} \in L^2(\omega), \end{cases} \tag{1.298}$$

where, in (1.298), operator Λ is defined as follows

$$\Lambda\hat{f} = \hat{\varphi}(T)\chi_\omega \quad \forall \hat{f} \in L^2(\omega), \tag{1.299}$$

with $\hat{\varphi}(T)$ obtained from $\hat{f}$ via (1.294) and

$$\frac{\partial\hat{\varphi}}{\partial t} + A\hat{\varphi} = \hat{\psi}\chi_{\mathcal{O}} \text{ in } Q, \quad \hat{\varphi}(0) = 0, \ \hat{\varphi} = 0 \text{ on } \Sigma. \tag{1.300}$$

Operator $\Lambda \in \mathcal{L}(L^2(\omega), L^2(\omega))$ and is *symmetric* and *positive definite* over $L^2(\omega)$. The numerical methods discussed in Section 1.8.8 can be easily modified in order to accommodate problem (1.298).

Consider, now, Example 1.3; the dual problem (1.297) can be written as

$$\begin{cases} \mathbf{f} \in \mathbb{R}^N, \\ (\mathbf{\Lambda f}, \hat{\mathbf{f}} - \mathbf{f})_{\mathbb{R}^N} + \beta\|\hat{\mathbf{f}}\|_{\mathbb{R}^N} - \beta\|\mathbf{f}\|_{\mathbb{R}^N} \geq (\mathbf{h}_T, \hat{\mathbf{f}} - \mathbf{f})_{\mathbb{R}^N} \ \forall \hat{\mathbf{f}} \in \mathbb{R}^N, \end{cases} \tag{1.301}$$

where, in (1.301), $\mathbf{\Lambda}$ is the $N \times N$ symmetric and *positive definite matrix* defined by

$$\mathbf{\Lambda} = (\lambda_{ij})_{1\leq i,j\leq N}, \quad \lambda_{ij} = \int_\Omega \varphi_i(T) g_j \, dx, \tag{1.302}$$

with, in (1.302), φ_i defined from g_i by

$$-\frac{\partial\psi_i}{\partial t} + A^*\psi_i = 0 \text{ in } Q, \quad \psi_i(T) = g_i, \ \psi_i = 0 \text{ on } \Sigma, \tag{1.303}$$

$$\frac{\partial\varphi_i}{\partial t} + A\varphi_i = \psi_i\chi_{\mathcal{O}} \text{ in } Q, \quad \varphi_i(0) = 0, \ \varphi_i = 0 \text{ on } \Sigma. \tag{1.304}$$

Remark 1.43 Problem (1.301) clearly has the 'flavour' of a *Galerkin method* (like those discussed in Section 1.8 to solve problems (1.115) and (1.116)).

To conclude this section we shall discuss a *solution method* for problem (1.301); this method is applicable when N *is not too large*, since it relies on the *explicit* construction of matrix $\mathbf{\Lambda}$. Our solution method is based on the fact that, according to, e.g., Glowinski, Lions and Trémolières (1976, Ch. 2) and (1981, Ch. 2 and Appendix 2), problem (1.301) is *equivalent* to the following *nonlinear system*

$$\begin{cases} \mathbf{\Lambda f} + \beta \mathbf{p} = \mathbf{h}_T \\ (\mathbf{p}, \mathbf{f})_{\mathbb{R}^N} = \|\mathbf{f}\|_{\mathbb{R}^N}, \|\mathbf{p}\|_{\mathbb{R}^N} \leq 1, \end{cases} \tag{1.305}$$

which has a *unique* solution since $\beta > 0$. System (1.305) is in turn *equivalent* to

$$\begin{cases} \mathbf{\Lambda f} + \beta \mathbf{p} = \mathbf{h}_T, \\ \mathbf{p} = P_B(\mathbf{p} + \rho \mathbf{f}) \ \forall \rho > 0, \end{cases} \tag{1.306}$$

where, in (1.306), $P_B : \mathbb{R}^N \to \mathbb{R}^N$ is the *orthogonal projector* from $\mathbb{R}^N$ on the *closed unit ball* B of $\mathbb{R}^N$; we clearly have $\forall \hat{\mathbf{f}} \in \mathbb{R}^N$,

$$P_B(\hat{\mathbf{f}}) = \begin{cases} \hat{\mathbf{f}} & \text{if } \hat{\mathbf{f}} \in B, \\ \hat{\mathbf{f}}/\|\hat{\mathbf{f}}\|_{\mathbb{R}^N} & \text{if } \hat{\mathbf{f}} \notin B. \end{cases}$$

Relations (1.306) suggest the following *iterative* method (of *fixed point* type):

$$\mathbf{p}^0 \in B \text{ is given (we can take, for example, } \mathbf{p}^0 = 0); \tag{1.307}$$

then for $n \geq 0$, *assuming that* $\mathbf{p}^n$ *is known, we compute* $\mathbf{f}^n$, *and then* $\mathbf{p}^{n+1}$, *by*

$$\mathbf{\Lambda f}^n = \mathbf{h}_T - \beta \mathbf{p}^n, \tag{1.308}$$

$$\mathbf{p}^{n+1} = P_B(\mathbf{p}^n + \rho \mathbf{f}^n). \tag{1.309}$$

Concerning the *convergence* of algorithm (1.307)–(1.309) we then have the following

Proposition 1.2 *Suppose that*

$$0 < \rho < 2\mu_1/\beta, \tag{1.310}$$

where μ_1 *is the smallest eigenvalue of matrix* $\mathbf{\Lambda}$. *Then,* $\forall \mathbf{p}^0 \in B$, *we have*

$$\lim_{n \to +\infty} \{\mathbf{f}^n, \mathbf{p}^n\} = \{\mathbf{f}, \mathbf{p}\}, \tag{1.311}$$

where $\{\mathbf{f}, \mathbf{p}\}$ *is the solution of (1.305).*

Proof. The convergence result (1.311) is a direct consequence of Glowinski *et al.* (1976, Ch. 2) and (1981, Ch. 2 and Appendix 2) (see also Ciarlet

(1989, Ch. 9)); however, we shall prove it here for the sake of completeness. Introduce therefore $\bar{\mathbf{f}}^n = \mathbf{f}^n - \mathbf{f}$ and $\bar{\mathbf{p}} = \mathbf{p}^n - \mathbf{p}$; we have

$$\mathbf{\Lambda}\bar{\mathbf{f}}^n = -\beta\bar{\mathbf{p}}^n. \tag{1.312}$$

We also have since P_B is a *contraction* $\|\bar{\mathbf{p}}^{n+1}\|_{\mathbb{R}^N} \leq \|\bar{\mathbf{p}}^n + \rho\bar{\mathbf{f}}^n\|_{\mathbb{R}^N}$, which implies in turn that

$$\begin{aligned} \|\bar{\mathbf{p}}^{n+1}\|^2_{\mathbb{R}^N} &\leq \|\bar{\mathbf{p}}^n\|^2_{\mathbb{R}^N} + 2\rho(\bar{\mathbf{p}}^n, \bar{\mathbf{f}}^n)_{\mathbb{R}^N} + \rho^2\|\bar{\mathbf{f}}^n\|^2_{\mathbb{R}^N} \\ &= \|\bar{\mathbf{p}}^n\|^2_{\mathbb{R}^N} - \frac{2\rho}{\beta}(\mathbf{\Lambda}\bar{\mathbf{f}}^n, \bar{\mathbf{f}}^n) + \rho^2\|\bar{\mathbf{f}}^n\|^2_{\mathbb{R}^N}. \end{aligned} \tag{1.313}$$

It follows from (1.313) that

$$\|\bar{\mathbf{p}}^n\|^2_{\mathbb{R}^N} - \|\bar{\mathbf{p}}^{n+1}\|^2_{\mathbb{R}^N} \geq \frac{2\rho}{\beta}(\mathbf{\Lambda}\bar{\mathbf{f}}^n, \bar{\mathbf{f}}^n)_{\mathbb{R}^N} - \rho^2\|\bar{\mathbf{f}}^n\|^2_{\mathbb{R}^N} \geq \rho\left(\frac{2\mu_1}{\beta} - \rho\right)\|\bar{\mathbf{f}}^n\|^2_{\mathbb{R}^N}, \tag{1.314}$$

where $\mu_1(>0)$ is the *smallest eigenvalue* of matrix $\mathbf{\Lambda}$. Suppose that (1.310) holds, then the sequence $\{\|\bar{\mathbf{p}}^n\|^2_{\mathbb{R}^N}\}_{n\geq 0}$ is *decreasing*; since it has 0 as a lower bound it converges to some (nonnegative) limit, implying that

$$\lim_{n\to+\infty}(\|\bar{\mathbf{p}}^n\|^2_{\mathbb{R}^N} - \|\bar{\mathbf{p}}^{n+1}\|^2_{\mathbb{R}^N}) = 0. \tag{1.315}$$

Combining (1.310), (1.314), (1.315) we obtain that $\lim_{n\to+\infty}\|\bar{\mathbf{f}}^n\|_{\mathbb{R}^N} = 0$; we have thus shown that $\lim_{n\to+\infty}\mathbf{f}^n = \mathbf{f}$. The convergence of $\{\mathbf{p}^n\}_{n\geq 0}$ to $\mathbf{p}$ follows from the convergence of $\{\mathbf{f}^n\}_{n\geq 0}$ and from (1.308) (or (1.312)).

Remark 1.44 If N is not too large, so that matrix $\mathbf{\Lambda}$ can be constructed (via (1.302)–(1.304)) at a reasonable cost, we shall use the *Cholesky factorization method* (see, e.g., Ciarlet (1989, Ch. 4)) to solve the various systems (1.308). If N is very large, we can expect $\mathbf{\Lambda}$ to be *ill conditioned* and expensive to construct and factorize; therefore, instead of using algorithm (1.307)–(1.309) we suggest solving problem (1.301) by simple variants of the methods used in Section 1.8.8 to solve problem (1.115).

1.10. Pointwise control

1.10.1. Generalities.

A rather natural question in the present framework is to consider situations where in (1.1) the open set $\mathcal{O}$ is replaced by a 'small' set, in particular a set of measure 0. One has then to consider 'functions' which are not in $L^2(\Omega)$ (for a given t).

Many situations can be considered. We confine ourselves here with the case where $\mathcal{O}$ *is reduced to a point*:

$$\mathcal{O} = \{b\}, \quad b \in \Omega. \tag{1.316}$$

Then, if $\delta(x-b)$ denotes the *Dirac measure* at b, the state function y is given by

$$\frac{\partial y}{\partial t} + Ay = v(t)\delta(x-b) \text{ in } Q, \quad y(0) = 0, \; y = 0 \text{ on } \Sigma. \tag{1.317}$$

In (1.317) the control v is now a function of t only. We shall assume that

$$v \in L^2(0,T). \tag{1.318}$$

Problem (1.317) has a *unique solution*, which is defined by *transposition*, as in Lions and Magenes (1968). It follows from this reference that, *if* $d \leq 3$ one has

$$y \in L^2(Q), \quad \frac{\partial y}{\partial t} \in L^2(0,T;H^{-2}(\Omega)), \tag{1.319}$$

so that

$$t \to y(t;v) \text{ is continuous from } [0,T] \text{ into } H^{-1}(\Omega). \tag{1.320}$$

When v spans $L^2(0,T), y(T;v)$ spans a subspace of $H^{-1}(\Omega)$. Let us look for the orthogonal of (the closure of) this subspace. Let f be given in $H_0^1(\Omega)$ such that

$$\langle y(T;v), f\rangle = 0 \tag{1.321}$$

(where $\langle \cdot\,, \cdot\rangle$ denotes the duality pairing between $H^{-1}(\Omega)$ and $H_0^1(\Omega)$). Let ψ be the solution of

$$-\frac{\partial \psi}{\partial t} + A^*\psi = 0 \text{ in } Q, \quad \psi(T) = f, \; \psi = 0 \text{ on } \Sigma. \tag{1.322}$$

Then

$$\langle y(T;v), f\rangle = \int_0^T \psi(b,t)v(t)\,\mathrm{d}t. \tag{1.323}$$

Remark 1.45 If $d = 1$, the 'function' $\{x,t\} \to v(t)\delta(x-b)$ belongs to

$$L^2(0,T;H^{-1}(\Omega));$$

this property implies in turn that

$$y \in L^2(0,T;H_0^1(\Omega)) \cap C^0([0,T];L^2(\Omega)), \quad \frac{\partial y}{\partial t} \in L^2(0,T;H^{-1}(\Omega)).$$

Remark 1.46 If $f \in H_0^1(\Omega)$ the solution ψ of (1.322) satisfies

$$\psi \in L^2(0,T;H^2(\Omega) \cap H_0^1(\Omega)),$$

so that, if $d \leq 3$, $\psi(b,t)$ makes sense and belongs to $L^2(0,T)$ (since the injection of $H^2(\Omega)$ into $C^0(\bar{\Omega})$ is continuous) and (1.323) is valid.

It follows from (1.321) and (1.323) that f belongs to the orthogonal of $\{y(T;v)\}$ iff

$$\psi(b,t) = 0. \tag{1.324}$$

Therefore

$$y(T;v) \text{ spans a dense subset of } H^{-1}(\Omega) \text{ when } v \text{ spans } L^2(0,T) \text{ iff } b \text{ is such that (1.324) implies } \psi \equiv 0. \quad (1.325)$$

This is a condition on b, as the following section shows.

1.10.2. On the concept of strategic point. Formulation of a control problem. We assume that

$$A^* = A, \quad A \text{ independent of } t. \quad (1.326)$$

We introduce the *eigenfunctions* and *eigenvalues* of A (we use here the fact that Ω is *bounded*), i.e.

$$Aw_j = \lambda_j w_j, \quad w_j = 0 \text{ on } \Gamma, \ w_j \neq 0. \quad (1.327)$$

Then (assuming in order to simplify the presentation that the spectrum is *simple*)

$$\psi = \sum_{j=1}^{\infty} (f, w_j) w_j \exp(-\lambda_j (T-t)), \quad (1.328)$$

where, in (1.328), $(\cdot\,,\cdot)$ denotes the scalar product of $L^2(\Omega)$.

We shall say that b is *a strategic point* in Ω if

$$w_j(b) \neq 0 \ \forall j. \quad (1.329)$$

Then (1.324) implies

$$(f, w_j) = 0 \ \forall j,$$

i.e. $f = 0$. *In this case (1.325) is true iff* b *is a strategic point.*

We assume from now on that (1.325) holds true. We are then looking for the solution of the following control problem

$$\inf_{v \in \mathcal{U}_f} \tfrac{1}{2} \int_0^T v^2 \, \mathrm{d}t, \quad (1.330)$$

with

$$\mathcal{U}_f = \{ v \mid v \in L^2(0,T), \ y(T;v) \in y_T + \beta B_{-1} \}, \quad (1.331)$$

where y_T is given in $H^{-1}(\Omega)$, where $\beta > 0$ and where B_{-1} denotes the unit ball of $H^{-1}(\Omega)$.

1.10.3. Duality results.

The *dual problem* is as follows. One looks (with obvious notation) for the solution of

$$\inf_{\hat{f} \in H_0^1(\Omega)} \left[\tfrac{1}{2} \int_0^T |\hat{\psi}(b,t)|^2 \, \mathrm{d}t - \langle y_T, \hat{f} \rangle + \beta \|\hat{f}\|_{H_0^1(\Omega)} \right]; \quad (1.332)$$

where, in (1.332), $\hat{\psi}$ is obtained from $\hat{f}$ via

$$-\frac{\partial\hat{\psi}}{\partial t} + A^*\hat{\psi} = 0 \text{ on } Q, \quad \hat{\psi}(T) = \hat{f}, \ \hat{\psi} = 0 \text{ on } \Sigma. \tag{1.333}$$

The minima in (1.330) and (1.332) are opposite.

If f is the solution of (1.332) then the *optimal control* u (i.e. the solution of (1.330)) is given by

$$u(t) = \psi(b, t), \tag{1.334}$$

where ψ is the solution of (1.333) corresponding to $\hat{f} = f$.

1.10.4. Iterative solution of the dual problem.

From a practical point of view it is convenient to introduce

$$\Lambda \in \mathcal{L}(H_0^1(\Omega), H^{-1}(\Omega))$$

defined by

$$\Lambda\hat{f} = \hat{\varphi}(T), \tag{1.335}$$

where, in (1.335), $\hat{\varphi}$ is obtained from $\hat{f}$ via (1.333) and

$$\frac{\partial\hat{\varphi}}{\partial t} + A\hat{\varphi} = \hat{\psi}(b, t)\delta(x - b) \text{ in } Q, \quad \hat{\varphi}(0) = 0, \quad \hat{\varphi} = 0 \text{ on } \Sigma. \tag{1.336}$$

We can easily show that

$$\langle \Lambda f_1, f_2 \rangle = \int_0^T \psi_1(b, t)\psi_2(b, t)\, dt \ \forall f_1, f_2 \in H_0^1(\Omega), \tag{1.337}$$

which implies that operator Λ is *self-adjoint* and *positive semi-definite*; operator Λ is positive definite if b is *strategic*.

Combining (1.332) and (1.337) we can rewrite (1.332) as follows

$$\inf_{\hat{f} \in H_0^1(\Omega)} \left[\tfrac{1}{2}\langle \Lambda\hat{f}, \hat{f} \rangle + \beta\|\hat{f}\|_{H_0^1(\Omega)} - \langle y_T, \hat{f} \rangle \right]. \tag{1.338}$$

The minimization problem (1.338) is *equivalent* to the following variational inequality

$$\begin{cases} f \in H_0^1(\Omega), \\ \langle \Lambda f, \hat{f} - f \rangle + \beta\|\hat{f}\|_{H_0^1(\Omega)} - \beta\|f\|_{H_0^1(\Omega)} \geq \langle y_T, \hat{f} - f \rangle \ \forall \hat{f} \in H_0^1(\Omega), \end{cases} \tag{1.339}$$

which can also be written as

$$y_T \in \mathbf{\Lambda} f + \beta\partial j(f), \tag{1.340}$$

where $\partial j(\cdot)$ is the *subgradient* of the convex functional $j : H_0^1(\Omega) \to \mathbb{R}$ defined by

$$j(\hat{f}) = \|\hat{f}\|_{H_0^1(\Omega)} \ \forall \hat{f} \in H_0^1(\Omega).$$

As seen in previous sections (particularly Section 1.8.8), to solve problem (1.340) we can associate the following *initial value problem* in $H_0^1(\Omega)$ (with $\Delta = \boldsymbol{\nabla}^2$ the *Laplace operator*)

$$\begin{cases} \dfrac{\partial}{\partial \tau}(-\Delta f) + \Lambda f + \beta \partial j(f) = y_T, \\ f(0) = f_0 \end{cases} \tag{1.341}$$

with it and integrate (1.341) from $\tau = 0$ to $\tau = +\infty$, to obtain the steady-state solution of (1.341), i.e. the solution of (1.340).

As in Section 1.8.8, the *Peaceman–Rachford scheme* is well suited to the solution of problem (1.341); we then obtain

$$f^0 = f_0, \tag{1.342}$$

then for $m \geq 0, f^m$ *being known we obtain* $f^{m+1/2}$ *and* f^{m+1} *from*

$$\frac{(-\Delta f^{m+1/2}) - (-\Delta f^m)}{\frac{1}{2}\Delta\tau} + \beta \partial j(f^{m+1/2}) + \Lambda f^m = y_T, \tag{1.343}$$

$$\frac{(-\Delta f^{m+1}) - (-\Delta f^{m+1/2})}{\frac{1}{2}\Delta\tau} + \beta \partial j(f^{m+1/2}) + \Lambda f^{m+1} = y_T. \tag{1.344}$$

Problem (1.343) is equivalent to the *minimization* problem

$$\min_{\hat{f} \in H_0^1(\Omega)} \left[\frac{1}{2} \int_\Omega |\boldsymbol{\nabla} \hat{f}|^2 \, dx + \beta \frac{\Delta\tau}{2} \left(\int_\Omega |\boldsymbol{\nabla} \hat{f}|^2 \, dx \right)^{1/2} - \int_\Omega \boldsymbol{\nabla} f^m \cdot \boldsymbol{\nabla} \hat{f} \, dx - \frac{\Delta\tau}{2} \langle y_T - \Lambda f^m, \hat{f} \rangle \right]. \tag{1.345}$$

Problem (1.345) has a unique solution $f^{m+1/2} \in H_0^1(\Omega)$, which is given by

$$f^{m+1/2} = \lambda^{m+1/2} f_*^{m+1/2}, \tag{1.346}$$

where, in (1.346),

(i) $f_*^{m+1/2}$ is *the* solution of the *Dirichlet* problem

$$\begin{cases} f_*^{m+1/2} \in H_0^1(\Omega), \\ \displaystyle\int_\Omega \boldsymbol{\nabla} f_*^{m+1/2} \cdot \boldsymbol{\nabla} \hat{f} \, dx = \int_\Omega \boldsymbol{\nabla} f^m \cdot \boldsymbol{\nabla} \hat{f} \, dx \\ \qquad + \dfrac{\Delta\tau}{2} \langle y_T - \Lambda f^m, \hat{f} \rangle \ \forall \hat{f} \in H_0^1(\Omega) \end{cases} \tag{1.347}$$

(i.e. $f_*^{m+1/2}$ satisfies $-\Delta(f_*^{m+1/2} - f^m) = \frac{1}{2}\Delta\tau(y_T - \Lambda f^m)$ in Ω, $f_*^{m+1/2} = 0$ on Γ).

(ii) $\lambda^{m+1/2} \geq 0$ and is *the* minimizer over $\mathbb{R}_+$ of the *quadratic polynomial*

$$\lambda \to (\tfrac{1}{2}\lambda^2 - \lambda)\|f_*^{m+1/2}\|^2_{H_0^1(\Omega)} + \beta\frac{\Delta\tau}{2}\lambda\|f_*^{m+1/2}\|_{H_0^1(\Omega)};$$

we then have

$$\lambda^{m+1/2} = \begin{cases} 1 - \beta\dfrac{\Delta\tau}{2}/\|f_*^{m+1/2}\|_{H_0^1(\Omega)}, & \text{if } \|f_*^{m+1/2}\|_{H_0^1(\Omega)} \geq \beta\dfrac{\Delta\tau}{2}, \\ 0, \quad \text{if } \|f_*^{m+1/2}\|_{H_0^1(\Omega)} \leq \beta\dfrac{\Delta\tau}{2}. \end{cases} \tag{1.348}$$

Now, to compute f^{m+1} we observe that (1.343), (1.344) imply

$$-\frac{\Delta(f^{m+1} - 2f^{m+1/2} + f^m)}{\frac{1}{2}\Delta\tau} + \Lambda f^{m+1} = \Lambda f^m,$$

i.e.

$$-\frac{2}{\Delta\tau}\Delta f^{m+1} + \Lambda f^{m+1} = \Lambda f^m - \frac{2}{\Delta\tau}\Delta(2f^{m+1/2} - f^m). \tag{1.349}$$

Problem (1.349) is a particular case of the *'generalized' elliptic problem*

$$-r^{-1}\Delta f + \Lambda f = g, \tag{1.350}$$

where $g \in H^{-1}(\Omega)$ and $r > 0$; the solution of problems like (1.350) will be discussed in the following section.

1.10.5. Solution of problem (1.350).

1.10.5.1. Generalities. From the properties of operators $-\Delta$ and Λ (ellipticity and symmetry) problem (1.350) can be solved by a *conjugate gradient algorithm* like the one discussed in Section 1.8.2. We think, however, that it may be instructive to discuss first a class of control problems closely related to problem (1.330) in which the *dual problems* are of the same form as in (1.350).

Let us consider therefore the following class of *approximate pointwise controllability problems*

$$\min_{v \in L^2(0,T)} \left[\frac{1}{2}\int_0^T v^2\,dt + \frac{k}{2}\|y(T) - y_T\|^2_{-1}\right], \tag{1.351}$$

obtained by *penalization* of the final condition $y(T) = y_T$. In (1.351)

(i) the *penalty* parameter k is *positive*;
(ii) the function y is obtained from v via (1.317);
(iii) the 'function' y_T belongs to $H^{-1}(\Omega)$;
(iv) the $H^{-1}(\Omega)$-norm $\|\cdot\|_{-1}$ is defined, $\forall g \in H^{-1}(\Omega)$, by

$$\begin{cases} \|g\|_{-1} = \|\tilde{g}\|_{H_0^1(\Omega)} \left(= \left(\int_\Omega |\boldsymbol{\nabla} \tilde{g}|^2 \, dx \right)^{1/2} \right) \text{ with } \tilde{g} \text{ the solution} \\ \text{of the Dirichlet problem } -\Delta \tilde{g} = g \text{ in } \Omega, \ \tilde{g} = 0 \text{ on } \Gamma. \end{cases} \tag{1.352}$$

Problem (1.351) has a *unique* solution y which is characterized by the existence of p belonging to $L^2(0,T; H^2(\Omega) \cap H_0^1(\Omega))$ such that the triple $\{u, y, p\}$ satisfies the following *optimality system:*

$$\frac{\partial y}{\partial t} + Ay = u\delta(x-b) \text{ in } Q, \quad y = 0 \text{ on } \Sigma, \ y(0) = 0, \tag{1.353}$$

$$-\frac{\partial p}{\partial t} + A^* p = 0 \text{ in } Q, \quad p = 0 \text{ on } \Sigma, \tag{1.354_1}$$

$$\begin{cases} p(T) \in H_0^1(\Omega), \\ -\Delta p(T) = k(y_T - y(T)) \text{ in } \Omega, \end{cases} \tag{1.354_2}$$

$$u(t) = p(b,t). \tag{1.355}$$

Let us define $f \in H_0^1(\Omega)$ by $f = p(T)$; it follows then from (1.353)–(1.355) that f is the solution of the (dual) problem

$$-k^{-1}\Delta f + \Lambda f = y_T. \tag{1.356}$$

Concerning the solution of problem (1.351) we have two options: we can use either the primal formulation (1.351) or the dual formulation (1.356). Both approaches will be discussed in the following two sections.

1.10.5.2. Direct solution of problem (1.351). Solving the control problem (1.351) *directly* (i.e. in $L^2(0,T)$) is worth considering for the following reasons:

(i) It can be generalized to pointwise control problems with *nonlinear* state equations.
(ii) The space $L^2(0,T)$ is a space of *one variable* functions, even for multidimensional domains Ω (i.e. $\Omega \subset \mathbb{R}^d$ with $d \geq 2$).
(iii) The structure of the space $L^2(0,T)$ is quite simple making the implementation of conjugate gradient algorithms operating in this space fairly easy.

Let us denote by $J(\cdot)$ the functional in (1.351); the solution u of problem (1.351) satisfies $J'(u) = 0$ where $J'(u)$ is the gradient of the functional $J(\cdot)$ at u. Let us consider $v \in L^2(0,T)$; we can identify $J'(v)$ with an element of $L^2(0,T)$ and we have

$$\int_0^T J'(v) w \, dt = \int_0^T (v(t) - p(b,t)) w(t) \, dt \ \forall v, w \in L^2(0,T), \tag{1.357}$$

where, in (1.357), p is obtained from v via (1.317) and the corresponding *adjoint equation*, namely

$$-\frac{\partial p}{\partial t} + A^*p = 0 \text{ in } Q, \quad p = 0 \text{ on } \Sigma, \tag{1.358$_1$}$$

$$\begin{cases} p(T) \in H_0^1(\Omega), \\ -\Delta p(T) = k(y_T - y(T)) \text{ in } \Omega. \end{cases} \tag{1.358$_2$}$$

Writing $J'(u) = 0$ in *variational form*, namely

$$\begin{cases} u \in L^2(0,T), \\ \int_0^T J'(u)v\,\mathrm{d}t = 0 \;\forall v \in L^2(0,T)), \end{cases}$$

and taking into account the fact that operator $v \to J'(v)$ is *affine* with respect to v (with a linear part associated with an $L^2(0,T)$-elliptic operator) we observe that problem (1.351) is a particular case of problem (1.121) (see Section 1.8.2); it can be solved therefore by the *conjugate gradient algorithm* (1.122)–(1.129). In the particular case considered here this algorithm takes the following form:

$$u^0 \in L^2(0,T) \textit{ is given;} \tag{1.359}$$

solve

$$\frac{\partial y^0}{\partial t} + Ay^0 = u^0\delta(x-b) \textit{ in } Q, \quad y^0 = 0 \textit{ on } \Sigma,\; y^0(0) = 0, \tag{1.360}$$

then

$$-\frac{\partial p^0}{\partial t} + A^*p^0 = 0 \textit{ in } Q, \quad p^0 = 0 \textit{ on } \Sigma, \tag{1.361$_1$}$$

$$\begin{cases} p^0(T) \in H_0^1(\Omega), \\ -\Delta p^0(T) = k(y_T - y^0(T)) \textit{ in } \Omega, \end{cases} \tag{1.361$_2$}$$

$$\begin{cases} g^0 \in L^2(0,T), \\ \int_0^T g^0(t)v(t)\,\mathrm{d}t = \int_0^T (u^0(t) - p^0(b,t))v(t)\,\mathrm{d}t \;\forall v \in L^2(0,T), \end{cases} \tag{1.362}$$

and set

$$w^0 = g^0. \quad \Box \tag{1.363}$$

Assuming that u^n, g^n, w^n are known, we obtain u^{n+1}, g^{n+1}, w^{n+1} as follows.

Solve

$$\frac{\partial \bar{y}^n}{\partial t} + A\bar{y}^n = w^n\delta(x-b) \textit{ in } Q, \quad \bar{y}^n = 0 \textit{ on } \Sigma,\; \bar{y}^n(0) = 0, \tag{1.364}$$

and then

$$-\frac{\partial \bar{p}^n}{\partial t} + A^* \bar{p}^n = \text{ in } Q, \quad \bar{p}^n = 0 \text{ on } \Sigma, \tag{1.365$_1$}$$

$$\begin{cases} \bar{p}^n(T) \in H_0^1(\Omega), \\ \Delta \bar{p}^n(T) = k\bar{y}^n(T) \text{ in } \Omega, \end{cases} \tag{1.365$_2$}$$

and

$$\begin{cases} \bar{g}^n \in L^2(0,T), \\ \int_0^T \bar{g}^n v \, dt = \int_0^T (w^n(t) - \bar{p}^n(b,t)) v(t) \, dt \ \forall v \in L^2(0,T). \end{cases} \tag{1.366}$$

Compute then

$$\rho_n = \int_0^T |g^n|^2 \, dt \Big/ \int_0^T \bar{g}^n w^n \, dt, \tag{1.367}$$

and update u^n and g^n by

$$u^{n+1} = u^n - \rho_n w^n, \tag{1.368}$$

and

$$g^{n+1} = g^n - \rho_n \bar{g}^n, \tag{1.369}$$

respectively. If $\|g^{n+1}\|_{L^2(0,T)}/\|g^0\|_{L^2(0,T)} \le \epsilon$, take $u = u^{n+1}$; if not, compute

$$\gamma_n = \|g^{n+1}\|^2_{L^2(0,T)}/\|g^n\|^2_{L^2(0,T)} \tag{1.370}$$

and update w^n by

$$w^{n+1} = g^{n+1} + \gamma_n w^n. \quad \square \tag{1.371}$$

Do $n = n + 1$ and go to (1.364).

A finite element/finite difference implementation of the above algorithm will be briefly discussed in Section 1.10.6.

1.10.5.3. A duality method for the solution of problem (1.351). Suppose that we can solve the *dual* problem (1.356), then from $f(= p(T))$ we can compute p, via $(1.354)_1$ and obtain the control u via (1.355). Problem (1.356) is equivalent to

$$\begin{cases} f \in H_0^1(\Omega), \\ k^{-1} \int_\Omega \nabla f \cdot \nabla \hat{f} \, dx + \langle \Lambda f, \hat{f} \rangle = \langle y_T, \hat{f} \rangle \ \forall \hat{f} \in H_0^1(\Omega). \end{cases} \tag{1.372}$$

From the *symmetry, positivity* and *continuity* of Λ (see Section 1.10.4) the *bilinear form* on the left-hand side of (1.372) is *continuous, symmetric* and $H_0^1(\Omega)$-*elliptic* (we have indeed

$$k^{-1} \int_\Omega |\nabla \hat{f}|^2 \, dx + \langle \Lambda \hat{f}, \hat{f} \rangle \ge k^{-1} \|\hat{f}\|^2_{H_0^1(\Omega)} \ \forall \hat{f} \in H_0^1(\Omega));$$

problem (1.372) (and therefore (1.351)) can be solved by a conjugate gradient algorithm operating this time in $H_0^1(\Omega)$. This algorithm – which is closely related to algorithm (1.134)–(1.148) – is given by

$$f^0 \in H_0^1(\Omega) \textit{ is given}; \tag{1.373}$$

solve

$$-\frac{\partial p^0}{\partial t} + A^* p^0 = 0 \textit{ in } Q, \quad p^0 = 0 \textit{ on } \Sigma,\ p^0(T) = f^0, \tag{1.374}$$

then

$$\frac{\partial y^0}{\partial t} + Ay^0 = p^0(b,t)\delta(x-b) \textit{ in } Q, \quad y^0 = 0 \textit{ on } \Sigma,\ y^0 = 0, \tag{1.375}$$

$$\begin{cases} g^0 \in H_0^1(\Omega), \\ \int_\Omega \nabla g^0 \cdot \nabla \hat{f}\, \mathrm{d}x = k^{-1} \int_\Omega \nabla f^0 \cdot \nabla \hat{f}\, \mathrm{d}x + \langle y^0(T) - y_T, \hat{f} \rangle \ \forall \hat{f} \in H_0^1(\Omega) \end{cases} \tag{1.376}$$

and set

$$w^0 = g^0. \quad \square \tag{1.377}$$

Assuming that f^n, g^n, w^n *are known, we obtain* u^{n+1}, g^{n+1}, w^{n+1} *as follows.*

Solve

$$-\frac{\partial \bar{p}^n}{\partial t} + A^* \bar{p}^n = 0 \textit{ in } Q, \quad \bar{p}^n = 0 \textit{ on } \Sigma,\ \bar{p}^n(T) = w^n, \tag{1.378}$$

then

$$\frac{\partial \bar{y}^n}{\partial t} + A\bar{y}^n = \bar{p}^n(b,t)\delta(x-b) \textit{ in } Q, \quad \bar{y}^n = 0 \textit{ on } \Sigma,\ \bar{y}^n(0) = 0 \tag{1.379}$$

and

$$\begin{cases} \bar{g}^n \in H_0^1(\Omega), \\ \int_\Omega \nabla \bar{g}^n \cdot \nabla \hat{f}\, \mathrm{d}x = k^{-1} \int_\Omega \nabla w^n \cdot \nabla \hat{f}\, \mathrm{d}x + \langle \bar{y}^n(T), \hat{f} \rangle \ \forall \hat{f} \in H_0^1(\Omega). \end{cases} \tag{1.380}$$

Compute then

$$\rho_n = \int_\Omega |\nabla g^n|^2\, \mathrm{d}x \Big/ \int_\Omega \nabla \bar{g}^n \cdot \nabla w^n\, \mathrm{d}x, \tag{1.381}$$

and update f^n *and* g^n *by*

$$f^{n+1} = f^n - \rho_n w^n, \tag{1.382}$$

and

$$g^{n+1} = g^n - \rho_n \bar{g}^n, \tag{1.383}$$

respectively. If $\|g^{n+1}\|_{H_0^1(\Omega)}/\|g^0\|_{H_0^1(\Omega)} \leq \epsilon$, *take* $f = f^{n+1}$; *if not compute*

$$\gamma_n = \int_\Omega |\nabla g^{n+1}|^2 \, dx \Big/ \int_\Omega |\nabla g^n|^2 \, dx \tag{1.384}$$

and update w^n *by*

$$w^{n+1} = g^{n+1} + \gamma_n w^n. \quad \square \tag{1.385}$$

Do $n = n+1$ *and go to* (1.378).

Remark 10.47 Concerning the speed of convergence of algorithm (1.373)–(1.385) we have (from Section 1.8.2, relation (1.130)) that the number of iterations necessary to achieve convergence verifies

$$n \leq n_0 \sim \ln \frac{1}{\epsilon} \Big/ \ln \left(\frac{\sqrt{\nu_k}+1}{\sqrt{\nu_k}-1} \right), \tag{1.386}$$

where

$$\nu_k = \|k^{-1}\mathbf{I} + \tilde{\Lambda}\| \; \|(k^{-1}\mathbf{I} + \tilde{\Lambda})^{-1}\| \quad (\text{with } \tilde{\Lambda} = (-\Delta)^{-1}\Lambda). \tag{1.387}$$

Since

$$\|k^{-1}\mathbf{I} + \tilde{\Lambda}\| = k^{-1} + \|\tilde{\Lambda}\|, \quad \|(k^{-1}\mathbf{I} + \tilde{\Lambda})^{-1}\| = k,$$

it follows from (1.386) and (1.387) that for *large values* of k we have

$$n \leq n_0 \sim \|\tilde{\Lambda}\|^{1/2} k^{1/2} \ln \epsilon^{-1/2}. \tag{1.388}$$

Similarly, we could have shown that the number of iterations of algorithm (1.359)–(1.371) necessary to obtain convergence also varies like $k^{1/2} \ln \epsilon^{-1/2}$ for large values of k.

From a practical point of view we shall implement finite-dimensional variants of the above algorithms; these variants will be discussed in the following section.

1.10.6. Spacetime discretizations of problems (1.330) and (1.351).

1.10.6.1. Generalities. We shall discuss in this section the *numerical solution* of the pointwise control problems addressed in Sections 1.10.2 to 1.10.5. The approximation methods to be discussed are closely related to those which have been employed in Section 1.8, namely they will combine *time discretizations* by *finite difference* methods to *space discretizations* by *finite element* methods. Since the solution to the control problem (1.330) can be reduced to a sequence of problems such as (1.351), we shall focus our discussion on this last problem.

1.10.6.2. Approximations of control problem (1.351). We now employ the *finite element* spaces V_h and V_{0h} defined as in Section 1.8.4 (the notation of which is mostly kept) we approximate control problem (1.351) as follows.

$$\min_{\mathbf{v}\in\mathbb{R}^N} J_h^{\Delta t}(\mathbf{v}), \tag{1.389}$$

where, in (1.389), we have $\Delta t = T/N, \mathbf{v} = \{v^n\}_{n=1}^N$ and

$$J_h^{\Delta t}(\mathbf{v}) = \frac{\Delta t}{2}\sum_{n=1}^{N}|v^n|^2 + \frac{k}{2}\int_\Omega |\boldsymbol{\nabla}\Phi_h^N|^2\,dx, \tag{1.390}$$

with Φ_h^N obtained from $\mathbf{v}$ via the solution of the following discrete parabolic problem:

$$y_h^0 = 0, \tag{1.391}$$

then for $n = 1, \dots, N$, assuming that y_h^{n-1} is known, we solve

$$\begin{cases} y_h^n \in V_{0h}, \\ \displaystyle\int_\Omega \frac{y_h^n - y_h^{n-1}}{\Delta t} z_h\,dx + a(y_h^n, z_h) = v^n z_h(b)\ \forall z_h \in V_{0h}, \end{cases} \tag{1.392}$$

and finally

$$\begin{cases} \Phi_h^N \in V_{0h}, \\ \displaystyle\int_\Omega \boldsymbol{\nabla}\Phi_h^H \cdot \boldsymbol{\nabla} z_h\,dx = \langle y_T - y_h^N, z_h\rangle\ \forall z_h \in V_{0h}. \end{cases} \tag{1.393}$$

Problems (1.392) (for $n = 1, \dots, N$) and (1.393) are *well-posed discrete Dirichlet problems* (we recall that $a(z_1, z_2) = \langle Az_1, z_2\rangle\ \forall z_1, z_2 \in H_0^1(\Omega)$).

The discrete control problem (1.389) is *well posed*; its unique solution – denoted by $\mathbf{u}_h^{\Delta t} = \{u^n\}_{n=1}^N$ – is *characterized by*

$$\boldsymbol{\nabla} J_h^{\Delta t}(\mathbf{u}_h^{\Delta t}) = \mathbf{0}, \tag{1.394}$$

where, in (1.394), $\boldsymbol{\nabla} J_h^{\Delta t}$ denotes the gradient of $J_h^{\Delta t}$.

Remark 1.48 The *convergence* of $\mathbf{u}_h^{\Delta t}$, and of the corresponding state vector, to their continuous counterparts is a fairly technical issue. It will not be addressed in this article. On the other hand, we shall address the solution of problem (1.389), via the solution of the equivalent *linear* problem (1.394); this will be the task of the following Sections 1.10.6.3 and 1.10.6.4.

Remark 1.49 The approximate control problem (1.389) relies on a time discretization by an *implicit Euler scheme.* Actually, we can improve accuracy by using, as in Section 1.8.6, a *second-order accurate two-step implicit time discretization scheme.* By merging the techniques described in the present section and in Section 1.8.6 we can easily derive a variant of the approximate problem (1.389) relying on the above second-order accurate time discretization scheme.

1.10.6.3. Iterative solution of the discrete control problem (1.389). I: Calculation of $\nabla J_h^{\Delta t}$. In order to solve via (1.394) the discrete control problem (1.389), by a *conjugate gradient* algorithm, we need to know $\nabla J_h^{\Delta t}(\mathbf{v})$ $\forall \mathbf{v} \in \mathbb{R}^N$. To compute $\nabla J_h^{\Delta t}(\mathbf{v})$, we observe that

$$\lim_{\substack{\theta \to 0 \\ \theta \neq 0}} \frac{J_h^{\Delta t}(\mathbf{v}+\theta\mathbf{w}) - J_h^{\Delta t}(\mathbf{v})}{\theta} = (\nabla J_h^{\Delta t}(\mathbf{v}), \mathbf{w})_{\Delta t} \ \forall \mathbf{v}, \mathbf{w} \in \mathbb{R}^N, \tag{1.395}$$

where

$$(\mathbf{v}, \mathbf{w})_{\Delta t} = \Delta t \sum_{n=1}^{N} v^n w^n \ \forall \mathbf{v}, \mathbf{w} \in \mathbb{R}^N \quad (\text{and } \|\mathbf{v}\|_{\Delta t} = (\mathbf{v}, \mathbf{v})_{\Delta t}^{1/2}).$$

Combining (1.390)–(1.393) and (1.395) we can prove that

$$(\nabla J_h^{\Delta t}(\mathbf{v}), \mathbf{w})_{\Delta t} = \Delta t \sum_{n=1}^{N} (v^n - p_h^n(b)) w^n \ \forall \mathbf{v}, \mathbf{w} \in \mathbb{R}^N, \tag{1.396}$$

where the family $\{p_h^n\}_{n=1}^N$ is obtained as the solution to the following *adjoint discrete parabolic problem*:

$$p_h^{N+1} = k\Phi_h^N; \tag{1.397}$$

then, for $n = N, \ldots, 1$, *assuming that* p_h^{N+1} *is known, solve (the well-posed discrete elliptic problem)*

$$\begin{cases} p_h^n \in V_{0h}, \\ \displaystyle\int_\Omega \frac{p_h^n - p_h^{n+1}}{\Delta t} z_h \, dx + a(z_h, p_h^n) = 0 \ \forall z_h \in V_{0h}. \end{cases} \tag{1.398}$$

Owing to the importance of relation (1.396), we shall give a short proof of it (of the engineer/physicist type) based on a (formal) *perturbation* analysis: Hence, let us consider a perturbation $\delta\mathbf{v}$ of $\mathbf{v}$; we have then, from (1.390),

$$\begin{aligned} \delta J_h^{\Delta t}(\mathbf{v}) &= (\nabla J_h^{\Delta t}(\mathbf{v}), \delta\mathbf{v})_{\Delta t} \\ &= \Delta t \sum_{n=1}^{N} v^n \delta v^n + k \int_\Omega \nabla\Phi_h^N \cdot \nabla\delta\Phi_h^N \, dx \end{aligned} \tag{1.399}$$

where, in (1.399), $\delta\Phi_h^N$ is obtained from $\delta\mathbf{v}$ via

$$\delta y_h^0 = 0, \tag{1.400}$$

then for $n = 1, \ldots, N$, we have

$$\begin{cases} \delta y_h^n \in V_{0h}, \\ \displaystyle\int_\Omega \frac{\delta y_h^n - \delta y_h^{n-1}}{\Delta t} z_h \, dx + a(\delta y_h^n, z_h) = \delta v^n z_h(b) \ \forall z_h \in V_{0h}, \end{cases} \tag{1.401}$$

and finally

$$\begin{cases} \delta\Phi_h^N \in V_{0h}, \\ \int_\Omega \nabla\delta\Phi_h^N \cdot \nabla z_h \,\mathrm{d}x = -\int_\Omega \delta y_h^N z_h \,\mathrm{d}x \ \forall z_h \in V_{0h}. \end{cases} \tag{1.402}$$

Taking $z_h = p_h^n$ in (1.401) we obtain, by summation from $n = 1$ to $n = N$,

$$\begin{aligned} \Delta t \sum_{n=1}^N p_h^n(b)\delta v^n &= \Delta t \sum_{n=1}^N \int_\Omega \frac{\delta y_h^n - \delta y_h^{n-1}}{\Delta t} p_h^n \,\mathrm{d}x + \Delta t \sum_{n=1}^N a(\delta y_h^n, p_h^n) \\ &= \int_\Omega p_h^{N+1} \delta y_h^N \,\mathrm{d}x \\ &\quad + \Delta t \sum_{n=1}^N \left[\int_\Omega \frac{p_h^n - p_h^{n+1}}{\Delta t} \delta y_h^n \,\mathrm{d}x + a(\delta y_h^n, p_h^n) \right]. \end{aligned} \tag{1.403}$$

Since $\{p_h^n\}_{n=1}^{N+1}$ satisfies (1.397), (1.398), it follows from (1.403) that

$$\int_\Omega p_h^{N+1} \delta y_h^N \,\mathrm{d}x = \Delta t \sum_{n=1}^N p_h^n(b)\delta v^n. \tag{1.404}$$

Taking $z_h = \Phi_h^N$ in (1.402), we obtain from (1.397)

$$k \int_\Omega \nabla\Phi_h^N \cdot \nabla\delta\Phi_h^N \,\mathrm{d}x = -k \int_\Omega \delta y_h^N \Phi_h^N \,\mathrm{d}x = -\int_\Omega p_h^{N+1} \delta y_h^N \,\mathrm{d}x,$$

which combined with (1.399) and (1.402) implies

$$(\nabla J_h^{\Delta t}(\mathbf{v}), \delta\mathbf{v})_{\Delta t} = \Delta t \sum_{n=1}^N (v^n - p_h^n(b))\delta v^n. \tag{1.405}$$

Since $\delta\mathbf{v}$ is 'arbitrary', relation (1.405) implies (1.396).

1.10.6.4. Iterative solution of the discrete control problem (1.389). II: Conjugate gradient solution of problem (1.389), (1.394). The discrete control problem (1.389) is equivalent to a linear system (namely (1.394)) which is associated with an $N \times N$ *symmetric* and *positive definite* matrix. Such a problem can therefore be solved by a conjugate gradient algorithm which is a particular case of algorithm (1.122)–(1.129) (see Section 1.8.2) and a variant of algorithm (1.169)–(1.183) (see Section 1.8.5). This algorithm takes the following form:

$$\mathbf{u}_0 = \{u_0^n\}_{n=1}^N \textit{ is given in } \mathbb{R}^N; \tag{1.406}$$

take then

$$y_0^0 = 0, \tag{1.407}$$

and, assuming that y_0^{n-1} *is known, solve for* $n = 1, \ldots, N$,

$$\begin{cases} y_0^n \in V_{0h}, \\ \displaystyle\int_\Omega \frac{y_0^n - y_0^{n-1}}{\Delta t} z_h \, dx + a(y_0^n, z_h) = u_0^n z_h(b) \ \forall z_h \in V_{0h}. \end{cases} \tag{1.408}$$

Solve next

$$\begin{cases} \Phi_0^N \in V_{0h}, \\ \displaystyle\int_\Omega \nabla \Phi_0^N \cdot \nabla z_h \, dx = \langle y_T - y_0^N, z_h \rangle \ \forall z_h \in V_{0h}. \end{cases} \tag{1.409}$$

Finally, take

$$p_0^{N+1} = k\Phi_0^N \tag{1.410}$$

and, assuming that p_0^{n+1} *is known, solve for* $n = N, \ldots, 1$

$$\begin{cases} p_0^n \in V_{0h}, \\ \displaystyle\int_\Omega \frac{p_0^n - p_0^{n+1}}{\Delta t} z_h \, dx + a(z_h, p_0^n) = 0 \ \forall z_h \in V_{0h}. \end{cases} \tag{1.411}$$

Set

$$\mathbf{g}_0 = \{u_0^n - p_0^n(b)\}_n^N = 1 \tag{1.412}$$

and

$$\mathbf{w}_0 = \mathbf{g}_0. \tag{1.413}$$

Then for $m \geq 0$, *assuming that* $\mathbf{u}_m, \mathbf{g}_m$ *and* $\mathbf{w}_m$ *are known compute* $\mathbf{u}_{m+1}, \mathbf{g}_{m+1}$ *and* $\mathbf{w}_{m+1}$ *as follows.*

Take

$$\bar{y}_m^0 = 0; \tag{1.414}$$

assuming that $\bar{y}_m^{n-1}$ *is known, solve for* $n = 1, \ldots, N$

$$\begin{cases} \bar{y}_m^n \in V_{0h}, \\ \displaystyle\int_\Omega \frac{\bar{y}_m^n - \bar{y}_m^{n-1}}{\Delta t} z_h \, dx + a(\bar{y}_m^n, z_h) = w_m^n z_h(b) \ \forall z_h \in V_{0h}. \end{cases} \tag{1.415}$$

Solve next

$$\begin{cases} \bar{\Phi}_m^N \in V_{0h}, \\ \displaystyle\int_\Omega \nabla \bar{\Phi}_m^N \cdot \nabla z_h \, dx = -\langle \bar{y}_m^N, z_h \rangle \ \forall z_h \in V_{0h}. \end{cases} \tag{1.416}$$

Finally, take

$$\bar{p}_m^{N+1} = k\bar{\Phi}_m^N, \tag{1.417}$$

and, assuming that $\bar{p}_m^{n+1}$ is known, solve for $n = N, \ldots, 1$

$$\begin{cases} \bar{p}_m^n \in V_{0h}, \\ \displaystyle\int_\Omega \frac{\bar{p}_m^n - \bar{p}_m^{n+1}}{\Delta t} z_h \, dx + a(z_h, \bar{p}_m^n) = 0 \ \forall z_h \in V_{0h}. \end{cases} \tag{1.418}$$

Set

$$\bar{\mathbf{g}}_m = \{w_m^n - \bar{p}_m^n(b)\}_{n=1}^N. \tag{1.419}$$

Compute

$$\rho_m = \frac{\|\mathbf{g}_m\|_{\Delta t}^2}{(\bar{\mathbf{g}}_m, \mathbf{w}_m)_{\Delta t}}, \tag{1.420}$$

and update $\mathbf{u}_m$ and $\mathbf{g}_m$ by

$$\mathbf{u}_{m+1} = \mathbf{u}_m - \rho_m \mathbf{w}_m, \tag{1.421}$$

$$\mathbf{g}_{m+1} = \mathbf{g}_m - \rho_m \bar{\mathbf{g}}_m, \tag{1.422}$$

respectively. If $\|\mathbf{g}_{m+1}\|_{\Delta t}/\|\mathbf{g}_0\|_{\Delta t} \leq \epsilon$ take $\mathbf{u}_h^{\Delta t} = \mathbf{u}^{m+1}$; else, compute

$$\gamma_m = \|\mathbf{g}_{m+1}\|_{\Delta t}^2 / \|\mathbf{g}_m\|_{\Delta t}^2, \tag{1.423}$$

and update $\mathbf{w}_m$ by

$$\mathbf{w}_{m+1} = \mathbf{g}_{m+1} + \gamma_m \mathbf{w}_m. \quad \square \tag{1.424}$$

Do $m = m + 1$ and go to (1.414).

Remark 1.50 Algorithm (1.406)–(1.424) is a discrete analogue of algorithm (1.359)–(1.371).

1.10.6.5. Approximation of the dual problem (1.356). It was shown in Section 1.10.5 that there is *equivalence* between the *primal* control problem (1.351) and its *dual* problem (1.356). We shall discuss now the approximation of problem (1.356). There is no difficulty in adapting problem (1.356) to the (backward Euler scheme based) approximation methods discussed in Sections 1.8.3 to 1.8.5 for the solution of problem (1.116). Therefore, to avoid tedious repetitions we shall focus our discussion on an approximation of problem (1.356) which is based on a time discretization by the two-step backward implicit scheme considered in Section 1.8.6 (whose notation is kept); for simplicity, we shall take $H = h$ and $E_{0h} = V_{0h}$.

We approximate the *dual problem* (1.356), (1.372) by

$$\begin{cases} f_h^{\Delta t} \in V_{0h}, \\ k^{-1} \displaystyle\int_\Omega \nabla f_h^{\Delta t} \cdot \nabla \hat{f}_h \, dx + \int_\Omega (\Lambda_h^{\Delta t} f_h^{\Delta t}) \hat{f}_h \, dx = \langle y_T, \hat{f}_h \rangle \ \forall \hat{f}_h \in V_{0h}, \end{cases} \tag{1.425}$$

where, in (1.425), $\Lambda_h^{\Delta t}$ denotes the *linear operator* from V_{0h} into V_{0h} defined as follows

$$\Lambda_h^{\Delta t}\hat{f}_h = 2\hat{\varphi}_h^{N-1} - \hat{\varphi}_h^{N-2}, \tag{1.426}$$

where, to obtain $\hat{\varphi}_h^{N-1}$ and $\hat{\varphi}_h^{N-2}$, we solve for $n = N-1, \ldots, 1$, the well-posed discrete elliptic problem

$$\begin{cases} \hat{\psi}_h^n \in V_{0h}, \\ \displaystyle\int_\Omega \frac{\frac{3}{2}\hat{\psi}_h^n - 2\hat{\psi}_h^{n+1} + \frac{1}{2}\hat{\psi}_h^{n+2}}{\Delta t} z_h \, dx + a(z_h, \hat{\psi}_h^n) = 0 \ \forall z_h \in V_{0h}, \end{cases} \tag{1.427}$$

with

$$\hat{\psi}_h^N = 2\hat{f}_h, \quad \hat{\psi}_h^{N+1} = 4\hat{f}_h, \tag{1.428}$$

then, with $\hat{\varphi}_h^0 = 0$,

$$\begin{cases} \hat{\varphi}_h^1 \in V_{0h}, \\ \displaystyle\int_\Omega \frac{\hat{\varphi}_h^1 - \hat{\varphi}_h^0}{\Delta t} z_h \, dx + a(\tfrac{2}{3}\hat{\varphi}_h^1 + \tfrac{1}{3}\hat{\varphi}_h^0, z_h) = \tfrac{2}{3}\hat{\psi}_h^1(b) z_h(b) \ \forall z_h \in V_{0h}, \end{cases} \tag{1.429}$$

and, finally, for $n = 2, \ldots, N-1$,

$$\begin{cases} \hat{\varphi}_h^n \in V_{0h}, \\ \displaystyle\int_\Omega \frac{\frac{3}{2}\hat{\varphi}_h^{n-1} - 2\hat{\varphi}_h^{n-1} + \frac{1}{2}\hat{\psi}_h^{n-2}}{\Delta t} z_h \, dx + a(\hat{\varphi}_h^n, z_h) = \hat{\psi}_h^n(b) z_h(b) \ \forall z_h \in V_{0h}. \end{cases} \tag{1.430}$$

It follows from (1.426)–(1.430) that (with obvious notation)

$$\int_\Omega (\Lambda_h^{\Delta t} f_1) f_2 \, dx = \Delta t \sum_{n=1}^{N-1} \psi_1^n(b)\psi_2^n(b) \ \forall f_1, f_2 \in V_{0h},$$

i.e. operator $\Lambda_h^{\Delta t}$ is *symmetric* and *positive semi-definite*, which implies in turn that the approximate dual problem (1.425) has a *unique* solution.

Remark 1.51 The discrete problem (1.425) is actually the *dual problem* of the following *discrete control problem* (a variant of problem (1.389); see Section 1.10.6.2):

$$\min_{\mathbf{v} \in \mathbb{R}^{N-1}} J_h^{\Delta t}(\mathbf{v}), \tag{1.431}$$

where, in (1.431), we have $\mathbf{v} = \{v^n\}_{n=1}^{N-1}$ and

$$J_h^{\Delta t}(\mathbf{v}) = \tfrac{1}{2}\Delta t \sum_{n=1}^{N-1} |v^n|^2 + \tfrac{1}{2}k \int_\Omega |\boldsymbol{\nabla}\Phi_h^N|^2 \, dx, \tag{1.432}$$

with Φ_h^N obtained from $\mathbf{v}$ via the solution of the following discrete parabolic problem

$$y_h^0 = 0, \tag{1.433}$$

$$\begin{cases} y_h^1 \in V_{0h}, \\ \int_\Omega \frac{y_h^1 - y_h^0}{\Delta t} + a(\tfrac{2}{3}y_h^1 + \tfrac{1}{3}y_h^0, z_h) = \tfrac{2}{3}v^1 z_h(b) \ \forall z_h \in V_{0h}, \end{cases} \tag{1.434}$$

then, for $n = 2, \ldots, N-1$, assuming that y_h^{n-1} is known we solve

$$\begin{cases} y_h^n \in V_{0h}, \\ \int_\Omega \frac{\tfrac{3}{2}y_h^n - 2y_h^{n-1} + \tfrac{1}{2}y_h^{n-2}}{\Delta t} z_h \, dx + a(y_h^n, z_h) = v^n z_h(b) \ \forall z_h \in V_{0h}, \end{cases} \tag{1.435}$$

and finally

$$\begin{cases} \Phi_h^N \in V_{0h}, \\ \int_\Omega \boldsymbol{\nabla}\Phi_h^N \cdot \boldsymbol{\nabla} z_h \, dx = \langle y_T - 2y_h^{N-1} + y_h^{N-2}, z_h\rangle \ \forall z_h \in V_{0h}. \end{cases} \tag{1.436}$$

Back to problem (1.425), it follows from the properties of operator $\Lambda_h^{\Delta t}$ that this problem can be solved by the following conjugate gradient algorithm (which is a discrete analogue of algorithm (1.373)–(1.385); see Section 1.10.5.3):

$$f_0 \in V_{0h} \textit{ is given;} \tag{1.437}$$

take

$$p_0^N = 2f_0, p_0^{N+1} = 4f_0, \tag{1.438}$$

and solve for $n = N-1, \ldots, 1$ *the following discrete elliptic problem*

$$\begin{cases} p_0^n \in V_{0h}, \\ \int_\Omega \frac{\tfrac{3}{2}p_0^n - 2p_0^{n+1} + \tfrac{1}{2}p_0^{n+2}}{\Delta t} z_h \, dx + a(z_h, p_0^n) = 0l \ \forall z_h \in V_{0h}. \end{cases} \tag{1.439}$$

Take now

$$y_0^0 = 0, \tag{1.440}$$

and solve

$$\begin{cases} y_0^1 \in V_{0h}, \\ \int_\Omega \frac{y_0^1 - y_0^0}{\Delta t} z_h \, dx + a(\tfrac{2}{3}y_0^1 + \tfrac{1}{3}y_0^0, z_h) = \tfrac{2}{3}p_0^1(b) z_h(b) \ \forall z_h \in V_{0h}; \end{cases} \tag{1.441}$$

solve next, for $n = 2, \ldots, N-1$,

$$\begin{cases} y_0^n \in V_{0h}, \\ \displaystyle\int_\Omega \frac{\frac{3}{2}y_0^n - 2y_0^{n-1} + \frac{1}{2}y_0^{n-2}}{\Delta t} z_h \,\mathrm{d}x + a(y_0^n, z_h) = p_0^n(b) z_h(b) \ \forall z_h \in V_{0h}. \end{cases} \tag{1.442}$$

Solve, next

$$\begin{cases} g_0 \in V_{0h}, \\ \displaystyle\int_\Omega \nabla g_0 \cdot \nabla \hat{f} \,\mathrm{d}x = k^{-1} \int_\Omega \nabla f_0 \cdot \nabla \hat{f} \,\mathrm{d}x \\ \qquad + \langle 2y_0^{N-1} - y_0^{N-2} - y_T, \hat{f} \rangle \ \forall \hat{f} \in V_{0h}, \end{cases} \tag{1.443}$$

and set

$$w_0 = g_0. \quad \square \tag{1.444}$$

Then for $m \geq 0$, *assuming that* f_m, g_m, w_m *are known compute* f_{m+1}, g_{m+1}, w_{m+1} *as follows.*

Take

$$\bar{p}_m^N = 2w_m, \bar{p}_m^{N+1} = 4w_m \tag{1.445}$$

and solve for $n = N-1, \ldots, 1$

$$\begin{cases} \bar{p}_m^n \in V_{0h}, \\ \displaystyle\int_\Omega \frac{\frac{3}{2}\bar{p}_m^n - 2\bar{p}_m^{n+1} + \frac{1}{2}p_m^{n+2}}{\Delta t} z_h \,\mathrm{d}x + a(z_h, \bar{p}_m^n) = 0 \ \forall z_h \in V_{0h}. \end{cases} \tag{1.446}$$

Take

$$\bar{y}_m^0 = 0; \tag{1.447}$$

solve

$$\begin{cases} \bar{y}_m^1 \in V_{0h}, \\ \displaystyle\int_\Omega \frac{\bar{y}_m^1 - \bar{y}_m^0}{\Delta t} z_h \,\mathrm{d}x + a(\tfrac{2}{3}\bar{y}_m^1 + \tfrac{1}{3}\bar{y}_m^0, z_h) = \tfrac{2}{3}\bar{p}_m^1(b) z_h(b) \ \forall z_h \in V_{0h}, \end{cases} \tag{1.448}$$

and then for $n = 2, \ldots, N-1$

$$\begin{cases} \bar{y}_m^n \in V_{0h}, \\ \displaystyle\int_\Omega \frac{\frac{3}{2}\bar{y}_m^n - 2\bar{y}_m^{n-1} + \frac{1}{2}\bar{y}_m^{n-2}}{\Delta t} z_h \,\mathrm{d}x + a(\bar{y}_m^n, z_h) = \bar{p}_m^n(b) z_h(b) \ \forall z_h \in V_{0h}. \end{cases} \tag{1.449}$$

Solve next

$$\begin{cases} \bar{g}_m \in V_{0h}, \\ \int_\Omega \nabla \bar{g}_m \cdot \nabla \hat{f}\,\mathrm{d}x = k^{-1} \int_\Omega \nabla w_m \cdot \nabla \hat{f}\,\mathrm{d}x \\ \qquad + \int_\Omega (2\bar{y}_m^{N-1} - \bar{y}_m^{N-2}) \hat{f}\,\mathrm{d}x \ \forall \hat{f} \in V_{0h}, \end{cases} \tag{1.450}$$

and compute

$$\rho_m = \int_\Omega |\nabla g_m|^2\,\mathrm{d}x \Big/ \int_\Omega \nabla \bar{g}_m \cdot \nabla w_m\,\mathrm{d}x; \tag{1.451}$$

then update f_m *and* g_m *by*

$$f_{m+1} = f_m - \rho_m w_m, \tag{1.452}$$

$$g_{m+1} = g_m - \rho_m \bar{g}_m, \tag{1.453}$$

respectively. If $\|g_{m+1}\|_{H_0^1(\Omega)}/\|g_0\|_{H_0^1(\Omega)} \leq \epsilon$, *take* $f_h^{\Delta t} = f_{m+1}$, *else compute*

$$\gamma_m = \int_\Omega |\nabla g_{m+1}|^2\,\mathrm{d}x \Big/ \int_\Omega |\nabla g_m|^2\,\mathrm{d}x \tag{1.454}$$

and update w_m *by*

$$w_{m+1} = g_{m+1} + \gamma_m w_m. \quad \square \tag{1.455}$$

Do $m = m + 1$ *and go to* (1.445).

Algorithm (1.437)–(1.455) is fairly easy to implement. It essentially requires *elliptic/finite element solvers* to compute the solutions to problems (1.439), (1.441)–(1.443), (1.446), (1.448)–(1.450); such solvers are easily available.

1.10.7. Numerical experiments

1.10.7.1. Generalities. Synopsis. In order to illustrate the results and methods from Sections 1.10.1 to 1.10.6 we shall discuss in this section the solution of some *pointwise control problems*; these problems will be particular cases and variants of the *penalized* problem (1.351). We suppose for simplicity that $d = 1$ (i.e. $\Omega \subset \mathbb{R}$); it follows then from Remark 1.45 that the solution of (1.317) satisfies

$$y \in C^0([0, T]; L^2(\Omega)),$$

which implies that, in (1.351), it makes sense to replace $\|y(T) - y_T\|_{-1}$ by $\|y(T) - y_T\|_{L^2(\Omega)}$. Also, for some of the test problems we shall replace $\frac{1}{2}\int_0^T |v|^2\,\mathrm{d}t$ by $(1/s)\int_0^T |v|^s\,\mathrm{d}t$, with $s > 2$, including some *very large* values of s for which the optimal control u (in fact, its discrete analogue) clearly has a *bang–bang behaviour*; this is expected from Section 1.7.

1.10.7.2. First test problems. What we have considered here is a family of test problems parametrized by T, y_T, k and by the 'support' b of the pointwise control. These test problems can be formulated as follows.

$$\min_{v \in L^2(0,T)} J_k(v), \tag{1.456}$$

where

$$J_k(v) = \tfrac{1}{2}\int_0^T v^2\,\mathrm{d}t + \frac{k}{2}\int_0^1 |y(T) - y_T|^2\,\mathrm{d}x, \tag{1.457}$$

with y the solution of the following *diffusion problem*

$$\frac{\partial y}{\partial t} - \nu\frac{\partial^2 y}{\partial x^2} = v(t)\delta(x-b) \text{ in } (0,1)\times(0,T), \tag{1.458}$$

$$y(0,t) = y(1,t) = 0 \text{ on } (0,T), \tag{1.459}$$

$$y(0) = 0. \tag{1.460}$$

In equation (1.458) we have $\nu > 0$ and $b \in (0,1)$. We clearly have $\Omega = (0,1)$.

We have considered, first, test problems where the *target function* y_T is *even* with respect to the variable $x - \frac{1}{2}$. These target functions are given by

$$y_T(x) = 4x(1-x), \tag{1.461}$$

$$y_T(x) = \begin{cases} 8(x-\frac{1}{4}) & \text{if } \frac{1}{4} \le x \le \frac{1}{2}, \\ 8(\frac{3}{4} - x) & \text{if } \frac{1}{2} \le x \le \frac{3}{4}, \\ 0 & \text{elsewhere on } (0,1), \end{cases} \tag{1.462}$$

$$y_T(x) = \begin{cases} 1 & \text{if } \frac{1}{4} \le x \le \frac{3}{4}, \\ 0 & \text{elsewhere on } (0,1), \end{cases} \tag{1.463}$$

respectively. We have taken $\nu = \frac{1}{10}$ in equation (1.458), and $T = 3$ for all the three target functions we have given above. The continuous problem (1.456)–(1.460) has been approximated using the methods described in Sections 1.10.6.2 to 1.10.6.4 (i.e. we have solved *directly* the control problems, taking into account the fact that for these test problems we use a penalty term associated with the L^2 norm, instead of the H^{-1} norm used in the general case). The *time discretization* has been obtained using the *backward Euler scheme* described in Section 1.10.6.2 with $\Delta t = 10^{-2}$, while the space discretization was obtained using a *uniform* mesh on (0,1) with $h = 10^{-2}$. The discrete control problems have been solved by a variant of the *conjugate gradient* algorithm (1.406)–(1.424); we have taken $\mathbf{u}_0 = \mathbf{0}$ as *initializer* for the above algorithm and $\|\mathbf{g}_m\|_{\Delta t}/\|\mathbf{g}_0\|_{\Delta t} \le 10^{-6}$ as the *stopping criterion* (for those cases for which this criterion could not be reached sufficiently quickly, we stopped iterating after a fixed number of iterations (300 or 500,

Table 1. *Summary of numerical results (target function defined by (1.461); $T = 3$, $h = \Delta t = 10^{-2}$).*

b	k	Number of iterations	$\|u^*\|_{L^2(0,T)}$	$\frac{\|y^*(T)-y_T\|_{L^2(0,1)}}{\|y_T\|_{L^2(0,1)}}$
$\sqrt{2}/3$	10^2	95	0.923	6×10^{-2}
	10^3	> 300	1.14	2.3×10^{-2}
	10^4	> 300	1.28	1.4×10^{-2}
$1/2$	10^2	93	0.909	5.5×10^{-2}
	10^3	> 300	1.09	2.1×10^{-2}
	10^4	> 300	1.20	1.3×10^{-2}
$\pi/6$	10^2	95	0.918	5.9×10^{-2}
	10^3	> 300	1.12	2.3×10^{-2}
	10^4	> 300	1.26	1.4×10^{-2}

depending on the test problem)). The corresponding numerical results have been summarized in Tables 1 to 3, where u^* and $y^*(T)$ denote the computed optimal control and the corresponding final state, respectively.

In Figures 1 to 9 we have visualized, for $k = 10^4$, the computed optimal control and compared the corresponding computed value of $y(T)$ (i.e. $y^*(T)$) to the target function y_T.

The above results deserve several comments:

(i) Since operator $A = -\nu\, \mathrm{d}^2/\mathrm{d}x^2$ is *self-adjoint* for the *homogeneous* Dirichlet boundary conditions we can apply the controllability results of Section 1.10.2. The eigenfunctions of operator A, i.e. the solutions of

$$-\nu \frac{\mathrm{d}^2}{\mathrm{d}x^2} w_j = \lambda_j w_j \text{ on } (0,1), \quad w_j(0) = w_j(1) = 0, \quad w_j \neq 0,$$

are clearly given by

$$w_j(x) = \sin j\pi x, \quad j = 1, 2, \ldots,$$

the corresponding *spectrum* being $\{\nu\pi^2 j^2\}_{j=1}^{+\infty}$. Since each eigenvalue is *simple*, b will be *strategic* if

$$\sin j\pi b \neq 0 \ \forall j = 1, 2, \ldots,$$

i.e. if

$$b \notin (0,1) \cap \mathbb{Q} \tag{1.464}$$

Table 2. *Summary of numerical results (target function defined by (1.462); $T = 3$, $h = \Delta t = 10^{-2}$).*

b	k	Number of iterations	$\|u^*\|_{L^2(0,T)}$	$\frac{\|y^*(T)-y_T\|_{L^2(0,1)}}{\|y_T\|_{L^2(0,1)}}$
$\sqrt{2}/3$	10^2	157	1.23	2.2×10^{-1}
	10^3	> 500	1.93	1.9×10^{-1}
	10^4	> 500	3.01	1.8×10^{-1}
$1/2$	10^2	113	1.27	1.1×10^{-1}
	10^3	> 500	1.74	6.3×10^{-2}
	10^4	> 500	2.03	5.6×10^{-2}
$\pi/6$	10^2	137	1.24	1.9×10^{-1}
	10^3	> 500	1.82	1.6×10^{-1}
	10^4	> 500	2.72	1.5×10^{-1}

Table 3. *Summary of numerical results (target function defined by (1.463); $T = 3$, $h = \Delta t = 10^{-2}$).*

b	k	Number of iterations	$\|u^*\|_{L^2(0,T)}$	$\frac{\|y^*(T)-y_T\|_{L^2(0,1)}}{\|y_T\|_{L^2(0,1)}}$
$\sqrt{2}/3$	10^2	279	0.94	3.47×10^{-1}
	10^3	> 500	2.2	3.13×10^{-1}
	10^4	> 500	3.0	3×10^{-1}
$1/2$	10^2	317	1.0	3.25×10^{-1}
	10^3	> 500	2.6	2.79×10^{-1}
	10^4	> 500	3.6	2.64×10^{-1}
$\pi/6$	10^2	291	0.96	3.4×10^{-1}
	10^3	> 500	2.3	3×10^{-1}
	10^4	> 500	3.2	2.9×10^{-1}

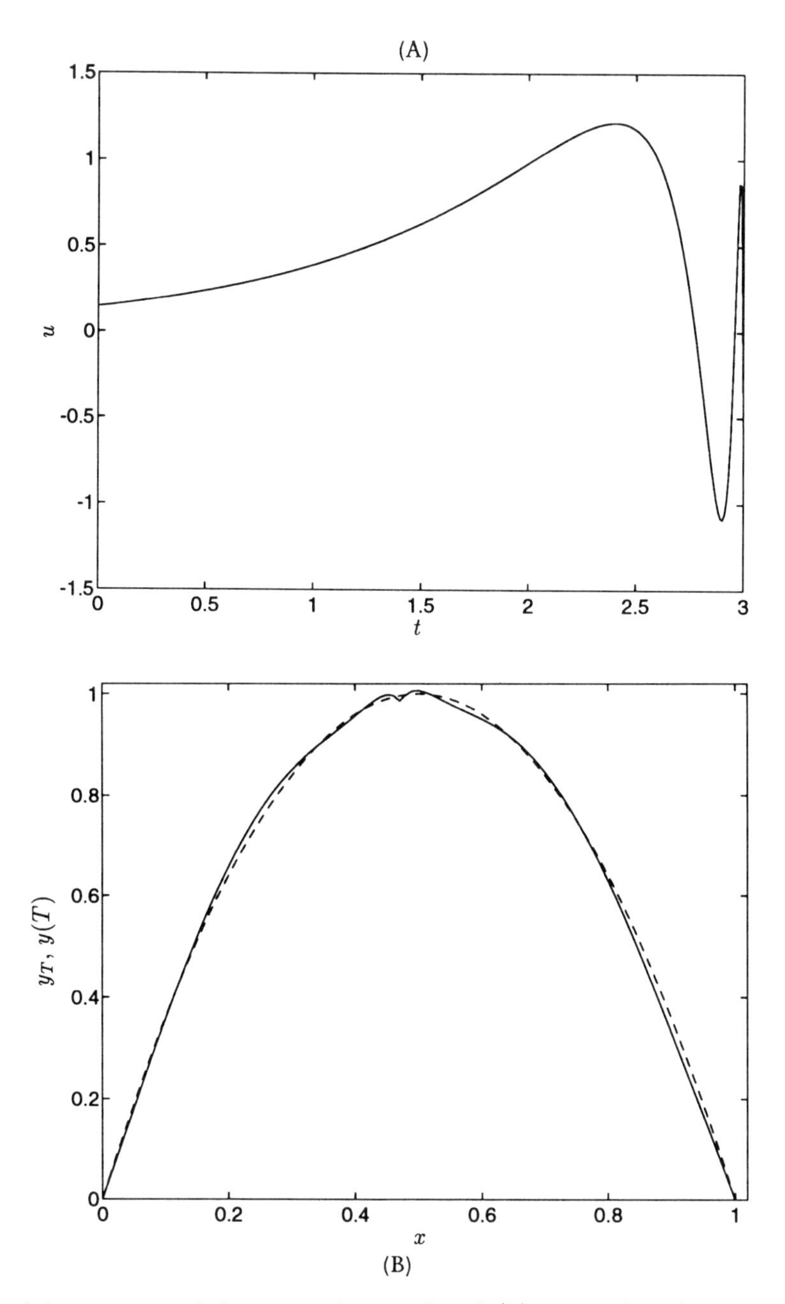

Fig. 1. (a) Variation of the optimal control and (b) comparison between y_T and $y^*(T)$ (target function (1.461): $T = 3$, $b = \sqrt{2}/3$, $k = 10^4$, $h = \Delta t = 10^{-2}$).

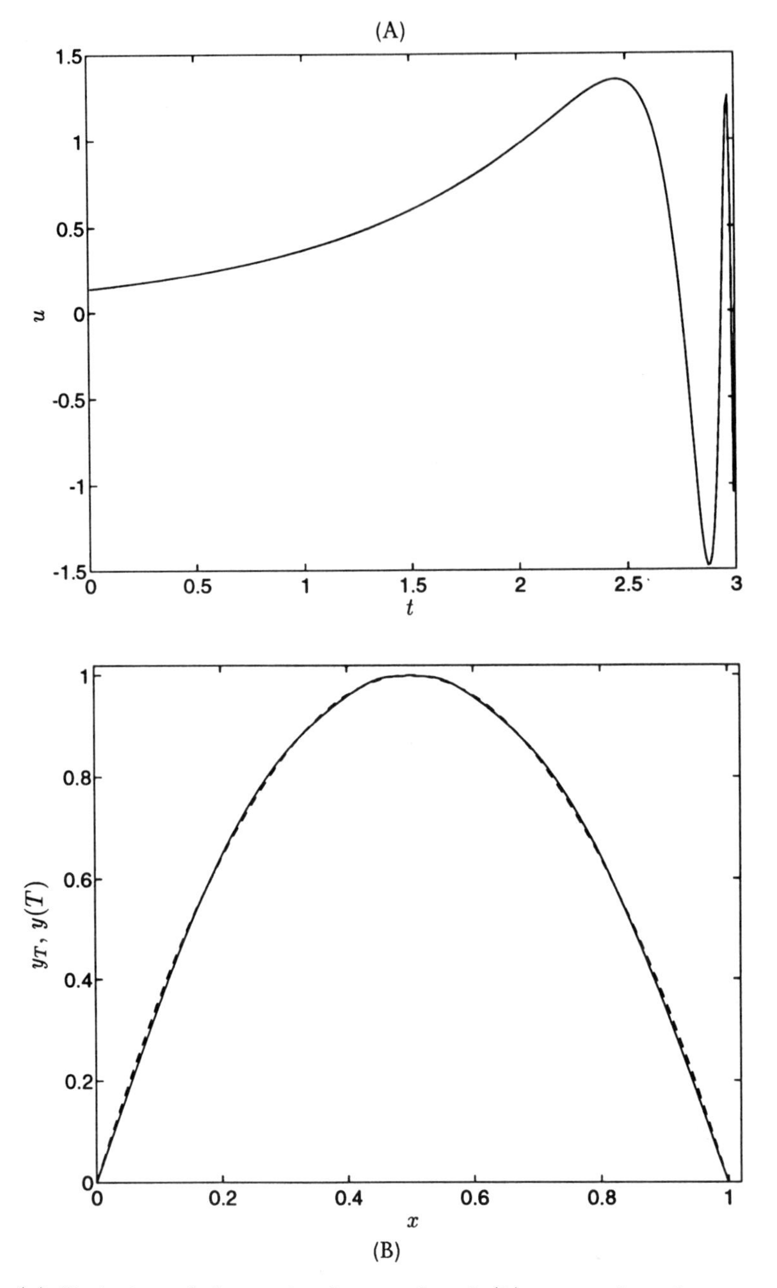

Fig. 2. (a) Variation of the optimal control and (b) comparison between y_T and $y^*(T)$ (target function (1.461): $T = 3$, $b = 1/2$, $k = 10^4$, $h = \Delta t = 10^{-2}$).

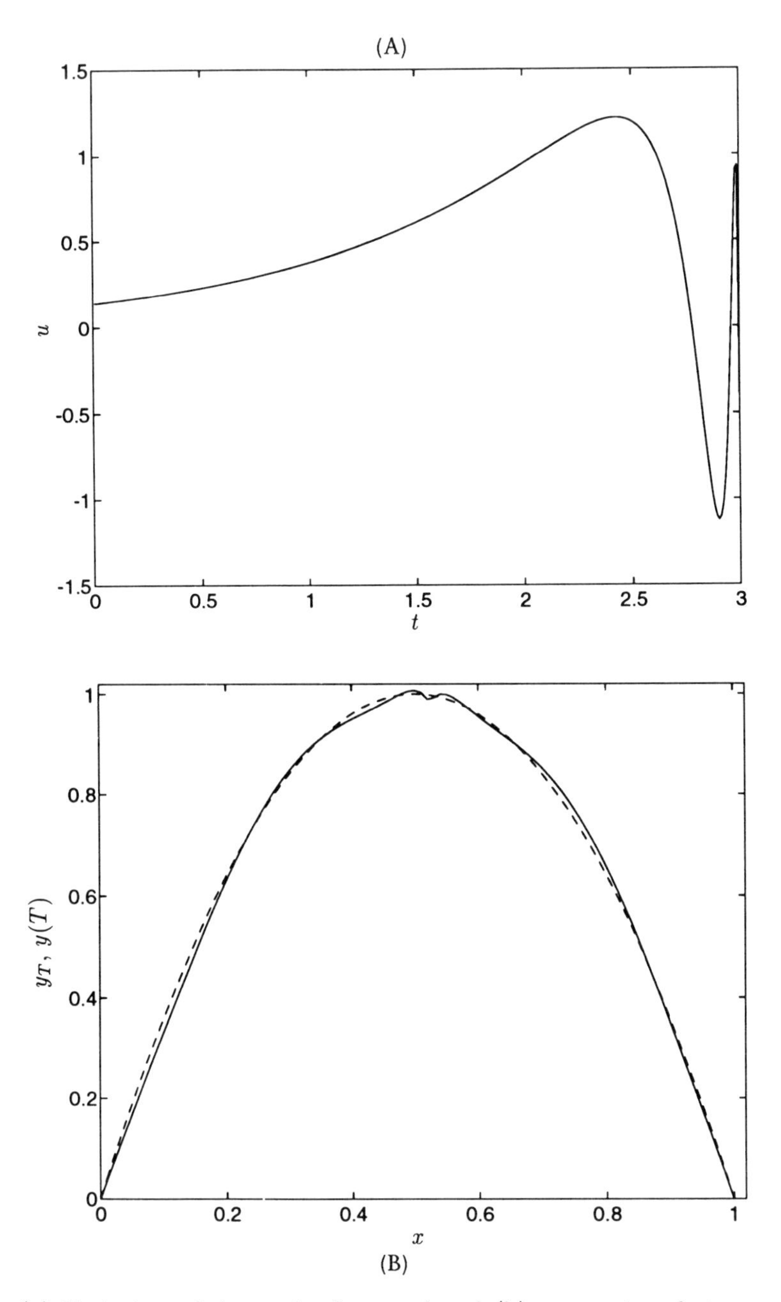

Fig. 3. (a) Variation of the optimal control and (b) comparison between y_T and $y^*(T)$ (target function (1.461): $T = 3$, $b = \pi/6$, $k = 10^4$, $h = \Delta t = 10^{-2}$).

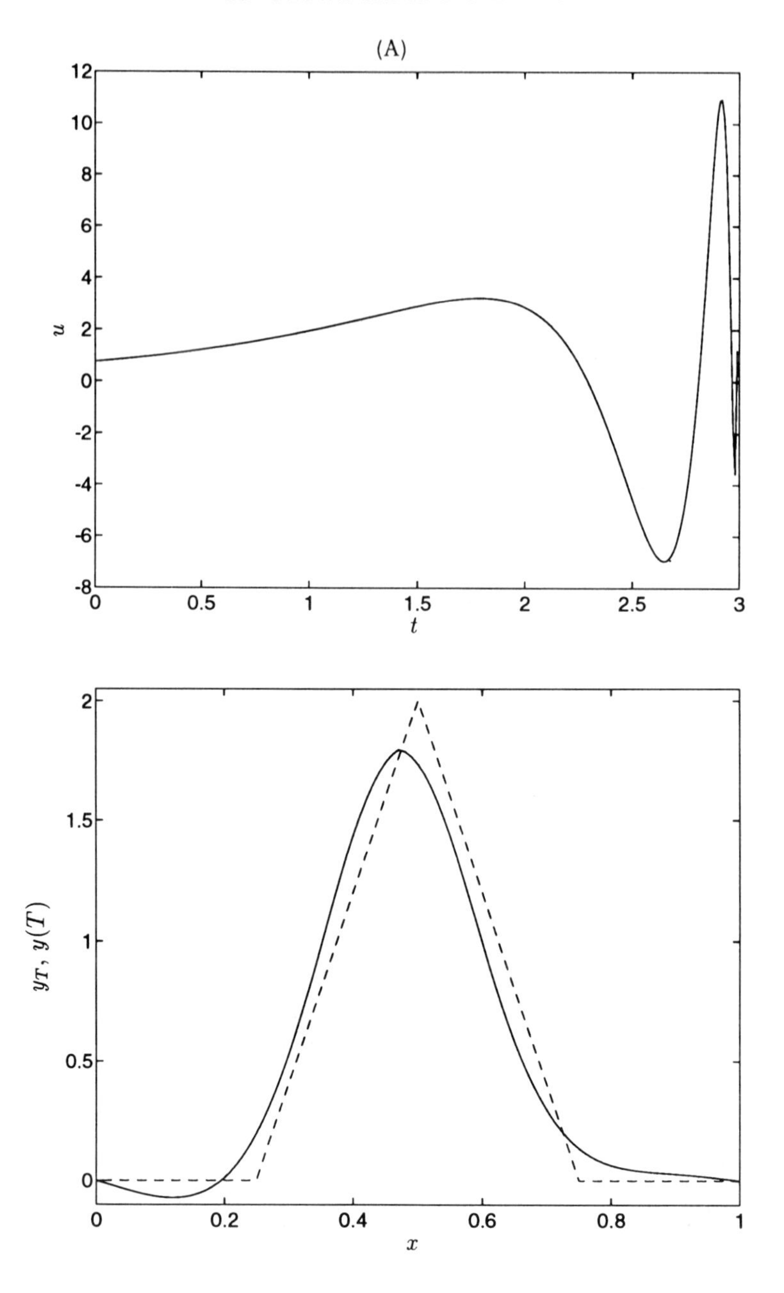

Fig. 4. (a) Variation of the optimal control and (b) comparison between y_T and $y^*(T)$ (target function (1.462): $T = 3$, $b = \sqrt{2}/3$, $k = 10^4$, $h = \Delta t = 10^{-2}$).

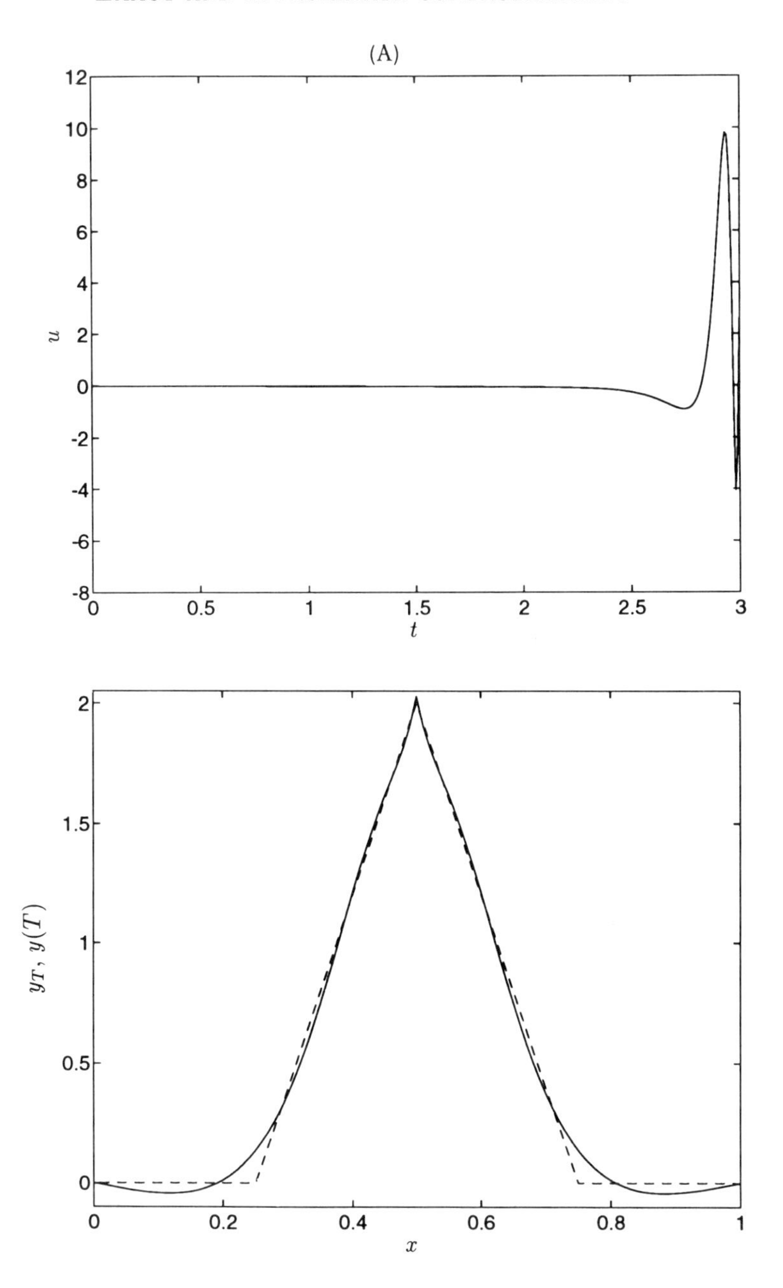

Fig. 5. (a) Variation of the optimal control and (b) comparison between y_T and $y^*(T)$ (target function (1.462): $T = 3$, $b = 1/2$, $k = 10^4$, $h = \Delta t = 10^{-2}$).

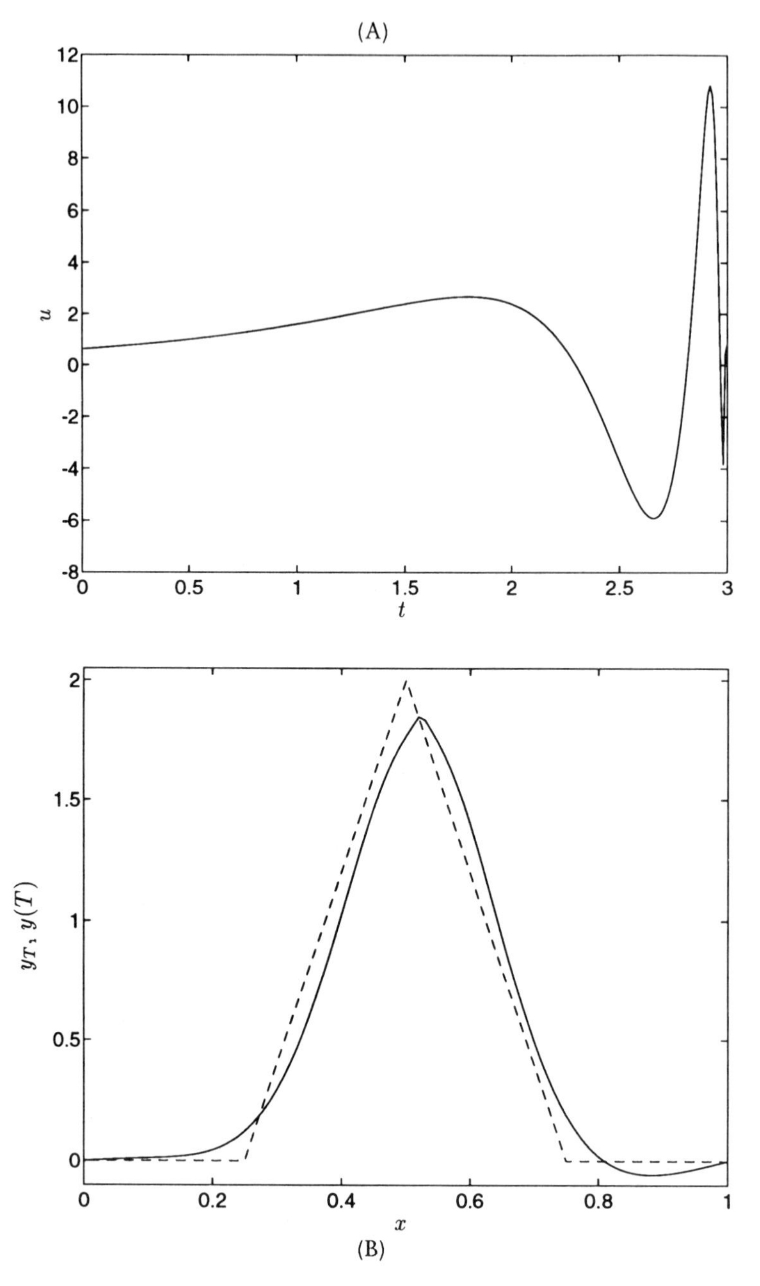

Fig. 6. (a) Variation of the optimal control and (b) comparison between y_T and $y^*(T)$ (target function (1.462): $T = 3$, $b = \pi/6$, $k = 10^4$, $h = \Delta t = 10^{-2}$).

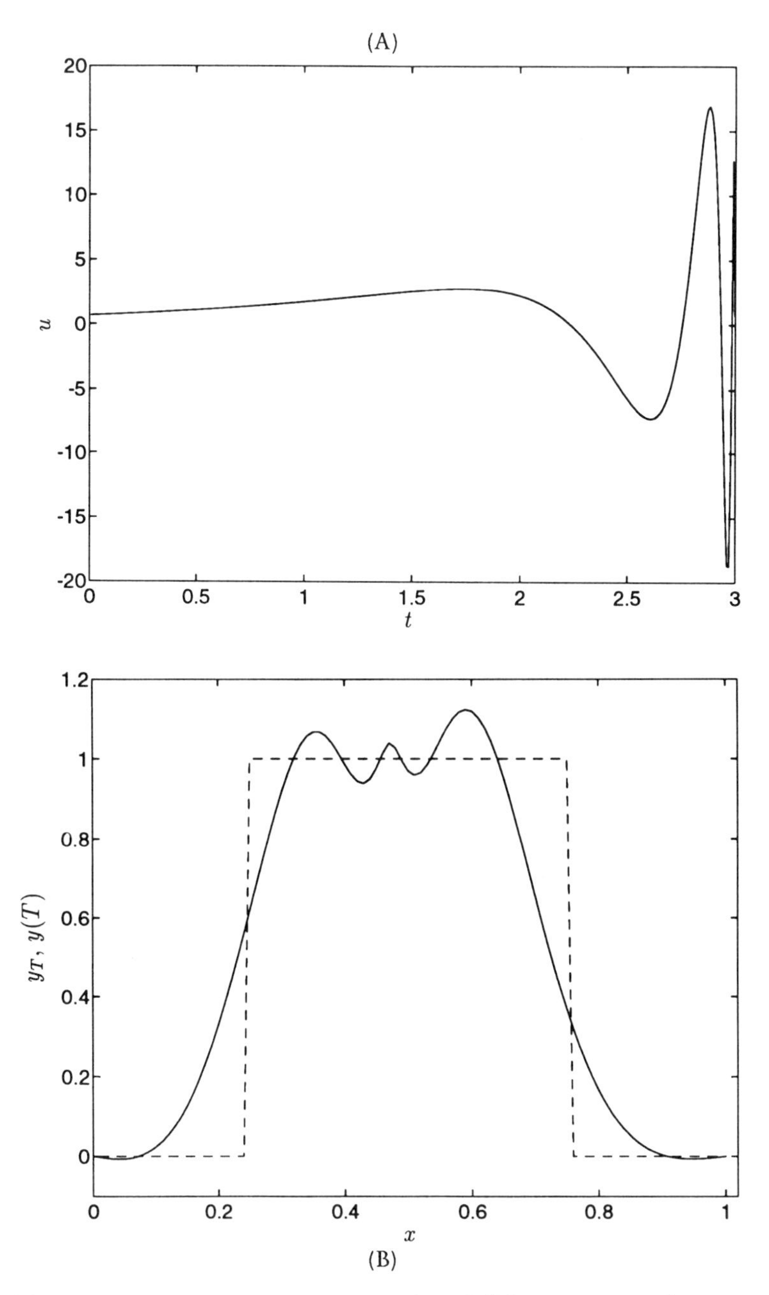

Fig. 7. (a) Variation of the optimal control and (b) comparison between y_T and $y^*(T)$ (target function (1.463): $T = 3$, $b = \sqrt{2}/3$, $k = 10^4$, $h = \Delta t = 10^{-2}$.)

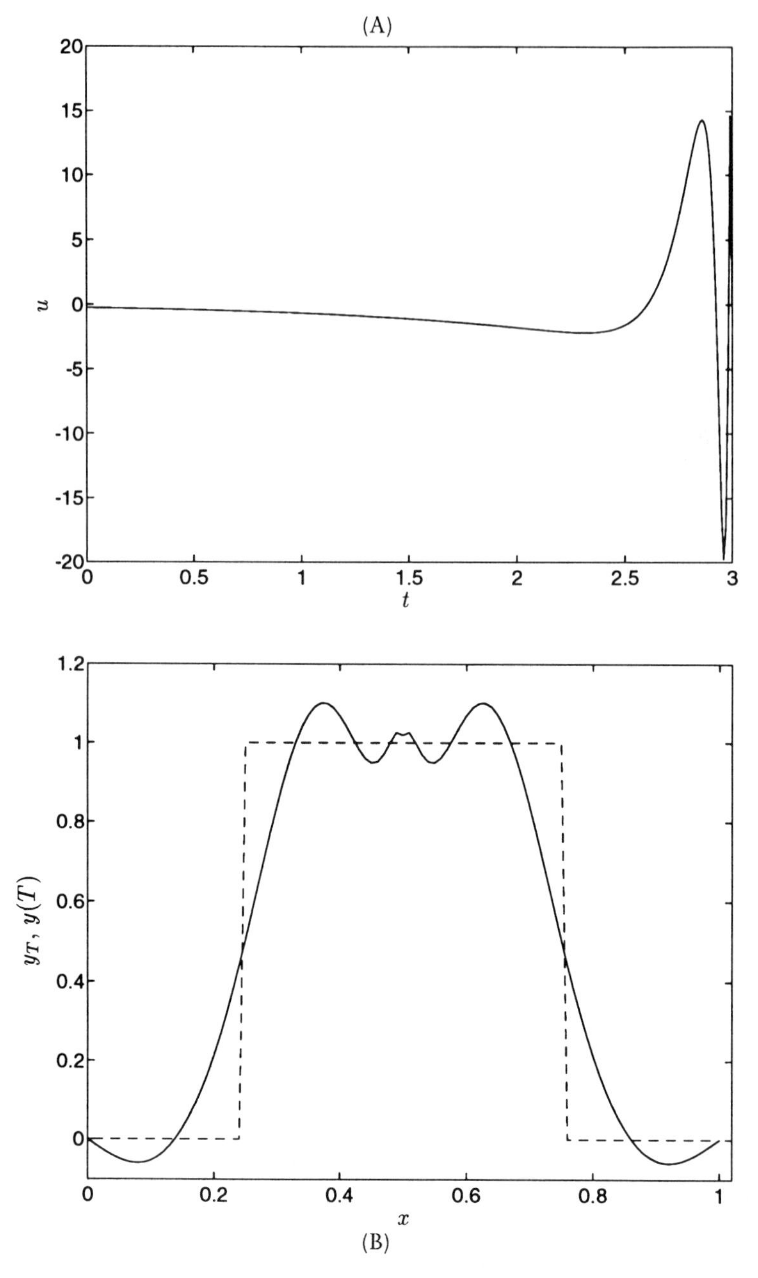

Fig. 8. Variation of optimal control and (b) comparison between y_T and $y^*(T)$ (target function (1.463): $T = 3$, $b = 1/2$, $k = 10^4$, $h = \Delta t = 10^{-2}$).

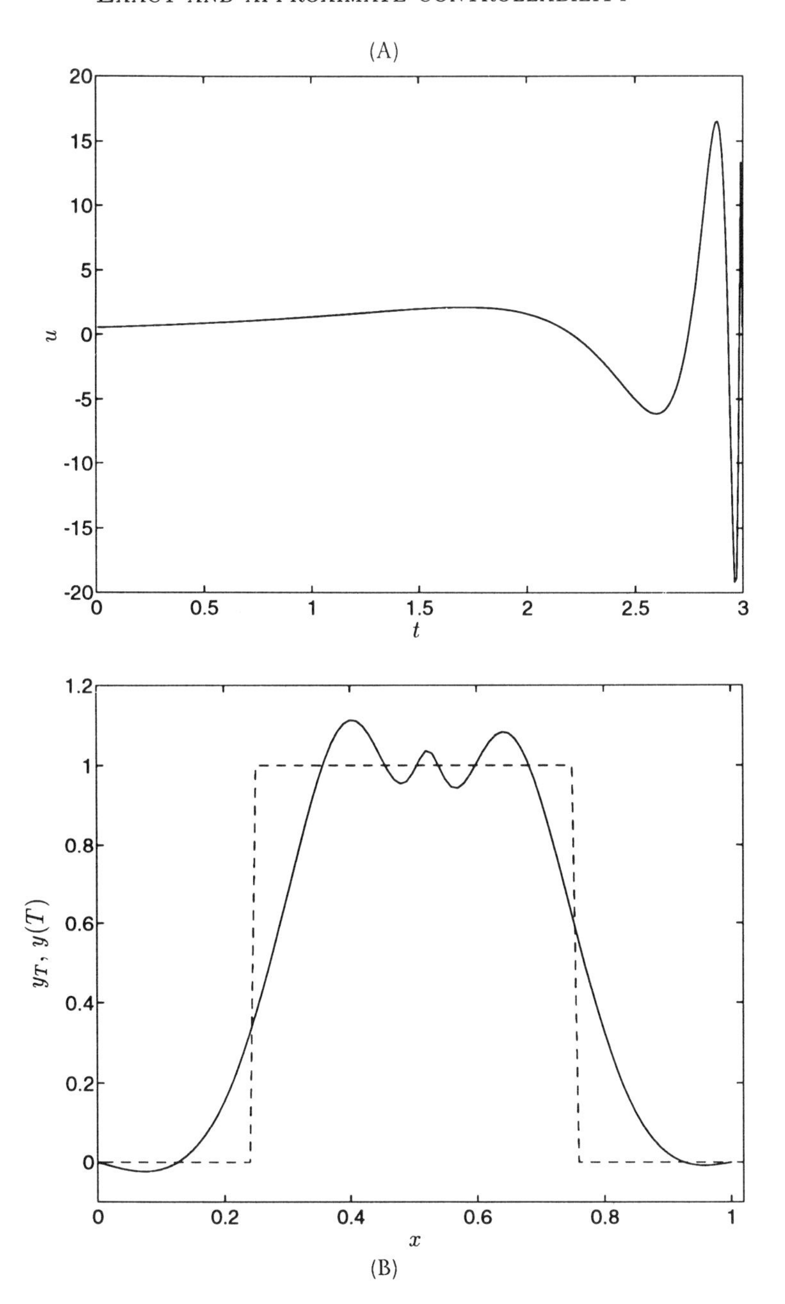

Fig. 9. Variation of the optimal control and (b) comparison between y_T and $y^*(T)$ (target function (1.463): $T = 3$, $b = \pi/6$, $k = 10^4$, $h = \Delta t = 10^{-2}$).

where, in (1.464), $\mathbb{Q}$ is the field of the *rational* real numbers. Clearly, $b = \sqrt{2}/3$ and $b = \pi/6$ being nonrational are strategic. On the other hand, $b = \frac{1}{2}$ is far from being strategic since $\sin j\pi/2 = 0$ for any *even integer* j; indeed, $b = \frac{1}{2}$ is *generically* the worst choice which can be made. However, if one takes $b = \frac{1}{2}$ the solution y of problem (1.458)–(1.460) satisfies

$$\forall t \in [0, T], y(t) \text{ is an even function of } x - \tfrac{1}{2}; \tag{1.465}$$

property (1.465) implies that the coefficients of w_j in the Fourier expansion of y are zero for j even. This property implies in turn that $b = \frac{1}{2}$ is strategic if y_T is also an even function of $x - \frac{1}{2}$; this is precisely the case for the target functions defined by (1.461)–(1.463). Actually, for target functions y_T which are even with respect to $x - \frac{1}{2}$, $b = \frac{1}{2}$ is the best strategic point; this appears clearly in Tables 1 to 3 where the smallest *control norms* and *controllability errors* are obtained for $b = \frac{1}{2}$. In Section 1.10.7.3 we shall consider target functions which are not even with respect to $x - \frac{1}{2}$; $b = \frac{1}{2}$ will not be strategic at all for these test problems.

(ii) A *digital computer* 'knows' only rational numbers; this means that for the particular test problems considered in this section, strictly speaking, there is no strategic point for pointwise control. However, if b is the computer approximation of a nonrational number, the integers j such that bj are also integers are *very large*. This last property implies that, unless h is extremely small and y_T quite pathological (i.e. its Fourier coefficients do not converge quickly to zero as $j \to +\infty$), such a b is strategic in practice.

(iii) The discrete control problems approximating (1.456) are equivalent to linear systems associated with an $N \times N$ symmetric and positive-definite matrix (we recall that $N = T/\Delta t$). These problems can be solved, therefore, by conjugate gradient algorithms. From the classical properties of conjugate gradient methods (see, e.g., Ciarlet (1989), Golub and Van Loan (1989)), we expect convergence in N iterations at most. Looking at Tables 1 to 3 we observe that for k sufficiently large this finite termination property does not hold. The main reason for this behaviour is that these discrete control problems are *badly conditioned* for large values of k, implying high sensitivity to round-off errors and, consequently, loss of the finite termination property.

An alternative to conjugate gradient methods is to construct the matrix and right-hand side of the equivalent linear system and to solve it by Cholesky's method. Let us briefly evaluate the cost of constructing the matrix and the right-hand side of this linear system. It follows from Section 1.10.6.3 that to construct the matrix (respectively the right-hand side) we need to solve N (respectively 1) discrete forward parabolic problems and then N (respectively 1) discrete backward parabolic problems, implying a total of $2(N + 1)$ parabolic problems. If one modifies y_T, with everything else staying the same, we only have to compute the corresponding new right-hand side at the cost of solving two discrete parabolic problems.

Table 4. *Summary of numerical results (target function defined by (1.466); $T = 3$, $h = \Delta t = 10^{-2}$).*

b	k	$\|u^*\|_{L^2(0,T)}$	$\frac{\|y^*(T)-y_T\|_{L^2(0,T)}}{\|y_T\|_{L^2(0,T)}}$
$\sqrt{2}/3$	10^4	42.4	1.5×10^{-1}
	10^5	73.6	4×10^{-2}
$1/2$	10^4	4.4	3.5×10^{-1}
	10^5	5.46	3.5×10^{-1}
$\pi/6$	10^4	40.4	1.8×10^{-1}
	10^5	84.5	5.6×10^{-2}

1.10.7.3. Further test problems The test problems in this section are still defined by (1.456)–(1.460) the main difference being that the target functions y_T are not even with respect to the variable $x - \frac{1}{2}$. Indeed, the two target functions considered here are defined by

$$y_T(x) = \tfrac{27}{4}x^2(1-x), \tag{1.466}$$

and

$$y_T(x) = \begin{cases} 0 & \text{on } [0, \frac{1}{2}], \\ 8(x - \frac{1}{2}) & \text{on } [\frac{1}{2}, \frac{3}{4}], \\ 8(1-x) & \text{on } [\frac{3}{4}, 1]; \end{cases} \tag{1.467}$$

we have taken $T = 3$ for both target functions. The approximation and solution methods being those of Section 1.10.7.2, still with $\nu = \frac{1}{10}$, $h = \Delta t = 10^{-2}$, we have obtained the results summarized in the Tables 4 and 5 and Figures 10–15 below (the notation is the same as in Section 1.10.7.2).

These results clearly show that $b = \frac{1}{2}$ is not strategic for the test problems considered here; this was expected since none of the functions y_T is even with respect to $x - \frac{1}{2}$. On the other hand, 'small' *irrational* shifts, either to the right or to the left of $\frac{1}{2}$, produce strategic values of b. The other comments made in Section 1.10.7.2 still hold for the examples considered here.

1.10.7.4. Test problems for nonquadratic cost functions. Motivated by Section 1.9 we have been considering pointwise control problems defined by

$$\min_{v \in L^s(0,T)} [\tfrac{1}{2}\|v\|^2_{L^s(0,T)} + \tfrac{1}{2}k\|y(T) - y_T\|^2_{L^2(0,1)}], \tag{1.468}$$

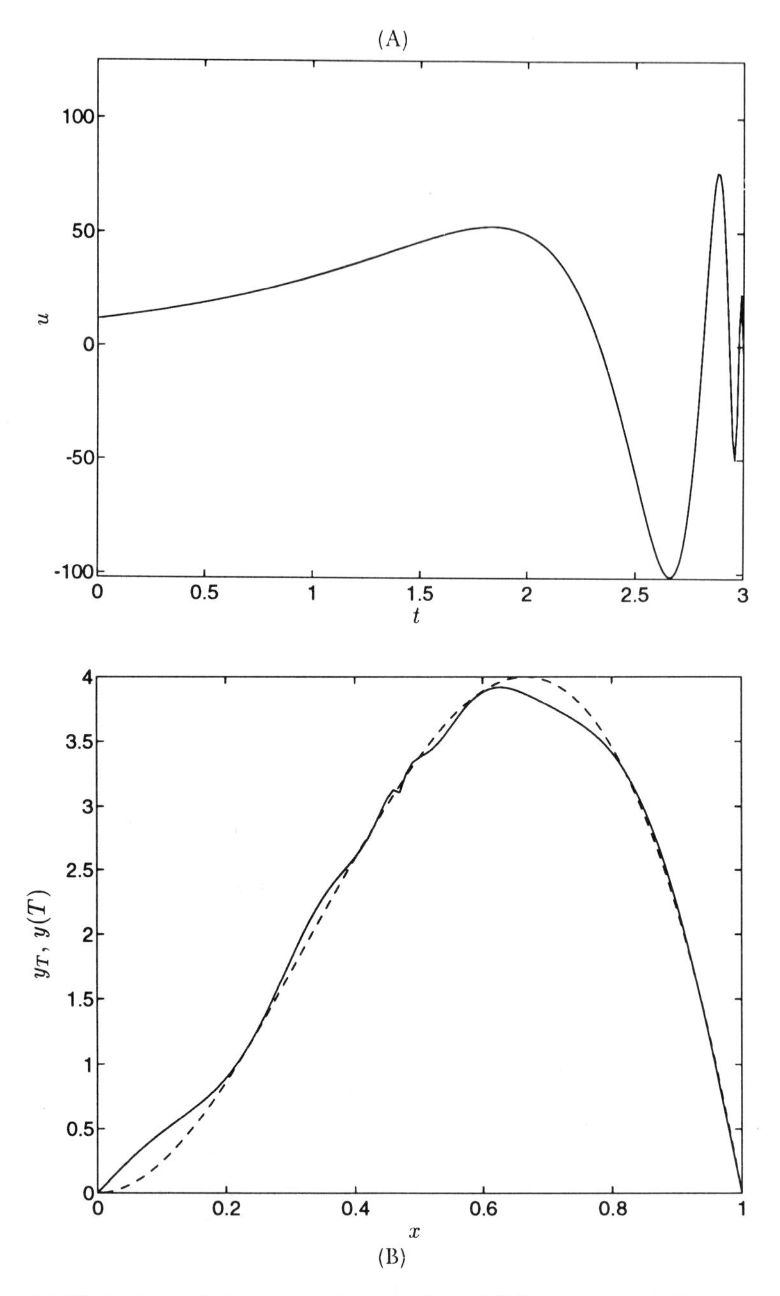

Fig. 10. (a) Variation of the optimal control and (b) comparison between y_T and $y^*(T)$ (target function (1.466): $T = 3$, $b = \sqrt{2}/3$, $k = 10^5$, $h = \Delta t = 10^{-2}$).

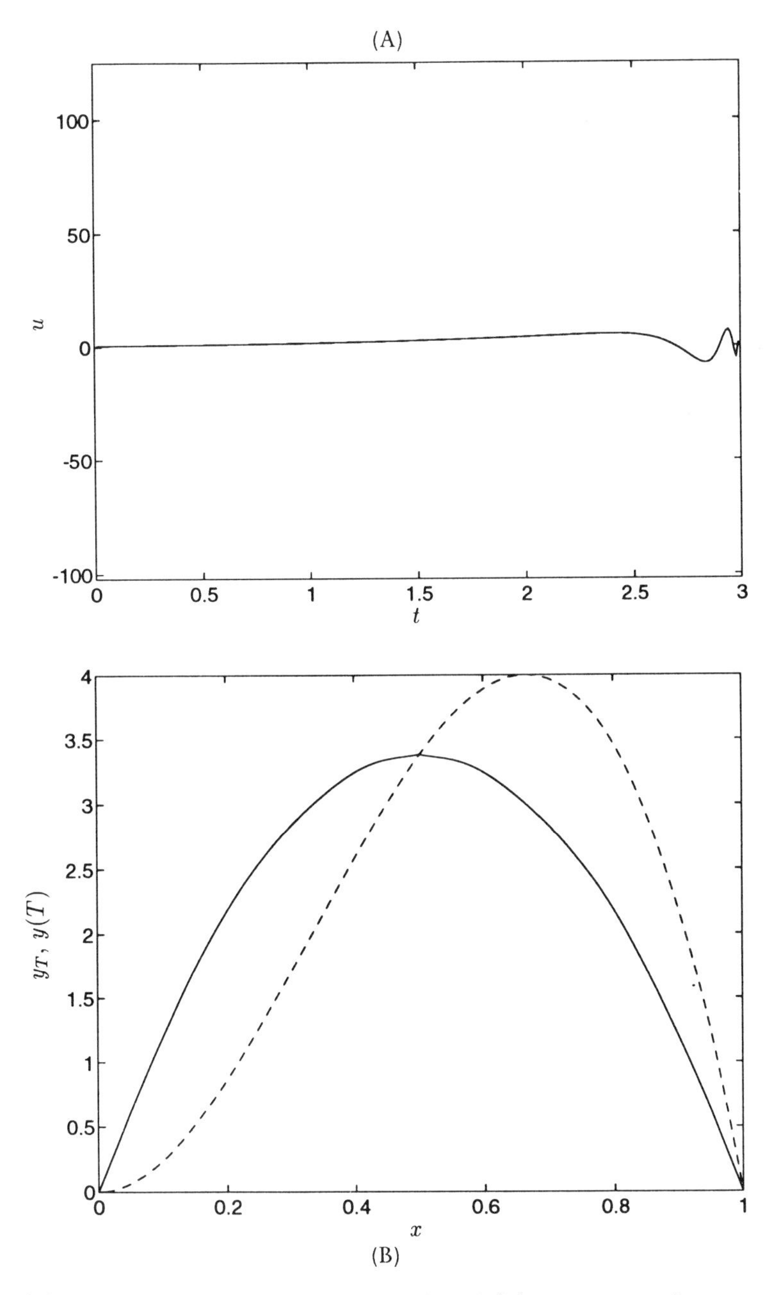

Fig. 11. (a) Variation of the optimal control and (b) comparison between y_T and $y^*(T)$ (target function (1.466): $T = 3$, $b = \frac{1}{2}$, $k = 10^5$, $h = \Delta t = 10^{-2}$).

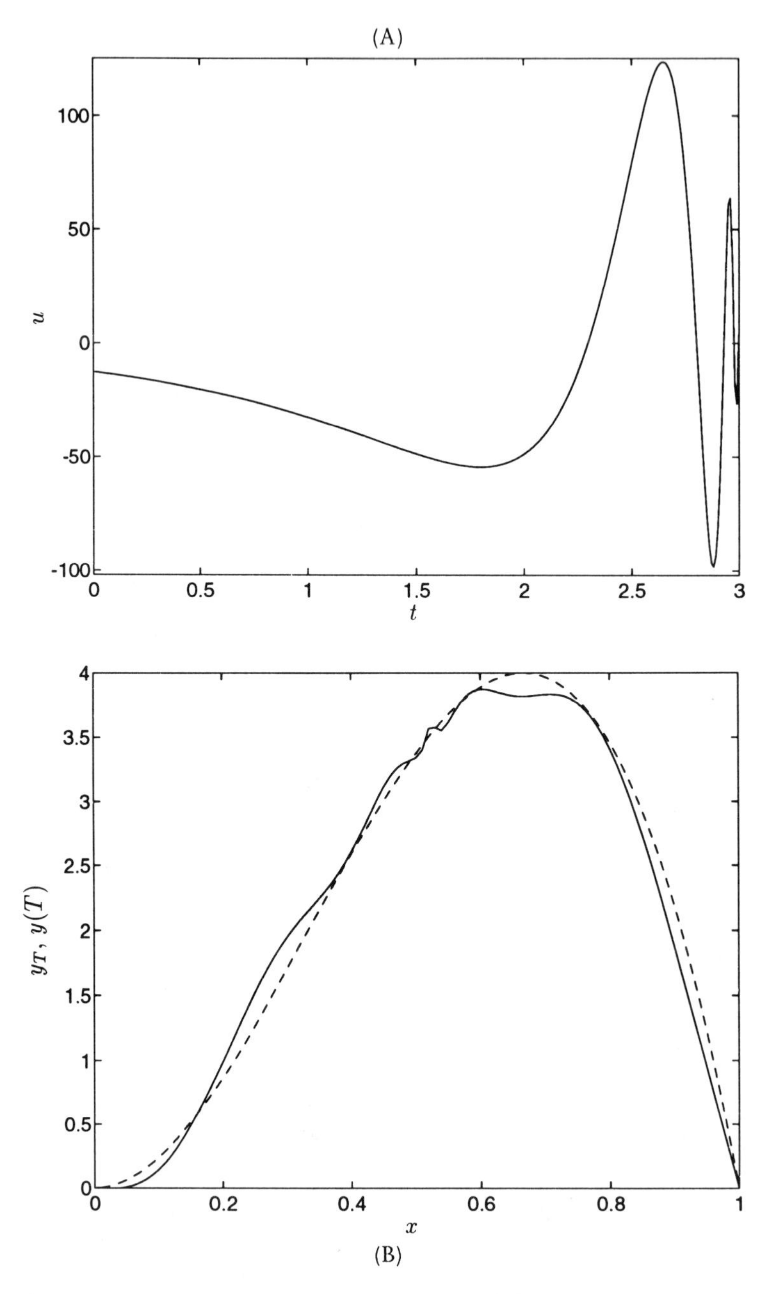

Fig. 12. (a) Variation of the optimal control and (b) comparison between y_T and $y^*(T)$ (target function (1.466); $T = 3$, $b = \pi/6$, $k = 10^5$, $h = \Delta t = 10^{-2}$).

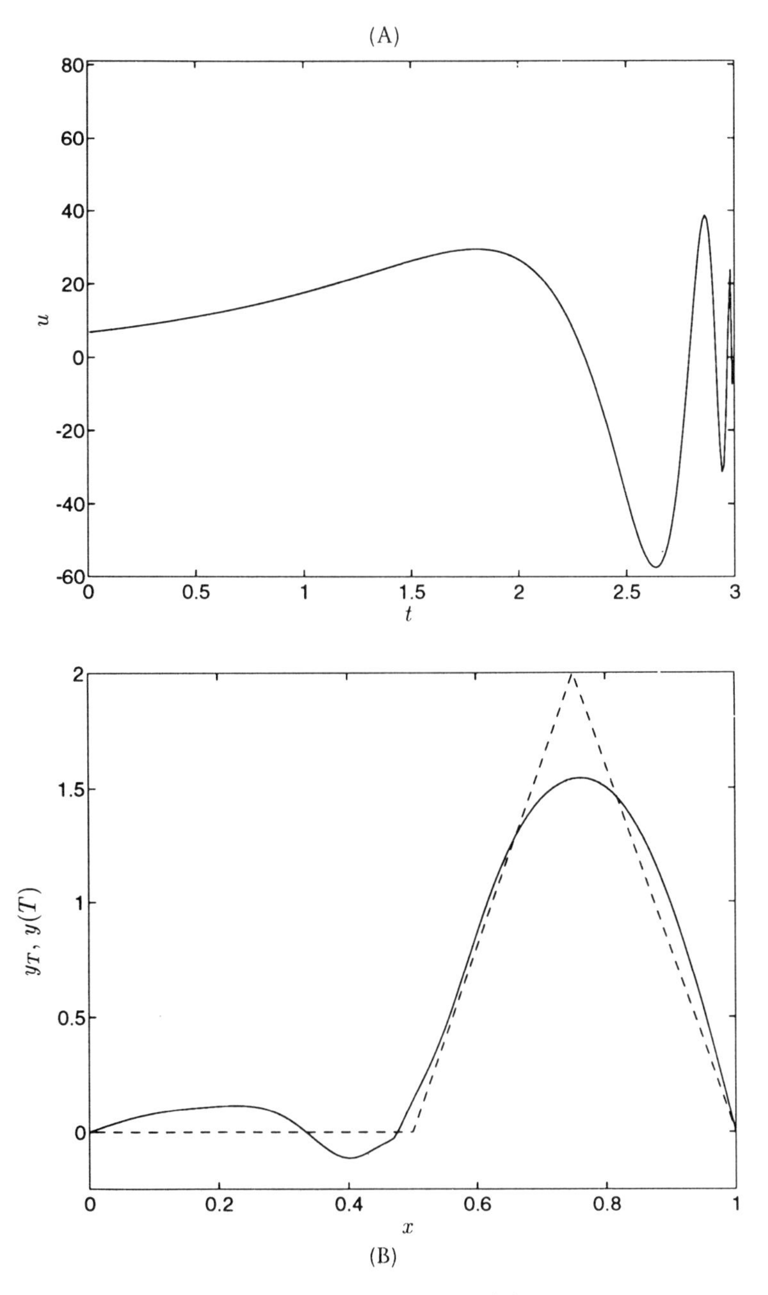

Fig. 13. (a) Variation of the optimal control and (b) comparison between y_T and $y^*(T)$ (target function (1.467): $T = 3$, $b = \sqrt{2}/3$, $k = 10^5$, $h = \Delta t = 10^{-2}$).

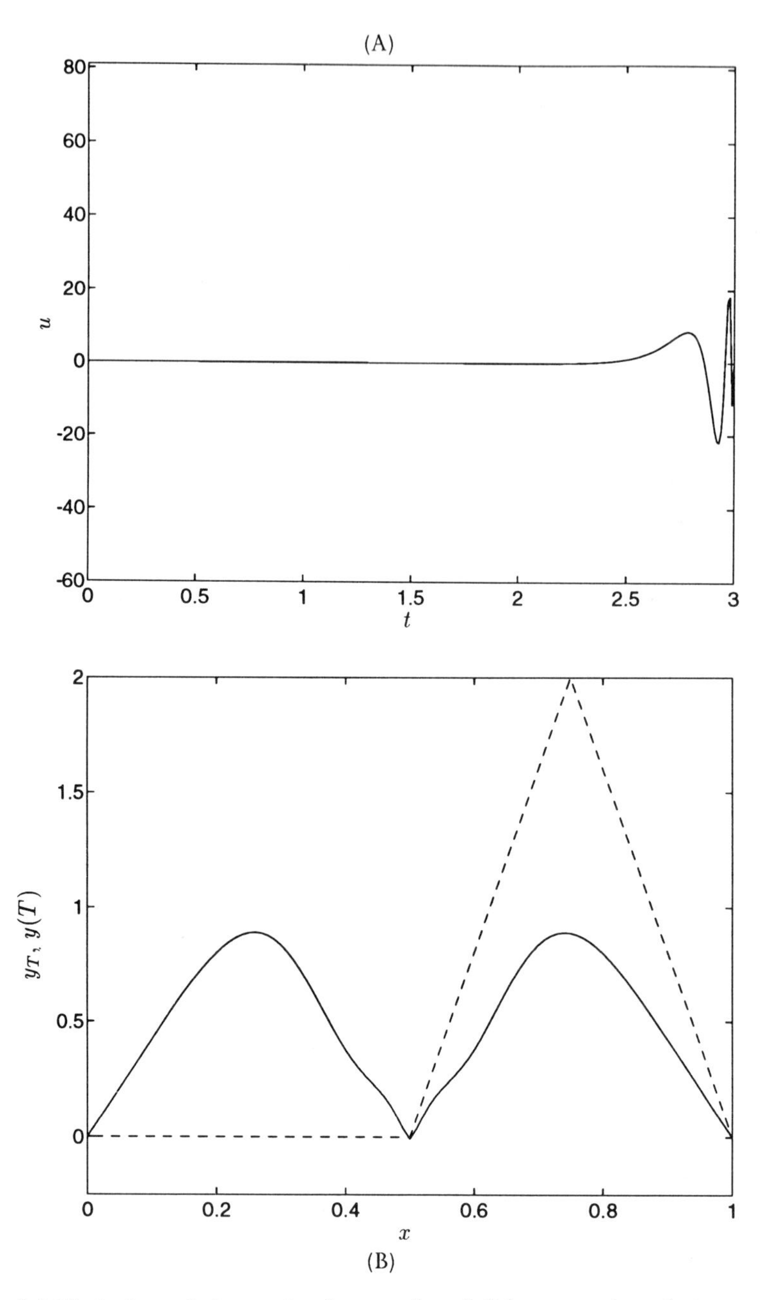

Fig. 14. (a) Variation of the optimal control and (b) comparison between y_T and $y^*(T)$ (target function (1.467): $T = 3$, $b = \frac{1}{2}$, $k = 10^5$, $h = \Delta t = 10^{-2}$).

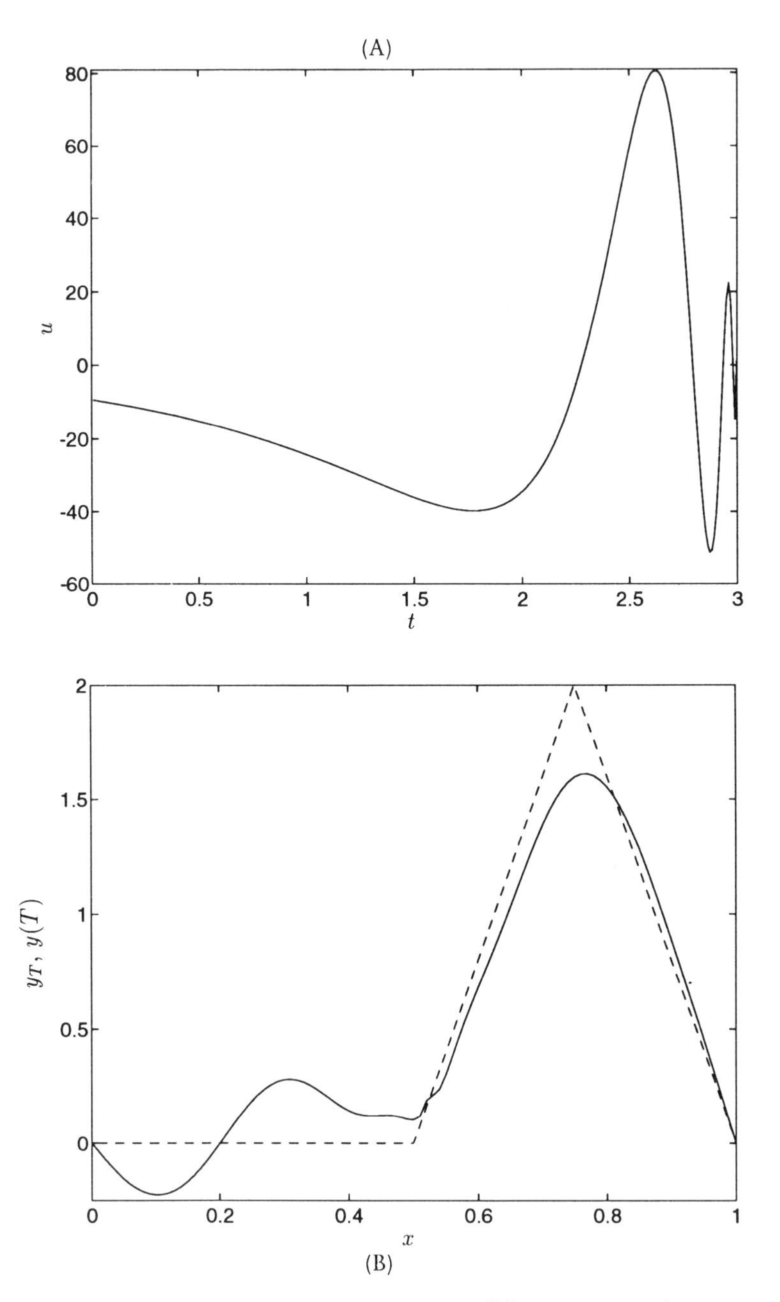

Fig. 15. (a) Variation of the optimal control and (b) comparison between y_T and $y^*(T)$ (target function (1.467): $T = 3$, $b = \pi/6$, $k = 10^5$, $h = \Delta t = 10^{-2}$).

Table 5. *Summary of numerical results (target function defined by (1.467); $T = 3,\ h = \Delta t = 10^{-2}$).*

b	k	$\|u^*\|_{L^2(0,T)}$	$\frac{\|y^*(T)-y_T\|_{L^2(0,T)}}{\|y_T\|_{L^2(0,T)}}$
$\sqrt{2}/3$	10^3	6.53	5.2×10^{-1}
	10^4	21.8	3×10^{-1}
	10^5	41.6	1.6×10^{-1}
$1/2$	10^3	2.05	7.1×10^{-1}
	10^4	2.67	7.1×10^{-1}
	10^5	6.26	7.1×10^{-1}
$\pi/6$	10^3	5.99	6.7×10^{-1}
	10^4	27.5	4.2×10^{-1}
	10^5	57.6	2×10^{-1}

with y still defined from v by (1.458)–(1.460), and s 'large'. It seems, unfortunately, that for $s > 2$, problem (1.468) is poorly conditioned implying that the various iterative methods we used to solve it (conjugate gradient, Newton and quasi-Newton methods) have failed to converge (or even worse, have stuck on some wrong solution). From these facts it is quite natural to consider the variation of problem (1.468) defined by

$$\min_{v \in L^s(0,T)} \left[\frac{1}{s} \int_0^T |v(t)|^s \, dt + \tfrac{1}{2} k \|y(T) - y_T\|^2_{L^2(0,1)} \right], \tag{1.469}$$

with y defined from v as above. The cost function in (1.469) has better differentiability properties than the one in (1.468).

Let us denote by u the solution of (1.469); assuming that b in (1.458) is strategic we can expect that for s fixed $y(u;T)$ will get closer to y_T as k increases. If, on the other hand, k is fixed we can expect the distance from $y(u;T)$ to y_T to increase with s, since in that case the relative importance of the term $s^{-1} \int_0^T |v|^s \, dt$ in the cost function increases with s. These predictions are fully confirmed by the numerical experiments whose results are shown below. For these experiments we have used essentially the same approximation methods as in Sections 1.10.7.2 and 1.10.7.3, with $h = \Delta t = 10^{-2}$, and taken $b = \sqrt{2}/3$, $T = 3$, $\nu = \frac{1}{10}$ and y_T defined by (1.466). The discrete control problems have been solved by *quasi-Newton's methods* à la BFGS, like those discussed, for example, in the classical text book by Dennis and Schnabel (1983) (see also Nocedal (1992)); for the problems considered here, these methods appear to be much more efficient than conjugate gradient methods.

On Figures 16 to 21 we have – for $k = 10^7$ and $s = 2, 4, 6, 10, 20, 30$ – visualized the computed optimal control u^* and compared the corresponding final state $y^*(T)$ with the target y_T. From these figures, we clearly see that the distance of $y^*(T)$ to y_T increases with s; we also see the *bang–bang* character of the optimal control for large values of s.

Finally, on Figures 22 to 25 we have shown some of the results obtained for large values of s and very large values of k; comparing these with Figures 16 to 21 we observe that if for a given s we increase k, then $y^*(T)$ gets closer to $y(T)$ and $\|u^*\|_{L^2(0,T)}$ increases, which makes sense. We observe again that for very large values of s the optimal control is very close to bang–bang.

Acknowledgments

The authors would like to acknowledge the help of the following individuals for friendly discussions and/or collaboration: A. Bamberger, C. Bardos, A. Bensoussan, M. Berggren, M.O. Bristeau, C. Carthel, E.J. Dean, C. Fabre, V. Fraysse, P. Grisvard, P. Joly, T. Lachand-Robert, J. Lagnese, I. Lasiecka, G. Lebeau, C.H. Li, J. Periaux, J.-P. Puel, J. Rauch, D.L. Russel, H. Steve, R. Triggiani, M.F. Wheeler, E. Zuazua.

The support of the following corporations or institutions is also acknowledged: AWARE, CERFACS, Collège de France, CRAY Research, Dassault Industries, INRIA, LCC-Rio de Janeiro, Rice University, University of Colorado, University of Houston, Université Pierre et Marie Curie. We also benefited from the support of DARPA (Contracts AFOSR F49620-89-C-0125 and AFOSR-90-0334), DRET, NSF (Grant INT 8612680), Texas Board of Higher Education (Grant ARP 003652156) and the Faculty Development Program, University of Houston (special thanks are due to Glenn Aumann, John Bear and Garret J. Etgen).

Finally, special thanks are due to the editors of *Acta Numerica* for suggesting to us that we write this article, and to D. Bidois, C. Demars, L. Ruprecht and J. Wilson for processing a preliminary version of it.

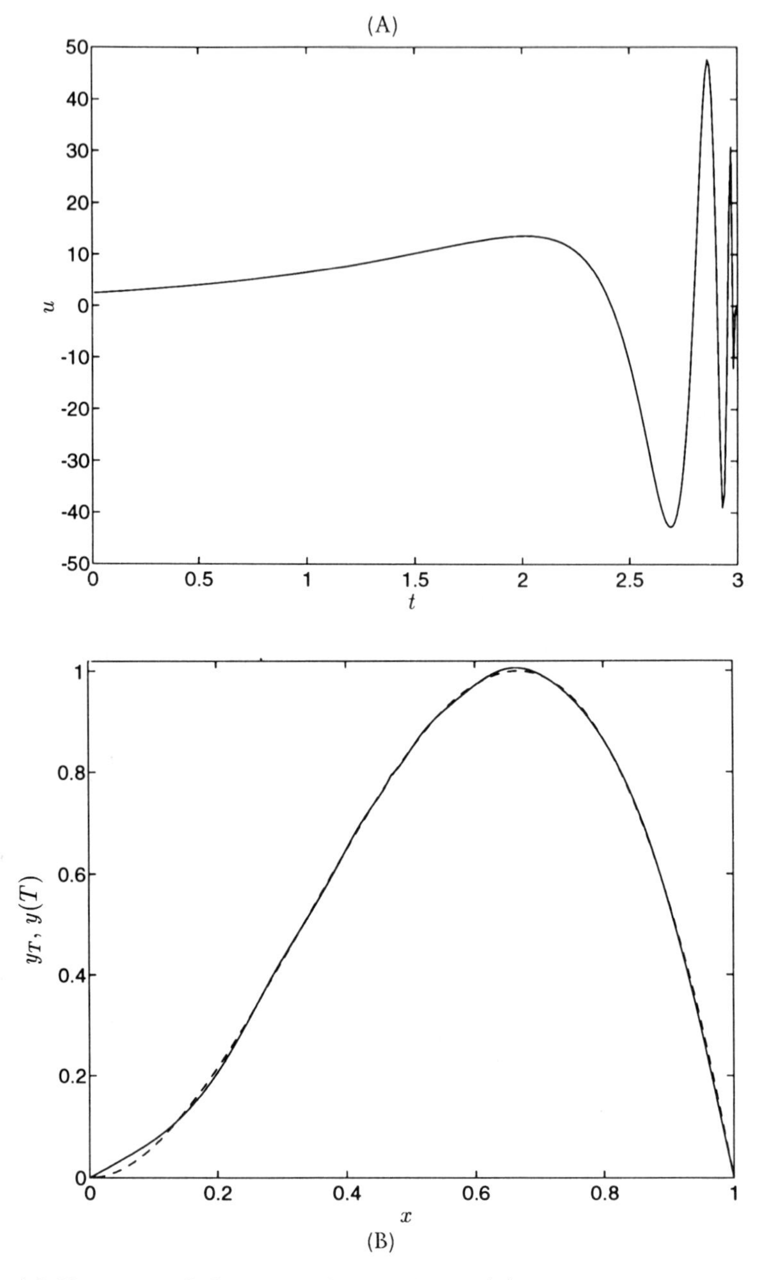

Fig. 16. (a) Variation of the optimal control and (b) comparison between y_T and $y^*(T)$ (target function (1.466): $s = 2$, $T = 3$, $b = \sqrt{2}/3$, $k = 10^7$, $h = \Delta t = 10^{-2}$).

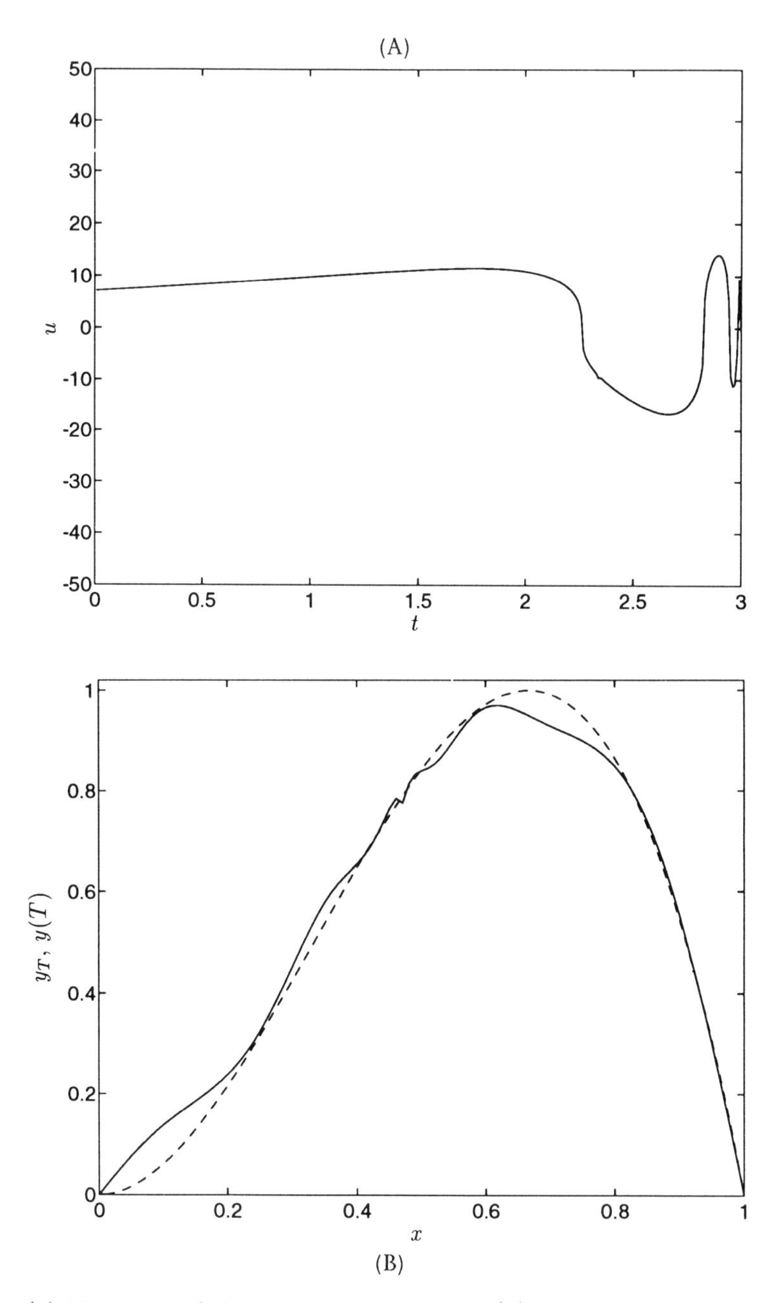

Fig. 17. (a) Variation of the optimal control and (b) comparison between y_T and $y^*(T)$ (target function (1.466): $s = 4$, $T = 3$, $b = \sqrt{2}/3$, $k = 10^7$, $h = \Delta t = 10^{-2}$).

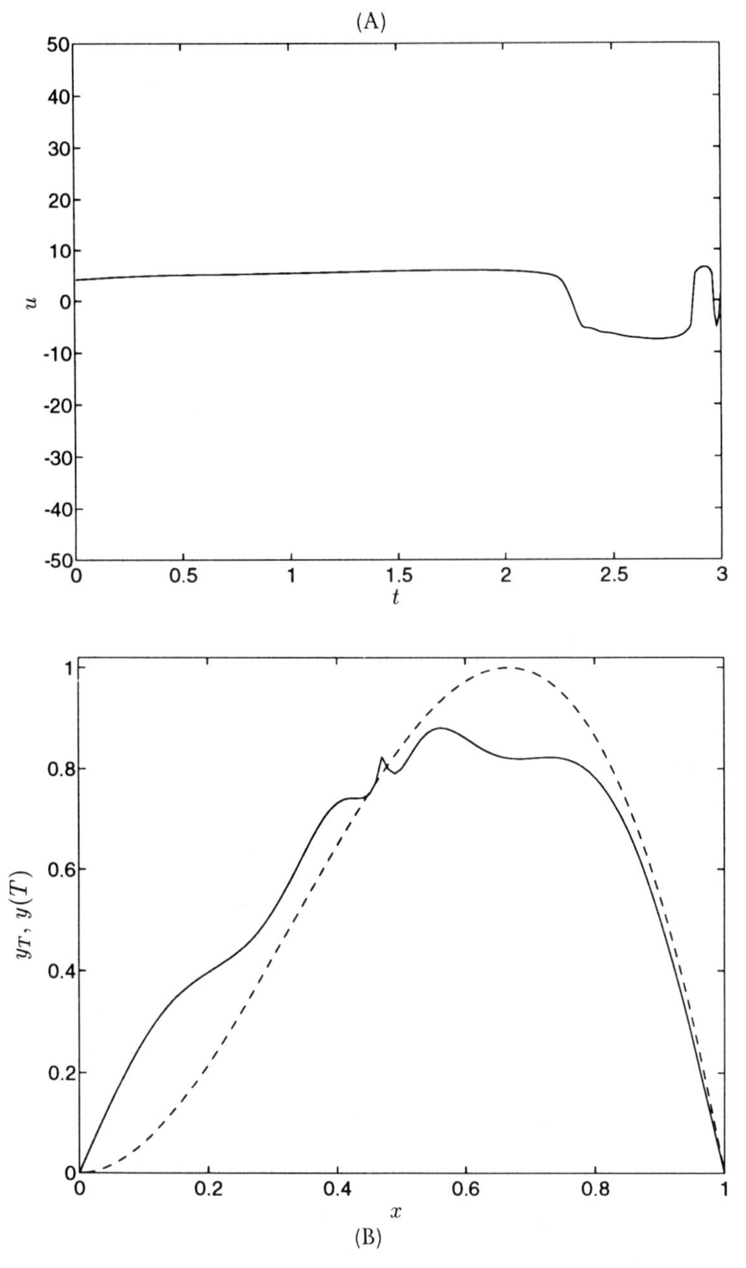

Fig. 18. (a) Variation of the optimal control and (b) comparison between y_T and $y^*(T)$ (target function (1.466): $s = 6$, $T = 3$, $b = \sqrt{2}/3$, $k = 10^7$, $h = \Delta t = 10^{-2}$).

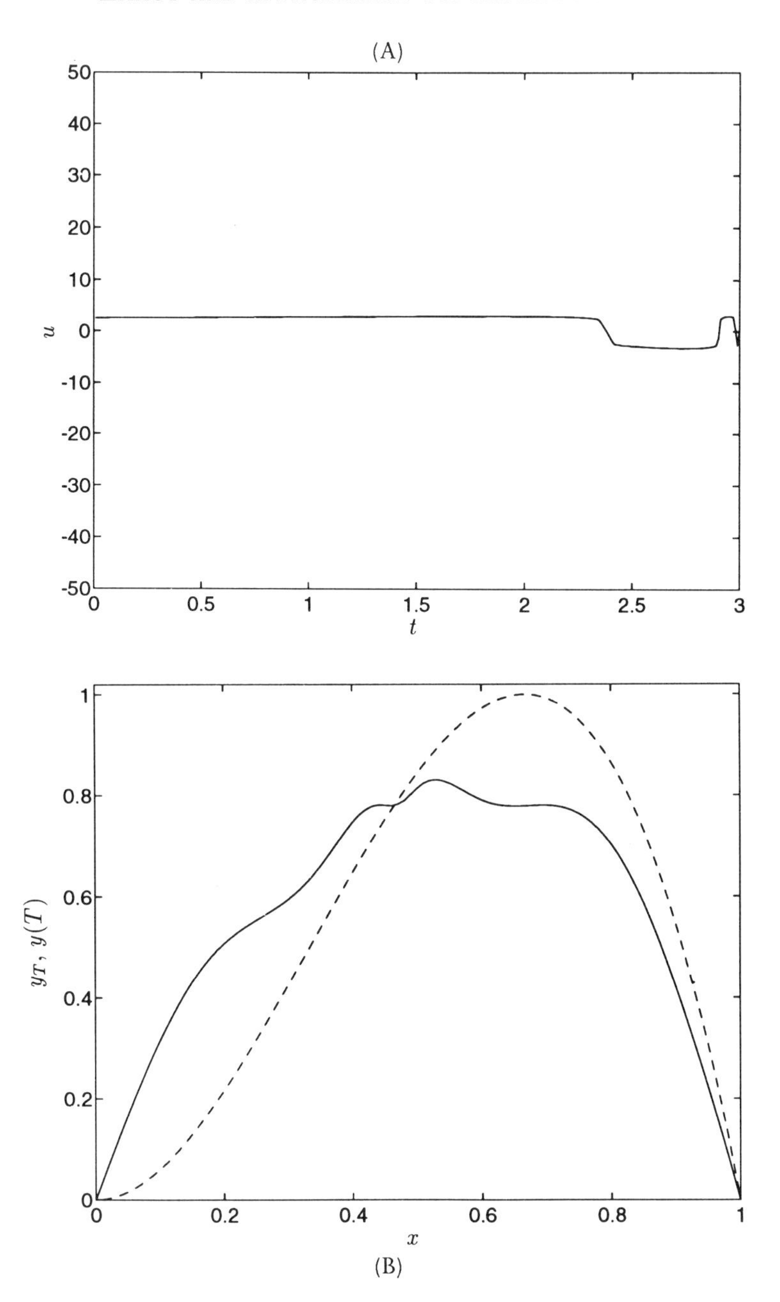

Fig. 19. (a) Variation of the optimal control and (b) comparison between y_T and $y^*(T)$ (target function (1.466): $s = 10$, $T = 3$, $b = \sqrt{2}/3$, $k = 10^7$, $h = \Delta t = 10^{-2}$).

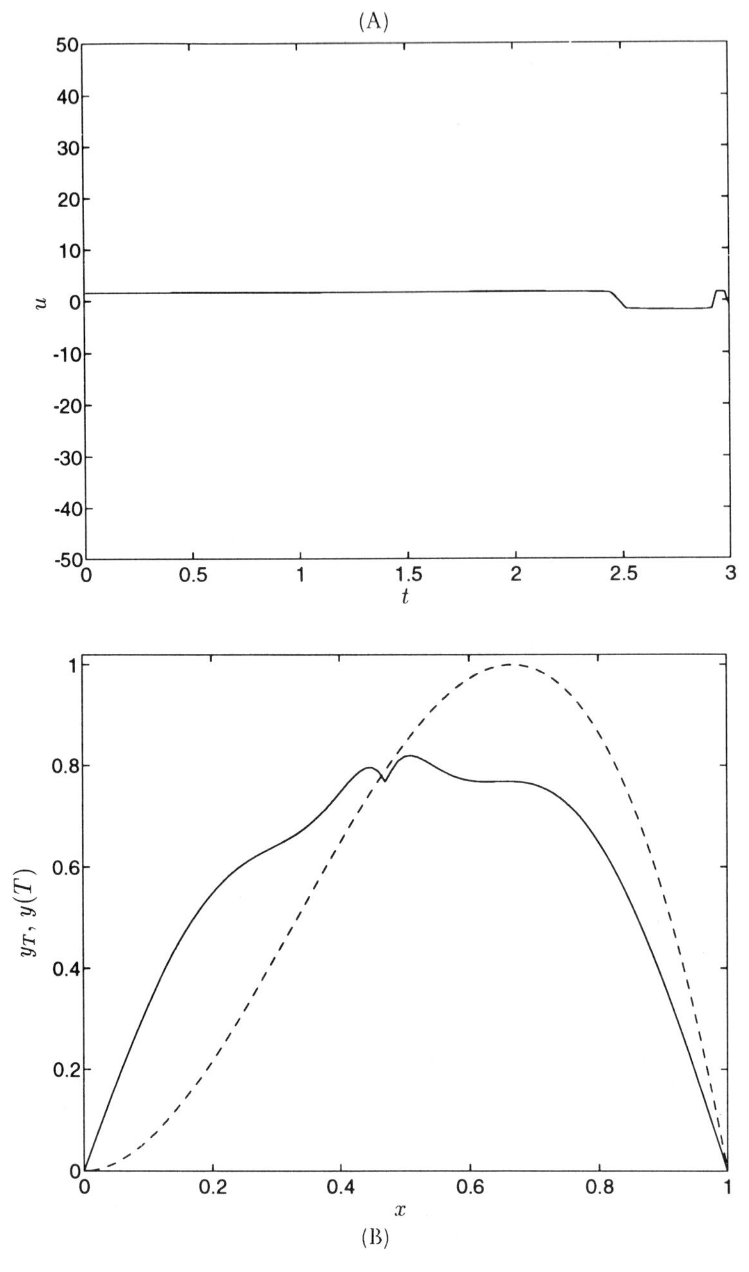

Fig. 20. (a) Variation of the optimal control and (b) comparison between y_T and $y^*(T)$ (target function (1.466): $s = 20$, $T = 3$, $b = \sqrt{2}/3$, $k = 10^7$, $h = \Delta t = 10^{-2}$).

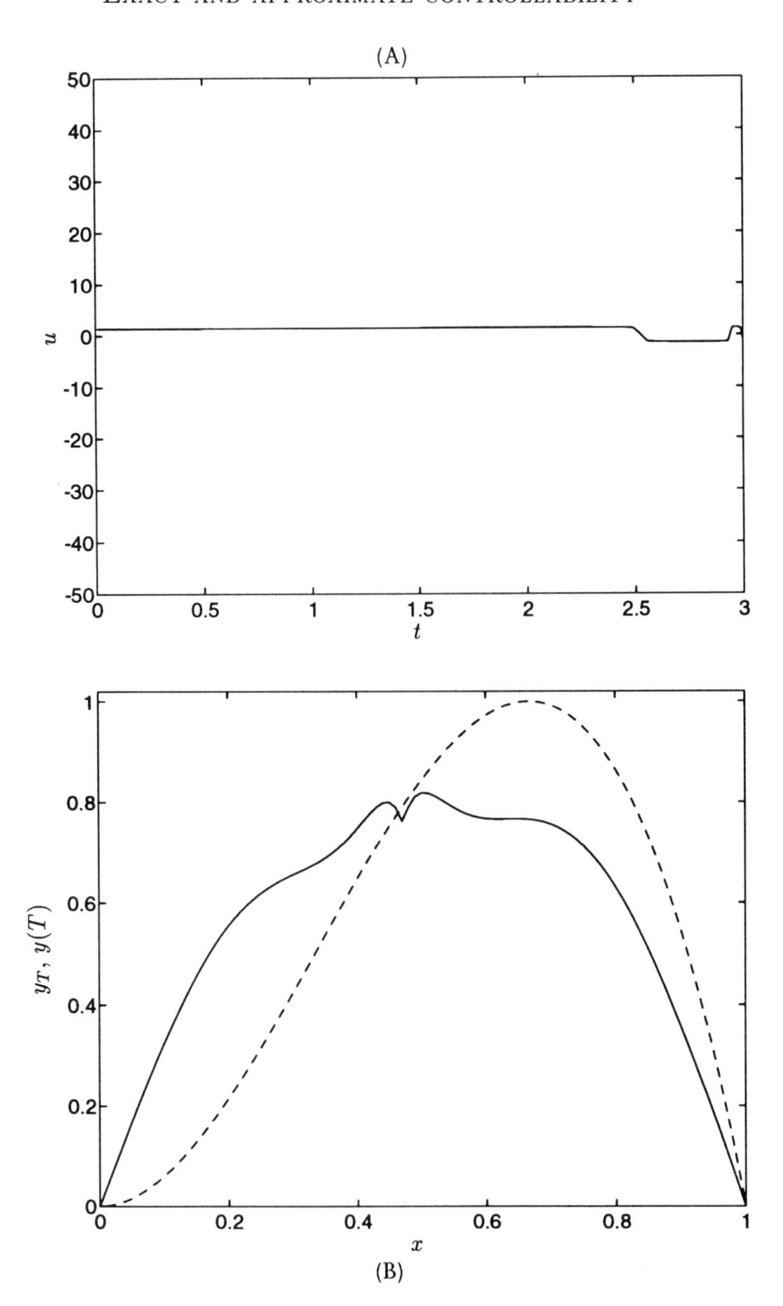

Fig. 21. (a) Variation of the optimal control and (b) comparison between y_T and $y^*(T)$ (target function (1.466): $s = 30$, $T = 3$, $b = \sqrt{2}/3$, $k = 10^7$, $h = \Delta t = 10^{-2}$).

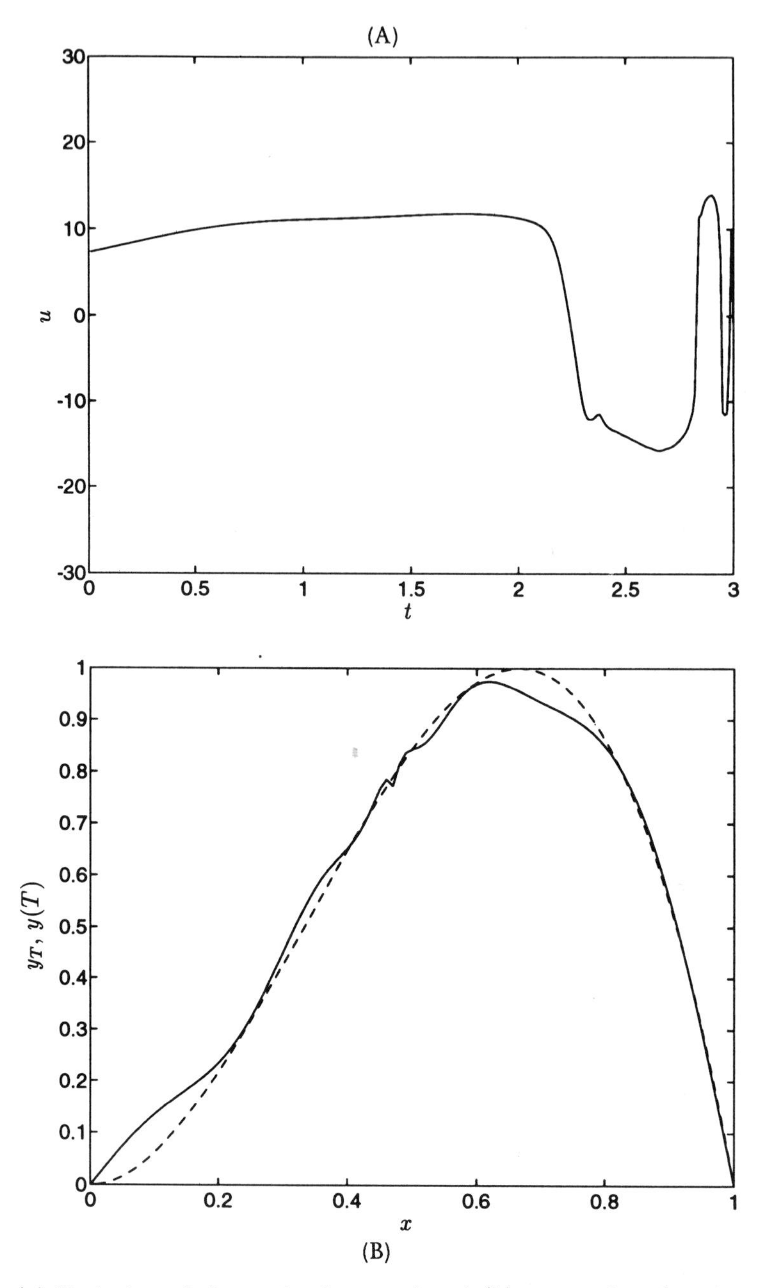

Fig. 22. (a) Variation of the optimal control and (b) comparison between y_T and $y^*(T)$ (target function (1.466): $s = 6$, $T = 3$, $b = \sqrt{2}/3$, $k = 2 \times 10^9$, $h = \Delta t = 10^{-2}$).

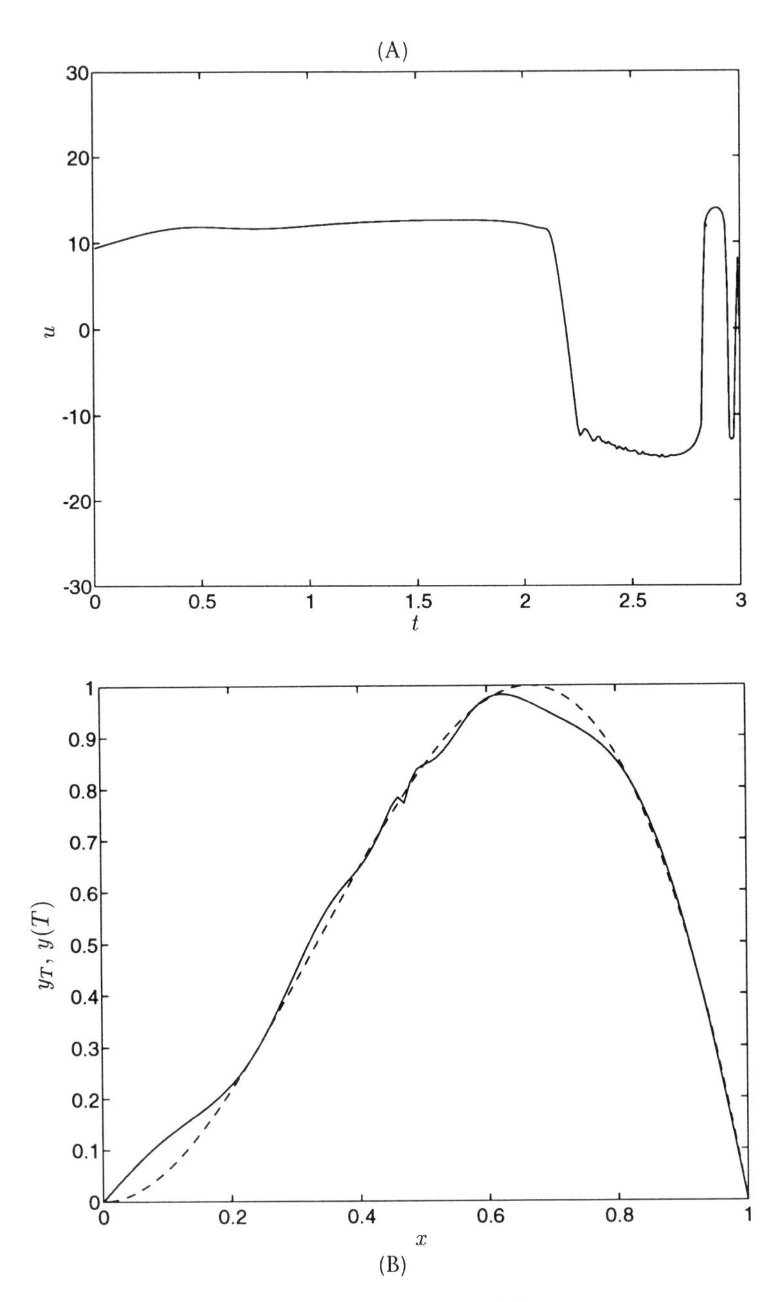

Fig. 23. (a) Variation of the optimal control and (b) comparison between y_T and $y^*(T)$ (target function (1.466): $s = 6$, $T = 3$, $b = \sqrt{2}/3$, $k = 2 \times 10^9$, $h = \Delta t = 10^{-2}$).

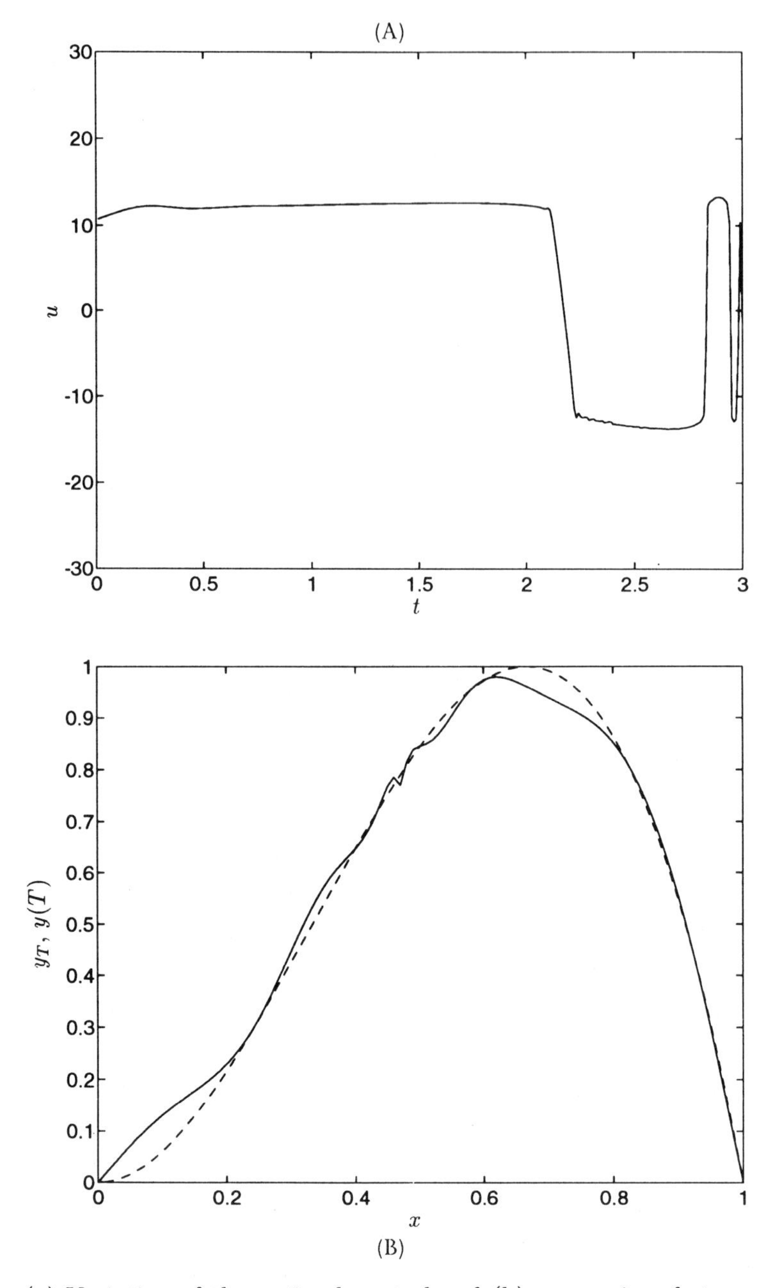

Fig. 24. (a) Variation of the optimal control and (b) comparison between y_T and $y^*(T)$ (target function (1.466): $s = 20, T = 3, b = \sqrt{2}/3, k = 10^{25}, h = \Delta t = 10^{-2}$).

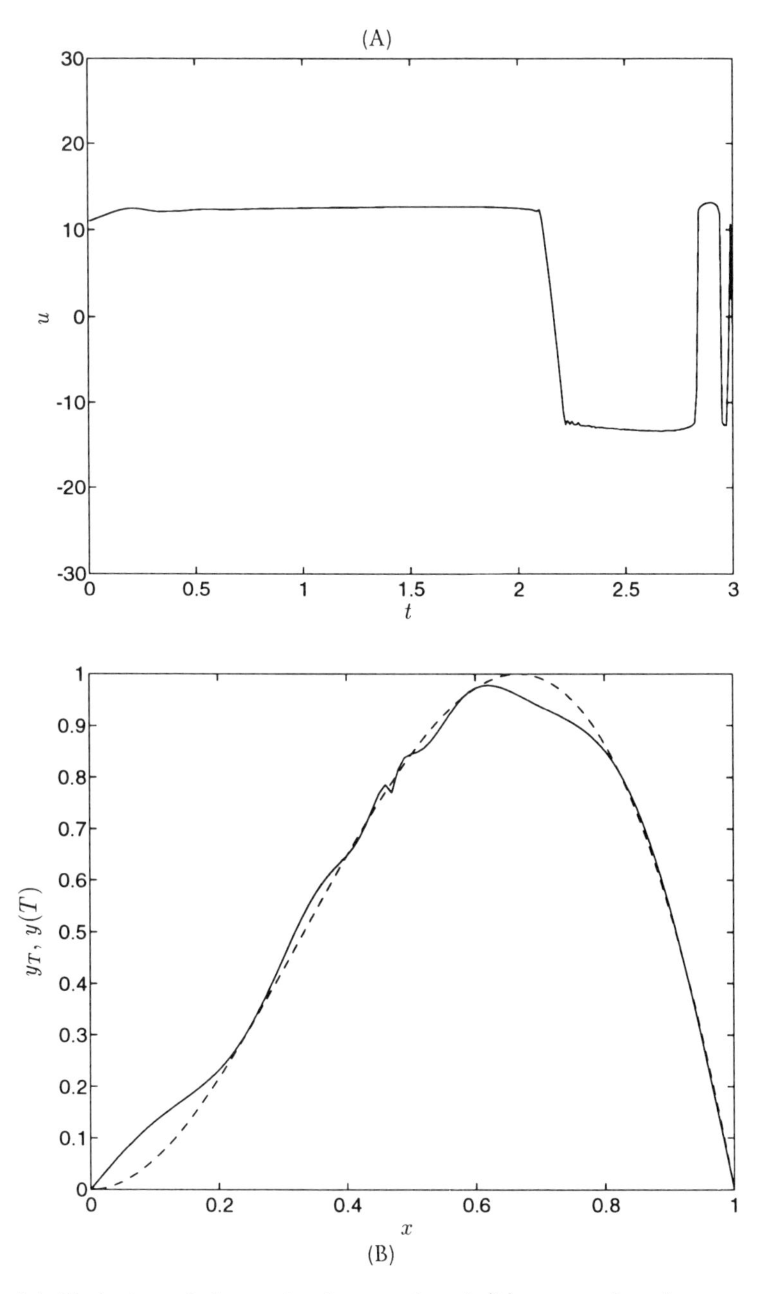

Fig. 25. (a) Variation of the optimal control and (b) comparison between y_T and $y^*(T)$ (target function (1.466): $s = 30, T = 3, b = \sqrt{2}/3, k = 10^{30}, h = \Delta t = 10^{-2}$).

REFERENCES

F. Abergel and R. Temam (1990), 'On some control problems in fluid mechanics', *Theoret. Comput. Fluid Dyn.* **1**, 303–326.

D.M. Buschnell and J.N. Hefner (1990), *Viscous Drag Reduction in Boundary Layers*, American Institute of Aeronautics and Astronautics (Washington, DC).

C. Carthel, R. Glowinski and J.L. Lions (1994), 'On exact and approximate boundary controllabilities for the heat equation: a numerical approach', *J. Opt. Theoret. Appl.* **82**, 3.

P.G. Ciarlet (1978), *The Finite Element Method for Elliptic Problems*, North-Holland (Amsterdam).

P.G. Ciarlet (1989), *Introduction to Numerical Linear Algebra and Optimization*, Cambridge University Press (Cambridge).

P.G. Ciarlet (1990a), 'A new class of variational problems arising in the modeling of elastic multi-structures', *Numer. Math.* **57**, 547–560.

P.G. Ciarlet (1990b), *Plates and Junctions in Elastic Multi-Structures: An Asymptotic Analysis*, Masson (Paris).

P.G. Ciarlet, H. Le Dret and R. Nzengwa (1989), 'Junctions between three-dimensional and two-dimensional linearly elastic structures', *J. Math. Pures et Appl.* **68**, 261–295.

J. Daniel (1970), *The Approximate Minimization of Functionals*, Prentice Hall (Englewood Cliffs, NJ).

E.J. Dean, R. Glowinski and C.H. Li (1989), 'Supercomputer solution of partial differential equation problems in computational fluid dynamics and in control', *Comput. Phys. Commun.* **53**, 401–439.

J. Dennis and R.B. Schnabel (1983), *Numerical Methods for Unconstrained Optimization and Nonlinear Equations*, Prentice-Hall (Englewood Cliffs, NJ).

J.I. Diaz (1991), 'Sobre la controlabilidad aproximada de problemas no lineales disipativos', *Jornadas Hispano-Francesas Sobre Control de Sistemas Distribuidos*, Universidad de Malaga 41–48.

J.D. Downer, K.C. Park and J.C. Chiou (1992), 'Dynamics of flexible beams for multibody systems: a computational procedure', *Comput. Meth. Appl. Mech. Engrg* **96**, 373–408.

I. Ekeland and R. Temam (1974), *Analyse Convexe et Problèmes Variationnels*, Dunod (Paris).

B. Engquist, B. Gustafsson and J. Vreeburg (1978), 'Numerical solution of a PDE system describing a catalytic converter', *J. Comput. Phys.* **27**, 295–314.

C. Fabre, J.P. Puel and E. Zuazua (1993), 'Contrôlabilité approchée de l'équation de la chaleur linéaire avec des contrôles de norme L^∞ minimale', *C. R. Acad. Sci.* I **316**, 679–684.

A. Friedman (1988), 'Modeling catalytic converter performance', in *Mathematics in Industrial Problems*, Part 4 (A. Friedman, ed.) Springer (New York) Ch. 7, 70–77.

C.M. Friend (1993), 'Catalysis and surfaces', *Scientific American* (April), 74–79.

H. Fujita and T. Suzuki (1991), 'Evolution problems', in *Handbook of Numerical Analysis* vol. II (P.G. Ciarlet and J.L. Lions, eds.) North-Holland (Amsterdam), 789–928.

D. Gabay (1982), 'Application de la méthode des multiplicateurs aux inéquations variationnelles', in *Méthodes de Lagrangien Augmenté* (M. Fortin and R. Glowinski, eds.) Dunod-Bordas, (Paris).

D. Gabay (1983) 'Application of the method of multipliers to variational inequalities', in *Augmented Lagrangian Methods* (M. Fortin and R. Glowinski, eds.) North-Holland (Amsterdam).

R. Glowinski (1984), *Numerical Methods for Nonlinear Variational Problems*, Springer (New York).

R. Glowinski (1992a), 'Ensuring well-posedness by analogy: Stokes problem and boundary control for the wave equation', *J. Comput. Phys.* **103**, 189–221.

R. Glowinski (1992b), 'Boundary controllability problems for the wave and heat equations', in *Boundary Control and Boundary Variation* (J.P. Zolesio, ed.) Lecture Notes in Control and Information Sciences, vol. 178, Springer (Berlin) 221–237.

R. Glowinski and P. Le Tallec (1989), *Augmented Lagrangian and Operator-Splitting Methods in Nonlinear Mechanics*, SIAM (Philadelphia, PA).

R. Glowinski and C.H. Li (1990), 'On the numerical implementation of the Hilbert uniqueness method for the exact boundary controllability of the wave equation', *C. R. Acad. Sci.* I **311**, 135–142.

R. Glowinski, C.H. Li and J.L. Lions (1990), 'A numerical approach to the exact boundary controllability of the wave equation (I) Dirichlet controls: description of the numerical methods', *Japan. J. Appl. Math.* **7**, 1–76.

R. Glowinski, J.L. Lions and R. Tremolières (1976), *Analyse Numérique des Inéquations Variationnelles*, Dunod (Paris).

R. Glowinski, J.L. Lions and R. Tremolières (1981), *Numerical Analysis of Variational Inequalities*, North-Holland (Amsterdam).

G.H. Golub and C. Van Loan (1989), *Matrix Computations*, Johns Hopkins Press (Baltimore, MD).

O.A. Ladyzenskaya, V.A. Solonnikov and N.N. Ural'ceva (1968), *Linear and Quasilinear Equations of Parabolic Type*, American Mathematical Society (Providence, RI).

J.E. Lagnese and G. Leugering. To appear.

T.A. Laursen and J.C. Simo (1993), 'A continuum-based finite element formulation for the implicit solution of multi-body, large deformation frictional contact problems', *Int. J. Numer. Meth. Engrg* **36**, 3451–3486.

J.L. Lions (1961), *Équations Différentielles Opérationnelles et Problèmes aux Limites*, Springer (Heidelberg).

J.L. Lions (1968), *Contrôle Optimal des Systèmes Gouvernés par des Équations aux Dérivées Partielles*, Dunod (Paris).

J.L. Lions (1988a), 'Exact controllability, stabilization and perturbation for distributed systems', *SIAM Rev.* **30**, 1–68.

J.L. Lions (1988b), *Controlabilité Exacte, Perturbation et Stabilisation des Systèmes Distribués*, vols. 1 and 2, Masson (Paris).

J.L. Lions (1990), *El Planeta Tierra*, Instituto de España, Espasa Calpe, S.A., Madrid.

J.L. Lions (1993), 'Quelques remarques sur la controlabilité en liaison avec des questions d'environnement', in *Les Grands Systémes des Sciences et de la Technologie* (J. Horwitz and J.L. Lions, eds.), Masson (Paris) 240–264.

J.L. Lions and E. Magenes (1968), *Problèmes aux Limites Non Homogènes*, vol. 1. Dunod (Paris).

P.L. Lions and B. Mercier (1979), 'Splitting algorithms for the sum of two nonlinear operators', *SIAM J. Numer. Anal.* **16**, 964–979.

S. Mizohata (1958), 'Unicité du prolongement des solutions pour quelques opérateurs différentiels paraboliques', *Mem. Coll. Sci. Univ. Kyoto* A **31**, 219–239.

J. Nocedal (1992), 'Theory of algorithms for unconstrained optimization', *Acta Numerica 1992*, Cambridge University Press, (Cambridge) 199–242.

K.C. Park, J.C. Chiou and J.D. Downer (1990), 'Explicit–implicit staggered procedures for multibody dynamics analysis', *J. Guidance Control Dyn.* **13**, 562–570.

D. Peaceman and H. Rachford (1955), 'The numerical solution of parabolic and elliptic differential equations', *J. Soc. Ind. Appl. Math.* **3**, 28–41.

P.A. Raviart and J.M. Thomas (1988), *Introduction à l'Analyse Numérique des Équations aux Dérivés Partielles*, Masson (Paris).

D.L. Russel (1978), 'Controllability and stabilizability theory for linear partial differential equations. Recent progress and open questions', *SIAM Rev.* **20**, 639–739.

J. Sanchez Hubert and E. Sanchez Palencia (1989), *Vibrations and Coupling of Continuous Systems (Asymptotic Methods)*, Springer (Berlin).

J.C. Saut and B. Scheurer (1987) 'Unique continuation for some evolution equations', *J. Diff. Eqns* **66**, 118–139.

R.H. Sellin and T. Moses (1989), *Drag Reduction in Fluid Flows*, Ellis Horwood (Chichester).

V. Thomee (1990), 'Finite difference methods for linear parabolic equations', in *Handbook of Numerical Analysis*, vol. I (P.G. Ciarlet and J.L. Lions, eds.) North-Holland (Amsterdam) 5–196.

Acta Numerica (1994), *pp.* 379–410

On the numerical evaluation of electrostatic fields in composite materials

Leslie Greengard* and Monique Moura†
Courant Institute of Mathematical Sciences
New York University
New York 10012, USA
E-mail: greengard@cims.nyu.edu

A classical problem in electrostatics is the determination of the effective electrical conductivity in a composite material consisting of a collection of piecewise homogeneous inclusions embedded in a uniform background. We discuss recently developed fast algorithms for the evaluation of the potential and electrostatic fields induced in multiphase composites by an applied potential, from which the desired effective properties may be easily obtained. The schemes are based on combining a suitable boundary integral equation with the Fast Multipole Method and the GMRES iterative method; the CPU time required grows linearly with the number of points in the discretization of the interface between the inclusions and the background material.

A variety of other questions in electrostatics, magnetostatics and diffusion can be formulated in terms of interface problems. These include the evaluation of electrostatic fields in the presence of dielectric inclusions, the determination of magnetostatic fields in media with variable magnetic permeability, and the calculation of the effective thermal conductivity of a composite material. The methods presented here apply with minor modification to these other situations as well.

CONTENTS

1 Introduction 2
2 Potential theory 8
3 Fast solution of integral equations 17
4 Numerical results 19
5 Conclusions 29
References 29

* The work of this author was supported by the Applied Mathematical Sciences Program of the U.S. Department of Energy under Contract DEFGO288ER25053, by a NSF Presidential Young Investigator Award and by a Packard Foundation Fellowship.
† The work of this author was supported in part by a Packard Foundation Fellowship to Leslie Greengard.

1. Introduction

Interface problems arise in a wide variety of areas of applied mathematics, including the determination of the electrostatic and magnetostatic fields in heterogeneous media, the calculation of effective transport properties, and the motion of multiphase fluids. The governing equation common to each of these problems is often the second-order elliptic partial differential equation

$$\nabla(\sigma \nabla u) = 0, \tag{1.1}$$

where σ is piecewise constant, subject to some appropriate boundary condition on u. In many situations, the dynamic range of σ can vary enormously and the geometry can be very complex. The reason for the use of the phrase 'interface' problem is that the differential equation (1.1) is often reformulated as follows: find a continuous function u which satisfies the Laplace equation in each phase (where σ is constant), and whose flux $\sigma \partial u / \partial n$ is continuous across each interface.

Definition 1.1 A function u which satisfies the above conditions will be referred to as a *total potential.*

For the sake of clarity, we will focus our attention on questions of electrical conductivity and leave the translation of our results to other application areas to the reader. We will restrict our attention for the most part to two-dimensional problems and consider primarily two issues. One is the determination of the electric field in the vicinity of a collection of inclusions in free space; the other is the determination of the effective conductivity of a composite material.

1.1. Inclusions in free space

The simplest problem of the first type is probably the determination of the electric field in an infinite plane with conductivity σ_e in which is embedded a disk D of conductivity σ_d. In the presence of a uniform applied field

$$\mathbf{E} = (E_a, 0),$$

corresponding to an applied potential $\Psi_a = -E_a x$, the interface problem takes the form

$$\begin{aligned}
\Delta u_e &= 0 \quad \text{in } \mathbb{R}^2 \setminus D \\
\Delta u_d &= 0 \quad \text{in } D \\
u_e &= u_d \quad \text{on } \partial D && (1.2) \\
\sigma_e \frac{\partial u_e}{\partial \nu} &= \sigma_d \frac{\partial u_d}{\partial \nu} \quad \text{on } \partial D, && (1.3)
\end{aligned}$$

where u_d denotes the restriction of the total potential u to D, u_e denotes the restriction of u to $\mathbb{R}^2 \setminus D$, and ν denotes the outward normal to ∂D. We also have the far field boundary condition

$$u_e(x,y) \to \Psi_a(x,y) \quad \text{as} \quad |(x,y)| \to \infty.$$

Without loss of generality, let us assume that the disk is of radius one, centred at the origin. The standard approach to solving this problem (Van Bladel, 1964; Jackson, 1975) is based on Fourier analysis. Making use of symmetry, we seek a solution of the form

$$\begin{aligned} u_d(r,\theta) &= \sum_{k=0}^{\infty} \alpha_k r^k \cos k\theta \\ u_e(r,\theta) &= -E_a r \cos\theta + \sum_{k=1}^{\infty} A_k r^{-k} \cos k\theta, \end{aligned}$$

where (r,θ) are the polar coordinates of a point in the plane. Imposition of the interface conditions (1.2) and (1.3) and a straightforward calculation yield

$$\begin{aligned} u_d(r,\theta) &= E_a \lambda\, r \cos\theta - E_a\, r \cos\theta \\ u_e(r,\theta) &= E_a \lambda \frac{\cos\theta}{r} - E_a\, r \cos\theta, \end{aligned}$$

where $\lambda = (\sigma_d - \sigma_e)/(\sigma_d + \sigma_e)$.

Definition 1.2 The difference between the total potential and the applied potential will be referred to as the *induced potential* or the *induced response*:

$$u_{\text{induced}} = u - \Psi_a.$$

In the preceding example, the induced potential is given by $u_{\text{induced}} = E_a \lambda r \cos\theta$ for $r \leq 1$ and $u_{\text{induced}} = E_a \lambda \cos\theta / r$ for $r \geq 1$. Contour plots of the total and induced potentials are shown in Figure 1.

Remark 1.1 In this article, the applied potential will always be assumed to be $\Psi_a = -E_a x$.

1.2. Periodic arrays

A more difficult problem is that of determining the electrostatic field in a simple composite consisting of a periodic array of disks in a uniform background (Figure 2). Rayleigh (1892) describes a method for solving this problem based on multipole expansions. Recent extensions and refinements have been developed by several groups, including Perrins *et al.* (1979b), McPhedran *et al.* (1988), and Sangani and Yao (1988).

Consider the plane to be tiled by unit squares with conductivity σ_e, each containing a disk of radius R_0 and conductivity σ_d.

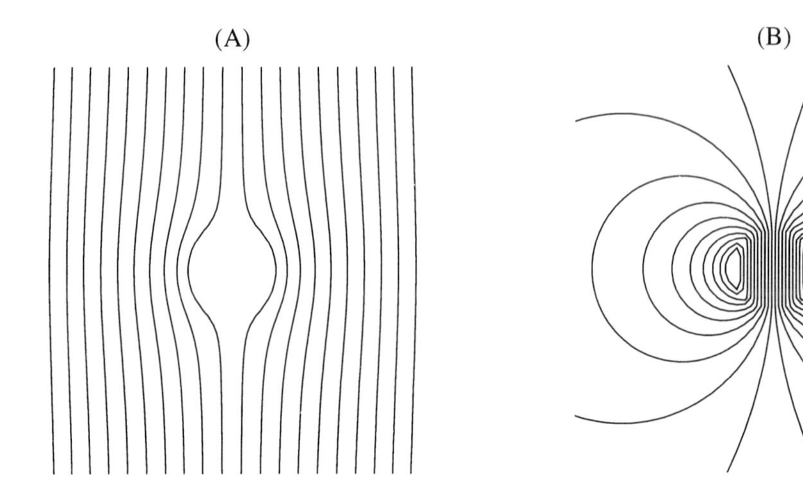

Fig. 1. The potential field in the vicinity of an inclusion in free space; the left-hand figure (A) is a contour plot of the total potential and the right-hand figure (B) is a contour plot of the induced potential. The x-axis is oriented in the horizontal direction.

Because of periodicity, it is sufficient to consider a single unit cell B centred at the origin and the governing equations

$$\begin{aligned} \nabla^2 u_e &= 0 \quad \text{in } B \setminus D, \\ \nabla^2 u_d &= 0 \quad \text{in } D, \\ u_e &= u_d \quad \text{on } \partial D, \qquad (1.4) \\ \sigma_e \frac{\partial u_e}{\partial \nu} &= \sigma_d \frac{\partial u_d}{\partial \nu} \quad \text{on } \partial D, \qquad (1.5) \end{aligned}$$

where u_d is the restriction of u to the disk D of radius R_0 and u_e is the restriction of u to $B \setminus D$. The boundary conditions on B are

$$u(x+1, y) - u(x, y) = -E_a \qquad (1.6)$$

$$u(x, y+1) - u(x, y) = 0. \qquad (1.7)$$

It is easy to see that outside the disk D, the potential can be represented as

$$u_e = A_0 + (A_1 r + B_1 r^{-1}) \cos\theta + (A_3 r^3 + B_3 r^{-3}) \cos 3\theta + \ldots \qquad (1.8)$$

and inside the disk D as

$$u_d = C_0 + C_1 r \cos\theta + C_3 r^3 \cos 3\theta + \ldots, \qquad (1.9)$$

where (r, θ) are the polar coordinates of a point with respect to the disk centre. (Note that we have had to introduce more unknown Fourier coeffi-

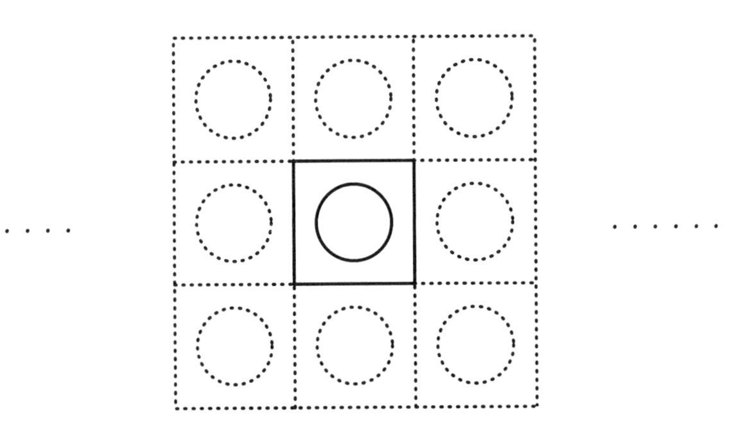

Fig. 2. A typical unit cell and its nearest neighbours in a periodic array of disks. Its length in the x-direction is α and its length in the y-direction is β. The disk is assumed to have radius R_0.

cients than in the earlier free-space example.) At the boundary of the disk, the potential must satisfy the interface conditions (1.4) and (1.5). Imposing these conditions on the above series, it is easy to derive a relation between the coefficients A_n and B_n, namely

$$B_n = -\lambda {R_0}^{2n} A_n \qquad \text{for } n \geq 1 \tag{1.10}$$

where, as before, $\lambda = (\sigma_d - \sigma_e)/(\sigma_d + \sigma_e)$. Letting A and B denote the vectors $(A_1, A_2, \cdots)$ and $(B_1, B_2, \cdots)$, respectively, and letting D be the infinite-dimensional diagonal matrix with entries $D_{nn} = -\lambda {R_0}^n$, the relations (1.10) can be written as

$$B = D\,A. \tag{1.11}$$

Since each square is subject to a constant potential drop $(-E_a)$, it is clear that the coefficient A_0 will depend on the particular location of the unit cell. It is also clear, however, that the coefficients $\{A_i\}$ and $\{B_i\}$ of the series expansions about the centre of each disk are translation invariant. In order to obtain the potential everywhere, it remains only to find another relation between the coefficient vectors A and B. For this, let (r, θ) be a point in $B \setminus D$ but close to the boundary of the disk. Then the part of the potential due to the applied field and to all disks other than D must be

$$A_0 + A_1 r \cos\theta + A_3 r^3 \cos 3\theta + \ldots. \tag{1.12}$$

If we now subtract Ψ_a, it is clear that the influence of all image disks is given by

$$\tilde{u} = E_0 r \cos\theta + A_0 + A_1 r \cos\theta + A_3 r^3 \cos 3\theta + \ldots \tag{1.13}$$

or, using complex notation,

$$\tilde{u} = \mathrm{Re}\,(A_0 + (E_0 + A_1)z + A_3 z^3 + \ldots), \tag{1.14}$$

where we identify the point $(r, \theta) \in \mathbb{R}^2$ with the complex number $z = re^{i\theta}$. Consider now a single distant disk centred at $w = m + in$. The influence it exerts at z, which we will denote by u_w, can be expressed in terms of a multipole expansion:

$$u_w = \mathrm{Re} \sum_{k=1}^{\infty} \frac{B_k}{(z-w)^k}. \tag{1.15}$$

Since the multipole coefficients B_k are the same for each image disk and $\tilde{u}$ is the field due to all image disks, we have

$$\tilde{u} = \mathrm{Re} \sum_{w \in \Lambda} \sum_{k=1}^{\infty} \frac{B_k}{(z-w)^k}, \tag{1.16}$$

where Λ denotes the set of lattice points (disk centres) excluding the origin,

$$\Lambda = \{(m,n) | m, n \in \mathbb{Z}, (m,n) \neq (0,0)\}.$$

Expanding each term as a Taylor series in z and using equation (1.13), the coefficients must satisfy the relations

$$A_n = \sum_{k=1}^{\infty} B_k \binom{n+k-1}{k-1} (-1)^k S_{n+k}, \quad n > 1 \tag{1.17}$$

$$A_1 + E_0 = \sum_{k=1}^{\infty} B_k \, k \, (-1)^k S_{k+1}, \tag{1.18}$$

where S_m denotes the lattice sum

$$S_m = \sum_{w \in \Lambda} \frac{1}{w^m}.$$

Letting P denote the matrix with entries

$$P_{nk} = \binom{n+k-1}{k-1} (-1)^k S_{n+k},$$

and letting $V^t = (E_0, 0, \cdots)$, the relations (1.17) and (1.18) can be rewritten as

$$P\,B = A - V. \tag{1.19}$$

Both (1.11) and (1.19) are infinite systems of equations. The error in truncating these systems depends on several factors, such as the distance between disks and the conductivity ratio σ_d/σ_e. Once truncated, however, it remains only to solve a finite-dimensional system for B

$$(P - D^{-1})B = V. \tag{1.20}$$

Rayleigh (1892) shows that the effective conductivity can be determined from the dipole moment B_1. This result is extended in a straightforward manner to arbitrary geometries and multiphase composites in Section 2.9.

1.3. Complex geometry

When the number of disks is large or the shapes of the inclusions are irregular, simple methods of the type described above are not available and other approaches need to be taken. A number of options have developed over the last century, including effective medium theory (Landauer, 1978; Bergman, 1978; Willis, 1981; Milton, 1985), variational methods (Hashin and Shtrikman, 1962; Beran, 1965; Prager, 1969; Phan-Tien and Milton, 1982; Torquato and Lado, 1988; Helsing, 1993), and asymptotic methods (Keller, 1987; McPhedran *et al.*, 1988; Bonnecaze and Brady, 1990).

The literature in each of these areas is vast and the references given above are by no means complete. Unfortunately, none of these approaches is suited to obtaining quantitatively precise evaluation of the field in complex geometries. For that purpose, one is obliged to consider direct solution of the governing partial differential equation.

Even within this category, there are several options including extensions of Rayleigh's method for systems of disks (Sangani and Yao, 1988), finite difference methods, finite element methods, and integral equation methods.

We will restrict our attention to the latter category since the problem we are interested in solving can be recast as a boundary integral equation. This reduces the dimensionality of the problem by one and greatly simplifies the discretization of the domain. Such methods are by no means new; integral equation techniques have been used to solve interface problems in electrostatics and magnetostatics for many years (Kellogg, 1953; Van Bladel, 1964; Jaswon and Symm, 1977; Lindholm, 1980; Brebbia *et al.*, 1983; Durand and Ungar, 1988; Hetherington and Thorpe, 1992; Nabors and White, 1992).

In the next section, we develop the mathematical apparatus of potential theory and construct second-kind Fredholm equations for interface problems by representing the solution as a single-layer potential. Numerical methods based on this formulation have been used previously for both smooth and polygonal inclusions (Jaswon and Symm, 1977; Hetherington and Thorpe, 1992). The dense linear systems which arise, however, will not be treated by standard factorization techniques which require $\mathcal{O}(N^3)$ operations where N is the number of points in the boundary discretization. Instead, the linear systems will be solved iteratively. This is by now standard in the integral equation community (see, for example, Atkinson (1976), Baker *et al.* (1982), Delves and Mohamed (1985), Rokhlin (1985), Nabors and White (1991)) and has been used for composite materials calculations as well (Gyure and Beale, 1992).

We have chosen to use the GMRES method of Saad and Schultz (1986), but a variety of other conjugate gradient type iterations are also acceptable. In addition, we will rely on the Fast Multipole Method (Rokhlin, 1985; Greengard and Rokhlin, 1987; Carrier *et al.*, 1988) to rapidly apply the integral operators at each step in the iterative process.* By combining these two schemes, the number of operations required is only $\mathcal{O}(N)$. Calculations with one hundred thousand boundary points, which have been viewed as intractable, require only minutes of CPU time on a workstation. Previous work which uses the Fast Multipole Method in this way includes that of Rokhlin (1985), Nabors and White (1991), and Greenbaum *et al.* (1992; 1993). A more thorough discussion of certain aspects of the present article and more extensive numerical experiments can be found in Moura (1993).

2. Potential theory

In this section, we review the basic properties of layer potentials involving a Green function for the Laplace equation. We then discuss two possible integral equation approaches to the solution of electrostatic problems in two phase composites. Both finite numbers of inclusions and periodic arrays will be treated. Our results will then be extended to multiphase systems.

2.1. Layer potentials

We will begin with some classical results from potential theory (Guenther and Lee, 1988; Jaswon and Symm, 1977; Kellogg, 1953; Mikhlin, 1957). Our treatment follows most closely that of Guenther and Lee (1988). We shall denote a two-dimensional domain by D and its boundary by S. S may consist of a number of disjoint components, but each is assumed to be a smooth curve with continuous curvature. If D is a bounded domain and $u, v \in C^2(D) \cap C^1_\nu(\overline{D})$, then Green identities state that

$$\int_D v\Delta u + \nabla v \nabla u = \int_S v\frac{\partial u}{\partial \nu}, \tag{2.1}$$

$$\int_D v\Delta u - u\Delta v = \int_S v\frac{\partial u}{\partial \nu} - u\frac{\partial v}{\partial \nu}, \tag{2.2}$$

$$\int_D \Delta u = \int_S \frac{\partial u}{\partial \nu}. \tag{2.3}$$

* Other fast algorithms could also be used (Anderson, 1986; Odlyzko and Schönhage, 1988; Hackbusch and Nowak, 1989; Van Dommelen and Rundensteiner, 1989; Brandt and Lubrecht, 1990).

Here ν denotes the unit outward normal vector and C^1_ν denotes the set of functions $u \in C^1(D) \cap C^0(\overline{D})$ such that

$$\frac{\partial u}{\partial \nu}(P) = \lim_{\substack{t \to 0 \\ t<0}} \nu(P) \cdot \nabla u(P + t\nu(P))$$

exists uniformly for all $P \in S$. For infinite regions, we require additional hypotheses on the functions involved. If we suppose, for example, that $u(P)$ is bounded at infinity and that $|\nabla u| = \mathcal{O}(1/|P|)$ as $|P| \to \infty$, then the Green identities hold.

One consequence of Green identities is that any function $u \in C^2(D) \cap C^1_\nu(\overline{D})$ can be expressed as the sum of three integrals.

Theorem 2.1 Every function $u \in C^2(D) \cap C^1_\nu(\overline{D})$ can be represented as

$$\begin{aligned} u(P) &= \int_S \frac{\partial G}{\partial \nu_Q}(P,Q) u(Q)\, \mathrm{d}s_Q - \int_S G(P,Q) \frac{\partial u}{\partial \nu}(Q)\, \mathrm{d}s_Q \\ &\quad + \int_D G(P,Q) \Delta u(Q)\, \mathrm{d}Q, \end{aligned} \tag{2.4}$$

where $G(P,Q) = (1/2\pi) \log |P - Q|$ is the fundamental solution of the Laplacian and ν_Q denotes the normal direction at Q. If D is unbounded, the additional hypotheses that u be bounded and that $|\nabla u| = \mathcal{O}(1/|P|)$ as $|P| \to \infty$ are assumed to hold.

Proof. In Green's second identity let $v = G(P,Q)$. Take the domain to be $D \setminus K_\epsilon(P)$, where $K_\epsilon(P)$ is the disk of radius ϵ centred at P, and consider the limit $\epsilon \to 0$. □

The first integral in (2.4) is referred to as a double-layer potential (with density u), the second is referred to as a single-layer potential (with density $\partial u/\partial \nu$) and the third is referred to as a volume potential.

2.2. Jump relations

In order to develop integral equation methods for the problems of potential theory, we need to study the analytic properties of layer potentials. It is well known and straightforward to prove that the double-layer potential defined on a smooth curve S maps continuous functions into harmonic, infinitely differentiable functions in $\mathbb{R}^2 \setminus S$. The resulting functions, however, are not continuous in $\mathbb{R}^2$.

Theorem 2.2 Let D be a bounded domain with boundary S and suppose that the function μ is continuous on S. Then the double-layer potential

$$u(P) = \int_S \frac{\partial G}{\partial \nu_Q}(P,Q)\, \mu(Q)\, \mathrm{d}s_Q$$

satisfies the jump relations at $P_0 \in S$

$$\lim_{\substack{P \to P_0 \\ P \in D}} u(P) = u(P_0) + \frac{1}{2}\mu(P_0) \tag{2.5}$$

$$\lim_{\substack{P \to P_0 \\ P \in {}^2 \setminus \overline{D}}} u(P) = u(P_0) - \frac{1}{2}\mu(P_0). \tag{2.6}$$

The single-layer potential maps continuous functions on S to continuous functions in $\mathbb{R}^2$. Although the result is infinitely differentiable and harmonic in $\mathbb{R}^2 \setminus S$, the normal derivative is discontinuous across S.

Theorem 2.3 Let D be a bounded domain with boundary S and suppose that the function ρ is continuous on S. Then the single-layer potential

$$u(P) = \int_S G(P, Q)\, \rho(Q)\, \mathrm{d}s_Q$$

satisfies the following jump relations. Let $P_0 \in S$ and let ν_{P_0} be the unit normal vector to S at P_0. Then

$$\frac{\partial u}{\partial \nu_-} \equiv \lim_{\substack{P \to P_0 \\ P \in D}} \frac{\partial u}{\partial \nu_{P_0}}(P) = \int_S \frac{\partial G}{\partial \nu_{P_0}}(P_0, Q)\rho(Q)\, \mathrm{d}s_Q - \frac{1}{2}\rho(P_0) \tag{2.7}$$

$$\frac{\partial u}{\partial \nu_+} \equiv \lim_{\substack{P \to P_0 \\ P \in {}^2 \setminus \overline{D}}} \frac{\partial u}{\partial \nu_{P_0}}(P) = \int_S \frac{\partial G}{\partial \nu_{P_0}}(P_0, Q)\rho(Q)\, \mathrm{d}s_Q + \frac{1}{2}\rho(P_0). \tag{2.8}$$

We may relate, as a result of the previous theorem, the source density and the normal derivative of the potential.

Corollary 2.1 Let $u(P) = \int_S G(P, Q)\, \rho(Q)\, \mathrm{d}s_Q$. Then

$$\rho = \frac{\partial u}{\partial \nu_+} - \frac{\partial u}{\partial \nu_-}.$$

Remark 2.1 In $\mathbb{R}^2$, $(\partial G/\partial \nu_Q)(P, Q)$ has a removable singularity at $P = Q \in S$. In fact

$$\lim_{P \to Q} \frac{\partial G}{\partial \nu_Q}(P, Q) = \frac{1}{2}\kappa(Q),$$

where κ is the curvature of S. Thus, the smoothness of the kernel of the double-layer potential is limited only by the smoothness of S. For infinitely differentiable curves, the kernel is infinitely differentiable.

2.3. The Fredholm alternative

Before proceeding with our investigation of interface problems, we state the Fredholm alternative, which allows us to investigate the solvability of a large class of second-kind integral equations. The theorem and its proof are well known (Guenther and Lee, 1988; Mikhlin, 1957).

Theorem 2.4 Consider the Fredholm equation

$$\rho(P) - \lambda \int_a^b K(P,Q)\rho(Q)\,\mathrm{d}Q = f(P), \tag{2.9}$$

where $K(P,Q)$ is an L_2 kernel, so that K is a compact operator on $L_2[a,b]$. Then (2.9) has a unique solution if and only if the equation

$$\rho(P) - \lambda \int_a^b K(P,Q)\rho(Q)\,\mathrm{d}Q = 0 \tag{2.10}$$

has only the trivial solution $\rho(P) = 0$. Furthermore, the adjoint equation to (2.10), defined by

$$\psi(P) - \overline{\lambda} \int_a^b \overline{K(Q,P)}\psi(Q)\,\mathrm{d}Q = 0 \tag{2.11}$$

has the same number of linearly independent solutions as (2.10). Finally, if (2.11) has at least one non-trivial solution, then (2.9) will have a solution only if

$$(f,\psi) = \int_a^b f(P)\overline{\psi(P)}\,\mathrm{d}P = 0$$

for all ψ satisfying (2.11). In this case, such a solution will not be unique.

2.4. Two-phase materials

Suppose that in the plane $\mathbb{R}^2$ with uniform conductivity σ_e we have embedded a finite number of smooth bounded inclusions each with conductivity σ_d. Let Ω denote the region occupied by the inclusions, let $c\Omega = \mathbb{R}^2 \setminus \Omega$, and let Γ denote the interface $\partial\Omega$. We will determine the electrostatic field in the plane in terms of a total potential function u, whose restriction to Ω and $c\Omega$ will be denoted by u_d and u_e, respectively. This corresponds to solving

$$\Delta u_d = 0 \quad \text{in } \Omega, \tag{2.12}$$

$$\Delta u_e = 0 \quad \text{in } c\Omega, \tag{2.13}$$

$$u_e = u_d \quad \text{on } \Gamma, \tag{2.14}$$

$$\sigma_e \frac{\partial u_e}{\partial \nu} = \sigma_d \frac{\partial u_d}{\partial \nu} \quad \text{on } \Gamma, \tag{2.15}$$

with the far-field boundary condition

$$u_e(P) \to \Psi_a(P) \quad \text{as} \quad P \to \infty. \tag{2.16}$$

We now look for a solution of equations (2.12) to (2.16) in the form of a single-layer potential

$$u(P) = \Psi_a(P) + \int_\Gamma G(P,Q)\rho(Q)\,\mathrm{d}s_Q, \tag{2.17}$$

where P is an arbitrary point in the plane and ρ is an unknown source density. It is convenient to write the solution as a single-layer potential because the continuity condition (2.14) is automatically satisfied. Physically, ρ represents the charge distribution on the interface which develops in response to the applied field (Jaswon and Symm, 1977; Hetherington and Thorpe, 1992).

In order to determine ρ, observe that using the jump relations (2.7) and (2.8), the interface condition can be rewritten in the form

$$\sigma_d \left[\frac{\partial \Psi_a}{\partial \nu_P}(P) - \frac{1}{2}\rho(P) + \int_\Gamma \frac{\partial G}{\partial \nu_P}(P,Q)\rho(Q)\,\mathrm{d}s_Q \right]$$
$$= \sigma_e \left[\frac{\partial \Psi_a}{\partial \nu_P}(P) + \frac{1}{2}\rho(P) + \int_\Gamma \frac{\partial G}{\partial \nu_P}(P,Q)\rho(Q)\,\mathrm{d}s_Q \right]. \tag{2.18}$$

Rearranging the previous equation we obtain the following second-kind Fredholm integral equation for ρ,

$$2\lambda \frac{\partial \Psi_a}{\partial \nu_P}(P) = \rho(P) - 2\lambda \int_\Gamma \frac{\partial G}{\partial \nu_P}(P,Q)\rho(Q)\,\mathrm{d}s_Q, \tag{2.19}$$

where $\lambda = (\sigma_d - \sigma_e)/(\sigma_d + \sigma_e)$.

It remains to determine for what values of λ the preceding equation can be solved. For this, we invoke the Fredholm alternative and make use of the following result.

Theorem 2.5 If the homogeneous integral equation

$$\rho(P) - 2\lambda \int_\Gamma \frac{\partial G}{\partial \nu_P}(P,Q)\rho(Q)\,\mathrm{d}s_Q = 0 \tag{2.20}$$

has a nontrivial solution, then $\lambda \in \mathbb{R}$ and lies on the rays $\lambda \geq 1$ or $\lambda < -1$.

Proof. See Kellogg (1953) or Mikhlin (1957). □

Corollary 2.2 As long as the ratio σ_d/σ_e is bounded and lies away from the negative real axis, the integral equation (2.19) has a unique solution.

2.5. Solution via the Green identity

Another, perhaps more common, integral formulation for the solution to the two-phase problem is based on Green's second identity (2.2). This is the approach taken, for example, by Van Bladel (1964) and Lindholm (1980). Letting $\phi_e = u - \Psi_a$ in $c\Omega$ and $\phi_d = u - \Psi_a$ in Ω and using Green's second identity, we have

$$\begin{aligned} u(P) &= \int_\Gamma \frac{\partial G}{\partial \nu_Q}(P,Q)u(Q) - G(P,Q)\frac{\partial u}{\partial \nu}(Q)\,\mathrm{d}s_Q, \quad P \in \Omega, \\ \Psi_a(P) &= \int_\Gamma \frac{\partial G}{\partial \nu_Q}(P,Q)\Psi_a(Q) - G(P,Q)\frac{\partial \Psi_a}{\partial \nu}(Q)\,\mathrm{d}s_Q, \quad P \in \Omega, \end{aligned}$$

$$\phi_e(P) = \int_\Gamma G(P,Q)\frac{\partial \phi_e}{\partial \nu}(Q) - \frac{\partial G}{\partial \nu_Q}(P,Q)\phi_e(Q)\,\mathrm{d}s_Q, \quad P \in c\Omega.$$

Taking the limit as P approaches a point on Γ, (2.5) and (2.6) yield

$$\frac{1}{2}u(P) = \int_\Gamma \frac{\partial G}{\partial \nu_Q}(P,Q)u(Q) - G(P,Q)\frac{\partial u}{\partial \nu}(Q)\,\mathrm{d}s_Q, \tag{2.21}$$

$$\frac{1}{2}\Psi_a(P) = \int_\Gamma \frac{\partial G}{\partial \nu_Q}(P,Q)\Psi_a(Q) - G(P,Q)\frac{\partial \Psi_a}{\partial \nu}(Q)\,\mathrm{d}s_Q, \tag{2.22}$$

$$\frac{1}{2}\phi_e(P) = \int_\Gamma G(P,Q)\frac{\partial \phi_e}{\partial \nu}(Q) - \frac{\partial G}{\partial \nu_Q}(P,Q)\phi_e(Q)\,\mathrm{d}s_Q, \tag{2.23}$$

repectively. Subtracting (2.22) from (2.23), we obtain

$$\frac{1}{2}u(P) - \Psi_a(P) = \int_\Gamma G(P,Q)\frac{\partial u}{\partial \nu}(Q) - \frac{\partial G}{\partial \nu_Q}(P,Q)u(Q)\,\mathrm{d}s_Q. \tag{2.24}$$

If we now multiply (2.21) by σ_d and (2.24) by σ_e, the flux interface condition and some algebra show that

$$u(P) - 2\lambda \int_\Gamma \frac{\partial G}{\partial \nu_Q}(P,Q)u(Q)\,\mathrm{d}s_Q = (\lambda - 1)\,\Psi_a(P). \tag{2.25}$$

The integral operator obtained this way is the adjoint of the operator obtained in equation (2.19), so that the analysis of solvability is the same. We have chosen to use the single-layer potential approach because several quantities of interest are computed more easily from the surface charge distribution than from values of the potential function u itself on the boundary.

2.6. Multiphase materials

Suppose now that a finite number of smooth bounded inclusions are embedded in a homogeneous background material with conductivity σ_e but that each inclusion is allowed a distinct conductivity. Ω_k will denote the region occupied by the kth inclusion with conductivity σ_k and its boundary will be denoted by Γ_k. Assuming there are M inclusions, the total interface is $\Gamma = \cup_{k=1}^M \Gamma_k$ and the total area occupied by the inclusions is $\Omega = \cup_{k=1}^M \Omega_k$. Let u denote the total potential, let u_k denote its restriction to the kth inclusion, and let u_e denote its restriction to the exterior domain $c\Omega$. Then

$$\Delta u_k = 0 \quad \text{in } \Omega_k,\ k = 1,\ldots,M, \tag{2.26}$$

$$\Delta u_e = 0 \quad \text{in } c\Omega, \tag{2.27}$$

$$u_e = u_k \quad \text{on } \Gamma_k,\ k = 1,\ldots,M, \tag{2.28}$$

$$\sigma_e\frac{\partial u_e}{\partial \nu} = \sigma_k\frac{\partial u_k}{\partial \nu} \quad \text{on } \Gamma_k,\ k = 1,\ldots,M \tag{2.29}$$

and

$$u_e(P) \to \Psi_a(P) \quad \text{as} \quad P \to \infty.$$

As for two-phase materials, we seek a solution in the form of a single-layer potential

$$u(P) = \Psi_a(P) + \int_\Gamma G(P,Q)\rho(Q)\,\mathrm{d}s_Q. \tag{2.30}$$

If we impose the condition (2.29) at each interface and let

$$\lambda_k = (\sigma_k - \sigma_e)(\sigma_k + \sigma_e),$$

we obtain

$$2\lambda_k \frac{\partial \Psi_a}{\partial \nu_P}(P) = \rho(P) - 2\lambda_k \int_\Gamma \frac{\partial G}{\partial \nu_P}(P,Q)\rho(Q)\,\mathrm{d}s_Q, \tag{2.31}$$

for $P \in \Gamma_k$, $k = 1, \ldots, M$. Since each of the λ_k may be distinct, this is actually a system of integral equations. Nevertheless, the Fredholm alternative can still be applied.

Theorem 2.6 Suppose λ_k is bounded for $k = 1, \ldots, M$ and that the homogeneous equation

$$\rho(P) - 2\lambda_k \int_\Gamma \frac{\partial G}{\partial \nu_P}(P,Q)\rho(Q)\,\mathrm{d}s_Q = 0 \tag{2.32}$$

has a nontrivial solution. Then at least one of the ratios σ_k/σ_e has negative real part.

Proof. Let $u = u_R + \mathrm{i}u_I$ denote the single-layer potential

$$u(P) = \int_\Gamma G(P,Q)\rho(Q)\,\mathrm{d}s_Q$$

corresponding to a complex-valued nontrivial solution of (2.32). Then

$$\frac{2}{\sigma_i + \sigma_e}\left(\sigma_e \frac{\partial u}{\partial \nu_+}(P) - \sigma_k \frac{\partial u}{\partial \nu_-}(P)\right) = \rho(P) - 2\lambda_k \int_\Gamma \frac{\partial G}{\partial \nu_P}(P,Q)\rho(Q)\,\mathrm{d}s_Q, \tag{2.33}$$

for $P \in \Gamma_k$, so that

$$\sigma_e \frac{\partial u}{\partial \nu_+}(P) - \sigma_k \frac{\partial u}{\partial \nu_-}(P) = 0. \tag{2.34}$$

To simplify notation, let us now define σ by

$$\sigma(P) = \sigma_k/\sigma_e$$

for $P \in \overline{\Omega_k}$, with $\sigma = \sigma_R + \subset \sigma_I$. Separating (2.34) into real and imaginary parts,

$$\frac{\partial u_R}{\partial \nu}_+ - \sigma_R \frac{\partial u_R}{\partial \nu}_- + \sigma_I \frac{\partial u_I}{\partial \nu}_- = 0, \tag{2.35}$$

$$\frac{\partial u_I}{\partial \nu}_+ - \sigma_R \frac{\partial u_I}{\partial \nu}_- - \sigma_I \frac{\partial u_R}{\partial \nu}_- = 0. \tag{2.36}$$

Multiplying equation (2.35) by u_R and (2.36) by u_I, adding them, and integrating along Γ yields

$$\int_\Gamma \left(u_R \frac{\partial u_R}{\partial \nu}_+ + u_I \frac{\partial u_I}{\partial \nu}_+ \right) - \int_\Gamma \sigma_R \left(u_R \frac{\partial u_R}{\partial \nu}_- + u_I \frac{\partial u_I}{\partial \nu}_- \right)$$

$$+ \int_\Gamma \sigma_I \left(u_R \frac{\partial u_I}{\partial \nu}_- - u_I \frac{\partial u_R}{\partial \nu}_- \right) = 0. \tag{2.37}$$

The third integral vanishes because of Green's second identity (2.2) and, from Green's first identity, we obtain

$$\iint_{c\Omega} (\nabla u_R)^2 + (\nabla u_I)^2 + \iint_\Omega \sigma_R((\nabla u_R)^2 + (\nabla u_I)^2) = 0.$$

Thus, if σ_R is nonnegative, u is identically zero and ρ is the trivial solution.
□

Corollary 2.3 As long as all the ratios σ_k/σ_e are bounded and have nonnegative real part, the integral equation (2.31) has a unique solution.

2.7. Periodic structures

Consider a two-phase composite medium in the plane consisting of a periodic array of inclusions embedded in a uniform background with conductivity σ_e. Let Ω_k denote the kth inclusion with conductivity σ_d and boundary Γ_k, and suppose that a square unit cell denoted by B contains M such inclusions. Let $\Gamma = \cup_{k=1}^M \Gamma_k$ and let $\Omega = \cup_{k=1}^M \Omega_k$. We are interested in calculating the induced electrostatic potential (and eventually the effective conductivity of the material). The potential equation to be solved can be written as follows:

$$\Delta u_d = 0 \quad \text{in } B \setminus \Omega, \tag{2.38}$$

$$\Delta u_e = 0 \quad \text{in } c\Omega, \tag{2.39}$$

$$u_e = u_d \quad \text{on } \Gamma, \tag{2.40}$$

$$\sigma_e \frac{\partial u_e}{\partial \nu} = \sigma_d \frac{\partial u_d}{\partial \nu} \quad \text{on } \Gamma, \tag{2.41}$$

where u_d and u_e are the restrictions of the total potential to the inclusions and the background, respectively. We also require that $u - \Psi_a$ be doubly periodic, that is

$$u(x+1,y) - u(x,y) = -E_a, \tag{2.42}$$

$$u(x,y+1) - u(x,y) = 0. \tag{2.43}$$

We again look for a solution of equations (2.38) to (2.43) in the form of a single-layer potential

$$u(P) = \Psi_a(P) + \int_\Gamma K(P,Q)\rho(Q)\, ds_Q, \tag{2.44}$$

but now $K(P,Q)$ is the doubly periodic Green function rather than the fundamental solution. There are a number of questions which arise in the evaluation of such a Green function that we will not review here in detail (Rayleigh, 1892; Perrins *et al.*, 1979a; Greengard and Rokhlin, 1987). We simply observe that periodicity can be imposed by considering the entire lattice of charge sources in the plane. These sources are translates of the density ρ to all image cells (see Section 3.1).

It is clear from our construction that the function $u(P)$ defined in (2.44) satisfies the conditions (2.42), (2.43) and (2.40). It remains only to satisfy the flux interface condition. By using the jump relations (2.7) and (2.8), we obtain

$$2\lambda \frac{\partial \Psi_a}{\partial \nu_P}(P) = \rho(P) - 2\lambda \int_\Gamma \frac{\partial K}{\partial \nu_P}(P,Q)\rho(Q)\, \mathrm{d}s_Q, \tag{2.45}$$

where $\lambda = (\sigma_d - \sigma_e)/(\sigma_d + \sigma_e)$. The analysis of solvability for this system is virtually identical to that of the two-phase composite in free space, so we simply state the result as

Theorem 2.7 If the homogeneous integral equation

$$\rho(P) - 2\lambda \int_\Gamma \frac{\partial K}{\partial \nu_P}(P,Q)\rho(Q)\, \mathrm{d}s_Q = 0 \tag{2.46}$$

has a nontrivial solution, then $\lambda \in \mathbb{R}$ and lies on the rays $\lambda \geq 1$ or $\lambda < -1$.

2.8. Periodic multiphase composites

It is a straightforward matter to extend our integral equation approach to multiphase composites using the same single-layer potential representation as for the two-phase system. The integral equation is solvable if the real parts of all the conductivities are positive, as in Section 2.6.

2.9. Computing the effective conductivity

Once the integral equation (2.45), or its multiphase analogue, has been solved, one of the important functionals one can extract from the source density ρ is the effective conductivity. We assume that the periodic cell is a unit square with vertices A, B, C, and D, listed counterclockwise from the lower left hand corner. We denote the boundary of the square by L. The effective conductivity matrix

$$\sigma_{\text{eff}} = \begin{pmatrix} \sigma_{11} & \sigma_{12} \\ \sigma_{21} & \sigma_{22} \end{pmatrix}$$

relates the current density vector J and the applied field E via

$$J = \sigma_{\text{eff}} E. \tag{2.47}$$

If we suppose that $E = (1, 0)$, then clearly

$$\sigma_{11} = \int_B^C \frac{\partial u}{\partial \nu} \, ds, \tag{2.48}$$

$$\sigma_{21} = \int_A^B \frac{\partial u}{\partial \nu} \, ds. \tag{2.49}$$

Consider first the quantity σ_{11} and apply Green's second identity (2.2) in the region $B \setminus \Omega$ to the functions u and v where $v(x,y) = x$, $u(x,y)$ is the computed total potential, and Ω denotes the region occupied by the inclusions. Since u and v are harmonic functions in this domain, we have

$$\int_{\Gamma \cup L} v \frac{\partial u}{\partial \nu_*} - u \frac{\partial v}{\partial \nu_*} \, ds = 0, \tag{2.50}$$

where ν_* denotes the unit inward normal to Γ. Green's third identity and a small amount of algebra show that

$$\int_\Gamma v \frac{\partial u}{\partial \nu_*} - u \frac{\partial v}{\partial \nu_*} \, ds = \int_\Gamma v \left(\frac{\partial u_e}{\partial \nu} - \frac{\partial u_d}{\partial \nu} \right) ds = \int_\Gamma x\rho \, ds, \tag{2.51}$$

so that it remains to analyse the integral in (2.50) along L. A straightforward calculation yields

$$\int_L v \frac{\partial u}{\partial \nu_*} - u \frac{\partial v}{\partial \nu_*} \, ds = \int_B^C \frac{\partial u}{\partial \nu} \, ds - 1 \tag{2.52}$$

and, therefore,

$$\sigma_{11} = 1 + \int_\Gamma x\rho \, ds. \tag{2.53}$$

Similarly,

$$\sigma_{21} = \int_\Gamma y\rho \, ds. \tag{2.54}$$

Finally, the components σ_{12} and σ_{22} can be computed by applying a field in the y-direction. If $E = (0, 1)$, then

$$\sigma_{12} = \int_B^C \frac{\partial u}{\partial \nu} \, ds = \int_\Gamma x\rho \, ds, \tag{2.55}$$

$$\sigma_{22} = \int_C^D \frac{\partial u}{\partial \nu} \, ds = 1 + \int_\Gamma y\rho \, ds. \tag{2.56}$$

3. Fast solution of integral equations

Consider now the numerical solution of the two-phase interface problem in free space using the integral equation (2.19), which we write explicitly as

$$\rho(P) - \frac{\lambda}{\pi} \int_\Gamma \frac{\partial}{\partial \nu_P} \log|P - Q| \rho(Q) \, dsQ = 2\lambda \frac{\partial \Psi_a}{\partial \nu_P}(P). \tag{3.1}$$

We select N_j points on the boundary Γ_j of the jth inclusion which are equispaced in arclength and define $h_j = |\Gamma_j|/N_j$, where $|\Gamma_j|$ denotes the length of boundary. The total number of discretization points is $N = \sum_{j=1}^{M} N_j$. Associated with each such point, denoted P_i^j, is an unknown charge density value ρ_i^j. Using the trapezoidal rule, we replace (3.1) by

$$\rho_i^j - \frac{\lambda}{\pi}\sum_{l=1}^{M} h_l \sum_{k=1}^{N_l} \frac{\partial}{\partial \nu_{P_i^j}} \log|P_i^j - P_k^l|\rho_k^l = -2\lambda\nu_i^j \cdot (1,0), \tag{3.2}$$

for $i = 1,\ldots,N_j$ and $j = 1,\ldots,M$. Care must be taken when $P_j^i = P_l^k$ to use the appropriate limit $\frac{1}{2}\kappa(P_l^k)$ in place of $(\partial/\partial\nu_{P_i^j})\log|P_i^j - P_k^l|$, where κ denotes curvature. The trapezoidal rule is used for quadrature since it achieves superalgebraic convergence on smooth contours.

We solve linear systems like (3.2) iteratively, using the generalized minimum residual method GMRES (Saad and Schultz, 1986). The reason for choosing a conjugate gradient type iterative method is that the integral operator in (3.1) is compact and well approximated by a finite rank operator. The eigenvalues of $I + K$ are bounded and cluster at one. As a result, the linear system (3.2) has a bounded condition number and the number of iterations required is independent of N (for a fixed physical problem). The amount of work required to solve the linear system, therefore, scales like $J \cdot f(N)$ where J is the number of iterations and $f(N)$ is the amount of work required to compute matrix–vector products. Since K (or its discrete version) is dense, naive methods require $\mathcal{O}(J \cdot N^2)$ work. The Fast Multipole Method, however, allows the cost to be reduced to $\mathcal{O}(N)$, so that the cost of solving the linear system is $\mathcal{O}(J \cdot N)$.

3.1. The fast multipole method

The Fast Multipole Method (FMM) is a hierarchical scheme for the evaluation of Coulombic interactions in both two and three space dimensions (Rokhlin, 1985; Greengard and Rokhlin, 1987; Carrier *et al.*, 1988; Greengard, 1988; Greengard and Rokhlin, 1989). Like the schemes of Van Dommelen and Rundensteiner (1989), Odlyzko and Schönhage (1988), Appel (1985), Barnes and Hut (1986) and others, it is based on using multipole expansions and/or Taylor series to compute far field interactions. For a system of N sources (charges, dipoles, etc.), the FMM requires $\mathcal{O}(N)$ work to evaluate all pairwise interactions, with the constant depending on the desired precision. With minor modification, the FMM allows for the calculation of electrostatic interactions in a periodic array as well (Greengard and Rokhlin, 1987). We refer the reader to the articles listed above for a complete description of the method.

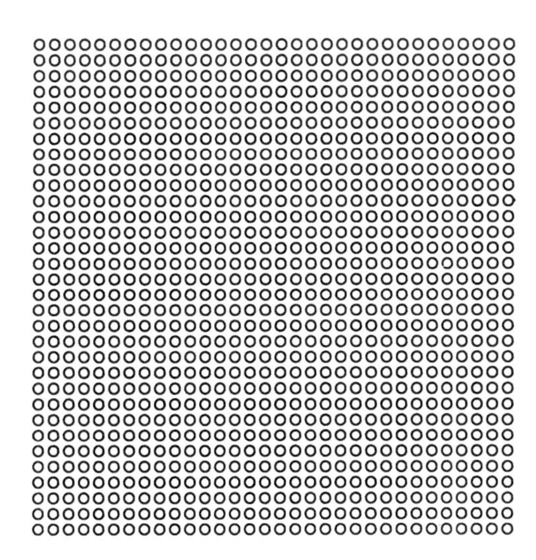

Fig. 3. A square array of 1,024 inclusions in the plane. This array is studied in free space and as a periodic system in Examples 1 and 2.

4. Numerical results

We have examined the behaviour of the integral equations (2.19), (2.31) and (2.45) over a wide range of geometries and conductivity ratios, and have selected three for the purpose of illustration. In each case, the GMRES method was used to reduce the Euclidean norm of the residual to below 10^{-6} and the FMM was used to compute matrix–vector products with a tolerance of 10^{-8}. All calculations were carried out on an IBM RS/6000 Model 580 in double precision. Evaluation of the potential off the boundary (once the integral equation was solved) was done using the FMM. Incorporation of Mayo's method (Mayo, 1984) would accelerate this part of the calculation and will be incorporated at a later date.

Example 1 The geometry in the first example consists of a square array of 1,024 disks in free space (Figure 3). The infinite medium is assumed to have conductivity one, while the inclusions have been assigned either a conductivity of 10^6 or a random number in the range $[10^{-7}, 10^2]$. Table 1 shows the type of problem being solved ($\sigma_i/\sigma_e = 10^6$ for the two-phase case and $\sigma_i/\sigma_e =$ 'Random' for the multiphase case), the number of boundary points used (N), the number of iterations required (Its), the time required for solving the integral equation (T) and the dipole moment

$$\mathbf{p} = \left(\int_\Gamma x\rho \, \mathrm{d}s, \int_\Gamma y\rho \, \mathrm{d}s \right)$$

induced in response to the applied field $E = (1, 0)$. Figure 4 shows contour plots of the total and induced potentials.

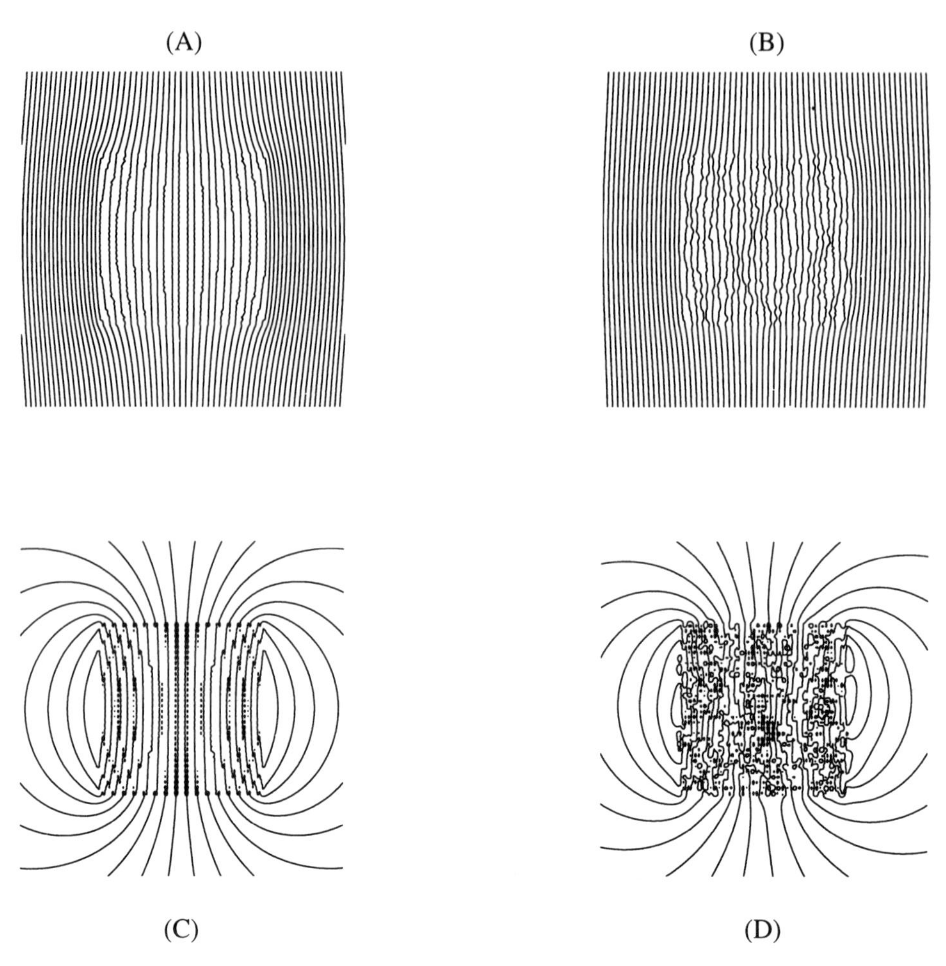

Fig. 4. Contour plots of the electrostatic potential for Example 1. (A) and (C) show the total and induced potential, respectively, for the two-phase case; (B) and (D) are the corresponding results for the (random) multiphase case.

Example 2 We now consider the same collection of 1,024 disks as in Example 1, but extended periodically. The background matrix is assumed to have conductivity one, while the inclusions have again been assigned either a conductivity of 10^6 or a random number in the range $[10^{-7}, 10^2]$. Table 2 summarizes our results, while Figure 5 shows contour plots of the total and induced potentials. (Note that in the two-phase case, we are solving a problem 1,024 times larger than necessary.)

Example 3 While materials with disk-like inclusions constitute an important class of composites, other geometries are clearly of interest as well. We therefore consider a collection of eleven slender inclusions in free space (Figure 6). Each inclusion is an ellipse which has been slightly perturbed in

Table 1. *Performance of the numerical method in Example 1. For insufficiently resolved problems, GMRES was unable to achieve the desired residual in less than 100 iterations. The iteration counts in parentheses indicate the number of GMRES steps allowed in such cases.*

σ_i/σ_e	N	Its	T (s)	$\mathbf{p}$
10^6	8,192	(7)	26.13	(−1.65231,0.0000000)
10^6	16,384	7	40.1	(−0.75483,0.0000000)
10^6	32,768	7	104.9	(−0.75484,0.0000000)
10^6	65,536	7	166.2	(−0.75484,0.0000000)
Random	8,192	(9)	32.1	(−0.396286,−0.000747)
Random	16,384	9	50.4	(−0.397286,−0.001183)
Random	32,768	9	131.5	(−0.397287,−0.001184)
Random	65,536	9	208.3	(−0.397287,−0.001184)

Table 2. *Performance of the numerical method in Example 2. For insufficiently resolved problems, GMRES was unable to achieve the desired residual in less than 100 iterations. The iteration counts in parentheses indicate the number of GMRES steps allowed in such cases.*

σ_i/σ_e	N	Its	T (s)	$\mathbf{p}$
10^6	8,192	2	5.6	(−1.09217,0.0000000)
10^6	16,384	3	31.3	(−1.08005,0.0000000)
10^6	32,768	3	36.63	(−1.08005,0.0000000)
10^6	65,536	3	106.2	(−1.08005,0.0000000)
Random	8,192	(9)	23.1	(−0.475469,−0.001001)
Random	16,384	10	82.9	(−0.458208,−0.001211)
Random	32,768	9	88.0	(−0.458201,−0.001211)
Random	65,536	9	262.1	(−0.458201,−0.001211)

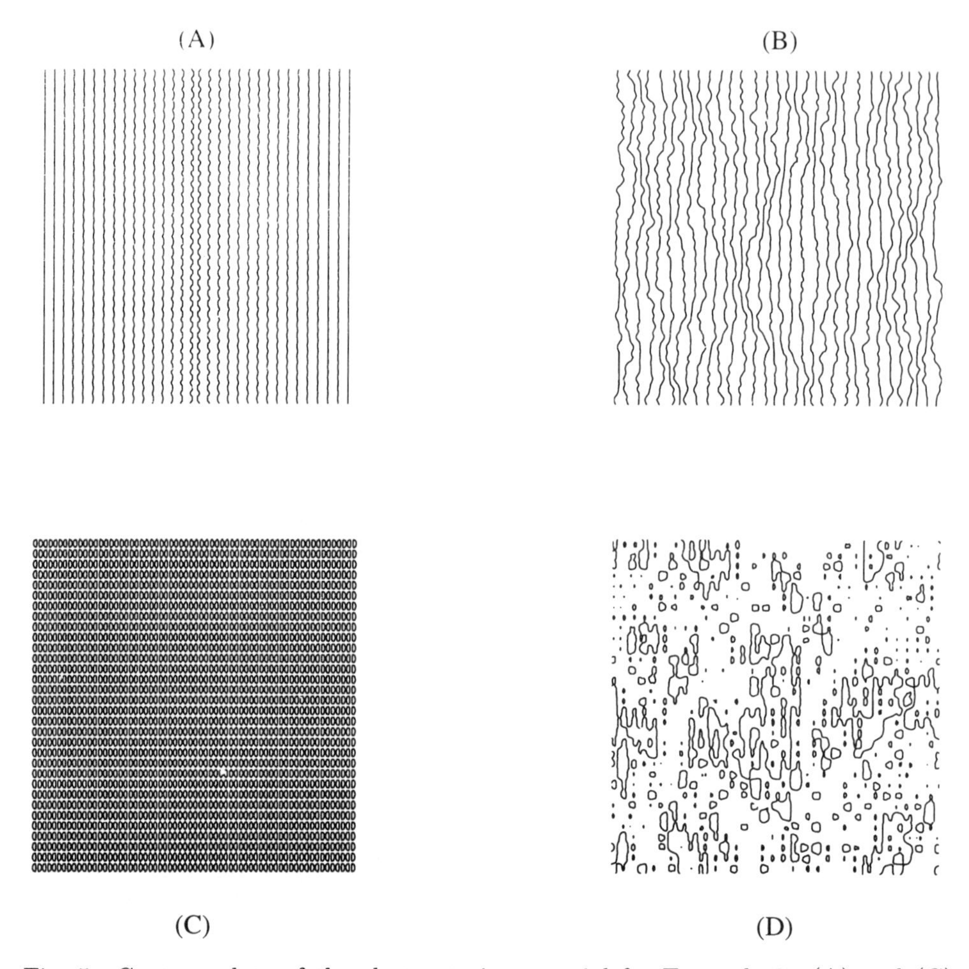

Fig. 5. Contour plots of the electrostatic potential for Example 2. (A) and (C) show the total and induced potential, respectively, for the two-phase case; (B) and (D) are the corresponding results for the (random) multiphase case.

order to make the geometry less regular. The background matrix is assumed to have conductivity one, while the inclusions have been assigned either a conductivity of 10^6 or 10^{-6}. Table 3 summarizes our results, while Figure 7 shows contour plots of the total and induced potentials.

Example 4 In this example, the slender inclusions of Example 3 are extended periodically. Table 4 summarizes our results, while Figure 8 shows contour plots of the total and induced potentials.

Example 5 The last geometry we consider is that of a fairly closely packed mixture of convex and nonconvex inclusions (Figure 9). In this free space calculation, the background matrix is assumed to have conductivity one,

Table 3. *Performance of the numerical method in Example 3. For insufficiently resolved problems, GMRES was unable to achieve the desired residual in less than 100 iterations. The iteration counts in parentheses indicate the number of GMRES steps allowed in such cases.*

σ_i/σ_e	N	Its	T (s)	p
10^6	1,100	(40)	15.5	(−0.688458,−0.387694)
10^6	2,200	(40)	25.7	(−0.874540,−0.251159)
10^6	4,400	40	45.2	(−0.874171,−0.251477)
10^6	8,800	40	87.6	(−0.874171,−0.251477)
10^{-6}	1,100	45	17.8	(0.860182,−0.252492)
10^{-6}	2,200	45	28.6	(0.855730,−0.251480)
10^{-6}	4,400	45	51.8	(0.855714,−0.251475)
10^{-6}	8,800	45	99.1	(−0.855714,−0.251475)

while the inclusions have been assigned either a conductivity of 10^6 or 10^{-6}. Table 5 summarizes our results, while Figure 10 shows contour plots of the total and induced potentials.

Example 6 In our last example, the inclusions of Example 5 are extended periodically. Table 6 summarizes our results, while Figure 11 shows contour plots of the total and induced potentials.

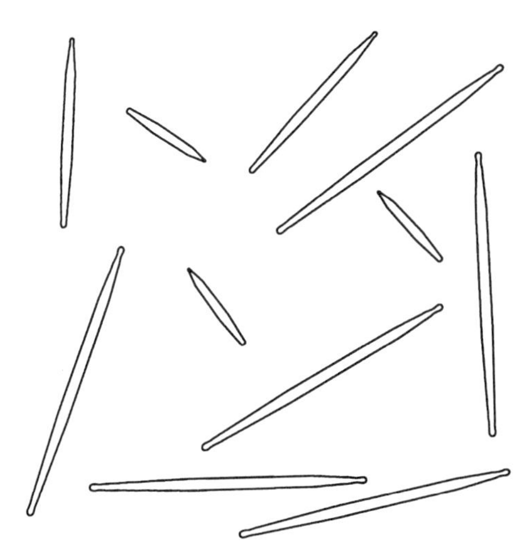

Fig. 6. Eleven slender inclusions in the plane. Each inclusion is a slightly perturbed ellipse. This geometry is studied in Examples 3 and 4.

Table 4. *Performance of the numerical method in Example 4. For insufficiently resolved problems, GMRES was unable to achieve the desired residual in less than 100 iterations. The iteration counts in parentheses indicate the number of GMRES steps allowed in such cases.*

σ_i/σ_e	N	Its	T (s)	p
10^6	1,100	(48)	18.1	(−3.22233,−3.67602)
10^6	2,200	(48)	25.6	(−1.65745,−0.75526)
10^6	4,400	48	63.8	(−1.64403,−0.73715)
10^6	8,800	48	98.9	(−1.64399,−0.73723)
10^6	17,600	48	256.9	(−1.64399,−0.73723)
10^{-6}	1,100	51	19.8	(0.563947,−0.124834)
10^{-6}	2,200	48	25.3	(0.554749,−0.124215)
10^{-6}	4,400	48	63.3	(0.554611,−0.124189)
10^{-6}	8,800	48	99.4	(0.554611,−0.124189)
10^{-6}	17,600	48	258.0	(0.554611,−0.124189)

Table 5. *Performance of the numerical method in Example 5. For insufficiently resolved problems, GMRES was unable to achieve the desired residual in less than 100 iterations. The iteration counts in parentheses indicate the number of GMRES steps allowed in such cases.*

σ_i/σ_e	N	Its	T (s)	p
10^6	450	(29)	4.6	(−1.48303,−0.07427)
10^6	900	(29)	9.6	(−1.38406,−0.01940)
10^6	1,800	29	17.5	(−1.37925,−0.01722)
10^6	3,600	29	30.9	(−1.37925,−0.01723)
10^6	7,200	29	51.7	(−1.37925,−0.01723)
10^{-6}	450	34	5.4	(1.4857,−0.05119)
10^{-6}	900	31	10.5	(1.45736,−0.017431)
10^{-6}	1,800	31	18.3	(1.45698,−0.01723)
10^{-6}	3,600	31	32.8	(1.45698,−0.01723)
10^{-6}	7,200	31	55.6	(1.45698,−0.01723)

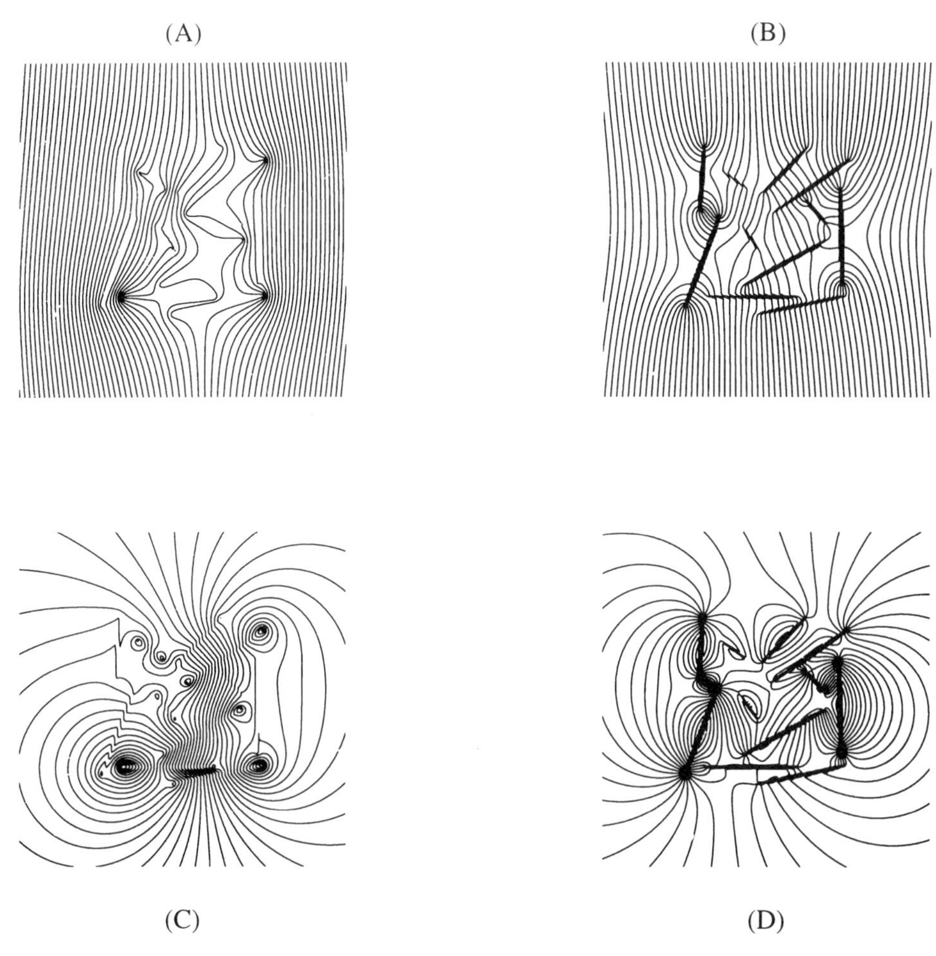

Fig. 7. Contour plots of the electrostatic potential for Example 3. (A) and (C) show the total and induced potential, respectively, for the case of highly conducting inclusions; (B) and (D) are the corresponding results for the case of poorly conducting inclusions.

Several observations can be made on the basis of the preceding examples.

1 For a fixed problem, the number of GMRES iterations required is constant, once sufficient resolution has been achieved.
2 The CPU time grows linearly with the number of discretization points.
3 The rate of convergence of the computed dipole moment is super-algebraic. (This rapid convergence can also be demonstrated for pointwise values of the charge density or other functionals of the solution.)
4 While the number of iterations required varies with the complexity of the geometry, it is remarkably insensitive to the conductivity ratio (in

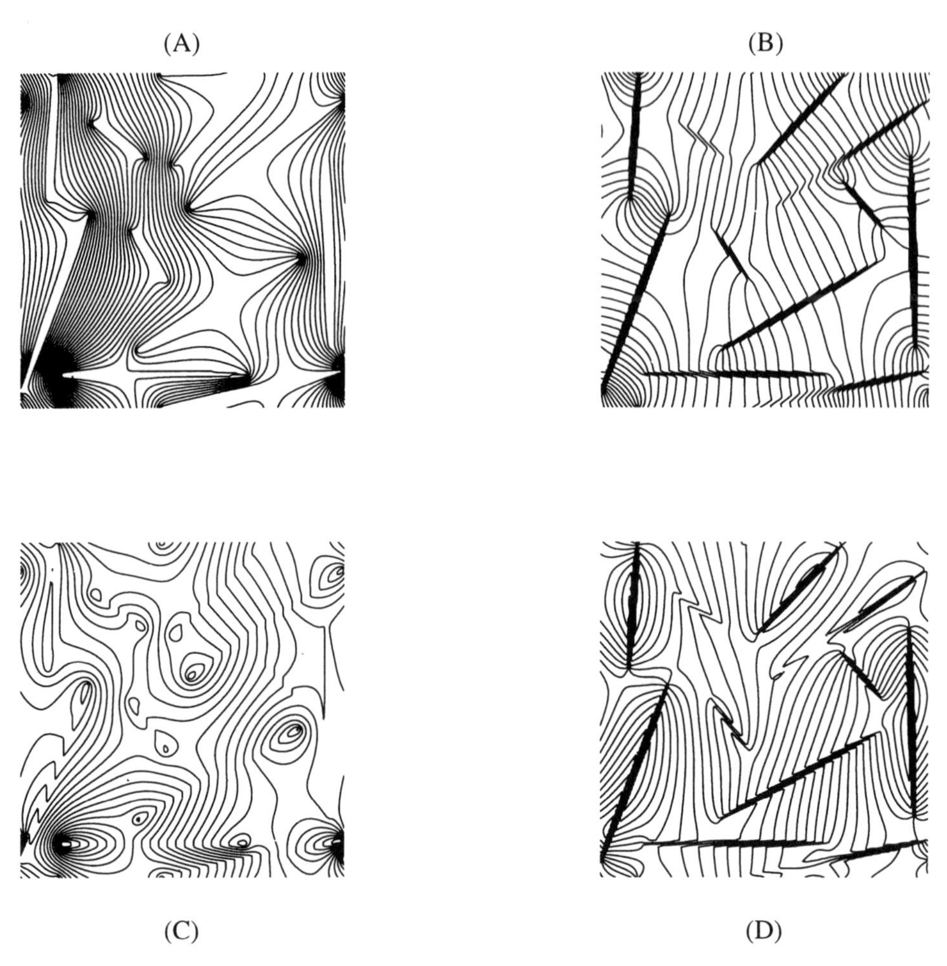

Fig. 8. Contour plots of the electrostatic potential for Example 4. (A) and (C) show the total and induced potential, respectively, for the case of highly conducting inclusions; (B) and (D) are the corresponding results for the case of poorly conducting inclusions.

marked contrast to the behavior of finite difference and finite element schemes).

As a final check on our calculations, we have computed the full effective conductivity tensor for the slender inclusion case (Example 4). With conductivity ratio 10^6,

$$\sigma^1_{\text{eff}} = \begin{pmatrix} 4.93128 & 0.16328 \\ 0.16328 & 6.22091 \end{pmatrix},$$

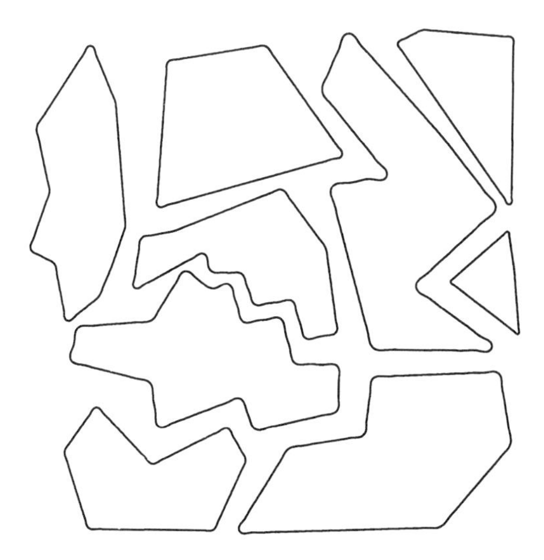

Fig. 9. A mixture of nine convex and nonconvex inclusions in the plane. This geometry is studied in Examples 5 and 6.

while with conductivity ratio 10^{-6},

$$\sigma^2_{\text{eff}} = \begin{pmatrix} 0.160888 & 0.005327 \\ 0.005327 & 0.202963 \end{pmatrix}.$$

Table 6. *Performance of the numerical method in Example 6. For insufficiently resolved problems, GMRES was unable to achieve the desired residual in less than 100 iterations. The iteration counts in parentheses indicate the number of GMRES steps allowed in such cases.*

σ_i/σ_e	N	Its	T (s)	$\mathbf{p}$
10^6	450	(30)	4.2	(−7.78783,2.2131022)
10^6	900	(30)	8.3	(−2.84366,0.74131)
10^6	1,800	30	14.9	(−3.93129,0.16328)
10^6	3,600	30	22.6	(−3.93128,0.16328)
10^6	7,200	30	49.6	(−3.93128,0.16328)
10^{-6}	450	30	4.1	(0.846530,−0.006531)
10^{-6}	900	29	7.8	(0.839114,0.005338)
10^{-6}	1,800	29	14.4	(0.839112,0.005327)
10^{-6}	3,600	29	22.8	(0.839112,0.005327)
10^{-6}	7,200	29	47.3	(0.839112,0.005327)

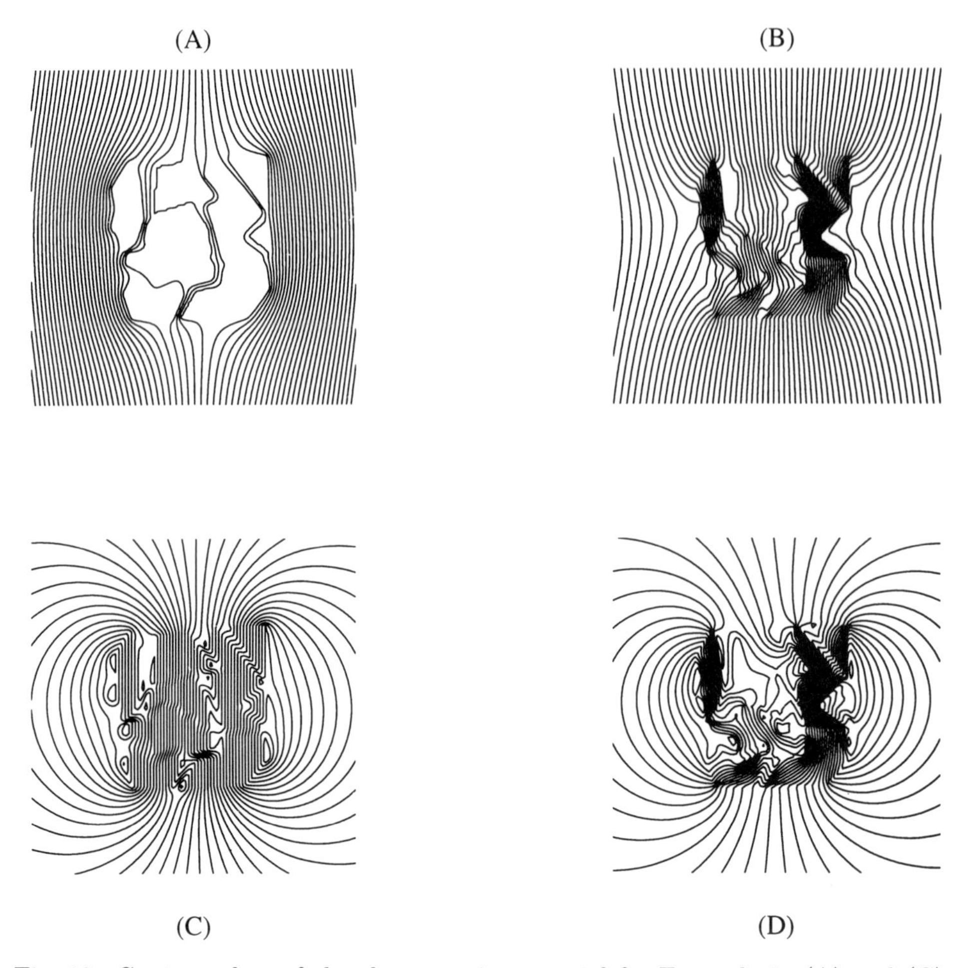

Fig. 10. Contour plots of the electrostatic potential for Example 5. (A) and (C) show the total and induced potential, respectively, for the case of highly conducting inclusions; (B) and (D) are the corresponding results for the case of poorly conducting inclusions.

These matrices satisfy the Keller–Dykhne–Mendelson relation (Keller, 1964; Dykhne, 1970; Mendelson, 1975)

$$\begin{pmatrix} 0 & 1 \\ -1 & 0 \end{pmatrix} \left(\sigma_{\text{eff}}^2\right)^{-1} = \sigma_{\text{eff}}^1 \begin{pmatrix} 0 & 1 \\ -1 & 0 \end{pmatrix}$$

to full accuracy.

5. Conclusions

We have presented an algorithm for the solution of the electrostatic field equations in composite media based on a fast multipole accelerated integral

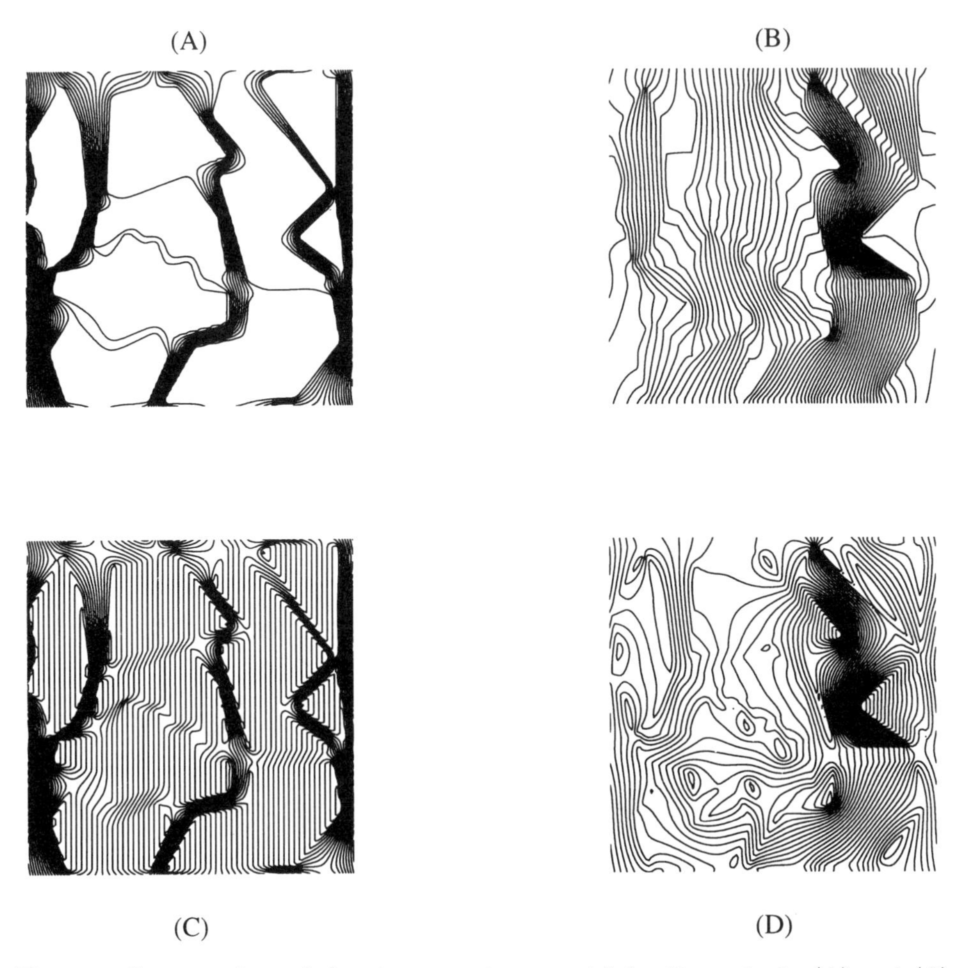

Fig. 11. Contour plots of the electrostatic potential for Example 6. (A) and (C) show the total and induced potential, respectively, for the case of highly conducting inclusions; (B) and (D) are the corresponding results for the case of poorly conducting inclusions.

equation solver. Large-scale problems, involving thousands of inclusions with perhaps one hundred thousand boundary points, can be solved in minutes using modest computational resources (such as a single workstation). Similar methods have been proposed for related problems by Rokhlin (1985), Nabors and White (1991), and Greenbaum *et al.* (1992, 1993). While we have only considered smooth inclusions, the method can be extended by modifying the quadrature to allow for the presence of corners (Hetherington and Thorpe, 1992). Extension to three dimensions is straightforward but requires incorporation of the three-dimensional FMM (Greengard and Rokhlin, 1988; 1989; Nabors and White, 1991). It is our hope that this method will provide researchers in materials science with a new tool – the ability to compute effective properties of systems with complex microstructure by direct solution of the governing equation.

Acknowledgements
We would like to thank V. Rokhlin for providing a subroutine which allowed us to obtain a smooth resampling in arclength from a user-specified curve.

REFERENCES

C. R. Anderson (1986), 'A method of local corrections for computing the velocity field due to a distribution of vortex blobs', *J. Comput. Phys.* **62**, 111–123.

A. W. Appel (1985), 'An efficient program for many-body simulation', *SIAM J. Sci. Stat. Comput.* **6**, 85–103.

K. Atkinson (1976), *A Survey of Numerical Methods for the Solution of Fredholm Integral Equations of the Second Kind*, SIAM (Philadelphia, PA).

G.R. Baker, D.I. Meiron and S.A. Orszag (1982), 'Generalized vortex methods for free-surface flow problems', *J. Fluid Mech.* **123**, 477–501.

J. Barnes and P. Hut (1986), 'A hierarchical $O(N \log N)$ force-calculation algorithm', *Nature* **324**, 446–449.

M.J. Beran (1965), 'Use of the variational approach to determine bounds for the effective permittivity in random media', *Nuovo Cimento* **38**, 771–782.

D. Bergman (1978), 'The dielectric constant of a composite material – a problem in classical physics', *Phys. Rep.* (Section C of *Phys. Lett.*) **43**(9), 377–407.

R.T. Bonnecaze and J.F. Brady (1990), 'A method for determining the effective conductivity of dispersions of particles', *Proc. R. Soc. London* A **369**, 207–225.

A. Brandt and A.A. Lubrecht (1990), 'Multilevel matrix multiplication and fast solution of integral equations', *J. Comput. Phys.* **90**, 348–370.

C.A. Brebbia, J.C.F. Telles and L.C. Wrobel (1983), *Boundary Element Techniques* Springer (Berlin).

J. Carrier, L. Greengard and V. Rokhlin (1988), 'A fast adaptive multipole algorithm for particle simulations', *SIAM J. Sci. Statist. Comput.* **9**, 669–686.

L.M. Delves and J.L. Mohamed (1985), *Computational Methods for Integral Equations*, Cambridge University Press (Cambridge).

P.P. Durand and L.H. Ungar (1988), 'Application of the boundary element method to dense dispersions', *Int. J. Numer. Methods Engrg* **26**, 2487–2501.

A.M. Dykhne (1970), 'Conductivity of a two dimensional two-phase system', *Zh. Eksp. Teor. Fiz.* **59**, 110–115.

A. Greenbaum, L. Greengard and A. Mayo (1992), 'On the numerical solution of the biharmonic equation in the plane', *Physica* D **60**, 216–225.

A. Greenbaum, L. Greengard and G.B. McFadden (1993), 'Laplace's equation and the Dirichlet–Neumann map in multiply connected domains', *J. Comput. Phys.* **105**, 267–278.

L. Greengard (1988), *The Rapid Evaluation of Potential Fields in Particle Systems*, MIT Press (Cambridge, MA).

L. Greengard and V. Rokhlin (1987), 'A fast algorithm for particle simulations', *J. Comput. Phys.* **73**, 325–348.
L. Greengard and V. Rokhlin (1988), 'Rapid evaluation of potential fields in three dimensions', in *Vortex Methods* (C. Anderson and C. Greengard, eds). Lecture Notes in Mathematics, vol. 1360, Springer (Berlin) 121–141.
L. Greengard and V. Rokhlin (1989), 'On the evaluation of electrostatic interactions in molecular modeling', *Chem. Scr.* A **29**, 139–144.
R.B. Guenther and J.W. Lee (1988), *Partial Differential Equations of Mathematical Physics and Integral Equations*, Prentice-Hall (Englewood Cliffs, NJ).
M. Gyure and P.D. Beale (1992), 'Dielectric breakdown in continuous models of metal-loaded dielectrics', *Phys. Rev.* B **46**, 3736–3746.
W. Hackbusch and Z.P. Nowak (1989), 'On the fast matrix multiplication in the boundary element method by panel clustering', *Numer. Math.* **54**, 463–491.
Z. Hashin and S. Shtrikman (1962), 'A variational approach to the theory of the effective magnetic permeability of multiphase materials', *J. Appl. Phys.* **33**(10), 3125–3131.
J. Helsing (1993), 'Bounds to the conductivity of some two-component composites', *J. Appl. Phys.* **73**(3), 1240–1245.
J. Hetherington and M.F. Thorpe (1992), 'The conductivity of a sheet containg inclusions with sharp corners', *Proc. R. Soc. London* A **438**, 591–604.
J.D. Jackson (1975), *Classical Electrodynamics*, Wiley (New York).
M.A. Jaswon and G.T. Symm (1977), *Integral Equation Methods in Potential Theory and Elastostatics*, Academic (New York).
J.B. Keller (1964), 'A theorem on the conductivity of a composite medium', *J. Math. Phys.* **5** 548–549.
J.B. Keller (1987), 'Effective conductivity of periodic composites composed of two very unequal conductors', *J. Math. Phys.* **28**, 2516–2520.
O.D. Kellogg (1953), *Foundations of Potential Theory*, Dover (New York).
R. Landauer (1978), 'Electrical conductivity in inhomogeneous media' in *Electrical Transport and Optical Properties of Inhomogeneous Media* (J.C. Garland and D.B. Tanner, eds), AIP (New York) 2–61.
D.A. Lindholm (1980), 'Notes on boundary integral equations for three-dimensional magnetostatics', *IEEE Trans. Magnetics* **16**, 1409–1413.
A. Mayo (1984), 'The fast solution of Poisson's and the biharmonic equation on irregular regions', *SIAM J. Numer. Anal.* **21**, 285–299.
R.C. McPhedran, L. Poladian and G.W. Milton (1988), 'Asymptotic studies of closely spaced, highly conducting cylinders', *Proc. R. Soc. London* A **415**, 185–196.
K.S. Mendelson (1975), 'Effective conductivity of two-phase materials with cylindrical boundaries', *J. Appl. Phys.* **46**, 917–918.
S.G. Mikhlin (1957), *Integral Equations*, Pergamon (London).
G.W. Milton (1985), 'The coherent potential approximation is a realizable effective medium scheme', *Comm. Math. Phys.* **99**, 463–500.
M. Moura (1993), 'On the numerical calculation of electrostatic fields in composite media', Ph.D. Thesis, New York University.
K. Nabors and J. White (1991), 'FastCap: a multipole accelerated 3-D capacitance extraction program', *IEEE Trans. Computer-Aided Design* **10**, 1447–1459.

K. Nabors and J. White (1992), 'Multipole-accelerated capacitance extraction algorithms for 3-D structures with multiple dielectrics', *IEEE Trans. Circuits and Systems* **39**, 946–954.

A.M. Odlyzko and A. Schönhage (1988), 'Fast algorithms for multiple evaluations of the Riemann zeta function', *Trans. Am. Math. Soc.* **309**, 797–809.

W.T. Perrins, R.C. McPhedran and D.R. McKenzie (1979a), 'Optical properties of dense regular cermets with relevance to selective solar absorbers', *Thin Solid Films* **57**, 321–326.

W.T. Perrins, D.R. McKenzie and R.C. McPhedran (1979b), 'Transport properties of regular arrays of cylinders', *Proc. R. Soc. London* A **369**, 207–225.

N. Phan-Thien and G.W. Milton (1982), 'New bounds on the effective thermal conductivity of N-phase materials', *Proc. R. Soc. London* A **380**, 333–348.

S. Prager (1969), 'Improved variational bounds on some bulk properties of a two-phase random medium', *J. Chem. Phys.* **50**, 4305–4312.

Lord Rayleigh (1892), 'On the influence of obstacles arranged in rectangular order upon the properties of a medium', *Phil. Mag.* **34**, 481–502.

V. Rokhlin (1985), 'Rapid solution of integral equations of classical potential theory', *J. Comput. Phys.* **60**, 187–207.

V. Rokhlin (1990), 'End-point corrected trapezoidal quadrature rules for singular functions', *Comput. Math. Applic.* **20**, 51–62.

Y. Saad and M.H. Schultz (1986), 'GMRES: a generalized minimum residual algorithm for solving nonsymmetric linear systems', *SIAM J. Sci. Stat. Comput.* **7**, 856–869.

A.S. Sangani and C. Yao (1988), 'Transport processes in random arrays of cylinders. I. thermal conduction', *Phys. Fluids* **31**, 2426–2434.

S. Torquato and F. Lado (1988), 'Bounds on the effective conductivity of a random array of cylinders', *Proc. R. Soc. London* A **417**, 59–80.

J. Van Bladel (1964), *Electromagnetic Fields*, McGraw-Hill (New York).

L. Van Dommelen and E.A. Rundensteiner (1989), 'Fast, adaptive summation of point forces in the two-dimensional Poisson equation', *J. Comput. Phys.* **83**, 126–147.

J.R. Willis (1981), 'Variational and related methods for the overall properties of composites', *Adv. Appl. Mech.* **21**, 1–78.

Acta Numerica (1994), *pp.* 411–466

Numerical geometry of surfaces

Malcolm Sabin
Department of Industrial Studies
University of Liverpool, England
E-mail: mal0r@liverpool.ac.uk

The mathematical techniques used within Computer Aided Design software for the representation and calculation of surfaces of objects are described. First the main techniques for dealing with surfaces as computational objects are described, and then the methods for enquiring of such surfaces the properties required for their assessment and manufacture.

CONTENTS

1	SURFACE DEFINITIONS	412
2	Analytic surfaces	413
3	Parametric surfaces	416
4	Recursive division surfaces	429
5	Conversions between different representations	431
6	SURFACE INTERROGATIONS	432
7	Nilvariate interrogations	435
8	Univariate interrogations	445
9	Bivariate interrogations	459
References		462

Notation Lower case letters are used to denote scalars. x, y and z are coordinates in $\mathbb{R}^3$, u and v are coordinates in parameter space. f, g and h are scalar functions. Scalar expressions are bracketted by round brackets ( ). Greek letters are usually scalar functions, except for α and θ, which are angles, and ρ which is a radius.

Upper case letters are used to denote vectors or, occasionally, matrices. P denotes a point, and N a surface normal vector. Point-valued or vector-valued expressions are bracketted by square brackets [ ]. Square brackets are also the convention for the vector triple product $[A, B, C] = [A \times B] \cdot C$

Unit vector expressions are denoted by the use of angle brackets $\langle\ \rangle$, or by the notation $\hat{N}$.

1. SURFACE DEFINITIONS

In data reduction and similar numerical activities, the word surface is sometimes used loosely to dramatize the behaviour of some function, and the surface equation with which most generalist mathematicians will be most comfortable is the simple form

$$z = f(x, y) \tag{1.1}$$

but numerical geometers working in the application of computing to design and manufacture soon found that this equation did not capture some of the most fundamental properties of manufactured artifacts.

The main problem was that equation (1.1) is firmly locked within one coordinate system. Rotating such a surface definition through a few degrees to a new position will in general not give a new definition of the same form as the old. We required surface descriptions which were closed (i.e. merely involved changes of coefficient values) under such operations as solid body rotations.

Those involved in computer graphics, who wanted to draw perspective pictures, preferred representations which were closed under perspective transformations.

In response to those needs, three generalizations have been used.

The first is the symmetric function of the coordinates.

$$f(x, y, z) = 0. \tag{1.2}$$

This form must have been discovered almost as soon as Descartes invented coordinates, and it has been the form in which the quadrics have been expounded ever since, in such text-books as Cohn (1961), McCrea (1960) and Eisenhart (1960).

The golden age of algebraic geometry in the mid- to late-1800s discovered all that was to be known about surfaces of this form when f was a polynomial function, and much of their knowledge still stands us in good stead today. See Book II of Coolidge (1963) for a good flavour of algebraic geometry.

This type of surface has been the main tool for those numerical geometers who have been representing machined artifacts, which are mainly bounded by surfaces which are plane or quadric, with the occasional torus. Such surfaces have fairly simple analytic equations.

The second is the parametric form

$$\begin{aligned} x &= f_1(u, v), \\ y &= f_2(u, v), \\ z &= f_3(u, v). \end{aligned} \tag{1.3}$$

This was devised by Gauss (1828) in his studies of mapping.

This form of surface has been the main tool used by those of us who have had to deal with smoothly flowing aesthetic shapes, such as those bounding aircraft, cars, ships, shoes or consumer articles such as hair-driers or electric shavers. Their main attraction is that it is fairly simple to make a piecewise definition, whereby one set of coefficients can be used in one part of the u, v domain, another set in another. You can change the shape of the rear fender without altering the front.

The third is a form which has appeared in own own generation. It does not have a single simple equation in terms of coordinates, but instead has a procedural definition, which says that a piece of surface lies within some piece of space. If you want to nail it down more tightly, an algorithm is available which divides the piece you have into smaller pieces, each of which lies within a smaller piece of space.

2. Analytic surfaces

This approach treats a bivariate point set definition as the set of zeros of some function of position

$$f(P) = 0.$$

Because we are concerned with real geometry in the colloquial sense, we deal with functions whose coefficients are real, and which map from $\mathbb{R}^3$ to $\mathbb{R}$.

A linear function gives a plane. Polynomial functions of higher order give, in general, curved surfaces. Quadratic functions give a family of surfaces called the *quadrics*, which includes the sphere, the cylinder, the cone, ellipsoids, and various paraboloids and hyperboloids. The first three of these are called the *natural quadrics*, and are important in that faces of machined objects are frequently of this form. In fact the whole technology of the representation of machined parts, whose complexity is primarily in the way large numbers of faces interact with each other, was built for at least its first ten years on natural quadrics only. In that context the faces are seen as boundaries between material and 'outside', and so the function is regarded as defining the half-space

$$f(P) < 0$$

rather than just the zero-set.

This section is concerned with faces in small numbers, and we do not distinguish particularly between the two sides of a surface.

Differential properties Consider a point P lying in a surface f. Because the surface is bivariate, there are directions in which we can move from P while remaining in the surface. Let such a direction be T, represented by a *tangent vector*.

Thus for small displacements such as $\mathrm{d}s$,

$$f(P + T\,\mathrm{d}s) = 0.$$

This can be expanded as a local Taylor series

$$f(P) + (\mathrm{d}f/\mathrm{d}x\, T_x + \mathrm{d}f/\mathrm{d}y\, T_y + \mathrm{d}f/\mathrm{d}z\, T_z)\, \mathrm{d}s = 0.$$

The term in brackets is just an inner product, which has to be zero

$$\mathrm{d}f/\mathrm{d}P \cdot T = 0.$$

Now T is any tangent vector, and $\mathrm{d}f/\mathrm{d}P$, which is a triple and can therefore be interpreted as a vector, must lie in the direction perpendicular to the surface, because its inner product with any tangent is zero. It is called the *surface normal*, and will usually be denoted by N.

Sometimes it is convenient to represent the direction rather than the magnitude of N. For this we take the vector in the same direction as N, but of unit magnitude. This is called the *unit surface normal*, and is denoted by $\hat{N}$.

If the function f is such that N is of unit magnitude near a point P of the surface, the value of f measures distance from the surface locally.

The second derivatives $\mathrm{d}^2 f/\mathrm{d}P^2$ form a matrix which encapsulates the local curvature behaviour. This is most obvious if we think of it as

$$\frac{\mathrm{d}}{\mathrm{d}P}\left(\frac{\mathrm{d}f}{\mathrm{d}P}\right),$$

which can be written as

$$\frac{\mathrm{d}N}{\mathrm{d}P}$$

so that the inner product with any displacement δP gives the corresponding change in surface normal.

2.1. *Definitions*

The most frequent form of analytic surface definition is by the specific geometric properties of the specific surface equation. For example, the centre and radius of a sphere, the vertex and semi-angle of a cone, a point on and the normal vector of a plane.

In a software system which is known to deal only with planes and quadrics, this initial data can be converted into the equivalent coefficient matrix of the homogeneous quadratic form.

$$P^i A_{ij} P^j = \begin{bmatrix} x & y & z & 1 \end{bmatrix} \begin{bmatrix} A & B \\ B^T & 1 \end{bmatrix} \begin{bmatrix} x \\ y \\ z \\ 1 \end{bmatrix} = 0. \tag{2.1}$$

Any more general system needs to provide a procedural interface, which will be described in more detail under Interrogations, below. There is a trade-off between holding data in the form in which it was first supplied,

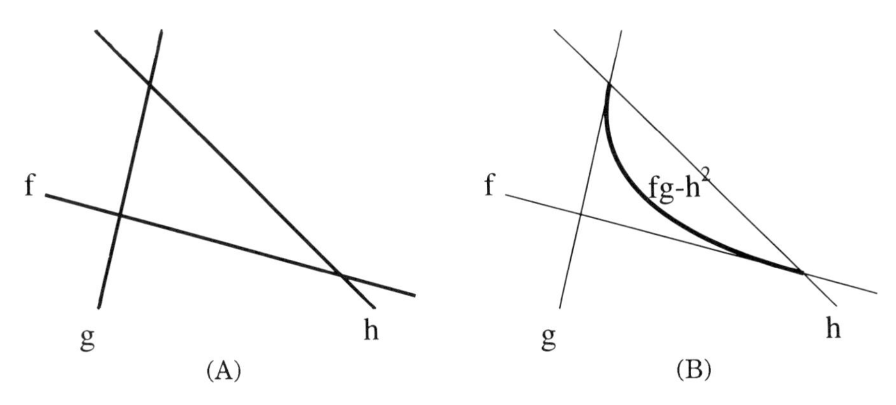

Fig. 1. Liming blends.

thus making editing more transparent, and doing as much pre-processing as possible to speed subsequent enquiries.

2.2. Blends and fillets

For the most part the analytic shapes which are actually defined are of low order, the highest being quartic for the torus. There is one exception to this: where blending surfaces are defined by an algebraic combination of the functions of the surfaces which they join.

The principle is a simple one, articulated by Liming in the late 1930s, but building on algebraic geometry ideas from the mid 19th century.

Let $f = 0$ and $g = 0$ be equations of two intersecting curves in two dimensions, and let $h = 0$ be the equation of a curve running close to their intersection. Then for any general value of a parameter λ, $fg - \lambda h^2 = 0$ is the equation of another curve, which lies tangent to each of f and g at its intersection with h.

If f, g and h are all of the same order, n, the order of the combination is $2n$. In Liming's work n was 1, and the blend curves conics.

If f, g and h are all normalized so that $\mathrm{d}f/\mathrm{d}P$ is of unit length locally, a value of λ of about 1 gives an intuitive fillet. Values nearer to 0 give a blend going closer to the intersection of f and g, and larger values give a blend which lies closer to h.

The same principle applies where f, g and h are surface equations in three dimensions. The blend, articulated by several groups of researchers almost simultaneously (Middleditch and Sears 1985; Rockwood and Owen, 1985; Hoffman and Hopcroft, 1986; 1987) is now a surface tangent to the base surfaces f and g all along their respective intersection curves with h.

Such surfaces are best described as blends, as the term fillet is usually taken to mean a surface generated by a ball rolling along the intersection (except by numerical geometers, who prefer to use the term fillet more generally).

The rolling ball fillet is a much harder shape to represent. We have to determine the locus of the ball's centre, which is the intersection of two surfaces offset from the base surfaces by the radius of the ball, and then construct the envelope of the ball as it travels along this locus. All these operations are in fact closed in the universe dealt with by algebraic geometry, but the orders of the objects involved escalate inconveniently. The equations also tend to have spurious parts to their solutions which do not correspond to the intended shape.

For example, the offset of an ellipse in 2D is a curve of order 8, which includes both inwards and outwards offsets. These two parts are not separable, in general, by factorization of the eighth-order equation into two fourth-order factors.

3. Parametric surfaces

Although most sculptured surface software deals with analytic surfaces to some extent, if only for planes, it is the parametric representation which is the workhorse.

In this form, the surface is regarded as a mapping from a *parameter plane* into R^3. The computational representation of the surface is the set of coefficients of the mapping.

The most general case is thus

$$P = F(C : u, v), \tag{3.1}$$

where u and v are the parameters, coordinates in the parameter plane, and C is the set of coefficients.

Differential properties The defining equation of a parametric surface can be differentiated with respect to the parameters. The two partial derivatives $\mathrm{d}P/\mathrm{d}u$ and $\mathrm{d}P/\mathrm{d}v$ lie tangent to the surface. This can be seen from the definition of the derivative as the limit of a difference.

$$\frac{\mathrm{d}P}{\mathrm{d}u} = \lim_{\delta u \to 0} \left[\frac{P(u + \delta u, v) - P(u, v)}{\delta u} \right]. \tag{3.2}$$

Both of the points in this equation lie in the surface, and the secant vector between them becomes a tangent in the limit.

The cross product of the two derivative vectors lies perpendicular to the surface, and is called the surface normal vector, N.

$$N = \mathrm{d}P/\mathrm{d}u \times \mathrm{d}P/\mathrm{d}v. \tag{3.3}$$

The unit vector in the same direction as N is called the unit surface normal, $\hat{N}$.

The second derivatives with respect to parameter provide information about the curvature of the surface. This is evident from the fact that the derivative of N with respect to parameter involves the second derivatives of P.

Definitions Although the general form of parametric surface equation is as quoted in equation (3.1), the subset actually used is much smaller. There are only two forms at all widely used for primary definition, the piecewise polynomials, and the transfinite interpolants.

3.1. Functions linear in the coefficients

The first limitation is to functions of u and v, linear in the coefficients.

$$P = \sum C_i \phi_i(u, v), \tag{3.4}$$

where the ϕ_i are scalar valued functions and the C_i are vector-valued coefficients.

There have been various attempts to use *shape parameters*, which are coefficients taking a nonlinear role, notably the superellipses of the Lockheed Master Dimensions system, and the ν-splines of Nielson (1974), but such coefficients are not found to be particularly easy to use, except possibly for generating academic dissertations.

3.2. Points and vectors

It is useful at this point to distinguish between *points* and *vectors*. Both are represented by triples of coordinates, but they transform differently under solid body transformations. If we change coordinate system, the coordinates of a point P are transformed to

$$P' = M \cdot P + O, \tag{3.5}$$

where M is a 3×3 orthonormal matrix and O is a point.

A vector, typified by the displacement from one point to another, is transformed according to

$$P' = M \cdot P. \tag{3.6}$$

We assume that the set of basis functions is independent, so that there is no set of nonzero scalar coefficients α_i such that $\sum \alpha_i \phi_i(u, v) = 0$.

There can then be at most one subset of the ϕ_i such that the sum of the functions within this subset is identically equal to 1. All the bases of interest do have such a subset. Call these functions the ρ_i and the remainder the ψ_i. Let the coefficients of the ρ_i be called R_i and of the ψ_i S_i.

Then the equation for a general point of the surface can be written

$$P(u,v) = \sum R_i \rho_i(u,v) + \sum S_i \psi_i(u,v). \tag{3.7}$$

Now suppose that we apply a point transformation to the R_i and a vector transformation to the S_i.

$$\begin{aligned} P'(u,v) &= \sum (M \cdot R_i + O)\rho_i(u,v) + \sum M \cdot S_i \psi_i(u,v) \\ &= M \cdot \sum R_i \rho_i(u,v) + O \sum \rho_i(u,v) + M \cdot \sum S_i \psi_i(u,v) \\ &= M \cdot \left(\sum R_i \rho_i(u,v) + S_i \psi_i(u,v)\right) + O \sum \rho_i(u,v) \\ &= M \cdot P(u,v) + O. \end{aligned} \tag{3.8}$$

This is exactly the transformation of $P(u,v)$ as a point, which is what we require for the surface representation to be closed under solid body transformation.

3.3. Tensor product surfaces

Even within the limitation to scalar functions with point and vector coefficients, the choice of basis functions ϕ is dominated by the tensor products where the coefficients are viewed as forming some kind of rectangular grid:

$$P = \sum C_{ij} \phi_i(u) \phi_j(v). \tag{3.9}$$

The mainstream of development can be followed by considering what univariate functions are applied within a tensor product. This has the great convenience of allowing the surface equations to be explained in terms of curves,

$$P = \sum C_i \phi_i(u) \tag{3.10}$$

because the surface equation can be regarded as two nested summations. The inner one describes a curve swept out by a point moving as u changes, the outer the way that this curve sweeps out a surface as v changes.

3.4. Piecewise polynomials

The first serious work used Hermite cubics within foursided patches considered as independent entities to be stitched together with various degrees of continuity. Ferguson (1965; 1993) built a system which enabled continuity of the first derivative to be achieved between patches which shared corner data. In fact he initially omitted some of the components of the full tensor product, but these were later added to the APTLFT-FMILL numerical control programming system very widely used in the 1970s and early 1980s.

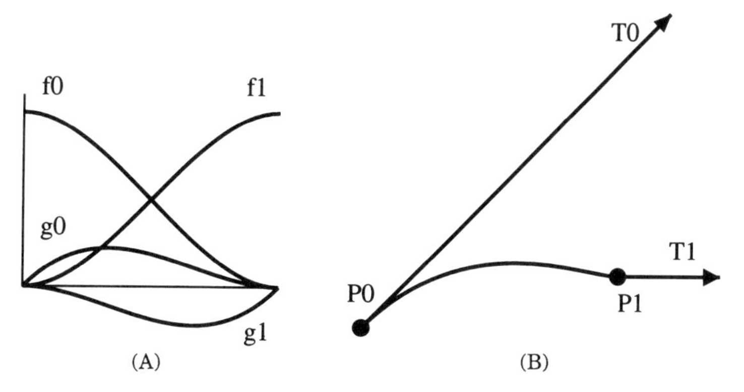

Fig. 2. Hermite functions and a Hermite curve.

The basis functions were

$$\begin{aligned} f_0(u) &= 1 - 3u^2 + 2u^3, \\ f_1(u) &= 3u^2 - 2u^3, \\ g_0(u) &= u(1-u)^2, \\ g_1(u) &= -(1-u)u^2. \end{aligned} \tag{3.11}$$

In this scheme, the coefficient of f_0 in a curve is the initial point, $u = 0$, that of f_1 the final point, $u = 1$. The coefficient of g_0 is the first derivative of position with respect to parameter at $u = 0$, of g_1 the first derivative at $u = 1$. The finite piece of curve defined in this way is the image of the interval from 0 to 1. In this context the distinction between open and closed intervals is not significant. Similarly, the piece of tensor product surface is the image of the unit square of parameter plane.

Clearly, with coefficients of this form, pieces of curve which will join with either continuity of position only or continuity of position and first derivative can easily be defined. This is achieved by having adjacent pieces share defining data.

In the surface context, the ff products have as coefficients the four corners of the patch, the fg products the corner values of the derivatives with respect to u, and the gf products the derivatives with respect to v. The gg products which Ferguson initially omitted have as coefficients the mixed partials $\mathrm{d}^2P/\mathrm{d}u\,\mathrm{d}v$.

The success of this scheme must not be understated. Hundreds of millions of dollars worth of aerospace parts have been machined successfully with this mathematics. However, it had two problems: using different magnitudes for the first derivative vector would give different curves, and it was not obvious how to choose the right magnitude first time; and that the mixed partials were even harder to choose.

One solution to both these problems was published by de Boor in 1962. He

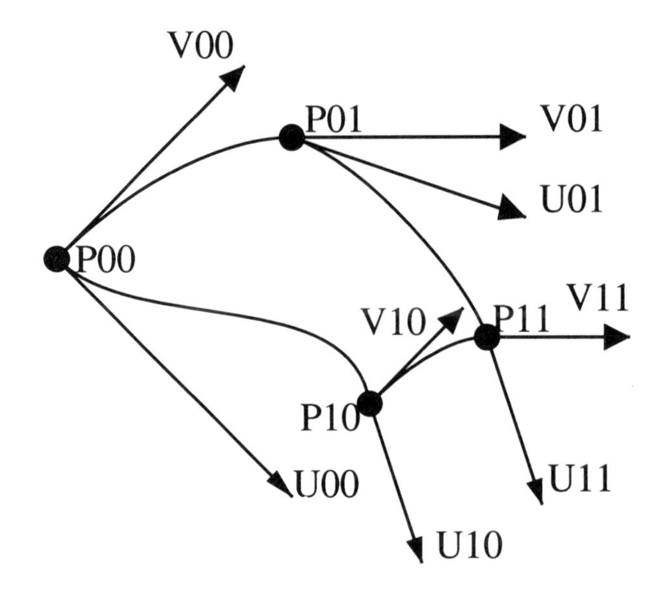

Fig. 3. Hermite surface definition.

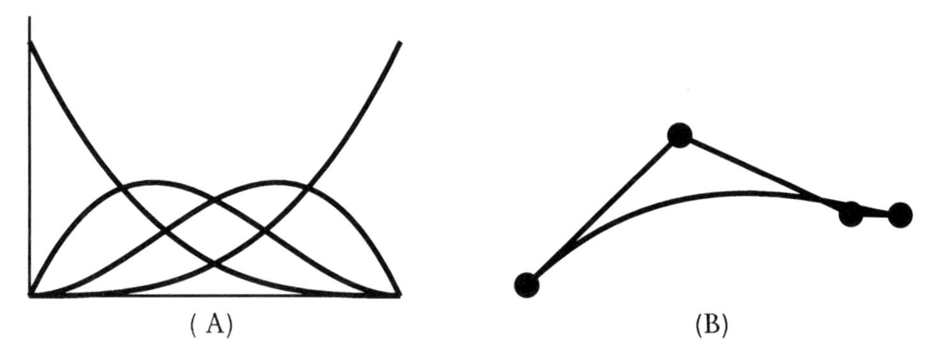

Fig. 4. Bernstein functions and a Bernstein curve.

used interpolating cubic splines as differentiation operators to determine all the required derivatives, starting from an array of points to be interpolated.

The other was devised independently by de Casteljau and by Bezier, both working in the French automobile industry (Bezier 1971).

They used, instead of the Hermite basis, the Bernstein basis

$$\phi_i(u) = \left(\frac{6}{i!(3-i)!}\right)(1-u)^i u^{3-i}, \quad i = 0, \ldots, 3. \tag{3.12}$$

The coefficients of these functions in the curve context are a sequence of points, usually thought of as forming an open polygon, usually called the control polygon. This has the same end-points as the curve itself, and is tangent to the curve at the end-points. Further, the curve has an inflexion

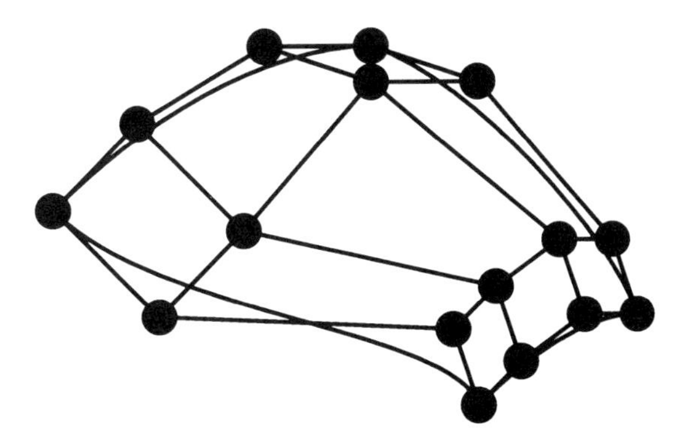

Fig. 5. Bernstein surface definition.

only if the polygon does. (It can also have two inflexions if the curve has a loop.)

In the surface context the coefficients form a rectangular network of points whose corners are the corners of the patch, and whose edges are the control polygons of the edges of the patch.

This approach solved both the vector magnitudes and the second derivatives problems. The obviously right vector magnitude was normally that which made all three of the polygon legs about the same length, but vectors could be shortened predictably to avoid inflexions; and a smoothly laid out control network would give good second derivatives.

It also made it possible for higher- or lower-order polynomials to be used with facility. The originators found useful transformations which permitted a part of a patch, trimmed by a cubic in uv-space to be expressed as a single patch, but of higher order.

However, the assembly of an array of patches to model, for example, a car door with a feature line running along it, was still a detailed task, which had to be performed with great accuracy if discontinuities of tangent plane were not to creep in.

3.5. B-splines

The next step in the development was a combination of the Bernstein and spline ideas. Bernstein brought legitimacy to the idea that control points need not be interpolated, splines the idea that a complete surface could be regarded as a single entity with piecewise basis functions, rather than a collection of independent pieces. The Ph.D. dissertation of Riesenfeld, supervised by Gordon (see Gordon and Riesenfeld, 1974) explored the use of the B-spline functions, first described much earlier by Schoenberg (1946), as analogues for the Bernstein polynomials.

B-spline functions are well explained in de Boor (1987). A univariate

B-spline function is a piecewise polynomial defined over a particular partitioning of the real line into segments over each of which the function is polynomial. The values at which these segments meet are called *knots.*

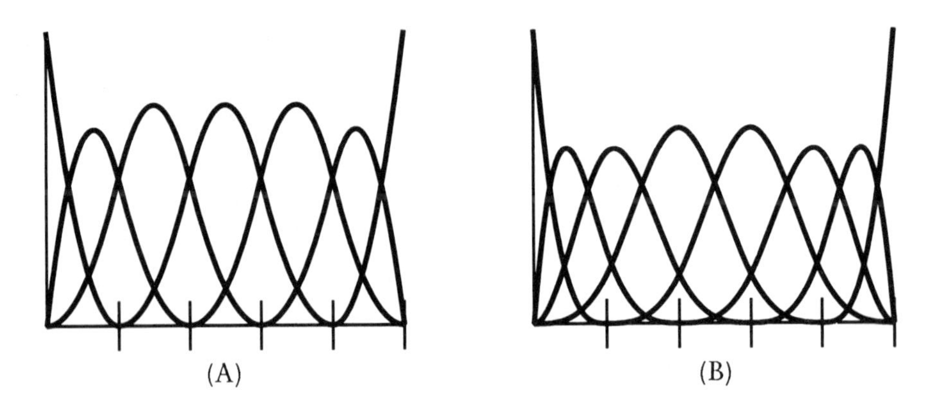

Fig. 6. Quadratic and cubic B-spline basis functions.

A B-spline function has finite support, i.e. it is nonzero only within a finite abscissa interval. It is strictly positive in the nonzero region. A nondegenerate B-spline has the maximum continuity possible with polynomial pieces of a given order, and minimum support. These properties are sufficient to define the function within a scaling factor, and the scaling factor for curve definition purposes is chosen so that the sum of all the nonzero B- splines at any given abscissa is exactly unity. As shown above, this partition of unity property is required so that when all the coefficients are transformed as points under a solid body transformation, the shape of the curve remains invariant.

By taking the limit as one of the pieces of abscissa becomes shorter and shorter, the concept of *coincident knots* (or even *multiple knots*) is obtained.

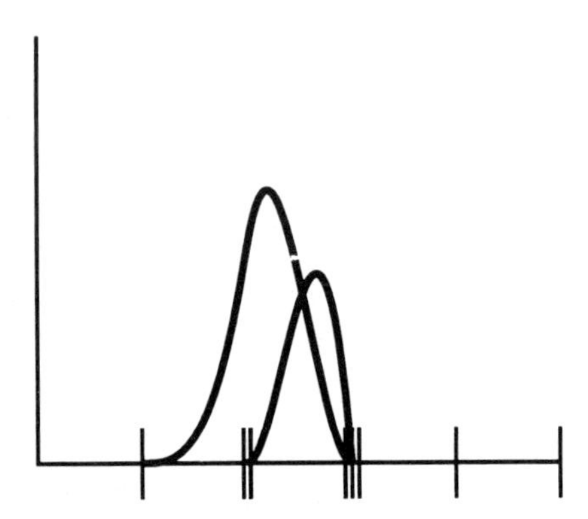

Fig. 7. Some degenerate B-splines.

B-splines with coincident knots can have discontinuities of lower than expected derivative, and narrower support regions.

A B-spline curve may be compared with a high-order Bernstein curve. The individual polynomial pieces for the B-spline are of lower order, but there are more of them. A change in the position of any one control point will, in general, affect only part of the B-spline curve, but all of the Bernstein curve.

A B-spline curve may also be compared with a collection of Bernstein curve segments. There are, in general, fewer control points for the same curve regarded as a B-spline. Moving one of the B-spline control points alters more of the shape, but maintains the continuity properties. Moving one of the Bernstein control points alters the shape, and in general reduces the degree of continuity.

In fact, a single Bernstein segment can be regarded as a special case of a B-spline, in which the domain is a single segment, with a sufficiently multiple knot at each end. Because B-spline curves can be concatenated to form another B-spline curve, a sequence of Berstein curves can also be regarded as a degenerate B-spline.

3.6. NURBS

Within the B-spline context, there are variations. If all the segments are of equal length in parameter space, the B-spline is termed *uniform*; if the software permits segments to differ in length, the B-spline is termed *non-uniform*.

The schemes described above all map from $\mathbb{R}^2$ to $\mathbb{R}^3$. It is also possible to map not into Euclidean space, but into projective space. It was mentioned above that the graphics community found it useful to have descriptions which were closed under perspective transformations.

This is achieved by mapping from $\mathbb{R}^2$ to $\mathbb{R}^4$, into the space of homogeneous coordinates.

Such an approach also makes it possible for the parametric form to represent a circle or ellipse exactly, using polynomial basis functions.

The conversion to Euclidean coordinates, which is necessary at some stage, is achieved by dividing the first three homogeneous coordinates by the fourth. This division gives the name *Rational* to the variant.

The full description Non-Uniform Rational B-Splines is universally truncated to the acronym *NURBS*.

Each control point now has four components. The extra degree of freedom is the *weight* by which the coordinates of a point are multiplied to give the first three components. These weights have the nature of nonlinear coefficients, and so are not easy to use when trying to achieve specific effects. Most designers use them only to match circular arcs; otherwise they just use unit weights. However, the ability to match conics, the closure under perspective transformations, and the better match to algebraic geometry assumptions make NURBS the scheme normally chosen for surface representation systems of the late 20th century.

3.7. Multivariate B-splines

There is a further extension of the B-spline ideas, described in de Boor (1993b), which should be usable as a set of basis functions for nontensor product surface description. The arrangement of control points need not be strictly rectangular. However, the definition of the partitioning of the domain is a nonlinear control, and they do not give the topological freedom which recursive division definitions offer. To the best of my knowledge no commercial surface definition software has yet applied them.

3.8. Transfinite interpolation

The previous sections have dealt with surfaces defined by a finite number of coefficients, usually the positions of control points. A second important stream defines a surface patch in terms of its bounding curves. Because each curve contains an infinite set of points, these surface definitions are referred to as *transfinite*. If the curves are themselves represented by a finite description, this can be mapped through to give a finite description surface as a special case of a transfinite one.

The first transfinite surface was devised by Coons (1967), and dealt with the problem of building a parametric surface defined over the unit square, meeting a pre-defined curve on each of its four sides.

There is a compatibility requirement that the curves should meet at the corners, and we assume that this is met.

Let the four edges be $U_0(v)$, $U_1(v)$, $V_0(u)$ and $V_1(u)$. The compatible corners are

$$\begin{aligned} U_0(0) = V_0(0) &= C_{00}, \\ U_0(1) = V_1(0) &= C_{10}, \\ U_1(0) = V_0(1) &= C_{01}, \\ U_1(1) = V_1(1) &= C_{11}. \end{aligned} \tag{3.13}$$

We require a surface $P(u, v)$, such that

$$\begin{aligned} P(u, 0) &= V_0(u), \\ P(u, 1) &= V_1(u), \\ P(0, v) &= U_0(v), \\ P(1, v) &= U_1(v). \end{aligned} \tag{3.14}$$

Coons approached this in two stages: first to match two of the edges and then to worry about the other two.

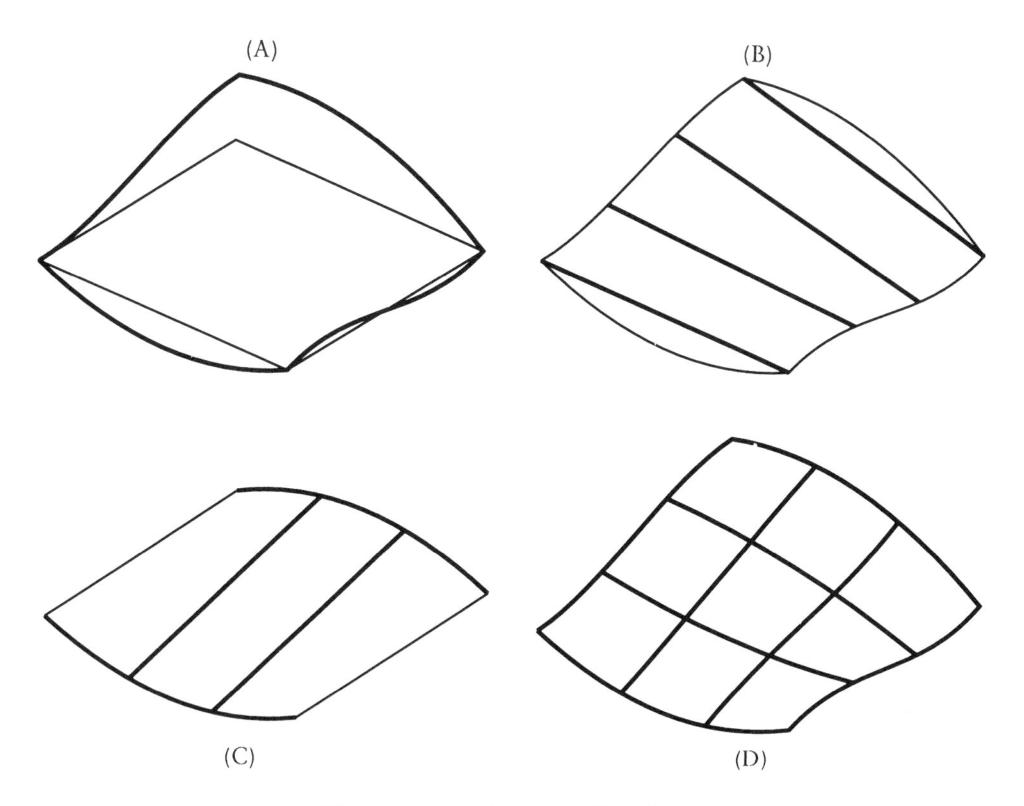

Fig. 8. Coons two-way blending.

The first two conditions above may be met by simple linear interpolation across the patch

$$P(u,v) = (1-v)V_0(u) + vV_1(u). \tag{3.15}$$

This does not match the other two conditions. There is a discrepancy of

$$U_0(v) - (1-v)V_0(0) - vV_1(0)$$

at the $u = 0$ edge and of

$$U_1(v) - (1-v)V_0(1) - vV_1(1)$$

at the $u = 1$ edge. So a second linear interpolation gives a correction function which can be added to the first try. Substituting the corner points by name, the correction becomes

$$(1-u)[U_0(v) - (1-v)C_{00} - vC_{01}] + u[U_1(v) - (1-v)C_{10} - vC_{11}] \tag{3.16}$$

giving the overall surface

$$\begin{aligned} P(u,v) \;=\; & (1-v)V_0(u) + vV_1(u) + (1-u)U_0(v) + uU_1(v) - (1-u)(1-v)C_{00} \\ & -(1-u)vC_{01} - u(1-v)C_{10} - uvC_{11}. \end{aligned} \tag{3.17}$$

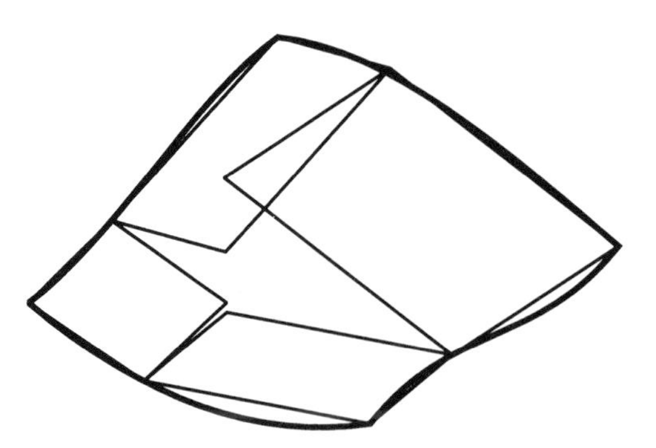

Fig. 9. Parallelogram construction.

Coons (1967) then took the 'add a correction' idea further, to achieve a match to given positions and cross edge derivatives, so that two adjacent patches, both being forced to match the same cross-derivatives, would land up tangent plane continuous. This used the f_0 and f_1 functions as described under Hermite surfaces above. However, it was not so successful, as, in the special case where the bounding edges were cubics, the twist terms were found to be zero.

Further, this approach gave only C^1 composite surfaces using cubic pieces. The splines gave C^1 using quadratics or C^2 using cubics.

Gordon (1969) made a much better generalization, by using interpolating spline functions as his generalization of the linear blending in the first-order Coons surface. This allowed a smooth surface to be passed through any compatible net of curves of constant u and curves of constant v.

The interpolation theorists formalized these ideas, and discovered that the Boolean sum of two interpolation or approximation operators was a powerful way of building new operators. Barnhill (1974) applied this technique to generate a transfinite surface in a triangle with given edges, and Gregory (1986) made one for n-sided regions with $n > 4$.

There is a second intuitive construction for the four-sided Coons patch.

Given a set of boundaries as before, and a value for the u, v pair at which a point is to be constructed, evaluate the edges at the appropriate points to give edge points $U_0(v)$, $V_1(u)$, $U_1(v)$, $V_0(u)$.

Now take each corner point, with the edge points on the adjacent edges, and construct a parallelogram. The fourth corner is an estimator for the required point. Take a weighted mean of these estimators, using weights inversely proportional to the logical area of the parallelogram.

As the required point nears one of the edges, the estimators from the corners at the ends of that edge have as their limit the edge point, and the weight of those two parallelograms dominate. Thus the surface interpolates the edges as required.

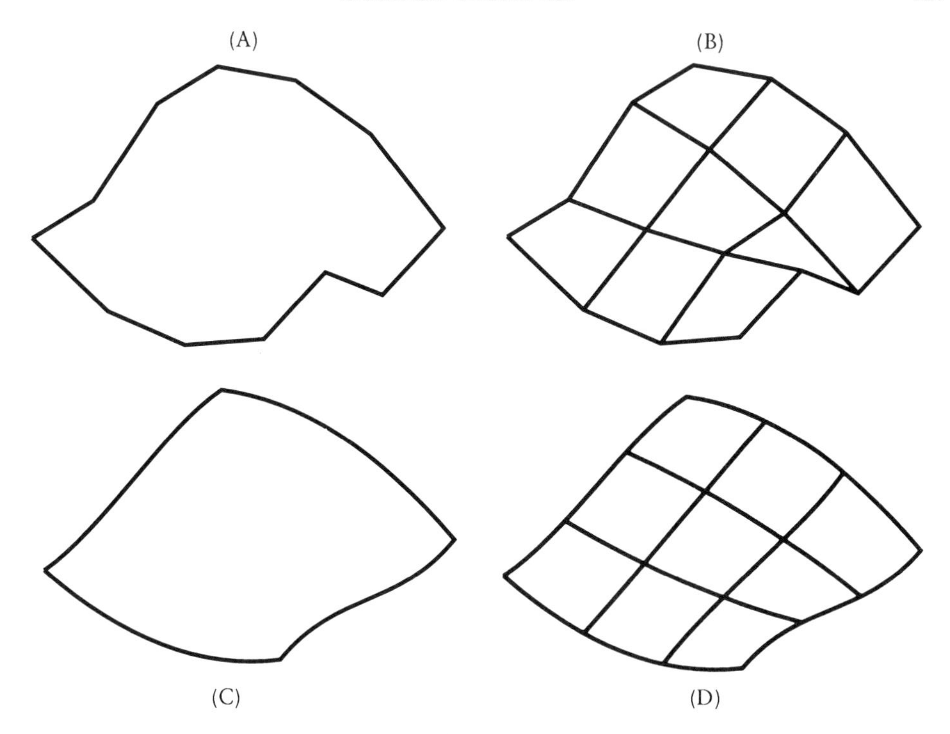

Fig. 10. Coons/Bernstein morphism.

This can be generalized to n-sided regions, and can also be applied to provide a way of estimating a surface when only three or even two of the sides are known.

Another intriguing property of the Coons first-order form is that if a cut is made across a patch, along a line of constant parameter, dividing the original patch into two, fitting each patch into its new boundaries gives exactly the same surface as the original.

Yet another is that if a first-order Coons patch is fitted to boundaries which are Bernstein curves, fitting the Coons surface to the curves gives exactly the same result as fitting a discrete interpolant to the control points of the boundary and then using Bernstein to interpolate the surface.

The transfinite techniques are extremely powerful, but more detailed in implementation than the B-spline ideas. They tend to be used in part-programming systems for numerically controlled machine tools, and in mesh generation for finite element analysis rather than in surface design systems.

A further method has been devised for fitting surfaces to desired boundaries. The Coons construction is actually the solution of a hyperbolic differential equation. Bloor and Wilson (1989; 1990) have tried using elliptic equations to determine the interior of a region with known boundary. There is much more computation in this case, because a point in the interior of an elliptic equation solution depends on all the boundary points, and so they now regard their technique as a way of generating the interior control points in a conventional NURBS representation.

3.9. Offset parametric surfaces

Offsets were mentioned above, as a step in the construction of rolling ball fillets. They are also extremely important in surface systems for two other purposes: the construction of the faces of objects which have a small but nonzero thickness, where a nominal surface may well lie in the centre of the object; and the calculation of tool centre paths for numerically controlled machining. (Bell *et al.*, 1974).

The latter is more demanding, and requires more generality.

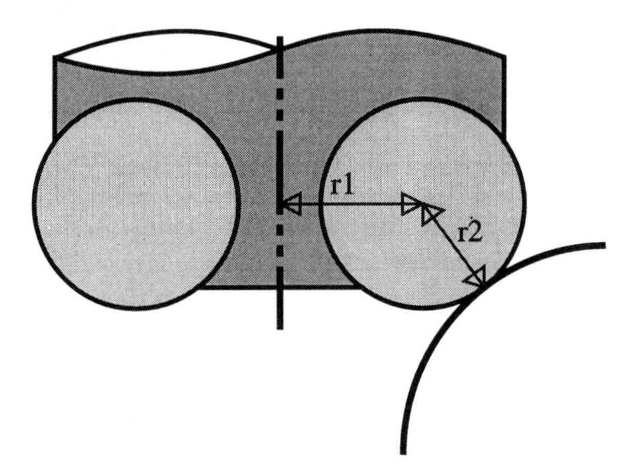

Fig. 11. Toroidal offset.

In the general case a machining cutter has teeth which sweep out a surface which is a surface of revolution made up of cylindrical, conical and toroidal pieces. The optimal cutting region is usually on just one of these pieces, and so the form of that piece is regarded as being the cuttter form for the purpose of computing a specific cut. It is usually a toroid.

Given a point at which the cutter envelope is to be tangential to the surface being machined, the question is where to put the tool datum point. Let this be the centre of the torus. Let the major and minor radii of the torus be r_1 and r_2 respectively.

Then moving up the surface normal from the cutting point by r_2 gives a point at the centre of a tooth. Moving perpendicular to the tool axis through r_1 gives the required tool centre.

$$P'(u,v) = P(u,v) + r_2\hat{N} + r_1\langle A \times N \times A\rangle. \qquad (3.18)$$

This is algebraically complex, because of the square roots implicit in the unit vectors, but computationally straightforward.

Tool paths can be computed in either of two ways, depending on circumstance. The first is to compute the path of the cutting point first, and then for each point along this path, correct it to give the tool centre. This is relevant when it is required to leave a particular trace on the surface.

The second is to treat the offset equation as the definition of an offset surface, and use the offset surface within some interrogation which defines the actual tool centre path. For example, the path where a cutter cleans out the intersection of two surfaces, leaving a fillet as side-effect, must be computed by intersecting the two offset surfaces. Efficient milling cutters, used for roughing, demand that the bottom of the cutter is never used for cutting; the tool centre path must lie in a plane perpendicular to the cutter axis, and so a plane section through the offset surface is used.

The offset surface can either be held for interrogation as an approximation using some other form, or the offset equation can be applied dynamically every time the evaluation of a point on it is required.

It should be noted that the offsetting context described above can introduce discontinuities of position where the surface normal lies in the direction of the cutter axis. A tiny disk around such a point maps into a thin annulus of radius r_1. This indicates that accurate approximation by one of the standard forms may be difficult.

4. Recursive division surfaces

One of the properties shared by the Bernstein surfaces and the B-splines is that their bases are variation diminishing. Any plane cuts a B-spline curve no more often than it cuts the control polygon. In the surface case, the equivalent property is that a piece of surface lies entirely within the convex hull of its control points.

This means that there is a simple test which may identify quickly that there is no intersection. This leads to a 'divide and conquer' style of interrogation based on the idea that if a piece of surface is simple enough it can be interrogated directly; if not, it can be divided into two, and a test applied to see whether each half contains any result of the interrogation. Examples of such interrogation algorithms will be presented later in this section.

The key to such an approach is the ability to produce cheaply the control networks for the halves of a subdivided surface. In the case of the Bernstein basis, the de Casteljau construction provides this, where the new control points are generated through a tableau of simple linear combinations. In the case of B-splines the knot insertion algorithms play a similar role, which pretend that a single span of the B-spline is actually two, although there is no discontinuity at the join.

It was observed that all of these constructions of new polygons and control nets consisted of taking linear combinations of existing control points, the coefficients in these linear combinations depending on the detail of the surface mathematics.

The structure of the interrogations did not depend on that detail at all. Provided that a new set of control points could be calculated, the interrogations would work.

This meant that it was possible to put forward a surface definition in terms of the division construction. Catmull and Clark (1978) generalized the bicubic B-spline surface, and Doo and Sabin (1978) the biquadratic B-spline surface. What both achieved was the ability to have a control network which matched the desired natural topology of the surface being defined, whereas the basic tensor product B-spline enforced that the control network had to be a regular rectangular grid of control points.

This was not just an aesthetic issue. A unit square of the real parameter plane can map nicely onto pieces of surface which are bounded and square, or onto a finite length of cylinder, or onto a complete torus. It can map onto all of a sphere except one point, but in the neighbourhood of that point the mapping is highly singular. It cannot map onto closed surfaces of higher topological type. The recursive division control networks have no such limitation, and oriented surfaces, open or closed, of any topology (see Griffiths, 1976) can be represented as unitary entities. There is only a small question of how to handle nonoriented surfaces.

The rules of this game were rapidly sorted out. If the control network were actually regular in some region, the generalized construction should be equivalent to one of the standard tensor product formulations. The irregularities should not multiply as division takes place, and then at each stage of division, more and more of the surface area of the surface can be identified as equivalent to a standard form. The unknown regions round isolated irregularities can then be analysed in terms of the eigenstructures of the operator which replaces the configuration of control points round the singularity by a new, smaller configuration.

Despite the topological freedom which this approach gives, there are some difficulties which have led to it not being taken up at all widely.

The first is that the quadratic, which behaves well in terms of mathematical analysis, and has continuity of tangent plane but not, of course, of curvature, at the singular points, tends to give rather bulgy surfaces noticeably different from what one expects. A control network in the form of a cube, for example, gives a closed surface significantly different from a sphere. In other words, the method introduces features of shorter spatial wavelength than the density of control points implies. These features are artifacts. Adding extra density of control points merely increases the spatial frequency of the artifacts.

The second is that the cubic, which suffers from the same problem but less severely, does not have a formulation giving continuity of curvature at singularities. The mathematical uncertainty associated with this may have discouraged system builders from basing their software on this approach.

A third is that it is the uniform B-spline which can be generalized in this way. The non-uniform capability has been found so useful that to lose it would be a backwards step.

5. Conversions between different representations

It is often the case that the same point-set can be represented in more than one way. As will be seen in the next section, on Interrogations, it may be advantageous to have more than one representation available. Indeed, recursive division interrogation is important because it can be used for parametric surfaces.

(i) The natural quadrics have a familiar parametrization in terms of trigonometric functions as well as the analytic form. So do tori.

Quadrics also have a parametric representation in terms of rational parametric quadratics, and the tori in terms of rational parametric biquadratics. The quadratic representation of the quadrics maps the entire parameter plane onto the surface of the quadric; if it is required to cover a complete quadric by the unit square a quartic representation is available, but there is always at least one singular point.

It is also true that nonsingular cubic analytic surfaces can be parametrized as rational polynomials, but it is not clear how much higher in order one can go and still have rational polynomial equivalents. Even for quadrics the degenerate case is difficult to parametrize (consider the case where the quadric consists of two planes), and the cubic cylinder whose generator is a plane cubic with no double point cannot be parametrized with rational polynomials.

Parametrization with elliptic functions or with functions including square roots will doubtless extend parametrizability, but eventually the fact that higher order surfaces can easily be of high topological type will get in the way, forcing the use of a complex parameter plane, which rather spoils any advantage of having both analytic and parametric forms available.

(ii) On the other hand, all rational parametric polynomials have an implicit (analytic) form, equal in order to the highest total degree in the polynomial.

This result applies only to a single piece of parametric surface. Where a surface has a piecewise definition, each piece has its own analytic equation, and deciding which piece is relevant at any given sample point is nontrivial.

(iii) The parametric bipolynomials can all be expresssed as B-splines, and thence as recursive division surfaces.

Transfinite surfaces can also be expressed as recursive division surfaces provided that:

— the curve cutting the surface in two can be evaluated;

— the bounding curves can be subdivided; and

— bounding boxes can be evaluated for all these curves.

The bounding box for a transfinite surface can then be expressed in terms of the boxes of its bounding curves and that of its corner points. The resultant box is rather slack, and so this kind of surface will not be very efficient for recursive division interrogation.

(iv) Recursive division surfaces which are of use all consist of parametric pieces. These pieces are of reasonable size where the control network is regular, but form an infinite regress around the points of singularity. The regress is terminated by the unresolved piece becoming small enough and flat enough to be approximated by a plane polygon. This is not really a conversion, since all recursive division interrogation falls back on some other method at the leaves of the division tree.

6. SURFACE INTERROGATIONS

A surface defined by any of the methods above is a barren thing indeed if we cannot make use of that definition in the production of real artifacts. To do this we need to ask questions of the surface such as

'What shape must I make the supporting members to fit inside the skin?'

'What path must a milling cutter take to machine the shape of this surface?'

'What shape does this piece of surface have to be in the flat, so that I can cut it out before I bend it into its final form?'

We also need to ask questions of a candidate surface to find out if its shape will be satisfactory for its purpose. Examples of these concerns are

'Make me a shaded image as seen from here.'

'Show me a set of closely spaced plane sections.'

'Calculate the lift and drag of this wing.'

All these questions are higher or lower level examples of interrogations, and the ability to provide the answers robustly and at acceptable computing cost are the important issues in the provision of surface software.

The list of possible enquiries is endless. Those described here include those most obviously required, and also others chosen to illustrate additional paradigms.

Within the scope of this section come the purely geometric enquiries; these need to be combined with application-specific knowledge of illumination models for rendering of shaded images, of feeds and speeds for machining, and of flow simulation calculations for the aerodynamics calculations, to provide such function.

This discussion of interrogation techniques starts with important properties of surfaces on which we can build. It then gives examples of those interrogations which return results consisting of isolated points, continues with those which return curves, and then proceeds to those which process complete surfaces.

Within each group one example is taken in reasonable depth; others are then dealt with by identifying their equations and mentioning any special features which need to be taken into account.

A summary follows each group of interrogations.

In most cases we consider all three of

- analytic surfaces, where all we can ask is the value and derivatives of the defining function at specific points in space;
- parametric ones where we assume that we know nothing about the surface except its boundary in parameter space and what we can determine by enquiring the position and derivatives of the surface point at specific parameter values;
- recursive division surfaces where all we can do is ask for a piece of surface to be split into a number of pieces, and enquire the extent of some hull round each piece.

Limiting the available knowledge about the surface form to this extent makes the techniques extremely general, and certainly not restricted to the specific surface equations described in Sections 2, 3 and 4.

6.1. Basic interrogation properties

This section defines the basic interrogations which must be supported for each of the three surface forms, if the methods described below are to be applied.

Basic properties of analytic surfaces An analytic surface is one whose point set is the set of zeros of some computable scalar function.

It is necessary to be able to compute the value of the function, f, at any point, P.

It is necessary to compute its first derivative, $\mathrm{d}f/\mathrm{d}P$, with respect to the coordinates, and, for some interrogations, particularly those concerned with curvature properties, the second derivative, $\mathrm{d}^2f/\mathrm{d}P^2$.

The first derivative is a vector which for points actually lying on the surface, is perpendicular to the local tangent plane. This we shall denote by N, and call the *normal vector.*

It is unusual for this to be of unit length. The vector of the same direction, but of unit length, is called the *unit normal vector* and is denoted by $\hat{N}$.

The second derivative is a symmetric tensor which may also be regarded as $\mathrm{d}N/\mathrm{d}P$.

Numerical differentiation is always available as a way of providing these derivatives, but for most surfaces found in CAD (computer aided design) systems, it is faster to differentiate the code which evaluates the function, and thereby produce accurate pointwise derivatives, using code which can often share comomon subexpressions with the basic evaluation.

Basic properties of parametric surfaces A parametric surface is a map from some part of $\mathbb{R}^2$, the *parameter plane* to $\mathbb{R}^3$, together with a description of the domain.

It is necessary to be able to evaluate this map at any point, (u, v), of the domain. Because the map is point-valued, we shall use the symbol P for both the map and the resulting point.

It is necessary to be able to evaluate the first derivatives, $\mathrm{d}P/\mathrm{d}u$ and $\mathrm{d}P/\mathrm{d}v$ of the map with respect to the parameters, and for some interrogations the second derivatives, $\mathrm{d}^2P/\mathrm{d}u^2$, $\mathrm{d}^2P/\mathrm{d}u\,\mathrm{d}v$ and $\mathrm{d}^2P/\mathrm{d}v^2$.

Again, specific code for these derivatives is preferable to numerical differentiation.

The two first derivatives are vectors which lie within the local tangent plane of the surface. The vector cross product $\mathrm{d}P/\mathrm{d}u \times \mathrm{d}P/\mathrm{d}v$ is a vector perpendicular to the local tangent plane.It is called the surface normal, and denoted by N. Again, it is not generally of unit length, and a unit surface normal is defined in the same way as for analytic surfaces.

If $||N|| = 0$ the surface is improperly parametrized, and the methods described below will probably fail. Note that $||N|| > 0$ implies $||\mathrm{d}P/\mathrm{d}u|| > 0$ and $||\mathrm{d}P/\mathrm{d}v|| > 0$.

It is necessary to be able to determine when a point (u, v) lies outside the domain.

It is also necessary to be able to scan round the boundary. The domain is a closed point set, so that parameter pairs lying on the boundary can be evauated as within the domain. The boundary will be treated here as a collection of curves in the parameter plane themselves parametrized by a scalar variable t.

Basic properties of recursive division surfaces A recursive division surface is the limit of the set of points in some collection of hulls, each hull in a collection corresponding to a piece of the total surface.

A hull is any convex point set guaranteed to contain all the points of its piece of surface. The convex hull is the minimal such, but slacker hulls, such as the convex hull of a set of B-spline control points, or the bounding box thereof, generally lead to faster algorithms because the hull tests become much simpler. The implicit assumption here is that a hull is the intersection of a number of support planar half-spaces of predetermined orientations. Exactly which orientations are used is decided by the interrogation software writer. The bounding box is not misleading as a mental image, though using more support directions can improve performance.

It is necessary to be able to evaluate the hull (i.e. the pedal distances of the support planes) for a piece of surface.

It is necessary to be able to divide the piece of surface into two, and then produce a hull for each part.

Recursive division interrogations will only be efficient if as this division proceeds, the hull shrinks in linear dimension proportionately with the piece of surface it represents. The total volume of the hulls of all the pieces of a surface must shrink as the division is made to deeper levels.

For interrogations relating to orientation and curvature properties it is also necessary to be able to evaluate a hull containing the unit surface normal vectors of all points in the piece.

It is necessary to be able to decide when a piece of surface is small or simple enough for direct methods to be applied. As a default, the condition can be used that the entire hull is within the geometric precision required, but this demands very deep division, which is expensive in computing time.

For the purposes of this section, it is assumed that the boundary of the recursive division surface is that implicit in its piece structure. This will not be the case when recursive division methods are used for robustness on trimmed parametric surfaces. In such cases resolving the boundary issues will have to use the methods described under the parametric surfaces paragraphs of each interrogation.

7. Nilvariate interrogations

Within this section we are looking for results which take the form of single points. In general there may be multiple solutions. The type of each interrogation is therefore

$$\text{surface} \times \text{auxiliary data} \mapsto \text{set of points}$$

where the auxiliary data are whatever is needed for the specific enquiry. There are no particular data-structure complications.

7.1. Raycasting

This is the problem of finding the intersection of a straight line with a surface. It can be used in graphics in order to render a shaded image of a surface, and it is also a useful constituent of other algorithms.

Straight lines have a number of alternative representations. We can convert between them, and we use this ability to use the most convenient form for each manifestation of this interrogation.

The most fundamental form for any algebraic curve is probably the Cayley form

$$l(P,Q) = 0 \text{ if the line through } P \text{ and } Q \text{ meets the line } L. \tag{7.1}$$

In the case of a straight line the function l is bilinear.

$$l(P,Q) = P^i \cdot L_{ij} \cdot Q^j. \tag{7.2}$$

By taking two arbitrary values for Q we reach the form where the line is defined as the intersection between two planes

$$\begin{aligned} P^i.F_i &= 0, \\ P^i.G_i &= 0. \end{aligned}$$

In Euclidean coordinates a plane, F, can represented by a point thereon, Q_F, together with a plane normal F.

$$\begin{aligned} [P - Q_F] \cdot F &= 0, \\ [P - Q_G] \cdot G &= 0. \end{aligned} \tag{7.3}$$

For good conditioning, the normals F and G should be orthogonal to each other.

Taking any two distinct points of the line (and in homogeneous coordinates the intersections of the line with the four coordinate planes will always contain two distinct points), P_1 and P_2, any linear combination $\alpha P_1^i + \beta P_2^i$ will also satisfy equations (7.3).

In Euclidean coordinates we need to apply the normalization $\alpha + \beta = 1$, and it is most convenient to express the general point of the line in terms of

$$P = P_1 + \alpha T, \text{ where } T = [P_2 - P_1]. \tag{7.4}$$

From a given point P, known to lie on the line, the value of α can be recovered by

$$\alpha := \frac{[P - P_1] \cdot T}{T \cdot T}. \tag{7.5}$$

For some purposes the normalization can be carried further. If T is a unit vector, α measures distance along the line. This is useful only when dealing with metric properties, or when (7.5) is being invoked really often for a single line.

Raycasting analytic surfaces Here we use the parametric form of the line $P = P_1 + \alpha T$. Substituting this into the analytic equation $f(P) = 0$ gives an equation in α, thus reducing the problem to root-finding.

Root-finding of general functions is a nontrivial exercise, but for polynomials there are several well-understood methods. The problems are essentially those of conditioning: ill conditioned geometry can exist in the form of almost-tangencies, and will lead to ill conditioned roots. The purely numerical ill conditioning can be minimized by use of the Bernstein basis for all polynomials (Farouki, 1987a,b; Farouki and Rajan, 1988a,b).

When this interrogation is being used for graphics, the interesting root is the smallest positive one, corresponding to the nearest leaf of the surface to the eye point in the direction of view.

Raycasting parametric surfaces If L is represented as the intersection of two planes, the problem becomes one of solving two simultaneous equations.

$$\begin{aligned} [P(u,v) - Q_F] \cdot F &= 0, \\ [P(u,v) - Q_G] \cdot G &= 0. \end{aligned} \tag{7.6}$$

From any given starting point (u_0, v_0) sufficiently close to the solution, Newton iteration is fast and effective.

Expanding the surface as a local Taylor series about P_0 $(= P(u_0, v_0))$ gives

$$P(u_0 + \delta u, v_0 + \delta v) = P_0 + (\mathrm{d}P/\mathrm{d}u)_0 \delta u + (\mathrm{d}P/\mathrm{d}v)_0 \delta v. \tag{7.7}$$

Substituting this into equation (7.3) gives

$$\begin{aligned} [P_0 - Q_F] \cdot F + (\mathrm{d}P/\mathrm{d}u)_0 \cdot F\delta u + (\mathrm{d}P/\mathrm{d}v)_0 \cdot F\delta v &= 0, \\ [P_0 - Q_G] \cdot G + (\mathrm{d}P/\mathrm{d}u)_0 \cdot G\delta u + (\mathrm{d}P/\mathrm{d}v)_0 \cdot G\delta v &= 0, \end{aligned} \tag{7.8}$$

which may be cast as a matrix equation for δu and δv

$$\begin{bmatrix} \mathrm{d}P/\mathrm{d}u \cdot F & \mathrm{d}P/\mathrm{d}v \cdot F \\ \mathrm{d}P/\mathrm{d}u \cdot G & \mathrm{d}P/\mathrm{d}v \cdot G \end{bmatrix} \begin{bmatrix} \delta u \\ \delta v \end{bmatrix} = \begin{bmatrix} -[P_0 - Q_F] \cdot F, \\ -[P_0 - Q_G] \cdot G. \end{bmatrix} \tag{7.9}$$

solution of which gives increments of u and v to the next estimate.

$$\begin{aligned} u_{i+1} &:= u_i + \delta u_i, \\ v_{i+1} &:= v_i + \delta v_i. \end{aligned}$$

Iteration proceeds until the point evaluated lies within the required tolerance of each of the planes. For this test to be most appropriate, the planes themselves should be close to orthogonal.

The straightfoward Newton method described above almost always works fast and efficiently. Experience suggests that in typical surface interrogation situations the average number of steps taken is two (three evaluations altogether).

However, in the worst case, it is possible for divergence to happen. It is also possible for the computed increment to lead to a point outside the domain of the surface, so that strictly it cannot be evaluated.

Enforcement of convergence Even though the actual step may not result in reduction of the residuals of both halves of (7.3), the increment direction is a descent direction, and so a small enough step in that direction will lead to a better estimate. This is exactly the line search issue in unconstrained optimization (Nocedal, 1992).

Because this situation occurs rarely, we need not hunt for an optimum algorithm. A simple fix-up considers each of the residuals separately.

Let the residuals of one equation in (7.3) at two successive estimates be r_i and r_{i+1}.

If these are of opposite sign, the appropriate fraction of the step to take is $r_i/(r_i - r_{i+1})$, derived by linear interpolation: if they are of the same sign, use $r_i/(2r_{i+1})$, derived by linear interpolation on estimates of first derivative of residual.

Each of the equations in (7.3) gives an estimated fraction of the step to take. (If $|r_{i+1}| < |r_i|$ then the appropriate fraction is 1.) Use whichever fraction of the estimated step is smaller.

This gives a new P_{i+1} for which the residuals must again be estimated, and if necessary the fix-up must be repeated. In order to be able to prove termination of this loop there has to be a counter, because geometric situations can occur where there is no solution. In the case of raycasting this happens when the surface we are dealing with is not C^0, (for example, the offset of a surface which is not C^1) but when applying the same approach to situations where the equations being solved involve derivatives of the surface it can happen when some higher continuity is lacking.

The refinement part of the interrogation then has to report back to its calling code that no solution exists in this locality.

Enforcement of boundary When the next estimate lies outside the domain, we need to find a local point within the domain, so that the iteration can continue, because although one estimate has gone outside, the final solution may lie inside the domain. Since the boundary is typically well behaved, at most two segments of boundary are involved. If two are involved we use the corner itself as the next estimate; if only one, the intersection of the increment with the boundary.

If the solution really does lie outside the boundary, the refinement step must again report absence of a solution to its calling code.

Multiple solutions Multiple solutions can occur. It is possible for a ray to cut even a single bicubic patch in as many as 18 distinct points.

The second case of figure 12 is rather more worrying, as there is no separation of the solutions by a locus where the surface normal is orthogonal to the ray axis.

Each starting point leads to at most one solution. Where there are multiple solutions we need multiple starting points. Different starting points may

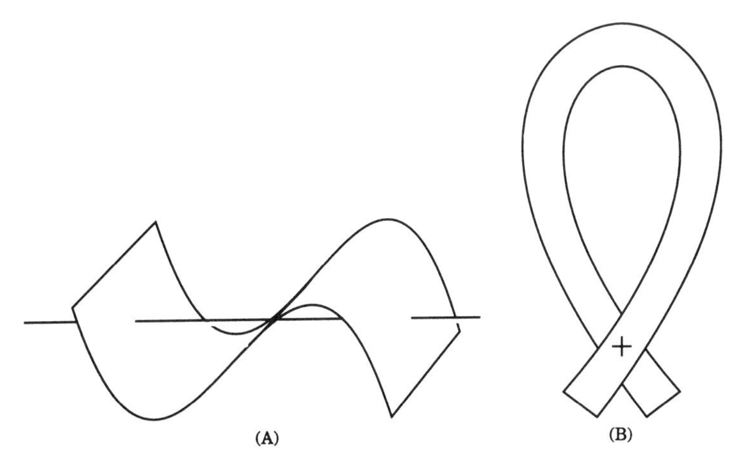

Fig. 12. Multiple solutions for raycasting.

converge to close estimates of a common solution: two different starting points may also converge to close but different solutions.

Handling this issue is thoroughly messy and *ad hoc*. Do not expect profound mathematics here.

The simplest approach is to use a dense spatter of starting points, and refine each of them. Finally, sort the solutions along the line, and combine those which are close in the parameter plane as well as in space.

This is unsatisfactory on two counts. First, there is no obvious way to decide how dense is dense enough: second, even a 10×10 grid means 100 evaluations; each refinement only three.

Reducing the number of start points considered is well worth while, reducing the number actually refined even more so.

System builders should consider associating information with each parametric surface, indicating how crinkly it is. This could be determined at the time the surface is defined, or over-ridden manually by the system operator. This information would then be used to decide the density of the spatter for all subsequent interrogations.

When the spatter is dense enough, the start points can be ranked according to the likelihood of their refining to a new solution. The first measure, in the case of raycasting, is the distance of the candidate from the line. However, there may be several starts close to one solution, while another solution may be some distance from the nearest start which leads to it. Starts which have a nearby start closer to the line should therefore be ranked low.

Coherence Rays are seldom cast in ones. A standard technique for improving performance is to use information from previous nearby instances of the same enquiry. For example, the places where the ray for an adjacent pixel cut the surface are likely to be close to where this one cuts it, and can act as excellent start points.

However, situations can change from one pixel to the next, and overuse of coherence can lead to being caught on the wrong leaf of a surface.

Two rays can have different sets of cut points if either a surface boundary or a silhouette come between them.

Raycasting recursive division surfaces In a recursive division of a surface, all those pieces whose hulls do not intersect the line can be eliminated. This is efficient, particularly when the hulls are aligned with the direction of the line. It is probably worth setting up a special set of hulls when a large array of parallel rays are to be cast.

If the hulls are not aligned, it may be simpler just to cull all pieces except those which cross both of the planes.

In the graphics situation, it is also possible to eliminate all pieces which lie further from the eye point than the nearest root found so far, and so some sorting of the pieces on the stack may be worth while, so that the nearest root is found early.

7.2. Maximum extent

This enquiry determines the places within a surface which might well be points of maximum extent in a direction given by a vector T. Its primary use is as an auxiliary interrogation for other enquiries, particularly plane sections. It is chosen for inclusion here to illustrate enquiries which depend on surface derivatives.

Maximum extent of parametric surfaces The equation which solution points satisfy is $T \times N = 0$, where N is the surface normal, given by $\mathrm{d}P/\mathrm{d}u \times \mathrm{d}P/\mathrm{d}v$.

When this is true, we have $\mathrm{d}P/\mathrm{d}u \cdot T = 0$ and $\mathrm{d}P/\mathrm{d}v \cdot T = 0$.

Each of these has a residual at any given surface point, and so we can solve for increments to u and v which reduce these residuals to zero in a way precisely analogous to the raycasting algorithm.

The Taylor expansions of $\mathrm{d}P/\mathrm{d}u$ and $\mathrm{d}P/\mathrm{d}v$ are

$$\frac{\mathrm{d}P}{\mathrm{d}u}(u_0+\delta u, v_0+\delta v) = \frac{\mathrm{d}P}{\mathrm{d}u}(u_0,v_0) + \left(\frac{\mathrm{d}^2P}{\mathrm{d}u^2}\right)_0 \delta u + \left(\frac{\mathrm{d}^2P}{\mathrm{d}u\,\mathrm{d}v}\right)_0 \delta v$$

$$\frac{\mathrm{d}P}{\mathrm{d}v}(u_0+\delta u, v_0+\delta v) = \frac{\mathrm{d}P}{\mathrm{d}v}(u_0,v_0) + \left(\frac{\mathrm{d}^2P}{\mathrm{d}u\,\mathrm{d}v}\right)_0 \delta u + \left(\frac{\mathrm{d}^2P}{\mathrm{d}v^2}\right)_0 \delta v$$

so that the matrix equation for δu and δv becomes

$$\begin{bmatrix} \mathrm{d}^2P/\mathrm{d}u^2\cdot T & \mathrm{d}^2P/\mathrm{d}u\,\mathrm{d}v\cdot T \\ \mathrm{d}^2P/\mathrm{d}u\,\mathrm{d}v\cdot T & \mathrm{d}^2P/\mathrm{d}v^2\cdot T \end{bmatrix} \begin{bmatrix} \delta u \\ \delta v \end{bmatrix} = \begin{bmatrix} -\mathrm{d}P/\mathrm{d}u\cdot T \\ -\mathrm{d}P/\mathrm{d}v\cdot T \end{bmatrix}. \tag{7.10}$$

Solution, including forced convergence and handling of the boundary proceeds exactly as in the raycasting case.

Maximum extent of recursive division surfaces Because we are basically solving the equations

$$\mathrm{d}P/\mathrm{d}u \cdot T = 0 \quad \text{and} \quad \mathrm{d}P/\mathrm{d}v \cdot T = 0,$$

recursive subdivision has to use culling based on the hulls of the first derivatives.

This is straightforward if we are using the recursive division form of a parametric surface, but if we are using a fully general recursive division surface the independent derivatives may not exist over pieces encountered early in the subdivision.

In this circumstance we need to recast the overall equation to use the surface normal hull.

Set up two auxiliary vectors, R and S, perpendicular to T and to each other.

We are now solving for

$$R \cdot N = 0, \quad S \cdot N = 0,$$

so if the surface normal hull of a piece does not cross the plane perpendicular to R passing through the origin, or if it does not cross the plane perpendicular to S, we can cull the piece.

7.3. Nearest point

We seek the nearest point, P, of a given surface to a given point Q.

This is a useful 'building-block' interrogation, second only to raycasting, since it is often useful for particular places on a surface to be indicated by the operator by specifying a nearby point, leaving the program to supply the precision.

It is also used in machining, where a path is traced until the tool centre meets a 'check surface'. This is checked efficiently by measuring the distance from a sample point to the nearest point on the check surface. The tool can then move that far along its trajectory before another test need be made.

Nearest point on an analytic surface Because we are not considering boundaries or discontinuities on analytic surfaces, the nearest point lies at the foot of a perpendicular.

A first estimate is derived very simply as

$$P_1 = Q - \left(\frac{f(Q)}{N_Q \cdot N_Q} \right) N_Q.$$

Typically this first estimate, when evaluated, will neither lie on the surface f nor have its normal passing through Q.

However, the Taylor expansion of f about P_1 gives a better estimate of the local tangent plane and the local perpendicular direction.

Unfortunately, if the distance of Q from the surface is larger than the local radius of curvature, the obvious iteration, intersecting the line from Q along N_1 with the Taylor plane, can oscillate and even diverge. It is necessary to take second derivatives into account.

The equation set to be solved is

$$\begin{aligned} f(P) &= 0, \\ [P-Q] \times N &= 0, \end{aligned}$$

where the second equation has rank 2, giving three equations in the three coordinates of P.

Let the residuals at P_1 of the equations be r_1 and R_1 respectively. Then for a step δP, the residuals will alter by amounts estimated by differentiating the above, and the Newton step is given by

$$\begin{aligned} \frac{\mathrm{d}f}{\mathrm{d}P}(P_1) \cdot \delta P &= -r_1 \\ -N(P_1) \times \delta P + [P-Q] \times \left[\frac{dN}{\mathrm{d}P} \cdot \delta P\right] &= -R_1. \end{aligned} \tag{7.11}$$

With this extra term, the iteration can converge on a point P whose distance from Q is a local maximum, rather than minimum. This is unlikely with surfaces of low order, but it is a wise precaution, having converged on an initial solution to set up a spatter of points at, for example, the vertices of a regular icosahedron centred on Q, with radius QP. Provided all of these points have a value of f of the same sign as Q there is confidence that the right solution has been reached. If not, a good starting point for a new convergence will be the point on the line between Q and the spatter point of most opposite sign of f, divided in the ratio of the two f-values.

Nearest point on a parametric surface The parametric surface case brings two complications: the first is finding a good place in the parameter plane from which to start; the second is the boundary.

Clearly this is a classical example of constrained optimization. Unfortunately, the mathematical interest in optimization has been in solving problems of high dimension in machines of limited memory. The dimension here is only two, and the ambition level significantly higher in terms of robustness and performance in the typical well-behaved cases.

Finding a good starting point can be addressed by spattering, just as in raycasting. In this case there is an argument for finding the nearest point on the boundary to Q, and using that as the first starting point. This is because scanning round the boundary is only a univariate haystack to find the needle in, and because the nearest point on the boundary may indeed be the solution. Unlike raycasting, there is always at least one nearest point of a surface, and it may lie on the boundary. Starting from such a point is easier than having to work out what to do when the normal process tries to cross the boundary.

The same Newton principle gives as an iterative solution method based on the equations

$$[P - Q] \cdot \mathrm{d}P/\mathrm{d}u = 0, \quad [P - Q] \cdot \mathrm{d}P/\mathrm{d}v = 0,$$

$$\begin{bmatrix} \phi_{uu} & \phi_{uv} \\ \phi_{vu} & \phi_{vv} \end{bmatrix} \begin{bmatrix} \delta u \\ \delta v \end{bmatrix} = \begin{bmatrix} -[P_1 - Q] \cdot \mathrm{d}P/\mathrm{d}u \\ -[P_1 - Q] \cdot \mathrm{d}P/\mathrm{d}u \end{bmatrix}, \tag{7.12}$$

where

$$\begin{aligned}
\phi_{uu} &= \mathrm{d}P/\mathrm{d}u \cdot \mathrm{d}P/\mathrm{d}u + [P_1 - Q] \cdot \mathrm{d}^2P/\mathrm{d}u^2 \\
\phi_{uv} &= \mathrm{d}P/\mathrm{d}u \cdot \mathrm{d}P/\mathrm{d}v + [P_1 - Q] \cdot \mathrm{d}^2P/\mathrm{d}u\,\mathrm{d}v \\
\phi_{vu} &= \mathrm{d}P/\mathrm{d}u \cdot \mathrm{d}P/\mathrm{d}v + [P_1 - Q] \cdot \mathrm{d}^2P/\mathrm{d}u\,\mathrm{d}v \quad \text{and} \\
\phi_{vv} &= \mathrm{d}P/\mathrm{d}v \cdot \mathrm{d}P/\mathrm{d}v + [P_1 - Q] \cdot \mathrm{d}^2P/\mathrm{d}v^2.
\end{aligned}$$

The problem with the straightforward Newton method converging on a furthest point occurs in the parametric case too. With a sparse initial spatter it is probably more likely. It can be detected by the Newton step not being a descent direction. When this happens an appropriate response is to use the steepest descent direction instead. The actual step to use is that given by equation (7.12) with the second derivative terms omitted. This fails only when an initial estimate falls exactly on a local maximum of distance.

Nearest point on a recursive division surface In a recursive division of a surface, all those pieces can be eliminated whose nearest point is further from Q than the nearest furthest-point of any piece met so far. More sophisticated tests using the hull of surface normals would risk discarding a nearest point which was not a foot of a perpendicular.

7.4. Umbilic points

An umbilic point is one at which the principal curvatures are equal, so that the principal directions are indeterminate. Calculating the positions of any umbilics is the first stage in dividing a surface into principal patches.

This enquiry is included here as an example of the use of second derivatives of a parametric surface.

The condition for an umbilic is that for some value of the local curvature k,

$$\begin{aligned}
\mathrm{d}^2P/\mathrm{d}u^2 \cdot N &= k\,\mathrm{d}P/\mathrm{d}u \cdot \mathrm{d}P/\mathrm{d}u, \\
\mathrm{d}^2P/\mathrm{d}u\,\mathrm{d}v \cdot N &= k\,\mathrm{d}P/\mathrm{d}u \cdot \mathrm{d}P/\mathrm{d}v, \\
\mathrm{d}^2P/\mathrm{d}v^2 \cdot N &= k\,\mathrm{d}P/\mathrm{d}v \cdot \mathrm{d}P/\mathrm{d}v.
\end{aligned}$$

Eliminating k from these three equations, and noting that a well behaved surface may easily have $\mathrm{d}P/\mathrm{d}u \cdot \mathrm{d}P/\mathrm{d}v = 0$, we can derive two equations of reasonable symmetry.

$$\begin{aligned} \frac{\mathrm{d}P}{\mathrm{d}u} \cdot \frac{\mathrm{d}P}{\mathrm{d}u}\frac{\mathrm{d}^2P}{\mathrm{d}v^2} \cdot N &= \frac{\mathrm{d}P}{\mathrm{d}v} \cdot \frac{\mathrm{d}P}{\mathrm{d}v}\frac{\mathrm{d}^2P}{\mathrm{d}u^2} \cdot N, \\ \left(\frac{\mathrm{d}P}{\mathrm{d}u} \cdot \frac{\mathrm{d}P}{\mathrm{d}u} + \frac{\mathrm{d}P}{\mathrm{d}v} \cdot \frac{\mathrm{d}P}{\mathrm{d}v}\right)\frac{\mathrm{d}^2P}{\mathrm{d}u\,\mathrm{d}v} \cdot N &= \frac{\mathrm{d}P}{\mathrm{d}u} \cdot \frac{\mathrm{d}P}{\mathrm{d}v}\left[\frac{\mathrm{d}^2P}{\mathrm{d}u^2} + \frac{\mathrm{d}^2P}{\mathrm{d}v^2}\right] \cdot N. \end{aligned}$$

Provided that the surface is C^2, these two equations can be solved using exactly the same techniques as raycasting and maximum extent above. Note that N is not a constant, but itself depends on the first derivatives of P, and so the four terms of the Newton matrix become fairly complicated expressions involving first, second and third derivatives.

7.5. Reflections of points

Suppose that we want to know at what point of a surface $P(u,v)$ the reflection appears of a point Q, as seen from a point E. This requirement appears in the checking of candidate surfaces for fairness in car design.

This problem is closely related to that of nearest points; the nearest point, P, to a given point, Q, if it lies within a surface, rather than on its boundary, is also a point at which a light ray from Q to Q bounces off the surface. The difference is that there may be no solutions, or many.

The equations to be solved are that

$$[\langle P - Q\rangle + \langle P - E\rangle] \times N = 0, \tag{7.13}$$

which may be resolved into

$$\begin{aligned} [\langle P - Q\rangle + \langle P - E\rangle] \cdot \mathrm{d}P/\mathrm{d}u &= 0, \\ [\langle P - Q\rangle + \langle P - E\rangle] \cdot \mathrm{d}P/\mathrm{d}v &= 0. \end{aligned}$$

We do require second derivatives for stability, but all zeros of the left-hand side function are valid solutions, and so, unlike nearest point, situations where the local Hessian is negative or mixed are required solutions.

7.6. Summary of nilvariate interrogations

We have seen examples above of enquiries which depend only on surface position, and those which depend on derivatives of the surface. Each has its own pair of equations which can normally be solved by an initial scan, followed by Newton iteration.

Generally speaking the biggest problem is finding the global solution when local ones also exist, or all the solutions if we want all of them. This demands fine scanning, except where recursive division methods can be applied.

The other problem is the interaction of the boundary, which may cut off solutions which the iteration would lead to, or which may require constrained solutions, satisfying different equations.

Recursive division interrogation avoids all of these problems, but the prospect of setting up hulls for many higher derivatives is a somewhat daunting one.

8. Univariate interrogations

Within this section we are looking for results which take the form of curves. In general there may be multiple pieces of curve within the required solution, and so the type of each interrogation is

$$\text{surface} \times \text{auxiliary data} \mapsto \text{set of curves.}$$

In the case of surface/surface intersection, the second surface is part of the auxiliary data, though the treatment is relatively symmetric between the two surfaces.

The data structure complication which does arise is the question *'What is a curve?'*.

A curve is a set of points satisfying some definitional equation, but this set, unlike the sets of points returned from univariate interrogations, contains a continuum of points. We therefore need a finite representation.

The most concise finite representation is the definition of the curve which forms the input to the interrogation, but that is not explicit enough. The whole point of the interrogation algorithms is to calculate an explicit form which can be drawn or printed or used in subsequent interrogations without repeating much calculation.

The ideal almost-explicit form is a set of coefficients of a parametric equation, so that positions along the curve can be mass-produced by just plugging values of parameters into some point-valued expression. This is seldom possible to do exactly.

What we can do is to provide the coefficients of a parametric *approximation* to the curve, and a highly convenient form is the set of coefficients of a first-order B-spline. These take the form of a sequence of points lying on the curve, sufficiently dense that points generated by linear interpolation between them lie within some operator-specified tolerance of the true curve.

The sound procedure is to hold, as well as this approximation, the original definition, so that if more accuracy is required, the approximation may be used as an initial starting point for further iteration. If this is done, the points on the curve need only be calculated densely enough for low-resolution applications, such as screen graphics. Higher precision can be obtained as and when it is required. To make this as fast as possible, every point needs to have, not just its coordinates, but also its parameter values in whatever surfaces are involved in its definition.

The convenient unit with which to deal is the connected piece. There will be a finite number of connected pieces in the curves we deal with, and so just collecting these into a set is quite acceptable.

8.1. Plane sections

Making a plane section through a surface is the most important single interrogation. It is required for effective visual evaluation of a surface's smoothness and also for the shaping of templates to assist in manufacture.

The process of calculating such curves is closely related to that of tracing contours in geographic systems.

The result of a particular cut may be one piece of curve, none, or several. A cut through a trimmed surface may meet the boundary or it may form a closed loop. The surface may be almost tangent to the sectioning plane, resulting in a tiny contour, or, in the limit, a single point.

A plane is the set of points whose coordinates satisfy some linear equation, $P^i F_i = 0$.

In Euclidean coordinates this takes the form $[P - P_F] \cdot F = 0$, where P_F is a point lying on the plane, and F is a vector perpendicular to the plane (the plane normal).

For some purposes it is also useful to have a parametric representation of the plane. This can be created by choosing two unit vectors, R and S, perpendicular to F and to each other. Then any point generated by $P := P_F + \alpha R + \beta S$ will satisfy the plane equation.

One way to find a pair such as R and S is to find which of the coordinate directions X, Y and Z has the smallest magnitude of dot product with F. Call it D. If F is nonzero, then $D \times F$ cannot be zero. Normalize it and call it R. Then take $T \times R$, normalize it and call it S.

Plane sections through analytic surfaces The best approach to this is to treat the plane as a parametric surface and use the methods described in surface/surface intersections below.

Plane sections through parametric surfaces There are two phases. The first is the identification of the topology of the required curve and isolation of one point lying on each of the pieces. Then from each of the start points, the curve can be traced around.

Finding start points Clearly, in the case of pieces of curve which meet the boundary it is most convenient to start from a point at the boundary, and so a scan round the boundary identifying points where the boundary cuts the plane is a good start.

Finding each start point itself has two stages: the identification that a start point exists, by noting the change of sign between two sample points; and the homing in. Along the boundary this is essentially a root-finding operation. Finding roots of general functions is a topic deserving a section in its own right. Polynomials are easier, since there is a known upper bound on the maximum number of roots, and derivatives are easily accessed.

If the boundary can be traversed in a consistent direction (for example, surface on the right if your head is pointing in the direction of the surface normal), it is possible to label each crossing of the plane depending on whether the curve is going from the positive side of the plane to the negative or *vice versa*. We only need to trace from the members of one set, not both, and the choice can be made consistent with the detail of the tracing code.

Finding start points within the interior of the surface has to be achieved by setting up some grid of sample points, and looking for sign changes between them. One approach is to determine all points where the surface normal is parallel to the plane normal, or where the boundary tangent is perpendicular to the plane normal, and set up a spanning tree between them. This uses the *maximum extent* enquiry described above.

Marching along the intersection The process of taking one step along the curve, from one known point to the next, currently unknown, is a univariate interrogation in its own right. The point we are seeking lies in the surface – that will be ensured by working in the parameter plane – and also in the cutting plane. The third condition that we need, to tie it down, is that the step along the curve is appropriate.

Suppose that the requirement is to take steps of some specific length, l_R. Then a good approximation is that the new point lies on a plane perpendicular to the tangent to the section at the start point and a distance l_R from it. We can now use the refinement step of the raycasting interrogation described above to home in on the new point, using the parameter values of the initial point as our initial estimate.

The same general strategy applies when more sophisticated step-length strategies are used, because each step-length strategy can be expressed as a locally linear condition equivalent to a second plane. Because the position of the step-length control plane typically moves when we know more about the surface near the solution, it is not worth taking the raycasting refinement more than one step at a time. Just as in constrained optimization by penalty functions (Wright, 1992), a single step gives enough precision to define the problem closer.

Step-length criteria Even in the case where equally spaced points are required, the normal to the step control plane can be updated at each step, to point along the chord from the last point to the current estimate of this one.

It is more typical, however, to require that generated points should be denser in regions of high curvature of the curve that they represent. There are a number of possible rules for this increase in density.

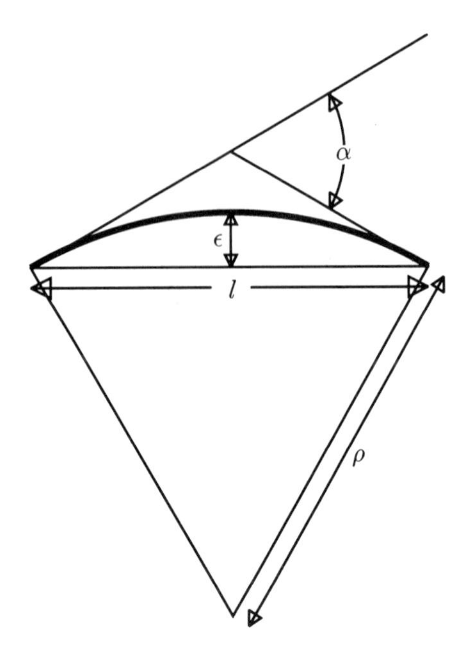

Fig. 13. Relationship between α, l, ϵ and ρ.

One of the simplest is that angles between the tangents at successive data points should make equal steps. If α_R is the required angle and ρ is the local radius of curvature, then the step length required is $2\rho \sin \alpha_R$, which for the angles likely to be used in this criterion is essentially $2\rho\alpha_R$. Now ρ will vary from point to point along the intersection, and so it needs to be estimated.

As soon as one step has been taken we can use the same equation in reverse. Let the tangent (given by $\langle F \times [\mathrm{d}P/\mathrm{d}u \times \mathrm{d}P/\mathrm{d}v]\rangle$) at the previous point be T_{i-1}, and the tangent at the current estimate be T_i. Then α_i, the actual angle between successive tangents, is close to $|T_{i-1} \times T_i|$. Similarly, l_i, the actual step taken, is $|P_i - P_{i-1}|$. We can therefore estimate $\rho = l_i/2\alpha_i$

Thus the required step length is this region is $l = l_i\alpha_R/\alpha_i$.

The new step control plane has $\langle P_i - P_{i-1}\rangle$ as its normal vector and $P_i + l\langle P_i - P_{i-1}\rangle$ as its sample point.

In situations where the sectioning plane is almost tangent to a surface (of negative Gaussian curvature), it is possible for two branches of the curve to come very close indeed to each other. It is essential to check that $T_i \cdot T_{i-1}$ is positive. A negative value is an indication that such a singularity is nearby, and that at very minimum the value of l_R should temporarily be reduced. A better response to this situation is to locate the singularity explicitly, model the situation round it and proceed using that information (Bajaj *et al.*, 1988).

The ideal step-length rule from the control of approximation point of view is that the error at the mid-chord should be constant from step to step, and

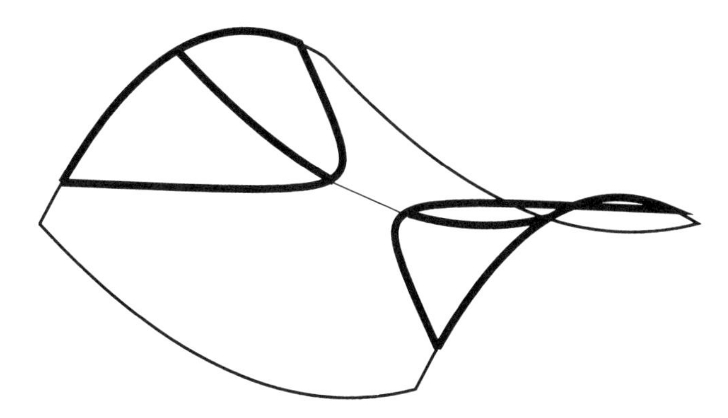

Fig. 14. Close-to-singular plane section.

equal to the desired tolerance. This corresponds to the length rule $l^2 = 8\epsilon\rho$, where ϵ is the required error. However, it is safer to temper this rule with a maximum step length for use in almost flat areas, and with a maximum angle for use in very tight corners.

A serendipitous combination is to require that rather than $l/l_R = 1$ or $\alpha/\alpha_R = 1$ conditions, we apply

$$l/l_R + \alpha/\alpha_R = 1.$$

At very small curvatures the actual α values will be small, and so the behaviour is dominated by the l term, and similarly when the curvatures are high, the angle term dominates.

For a range of curvatures about an order of magnitude wide, centred on $2\alpha_R/l_R$, the chord height error is close to $\alpha_R l_R/4$, and so it is possible to determine appropriate α_R and l_R from the mean principal curvature (determined when the surface is defined) and the required tolerance.

Handling the boundary What happens when the raycasting refinement step reports that it has hit a boundary?

Because surfaces are sometimes defined by plane cross-sections, and it is unreasonable to prevent those sections from being calculated, the first response is that if the next estimate returned is an acceptable next point for the curve (within tolerance of the sectioning plane, and with a step length within, say, 10% of the current ideal), it should be accepted, and the section continued as normal.

If the section piece started from a boundary, it is likely that the end of the piece has been reached. The accurate final point of the piece should be determined by homing in along the boundary in the neighbourhood of the point just found.

Handling failure What happens when the raycasting refinement step reports that it has failed to find a point which reduces the initial error?

This is almost certainly due to a C^0 discontinuity in the surface, and so the appropriate response is to locate the discontinuity and treat it as a local piece of boundary. If knowledge of the structure of the surface is accessible (for example, a knot-vector in the case of B-spline surfaces or the offset parameters in the case of a toroidal offset surface with a surface normal near the axis of the toroid) this can be used to define the local boundary. If not, special code may have to be written to explore the vicinity and deduce what the boundary should be.

Handling a closed loop When a curve piece starts at the boundary it also finishes at the boundary, and the mechanism described above covers terminating the piece. If it starts at a point in the interior, it may reach a boundary, (in which case work has been duplicated, because the same piece should have been found from a boundary start point), or else it returns to its own start point. However, it is most unlikely to return exactly to the start point, or even to within precision tolerance of it. What will typically happen is that at some stage it will overshoot. Let S be the start point, and P_i and P_{i+1} be the previous and current points. The simplest test to make is to see whether S lies in the sphere whose diameter is P_iP_{i+1} which is just

$$[P_i - S] \cdot [P_{i+1} - S] < 0$$

If this test indicates that S is inside the sphere, it should next be checked that the step from P_i to S would have a step-length criterion less than the step from P_i to P_{i+1}. If it does, then the loop has closed, and S can be accepted as the final point. If not, then we were unlucky, and started the curve piece at a place where two leaves approached each other closely.

Plane sections through recursive division surfaces In a recursive subdivision of a surface, any piece whose hull does not intersect the plane may be eliminated. This is reasonably efficient, particularly if the hulls are aligned with a face parallel to the plane.

A fuller elaboration of the detail of a recursive division bivariate interrogation follows in the next section (8.2), on surface/surface intersections. The plane section case is somewhat simpler.

8.2. Surface/surface intersection

This is normally regarded as the 'difficult' interrogation, but in fact it is very much of a kind with the others.

The main complication from the point of view of this section is that there are six possible combinations of the types of surface we consider. To reduce this complication we shall ignore the two most awkward combinations, of recursive division surfaces with other kinds.

Intersection of a parametric surface with an analytic This is the easiest case to deal with, because it is an almost trivial generalization of the plane section algorithm. Wherever we evaluate a parametric surface point we can immediately carry out a Taylor series expansion of the function which defines the analytic surface. The first two terms of this Taylor series define a local plane, which is used during that step as the local view of the analytic surface.

Intersection of two analytic surfaces This case can follow the same general principle, but has to be carried through in three variables, the space coordinates, instead of two.

Once a start point is found, the stepping process works by setting up a stepping direction vector as the cross product of the two surface normals. A step is made along this of the required amount, and the surfaces re-evaluated. The stepping plane and the two local linearizations are now three planes, the intersection point of which is the next candidate curve point. This iteration continues until the point found lies within the computational tolerance of both surfaces.

The starting point can be found by a slight variation on this. An arbitrary point is taken, and the Taylor planes evaluated there. Also evaluated is a third plane, whose normal is the cross product of those of the Taylor planes, and which passes through the arbitrary starting point. The intersection of these three planes gives a better approximation to a starting point. In typical situations where analytic shapes are encountered, the start point for a particular intersection will be known to lie on a third surface: in such cases, the intersection of three Taylor planes will converge to one of the intersections of the three surfaces.

The main problem is detecting the closure of a closed loop intersection without running the risk of aborting too early when two leaves of the intersection come close together. Related to this is taking the correct trace through singular situations where two surfaces are tangent to each other.

While this approach may be necessary if the analytic surfaces are fully general, there is another approach which may be applied if it is known that the only surfaces which will be met are planes, quadrics and tori. This uses the algebraic geometry result that the intersection curve of any two such surfaces may be parametrized by a curve of genus at most three and order at most eight. The genus three property means that for each value of curve parameter we can find four points on the curve by solving a quartic equation. The problem reduces to stepping along the curve by solving a quartic at each step. This could use the Tartaglia closed form, or else could just use the previous roots as first approximations for an iterative root finder.

This means tracing complex roots, but, although I have not implemented this myself, I do not regard it as particularly difficult. An interesting point is that from the algebraic geometry point of view, the more singular the configuration becomes, with points of common tangency between the surfaces, the lower the genus gets, and therefore the simpler the algebraic solution. See Farouki (1987a,b) for further elaboration on this point.

Intersection of two parametric surfaces This case takes us into even higher dimensionalities, as we have to trace a path in both parametric spaces. This is not a problem. Hoffman (1989) argues strongly that numerical tracing in spaces of many dimensions gives simpler, more reliable code than trying to eliminate the bulk of them algebraically. A useful tactic here, however, breaks down the system of four equations into two pairs, by resolving the equations

$$P_1(u_1, v_1) = P_2(u_2, v_2)$$

along the two surface normals and their mutual perpendicular.

This can be viewed geometrically as finding the intersection of one parametric surface with the local tangent plane of the other and the stepping plane, and then repeating the process for the other.

This means that the code for the refinement step of ray casting can be used yet again.

The main problem here is finding a start point which lies in (or on the boundary of) both surfaces. If we merely trail round one boundary we are likely to find a point which is outside the boundary of the other.

Intersection of two recursive division surfaces This case shows the recursive division method at its most elegant. In the outer product of two recursive division surfaces we may eliminate any pair of pieces whose hulls do not intersect.

This algorithm needs care in implementation to achieve efficiency. Specific points are:

- not to throw away any information;
- to use simple hulls, such as bounding boxes;
- to rebuild the hulls bottom up rather than just top down;
- to split only one surface, in only one direction, at each stage;
- to choose which surface to split;
- to choose which direction to split it in;
- to rejoin the constructed pieces of curve as soon as possible on the way out of the recursion.

The algorithm applying these tenets (based on Nasri (1987)) is no more complicated than some of the baroque methods which have been developed for making sense of the heap of small pieces of curve which results from not obeying the last one.

It requires as a data structure for each surface a binary tree in which each node holds the local data from which subdivision takes place, a hull, a flag which labels the node as primitive, subdivided or not subdivided, and if subdivided, the direction of subdivision and the parameter value at which the subdivision takes place.

It requires as a data structure for the pieces of curve a linked list structure within each curve. Each piece also has data identifying the logical place in the subdivision hierarchy of each of its ends, and these ends are linked together into chains accessible from the appropriate nodes of the surface trees.

Algorithm

```
Procedure Intersect(s1:surface, s2:surface):curveset
begin
        if     disjoint(s1,s2)
        then nil
        elseif primitive(s1) and primitive(s2)
        then   primitiveintersection(s1,s2)
        else   choosewhich(s1,s2,which)
               choose direction(which, dirn)
               if     which = s1
               then Split( s1, dirn, s1a, s1b)
                    Combine( Intersect(s1a, s2),
                             Intersect(s1b, s2) )
               else Split( s2, dirn, s2a, s2b)
                    Combine( Intersect(s1, s2a),
                             Intersect(s1, s2b) )
               endif
        endif
end
```

The procedure *disjoint* simply tests whether hulls overlap.

The procedure *primitive* returns TRUE if is argument is labelled as primitive.

The procedure *choosewhich* decides which of the two surfaces to split. This is based on whether either is primitive (they will not both be) and if neither is, on the relative sizes of the two hulls. The nonprimitive with the largest extent in any direction will be chosen for splitting.

The procedure *choosedirection* decides which way to split. Again this will be based on which direction of split does most to reduce the largest hull dimension. If the surface is already split, this procedure merely reports the direction previously chosen.

The procedure *Split* implements the splitting if it has not already taken place. As soon as the two new hulls are available, the hull of the surface node being split is updated, to bring the bounds in, if possible. If this results in tighter bounds they are propagated upwards until either the top of the tree is reached or no narrowing takes place. The two halves are examined for simplicity, interpreted simply as planarity, and if either half is found to be close to planar, it is marked as primitive, otherwise as undivided.

The procedure *Combine* takes those pieces of curve which are linked to the common edge of the two halves of the split surface, and sews them together into larger pieces. Where one side has been split to a deeper level than the other, its endpoint is chosen to give the coordinates of the common point.

To gain most advantage from this approach, the subdivisions will not be thrown away when a single intersection has been calculated. A great deal of work can be saved by keeping all the surface nodes for the intersection of $s1$ with $s3$ which is likely to be required next.

Clearly this arrangement is highly demanding of memory, but sufficient memory for reasonable problems is now available on current workstations. Further, if memory does become exhausted, surface nodes can be thrown away and their memory reused, provided that the immediate parent is flagged as no longer subdivided. If the data are required again, it will be recomputed, and in the meantime the tighter bounds on the parent are still having their good effect.

The only improvement available in the literature (Sederberg and Meyers, 1988) and (de Montaudouin, 1989) on the above scheme uses the concept of *co-simplicity* to fine-tune and accelerate the point at which a nonrecursive method can be used. If the surface normals of $s1$ can be bounded into a cone, $c1$, and those of $s2$ into a cone $c2$, and the cones $c1$ and $c2$ are disjoint (except at the origin), then there can at most be one piece of intersection curve between $s1$ and $s2$, and it will not form a closed loop. It is therefore safe to take recourse to marching methods to evaluate the detail of the piece of intersection.

This refinement requires more data per node, to hold the bounds on the surface normal, but it should permit the recursion to be truncated much earlier, and also give better resolution when two surfaces are almost tangent. It affects the criteria used for deciding which surface to split and which way, because it becomes just as important to reduce the surface normal spread quickly as to reduce the physical extent.

The two awkward cases which we have omitted are best handled by multiple representations. The analytic surfaces which are found in geometric modelling systems also have parametric forms, and can then be converted on, to a recursive division representation.

8.3. Reflection lines in a parametric surface

This interrogation determines the curves on a surface at which light rays are reflected from some curve C in space to an eye at point E.

It is used in validation of surfaces in car styling. It is also an example of another paradigm for univariate interrogation, in which nilvariate enquiries are made recursively until the necessary density is reached.

Suppose that C is a curve stored in exactly the same form as all the results of our univariate enquiries described so far. It has a start point, which can be used as auxiliary data for the nilvariate calculation of a reflection point. Each point in the resulting set of points is a start point for a reflection curve, and these should be taken one at a time.

Imagine a point moving along C from that start point. In fact C is parametrized, and so we can formalize this movement by differentiating C with respect to parameter. For high precision we should differentiate the true curve underlying C, but for graphics purposes the piecewise linear approximation will be quite adequate. The derivative is $\mathrm{d}C(t)/\mathrm{d}t$

As the point on C moves, the result of carrying out the nilvariate reflection also moves, and differentiating the reflection equation (equation (7.13) above) gives $\mathrm{d}P/\mathrm{d}C(t)$, and the chain rule then gives $\mathrm{d}P/\mathrm{d}t$. This is the tangent to the reflection curve.

We can judge an appropriate increment in t from the magnitude of $\mathrm{d}P/\mathrm{d}t$, and can then step along C using a bracketting in t to find points on the reflection curve which satisfy the step-length rules. At each step we can use the previous point on the curve as an initial estimate of the next, thus using coherence to simplify the process.

This method is probably adequate for judging car body panels, where the reflection curves do not hit the boundary, and where there is only one reflection of each C.

However, a full algorithm needs to handle correctly the cases where reflections fall off the edges of the surface, and also those cases where the reflection falls back on again. It also needs to cater for the bifurcation which can happen when the surface has an inflexion.

The first problem is fairly straightforward, since tracing can just stop when a boundary is reached. The second needs to be addressed by scanning the boundary and identifying the points and t-values at which reflections of C-points lie on the boundary.

The third requires the explicit evaluation of the parabolic lines of the surface, which can then be treated in the same way as the boundary as a source of start points for new bits of reflection curve.

Other interrogations using the same paradigm will require some other specific internal 'boundaries' depending on the detail of their equations.

8.4. Geodesics across a surface

A geodesic is the path between two points of a surface which lies entirely within the surface and whose length is least. Calculus of variations shows that it is the solution of a differential equation which states that the component of curvature within the tangent plane (the geodesic curvature) is zero.

Geodesics map into straight lines during the flattening of a developable, and so this calculation is applied in the determination of development. It is also a good example of a differential equation interrogation.

Geodesics across an analytic surface Let the two ends be P_1 and P_2. We can determine the surface normals N_1 and N_2 at these points from the gradient of the function defining the surface.

A surprisingly accurate cubic approximation to a geodesic is constructed by postulating that the Bezier control points of this cubic are given by

$$B_1 = (2P_1 + P_2)/3 + \alpha N_1 + \beta N_2, \tag{8.1}$$

$$B_2 = (P_1 + 2P_2)/3 + \gamma N_1 + \delta N_2, \tag{8.2}$$

and choosing the unknowns, α, β, γ and δ from the conditions that the tangent at each end must be perpendicular to the normal there, and that the local osculating plane at each end contains the surface normal. From these we can deduce that $\delta = 2\beta$ and $\alpha = 2\gamma$, and thence that

$$\begin{bmatrix} 2N_1 \cdot N_1 & N_1 \cdot N_2 \\ N_1 \cdot N_2 & 2N_2 \cdot N_2 \end{bmatrix} \begin{bmatrix} \gamma \\ \beta \end{bmatrix} = \begin{bmatrix} -D \cdot N_1/3 \\ D \cdot N_2/3 \end{bmatrix}. \tag{8.3}$$

The matrix on the left is always nonsingular, and so this equation can be solved to give the required control points.

The cubic may well depart from the surface in its interior. Evaluation of the distance of points evaluated on it from the surface gives the first measure of error.

The second measure of error is to evaluate a point at the parametric centre of the curve, and project it back on to the surface using the nearest-point algorithm. Then apply the same process as above, to determine an approach direction and a departure direction at this central point. The angle between these directions gives another measure of error. Let this angle be θ, and the distance between the two ends be l. A lateral movement of the midpoint (that is, a movement in the tangent plane there and in a plane perpendicular to the line joining P_1 and P_2) of $l\theta/4$ will approximately correct this mismatch.

However, inserting additional points at the quarter points would probably discover the need for a further correction: the best estimate of the correction needed if we inserted really densely is $l\theta/2$. The number of points necessary for insertion to give a solution accurate to a given tolerance ϵ is the square root of (estimated error/required error). These can be inserted either all at once or in binary stages, with a relaxation at every step correcting the positions.

Geodesics across a parametric surface An exactly similar approach can be taken over a parametric surface; the only additional stage is resolving back the components of the vector $B_1 - P_1$ into components in the parameter plane.

Because $B_1 - P_1$ lies in the tangent plane at P_1, it must be of the form

$$\mathrm{d}P/\mathrm{d}u\,\delta u + \mathrm{d}P/\mathrm{d}v\,\delta v.$$

To evaluate δu and δv take the triple product of $B_1 - P_1$ first with $\mathrm{d}P/\mathrm{d}v$ and N_1, then with $\mathrm{d}P/\mathrm{d}u$ and N_1. These give scalar equations

$$\delta u = [B_1 - P_1, \mathrm{d}P/\mathrm{d}v, N]/[\mathrm{d}P/\mathrm{d}u, \mathrm{d}P/\mathrm{d}v, N],$$
$$\delta v = [\mathrm{d}P/\mathrm{d}u, B_1 - P_1, N]/[\mathrm{d}P/\mathrm{d}u, \mathrm{d}P/\mathrm{d}v, N].$$

The cubic is now set up in parameter space, and a similar procedure of interpolating a central point and checking the accuracy is applied.

8.5. Lines of curvature

At any point of a surface (except an umbilic) there are two directions of *principal curvature.* These are the directions in which the curvature of a normal section is largest or least, and they are also directions in which there is no twist.

A line of curvature is a curve whose tangent is always along a local direction of curvature. Tracing such a curve is essentially the numerical solution of a first-order differential equation, and the obvious procedure is marching along it.

The most convenient equation to solve is that corresponding to the no-twist condition. Let T be a tangent to the curve at P, locally parametrized by t, and N the surface normal. As a point moves along T, the surface normal becomes

$$N(t + \delta t) = N + \mathrm{d}N/\mathrm{d}t\delta t,$$

and the condition of no twist is that $\mathrm{d}N/\mathrm{d}t$ lies in the plane of N and T. (N and T are always perpendicular, and so they always span a plane.)

$$[N, T, \mathrm{d}N/\mathrm{d}t] = 0. \tag{8.4}$$

Lines of curvature across an analytic surface At the start point, P, of the curve we evaluate the local surface normal, N, and the tensor $\mathrm{d}^2 f/\mathrm{d}P^2$ which may be viewed as $\mathrm{d}N/\mathrm{d}P$.

The direction, T, we need to march in satisfies the equations

$$T \cdot N = 0, \tag{8.5}$$
$$[N, T, dN/\mathrm{d}P \cdot T] = 0. \tag{8.6}$$

This is homogeneous in T, and so any multiple of a solution is also a solution. In particular, the exact reverse is also a solution. It is also quadratic in T, and so we expect there to be two distinct solutions.

Unless the context gives a reason for choosing one of the four rather than another, all four need to be treated equally, as start directions for lines of curvature.

However, once a particular path has been chosen, further steps can always be made in the direction most coherent with that of arrival at the current point.

The actual solution of the equations is easily carried out by constructing two vectors spanning the plane perpendicular to N and using coordinates in this plane as two freedoms, thus automatically satisfying the first equation. The second becomes a homogeneous quadratic equation in the two coordinates, whose solution ratios give the possible directions for T.

Once a direction is determined, the same methods of step length control apply as in previous curve interrogations.

The difference here is that there is no way of correcting back onto the true curve in any absolute sense, as there was when intersecting with other surfaces. Now the effect of making steps along the tangent to the curve at each point is to drift away from the true curve fairly rapidly. This is a familiar effect in solution of ordinary differential equations, and the appropriate response is the predictor–corrector paradigm (see Allgower and Georg (1993)). Because we have to iterate at each point to get back on to the surface itself, the corrector step gives no extra computing.

Lines of curvature across a parametric surface In this case we have our vectors spanning the tangent plane ready-made in $\mathrm{d}P/\mathrm{d}u$ and $\mathrm{d}P/\mathrm{d}v$. The equivalent equation to (8.5) is

$$\begin{bmatrix} \delta u & \delta v \end{bmatrix} \begin{bmatrix} \phi_{uu} & \phi_{uv} \\ \phi_{vu} & \phi_{vv} \end{bmatrix} \begin{bmatrix} \delta u \\ \delta v \end{bmatrix} = 0 \tag{8.7}$$

where

$$\begin{aligned}
\phi_{uu} &= N \cdot \frac{\mathrm{d}^2 P}{\mathrm{d}u\,\mathrm{d}v}\frac{\mathrm{d}P}{\mathrm{d}u} \cdot \frac{\mathrm{d}P}{\mathrm{d}u} - N \cdot \frac{\mathrm{d}^2 P}{\mathrm{d}u^2}\frac{\mathrm{d}P}{\mathrm{d}u} \cdot \frac{\mathrm{d}P}{\mathrm{d}v}, \\
\phi_{uv} &= N \cdot \frac{\mathrm{d}^2 P}{\mathrm{d}u\,\mathrm{d}v}\frac{\mathrm{d}P}{\mathrm{d}u} \cdot \frac{\mathrm{d}P}{\mathrm{d}v} - N \cdot \frac{\mathrm{d}^2 P}{\mathrm{d}u^2}\frac{\mathrm{d}P}{\mathrm{d}v} \cdot \frac{\mathrm{d}P}{\mathrm{d}v}, \\
\phi_{vu} &= N \cdot \frac{\mathrm{d}^2 P}{\mathrm{d}v^2}\frac{\mathrm{d}P}{\mathrm{d}u} \cdot \frac{\mathrm{d}P}{\mathrm{d}u} - N \cdot \frac{\mathrm{d}^2 P}{\mathrm{d}u\,\mathrm{d}v}\frac{\mathrm{d}P}{\mathrm{d}u} \cdot \frac{\mathrm{d}P}{\mathrm{d}v}, \\
\phi_{vv} &= N \cdot \frac{\mathrm{d}^2 P}{\mathrm{d}u^2}\frac{\mathrm{d}P}{\mathrm{d}u} \cdot \frac{\mathrm{d}P}{\mathrm{d}v} - N \cdot \frac{\mathrm{d}^2 P}{\mathrm{d}u\,\mathrm{d}v}\frac{\mathrm{d}P}{\mathrm{d}v} \cdot \frac{\mathrm{d}P}{\mathrm{d}v}.
\end{aligned}$$

Correction of each step to minimize drift is still necessary even though evaluation from the parameters ensures that generated points lie on the surface.

8.6. Summary of univariate interrogations

We have seen above examples of both algebraic equations, where accuracy can be maintained at each step, independent of what has come before, and of differential equations, whose tracing is possible from any start point, and in which numerical errors will inexorably, if slowly, build up.

Marching methods work well for both in situations where the geometry is actually well conditioned, and normal applications in manufacturing tend to give these situations.

In marching methods we first need to identify the topology of the solution; how many pieces the solution has, and how they interact with the boundary of the surface(s). Then we step along each piece, choosing the step length to match the local curvature of the result. Within each step we first find a first approximation for the next point and then refine it. Refinement can often use the refinement step from raycasting.

If, however, robustness is important the recursive subdivision techniques are more appropriate to algebraic equations and the differential equations should be cast ideally as boundary value, rather than initial value questions.

9. Bivariate interrogations

Within this section we are looking for results which map an entire surface into a new form. The examples are concerned with graphics.

9.1. Rendering a surface

With the advent of workstations capable of displaying a wide range of colours, it has become desirable to be able to compute exactly what colour each pixel should be in an image of the objects our software deals with.

One technique for this is raytracing, which uses geometric optics to determine what can be seen at each pixel.

The first level of this technique merely uses the nilvariate raycasting algorithm to find out what point on a given surface lies at the intersection of the surface with a ray through the eye and a given pixel.

In a typical model, with many surfaces, there is also the process of choosing the nearest surface, as well as that of choosing the nearest intersection on the given surface.

Once the nearest intersection has been found, its illumination is determined by applying an illumination model, which uses the positions and colours of the lights, the positions of the surface point and the eye, the surface normal, and the various reflection coefficients of the material of which the surface purports to be made. Such models typically use separate models of diffuse and specular reflection, and, if the coefficients are well-chosen, can give surprisingly recognizable impressions of different materials.

The process may be taken further, by starting either a reflection ray or a refraction ray (or both) from the visible surface point, and repeating the process to find out what can be seen reflected (or refracted) in the surface.

This capability is usually shown off in images of glass or shiny metal objects. It just uses the same raycasting code again.

Finally, for duller materials, shadows can also be determined using the same code. A point on a surface is illuminated if it is visible from the light. A ray can be fired from the light towards each point visible from the eye. If the first intersection is the visible point it is illuminated, if not, not.

Multiple lights just means lots more calls to the raycaster.

This technology runs out of steam when diffuse reflections from one surface to another dominate the illumination, as in images of domestic interior scenes. A simultaneous solution for the illumination levels on all surfaces then needs to be invoked. This technique is called *'luminosity'* (Hall, 1990).

Nor is raycasting the fastest technique with modern workstations which have significant amounts of special purpose hardware designed for rendering. These displays are driven, not pixel by pixel, but facet by facet, where facets are small pieces of surface, small enough to be treated as plane.

In order to keep the number of facets small, while still giving the impression of a smooth surface, there are two ways in which graphics workstations typically cheat. The first is called Gouraud shading, in which a true surface normal is evaluated at each vertex of each facet; the illumination model is applied there, and the illumination value is interpolated linearly across the facet. The second is called Phong shading. Again a true surface normal is evaluated at every vertex, and then an effective surface normal vector is interpolated across the facet, with the illumination model being applied at every pixel. The application of such methods can be detected in images which have smooth surfaces with polygonal edges.

In order to drive such powerful displays, it is necessary to split each surface up into facets.

9.2. Facetting a surface

One technique, applicable to both parametric and recursive division surfaces, is to base the facetting on the parametric structure of the surface.

In the simplest parametric case, a regular subdivision is made on a regular parametric grid, giving four-sided facets. If any facet is too far out of plane, it is just subdivided further.

In the recursive subdivision case, each piece of surface is examined for planarity. If it is flat enough it is issued as a facet, if not it is subdivided further. The planarity test can use surface normal hulls or else it can set up an approximate surface normal for an entire piece and just measure the upper bound on thickness along this direction.

Another approach is to divide the surface into triangular pieces, triangles being flat by definition. The problem here is that some parts of the surface will require small triangles, others can accept large ones, if the criterion on triangle size is economically matching the true surface within a stated tolerance.

Such a triangulation can be used for communicating models to certain 'rapid prototyping' machines, and also as a discretization for certain types of aerodynamic or electromagnetic analyses, as well as for graphics.

A method which has been used in mesh generation for finite-element analysis (Cavendish, 1974) is first to create vertices all round the boundary at an appropriate density, then to create a Delaunay triangulation of those vertices, and finally to insert additional vertices, always updating the Delaunay triangulation, until every triangle is small enough.

The interesting part of this process is that, while many different optimality criteria give the same triangulation of a set of points in two dimensions, there is no clean equivalent in three dimensions. It is better therefore to do the triangulation in the parameter plane.

However, the distortion introduced by the local affineness of the mapping from parameter space to real space means that a triangulation which is Delaunay in parameter space is likely to be a very poor one when mapped into object space. It is necessary to compensate for this distortion.

There are just three places within the overall process where compensation is necessary. Within the Delaunay process itself there is a 'swap test' which decides whether a pair of triangles forming a quadrilateral should be swapped to split the quadrilateral by its other diagonal.

Then there is the decision as to whether a triangle is acceptable in the final tesselation, or whether it should have an extra vertex inserted in it, and, finally, if an extra vertex is to be created, there is the computing of where it should go.

In each case, the local configuration can be mapped from parameter space into an orthonormal coordinate system which can be thought of as being of the tangent plane at the centroid of the three or four points concerned.

This gives well proportioned triangles, even on surfaces where $\mathrm{d}P/\mathrm{d}u$ and $\mathrm{d}P/\mathrm{d}v$ vary widely over the surface and are far from orthonormal.

The final step, of making the triangles' proportions and densities fit the local surface curvatures, is achieved by using, not an orthonormal system in the tangent plane, but a system which is orthonormal with respect to the sign-corrected second derivative matrix.

Take the principal curvatures and principal directions of curvature at a point on the surface, and scale the vectors by the square roots of the magnitudes of the radii of curvature. These two vectors are now conjugate with respect to an ellipse which is either the Dupin indicatrix, or else has the same maximum deflection from the tangent plane on both sides. Under the necessary final transformation, this ellipse is the image of a circle. If the maximum deflection from the tangent plane is the required tolerance, it is the image of a unit circle. Any triangle inscribed in it has an error no greater than the tolerance, and if the triangle is well enough shaped to include its (mapped) circumcentre, the error is equal to the specified tolerance.

REFERENCES

E.L. Allgower and K. Georg (1993), 'Continuation and path following', *Acta Numerica 1993*, Cambridge University Press (Cambridge), 1–64.

S. Aomura and T. Uehara (1990), 'Self-intersection of an offset surface', *Comput. Aided Des.* **22**, 417–422.

A.P. Armit (1971), 'Curve and surface design: using multipatch and multiobject design systems', *Comput. Aided Des.* **3**, 3–12.

C.L. Bajaj (1989), 'Geometric modelling with algebraic surfaces', in *The Mathematics of Surfaces III* (D.C. Handscomb, ed.), Clarendon (Oxford), 3–48.

C.L. Bajaj, C.M. Hoffman, J.E. Hopcroft and R.E. Lynch (1988), 'Tracing surface intersections', *Comput. Aided Geom. Des.* **5**, 285–308.

R.E. Barnhill (1974), 'Smooth interpolation over triangles', in *Computer Aided Geometric Design* (R.E. Barnhill and R.F. Riesenfeld, eds), Academic Press (New York), 45–70.

R.E. Barnhill, G. Farin, M. Jordan and B.R. Piper (1987), 'Surface/surface intersection', *Comput. Aided Geom. Des.* **4**, 3–16.

R.E. Barnhill and R.F. Riesenfeld, eds (1974), *Computer Aided Geometric Design*, Academic Press (New York).

C. Bell, B. Landi and M. Sabin (1974), 'The programming and use of numerical control to machine sculptured surfaces', in *Proceedings 14th MTDR Conference*, Macmillan (London), 233–238.

P. Bezier (1971) 'An existing system in the automobile industry', *Proc. R. Soc. London* A **321**, 207–218.

L. Biard and P. Chenin (1991), 'Ray tracing rational parametric surfaces', in *Curves and Surfaces*, (J. Laurent, A. le Mehaute and L.L. Schumaker, eds), Academic Press (New York), 37–42.

M.I.G. Bloor and M.J. Wilson (1989), 'Generating blend surfaces using partial differential equations', *Comput. Aided Des.* **21**, 165–171.

M.I.G. Bloor and M.J. Wilson (1990), 'Representing PDE surfaces in terms of B-splines', *Comput. Aided Des.* **22**, 324–331.

C. de Boor (1962), 'Bicubic Spline Interpolation', *J. Math. Phys.* **41**, 212–273.

C. de Boor (1987), 'B-form basics', in *Geometric Modelling: Algorithms and New Trends* (G. Farin, ed.), SIAM (Philadelphia), 131–148.

C. de Boor (1993a), *B-Spline Basics* in *Fundamental Developments of Computer Aided Geometric Modelling* (L. Piegl, ed.), Academic Press (New York), 327–350.

C. de Boor (1993b), 'Multivariate piecewise polynomials', *Acta Numerica 1993*, Cambridge University Press (Cambridge), 65–110.

H. Burger and R. Schaback (1993), 'A parallel multistage method for surface/surface intersection', *Comput. Aided Geom. Des.* **10**, 277–292.

E. Catmull and J. Clark (1978), 'Recursively generated B-spline surfaces on arbitrary topological meshes', *Comput. Aided Des.* **10**, 350–355.

J.C. Cavendish (1974), 'Automatic triangulation of arbitrary planar domains for the FE method', *Int. J. Numer. Meth. Engrg* **8**, 679–696.

P.M. Cohn (1961), *Solid Geometry*, Routledge and Kegan Paul (London).

J.L. Coolidge (1963), *A History of Geometrical Methods*, rep. 1963 of 1938 book, Dover (New York).

S.A. Coons (1967), 'Surfaces for computer aided design of space forms', MAC–TR–41, Massachusetts Institute of Technology.

W. Dahmen (1989) 'Smooth piecewise quadric surfaces', in *Mathematical Methods in Computer Aided Geometric Design* (T. Lyche and L.L. Schumaker, eds), Academic Press (New York), 181–194.

D. Doo and M.A. Sabin (1978), 'Behaviour of recursive division surfaces near extraordinary points', *Comput. Aided Des.* **10**, 356–362.

J.P. Duncan and S.G. Mair (1980), *Sculptured Surfaces in Engineering and Medicine*, Cambridge University Press (Cambridge, UK).

R.A. Earnshaw, ed. (1985), *Fundamental Algorithms for Computer Graphics*, NATO ASI F17, Springer (Berlin).

H. Einar and E. Skappel (1973), 'FORMELA: A general design and production system for sculptured products', *Comput. Aided Des.* **5**, 68–76

L.P. Eisenhart (1960), *Coordinate Geometry*, 1938 repub 1960 Dover (New York).

G. Farin (1982), 'Designing C1 surfaces consisting of triangular cubic patches', *Comput. Aided Des.* **14**, 253–256.

G. Farin, ed. (1987), *Geometric Modelling: Algorithms and New Trends*, SIAM (Philadelphia).

G. Farin (1988), *Curves and Surfaces for Computer Aided Geometric Design*, Academic Press (New York).

R.T. Farouki (1987a), 'Direct surface section evaluation', in *Geometric Modelling: Algorithms and New Trends* (G. Farin, ed.), SIAM (Philadelphia), 319–334.

R.T. Farouki (1987b), 'Numerical stability on geometric algorithms and representations', in *Mathematics of Surfaces III* (D. C. Handscomb, ed.), 83–114.

R.T. Farouki and V.T. Rajan (1987), 'On the numerical conditioning of polynomials in Bernstein form', *Comput. Aided Geom. Des.* **4**, 191–216.

R.T. Farouki and V.T. Rajan (1988a), 'Algorithms for polynomials "in Bernstein form" ', *Comput. Aided Geom. Des.* **5**, 1–26.

R.T. Farouki and V.T. Rajan (1988b), 'On the numerical condition of algebraic curves and surfaces (part 1)', *Comput. Aided Geom. Des.* **5**, 215–252.

I.D. Faux and M.J. Pratt (1979), *Computational Geometry for Design and Manufacture*, Ellis Horwood (New York).

J. Ferguson (1964) 'Multivariable curve interpolation', *JACM* **11**, 221–228.

J. Ferguson (1993), 'F-methods for freeform curve and hypersurface definition', in *Fundamental Developments of Computer Aided Geometric Modelling* (L. Piegl, ed.), Academic Press (New York), 99–116.

D. Filip, R. Magedson and R. Markot (1986), 'Surface algorithms using bounds on derivatives', *Comput. Aided Geom. Des.* **3**, 295–312.

T. Garrity and J. Warren (1989), 'On computing the intersection of a pair of algebraic surfaces', *Comput. Aided Geom. Des.* **6**, 137–154.

K.F. Gauss (1828), *General Investigation of Curved Surfaces* (J.C. Morehead and A.M. Hiltebeitel, trans.), reprinted Raven Press, 1965.

W.J. Gordon (1969), 'Spline blended surface interpolation through curve networks', *J. Math. Mech.* **18**, 10, 931–952.

W.J. Gordon and R.F. Riesenfeld (1974), 'B-spline curves and surfaces', in *Computer Aided Geometric Design* (R.E. Barnhill and R.F. Riesenfeld, eds), Academic Press (New York), 95–126.

J.A. Gregory, ed. (1986), *The Mathematics of Surfaces*, Clarendon (Oxford), 217–232.

H.B. Griffiths (1976), *Surfaces*, Cambridge University Press (Cambridge).

R. Hall (1990), 'Algorithms for realistic image synthesis', *Computer Graphics Techniques*, (D.F. Rogers and R.A. Earnshaw, eds), Springer (Berlin), 189–231.

D.C. Handscomb, ed. (1989), *The Mathematics of Surfaces III*, Clarendon Press (Oxford).

C. Hoffman (1989), *Geometric and Solid Modelling*, Morgan Kaufmann (San Mateo, CA).

C. Hoffman and J. Hopcroft (1986), 'Quadratic blending surfaces', *Comput. Aided Des.* **18**, 301–306.

C. Hoffman and J. Hopcroft (1987), 'The potential method for blending surfaces and corners', in *Geometric Modelling: Algorithms and New Trends* (G. Farin, ed.), SIAM (Philadelphia), 347–366.

J. Hoschek (1988), 'Spline approximation of offset curve', *Comput. Aided Geom. Des.* **5**, 33–40.

J.K. Johnstone (1993), 'A new intersection algorithm for cyclides and swept surfaces using circle decomposition', *Comput. Aided Geom. Des.* **10**, 1–24

S. Katz and T.W. Sederberg (1988), 'Genus of the intersection curve of two rational surface patches', *Comput. Aided Geom. Des.* **5**, 253–258.

A. Kaufmann (1991), 'A distributed algorithm for surface/plane intersection', in *Curves and Surfaces* (J. Laurent, A. le Mehaute and L.L. Schumaker, eds), Academic Press (New York) 251–254.

R. Klass (1980), 'Correction of local surface irregularities using reflection lines', *Comput. Aided Des.* **12**, 73–78.

P.A. Koparkar and S.P. Mudur (1985), 'Subdivision techniques for processing geometric objects', *Fundamental Algorithms for Computer Graphics* (R.A. Earnshaw, ed.), NATO ASI F17, Springer (Berlin) 751–801.

P.A. Koparkar and S.P. Mudur (1986), 'Generation of continuous smooth curves resulting from operations on parametric surface patches', *Comput. Aided Des.* **18**, 193–206.

G.A. Kriezis, P.V. Prakash and N.M. Patrikalakis (1990), 'A method for intersecting algebraic surfaces with rational polynomial patches', *Comput. Aided Des.* **22**, 645–654.

G.A. Kriezis, N.M. Patrikalakis and F-E. Wolter (1992), 'Topological and differential-equation methods for surface intersections', *Comput. Aided Des.* **24**, 41–55.

D. Lasser (1986), 'Intersection of parametric surfaces in the Bernstein–Bezier representation', *Comput. Aided Des.* **18**, 186–192.

P-J. Laurent, A. le Mehaute and L.L. Schumaker (1991), *Curves and Surfaces*, Academic Press (New York).
R.A. Liming (1979) *Mathematics for Computer Graphics*, Aero.
T. Lyche and L. L. Schumaker (1989), *Mathematical Methods in Computer Aided Geometric Design* Academic Press (New York).
R.P. Markot and R.L. Magedson (1989), 'Solutions of tangential surface and curve intersections', *Comput. Aided Des.* **21**, 421–429.
R.P. Markot and R.L. Magedson (1991), 'Procedural method for evaluating the intersection curves of two parametric surfaces', *Comput. Aided Des.* **23**, 395–404.
R.R. Martin, ed. (1987), *The Mathematics of Surfaces II*, Clarendon (Oxford).
W.H. McCrea (1960), *Analytic Geometry of Three Dimensions*, University Mathematical Texts, Oliver and Boyd (Edinburgh).
A. Middleditch and K. Sears (1985), 'Blend surfaces for set-theoretic volume modelling systems', *SIGRAPH Comput. Graphics* **19**, 161–170.
Y. de Montaudouin (1989), 'Cross product of cones of revolution', *Comput. Aided Des.* **21**, 404.
Y. de Montaudouin (1991), 'Resolution of $P(x, y) = 0$', *Comput. Aided Des.* **23**, 653–654.
G. Mullenheim (1990), 'Convergence of a surface/surface intersection algorithm', *Comput. Aided Geom. Des.* **7**, 415–424.
G. Mullenheim (1991), 'On determining start points for a surface/surface intersection', *Comput. Aided Geom. Des.* **8**, 401–408.
A.H. Nasri (1987), 'A polyhedral subdivision method for free-form surfaces', *ACM ToG* **6**, 29–73.
A.H. Nasri (1991), 'Boundary-corner control in recursive subdivision surfaces', *Comput. Aided Des.* **23**, 405–410.
G.M. Nielson (1974), 'Some piecewise polynomial alternatives to splines under tension', in *Computer Aided Geometric Design* (R.E. Barnhill and R.F. Riesenfeld, eds), Academic Press (New York) 209–235.
J. Nocedal (1992), 'Theory of algorithms for unconstrained optimization', *Acta Numerica 1992*, Cambridge University Press (Cambridge), 199–242
K. Ohkura and Y. Kakazu (1992), 'Generalization of the potential method for blending three surfaces', *Comput. Aided Des.* **24**, 599–610.
J.C. Owen and A.P. Rockwood (1987), 'Intersection of general implicit surfaces', in *Geometric Modelling: Algorithms and New Trends* (G. Farin, ed.), SIAM (Philadelphia), 335–346.
Q.S. Peng (1984), 'An algorithm for finding the intersection lines between two B-spline surfaces', *Comput. Aided Des.* **16**, 191–196.
J. Peters (1990), 'Smooth mesh interpolation with cubic patches', *Comput. Aided Des.* **22**, 109–120.
C.S. Petersen (1984), 'Adaptive contouring of three-dimensional surfaces', *Comput. Aided Geom. Des.* **1**, 61–74.
B. Pham (1992), 'Offset curves and surfaces: a brief survey', *Comput. Aided Des.* **24**, 223–229.
L. Piegl (1993), *Fundamental Developments of Computer Aided Geometric Modelling*, Academic Press (New York).

E. Polak (1971), *Computational Methods in Optimization*, Academic Press (New York).

M.J. Pratt and A.D. Geisow (1986), 'Surface/surface intersection problems', in *The Mathematics of Surfaces* (J.A. Gregory, ed.) Clarendon (Oxford), 117–142.

R.F. Riesenfeld (1975), 'On Chaikin's algorithm', *Comput. Graph. Image Proc.* **4**, 304–310.

R.G. Robertson (1966), *Descriptive Geometry*, Pitman (London).

A. Rockwood and J. Owen (1985), 'Blending surfaces in solid geometric modelling', in *Geometric Modelling: Algorithms and New Trends* (G. Farin, ed.), SIAM (Philadelphia), 367–384.

D.F. Rogers and R.A. Earnshaw, eds, (1990), *Computer Graphics Techniques*, Springer (Berlin).

P.K. Scherrer and B.M. Hilberry (1978), 'Determining distance to a surface represented in piecewise fashion with surface patches', *Comput. Aided Des.* **10**, 320–324.

I.J. Schoenberg (1946), 'Contributions to the problem of approximation of equidistant data by analytic functions', *Quart. Appl. Math.* **4**, 45–99.

T.W. Sederberg (1987), 'Algebraic geometry for surface and solid modelling', in *Geometric Modelling: Algorithms and New Trends* (G. Farin, ed.), SIAM (Philadelphia), 29–42.

T.W. Sederberg and R.J. Meyers (1988), 'Loop detection in surface patch intersections', *Comput. Aided Geom. Des.* **5**, 161–172.

T.W. Sederberg, H. Christiansen and S. Katz (1989), 'Improved test forclosed loops in surface intersections', *Comput. Aided Des.* **21**, 505–508.

T.W. Sederberg and X. Wang (1987), 'Rational hodographs', *Comput. Aided Geom. Des.* **4**, 333–336.

D.J.T. Storry and A.A. Ball (1989), 'Design of an n-sided patch from Hermite boundary data', *Comput. Aided Geom. Des.* **6**, 111–120.

I.E. Sutherland, R.F. Sproull and R.A. Schumaker (1974), 'A characterisation of ten hidden surface algorithms', *Comput. Surv.* **6**, 1–55.

G.W. Vickers (1977), 'Computer-aided manufacture of marine propellors', *Comput. Aided Des.* **9**, 267–274

J. Woodwark, ed., (1989), *Geometric Reasoning*, Clarendon (Oxford).

M.H. Wright (1992), 'Interior methods for constrained optimization', *Acta Numerica 1992*, Cambridge University Press (Cambridge), 341–407.

C-G. Yang (1987), 'On speeding up ray tracing of B-spline surfaces', *Comput. Aided Des.* **19**, 122–130.

Acta Numerica (1994), *pp.* 467–572

Numerical analysis of dynamical systems

Andrew M. Stuart
Program in Scientific Computing and Computational Mathematics
Division of Applied Mechanics
Stanford University
California, CA94305-4040, USA
E-mail: stuart@sccm.stanford.edu

This article reviews the application of various notions from the theory of dynamical systems to the analysis of numerical approximation of initial value problems over long-time intervals. Standard error estimates comparing individual trajectories are of no direct use in this context since the error constant typically grows like the exponential of the time interval under consideration.

Instead of comparing trajectories, the effect of discretization on various sets which are invariant under the evolution of the underlying differential equation is studied. Such invariant sets are crucial in determining long-time dynamics. The particular invariant sets which are studied are equilibrium points, together with their unstable manifolds and local phase portraits, periodic solutions, quasi-periodic solutions and strange attractors.

Particular attention is paid to the development of a unified theory and to the development of an existence theory for invariant sets of the underlying differential equation which may be used directly to construct an analogous existence theory (and hence a simple approximation theory) for the numerical method.

CONTENTS

1	Introduction	468
2	Background and motivation	470
3	Semigroups and their approximation	481
4	Neighbourhood of an equilibrium point	494
5	Periodic solutions and invariant tori	520
6	Uniform asymptotic stability and attractors	535
7	Conclusions	562
	References	565

1. Introduction

In this article we study the numerical approximation of the ordinary differential equation

$$u_t = f(u), \quad u(0) = U, \tag{1.1}$$

for $u(t) \in C^1(\mathbb{R}^+, \mathbb{R}^p)$. We introduce a time discretization through the points $t_n = n\Delta t$ and study the approximation of (1.1) by one-step numerical methods of the form

$$U_{n+1} = \mathcal{F}(U_n; \Delta t), \quad U_0 = U. \tag{1.2}$$

Here $U_n \in \mathbb{R}^p$ is considered as an approximation to $u(t_n)$. All Runge–Kutta methods, for example, can be considered in this form, provided solvability of the defining equations has been established.

The classical error bound for the approximation of (1.1) by (1.2) is of the form

$$\|U_n - u(t_n)\| \leq ce^{k\mathrm{T}}\Delta t^r \tag{1.3}$$

for $0 \leq t_n \leq T$. Such error bounds can be derived purely under the assumptions which yield existence, uniqueness, smoothness and continuous dependence of a solution to (1.1) and no further understanding of the behaviour of solutions is required. Typically $k > 0$ reflecting the fact that different solutions of (1.1) may diverge exponentially over certain parts of phase space. Consequently the error bound (1.3) is of little direct use in studying the long-time behaviour of approximations (1.2) for the equation (1.1) since it yields no information for fixed Δt as $T \to \infty$.

To understand the behaviour of the approximation (1.2) of (1.1) over long-time intervals requires a deeper knowledge of the behaviour of solutions of (1.1) and, in particular, an understanding of how these solutions behave over long-time intervals. In particular, the study of a variety of sets invariant under the evolution generated by the equation (1.1) is crucial. Such knowledge, combined with the standard error estimate (1.3) or a truncation error bound, can provide very powerful results about the long-time behaviour of numerical methods. The purpose of this review is to describe such results within the unified framework of dynamical systems.

Section 2 contains background and motivational material. In particular we state what we aim to show in this article and describe the types of problems that we have in mind. In so doing we also make it clear that a number of important issues will *not be covered* and give appropriate references to existing literature in these areas. We also describe, by means of simple examples, the types of theorems whose statements and detailed proofs we consider in the remainder of the article.

In Section 3 we formulate the basic notions from the theory of dynamical systems that are relevant to this article; in particular the concept of semigroup $S(t)$ for (1.1) is introduced. In addition, the basic assumptions concerning relationships between the true semigroup $S(t)$ and approximate semigroup $S^n_{\Delta t}$ for (1.2) are spelled out and certain convergence results proved for individual trajectories.

In the remaining three sections we study the existence and convergence properties of a variety of objects under discretization. Since the focus of the article is on convergence, and a rather general framework for this question is considered, we do not distinguish between the practical value of different methods. Rather we address the question 'what meaning can be attached to computations performed with arbitrary finite time convergent numerical methods when used over very long-time intervals?'

In all three sections the basic format is the same: the introduction is concerned with the development of an existence theory for the invariant sets of (1.1) itself, whilst the remaining sections contain modifications of this theory for the numerical approximation and the derivation of error bounds. Wherever possible, the existence theory is developed in such a way as to be directly applicable to both the equation (1.1) and its approximation (1.2). For this reason the existence theory is formulated in terms of the time Δt evolution of the equation (1.1).

Consequently it is true that, in many cases, much of the work involved in proving results about numerical approximation is concerned with formulating existing theories from continuous dynamical systems in a form amenable to the study of discrete maps arising in numerical analysis; this involves a fair amount of rehashing of well known theories in dynamical systems but is a fruitful process since the approximation theory for the numerical method then falls out in a relatively straightforward manner. Note that, in some cases, we will develop several approaches to the same question. In particular we provide two alternative constructions and convergence proofs for the stable and unstable manifolds, for uniformly asymptotically stable sets and for attractors.

In Section 4 we examine the behaviour of approximations $S^n_{\Delta t}$ in the neighbourhood of an equilibrium point of $S(t)$. We are led to study the existence and convergence of an approximate equilibrium point, the convergence of stable and unstable manifolds of the equilibrium point and the convergence of phase portraits near to the equilibrium point.

In Section 5 we study periodic solutions of $S(t)$ under approximation by $S^n_{\Delta t}$. We show that, under a condition which ensures that the periodic solution is isolated, the semigroup $S^n_{\Delta t}$ has a closed invariant curve which converges, in the sense of sets, to the periodic solution of $S(t)$. We also discuss briefly the effect of discretization on quasi-periodic solutions (the sum of two irrationally related periodic solutions).

In Section 6 we study uniformly asymptotically stable sets and attractors; these objects may include, for example, strange attractors such as those observed in the Lorenz equations. Again we study existence and convergence of approximations of these objects found in the numerical scheme.

2. Background and motivation

In order to motivate the material in the remainder of the article, we start by relating the approach taken here to the classical theory of numerical analysis of initial value problems. Broadly speaking, the two fundamental issues in the classical study of the approximation of (1.1) by (1.2) are *convergence* and *stability*; we consider these two issues in turn and discuss how they might be generalized to the consideration of nonlinear dynamical systems over long-time intervals.

2.1. Convergence

As mentioned in the introduction, standard error bounds relating (1.1) and (1.2) are of the form (1.3). This bound reflects the exponential divergence of trajectories that may be present in well posed problems of the form (1.1). Since most problems do not exhibit exponential divergence throughout the whole of phase space, the bound (1.3) can sometimes be improved upon in a number of ways: (a) for equations (1.1) exhibiting exponential contraction of trajectories throughout phase space, or asymptotically as $t \to \infty$, k may be negative yielding uniform convergence of trajectories for $t \in [0, \infty)$; (b) for equations with conserved quantities, such as Hamiltonian systems, the error bound (1.3) can sometimes be weakened to

$$\|U_n - u(t_n)\| \leq cT^{\alpha}\Delta t^r$$

for $0 \leq t_n \leq T$, for some $\alpha > 0$; (c) for equations whose solution and approximation ultimately lie in a bounded set B the error estimate (1.3) is clearly pessimistic as $t \to \infty$ and can trivially be replaced by

$$\|U_n - u(t_n)\| \leq \mathrm{diam}(B)$$

for t sufficiently large, where $\mathrm{diam}(B)$ denotes the largest distsnce between any two points in B.

Possibility (a) is of *minor interest* since it admits only convergence to a stable equilibrium point as $t \to \infty$ and therefore rules out many applications involving interesting dynamical behaviour; it is discussed briefly in Section 3. Possibility (b) *is of interest* and some results in this direction are described in Calvo and Sanz-Serna (1992; 1993a,b); however, the techniques of use in that case are rather specialized to Hamiltonian and other conservative systems, an area which is extensively reviewed in, for example, Sanz-Serna (1992a). It is possibility (c) with which we shall concern ourselves in this article.

Of course, there are important application areas where unbounded solutions are of relevance. Furthermore, in many cases it is not just the asymptotic behaviour which is of interest but also the transient behaviour. However, there *are many applications* in which bounded asymptotic behaviour is of paramount importance and here we concentrate on such situations; we do not study the approximation of unbounded trajectories nor do we study the approximation of transients in any detail. Thus we suppose that solutions of both (1.1) and (1.2) ultimately lie in some bounded set B. Within B the solution will typically approach an *ω-limit set* as $t \to \infty$. This might be, for example, an equilibrium point, a periodic solution, a quasi-periodic solution or a strange attractor. Examples of the four possibilities are given in Figure 1. The four objects observed as $t \to \infty$ are all examples of ω-limit sets; a precise definition is given in Section 3 but, roughly, they are objects observed for large t in (1.1).

A natural generalization of the standard convergence question to this situation is to ask *are ω-limit sets of* (1.1) *well approximated by ω-limit sets of* (1.2)? It is predominantly this question, and others closely related to it, that we address in this article. To introduce ideas concerned with convergence of limit sets, and other related sets, we describe six examples.

Examples

(i) *Equilibrium points.* The explicit Euler scheme for the approximation of (1.1) is

$$U_{n+1} = U_n + \Delta t f(U_n).$$

Its fixed points satisfy

$$\bar{U} = \bar{U} + \Delta t f(\bar{U}) \Leftrightarrow f(\bar{U}) = 0, \quad \forall \Delta t > 0.$$

Hence they coincide with the equilibrium points of (1.1). Thus convergence of these limit sets (equilibrium solutions) as $\Delta t \to 0$ is trivial. It is worth noting, however, that general Runge–Kutta methods may produce *spurious fixed points* which are not close to the true equilibria as $\Delta t \to 0$ – this point was first observed in Iserles (1990); see Theorem 4.11 and the example preceding it. In Section 4 we shall consider the questions of convergence of fixed points to true equilibria, and the existence of spurious solutions, under fairly weak hypotheses on the nature of the approximation – see Theorems 4.10 and 4.11.

(ii) *Unstable manifolds.* Consider the pair of equations

$$p_t = p, \quad q_t = -q + p^2. \tag{2.1}$$

These equations have the equilibrium point $p = q = 0$. Now consider the curve $q = p^2/3$. If we define the variable $z = q - p^2/3$ then

$$z_t = q_t - \tfrac{2}{3}pp_t = -q + p^2 - \tfrac{2}{3}p^2 = -z. \tag{2.2}$$

Hence, if $z(0) = 0$ then $z(t) = 0\ \forall t \in \mathbb{R}$. Thus the curve

$$q = \tfrac{1}{3}p^2 \tag{2.3}$$

is invariant for the equations – solutions starting on the curve remain on it. Furthermore, using (2.1), we find that if (2.3) holds then

$$p(t) = Ae^t, \quad q = \tfrac{1}{3}A^2e^{2t};$$

thus $p(t), q(t) \rightarrow 0$ as $t \rightarrow -\infty$ for solutions on the invariant curve; this curve is referred to as *the unstable manifold* of the origin.

Now consider the Euler approximation

$$p_{n+1} = (1 + \Delta t)p_n, \quad q_{n+1} = (1 - \Delta t)q_n + \Delta t p_n^2. \tag{2.4}$$

It is natural to seek an invariant curve of the same form as for the differential equation. Specifically, we seek an $a \in \mathbb{R}$ such that

$$q_n = ap_n^2 \Leftrightarrow q_{n+1} = ap_{n+1}^2.$$

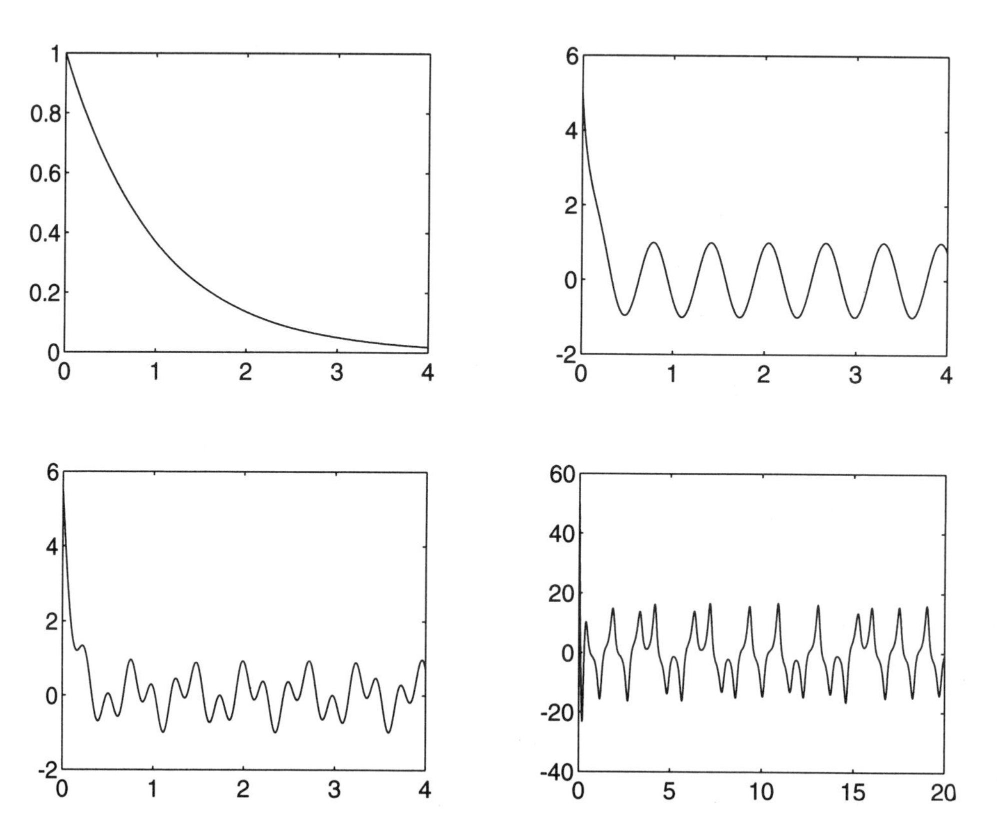

Fig. 1. Plot of a component of $u(t)$ (vertical axis) against t illustrating four different limit sets; from left to right, top to bottom: an equilibrium point, a periodic solution, a quasi-periodic solution, a chaotic solution.

From (2.4) we see that this implies that

$$(1-\Delta t)ap_n^2 + \Delta t p_n^2 = q_{n+1} = ap_{n+1}^2 = a(1+\Delta t)^2 p_n^2.$$

Hence, since this must be satisfied for all p_n, we obtain

$$(1-\Delta t)a + \Delta t = a(1 + 2\Delta t + \Delta t^2)$$

which yields $a = (3+\Delta t)^{-1}$. Thus the curve

$$q = \frac{1}{3+\Delta t} p^2 \tag{2.5}$$

is invariant for the numerical method. Furthermore, on (2.5),

$$p_n = (1+\Delta t)^n p_0, \quad q_n = \frac{(1+\Delta t)^{2n}}{3+\Delta t} p_0^2$$

so that $p_n, q_n \to 0$ as $n \to -\infty$. This demonstrates that (2.5) is the unstable manifold of the origin for the Euler approximation.

The important observation to make from this example is that both the underlying equations and their approximation have an invariant manifold and that, furthermore, by virtue of (2.3) and (2.5), these manifolds are close as $\Delta t \to 0$. Generalizations of this idea will be proved in Theorem 4.12 and Corollary 4.13.

(iii) *Phase portraits.* Consider approximation of the equation

$$x_t = -x, \quad x(0) = x_0,$$

by the explicit Euler scheme

$$X_{n+1} = (1-\Delta t)X_n, \quad X_0 = x_0.$$

It is straightforward to show that there exists $C > 0$ so that the error satisfies

$$|X_n - x(t_n)| \le C\Delta t[1 - (1-\Delta t)^n] \le C\Delta t \quad \forall n, \Delta t : 0 \le n\Delta t < \infty,$$

for Δt sufficiently small. Thus, due to the exponential contraction of the true solution, this error bound is uniform in T for $0 \le t_n \le T$.

The equations

$$x_t = -x, \quad y_t = y,$$

are illustrative of the behaviour of trajectories of (1.1) in the neighbourhood of an equilibrium point. Because of the exponential divergence present in the y-component of the solution, standard error bounds are not uniform with respect to T on the time interval $0 \le t \le T$. However, if we consider numerical trajectories with different initial conditions from the true trajectories *it is possible* to find approximations of solutions which are uniform with respect to the length of time interval T. The key point is that the y-equation *is contractive backwards in time.* We choose the initial condition in y so that the true and numerical solutions agree at $t_n = T$ and then exploit backwards contractivity in y. As an example, consider solving the differential equations subject to the boundary conditions

$$x(0) = \eta, \quad y(T) = \xi, \quad T = N\Delta t.$$

Thus the true solution to the problem is

$$x(0) = \eta e^{-t}, \ y(t) = \xi e^{t-T}.$$

The backward Euler scheme scheme

$$X_{n+1} = (1 - \Delta t)X_n, \quad X_0 = \eta$$

$$Y_{n+1} = (1 + \Delta t)Y_n, \quad Y_N = \xi$$

gives the solution

$$X_n = \eta(1 - \Delta t)^{-n}, \ Y_n = (1 + \Delta t)^{n-N}\xi.$$

It can then be shown that there exists $C > 0$ such that the errors satisfy

$$|X_n - x(t_n)| \leq C\Delta t[1 - (1 - \Delta t)^n] \leq C\Delta t \quad \forall n, \Delta t : 0 \leq n\Delta t \leq T$$

and

$$|Y_n - y(t_n)| \leq C\Delta t[1 - (1 + \Delta t)^{n-N}] \leq C\Delta t \quad \forall n, \Delta t : 0 \leq n\Delta t \leq T$$

for Δt sufficiently small. Again C is independent of T. Thus, by comparing suitably chosen solutions, it is possible to find error bounds which *do not depend on the length of the time interval* for problems exhibiting exponential divergence of trajectories. The basic idea described here was described for a general class of linear problems in Enquist (1969). The idea can be generalized to a wide class of nonlinear problems in the neighbourhood of an equilibrium point; this we show in Section 4 – see Theorem 4.14 and Corollary 4.15.

(iv) *Periodic solutions.* Consider the complex equation ($i^2 = -1$)

$$z_t = (\alpha i + 1 - |z|^2)z$$

with periodic solution $z(t) = e^{\alpha i t}$. As a set of points in $\mathbb{C}$ the periodic solution is given by

$$\mathcal{P} := \{z \in \mathbb{C} : |z| = 1\}. \tag{2.6}$$

The explicit Euler approximation yields the map

$$Z_{n+1} = Z_n + \Delta t(\alpha i + 1 - |Z_n|^2)Z_n.$$

The analogue of the periodic solution of the differential equation is to seek a circle in the complex plane which is *invariant* under the maps – that is a circle with the property that points starting on the circle remain on the circle. (A precise definition of invariant will be given in Section 3.) Thus we seek fixed points of the map $|Z_n|^2 \mapsto |Z_{n+1}|^2$. A little algebra shows that there is a circle of fixed points with the form

$$|Z_n| = R_-(\alpha)$$

where

$$R_\pm(\alpha)^2 = 1 + \frac{1 \pm [1 - \alpha^2 \Delta t^2]^{1/2}}{\Delta t}. \tag{2.7}$$

Hence the mapping has the invariant circle

$$\mathcal{P}_{\Delta t} := \{z \in \mathbb{C} : |z| = R_-(\alpha)\}. \tag{2.8}$$

Noting that $R_-(\alpha) = 1 + \mathcal{O}(\Delta t)$, we deduce from (2.6) and (2.8) that the set $\mathcal{P}_{\Delta t}$ converges to $\mathcal{P}$ as $\Delta t \to 0$. Such a convergence result for periodic solutions is true under much more general circumstances and this is investigated further in Section 5 – see Theorem 5.7 and Corollary 5.8. Note, however, that the numerical method also has a spurious limit set in the form of the invariant circle $|Z_n| = R_+(\alpha) = \mathcal{O}(\Delta t^{-1/2})$. The example constructed here was introduced in Brezzi *et al.* (1984).

We introduce a brief note of caution concerning the approximation of periodic solutions by numerical methods. In order to generalize the example considered to other periodic solutions and other numerical methods, it is necessary to assume that the periodic solution is isolated in phase space (no other periodic solutions arbitrarily close to it). To illustrate why this is necessary, consider the equations

$$x_t = -y, \quad y_t = x$$

with periodic solutions

$$x(t) = A\cos(t), \;\; y(t) = A\sin(t), \quad A \in \mathbb{R}$$

Since A is arbitrary these solutions are not isolated. Furthermore, for the explicit Euler scheme

$$X_{n+1} = X_n - \Delta t Y_n, \quad Y_{n+1} = Y_n + \Delta t X_n,$$

all solutions satisfy

$$[X_n^2 + Y_n^2] = (1 + \Delta t^2)^n [X_0^2 + Y_0^2];$$

thus, unless the initial data are at the origin,

$$X_n^2 + Y_n^2 \to \infty, \quad n \to \infty.$$

Thus no closed invariant curves approximating a periodic solution can exist. Issues of this nature are encountered frequently in the approximation of Hamiltonian and other conservative systems; as mentioned earlier, this is a somewhat separate subject area which we will not address in this article.

(v) *Quasi-periodic solutions.* Consider the coupled complex equations

$$z_t = (\mathrm{i} + 1 - |w|^2)z,$$

$$w_t = (\sqrt{2}\,\mathrm{i} + 1 - |z|^2)w.$$

Note that the equations admit the solution $z(t) = \mathrm{e}^{\mathrm{i}t}, w(t) = \mathrm{e}^{\sqrt{2}\,\mathrm{i}t}$. This is a quasi-periodic solution of the coupled system and, as a set, it may be written

$$\mathcal{Q} = \{z, w \in \mathbb{C} \colon |z| = |w| = 1\}. \tag{2.9}$$

The explicit Euler scheme for these equations is

$$Z_{n+1} = Z_n + \Delta t(\mathrm{i} + 1 - |W_n|^2)Z_n,$$

$$W_{n+1} = W_n + \Delta t(\sqrt{2}\,\mathrm{i} + 1 - |Z_n|^2)W_n.$$

By analogy with the continuous solution, we seek an invariant set with the form

$$|Z_{n+1}| = |Z_n|, \quad |W_{n+1}| = |W_n|, \quad \forall n \geq 0.$$

A calculation shows that such an invariant set may be found with the form

$$|Z_n| = R_-(\sqrt{2}), \quad |W_n| = R_-(1),$$

where $R_-(\alpha)$ is given by (2.7). Noting that $R_-(\alpha) = 1 + \mathcal{O}(\Delta t)$ we deduce that the numerical method has an invariant set

$$\mathcal{Q}_{\Delta t} = \{z, w \in \mathbb{C} \colon |z| = R_-(\sqrt{2}), |w| = R_-(1)\},$$

which converges to the true invariant set $\mathcal{Q}$ given by (2.9) as $\Delta t \to 0$. Note that the example constructed here is simply a modification of that described for periodic solutions in the previous example. Spurious invariant sets can be constructed by choosing the root $R_+(\cdot)$ in the construction of the invariant set.

Once again, in order to generalize this example, it is crucial to require that the quasi-periodic solution be isolated. The convergence of quasi-periodic solutions is discussed in Section 5.

(vi) *Strange attractors.* Consider the Lorenz equations (Lorenz, 1963)

$$\begin{aligned} x_t &= \sigma(y - x), \\ y_t &= rx - y - xz, \\ z_t &= xy - bz. \end{aligned} \tag{2.10}$$

Figure 2 shows solutions of the equations, with parameters set at $\sigma = 10, r = 28$ and $b = \frac{8}{3}$, for four entirely different initial conditions. Note that, in all cases, the solutions are attracted to a very complicated set in $\mathbb{R}^3$ and this set is an example of a *strange attractor.* The same set is observed in all four cases.

This complicated set is observed for almost all initial conditions chosen and forms the ω-limit set for equations (2.10) for almost all initial data. The issues involved in proving convergence of such strange attractors are far more complicated than for equilibrium points and periodic solutions; no simple illustrations can be constructed. Indeed one of the stumbling blocks in the numerical analysis of such objects is that the existing theory of perturbations to such strange attractors is itself far from fully developed. The convergence of such attractors, and other related objects, is considered in Section 6. See Theorems 6.12, 6.20, 6.21, 6.22, 6.26 and Corollaries 6.18 and 6.30.

2.2. Stability

In the previous section we discussed the notion of convergence and described a particular generalization that is useful in the study of dynamical systems – namely to look at the existence of limit sets (and other related objects) and then study their convergence as $\Delta t \to 0$. Such a convergence study will

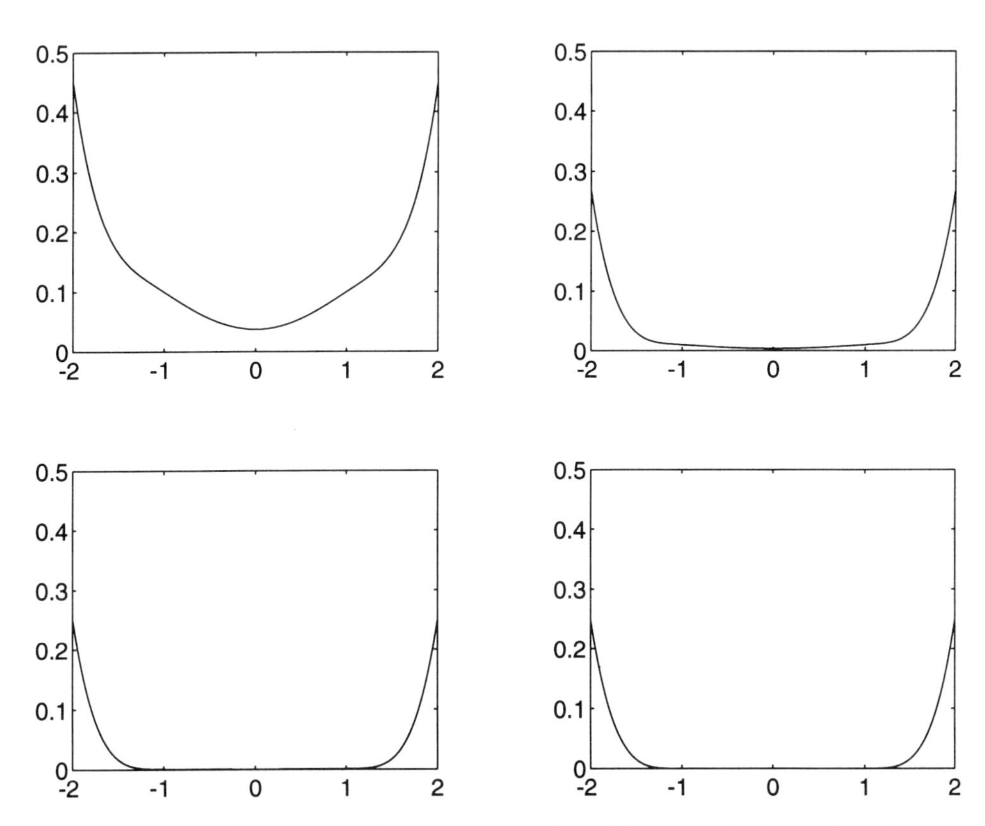

Fig. 2. Plot of x (vertical axis) against y for four different solutions of the Lorenz equations, with $\sigma = 10, r = 28, b = \frac{8}{3}$.

form the majority of the remainder of the article. However, to set things in context, we briefly discuss notions of stability appropriate to the study of dynamical systems.

Classical linear stability theory is concerned with the analysis of approximating the equation

$$u_t = \lambda u, \quad u(0) = U, \quad \mathrm{Re}\,(\lambda) < 0. \tag{2.11}$$

(For the study of linear conservative or Hamiltonian problems it is important to include the effect of approximating this problem for purely imaginary values of λ – the equation $u_t = iu$ is the archetypal example of a Hamiltonian equation. However, such problems are not our concern here and so we restrict attention to the case where $\mathrm{Re}\,(\lambda) < 0$.)

The approximation (1.2) to (2.11) is typically a rational function of $\lambda\Delta t$; for a given numerical method applied to (2.11), the *region of absolute stability* $\mathcal{S} \subseteq \mathbb{C}$ is defined to be the set with the property that

$$\lambda\Delta t \in \mathcal{S} \Leftrightarrow |U_n| \to 0 \text{ as } n \to \infty.$$

Thus, if $\lambda\Delta t \in \mathcal{S}$, the numerical solution replicates the behaviour of the underlying equation (2.11). In many circumstances it is important that this occurs without restriction on Δt. Hence the following definition is useful: the numerical method is said to be *A-stable* if

$$\{z \in \mathbb{C}\colon \mathrm{Re}\,(z) < 0\} \subseteq \mathcal{S}.$$

One abstraction of the above concepts of stability (which we can generalize to other problems) is that they yield *conditions under which an important qualitative property of the equation is inherited by the discretization.* (This abstraction misses the important connection between such practical stability conditions and error propagation in stiff problems, but is nonetheless a useful notion.) With this abstraction in mind, let us consider nonlinear problems.

A fundamental difference between linear and nonlinear problems is that, in the latter case, the stability notions become *initial data dependent.* Consider the approximation (1.2) of a nonlinear problem of the form (1.1) with the property that all solutions tend to the origin as $t \to \infty$. In this case we could define the *region of absolute stability* of a given method to be the set $\mathcal{S} \subseteq \mathbb{R} \times \mathbb{R}^p$ with the property that

$$(\Delta t, U_0) \in \mathcal{S} \Leftrightarrow |U_n| \to 0 \text{ as } \quad t \to \infty.$$

Defining the *basin of attraction* of the origin to be the set of initial data which yield an asymptote at the origin as t or n tends to infinity, we see that this notion of stability is nothing more than seeking conditions on Δt which preserve the basin of attraction. We have assumed that the basin of attraction of the origin for the differential equation is the whole of $\mathbb{R}^p$. Thus it might be natural to seek numerical methods which replicate this property for an interval of Δt; that is to seek methods with the property that there exists $\Delta t_c > 0$ such that

$$\{(\Delta t, U) \in \mathbb{R} \times \mathbb{R}^p : \Delta t \in (0, \Delta t_c)\} \subseteq \mathcal{S}.$$

Under such conditions, all choices of $\Delta t \in (0, \Delta t_c)$ will yield a numerical solution with the correct asymptotic behaviour, *independently of initial data.* An even stronger constraint on the numerical method would be to ask that no upper bound on Δt is required either; that is to seek methods with the property that

$$\{(\Delta t, U) \in \mathbb{R} \times \mathbb{R}^p : \Delta t \in (0, \infty)\} \subseteq \mathcal{S}.$$

Such notions of nonlinear stability are contained in the literature although they are not framed in this way. In particular, there has been a great deal of work devoted to problems of the form (1.1) where $f(\cdot)$ satisfies

$$\langle f(u) - f(v), u - v \rangle \leq -\alpha \|u - v\|^2 \tag{2.12}$$

for some $\alpha > 0$. Under (2.12), equation (1.1) has a unique steady solution (without loss of generality at the origin) which all solutions approach exponentially as $t \to \infty$. The study of numerical stability for such problems was initiated in Dahlquist (1975; 1978) where linear multistep methods were considered and generalized to Runge–Kutta methods by Burrage and Butcher (Butcher, 1975; Burrage and Butcher, 1979).

Whilst the nonlinear stability theory developed for (1.1) satisfying (2.12) has been very important in unifying the linear and nonlinear theories of error propagation, its range of applicability is somewhat limited since the condition (2.12) rules out nontrivial dynamical behaviour. Nonetheless, analogous stability theories can be developed under other hypotheses on $f(\cdot)$. For instance, the assumption that

$$\exists a, b > 0 : \langle f(u), u \rangle \leq a - b\|u\|^2 \tag{2.13}$$

is of interest in this context. Under (2.13) all solutions of (1.1) eventually enter the ball

$$\{u \in \mathbb{R}^p : \|u\|^2 \leq (a + \epsilon)/b\} \text{ for any } \epsilon > 0.$$

Hence this asymptotic bound on the solution is *independent of the initial data.* It is natural to examine numerical methods which replicate this property and this is done in Humphries and Stuart (1994) for Runge–Kutta methods and in Hill (1994) for linear multistep methods. It is interesting that the stability conditions required to make numerical methods replicate the qualitative behaviour of the underlying equation (1.1) under (2.11), (2.12) or (2.13) are very closely related. An overview of the area of nonlinear stability for (1.1) under a variety of different structural assumptions, including those described here, may be found in Stuart and Humphries (1992a). Since that article is self-contained we will not pursue stability issues any further in this article.

2.3. *Summary*

There are two important obervations to make concerning the discussion in this section. First *one possible* generalization of the notion of convergence to include dynamical systems is to consider the convergence of ω-limit sets – the objects observed for large time in (1.1). Second *one possible* generalization of the notion of stability to dynamical systems is to ask for preservation of certain qualitative properties of the underlying differential equation (1.1) under numerical approximation; in particular it is desirable to have such preservation occurring for a wide range of time step Δt and initial data U.

These two separate questions of convergence and stability are bridged by the question of the convergence of basins of attraction of ω-limit sets. This is because numerical instability is often manifested in a blow-up of the scheme so that the basins of attraction of limit sets are affected. Consider the following example.

Example This example shows relationships between the convergence of limit sets and basins of attraction, and numerical stability. All solutions of the equation

$$u_t = -u^3, \quad u(0) = U \tag{2.14}$$

tend to the origin 0 as $t \to \infty$. Thus $\{0\}$ is the only ω-limit set and its basin of attraction is $\mathbb{R}$ Now consider the Euler approximation

$$U_{n+1} = (1 - \Delta t U_n^2)U_n, \quad U_0 = U.$$

Analysis of this map given in Stuart (1991) shows that the origin $\{0\}$ is an ω-limit point (which trivially converges to the true limit set $\{0\}$ as $\Delta t \to 0$.) The origin has basin of attraction $(-\sqrt{(}2/\Delta t), \sqrt{(}2/\Delta t))$ and thus the basin of attraction converges to the true basin of attraction $\mathbb{R}$ as $\Delta t \to 0$. To see the connection with numerical stability, consider initial data outside the basin of attraction: if $|U|^2 > \sqrt{(}2/\Delta t)$ then it may be shown that $|U_n| \to \infty$ as $n \to \infty$, a form of numerical instability. Thus the convergence of basins of attraction is closely related to the determination of conditional numerical stability questions.

The question of convergence of basins of attraction is little studied and there are many open questions in the area; see Humphries (1994) for some analysis in this direction.

As stated before, in this article we concentrate solely on the convergence of ω-limit sets and other related objects. This in itself is an enormous subject area but, as we hope to show, comprises a cohesive body of knowledge. In particular we have striven to put the results already contained in the literature in a unified framework, paying particular attention to the development of an existence theory for the objects of interest which can also be used to study analogous questions for numerical approximations.

2.4. Bibliography

Interest in the subject of the interaction between numerical analysis and dynamical systems has been growing steadily over the past decade. In particular two major international conferences have been held concerning the subject – the first was in Bristol, UK, in 1990 (see Broomhead and Iserles, (1992), Budd (1990) and Sanz-Serna (1992a)) and the second in Geelong, Australia, in 1993 (see Kloeden and Palmer (1994)). Furthermore, a series of lectures at the IVth SERC Numerical Analysis Summer School in Lancaster, UK, was given in 1991 – see Beyn (1992).

3. Semigroups and their approximation

3.1. Notation

We shall not define a specific norm in this article except in a few special circumstances. However, the norm should be taken as fixed throughout any given argument used and, furthermore, all matrix norms are those subordinate to the underlying vector norm. For simplicity it is sufficient to consider the Euclidean norm unless otherwise stated.

It will be important to have an appropriate definition of the distance between sets. Let A and B be sets in $\mathbb{R}^p$ and u a point in $\mathbb{R}^p$; we introduce the following notation:

$$\begin{aligned} \mathrm{dist}(u, A) &= \inf_{v \in A} \|u - v\|, \\ \mathrm{dist}(B, A) &= \sup_{u \in B} \mathrm{dist}(u, A), \\ \mathcal{N}(A, \epsilon) &= \{u \in \mathbb{R}^p : \mathrm{dist}(u, A) < \epsilon\}, \\ \partial\mathcal{N}(A, \epsilon) &= \{u \in \mathbb{R}^p : \mathrm{dist}(u, A) = \epsilon\}. \end{aligned} \tag{3.1}$$

Notice that, if $\mathrm{dist}(B, A) < \epsilon$ it follows that $\bar{B} \subseteq \mathcal{N}(\bar{A}, \epsilon)$ so that

$$\mathrm{dist}(B, A) = 0 \Longrightarrow \bar{B} \subseteq \bar{A}.$$

Hence 'dist' only defines a *semidistance* – the asymmetric Hausdorff semidistance as distinct from the *Hausdorff distance* between two sets A and B which is defined by

$$d_{\mathrm{H}}(A, B) = \max\{\mathrm{dist}(A, B), \mathrm{dist}(B, A)\}. \tag{3.2}$$

We also employ the following notation for open balls:

$$\begin{aligned} B(v, \epsilon) &:= \{u \in \mathbb{R}^p : \|u - v\| < \epsilon\}, \\ \partial B(v, \epsilon) &:= \{u \in \mathbb{R}^p : \|u - v\| = \epsilon\}. \end{aligned} \tag{3.3}$$

Thus $B(v, \epsilon) = \mathcal{N}(v, \epsilon)$ and $\partial B(v, \epsilon) = \partial\mathcal{N}(v, \epsilon)$.

3.2. The differential equation

Let us assume for the moment that a unique solution of (1.1) exists for all $t \geq 0$ and any $U \in \mathbb{R}^p$. Consequently we may define a semigroup $S(t) : \mathbb{R}^p \to \mathbb{R}^p$ in such a way that the solution $u(t)$ of (1.1) is given by

$$u(t) = S(t)U.$$

The one-parameter mapping $S(t)$ satisfies the usual semigroup properties

(i) $S(0) = I$, the identity on $\mathbb{R}^p$;
(ii) $S(t+s) = S(t)S(s) \quad \forall t, s \in \mathbb{R}^+$.

Under the assumption that f is differentiable on $\mathbb{R}^p$ it follows that the semigroup $S(t)U$ is continuous in both $t \in \mathbb{R}^+$ and $U \in \mathbb{R}^p$. We make this assumption throughout. We denote the Jacobian of $S(t)U$ with respect to $U \in \mathbb{R}^p$, evaluated at a point $V \in \mathbb{R}^p$, by $\mathrm{d}S(V;t)$.

Example To illustrate the semigroup $S(t)$ we consider the equation (2.14). Since solutions exist for all positive time, a semigroup may be defined. Solving the equation explicitly gives

$$S(t)U = \frac{U}{(1 + 2tU^2)^{1/2}}. \tag{3.4}$$

Properties (i) and (ii) are easily verified. Differentiation with respect to U shows that

$$dS(V;t) = \frac{1}{(1 + 2tV^2)^{3/2}}. \quad \square \tag{3.5}$$

Frequently we shall we require the action of $S(t)$ on a set of points $E \subset \mathbb{R}^p$. We define

$$S(t)E = \bigcup_{x \in E} S(t)x. \tag{3.6}$$

When analysing dynamical systems it is of great importance to study sets with the property that trajectories starting in a given set remain within that set. This motivates the following definition:

Definition 3.1 A set E is said to be *invariant* (respectively *positively invariant*) if, for any $t \geq 0$, $S(t)E \equiv E$ (respectively $S(t)E \subseteq E$).

Example Consider the equation

$$u_t = u(1 - u^2), \quad u(0) = U. \tag{3.7}$$

This has solution

$$u(t) = \frac{U}{[U^2 + (1 - U^2)\mathrm{e}^{-2t}]^{1/2}}. \tag{3.8}$$

Let $E = [-1, 1]$. We show that $S(t)E = E$. If $U \in [-1, 1]$ then it follows from (3.8) that $S(t)U \in [-1, 1]$ and hence that $S(t)E \subseteq E$. Furthermore reversing time in (3.8) shows that, if $V = S(t)U$, then

$$U = \frac{V}{[V^2 + (1 - V^2)e^{2t}]^{1/2}}.$$

Thus, if $V \in [-1, 1]$ then $U \in [-1, 1]$ and hence $E \subseteq S(t)E$. Thus we have shown that $S(t)E \equiv E$ and so E is invariant.

Now consider the interval $B = [-a, a], a > 1$. From (3.7) we have

$$\frac{1}{2}\frac{\mathrm{d}}{\mathrm{d}t}|u|^2 = u^2 - u^4 < 0, \quad |u| > 1.$$

Thus solutions starting on the boundary of B enter B and hence no solutions can leave B. Thus $S(t)B \subseteq B$ and B is positively invariant. □

The behaviour as $t \to \infty$ of the dynamical system defined by $S(t)$ is captured by its ω-limit sets. Roughly these are sets of accumulation points at $t = \infty$ for subsequences in time extracted from a solution $u(t)$, $t \geq 0$. Precisely we have:

Definition 3.2 *The* ω-limit set *of a point* U *is defined by*

$$\omega(U) = \{x \in \mathbb{R}^p | \exists \{t_i\}, t_i \to \infty : S(t_i)U \to x \text{ as } t_i \to \infty\}.$$

An equivalent definition is

$$\omega(U) = \bigcap_{s \geq 0} \overline{\bigcup_{t \geq s} S(t)U}. \tag{3.9}$$

Similary we may define the ω-limit set *of a set* E *by*

$$\omega(E) = \{x \in \mathbb{R}^p | \exists \{t_i\}, \{U_i\}, t_i \to \infty, U_i \in E : S(t_i)U_i \to x \text{ as } t_i \to \infty\}.$$

An equivalent definition is

$$\omega(E) = \bigcap_{s \geq 0} \overline{\bigcup_{t \geq s} S(t)E}. \tag{3.10}$$

Examples Typical examples of ω-limit sets of individual points are equilibrium points, periodic solutions, quasi-periodic solutions and strange attractors. We illustrate these objects by example.

(i) *Equilibrium point.* Consider the equation

$$u_t = -u, \quad u(0) = U. \tag{3.11}$$

Since $u(t) = e^{-t}U$ it follows that $\omega(U) = \{0\}$ for all $U \in \mathbb{R}^p$. The point $\{0\}$ is simply the equilibrium point for the equation.

(ii) *Periodic solutions.* Consider the equations

$$\begin{aligned} x_t &= x + y - x(x^2 + y^2), \quad x(0) = x_0, \\ y_t &= -x + y - y(x^2 + y^2), \quad y(0) = y_0. \end{aligned} \tag{3.12}$$

If we change variables to polar coordinates by introducing R and ϕ given by

$$x = R\cos\phi, \quad y = R\sin\phi$$

then we obtain the equations

$$R_t = R(1 - R^2), \quad \phi_t = -1. \tag{3.13}$$

Thus $\phi(t) = \phi(0) - t$. Explicit use of the solution (3.8) shows that, for all $R(0) > 0$, $R(t) \to 1$ as $t \to \infty$. Thus

$$x(t) \to \cos(\phi(0) - t), \quad y(t) \to \sin(\phi(0) - t).$$

The solution rotates clockwise in the plane and asymptotically approaches a solution with radius 1; we have found a periodic solution of equations (3.12). Thus, for every solution with $R(0) > 0$ the ω-limit set is simply

$$\mathcal{P} = \{(x, y)^T \in \mathbb{R}^2 : x^2 + y^2 = 1\},$$

that is the set of all points on the periodic solution.

(iii) *Quasi-periodic solutions.* Consider equation (1.1) in $\mathbb{R}^p$, $p \geq 2$, with solution $u(t) = (u_1(t), u_2(t), \ldots, u_p(t))^T$. Assume that the first two solution components satisfy

$$u_1(t) = \mathrm{e}^{-t} + \sin(t), \quad u_2(t) = \mathrm{e}^{-t} + \cos(\delta t)$$

and that the remaining solution components approach 0 as $t \to \infty$. Then, if δ is irrational, as $t \to \infty$ the limiting solution is quasi-periodic. The ω-limit set is given by

$$\mathcal{Q} = \{(u \in \mathbb{R}^p : -1 \leq u_1 \leq 1, -1 \leq u_2 \leq 1, u_j = 0, \ j = 3, \ldots, p\}.$$

(iv) *Strange attractors.* Consider the *Lorenz equations* given by (2.10). Figure 2 shows solutions of the equations, with parameters set at $\sigma = 10, r = 28$ and $b = \frac{8}{3}$, for four entirely different initial conditions. As described in Section 2, the solutions eventually lie on a very complicated set in $\mathbb{R}^3$ and this set is an example of a *strange attractor*. It is the ω-limit set for equations (2.10) for almost all initial data. □

It is important to note that, in general,

$$\omega(E) \neq \bigcup_{x \in E} \omega(x).$$

The following example illustrates this.

Example Consider the equation (3.7) with solution (3.8). From the explicit solution it is clear that

$$\omega(U) = 1, \ U > 0; \quad \omega(U) = -1, \ U < 0; \quad \omega(0) = 0.$$

Thus for any interval $E = [-a, a]$, $a > 0$ we have

$$\bigcup_{x \in E} \omega(x) = \{-1, 0, 1\}.$$

It may also be shown that

$$\omega(E) = [-1, 1];$$

this follows from showing that

$$S(t)E = [-S(t)a, S(t)a],$$

with $S(t)$ determined by (3.8), and noting that $S(t)a \to 1$ as $t \to \infty$.

The following property of ω limit sets is very useful:

Theorem 3.3 The ω-limit set of any bounded set $E \subset \mathbb{R}^p$, $\omega(E)$, is a closed positively invariant set. Furthermore, if $\exists T > 0 : S(t)E$ is bounded for $t \geq T$, then $\omega(E)$ is invariant. Finally, if $\omega(U)$ is bounded for some $U \in \mathbb{R}^p$ then it is connected.

Proof. We first establish closure: consider a sequence of ω-limit points $x_k \to x$, with each $x_k \in \omega(E)$. We wish to show that $x \in \omega(E)$. By Definition 3.2, for each k there are sequences $\{v_i^k\}, \{t_i^k\}$ such that

$$S(t_i^k)v_i^k \to x_k \text{ as } i \to \infty.$$

Hence, without loss of generality, let

$$\|x_k - S(t_i^k)v_i^k\| \leq k^{-1} \quad \text{and} \quad t_i^k \geq k \text{ for } i \geq k.$$

Now define $v_i^* := v_i^i$ and $t_i^* := t_i^i$. By construction $t_i^* \to \infty$ as $i \to \infty$. Now

$$\begin{aligned} \|x - S(t_i^*)v_i^*\| &\leq \|x - x_i\| + \|S(t_i^*)v_i^* - x_i\| \\ &\leq \|x - x_i\| + i^{-1}. \end{aligned}$$

Taking $i \to \infty$ we find $S(t_i^*)v_i^* \to x$. Hence, by Definition 3.2, $x \in \omega(E)$.
We now show positive invariance. Assume that $x \in \omega(E)$. If

$$S(t_i)v_i \to x$$

then by continuity of $S(t)\cdot$,

$$S(t + t_i)v_i = S(t)S(t_i)v_i \to S(t)x \quad \forall t \geq 0.$$

Thus $S(t + t_i)v_i \to S(t)x$ and, hence $S(t)x \in \omega(E)$ by Definition 3.2. Thus we deduce positive invariance of $\omega(E)$.

Now we establish negative invariance. We assume that $S(t)E$ is bounded for $t \geq T$ and assume that $x \in \omega(E)$. We aim to show that, for any $t > 0$, $\exists y$: $S(t)y = x$ and $y \in \omega(E)$. Let $S(t_i)v_i \to x$, where, without loss of generality, we may choose $t_1 \geq 1 + T + t$. Now consider the sequence $S(t_i - t)v_i$. Since

$S(t)E$ is bounded for $t \geq T$ and $t_i - t \geq T+1$ for all $i \geq 1$ it follows that there exists a convergent subsequence

$$S(t_{i_j} - t)v_{i_j} \to y.$$

Now

$$\begin{aligned} x &= \lim_{j\to\infty} S(t_{i_j})v_{i_j} = \lim_{j\to\infty} S(t)S(t_{i_j} - t)v_{ij} \\ &= S(t) \lim_{j\to\infty} S(t_{i_j} - t)v_{i_j} = S(t)y. \end{aligned}$$

Hence the negative invariance is proved.

Finally we show that boundedness of $\omega(U)$ implies connectedness. Assume for contradiction that $\omega(U)$ comprises two disjoint components P and Q with $\mathcal{N}(P,\epsilon) \cap \mathcal{N}(Q,\epsilon) = \emptyset$, for some $\epsilon > 0$. Then there exist sequences $t_i \to \infty$ and $\tau_i \to \infty$ such that $S(t_i)U \to x \in P$ and $S(\tau_i)U \to y \in Q$. Without loss of generality we may assume that $t_i < \tau_i$ and that $S(t_i)U \in \mathcal{N}(P,\epsilon), S(\tau_i) \in \mathcal{N}(Q,\epsilon) \quad \forall i \geq 1$. By continuity of $S(\cdot)U$ it follows that there exists $T_i \in (t_i, \tau_i)$ such that $S(T_i)U \in \partial\mathcal{N}(P,\epsilon)$. But the set $\partial\mathcal{N}(P,\epsilon)$ is closed and bounded, since P is bounded, and hence compact. Thus there exists a convergent subsequence $S(T_{i_j})U \to z \in \partial\mathcal{N}(P,\epsilon)$. But this is a contradiction since then, by definition, $z \in \omega(U)$ but $z \notin P \cup Q$. This completes the proof. □

The behaviour of a dynamical system is very well understood if all the ω-limit sets can be determined together with knowledge of which initial data are associated with a given limit set; this motivates the following:

Definition 3.4 *An ω-limit set $\mathcal{W}$ has* basin of attraction $\mathcal{B}$ *if*

$$\{\omega(U) = \mathcal{W}\} \Leftrightarrow \{U \in \mathcal{B}\}.$$

Examples For equation (3.11) it is is clear that the basin of attraction of the ω-limit set $\mathcal{Q} = \{0\}$ is the whole of $\mathbb{R}$. For equation (3.12) the basin of attraction of the periodic solution $\mathcal{P}$ is $\mathbb{R}^2\backslash\{0\}$. □

3.3. Approximating semigroups

The numerical method (1.2) generates a semigroup $S^n_{\Delta t} : \mathbb{R}^p \to \mathbb{R}^p$ in such a way that the solution U_n of (1.2) is given by

$$U_n = S^n_{\Delta t} U_0.$$

Here the subscript Δt is used simply to emphasize the dependence of the numerical method on Δt. This semigroup satisfies properties analogous to those for $S(t)$:

(i) $S^0_{\Delta t} = I$, the identity on $\mathbb{R}^p$;

(ii) $S^{n+m}_{\Delta t} = S^n_{\Delta t} S^m_{\Delta t} \quad \forall n,m \in \mathbb{Z}^+$.

If f is differentiable then standard numerical methods yield a differentiable semigroup $S^1_{\Delta t}U$. Throughout we denote the Jacobian of $S^1_{\Delta t}U$ with respect to $U \in \mathbb{R}^p$ evaluated at a point $V \in \mathbb{R}^p$ by $\mathrm{d}S^1_{\Delta t}(V)$.

Example We consider equation (2.14) under approximation by the explicit Euler scheme; this yields the mapping

$$U_{n+1} = U_n - \Delta t U_n^3, \tag{3.14}$$

so that

$$S^1_{\Delta t}U = U - \Delta t U^3 \tag{3.15}$$

and $S^n_{\Delta t}$ is an n-fold composition of $S^1_{\Delta t}$. Again Properties (i) and (ii) are straightforward to check. Differentiation with respect to U gives

$$\mathrm{d}S^1_{\Delta t}(V) = 1 - 3\Delta t V^2. \quad \square \tag{3.16}$$

Important remark It is straightforward to define concepts of invariance, ω-limit set and basin of attraction for the semigroup $S^n_{\Delta t}$. Indeed the only change necessary to Definitions 3.1, 3.2 and 3.4 is to replace t_i by a sequence of integers $n_i \to \infty$. Furthermore, Theorem 3.3 has a discrete analogue for $S^n_{\Delta t}$ with the caveat that the last part of the theorem, concerning connectedness, does not hold.

Example Consider the explicit Euler approximation of (3.11). This yields the map

$$U_{n+1} = (1 - \Delta t)U_n, \quad U_0 = U,$$

so that $S^1_{\Delta t}U = (1 - \Delta t)U$. If $\Delta t \in (0, 2)$ then $\omega(U) = 0$ for any $U \in \mathbb{R}$, replicating the behaviour of the differential equation. Thus $\{0\}$ has basin of attraction $\mathbb{R}$ if $\Delta t \in (0, 2)$. However, if $\Delta t = 2$ then $\omega(U) = \{U, -U\}$ and the basin of attraction of $\omega(U)$ is $\omega(U)$ itself. Note that $\omega(U)$ is not connected in this case. $\square$

The truncation error is now defined to be the error committed by the approximation (1.2) over one time step of length Δt.

Definition 3.5 The *truncation error* for the map (1.2) as an approximation to the ordinary differential equation (1.1) at a point $U \in \mathbb{R}^p$ is defined by

$$T(U; \Delta t) := S(\Delta t)U - S^1_{\Delta t}U.$$

The Jacobian of $T(U; \Delta t)$ with respect to $U \in \mathbb{R}^p$ evaluated at $V \in \mathbb{R}^p$ is denoted by

$$\mathrm{d}T(V; \Delta t) := \mathrm{d}S(V; \Delta t) - \mathrm{d}S^1_{\Delta t}(V).$$

Example We illustrate this definition by considering the approximation of (2.14) by (3.14). First notice that from (3.15)

$$|S^1_{\Delta t}U - S^1_{\Delta t}V| \le |U - V| + \Delta t|U^3 - V^3| \le [1 + \Delta t K_1(U, V)]|U - V|,$$

where $K_1(U,V) = |U^2+UV+V^2|$. This shows the local Lipschitz continuity of $S^1_{\Delta t}$. By (3.4) and (3.15) we see that the truncation error satisfies

$$T(U;\Delta t) = \frac{U}{(1+2\Delta tU^2)^{1/2}} - (U - \Delta tU^3).$$

Taylor expansion of $S(\Delta t)U$ shows that there exists $\Delta t_c = \Delta t_c(U) > 0$ and $K_2 = K_2(U) > 0$ such that

$$|T(U;\Delta t)| \le K_2\Delta t^2 \quad \forall \Delta t \in (0, \Delta t_c].$$

This shows that the truncation error is bounded above by a quantity proportional to Δt^2. Furthermore

$$\mathrm{d}T(U;\Delta t) = \frac{1}{(1+2\Delta tU^2)^{3/2}} - (1 - 3\Delta tU^2).$$

Again, a Taylor series expansion shows that there exists $\Delta t_c = \Delta t_c(U) > 0$ and $K_3 = K_3(U) > 0$ such that

$$|\mathrm{d}T(U;\Delta t)| \le K_3\Delta t^2 \quad \forall \Delta t \in (0, \Delta t_c],$$

possibly by further reduction of Δt_c. This shows that the Jacobian of the truncation error is also bounded above by a constant of $\mathcal{O}(\Delta t^2)$.

There are two remaining properties of $S^1_{\Delta t}$ worth emphasizing by example. Note from (3.16) that

$$|\mathrm{d}S^1_{\Delta t}(V)| \le 1 + 3V^2\Delta t$$

and

$$|\mathrm{d}S^1_{\Delta t}(U) - \mathrm{d}S^1_{\Delta t}(V)| \le 3\Delta t|U+V||U-V|. \quad \square$$

The example illustrates a number of basic properties of the approximate semigroup that we will need throughout this article. It is clear that four important properties of the approximate semigroup are: (i) a Lipschitz condition on $S^1_{\Delta t}$ (or, relatedly, a bound on the Jacobian $\mathrm{d}S^1_{\Delta t}$) of size $1+\mathcal{O}(\Delta t)$; (ii) a Lipschitz condition on $\mathrm{d}S^1_{\Delta t}$ of size $\mathcal{O}(\Delta t)$; (iii) an $\mathcal{O}(\Delta t^2)$ closeness of $S^1_{\Delta t}$ to $S(\Delta t)$ uniformly in any bounded set $B(0,R)$; (iv) an $\mathcal{O}(\Delta t^2)$ closeness of $\mathrm{d}S^1_{\Delta t}(U)$ to $\mathrm{d}S(U;\Delta t)$ uniformly in any bounded set $B(0,R)$.

We will assume that $S^1_{\Delta t}$ satisfies generalizations of these four conditions throughout the remainder of the article. However, our analysis will be greatly streamlined if estimates for the size of the truncation error in terms of Δt are uniform across the whole of $\mathbb{R}^p$. Since our interest is primarily in the local behaviour of $S(t)$ and $S^n_{\Delta t}$ near to bounded limit sets it is sufficient to consider vector fields which are globally bounded. Specifically we make the following assumption concerning the vector field f :

Assumption 3.6 The vector field f in (1.1) satisfies $f \in C^\infty(\mathbb{R}^p, \mathbb{R}^p)$ and f and all of its derivatives are uniformly bounded for all $u \in \mathbb{R}^p$.

In fact this assumption also yields both forwards and backwards in time global existence and uniqueness for the equation (1.1) so that the semigroup may be extended to a group, but we do not pursue this further. Assumption 3.6 will be made throughout the remainder of this section and throughout Sections 4–6; it will not be stated explicitly in the theorems. The assumption is made for simplicity and, since the results described are local in nature, is not necessary – the vector field $f(\cdot)$ can always be modified outside a compact set to yield Assumption 3.6.

From Assumption 3.6 it is possible to prove that many one-step approximations $S^1_{\Delta t}$ to the semigroup $S(t)$ (including all Runge–Kutta methods) generated by (1.1) satisfy the following uniform continuity and approximation properties:

Assumption 3.7 There exists constants $K > 0, \Delta t_c > 0$ and an integer $r \geq 1$ such that for all $\Delta t \in [0, \Delta t_c)$ the semigroups $S(t)$ and $S^1_{\Delta t}$ satisfy

(i) $\|\mathrm{d}S^1_{\Delta t}(U)\| \leq (1 + K\Delta t) \quad \forall U \in \mathbb{R}^p$;
(ii) $\|\mathrm{d}S^1_{\Delta t}(U) - \mathrm{d}S^1_{\Delta t}(V)\| \leq K\Delta t\|U - V\| \quad \forall U, V \in \mathbb{R}^p$;
(iii) $\|T(U; \Delta t)\| \leq K\Delta t^{r+1} \quad \forall U \in \mathbb{R}^p$;
(iv) $\|\mathrm{d}T(U; \Delta t)\| \leq K\Delta t^{r+1} \quad \forall U \in \mathbb{R}^p$.

We will not make this assumption explicit in the statement of the theorems but will assume it throughout the remainder of this section and throughout Sections 4–6. Assumption 3.7 is satisfied by standard Runge–Kutta methods applied to (1.1) under Assumption 3.6.

We now prove certain results concerning the closeness of the semigroups $S^n_{\Delta t}$ and $S(t_n)$ over fixed time intervals $0 \leq n\Delta t = t_n \leq T$. These results follow directly from Assumption 3.7. We need one preliminary observation. Note that, if $u(t) = S(t)U$ then $w(t) = \mathrm{d}S(U; t)v$ satisfies the equation

$$w_t = \mathrm{d}f(u(t))w, \quad w(0) = v$$

where $\mathrm{d}f(\cdot)$ denotes the Jacobian of f. By Assumption 3.6 we may assume, without loss of generality, that $\|\mathrm{d}f(u)\| \leq K \; \forall u \in \mathbb{R}^p$ so that

$$\frac{1}{2}\frac{\mathrm{d}}{\mathrm{d}t}\|w\|^2 = \langle w, w_t\rangle = \langle w, \mathrm{d}f(u)w\rangle \leq K\|w\|^2.$$

Hence

$$\|w(t)\| \leq \mathrm{e}^{Kt}\|v\| \Rightarrow \|\mathrm{d}S(U,t)\| \leq \mathrm{e}^{Kt}. \tag{3.17}$$

We can now prove the following theorem concerning the closeness of both the semigroups $S(t)$ and $S^1_{\Delta t}$ and their derivatives.

Theorem 3.8 Consider the semigroups $S^1_{\Delta t}U$ and $S(\Delta t)U$. It follows that, if $t_n = n\Delta t, 0 \leq t_n \leq T$ and $\Delta t \in (0, \Delta t_c]$, then

$$e_n := \|S^n_{\Delta t}U - S(t_n)U\| \leq [\mathrm{e}^{KT} - 1]\Delta t^r$$

and

$$E_n := \|\mathrm{d}S^n_{\Delta t}(U) - \mathrm{d}S(U;t_n)\| \le [\mathrm{e}^{3KT} - \mathrm{e}^{2KT}]\Delta t^r.$$

Furthermore

$$\|S^n_{\Delta t}U - S(t_n)V\| \le (\mathrm{e}^{KT} - 1)\Delta t^r + \mathrm{e}^{KT}\|U - V\|.$$

Proof. Clearly Assumption 3.7(i) yields

$$\|S^1_{\Delta t}V - S^1_{\Delta t}W\| \le (1 + K\Delta t)\|V - W\| \quad \forall V, W \in \mathbb{R}^p. \tag{3.18}$$

Thus, by (i) and (iii) of Assumption 3.7, we have

$$\begin{aligned} e_{m+1} &= \|S^{m+1}_{\Delta t}U - S(t_{m+1})U\| = \|S^1_{\Delta t}S^m_{\Delta t}U - S(\Delta t)S(t_m)U\| \\ &\le \|S^1_{\Delta t}S^m_{\Delta t}U - S^1_{\Delta t}S(t_m)U\| + \|S(\Delta t)S(t_m)U - S^1_{\Delta t}S(t_m)U\| \\ &\le (1 + K\Delta t)e_m + K\Delta t^{r+1}. \end{aligned}$$

Thus, by induction, we obtain

$$e_n \le [(1 + K\Delta t)^n - 1]\Delta t^r + (1 + K\Delta t)^n e_0. \tag{3.19}$$

Since $e_0 = 0$ and $(1 + K\Delta t)^n \le \mathrm{e}^{KT}$ the desired result follows.

Now we consider E_n. Note that from (3.17) we have

$$\mathrm{d}S(U;t_m) \le \mathrm{e}^{KT}, \quad \forall t_m \in [0, T]. \tag{3.20}$$

The quantity E_m satisfies

$$\begin{aligned} E_{m+1} &= \|\mathrm{d}S^{m+1}_{\Delta t}(U) - \mathrm{d}S(U;t_{m+1})\| \\ &= \|\mathrm{d}\{S^1_{\Delta t}S^m_{\Delta t}(U)\} - \mathrm{d}\{S(\Delta t)S(t_m)U\}\| \\ &= \|\mathrm{d}S^1_{\Delta t}(S^m_{\Delta t}U)\,\mathrm{d}S^m_{\Delta t}(U) - \mathrm{d}S(S(t_m)U;\Delta t)\,\mathrm{d}S(U;t_m)\| \\ &\le \|\mathrm{d}S^1_{\Delta t}(S^m_{\Delta t}U)\,\mathrm{d}S^m_{\Delta t}(U) - \mathrm{d}S^1_{\Delta t}(S^m_{\Delta t}U)\,\mathrm{d}S(U;t_m)\| \\ &\quad + \|\mathrm{d}S^1_{\Delta t}(S^m_{\Delta t}U)\,\mathrm{d}S(U;t_m) - \mathrm{d}S(S(t_m)U;\Delta t)\,\mathrm{d}S(U;t_m)\|. \end{aligned}$$

Hence, by Assumption 3.7(i),

$$\begin{aligned} E_{m+1} \le\ & (1 + K\Delta t)E_m \\ & + \|\mathrm{d}S^1_{\Delta t}(S^m_{\Delta t}U)\,\mathrm{d}S(U;t_m) - \mathrm{d}S^1_{\Delta t}(S(t_m)U)\,\mathrm{d}S(U;t_m)\| \\ & + \|\mathrm{d}S^1_{\Delta t}(S(t_m)U)\,\mathrm{d}S(U;t_m) - \mathrm{d}S(S(t_m)U;\Delta t)\,\mathrm{d}S(U;t_m)\|. \end{aligned}$$

Assumption 3.7(ii),(iv) and (3.19), (3.20) thus give us

$$E_{m+1} \le (1 + K\Delta t)E_m + K\mathrm{e}^{KT}\Delta t e_m + K\mathrm{e}^{KT}\Delta t^{r+1}.$$

Using the known bound on e_m we obtain

$$E_{m+1} \le (1 + K\Delta t)E_m + K\mathrm{e}^{KT}\Delta t(\mathrm{e}^{KT} - 1)\Delta t^r + K\mathrm{e}^{KT}\Delta t^{r+1}.$$

Hence

$$E_{m+1} \le (1 + K\Delta t)E_m + K\mathrm{e}^{2KT}\Delta t^{r+1}.$$

By induction we obtain

$$E_n \leq [(1+K\Delta t)^n - 1]e^{2KT}\Delta t^r + (1+K\Delta t)^n E_0.$$

Since $E_0 = 0$ and $(1+K\Delta t)^n \leq e^{KT}$ the desired result follows.

To prove the final result, note that

$$\|S^n_{\Delta t}U - S(t_n)V\| \leq \|S^n_{\Delta t}V - S(t_n)V\| + \|S^n_{\Delta t}U - S^n_{\Delta t}V\|.$$

Using the error bound for e_n and (3.18) we obtain

$$\|S^n_{\Delta t}U - S(t_n)V\| \leq [e^{KT} - 1]\Delta t^r + (1+K\Delta t)^n\|U - V\|$$

and, since $0 \leq n\Delta t \leq T$, the required result follows. □

Remarks

(i) Note that the derivation of the error e_n is obtained by using two facts: the Lipshitz continuity of the numerical method together with a uniform truncation error bound. An alternative method of proof is to exploit the Lipschitz continuity of the underlying differential equation. Assume that

$$\|S(t)V - S(t)W\| \leq e^{Kt}\|V - W\|. \tag{3.21}$$

(The choice of a constant K in this bound can be made without loss of generality.) The error equation can now be studied thus:

$$\begin{aligned} e_{m+1} &= \|S^{m+1}_{\Delta t}U - S(t_{m+1})U\| = \|S^1_{\Delta t}S^m_{\Delta t}U - S(\Delta t)S(t_m)U\| \\ &\leq \|S(\Delta t)S^m_{\Delta t}U - S(\Delta t)S(t_m)U\| + \|S(\Delta t)S^m_{\Delta t}U - S^1_{\Delta t}S^m_{\Delta t}U\|. \end{aligned}$$

Using the Lipschitz continuity of $S(t)$ and Assumption 3.7(iii) we obtain

$$e_{m+1} \leq e^{K\Delta t}e_m + K\Delta t^{r+1}. \tag{3.22}$$

By induction we obtain

$$e_n \leq [e^{Kt_n} - 1]\frac{K\Delta t^{r+1}}{e^{K\Delta t} - 1} + e^{Kt_n}e_0.$$

Noting that

$$e^x - 1 \geq x/2$$

and that $e_0 = 0$ we obtain

$$e_n \leq 2[e^{KT} - 1]K\Delta t^r, \quad 0 \leq t_n \leq T$$

an analogous bound to that obtained in Theorem 3.8.

(ii) It is worth observing that the constant K appearing in the error bounds derived here is far from optimal. This is since the straightforward bound on the Lipschitz constant used to obtain (3.17) is often pessimistic in estimating the divergence between solutions of the differential equation. Smaller constants can be obtained by working with the so-called logarithmic norm which, roughly speaking, captures the possible rate of divergence more accurately. See Dekker and Verwer (1984).

As briefly mentioned in Section 2 it is possible to obtain uniformly valid error bounds in the case where $u(t)$ approaches an exponentially stable equilibrium point. Specifically we assume that (2.12) holds in a neighbourhood of an equilibrium point $\bar{u}$ from which it follows that

$$\exists \alpha, R > 0: \quad \|S(t)v - S(t)w\| \leq e^{-\alpha t}\|v - w\| \quad \forall v, w \in B(\bar{u}, R). \tag{3.23}$$

The following result is a precursor for the remainder of the article: the proof contains two essential components namely (i) the use of a standard truncation error estimate or resulting finite-time error bound together with (ii) exploitation of a structural property of the underlying solutions of (1.1). In this case the structural property is the exponential stability of $\bar{u}$ manifest in (3.23).

Theorem 3.9 Consider the semigroups $S^1_{\Delta t}U$ and $S(\Delta t)U$ under Assumption 3.7. Assume further that $S(t)U \to \bar{u}$ as $t \to \infty$ and that (3.23) holds. It follows that, if $t_n = n\Delta t, 0 \leq t_n \leq T$ and $\Delta t \in (0, \Delta t_c]$, then there exists $C > 0$ such that

$$e_n := \|S^n_{\Delta t}U - S(t_n)U\| \leq C\Delta t^r \quad \forall n, \Delta t : 0 \leq n\Delta t < \infty.$$

Proof. Assume that $T = N\Delta t$ is a time chosen so that $\|S(T)U - \bar{u}\| < R/2$. By assumption it follows that, for $\tau > 0$ and whilst $S(\tau + T)U \in B(\bar{u}, R)$,

$$\begin{aligned} \|S(\tau + T)U - S(\tau + T)\bar{u}\| &= \|S(T)S(\tau)U - S(T)S(\tau)\bar{u}\| \\ &= \|S(T)S(\tau)U - S(\tau)\bar{u}\| \\ &\leq e^{-\alpha\tau}\|S(T)U - \bar{u}\| < R/2. \end{aligned}$$

Hence $S(t)U \in B(\bar{u}, R/2)$ for all $t \geq T$. From Theorem 3.8 we have that, for $\Delta t \in (0, \Delta t_c]$,

$$e_n \leq [e^{KT} - 1]\Delta t^r, \quad 0 \leq t_n \leq T. \tag{3.24}$$

Thus we can ensure that (possibly by further reduction of Δt_c) that

$$\|S^n_{\Delta t}U - S(t_n)U\| \leq R/2 \quad 0 \leq t_n \leq T$$

and hence that $S^N_{\Delta t}U \in B(\bar{u}, R)$. Now assume, for the purposes of induction, that for some $m \geq N$,

$$e_m \leq \frac{[1 - e^{-\alpha(t_m - T)}]}{1 - e^{-\alpha\Delta t}} K\Delta t^{r+1} + e^{-\alpha(t_m - T)}[e^{KT} - 1]\Delta t^r. \tag{3.25}$$

We also assume that, again possibly by further reduction of Δt_c,

$$(1 - e^{-\alpha\Delta t})^{-1} \leq \frac{2}{\Delta t} \quad \text{and} \quad (2K + e^{KT} - 1)\Delta t^r \leq \tfrac{1}{2}R. \tag{3.26}$$

Note that (3.26) and (3.25) yield $e_m \leq R/2$ and hence $S^m_{\Delta t}U \in B(\bar{u}, R)$.

Clearly (3.25) holds for $m = N$ by (3.24). By an identical argument to that yielding (3.22) but with $K = -\alpha$ in (3.21) we have

$$e_{m+1} \leq \mathrm{e}^{-\alpha\Delta t} e_m + K\Delta t^{r+1}.$$

If (3.25) holds then

$$\begin{aligned} e_{m+1} &\leq \frac{\mathrm{e}^{-\alpha\Delta t}}{1-\mathrm{e}^{-\alpha\Delta t}}[1-\mathrm{e}^{-\alpha(t_M-T)}]K\Delta t^{r+1} + K\Delta t^{r+1} \\ &\quad + \mathrm{e}^{-\alpha\Delta t}\mathrm{e}^{-\alpha(t_M-T)}[\mathrm{e}^{KT}-1]\Delta t^r \\ &= \frac{1-\mathrm{e}^{-\alpha(t_{M+1}-T)}}{1-\mathrm{e}^{-\alpha\Delta t}}K\Delta t^{r+1} + \mathrm{e}^{-\alpha(t_{M+1}-T)}[\mathrm{e}^{KT}-1]\Delta t^r. \end{aligned}$$

Thus (3.25) holds with $m \mapsto m+1$ and the induction is complete. Hence (3.25) and (3.26) give

$$e_n \leq [2K + \mathrm{e}^{KT} - 1]\Delta t^r \quad \forall n, \Delta t : T \leq n\Delta t < \infty.$$

Combining this estimate and (3.24) we have the desired result by choosing $C = 2K + \mathrm{e}^{KT} - 1$. □

3.4. Bibliography

The subject of dynamical systems is discussed in numerous text books and monographs. In the context of ordinary differential equations see, for example, Arrowsmith and Place (1990), Bhatia and Szego (1970), Devaney (1989), Drazin (1992), Guckenheimer and Holmes (1983) and Hale and Kocak (1991); in the context of partial differential equations see, for example, Babin and Vishik (1992), Hale *et al.* (1988), Hale (1984), Ladyzhenskaya (1991) and Temam (1988). The text by Bhatia and Szego (1970) is closest in outlook to the presentation of results in Section 3.1.

The pivotal point in Section 3.2 is Assumption 3.7. It is worth noting that in Section 4 we only require (i), (iii) and (iv) of Assumption 3.7, in Section 5 we require all four points and in Section 6 we use only (i) and (iii). Assumption 3.7 hold for all consistent Runge–Kutta methods: point (iii) is a standard truncation error bound proved in, for example, Butcher (1987) whilst point (iv) is proved in Stoffer (1994); points (i) and (ii) are readily established and, indeed, use of point (i) is implicit in all convergence proofs for Runge–Kutta methods. The situation for s-step multistep methods considered as dynamical systems is somewhat more complicated since the natural phase space for the problem is $\mathbb{R}^{ps}$. However, it has been shown in Kirchgraber (1986) that, under Assumption 3.6, for all strictly stable multistep methods there exists a consistent one-step method which is an attractive invariant manifold for the multistep method. In essence this means that there is a linear combination of $s-1$ successive steps of the method whose behaviour is governed by a one-step method – after a large number of iterations. This work has been generalized in Stoffer (1993) and a related question considered in Eirola and Nevanlinna (1988). These results enable the use of Assumption 3.7 in the study of multistep methods as well as one-step methods.

The first error bound in Theorem 3.8 is standard for Runge–Kutta methods and proofs may be found in (for example) Butcher (1987), Hairer *et al.* (1987), Lambert (1991) and Stetter (1973). The second error bound, concerning the C^1 closeness of the true and approximate semigroups, is not in the literature to the best of our knowledge in this general form; however, for the specific case of approximations of reaction-diffusion equations, such results are proved in Alouges and Debussche (1991) and Hale *et al.* (1988). The uniform in time error bound of Theorem 3.9 may be found in Stetter (1973). Generalizations of this idea may be found in Heywood and Rannacher (1986) for finite-element approximations of the Navier–Stokes equation, in Larsson (1989) for finite element approximations of nonsmooth solutions to reaction-diffusion equations and in Sanz-Serna and Stuart (1992) for finite difference approximations of smooth solutions to reaction-diffusion equations.

4. Neighbourhood of an equilibrium point

4.1. Background theory

In this section we study the affect of approximation on equilibrium points, their stable and unstable manifolds and their local phase portraits. In all cases we employ the contraction mapping theorem to develop an existence theory and exploit this to prove convergence results. This means that the basic existence theory for $S(t)$ takes the longest to develop whilst the existence and approximation theory for $S^n_{\Delta t}$ follows simply.

An *equilibrium point* for (1.1) is a point $\bar{u} \in \mathbb{R}^p$ satisfying

$$f(\bar{u}) = 0. \tag{4.1}$$

Consequently such a point $\bar{u}$ also satisfies the defining equation

$$\bar{u} = S(t)\bar{u} \quad \forall t \in \mathbb{R} \tag{4.2}$$

Thus $\bar{u}$ is a *fixed point* of the mapping $S(t)$ for every $t \in \mathbb{R}^p$ – see (4.5) below. The equilibrium point is said to be *hyperbolic* if none of the eigenvalues of the Jacobian of f at $\bar{u}$, $\mathrm{d}f(\bar{u})$, lies on the imaginary axis. A hyperbolic equilibrium point is said to be *stable* if all eigenvalues of $\mathrm{d}f(\bar{u})$ lie in the left-half plane. It is *unstable* if at least one eigenvalue of $\mathrm{d}f(\bar{u})$ lies in the right-half plane.

It may be shown that $\mathrm{d}S(\bar{u};t) = \exp[\mathrm{d}f(\bar{u})t]$ and hence, that $\mathrm{d}S(\bar{u};t)$ has no eigenvalues on the unit circle if $\bar{u}$ is hyperbolic. Thus we can define

$$D(t) = [I - \mathrm{d}S(\bar{u};t)]^{-1}. \tag{4.3}$$

Using the fact that 0 is not in the spectrum of $\mathrm{d}f(\bar{u})$ and writing $\mathrm{d}S(\bar{u};t)$ as an infinite series in $\mathrm{d}f(\bar{u})$ it may be shown that $\exists \beta > 0, t_c > 0$ such that

$$\|D(t)\| \leq \beta/t \quad \forall t \in (0, t_c). \tag{4.4}$$

Note that stability may also be formulated in terms of the eigenvalues of $\mathrm{d}S(\bar{u};t)$; this can be done by modifying the definitions that we are about to make for fixed points of $S^1_{\Delta t}$.

A *fixed point* $\bar{U}$ of the semigroup $S^1_{\Delta t}$ satisfies the equation

$$\bar{U} = S^1_{\Delta t}\bar{U} \tag{4.5}$$

or, equivalently for (1.2),

$$\bar{U} = \mathcal{F}(\bar{U}, \Delta t).$$

The fixed point is *hyperbolic* if $\mathrm{d}S^1_{\Delta t}(\bar{U})$ has no eigenvalues on the unit circle; such a hyperbolic fixed point is said to be stable if all eigenvalues of $\mathrm{d}S^1_{\Delta t}(\bar{U})$ lie inside the unit circle and unstable if at least one eigenvalue lies outside the unit circle.

Throughout this article we will use the following notation for the set of fixed points of $S(t)$ and $S^1_{\Delta t}$ together with their neighbourhoods:

$$\begin{aligned} \mathcal{E} &= \{v \in \mathbb{R}^p : f(v) = 0\}, \\ \mathcal{E}(\epsilon) &= \{v \in \mathbb{R}^p : \|f(v)\| \leq \epsilon\}, \\ \mathcal{E}_{\Delta t} &= \{v \in \mathbb{R}^p : v = S^1_{\Delta t}v\}. \end{aligned} \tag{4.6}$$

We will need the following definitions; illustrative examples will be given later on.

Definition 4.1 The *unstable manifold* of an equilibrium point $\bar{u}$ of (1.1) is the set

$$W^u(\bar{u}) := \{U \in \mathbb{R}^p : u(t) \to \bar{u} \ \ as \ \ t \to -\infty\}.$$

The *local unstable manifold* of $\bar{u}$ is the set

$$W^{u,\epsilon}(\bar{u}) := \{U \in W^u(\bar{u}) : \|u(t) - \bar{u}\| \leq \epsilon \, \forall t \leq 0\}.$$

The *stable manifold* of an equilibrium point $\bar{u}$ of (1.1) is the set

$$W^s(\bar{u}) := \{U \in \mathbb{R}^p : u(t) \to \bar{u} \ \ as \ \ t \to \infty\}.$$

The *local stable manifold* of $\bar{u}$ is the set

$$W^{s\epsilon}(\bar{u}) := \{U \in W^s(\bar{u}) : \|u(t) - \bar{u}\| \leq \epsilon \, \forall t \geq 0\}$$

The following facts concerning unstable manifolds are of interest:

Lemma 4.2 The unstable manifold $W^u(\bar{u})$ is invariant and, furthermore,

$$W^u(\bar{u}) = W^{u,\epsilon}(\bar{u}) \cup \bigcup_{t>0} S(t)\Gamma$$

where

$$\Gamma = W^{u,\epsilon}(\bar{u}) \cap \partial B(\bar{u}, \epsilon). \tag{4.7}$$

Proof. It follows from the definition that, if $u \in W^u(\bar{u})$ then, for every $\tau > 0$ there exists $v^\tau \in \mathbb{R}^p$ such that

$$S(\tau)v^\tau = u \quad v^\tau \to \bar{u} \quad \text{as} \quad \tau \to \infty. \tag{4.8}$$

The converse is also true: if (4.8) holds for every $\tau > 0$ then $u \in W^u(\bar{u})$.

Thus $S(\tau + t)v^\tau = S(t)u$ and, since (4.8) holds, we deduce that $S(t)u \in W^u(\bar{u})$ so that $S(t)W^u(\bar{u}) \subseteq W^u(\bar{u})$. Furthermore, since $S(t)v^t = u$ we have that, for every $t > 0$, $S(\tau - t)v^\tau = v^t$. Thus, from (4.8), we deduce that $v^t \in W^u(\bar{u})$. Thus $W^u(\bar{u}) \subseteq S(t)W^u(\bar{u})$ and the first part of the proof is complete.

We now establish (4.7). First we show that

$$W^u(\bar{u}) \subseteq W^{u,\epsilon}(\bar{u}) \cup \bigcup_{t>0} S(t)\Gamma.$$

Let $u \in W^u(\bar{u}) \backslash W^{u,\epsilon}(\bar{u})$. If $u \notin B(\bar{u}, \epsilon)$ then, since (4.8) holds it follows that $v^0 = u$ and, by continuity, there exists $t > 0$ such that $S(t)v^t = u$ and $v^t \in \Gamma$. On the other hand, if $u \in B(\bar{u}, \epsilon)$ then $\exists t > 0, v^t \in \mathbb{R}^p : S(t)v^t = u$ with $v^t \in \Gamma$ since otherwise we have $u \in W^u, \epsilon(\bar{u})$.

Now we show that

$$W^u(\bar{u}) \supset W^{u,\epsilon}(\bar{u}) \cup \bigcup_{t>0} S(t)\Gamma.$$

If

$$u \in \bigcup_{t \geq 0} S(t)\Gamma$$

then there exists $w \in \Gamma$ such that $S(t)w = u$. Furthermore, since $w \in W^{u,\epsilon}(\bar{u})$ it follows that $u \in W^{u,\epsilon}(\bar{u})$ and the result is proved. □

Important remark It is straightforward to generalize Definition 4.1 to the discrete semigroup $S^1_{\Delta t}$. We employ the notation

$$W^u_{\Delta t}(\bar{U}), \quad W^{u,\epsilon}_{\Delta t}(\bar{U}) \tag{4.9}$$

to denote the unstable and local unstable manifolds of a fixed point $\bar{U}$ respectively. Similar notation is employed for the stable manifold. With this notation it is also straightforward to generalize Theorem 4.2 to the discrete semigroup.

We now discuss the behaviour of trajectories in the neighbourhood of hyperbolic equilibrium points. If we introduce the variable

$$v(t) = u(t) - \bar{u}$$

and change variables in (1.1) then we find that $v(t)$ satisfies the equation

$$\begin{aligned} v_t &= Av + g(v), v(0) = V := U - \bar{u}, \\ A &= \mathrm{d}f(\bar{u}), g(v) = [f(v+\bar{u}) - \mathrm{d}f(\bar{u})v]. \end{aligned} \tag{4.10}$$

Furthermore we introduce the notation $\bar{S} : \mathbb{R}^p \to \mathbb{R}^p$ to denote the semi-group constructed so that $v(t) = \bar{S}v_0$. Hence

$$\bar{S}(t)v = S(t)(\bar{u}+v) - \bar{u}. \tag{4.11}$$

Since $f(\bar{u}) = 0$ it is clear that $g(v) = \mathcal{O}\|v\|^2$ and hence it is reasonable to expect that, for hyperbolic equilibria, the properties of the linear equation $w_t = Aw$ describe the dynamics of solutions to (4.10) in the neighbourhood of $v = 0$. Our aim is to put this intuition on a firm mathematical basis and then to understand analogous behaviour for the numerical method. For the purposes of comparison with the numerical method we now formulate the solution of (4.10) as a mapping over time interval Δt. Since $\bar{u}$ is hyperbolic we can split the space

$$\mathbb{R}^p = X \oplus Y$$

where X (respectively Y) is an l (respectively m) dimensional subspace of $\mathbb{R}^p$ spanned by the generalized eigenspace of A corresponding to eigenvalues with positive (respectively negative) real parts so that $p = l + m$.

We denote by $\mathcal{P}$ and $\mathcal{Q}$ the spectral projections $\mathcal{P} : \mathbb{R}^p \to X$ and $\mathcal{Q} : \mathbb{R}^p \to Y$. Using the variation of constants formula we write the solution of (4.10) as

$$v(t) = \mathrm{e}^{At}v(0) + \int_0^t \mathrm{e}^{A(t-s)}g(v(s))\,\mathrm{d}s. \tag{4.12}$$

Hence we may write

$$v(t) = L(t)v(0) + G(v(0), t) \tag{4.13}$$

where

$$L(t) := \mathrm{e}^{At}, \quad G(v,t) := \int_0^t L(t-s)g(\bar{S}(s)v)\,\mathrm{d}s. \tag{4.14}$$

Now we define $t_n = n\Delta t$, $v_n = v(t_n)$ and write (4.13) as

$$v_{n+1} = Lv_n + G(v_n) \tag{4.15}$$

where $L := L(\Delta t)$ and $G(\cdot) := G(\cdot, \Delta t)$. Using the spectral projections $\mathcal{P}$ and $\mathcal{Q}$ we can decompose v_n as $v_n = p_n + q_n$ where $p_n = \mathcal{P}v_n$ and $q_n = \mathcal{Q}v_n$ to obtain

$$\begin{aligned} p_{n+1} &= Lp_n + \mathcal{P}G(p_n + q_n), \\ q_{n+1} &= Lq_n + \mathcal{Q}G(p_n + q_n). \end{aligned} \tag{4.16}$$

This splitting of the variable v will be particularly useful to us both in studying stable and unstable manifolds and in the examination of phase portraits. Finally note that because of the spectral properties of A on X and Y it follows that there exist norms $\|\cdot\|_u$ and $\|\cdot\|_s$ on $\mathbb{R}^p$ and an $\alpha > 0$ such that, for all $t \geq 0$

$$\begin{aligned} \|L(-t)v\|_u &\leq e^{-\alpha t}\|v\|_u \quad \forall v \in X, \\ \|L(t)v\|_s &\leq e^{-\alpha t}\|v\|_s \quad \forall v \in Y. \end{aligned} \tag{4.17}$$

Equivalently

$$\begin{aligned} \|L^{-1}v\|_u &\leq e^{-\alpha\Delta t}\|v\|_u \quad \forall v \in X, \\ \|Lv\|_s &\leq e^{-\alpha\Delta t}\|v\|_s \quad \forall v \in Y. \end{aligned} \tag{4.18}$$

For the remainder of this section, whenever estimating v or its numerical counterpart, we employ the norm on $\mathbb{R}^p$ given by

$$\|v\| = \max\{\|\mathcal{P}v\|_u, \|\mathcal{Q}v\|_s\}. \tag{4.19}$$

Here the subscripts 'u' and 's' denote 'unstable' and 'stable' respectively. This choice of norm simplifies the exposition considerably since the majority of the estimation takes place either in X (respectively Y) where $\|\cdot\| \equiv \|\cdot\|_u$ (respectively $\|\cdot\| \equiv \|\cdot\|_s$.)

Using the formulation (4.16) we prove several results concerning the behaviour of solutions of equation (1.1) in the neighbourhood of a hyperbolic equilibrium point $\bar{u}$ – we study stable and unstable manifolds and phase portraits. In subsequent sections we examine the effect of discretization error on these objects. The following two examples serve to illustrate the type of results which interest us here.

Example (Unstable manifolds) Consider the equations (2.1) with the equilibrium point $p = q = 0$. Linearizing about the equilibrium point gives the system

$$p_t = p, \quad q_t = -q. \tag{4.20}$$

Thus, in the notation (4.10), the matrix A has eigenvalues ± 1.

The system (4.20) has an *unstable manifold* given by $q = 0$: on the unstable manifold solutions tend to the origin as $t \to -\infty$ since $q = 0$ and $p(t) = \exp(t)p(0)$. Notice also that the unstable manifold is attractive in the sense that solutions approach it as $t \to \infty$.

Since (2.1) is, in a neighbourhood of the origin, a small perturbation of the linear system (4.20) we expect that it should have an unstable manifold (on which solutions tend to the origin as $t \to -\infty$) close to $q = 0$. This is indeed the case – recall that in Section 2 we showed that the curve $q = p^2/3$ is the unstable manifold for the nonlinear system.

The important point to take away from this example is that the linear system has unstable manifold $q = \Phi_l(p) := 0$ whilst the true equation has unstable manifold $q = \Phi(p) := p^2/3$. It follows that

$$\sup_{|p|\le\epsilon} |\Phi(p) - \Phi_l(p)| \le \tfrac{1}{3}\epsilon^2$$

so that the true and linearized unstable manifolds are close in a neighbourhood of the equilibrium point. Indeed they are tangential at the equilibrium point itself. Our aim is to generalize this to the system (4.10).

The second example broadens our study from the unstable manifold, which comprises certain subclass of trajectories in the neighbourhood of equilibria, to the complete local phase portrait, which comprises all trajectories in the neighbourhood of equilibria.

Example (Phase portrait) We return to the system (2.1) and its linearization (4.20). It is clear that, for any $T > 0$, equations (4.20) can be solved subject to the boundary conditions

$$p(T) = \xi, \quad q(0) = \eta \tag{4.21}$$

and this yields a solution of the form

$$p(t) = \mathrm{e}^{t-T}\xi, \quad q(t) = \mathrm{e}^{-t}\eta.$$

By choosing different ξ and η and different values of T the complete phase portrait for the equation can be constructed. Thus we might expect that a local phase portrait for the nonlinear system can be constructed analogously. We show this to be so.

Explicit solution of (2.1) subject to (4.21) yields

$$p(t) = \mathrm{e}^{t-T}\xi, \quad q(t) = \mathrm{e}^{-t}\eta + \tfrac{1}{3}(\mathrm{e}^{-2T}\xi^2)(\mathrm{e}^{2t} - \mathrm{e}^{-t}).$$

Notice that, if ξ and η are small, then the perturbation from the linear solution is uniformly small in $(0, T)$. Specifically

$$\sup_{0\le t\le T} |\tfrac{1}{3}(\mathrm{e}^{-2T}\xi^2)(\mathrm{e}^{2t} - \mathrm{e}^{-t})| \le \tfrac{1}{3}\xi^2.$$

Thus the perturbation is bounded independently of the time of flight T between boundary conditions.

We now proceed to extend what we have observed in these two examples to the more general case. In the following note that, from (4.14) and the properties of projections, there exists a constant $\kappa > 1$ and $\beta > 0$ such that

$$\|L(t)\| \le \kappa \mathrm{e}^{\beta t} \quad \forall t \ge 0 \quad \text{and} \quad \|\mathcal{P}\|, \|\mathcal{Q}\| \le \kappa. \tag{4.22}$$

Since the function $g(v)$ defined in (4.10) is $\mathcal{O}(\|v\|^2)$ it follows that $\exists C > 0$:

$$\begin{aligned} \|\mathcal{R}g(v)\| &\le (C/2\kappa)\epsilon^2, \quad \mathcal{R} = I, \mathcal{P}, \mathcal{Q} \\ \|\mathcal{R}(g(v) - g(w))\| &\le (C/2\kappa)\epsilon\|v - w\|, \quad \mathcal{R} = I, \mathcal{P}, \mathcal{Q} \end{aligned} \tag{4.23}$$

for all $v, w \in B(0, \epsilon)$. Since our interest in this section is focused on a small neighbourhood of the equilibrium point, smooth modification of the function g outside such a neighbourhood will not affect the results. Thus we assume:

Assumption 4.3 The function g in equation (4.10) satisfies (4.23) for all $v, w \in \mathbb{R}^p$.

After proving results under Assumption 4.3 we will derive corollaries concerning the original, unmodified flow generated by (4.10). Under Assumption 4.3 we have

$$\begin{aligned} \|G(v)\| &\leq (1/2\kappa)C\epsilon^2 \int_0^{\Delta t} \kappa \mathrm{e}^{\beta(\Delta t - s)}\,\mathrm{d}s \\ &\leq C\epsilon^2[-\mathrm{e}^{\beta(\Delta t-s)}/2\beta]_0^t \\ &= [(\mathrm{e}^{\beta\Delta t}-1)/2\beta]C\epsilon^2 \leq \Delta t C\epsilon^2 \end{aligned}$$

for Δt sufficiently small.

Using similar analysis, it is possible to show that $\exists \Delta t_c > 0$ such that

$$\begin{aligned} \|\mathcal{R}G(v)\| \leq \Delta t C\epsilon^2, \quad \mathcal{R} = I, \mathcal{P}, \mathcal{Q} \\ \|\mathcal{R}(G(v) - G(w))\| \leq \Delta t C\epsilon\|v - w\|, \quad \mathcal{R} = I, \mathcal{P}, \mathcal{Q} \end{aligned} \tag{4.24}$$

for all $v, w \in \mathbb{R}^p$ and all $\Delta t \in (0, \Delta t_c]$.

Important remark To simplify notation we denote the norms in the v-coordinates and the u-coordinates in the same way ($\|\cdot\|$). However, the norm in v is always defined through (4.19) whereas the norm in u is defined differently (e.g. the Euclidean norm). This should not cause confusion but it is important to note that, in the remainder of this section, $B(0, \delta)$ denotes a ball in the v-coordinate with norm (4.19) whilst $B(\bar{u}, \delta)$ denotes a ball in the u-coordinates with a different norm.

We start by considering unstable manifolds. Our aim now is to prove the existence of an invariant manifold for the mapping (4.16). Specifically we seek a function $\Phi : X \to Y$ which satisfies the following:

$$q_n = \Phi(p_n) \Leftrightarrow q_{n+1} = \Phi(p_{n+1}). \tag{4.25}$$

We shall look for Φ lying in the space

$$\begin{aligned} \Gamma &= \{\Phi \in C(X,Y) : \|\Phi\|_C = \sup_{p\in X} \|\Phi(p)\| \leq \epsilon, \\ &\quad \|\Phi(p_1) - \Phi(p_2)\| \leq \|p_1 - p_2\| \quad \forall p_1, p_2 \in X\}. \end{aligned}$$

The subscript C in the norm on $C(X, Y)$ is simply to denote the space of functions in which Φ lies.

In the following note that, if $16C\epsilon \leq \alpha$ and $8C\epsilon\Delta t \leq 1$, we have that, for all $\gamma \in [1,2]$,

$$\begin{aligned}(1-\tfrac{1}{2}\Delta t\alpha)\epsilon + \gamma\Delta t C\epsilon^2 \le \epsilon;\\ 1-\tfrac{1}{2}\Delta t\alpha + 2\gamma\Delta t C\epsilon \le (1-2\gamma\Delta t C\epsilon);\\ (1-\tfrac{1}{2}\Delta t\alpha) + 2\gamma\Delta t C\epsilon \le 1-\tfrac{1}{4}\Delta t\alpha\\ 2\gamma\Delta t C\epsilon \le \tfrac{1}{2}.\end{aligned} \tag{4.26}$$

Furthermore, note that if $\Phi \in \Gamma$ then

$$\|\mathcal{R}[G(\xi_1+\Phi(\xi_1)) - G(\xi_2+\Phi(\xi_2))]\| \le 2\Delta t C\epsilon\|\xi_1-\xi_2\|, \mathcal{R} = I, \mathcal{P}, \mathcal{Q}, \tag{4.27}$$

by (4.24). That the estimates (4.26) are robust for $\gamma \in [1,2]$ will be used in Section 4.3 where the same method of proof used to construct an invariant manifold for (4.15) will be employed to construct an invariant manifold for the approximate semigroup; hence C will be enlarged by a factor of 2 to incorporate the effect of the truncation error.

Theorem 4.4 Assume that Δt_c is chosen so that

$$\mathrm{e}^{-\alpha\Delta t} \le 1-\tfrac{1}{2}\alpha\Delta t \,\forall \Delta t \in (0, \Delta t_c],$$

that ϵ is chosen so that $16C\epsilon \le \alpha$ and that $8C\epsilon\Delta t \le 1$. Then, under Assumption 4.3, there exists a function $\Phi \in \Gamma$ so that solutions of (4.16) satisfy (4.25). Furthermore, the graph of Φ is attractive in the sense that

$$\|q_{n+1} - \Phi(p_{n+1})\| \le (1-\tfrac{1}{4}\alpha\Delta t)\|q_n - \Phi(p_n)\|. \tag{4.28}$$

Proof. We use the contraction mapping theorem. Given a function $\Phi \in \Gamma$ consider the construction of a new function $M\Phi : X \mapsto Y$ defined by

$$\begin{aligned}p = L\xi + \mathcal{P}G(\xi+\Phi(\xi))\\ (M\Phi)(p) = L\Phi(\xi) + \mathcal{Q}G(\xi+\Phi(\xi)).\end{aligned} \tag{4.29}$$

We show that this new function $M\Phi$ is well defined, lies in Γ and that M contracts on Γ. Thus we construct a fixed point of M; comparison with (4.16) shows that this fixed point is an invariant manifold for (4.16) so that (4.25) is satisfied. Exponential attractivity will then be shown.

To show that $M\Phi$ is well defined we must show that, for every $p \in X$ $\exists \xi \in X$ such that (4.29) is satisfied. To do this consider the mapping

$$\xi^{k+1} = L^{-1}p - L^{-1}\mathcal{P}G(\xi^k + \Phi(\xi^k)).$$

A fixed point of this mapping satisfies (4.29) and will provide the requisite ξ. If η^k also satisfies

$$\eta^{k+1} = L^{-1}p - L^{-1}\mathcal{P}G(\eta^k + \Phi(\eta^k))$$

then, provided that $\Phi \in \Gamma$, (4.18), (4.26) and (4.27) give

$$\|\xi^{k+1} - \eta^{k+1}\| \le 2\mathrm{e}^{-\alpha\Delta t}C\epsilon\Delta t\|\xi^k - \eta^k\| \le \tfrac{1}{2}\|\xi^k - \eta^k\|.$$

Thus $\exists \xi \in X$:

$$\xi = L^{-1}p - L^{-1}\mathcal{P}G(\xi + \Phi(\xi)).$$

Since L is invertible we deduce that $\exists \xi \in X$ so that the first equation in (4.29) can be satisfied for any $p \in X$; hence $M\Phi : X \mapsto Y$ is well defined if $\Phi \in \Gamma$.

Now we show that $M : \Gamma \to \Gamma$. From (4.29) we obtain, using (4.18), (4.24) and (4.26)

$$\begin{aligned} \|M\Phi(p)\| &\le e^{-\alpha\Delta t}\|\Phi(\xi)\| + \|\mathcal{Q}G(\xi + \Phi(\xi))\| \\ &\le (1 - \tfrac{1}{2}\alpha\Delta t)\epsilon + \Delta t C\epsilon^2 \le \epsilon \end{aligned}$$

as required. Since this is true for every $p \in X$ we have

$$\|M\Phi\|_C \le \epsilon.$$

Also, by considering (4.29) with $\xi \to \{\xi_i\}_{i=1}^2$ and $p \to \{\mathbf{p}_i\}_{i=1}^2$ we obtain, using (4.18) and (4.27),

$$\begin{aligned} \|(M\Phi)(p_1) - (M\Phi)(p_2)\| &\le e^{-\alpha\Delta t}\|\Phi(\xi_1) - \Phi(\xi_2)\| + 2\Delta t C\epsilon\|\xi_1 - \xi_2\| \\ &\le [e^{-\alpha\Delta t} + 2\Delta t C\epsilon]\|\xi_1 - \xi_2\|. \end{aligned}$$

But, also from (4.29),

$$\xi_1 - \xi_2 = L^{-1}(p_1 - p_2) - L^{-1}[\mathcal{P}G(\xi_1 + \Phi(\xi_1)) - \mathcal{P}G(\xi_2 + \Phi(\xi_2))]$$

so that by (4.18) and (4.27)

$$\|\xi_1 - \xi_2\| \le \|p_1 - p_2\| + 2\Delta t C\epsilon\|\xi_1 - \xi_2\|.$$

Combining these two estimates we obtain, using (4.26)

$$\|(M\Phi)(p_1) - (M\Phi)(p_2)\| \le \frac{(1 - \frac{1}{2}\alpha\Delta t) + 2\Delta t C\epsilon}{1 - 2\Delta t C\epsilon}\|\xi_1 - \xi_2\| \le \|\xi_1 - \xi_2\|$$

concluding the proof that $M : \Gamma \to \Gamma$.

We now show that $M : \Gamma \to \Gamma$ contracts. Consider (4.29) with $\Phi \to \{\Phi_i\}_{i=1}^2$ and $\xi \to \{\xi_i\}_{i=1}^2$. Now

$$\|M\Phi_1 - M\Phi_2\|_c = \sup_{p\in X}\|M\Phi_1(p) - M\Phi_2(p)\|.$$

Using (4.18) and (4.24) we obtain from (4.16) that, for any $p \in X$

$$\begin{aligned} &\|M\Phi_1(p) - M\Phi_2(p)\| \\ &\le e^{-\alpha\Delta t}\|\Phi_1(\xi_1) - \Phi_2(\xi_2)\| + \Delta t C\epsilon\|\xi_1 - \xi_2 + \Phi_1(\xi_1) - \Phi_2(\xi_2)\| \\ &\le (e^{-\alpha\Delta t} + \Delta t C\epsilon)\|\Phi_1(\xi_1) - \Phi_2(\xi_1) + \Phi_2(\xi_1) - \Phi_2(\xi_2)\| + \Delta t C\epsilon\|\xi_1 - \xi_2\| \\ &\le (e^{-\alpha\Delta t} + \Delta t C\epsilon)\|\Phi_1(\xi_1) - \Phi_2(\xi_1)\| + [e^{-\alpha\Delta t} + 2\Delta t C\epsilon]\|\xi_1 - \xi_2\| \\ &\le (e^{-\alpha\Delta t} + \Delta t C\epsilon)\|\Phi_1 - \Phi_2\|_C + [e^{-\alpha\Delta t} + 2\Delta t C\epsilon]\|\xi_1 - \xi_2\|. \end{aligned}$$

But we also know, by similar reasoning, that

$$\|\xi_1 - \xi_2\| \leq \Delta t C\epsilon \|\Phi_1(\xi_1) - \Phi_2(\xi_1)\| + 2\Delta t C\epsilon \|\xi_1 - \xi_2\|$$

so that

$$\|\xi_1 - \xi_2\| \leq \frac{\Delta t C\epsilon \|\Phi_1(\xi_1) - \Phi_2(\xi_1)\|}{1 - 2C\epsilon\Delta t} \leq \frac{\Delta t C\epsilon \|\Phi_1 - \Phi_2\|_C}{1 - 2C\epsilon\Delta t}.$$

Combining the two estimates and using (4.26) we obtain

$$\begin{aligned}
\|M\Phi_1(p) - M\Phi_2(p)\| &\leq (\mathrm{e}^{-\alpha\Delta t} + \Delta t C\epsilon)\|\Phi_1 - \Phi_2\|_C \\
&\quad + \frac{\mathrm{e}^{-\alpha\Delta t} + 2\Delta t C\epsilon}{1 - 2\Delta t C\epsilon}\Delta t C\epsilon \|\Phi_1 - \Phi_2\|_C \\
&\leq (\mathrm{e}^{-\alpha\Delta t} + 2\Delta t C\epsilon)\|\Phi_1 - \Phi_2\|_C \\
&\leq (1 - \tfrac{1}{4}\alpha\Delta t)\|\Phi_1 - \Phi_2\|_C.
\end{aligned}$$

Since this is true for any $p \in X$ it follows that

$$\|M\Phi_1 - M\Phi_2\|_C \leq (1 - \tfrac{1}{4}\alpha\Delta t)\|\Phi_1 - \Phi_2\|_C.$$

Thus the mapping is a contraction and the existence of an invariant manifold satisfying (4.25) follows.

Finally we show that the manifold is attracting. Let

$$\begin{aligned}
p &= Lp_n + \mathcal{P}G(p_n + \Phi(p_n)), \\
\Phi(p) &= L\Phi(p_n) + \mathcal{Q}G(p_n + \Phi(p_n)).
\end{aligned}$$

Subtracting this from (4.16) yields, by (4.26),

$$\begin{aligned}
\|q_{n+1} - \Phi(p_{n+1})\| &\leq \|q_{n+1} - \Phi(p)\| + \|\Phi(p) - \Phi(p_{n+1})\| \\
&\leq (\mathrm{e}^{-\alpha\Delta t} + \Delta t C\epsilon)\|q_n - \Phi(p_n)\| + \|p - p_{n+1}\| \\
&\leq (\mathrm{e}^{-\alpha\Delta t} + 2\Delta t C\epsilon)\|q_n - \Phi(p_n)\| \\
&\leq (1 - \tfrac{1}{2}\alpha\Delta t + 2\Delta t C\epsilon)\|q_n - \Phi(p_n)\| \\
&\leq (1 - \tfrac{1}{4}\alpha\Delta t)\|q_n - \Phi(p_n)\|
\end{aligned}$$

and the desired result follows. □

Using Theorem 4.4 we may prove:

Corollary 4.5 (**Local unstable manifolds**) Assume that ϵ is chosen so that $16C\epsilon \leq \alpha$. Then, there exists a function $\Phi \in \Gamma$ such that, if $v(t)$ satisfies (4.10) and $v(t) \in B(0,\epsilon)$ for $t \in [t_n, t_{n+1}]$ then $v_n = v(t_n)$ satisfies (4.25) and (4.28). Furthermore, there exists $c > 0$ such that the set of points

$$\{u \in \mathbb{R}^p | \mathcal{P}(u - \bar{u}) = p, \mathcal{Q}(u - \bar{u}) = \Phi(p)),\ p \in X\} \cap B(\bar{u}, c\epsilon)$$

is the local unstable manifold of the equilibrium point $\bar{u}$ of (1.1).

Proof. Note that if we are considering a solution $v(t) \in B(0,\epsilon)$ then we can modify g outside $B(0, \epsilon)$ to make Assumption 4.3 valid, without affecting the solution. Thus to establish the first part of the corollary it is sufficient to show that, under Assumption 4.3, Φ is indeed an invariant manifold for the equation (4.10) as well as for the map (4.16). To do this it is sufficient to show that the function Φ constructed in Theorem 4.4 is independent of the choice of $\Delta t \in (0, \Delta t_c]$ used in the construction. To this end we denote Φ by $\Phi(\Delta t)$ and the mapping $M : \Gamma \mapsto \Gamma$ by $M(\Delta t)$ defined by (4.29).

A little work shows that, since M is constructed using the semigroup $S(t)$,

$$M(t) \cdot M(s) \equiv M(s) \cdot M(t).$$

Now consider $t, s \in (0, \Delta t_c]$ and assume that $8C\epsilon\Delta t_c \leq 1$ and that $e^{-\alpha\Delta t_c} \leq 1 - \alpha\Delta t_c/2$. Then, by Theorem 4.4, both $\Phi(s)$ and $\Phi(t)$ lie in Γ. Furthermore, $M(s)\Phi(t) \in \Gamma$. Now, by definition

$$M(t)\Phi(t) = \Phi(t).$$

Thus

$$M(s) \cdot M(t)\Phi(t) = M(s)\Phi(t) \Rightarrow M(t) \cdot M(s)\Phi(t) = M(s)\Phi(t).$$

Hence, since $M(s)\Phi(t) \in \Gamma$, and since $M(s)\Phi(t)$ is a fixed point of $M(t)$ we deduce that $M(s)\Phi(t) = \Phi(t)$. But this shows that $\Phi(t)$ is a fixed point of $M(s)$ and, since it lies in Γ, we deduce that $\Phi(t) \equiv \Phi(s)$. Hence the manifold $\Phi(\Delta t)$ is independent of Δt and the result follows.

Now we show that the set

$$\mathcal{M} := \{v \in \mathbb{R}^p | \mathcal{P}v = p, \mathcal{Q}v = \Phi(p)\; p \in X\} \cap B(0, \epsilon)$$

defines the local unstable manifold of the equilibrium point 0 of (4.10). We must show that if $v(0) \in \mathcal{M}$ then $v(t) \in \mathcal{M}$ $\forall t \leq 0$ and that $v(t) \to 0$ as $t \to -\infty$.

First note that, since $v = 0$ is an equilibrium point for (4.13), analysis of the mapping (4.29) shows that the fixed point Φ satisfies $\Phi(0) = 0$. Now, on the invariant manifold we have from (4.16), that

$$\begin{aligned} p_{n+1} &= Lp_n + \mathcal{P}G(p_n + \Phi(p_n)), \\ 0 &= 0 + \mathcal{P}G(0 + \Phi(0)). \end{aligned}$$

Hence, by (4.18) and (4.27),

$$\begin{aligned} \|p_n\| &\leq e^{-\alpha\Delta t}\|p_{n+1}\| + \|\mathcal{P}[G(p_n + \Phi(p_n)) - G(0 + \Phi(0))]\| \\ &\leq (1 - \tfrac{1}{2}\alpha\Delta t)\|p_{n+1}\| + 2C\Delta t\epsilon\|p_n\|. \end{aligned}$$

Thus

$$\|p_n\| \leq \frac{1 - \frac{1}{2}\alpha\Delta t}{1 - 2C\Delta t\epsilon}\|p_{n+1}\|$$

$$\leq \frac{1-\frac{1}{2}\alpha\Delta t}{1-\frac{1}{4}\alpha\Delta t}\|p_{n+1}\|$$
$$\leq (1-\tfrac{1}{4}\alpha\Delta t)\|p_{n+1}\|$$

by (4.26).

From this it is clear that, if $v_0 \in \mathcal{M}$ so that $\|p_0\| \leq \epsilon$ then $\|p_n\| \leq \epsilon\ \forall n \leq 0$. Since $\|q_n\| = \|\Phi(p_n)\| \leq \epsilon$ it follows that $v_n \in \mathcal{M}\ \forall n \leq 0$. Since $\Delta t \in (0, \Delta t_c]$ is arbitrary this shows that $v(t) \in \mathcal{M}\ \forall t \leq 0$ as required. Furthermore it is clear that $\|p_n\| \to 0$ as $n \to -\infty$ so that, since $q_n = \Phi(p_n)$ for $v_n \in \mathcal{M}$ and $\Phi(0) = 0$ it follows that $q_n \to 0$ as $n \to -\infty$. Thus $v_n \to 0$ as $n \to -\infty$ and hence that $v(t) \to 0$ as $t \to -\infty$. Thus we have constructed the local unstable manifold for (4.10). Converting back to the u variables from the v variables and changing norms introduces the constant c and completes the proof. □

It should be noted that the methodology used here can be extended to construct the stable manifold of (4.10).

Motivated by the example described above concerning local phase portraits, we now examine the phase portrait of the nonlinear equation (4.10) near to the origin, again using the mapping formulation (4.16). In the particular example above, the result is established easily because the p and q equations decouple. In the general case they do not decouple but, nonetheless, a result of this type still holds. Specifically we seek a solution of (4.16) which satisfies the boundary conditions

$$p_N = \xi \in X, \quad q_0 = \eta \in Y \tag{4.30}$$

where $\|\xi\|, \|\eta\| \leq \epsilon/2$. Noting that $v_n = p_n + q_n$ induction on (4.16) yields

$$\begin{aligned} p_n &= L^{n-N}p_N - \textstyle\sum_{j=n}^{N-1} L^{n-j-1}\mathcal{P}G(v_j), \\ q_n &= L^n q_0 + \textstyle\sum_{j=0}^{n-1} L^{n-1-j}\mathcal{Q}G(v_j). \end{aligned} \tag{4.31}$$

Thus it is our purpose to solve (4.31) subject to (4.30) for arbitrary $N > 0$. Such a solution corresponds to solving the equation (4.10) with boundary conditions specified in X at $t = N\Delta t$ and in Y at $t = 0$ rather than the initial condition $v(0) = V$; since N is arbitrary, the time of flight between these points is arbitrary. Finding all such solutions with ϵ small corresponds to constructing the local phase portrait near $v = 0$.

In the following we let $\mathcal{V} = \{v_n\}_{n=0}^N$ denote an element of the product space $\Psi = \{\mathbb{R}^p\}^N$ and define

$$\|\mathcal{V}\|_\infty = \max_{0\leq n\leq N} \|v_n\|.$$

We consider the set

$$\Psi_\epsilon = \{\mathcal{V} \in \Psi : \|\mathcal{V}\|_\infty \leq \epsilon\}.$$

To study (4.31), (4.30) we use the contraction mapping theorem in Ψ_ϵ. We generate iterates $V^k = \{v_n^k\}_{n=0}^N$ through the definition

$$M\mathcal{V} = \{Mp_n + Mq_n\}_{n=0}^N$$

where $Mp_n \in X$ and $Mq_n \in Y$ are defined by

$$\begin{aligned} Mp_n &= L^{n-N}\xi - \textstyle\sum_{j=n}^{N-1} L^{n-j-1}\mathcal{P}G(v_j), \\ Mq_n &= L^n\eta + \textstyle\sum_{j=0}^{n-1} L^{n-1-j}\mathcal{Q}G(v_j). \end{aligned} \tag{4.32}$$

Clearly a fixed point of M is a solution of (4.30), (4.31).

Now it is straightforward to show that, under the conditions on Δt imposed in Theorem 4.4 and as a result of the bounds (4.18),

$$\begin{aligned} \sum_{j=n}^{N-1} \|L^{n-j-1}v_j\| &\leq \sum_{j=n}^{N-1} \mathrm{e}^{\alpha(n-j-1)\Delta t}\|v_j\| \\ &\leq \left[\frac{1-\mathrm{e}^{-\alpha(N-n)\Delta t}}{1-\mathrm{e}^{-\alpha\Delta t}}\right]\|\mathcal{V}\|_\infty \leq [2/\alpha\Delta t]\|\mathcal{V}\|_\infty \quad \forall v \in X, \\ \sum_{j=0}^{n-1} \|L^{n-1-j}v\| &\leq \sum_{j=0}^{n-1} \mathrm{e}^{-\alpha(n-j-1)\Delta t}\|v_j\| \\ &\leq \left[\frac{1-\mathrm{e}^{-\alpha n\Delta t}}{1-\mathrm{e}^{-\alpha\Delta t}}\right]\|\mathcal{V}\|_\infty \leq [2/\alpha\Delta t]\|\mathcal{V}\|_\infty \quad \forall v \in Y. \end{aligned} \tag{4.33}$$

We may now prove:

Theorem 4.6 Assume that Δt_c is chosen so that

$$\mathrm{e}^{-\alpha\Delta t} \leq 1 - \tfrac{1}{2}\alpha\Delta t \; \forall \Delta t \in (0, \Delta t_c]$$

and that ϵ is chosen so that $8C\epsilon \leq \alpha$. Then, under Assumption 4.3, for any $N > 0$ and any $\xi \in X$, $\eta \in Y$ with $\|\xi\|, \|\eta\| \leq \frac{1}{2}\epsilon$ $\exists$ a solution of (4.16) subject to (4.30) satisfying

$$\max_{0\leq n\leq N} \|\mathbf{v}_n\| \leq \epsilon.$$

Proof. Since (4.16) implies (4.31) we examine (4.31), (4.30). In the following we will use the fact that

$$2\gamma C\epsilon/\alpha \leq \tfrac{1}{2} \tag{4.34}$$

for all $\gamma \in [1, 2]$. Again the factor of γ is incorporated to allow an analogous proof for the numerical method where C is enlarged by a factor of 2 to incorporate the truncation error.

To prove the result we show that $M : \Psi_\epsilon \mapsto \Psi_\epsilon$ and is a contraction. From

(4.32) we have, using (4.18), (4.24) and (4.33), that

$$\begin{aligned}\|Mp_n\| &\leq \|L^{n-N}\xi\| + \sum_{j=n}^{N-1} \|L^{n-j-1}\mathcal{P}G(v_j)\| \\ &\leq \|\xi\| + \frac{2}{\alpha\Delta t}\Delta t C\epsilon^2.\end{aligned}$$

Hence, by the assumptions on ϵ and (4.34) it follows that

$$\|Mp_n\| \leq \epsilon \quad \forall n : 0 \leq n \leq N.$$

Likewise it may be shown that

$$\|Mq_n\| \leq \epsilon \quad \forall n : 0 \leq n \leq N$$

and hence that $M\mathcal{V} \in \Psi_\epsilon$.

To show that the mapping contracts, consider (4.32) with $p_n \mapsto x_n$, $q_n \mapsto y_n$, $v_n \mapsto w_n$, define $w_n = x_n + y_n$ and set $\Omega = \{w_n\}_{n=0}^N$. Then, using (4.18), (4.24) and (4.33) we obtain from (4.32)

$$\|Mp_n - Mx_n\| \leq \frac{2}{\alpha\Delta t}\Delta t C\epsilon\|\mathcal{V} - \Omega\|_\infty \quad \forall n : 0 \leq n \leq N$$

and

$$\|Mq_n - Mz_n\| \leq \frac{2}{\alpha\Delta t}\Delta t C\epsilon\|\mathcal{V} - \Omega\|_\infty \quad \forall n : 0 \leq n \leq N.$$

Thus it follows from (4.34) that

$$\|\mathcal{V}^{k+1} - \Omega^{k+1}\|_\infty \leq \frac{2C\epsilon}{\alpha}\|\mathcal{V}^k - \Omega^k\|_\infty \leq \frac{1}{2}\|\mathcal{V}^k - \Omega^k\|_\infty.$$

Hence $M : \Psi_\epsilon \mapsto \Psi_\epsilon$ is a contraction and the result follows. □

We may now remove the Assumption 4.3 from Theorem 4.6. Our aim is to solve the equation (4.10) subject to specified boundary conditions:

$$v_t = Av + g(v), \quad \mathcal{P}v(T) = \xi, \ \mathcal{Q}v(0) = \eta. \tag{4.35}$$

This is equivalent to solving

$$u_t = f(u), \quad \mathcal{P}(u(T) - \bar{u}) = \xi, \ \mathcal{Q}(u(0) - \bar{u}) = \eta. \tag{4.36}$$

Recall the constant $\kappa > 1$ from (4.22).

Corollary 4.7 (**Phase portraits**) Assume that ϵ is chosen so that $8C\epsilon \leq \alpha$. Then for any $T > 0$ and any $\xi \in X, \eta \in Y$ with $\|\xi\|, \|\eta\| \leq \frac{1}{2}\epsilon$ there exists a constant $c > 0$ and a unique solution $u(t)$ of (4.36) satisfying $u(t) \in B(\bar{u}, c\epsilon)$ for all $t \in [0, T]$.

Proof. We consider (4.35). The simple change of variable $u(t) = \bar{u} + v(t)$ will then yield the required result; the constant c is introduced since the norms used to measure $u \in \mathbb{R}^p$ and $v \in \mathbb{R}^p$ may differ. We first show that if (4.35) has a solution in $B(0, \epsilon)$ then it is unique. Let $v^i(t), i = 1, 2$ denote two solutions of (4.35) and decompose them as

$$v^i(t) = p^i(t) + q^i(t), \quad p^i(t) \in X, \; q^i(t) \in Y, \; i = 1, 2.$$

Projecting the solution appropriately and using the variation of constants formula we obtain

$$\begin{aligned} p^i(t) &= L(t - T)\xi + \textstyle\int_T^t \mathcal{P}L(t - s)g(v^i(s))\,\mathrm{d}s, \\ q^i(t) &= L(t)\eta + \textstyle\int_0^t \mathcal{Q}L(t - s)g(v^i(s))\,\mathrm{d}s. \end{aligned}$$

Thus, by subtracting and using (4.17), (4.23) we obtain

$$\|p^1(t) - p^2(t)\| \le \int_t^T \mathrm{e}^{-\alpha(s-t)} \frac{C}{2\kappa} \epsilon \|v^1(s) - v^2(s)\|\,\mathrm{d}s$$

and

$$\|q^1(t) - q^2(t)\| \le \int_0^t \mathrm{e}^{-\alpha(t-s)} \frac{C}{2\kappa} \epsilon \|v^1(s) - v^2(s)\|\,\mathrm{d}s.$$

Thus it follows that, since $\kappa > 1$,

$$\sup_{0 \le t \le T} \|v^1(t) - v^2(t)\| \le \frac{C\epsilon}{2\alpha} \sup_{0 \le s \le T} \|v^1(s) - v^2(s)\|.$$

Since $8C\epsilon \le \alpha$ it follows that $\|v^1(t) - v^2(t)\| = 0$ for $t \in [0, T]$ as required.

To establish the existence of a solution in $B(0, \epsilon)$ we simply use Theorem 4.6. For any choice of $N, \Delta t$ such that $N\Delta t = T$ this gives a solution of (4.35) if Assumption 4.3 holds and, by uniqueness, this solution is independent of the choice of $\Delta t \in (0, \Delta t_c]$. It remains to establish that the solution is in $B(0, \epsilon)$ for all $t \in [0, T]$ so that Assumption 4.3 is not needed. Assume to the contrary that $\exists \tau \in [0, T]$ such that $\|v(\tau)\| = \epsilon + \eta, \eta > 0$. Clearly, for any $\Delta t > 0$ $\exists m \in \mathbb{Z}^+ : \tau \in [m\Delta t, (m + 1)\Delta t]$ and, by Theorem 4.6, $\|v_m\|, \|v_{m+1}\| \le \epsilon$. Now, by the boundeness of f it follows that $\exists L > 0 :$ $\|v(t)\| \le \epsilon + L\Delta t \; \forall t \in [m\Delta t, (m+1)\Delta t]$. Since Δt may be chosen arbitrarily small the choice of Δt so that $L\Delta t < \eta$ yields a contradiction. □

It is possible to modify the analysis of phase portraits to prove the existence of stable and unstable manifolds. For brevity we consider stable manifolds. Essentially the stable manifold is constructed by solving (4.30) and (4.31) in the limit $N \to \infty$ whilst asking that $\|p_n\|$ remain uniformly bounded in $n \ge 0$; this yields the problem

$$\begin{aligned} p_n &= -\textstyle\sum_{j=n}^{\infty} L^{n-j-1} \mathcal{P}G(v_j), \\ q_n &= L^n q_0 + \textstyle\sum_{j=0}^{n-1} L^{n-1-j} \mathcal{Q}G(v_j), \\ q_0 &= \eta, \quad \exists \delta > 0 : \|p_n\| \le \delta \; \forall n \ge 0. \end{aligned} \tag{4.37}$$

Recall that $v_n = p_n + q_n$. The following theorem may be proved identically to Theorem 4.6.

Theorem 4.8 Assume that Δt_c is chosen so that $e^{-\alpha \Delta t} \leq 1 - \frac{1}{2}\alpha \Delta t \, \forall \Delta t \in (0, \Delta t_c]$ and that ϵ is chosen so that $8C\epsilon \leq \alpha$. Then, under Assumption 4.3, any $\eta \in Y$ with $\|\eta\| \leq \frac{\epsilon}{2}$ $\exists$ a solution of (4.37) satisfying

$$\max_{0 \leq n < \infty} \|\mathbf{v}_n\| \leq \epsilon.$$

□

We are now in a position to show that the stable manifold has been constructed.

Corollary 4.9 **(Stable manifolds)** Assume that ϵ is chosen so that

$$8C\epsilon \leq \alpha \quad \text{and} \quad \epsilon < \alpha.$$

Then there exists a function $\Phi : X \mapsto Y$ and $c > 0$ such that the set of points

$$\{u \in \mathbb{R}^p | \mathcal{P}(u - \bar{u}) = \Phi(q), \ \mathcal{Q}(u - \bar{u}) = q, \ q \in B(0, \epsilon/2)\}$$

is the local unstable manifold $W^{s,c\epsilon}(\bar{u})$ of the equilibrium point $\bar{u}$ of (1.1).

Proof. It is possible to prove that the solution of (4.37) is independent of Δt as in the proof of Corollary 4.7. To construct Φ solve (4.37) for all $\eta : \|\eta\| \leq \epsilon/2$ and set $\Phi(\eta) = p_0$. It is straightforward to show that

$$\|\Phi(\eta^1) - \Phi(\eta^2)\| \leq 2\|\eta^1 - \eta^2\| \tag{4.38}$$

by considering the Lipschitz properties of solutions to (4.37) with respect to the data η.

Now note that the graph of Φ is positively invariant: given any solution p_n^1, q_n^1 of (4.37) with $p_0^1 = \Phi(\eta_1)$ we can construct another solution p_n^2, q_n^2 of (4.37) by setting $p_n^2 = p_{n+m}^1$, $q_n^2 = q_{n+m}^1$ and imposing the boundary condition that $q_0^2 = \eta^2 = q_m^1$. Hence $\Phi(\eta^2) = p_0^2$; this construction can be done for any $m > 0$ and, since $p_0^2 = p_m^1$, we deduce that $p_m^1 = \Phi(q_m^1)$ so that the graph of Φ is positively invariant.

Thus any solution of (4.37) satisfies

$$q_n = L^n q_0 + \sum_{j=0}^{n-1} L^{n-j-1} \mathcal{Q}G(\Phi(q_j) + q_j).$$

Hence, by (4.33) and (4.38), we have

$$\|q_n\| \leq e^{-\alpha n \Delta t}\|q_0\| + \sum_{j=0}^{n-1} 2C\epsilon \Delta t e^{-\alpha(n-j-1)\Delta t}\|q_j\|.$$

Application of the Gronwall lemma gives

$$\|q_n\| \leq \|q_0\| e^{2(\epsilon - \alpha)n\Delta t}$$

so that, since $\epsilon < \alpha$, $\|q_n\| \to 0$ as $n \to \infty$. Hence $\|p_n\| = \|\Phi(q_n)\| \to 0$ as $n \to \infty$ and the proof is complete. □

This concludes our analysis of equation (1.1) in the neighbourhood of a hyperbolic equilibrium point. We now proceed to study the effect of numerical approximation.

4.2. Equilibrium points and stability

A natural first question to ask about the approximation (1.2) under Assumption 3.7 is whether or not the equilibrium points of (1.1) are inherited by (1.2) and, furthermore, to study the stability of the approximate fixed points. Recall $D(t)$ given by (4.3) and satisfying (4.4).

Theorem 4.10 **(Equilibrium points under approximation)** Let $\bar{u}$ be a hyperbolic equilibrium point of (1.1). Then there exists $\Delta t_c > 0$ such that the numerical approximation (1.2) has a fixed point $\bar{U} \in B(\bar{u}; 2K\beta\Delta t^r)$ for all $\Delta t \in (0, \Delta t_c]$. Furthermore $\bar{U}$ is stable (respectively unstable) if $\bar{u}$ is stable (respectively unstable).

Proof. For simplicity we consider initially the case where $r > 1$ in Assumption 3.7, returning to $r = 1$ at the end of the proof. The proof is a modification of the proof of the implicit function theorem. Consider the mapping

$$W^{k+1} = W^k - D[W^k - S^1_{\Delta t} W^k] \tag{4.39}$$

where $D = D(\Delta t)$ is given by (4.3). To prove existence of a fixed point of (4.39) we show that the iteration maps $B(\bar{u}; 2K\beta\Delta t^r)$ into itself and is a contraction on that set. Clearly a fixed point of this mapping is necessarily a fixed point of $S^1_{\Delta t}$ and hence of (1.2).

To show that the mapping is into, note that by Definition 3.5, (4.39) may be written as

$$W^{k+1} = W^k - D[W^k - S(\Delta t)W^k + T(W^k; \Delta t)]. \tag{4.40}$$

Also, from (4.2) it follows that

$$\bar{u} = \bar{u} - D[\bar{u} - S(\Delta t)\bar{u}].$$

Let $W^k \in B(\bar{u}; 2K\beta\Delta t^r)$ and set $\mathrm{e}^k = W^k - \bar{u}$. Then (4.40) yields, upon appplication of the mean value theorem,

$$\|\mathrm{e}^{k+1}\| = \|\mathrm{e}^k - D[(I - \mathrm{d}S(\bar{u}, t))\mathrm{e}^k + Q_1 + T(W^k; \Delta t)]\|$$

where

$$\|Q_1\| \le C_1 \|\mathrm{e}^k\|^2,$$

for some $C_1 > 0$. Thus, using (4.4) and Assumption 3.7(iii),

$$\|\mathrm{e}^{k+1}\| \le \|D\|\|Q_1\| + \|D\|\|T(W^k; \Delta t)\|$$

$$\begin{aligned} &\leq \frac{\beta C_1 4K^2\beta^2\Delta t^{2r}}{\Delta t} + \beta K\Delta t^r \\ &\leq 2\beta K\Delta t^r \end{aligned}$$

for Δt sufficiently small. Hence the mapping takes $B(\bar{u}, 2K\beta\Delta t^r)$ into itself for Δt sufficiently small.

To show that the mapping is a contraction, let V^k satisfy (4.40) with $W^k \to V^k$ and and define $d^k = W^k - V^k$. A similar manipulation to that used in showing that the mapping is 'into' yields

$$\|d^{k+1}\| \leq \|d^k - D[(I - \mathrm{d}S(\bar{u}, t))d^k + Q_2 + T(W^k; \Delta t) - T(V^k; \Delta t)]\|$$

where

$$\|Q_2\| \leq C_2\|d^k\|^2,$$

for some constant $C_2 > 0$. Hence, by (4.4) and Assumption 3.7(iv),

$$\begin{aligned} \|d^{k+1}\| &\leq \|D\|\|Q_2\| + \|D\|\|T(W^k; \Delta t) - T(V^k; \Delta t)\| \\ &\leq \frac{\beta C_2}{\Delta t}\|d^k\|^2 + \beta K\Delta t^r\|d^k\|. \end{aligned}$$

Since $V^k, W^k \in B(\bar{u}; 2K\beta\Delta t^r)$ it follows that $\|d^k\| \leq 4K\beta\Delta t^r$ and hence that

$$\|d^{k+1}\| \leq \tfrac{1}{2}\|d^k\|$$

for Δt sufficiently small. The existence of a fixed point $\bar{U}$ of $S^1_{\Delta t}$ follows for Δt sufficiently small.

To deal with the case $r = 1$ it is sufficient to show that $C_1, C_2 \to 0$ as $\Delta t \to 0$: this holds since $S(\Delta t)$ and $S^1_{\Delta t}$ yield the identity for $\Delta t = 0$.

The stability of $\bar{U}$ follows from the spectral properties of $\mathrm{d}S^1_{\Delta t}\bar{U}$. By Assumption 3.7(iv) the eigenvalues of $\mathrm{d}S^1_{\Delta t}\bar{U}$ converge to those of $\mathrm{d}S(\bar{U}; \Delta t)$ as $\Delta t \to 0$; furthermore, the eigenvalues of $\mathrm{d}S(\bar{U}; \Delta t)$ converge to those of $\mathrm{d}S(\bar{u}; \Delta t)$ as $\Delta t \to 0$ by standard finite-dimensional spectral theory since $\bar{U} \to \bar{u}$. Since $\mathrm{d}S(\bar{u}; \Delta t)$ has no eigenvalues on the unit circle it follows that, for Δt sufficiently small, $\mathrm{d}S^1_{\Delta t}\bar{U}$ has the same number of eigenvalues inside and outside the unit circle as $\mathrm{d}S(\bar{u}; \Delta t)$ and the result follows. □

Consideration of standard Runge–Kutta methods shows that all equilibrium points of (1.1) become fixed points of the Runge–Kutta method for any $\Delta t > 0$. However, not all fixed points of the Runge–Kutta method are equilibrium points of (1.1) as the following example shows:

Example Consider the scalar equation (1.1) with

$$f(u) = -\lambda u/(1 + u^2)$$

and the Runge–Kutta method

$$\eta = U_n + \Delta t f(U_n), \quad U_{n+1} = U_n + \Delta t f(\eta).$$

Notice that the differential equation has a single equilibrium solution $\bar{u} = 0$. If $\Delta t > 1/\lambda$ then the Runge–Kutta method has the fixed points

$$U = \pm(\lambda\Delta t - 1)^{1/2}$$

in addition to the true fixed point $U = 0$. □

However, it is possible to show that such spurious fixed points cannot exist for Δt sufficiently small; recall (4.6):

Theorem 4.11 **(Spurious solutions as $\Delta t \to 0$)** For any $\epsilon > 0$ $\exists \Delta t_c > 0$ such that $\mathcal{E}_{\Delta t} \in \mathcal{E}(\epsilon)$ $\forall \Delta t \in (0, \Delta t_c]$.

Proof. Let $v \in \mathbb{R}^p \backslash \mathcal{E}(\epsilon)$ so that $\|f(v)\| > \epsilon$. We prove that there exists $\Delta t_c > 0$ such that $v \notin \mathcal{E}_{\Delta t}$ for $\Delta t \in (0, \Delta t_c]$. Note that, from (1.1),

$$S(t)v = v + \int_0^t f(S(s)v)\, ds. \tag{4.41}$$

Thus, by Assumption 3.6 $\exists L > 0$ such that $\|S(t)v - v\| \leq tL$ and hence, for any $\delta > 0$, $\exists \Delta t_c > 0$:

$$\|S(t)v - v\| \leq \delta \quad \forall t \in (0, \Delta t_c].$$

Thus, by (4.41) and continuity of f,

$$\|S(\Delta t)v - v\| \geq \|\int_0^{\Delta t} f(v)ds\| - \|\int_0^{\Delta t}[f(S(s)v) - f(v)]ds\| \geq \frac{\Delta t\epsilon}{2} \quad \forall \Delta t \in (0, \Delta t_c]$$

possibly by further reduction of Δt_c. Now, by Assumption 3.7(iii),

$$\begin{aligned} \|S^1_{\Delta t}v - v\| &= \|S(\Delta t) - v + S^1_{\Delta t}v - S(\Delta t)v\| \\ &\geq \|S(\Delta t)v - v\| - K\Delta t^{r+1} \\ &\geq \tfrac{1}{2}\Delta t\epsilon - K\Delta t^{r+1} \geq \tfrac{1}{4}\Delta t\epsilon, \end{aligned}$$

possibly by further reduction of Δt_c. Hence $v \notin \mathcal{E}_{\Delta t}$ and the result follows. □

The strength of this result relies heavily on Assumption 3.6. Without Assumption 3.6 Theorem 4.11 can be used to show that, given any ball $B(0, R)$ there exists $\Delta t_c = \Delta t_c(R) > 0$ such that no spurious steady solutions can be found in $B(0, R)$ for all $\Delta t \in (0, \Delta t_c]$.

4.3. Unstable manifolds

In this section we show that the unstable manifolds for (1.1) constructed in Corollary 4.5 persist under numerical approximation and that, furthermore, the numerical unstable manifold is close to the true unstable manifold.

Recall that $U_n \approx u(t_n)$ and define

$$V_n = U_n - \bar{u}.$$

Thus V_n is our numerical approximation to $v(t_n) = u(t_n) - \bar{u}$. Recall also (4.11) and define

$$\bar{S}^1_{\Delta t}v = S^1_{\Delta t}(\bar{u} + v) - \bar{u}.$$

Thus

$$\bar{S}^1_{\Delta t}v = \bar{S}(t)v - S(t)(\bar{u} + v) + S^1_{\Delta t}(\bar{u} + v).$$

Hence

$$\bar{S}^1_{\Delta t}v - \bar{S}(t)v = -T(\bar{u} + v; \Delta t).$$

Using the Definition 3.5 of truncation error and (4.15) we deduce that

$$V_{n+1} = LV_n + G(V_n) - T(\bar{u} + V_n; \Delta t).$$

Defining

$$\tilde{G}(v) = G(v) - T(\bar{u} + v; \Delta t) \tag{4.42}$$

we obtain

$$V_{n+1} = LV_n + \tilde{G}(V_n). \tag{4.43}$$

If we let $P_n = \mathcal{P}V_n$ and $Q_n = \mathcal{Q}V_n$ then (4.43) can be written as

$$\begin{aligned} P_{n+1} &= LP_n + \mathcal{P}\tilde{G}(V_n), \\ Q_{n+1} &= LQ_n + \mathcal{Q}\tilde{G}(V_n). \end{aligned} \tag{4.44}$$

Our aim is to prove that, as for (4.15), the mapping (4.43) has an attractive invariant manifold $\Phi_{\Delta t} : X \mapsto Y$ satisfying

$$Q_n = \Phi_{\Delta t}(P_n) \Leftrightarrow Q_{n+1} = \Phi_{\Delta t}(P_{n+1}). \tag{4.45}$$

and, in addition, to show that Φ and $\Phi_{\Delta t}$ are close.

Using Assumption 3.7(iii) and (iv) it follows from (4.42) that

$$\begin{aligned} \|\tilde{G}(v) - G(v)\| &\le K\Delta t^{r+1}, \\ \|\tilde{G}(v) - \tilde{G}(w)\| &\le \|G(v) - G(w)\| + K\Delta t^{r+1}\|v - w\| \end{aligned} \tag{4.46}$$

and hence from (4.24) that under Assumption 4.3,

$$\begin{aligned} \|\mathcal{R}\tilde{G}(v)\| &\le 2\Delta t C\epsilon^2, \quad \mathcal{R} = I, \mathcal{P}, \mathcal{Q} \\ \|\mathcal{R}(\tilde{G}(v) - \tilde{G}(w))\| &\le 2\Delta t C\epsilon\|v - w\|, \quad \mathcal{R} = I, \mathcal{P}, \mathcal{Q} \end{aligned} \tag{4.47}$$

for all $v, w \in \mathbb{R}^p$ and all $\Delta t \in (0, \Delta t_c]$. We now exploit this to prove:

Theorem 4.12 Assume that Δt_c is chosen so that

$$\mathrm{e}^{-\alpha\Delta t} \le 1 - \tfrac{1}{2}\alpha\Delta t \; \forall \Delta t \in (0, \Delta t_c]$$

and that ϵ is chosen so that $16C\epsilon \le \alpha$ and that $8C\epsilon\Delta t \le 1$. Then, under Assumption 4.3, there exists a function $\Phi_{\Delta t} \in \Gamma$ so that solutions of (4.44) satisfy (4.45). Furthermore, the graph of $\Phi_{\Delta t}$ is attractive in the sense that

$$\|Q_{n+1} - \Phi_{\Delta t}(P_{n+1})\| \le (1 - \tfrac{1}{4}\alpha\Delta t)\|Q_n - \Phi_{\Delta t}(P_n)\|. \tag{4.48}$$

Finally the graph $\Phi_{\Delta t}$ is close to Φ given in Theorem 4.4 in the sense that

$$\|\Phi - \Phi_{\Delta t}\|_c \leq 8K\Delta t^r/\alpha.$$

Proof. The existence of $\Phi_{\Delta t}$ is proved precisely as for Φ in Theorem 4.4, except that $C \mapsto 2C$ since conditions (4.24) have been replaced by (4.47), by considering the fixed point mapping

$$\begin{aligned} P &= L\xi + \mathcal{P}\tilde{G}(\xi + \Phi_{\Delta t}(\xi)), \\ (M_{\Delta t}\Phi_{\Delta t})(P) &= L\Phi_{\Delta t}(\xi) + \mathcal{Q}\tilde{G}(\xi + \Phi_{\Delta t}(\xi)). \end{aligned} \tag{4.49}$$

Note that the conditions (4.26) employed in the proof of Theorem 4.4 are sufficiently robust to admit essentially the same proof with C enlarged by a factor of 2; indeed this is why they were constructed that way. Hence the existence and attractivity of $\Phi_{\Delta t} \in \Gamma$ follows. It remains to estimate the closeness of $\Phi_{\Delta t}$ to Φ. To do this we use an argument which is essentially the *uniform contraction principle.*

Now, since Φ and $\Phi_{\Delta t}$ both lie in Γ, are fixed points of M and $M_{\Delta t}$ and M has contraction constant $(1 - \frac{1}{4}\alpha\Delta t)$ on Γ it follows that

$$\begin{aligned} \|\Phi - \Phi_{\Delta t}\| &= \|M\Phi - M_{\Delta t}\Phi_{\Delta t}\| \\ &\leq \|M\Phi - M\Phi_{\Delta t}\| + \|M\Phi_{\Delta t} - M_{\Delta t}\Phi_{\Delta t}\|, \\ &\leq (1 - \tfrac{1}{4}\alpha\Delta t)\|\Phi - \Phi_{\Delta t}\| + \|M\Phi_{\Delta t} - M_{\Delta t}\Phi_{\Delta t}\|. \end{aligned}$$

Hence

$$\|\Phi - \Phi_{\Delta t}\| \leq \frac{4}{\alpha\Delta t}\|M\Phi_{\Delta t} - M_{\Delta t}\Phi_{\Delta t}\|.$$

Thus it remains to estimate $M - M_{\Delta t}$.

Clearly

$$\begin{aligned} \|(M_{\Delta t}\Phi_{\Delta t})(P) - (M\Phi_{\Delta t})(P)\| \leq\; & \|(M_{\Delta t}\Phi_{\Delta t})(P) - (M\Phi_{\Delta t})(p)\| \\ & + \|(M\Phi_{\Delta t})(p) - (M\Phi_{\Delta t})(P)\|. \end{aligned}$$

Since $M\Phi_{\Delta t} \in \Gamma$ we deduce that

$$\|(M_{\Delta t}\Phi_{\Delta t})(P) - (M\Phi_{\Delta t})(P)\| \leq \|(M_{\Delta t}\Phi_{\Delta t})(P) - (M\Phi_{\Delta t})(p)\| + \|p - P\|.$$

Now consider (4.29) with $\Phi \mapsto \Phi_{\Delta t}$ and (4.49); subtracting and using (4.46) we obtain

$$\|(M_{\Delta t}\Phi_{\Delta t})(P) - (M\Phi_{\Delta t})(P)\| \leq 2K\Delta t^{r+1}.$$

Thus, in summary we have that

$$\|\Phi - \Phi_{\Delta t}\| \leq 8K\Delta t^r/\alpha$$

and the proof is complete. □

We can use this result to prove convergence of the local unstable manifold of the map (1.2) to the local unstable manifold of the equation (1.1). Recall that by Theorem 4.10 the map (1.2) has a fixed point $\bar{U}$ close to the equilibrium solution $\bar{u}$ of (1.1). Recall Definition 4.1 and the analogous notation (4.9) for the unstable manifolds of the map (1.2). We may now prove:

Corollary 4.13 (Local unstable manifolds under approximation) Let $W^{u,\epsilon}(\bar{u})$ denote the local unstable manifold of an equilibrium point $\bar{u}$ of (1.1), and $W^{u,\epsilon}_{\Delta t}(\bar{U})$, the local unstable manifold of the fixed point $\bar{U}$ of (1.2) given by Theorem 4.10. Then there exists $C, \Delta t_c, \epsilon_c > 0$ such that for any $\epsilon \in (0, \epsilon_c]$ and $u \in W^{u,\epsilon}(\bar{u})$ there exists $\epsilon' > 0$ and $U \in W^{u,\epsilon'}_{\Delta t}(\bar{U})$ such that

$$\|u - U\| \leq C\Delta t^r \quad \forall \Delta t \in (0, \Delta t_c].$$

That is

$$\mathrm{dist}(W^{u,\epsilon}(\bar{u}), W^{u,\epsilon'}_{\Delta t}(\bar{U})) \leq C\Delta t^r \quad \forall \Delta t \in (0, \Delta t_c].$$

Proof. The existence of a local invariant manifold for (1.2) follows from Theorem 4.12. That it is in fact the unstable manifold of $\bar{U}$ may be proved analogously to the proof of Corollary 4.5; it is necessary to use the fact that $\Phi_{\Delta t}(\mathcal{P}\bar{U}) = \mathcal{Q}\bar{U}$, that is that the fixed point lies on the invariant manifold. The closeness of the two local unstable manifolds follows from the closeness of the graphs Φ and $\Phi_{\Delta t}$ given in Theorem 4.12. The fact that ϵ' may differ from ϵ occurs when changing from a global to a local result since nearby points on the global graphs constructed under Assumption 4.3 may lie in balls of slightly different radii when localizing the result and removing Assumption 4.3. The change in norms yields the constant $C > 0$. □

4.4. Phase portraits and stable manifolds

Our aim in this section is to show the existence of a solution to (4.44) subject to the boundary conditions

$$P_N = \xi, \quad Q_0 = \eta \tag{4.50}$$

for any $N > 0$ and sufficiently small ξ, η. Furthermore, we then show that this solution is $\mathcal{O}(\Delta t^r)$ close to the analogous solution of (4.16) subject to (4.30). Since N is arbitrary this result yields convergence of the approximate trajectories to true trajectories over arbitrarily long-time intervals in the neighbourhood of equilibrium points and *could not be obtained by standard error analysis.* As we shall see, the key to the uniform in time convergence result is that the initial condition for the two trajectories is not the same.

Let $\mathcal{V}_{\Delta t}$ denote the sequence $\{V_n\}_{n=0}^N$ where $V_n = P_n + Q_n$, as in Section 4.3. To solve (4.44), (4.50) we consider finding fixed points $\mathcal{V}_{\Delta t}$ of the mapping $M_{\Delta t} : \Psi \mapsto \Psi$ defined by setting $M_{\Delta t}\mathcal{V}_{\Delta t} = \{M_{\Delta t}P_n + M_{\Delta t}Q_n\}_{n=0}^N$ where

$$\begin{aligned} M_{\Delta t}P_n &= L^{n-N}\xi - \textstyle\sum_{j=n}^{N-1} L^{n-j-1}\mathcal{P}\tilde{G}(V_j), \\ M_{\Delta t}Q_n &= L^n\eta + \textstyle\sum_{j=0}^{n-1} L^{n-1-j}\mathcal{Q}\tilde{G}(V_j). \end{aligned} \tag{4.51}$$

This map should be compared with its continuous counterpart (4.32). To prove existence of a solution to (4.50), (4.51) we follow the method of proof employed for the differential equation. To prove closeness of the solution $\mathcal{V}_{\Delta t}$ to $\mathcal{V}$ we use the uniform contraction principle in a similar manner to the proof of Theorem 4.12.

Theorem 4.14 Assume that Δt_c is chosen so that

$$e^{-\alpha\Delta t} \leq 1 - \tfrac{1}{2}\alpha\Delta t \forall \Delta t \in (0, \Delta t_c]$$

and that ϵ is chosen so that $8C\epsilon \leq \alpha$. Then, under Assumption 4.3, for any $N > 0$ and any $\xi \in X$, $\eta \in Y$ with $\|\xi\|, \|\eta\| \leq \frac{1}{2}\epsilon$ $\exists$ a solution of (4.44) subject to (4.50) satisfying

$$\max_{0\leq n\leq N} \|V_n\| \leq \epsilon.$$

Furthermore the following error estimate holds between the solution v_n of (4.16) and (4.30) and V_n:

$$\max_{0\leq n\leq N} \|\mathbf{v}_n - V_n\| \leq 4K\Delta t^r/\alpha.$$

Proof. To show the existence of $\mathcal{V}_{\Delta t} \in \Psi_\epsilon$ we show that $M_{\Delta t} : \Psi_\epsilon \mapsto \Psi_\epsilon$ is a contraction. This may be achieved by the following the proof of Theorem 4.6; note that (4.47) holds and so it is sufficient to enlarge C by a factor of 2. Since the estimate (4.34) was constructed to be robust under enlargement of C by a factor of 2 the existence of $\mathcal{V}_{\Delta t} \in \Psi_\epsilon$ follows, giving the bound on $\|V_n\|$.

To show convergence of $\mathcal{V}$ to $\mathcal{V}_{\Delta t}$ note that

$$\begin{aligned}\|\mathcal{V} - \mathcal{V}_{\Delta t}\| &= \|M\mathcal{V} - M_{\Delta t}\mathcal{V}_{\Delta t}\| \\ &\leq \|M\mathcal{V} - M\mathcal{V}_{\Delta t}\| + \|M\mathcal{V}_{\Delta t} - M_{\Delta t}\mathcal{V}_{\Delta t}\| \\ &\leq \tfrac{1}{2}\|\mathcal{V} - \mathcal{V}_{\Delta t}\| + \|M\mathcal{V}_{\Delta t} - M_{\Delta t}\mathcal{V}_{\Delta t}\|.\end{aligned}$$

Hence

$$\|\mathcal{V} - \mathcal{V}_{\Delta t}\| \leq 2\|M\mathcal{V}_{\Delta t} - M_{\Delta t}\mathcal{V}_{\Delta t}\|.$$

Now, by consideration of (4.32) with $\mathcal{V} \mapsto \mathcal{V}_{\Delta t}$ (that is $v_n \mapsto V_n$) and (4.51) we deduce that $\|M\mathcal{V}_{\Delta t} - M_{\Delta t}\mathcal{V}_{\Delta t}\|$ can be bounded above by

$$\max\left\{\sum_{j=n}^{N-1} L^{n-j-1}[\mathcal{P}G(V_j) - \mathcal{P}\tilde{G}(V_j)], \sum_{j=0}^{n-1} L^{n-1-j}[\mathcal{Q}G(V_j) - \mathcal{Q}\tilde{G}(V_j)]\right\}.$$

Using (4.33) and (4.46) we thus find that

$$\|M\mathcal{V}_{\Delta t} - M_{\Delta t}\mathcal{V}_{\Delta t}\| \leq \frac{2}{\alpha\Delta t}K\Delta t^{r+1}$$

and hence that

$$\|\mathcal{V} - \mathcal{V}_{\Delta t}\| \leq 4K\Delta t^r/\alpha. \quad \square$$

□

Using Theorem 4.14 we are able to state an interesting result concerning error bounds for approximate solutions of (1.1) near to an equilibrium point.

Consider the boundary value problems (4.36) and the discrete analogue

$$U_{n+1} = \mathcal{F}(U_n; \Delta t), \quad \mathcal{P}(U_N - \bar{u}) = \xi, \ \mathcal{Q}(U_0 - \bar{u}) = \eta. \qquad (4.52)$$

We can now prove:

Corollary 4.15 **(Phase portrait under approximation)** There exist $C, \Delta t_c, \epsilon_c > 0$ such that, if $\epsilon \in (0, \epsilon_c]$ then for any $T > 0$ and any $\xi \in X$, $\eta \in Y$ with $\|\xi\|, \|\eta\| \leq \frac{1}{2}\epsilon$ $\exists$ a solution of (4.36) and, if $N\Delta t = T$, a solution of (4.52) satisfying

$$\max_{0 \leq n \leq N} \|U_n - u(t_n)\| \leq C\Delta t^r \quad \forall \Delta t \in (0, \Delta t_c].$$

Proof. This is simply a restatement of Theorem 4.14 in the original variables $u(t)$ and U_n. The change of variables introduces a change in the error constant through the change of norms. □

The important point here is that the error bound is independent of T. Thus it improves upon the standard estimate (1.3) which contains a constant growing exponentially with T. Note that this is achieved by comparing two solutions of (1.1) and (1.2) which *do not share the same initial condition*; specifically, only the projection of the initial condition into the subspace Y is identical at $t = 0$.

We now consider the existence and convergence of stable manifolds under approximation. The local stable manifold for the map (1.2) can be constructed by solving

$$\begin{aligned} P_n &= -\textstyle\sum_{j=n}^{\infty} L^{n-j-1}\mathcal{P}\tilde{G}(V_j), \\ Q_n &= L^n Q_0 + \textstyle\sum_{j=0}^{n-1} L^{n-1-j}\mathcal{Q}\tilde{G}(V_j), \\ Q_0 &= \eta, \quad \exists \delta > 0 : \|P_n\| \leq \delta \ \forall n \geq 0. \end{aligned} \qquad (4.53)$$

where $V_n = P_n + Q_n$ and P_n, Q_n and $\tilde{G}(\cdot)$ are defined by (4.44), (4.42). The stable manifold is formed from solutions of (4.53) as the graph $\Theta : Y \mapsto X$ given by $\Theta(\eta) = P_0$.

Since the proof of Theorem 4.8 is robust to enlargement of C by a factor of 2 it follows that

Theorem 4.16 Assume that Δt_c is chosen so that

$$\mathrm{e}^{-\alpha\Delta t} \leq 1 - \tfrac{1}{2}\alpha\Delta t \, \forall \Delta t \in (0, \Delta t_c]$$

and that ϵ is chosen so that $8C\epsilon \leq \alpha$. Then, under Assumption 4.3, for any $\eta \in Y$ with $\|\eta\| \leq \frac{1}{2}\epsilon$ $\exists$ a solution of (4.53) satisfying

$$\max_{0 \leq n < \infty} \|\mathbf{V}_n\| \leq \epsilon.$$

We are now in a position to show that the stable manifold is well approximated numerically.

Corollary 4.17 **(Local stable manifolds under approximation)** Let $W^{s,\epsilon}(\bar{u})$ denote the local unstable manifold of an equilibrium point $\bar{u}$ of (1.1), and $W^{s,\epsilon}_{\Delta t}(\bar{U})$, the unstable manifold of the fixed point $\bar{U}$ of (1.2) given by Theorem 4.10. Then there exists $C, \Delta t_c, \epsilon_c > 0$ such that for any $\epsilon \in (0, \epsilon_c]$ and $u \in W^{s,\epsilon}(\bar{u})$ there exists $\epsilon' > 0$ and $U \in W^{s,\epsilon'}_{\Delta t}(\bar{U})$ such that

$$\|u - U\| \leq C\Delta t^r \quad \forall \Delta t \in (0, \Delta t_c].$$

That is

$$\text{dist}(W^{s,\epsilon}(\bar{u}), W^{s,\epsilon'}_{\Delta t}(\bar{U})) \leq C\Delta t^r \quad \forall \Delta t \in (0, \Delta t_c].$$

Proof. The existence of a local invariant manifold for (1.2) follows from Theorem 4.16 by setting $\Phi_{\Delta t}(\eta) = P_0$ for every $\eta : \|\eta\| \leq \epsilon/2$. That it is in fact the stable manifold of $\bar{U}$ may be proved analogously to the proof of Corollary 4.9; it is necessary to use the fact that $\Phi_{\Delta t}(\mathcal{Q}\bar{U}) = \mathcal{P}\bar{U}$, that is that the fixed point lies on the invariant manifold. The closeness of the two local stable manifolds follows from the closeness of the solution v_n and V_n of (4.37) and (4.53) given in Theorems 4.8 and 4.16. The change in norms yields the constant $C > 0$. □

4.5. Bibliography

For background material concerning equilibria, fixed points, hyperbolicity and stability see Hale and Kŏcak (1991) and Wiggins (1990). For discussion of unstable manifolds see Babin and Vishik (1992), Hale (1988), Hale and Kŏcak (1991) and Wiggins (1990). The construction of unstable manifolds described in Section 4.1 is based on an approach known as the *Hadamard graph transform*; this transform technique can also be used to construct stable and centre manifods. The construction of the phase portrait in Section 4.1 is closely related to the *Hartman–Grobman* theorem which states that there is a 1:1 correspondence between solutions of (1.2) and its linearization in the neighbourhood of a hyperbolic equilibrium point. See Hartman (1982). The construction of the stable manifold in Section 4.1 is based on the *Lyapunov–Perron* technique, here modified from differential equations to mappings. For a discussion of this technique see, for example, Carr (1982) and Medved (1991). Again, this technique can be modifed to construct unstable and centre manifolds.

Since most numerical methods for ordinary differential equations replicate exactly all the equilibria of the underlying equation as fixed points of the numerical method, Theorem 4.10 may seem a little pointless. However, the method of proof employed there can be used to study approximation of partial differential equations where exact preservation of equilibria under spatial approximation does not occur; see Crouziex and Rappaz (1990). Analysis of the existence of spurious solutions introduced by discretization using techniques from dynamical systems can be traced back to the article by Newell (1977) and the subsequent related work undertaken in Mitchell and Griffiths (1986) and Stuart (1989a,b); all these articles concerned spurious solutions oscillating on a grid scale in time. The article by Brezzi *et al.* (1984) considered the existence of spurious invariant curves introduced by time discretization. However, it was not until the work of Iserles (1990) that the more interesting question of the possible existence of spurious equilibrium solutions was investigated. He showed that Runge–Kutta methods could admit spurious equilibria whilst linear multistep methods could not. Subsequent analysis of this phenomena can be found in Hairer *et al.* (1990), Yee *et al.* (1991) and Griffiths *et al.* (1992). It is fair to say that the area of spurious solutions introduced by time discretization is now very well understood – the article by Iserles *et al.* (1991) puts the subject in a unified framework whilst in Humphries (1993) it is proved that spurious solutions must either converge to true solutions or become unbounded as $\Delta t \to 0$; such a result can also be deduced from Theorem 4.11. There is probably little of interest remaining to do in the area of spurious solutions introduced by fixed time-step time discretization. Note also that it is reasonable to expect that, under many circumstances, codes which vary the time-step to control the local error will also prevent spurious solutions. Such a result was conjectured in Sanz-Serna (1992b) and is proved for certain error control schemes applied to (1.1) under a variety of a structural assumptions in Stuart and Humphries (1992b). The effect of spurious solutions introduced by spatial discretization is an area in which there are still many open questions. For representative work in this area see Beyn and Doedel (1981), Budd (1991), Murdoch and Budd (1990), Elliott and Stuart (1993) and Stephens and Shubin (1987).

The first proof of convergence of local stable and unstable manifolds, together with phase portraits, was contained in the article by Beyn (1987b). This article employed a very clean presentation, involving use of a Lipschitz inverse mapping theorem. The approach presented there can be extended to multistep methods. We have chosen to present a more transparent, if lengthier, proof of the convergence of local unstable manifolds; it is based on a similar proof for centre-unstable manifolds contained in Beyn and Lorenz (1987) involving the Hadamard graph transform. It should be noted that Corollary 4.13 can easily be extended to show that the distance $d_{\mathrm{H}}(\cdot,\cdot)$ between the local unstable manifolds is small, rather than just the semi-distance described. A thorough study of the behaviour of discretizations near equilibria may be found in Garay (1993) which unifies and extends much of the work described here.

The proof we employ to construct the phase portrait follows the approach taken in Larrson and Sanz-Serna (1993) very closely (see also Sanz-Serna and Larsson (1993)) where finite-element approximations of reaction-diffusion equations are studied; their approach is closely related to the construction of stable and unstable manifold by the Lyapunov–Perron method in Henry (1981). Hence the Lyapunov–Perron method underlies our result concerning the convergence of stable manifolds. The approach of Beyn (1987b) has been generalized to study the numerical approximation of certain partial differential equations in Alouges and Debussche (1991).

Whilst on the subject of unstable manifolds, it is relevant to mention the literature concerning the effect of numerical approximation on *inertial manifolds*. These attractive invariant manifolds for partial differential equations on a Hilbert space $\mathcal{H}$ may be represented as graphs relating a certain projection of the space $\mathcal{H}$ to its complement. See Foais *et al.* (1988), Mallet-Paret and Sell (1988) and Constantin *et al.* (1989) for the background theory. The original construction of the inertial manifold in Foais *et al.* (1988) uses the Lyapunov–Perron approach and contains a convergence result concerning the effect of Galerkin approximation on the inertial manifold; a related method of analysis was employed in Demengel and Ghidaglia (1989) to study the effect of a particular time discretization on the problem. In Jones and Stuart (1993) the inertial manifold is constructed by use of a technique similar to that employed to prove Theorem 4.4 (the Hadamard graph transform) and a convergence proof, sufficiently general to include a variety of numerical approximations and similar to the proof of Corollary 4.13, is given.

5. Periodic solutions and invariant tori

5.1. Background theory

In this section we study the effect of discretization on periodic solutions of (1.1); we shall not describe the theory for quasi-periodic solutions but give some references to the literature in the final section. The methods employed are very similar to those we describe for the study of periodic solutions.

For simplicity we assume that the periodic solution is stable and hyperbolic – we shall be precise about the meaning of this later on – see (5.6). Let us assume that (1.1) has the periodic solution $\bar{u}(t)$ with period T :

$$\{\bar{u}(t) \in C^1(\mathbb{R}, \mathbb{R}^p) | \bar{u}(t+T) = \bar{u}(t) \; \forall t \in \mathbb{R}\}. \tag{5.1}$$

In order to facilitate study of the periodic solution, we introduce new coordinates $r \in \mathbb{R}^{p-1}$ and $\theta \in \mathbb{R}$ where, roughly, r measures the coordinates normal to the tangent space of the periodic solution and θ measures an angular coordinate in the tangent space of the periodic solution. Letting $v = (r^T, \theta)^T \in \mathbb{R}^p$ it may be shown that there exists a C^3 diffeomorphism $\chi : \mathbb{R}^p \mapsto \mathbb{R}^p$ under which the transformation $u = \chi(v)$ renders (1.1) in a very useful form. Specifically it may be shown that we obtain

$$\begin{aligned} r_t &= A(\theta)r + g(r,\theta), \quad r(0) = \xi, \\ \theta_t &= 1 + h(r,\theta), \quad \theta(0) = \phi. \end{aligned} \tag{5.2}$$

Here A, g and h are C^2 in a neighbourhood of the periodic solution and satisfy the following conditions for all $\theta \in \mathbb{R}$:

$$g(0,\theta) = 0, \quad g_r(0,\theta) = 0, \quad h(0,\theta) = 0, \tag{5.3}$$

$$A(\theta + T) = A(\theta), \quad g(r,\theta + T) = g(r,\theta), \quad h(r,\theta + T) = h(r,\theta). \tag{5.4}$$

Thus the periodic solution is simply $r = 0$ and $\theta = t$ in this coordinate system and A, g and h are defined for all $\theta \in \mathbb{R}$ and all $r \in B(0,\epsilon)$, for some ϵ sufficiently small.

For simplicity consider the case $p = 2$ so that $r(t) \in \mathbb{R}$. If we define

$$\begin{aligned} B(s,t;\phi) &:= \exp[\int_s^t A(\phi + \tau)\mathrm{d}\tau], \\ B &:= B(0,T;\phi), \end{aligned} \tag{5.5}$$

then, using the fact that the periodic solution is hyperbolic and stable, it follows that there exists a norm on $\mathbb{R}^{p-1}$ in which the following property holds for B:

$$\|B\| \le \alpha < 1 \quad \forall \phi \in \mathbb{R} \tag{5.6}$$

In dimension $p > 2$ a bound similar to (5.6) holds where $B(s,t;\phi)$ is replaced by the solution operator for the non-autonomous equation

$$r_t = A(\phi + t)r, \quad r(s) = \psi.$$

Whenever we require use of $B(s,t;\phi)$ in this chapter we will consider the case $p = 2$ and refer to the representation (5.5) for $B(s,t;\phi)$; however, by using the more general definition of $B(s,t;\phi)$ instead of (5.5) arbitrary $p > 2$ may be considered similarly. We will employ the norm on $\mathbb{R}^{p-1}$ given in (5.6) throughout the remainder of our discussion of periodic solutions. The following example illustrates the transformation of variables just described.

Example Consider equations (3.12). We modify the change to polar coordinates used to study these equations and introduce the variables

$$r = R - 1, \quad \theta = -\phi.$$

Thus

$$x = (1 + r)\cos\theta, \quad y = -(1 + r)\sin\theta. \tag{5.7}$$

Then, from (3.13), we obtain

$$r_t = -2r - (3r^2 + r^3), \quad \theta_t = 1. \tag{5.8}$$

Hence $A(\theta) = -2$, $g(r, \theta) = -(3r^2 + r^3)$ and $h(r, \theta) = 0$. Note that (5.3) and (5.4) are trivially satisfied. Furthermore

$$B(s, t; \phi) = \mathrm{e}^{-2(t-s)}, \quad B = \mathrm{e}^{-4\pi}$$

since the period $T = 2\pi$. This shows that (5.6) is satisfied. □

As we have observed, the existence of a periodic solution $r = 0, \theta = t$ in (5.2) is trivial under (5.3), (5.4). However, since our aim is to develop an existence theory which is sufficiently robust to incorporate the effect of numerical approximation at a later point, we must relax (5.3). The crucial consequence of (5.3) (which is also shared by equations generated by applying $u = \chi(v)$ to equations (1.1) found from smooth perturbations of a vector field yielding (5.2)) is that $\exists C_1 > 0$:

$$\begin{gathered} \|B(0,t;\theta)\| \leq C_1 \; \forall t \in [0,T], \quad \|g(r,\theta)\|, \|g_\theta(r,\theta)\| \leq C_1\epsilon^2, \\ \|h(r,\theta)\|, \|h_\theta(r,\theta)\|, \|g_r\| \leq C_1\epsilon, \quad \|h_r(r,\theta)\| \leq C_1 \end{gathered} \tag{5.9}$$

for all $r \in B(0, \epsilon)$, $t \in \mathbb{R}$, where the subscripts r and θ denote appropriate derivatives. Thus, by considering (5.9) instead of (5.3) we are considering the effect of small perturbations of the vector field $f(\cdot)$ on the periodic solution $\bar{u}(t)$.

The mappings generated by (5.2) under (5.9) are sufficiently general to enable us to incorporate the effect of numerical approximations within the same framework.

Since our interest in now focused on a small neighbourhood of the periodic solution, smooth modification of the functions g and h outside such a neighbourhood will not affect the results. Thus we assume:

Assumption 5.1 The functions

$$A(\theta) \in C^2(\mathbb{R}^{p-1}, \mathbb{R}^{p-1}), \quad g(r,\theta) \in C^2(\mathbb{R}^{p-1} \times \mathbb{R}, \quad \mathbb{R}^{p-1})$$

and

$$h(r, \theta) \in C^2(\mathbb{R}^{p-1} \times \mathbb{R}, \mathbb{R}).$$

Furthermore, (5.9) and (5.4) are satisfied for all $r \in \mathbb{R}^{p-1}$ and $\theta \in \mathbb{R}$

Under Assumption 5.1 it is straightforward to prove that (5.2) has a unique solution for all $t \geq 0$. The following example illustrates that periodic solutions persist under perturbations to the vector field $f(\cdot)$ in (1.1) such that Assumption 5.1 holds.

Example If, instead of considering equations (5.8) we modify g and h to obtain

$$\begin{gathered} r_t = -2r + \epsilon^2\{\cos((1+\epsilon)\theta) + 2\sin\theta\}, \\ \theta_t = 1 + \epsilon, \end{gathered} \tag{5.10}$$

then we obtain a solution in the form

$$r(t) = \epsilon^2 \sin((1+\epsilon)t), \quad \theta(t) = (1+\epsilon)t. \tag{5.11}$$

Recalling the transformation (5.7) it is clear that the solution (5.11) yields a periodic solution in x, y coordinates. □

After proving results under Assumption 5.1 we will derive a corollary concerning the original, unmodified flow generated by (1.1). We shall employ the theory of attractive invariant manifolds to construct a periodic solution for (5.2) under Assumption 5.1 and to study the effect of numerical approximation. The basic idea of the proof is that if $g, h \equiv 0$ then (5.2) has solution

$$r(t) = B(0, t; \phi)\xi, \quad \theta(t) = \phi + T.$$

Hence $\|r(mT)\| \leq \alpha^m \xi$ by (5.6). It is this contractivity that we wish to exploit in the case where g and h are small, but not identically zero. Hence we integrate (5.2), using the integrating factor $\{B(0, t; \phi)\}^{-1}$ and the variation of constants formula to obtain

$$\begin{aligned} r(t) &= B(0, t; \phi)\xi + \int_0^t B(s, t; \phi) Z(s)\, \mathrm{d}s, \\ \theta(t) &= \phi + t + \int_0^t h(r(s), \theta(s))\, \mathrm{d}s, \end{aligned} \tag{5.12}$$

where

$$Z(s) = [A(\theta(s)) - A(\phi + s)]r(s) + g(r(s), \theta(s)).$$

In order to exploit the contractivity induced by (5.6) it will be convenient to consider the solution of (5.12) at time $t = T$. Denoting the solutions of (5.2) by $r(s)$ and $\theta(s)$ and noting that these vectors are functions of ξ and ϕ we obtain from (5.12)

$$\begin{aligned} r(T) &= B\xi + G(\xi, \phi), \\ \theta(T) &= \phi + T + H(\xi, \phi), \end{aligned} \tag{5.13}$$

where $G : \mathbb{R}^{p-1} \times \mathbb{R} \mapsto \mathbb{R}^{p-1}$ and $H : \mathbb{R}^{p-1} \times \mathbb{R} \mapsto \mathbb{R}$ are defined in the following way

$$\begin{aligned} G(\xi, \phi; \tau) &:= \int_0^\tau B(s, \tau; \phi) Z(s)\, \mathrm{d}s, \\ H(\xi, \phi; \tau) &:= \int_0^\tau h(r(s), \theta(s))\, \mathrm{d}s, \\ G(\xi, \phi) &:= G(\xi, \phi; T), \\ H(\xi, \phi) &:= H(\xi, \phi; T). \end{aligned} \tag{5.14}$$

We now give an example to illustrate the preceding definitions.

Example Consider the equations

$$r_t = -2r + \epsilon^2, \quad \theta_t = 1 + \epsilon.$$

Then it follows that

$$G(\xi,\phi)=\epsilon^2(1-\mathrm{e}^{-2T})/2,\quad H(\xi,\phi)=\epsilon T.$$

The following space of functions will be useful in the succeeding analysis:

$$\begin{aligned}\Gamma=\{\Psi\in C(\mathbb{R},\mathbb{R}^{p-1}):\|\Psi\|_P=\sup\nolimits_{\theta\in[0,T)}\|\Psi(\theta)\|\le K\epsilon^2,\\ \|\Psi(\theta_1)-\Psi(\theta_2)\|\le\epsilon|\theta_1-\theta_2|,\ \Psi(\theta_1+T)=\Psi(\theta_1)\ \forall\theta_1,\theta_2\in\mathbb{R}\}.\end{aligned}\tag{5.15}$$

The subscript P on the norm simply denotes the space of periodic functions in $C(\mathbb{R},\mathbb{R}^{p-1})$ with period T. After Lemma 5.2 we will require a particular value for K and use

$$K=4C_1T/(1-\alpha).\tag{5.16}$$

Our aim is to find an invariant manifold for (5.13), namely $\Phi\in\Gamma$ such that

$$\xi=\Phi(\phi)\Leftrightarrow r(T)=\Phi(\theta(T)).\tag{5.17}$$

Comparison with (5.13) shows that to do this is equivalent to finding a fixed point of the mapping $\mathcal{T}$ defined by

$$\begin{aligned}(\mathcal{T}\Phi)(\theta)=B\Phi(\phi)+G(\Phi(\phi),\phi),\\ \theta=\phi+T+H(\Phi(\phi),\phi).\end{aligned}\tag{5.18}$$

Our proof of existence of such a fixed point is closely related to our proof of the existence of an unstable manifold of a fixed point, given in Section 4. We first establish certain 'smallness' properties of G, H and their Lipschitz constants; we then show that $\mathcal{T}$ maps Γ into itself and finally that $\mathcal{T}$ is a contraction on Γ. The details are more complicated than for the unstable manifold and hence we break up the proof into a sequence of lemmas. The proof of Lemma 5.2, in particular, is very technical and may be omitted without disrupting the flow.

Lemma 5.2 Assume that Assumption 5.1 holds. Let $\Phi^i\in\Gamma$, $i=1,2$. Then, for all $\phi^i\in\mathbb{R}$, $i=1,2$ it follows that there exists $C_2=C_2(T)$ and $\epsilon^*>0$ such that, for all $\epsilon\in(0,\epsilon^*]$,
(i) $|H(\Phi^1(\phi^1),\phi^1)|\le C_1T\epsilon$;
(ii) $\|G(\Phi^1(\phi^1),\phi^1)\|\le 2C_1T\epsilon^2$;
(iii) $|H(\Phi^1(\phi^1),\phi^1)-H(\Phi^1(\phi^2),\phi^2)|\le C_2\epsilon|\phi^1-\phi^2|$;
(iv) $\|G(\Phi^1(\phi^1),\phi^1)-G(\Phi^1(\phi^2),\phi^2)\|\le C_2\epsilon^2|\phi^1-\phi^2|$;
(v) $|H(\Phi^1(\phi),\phi)-H(\Phi^2(\phi),\phi)|\le C_2\|\Phi^1-\Phi^2\|_P$;
(vi) $\|G(\Phi^1(\phi),\phi)-G(\Phi^2(\phi),\phi)\|\le C_2\epsilon\|\Phi^1-\Phi^2\|_P$.

Proof. Throughout the proof we use $C(T)$ to denote a global constant independent of ϵ.
(i), (ii) Consider equations (5.12) with $\xi=\Phi(\phi)$ and let Assumption 5.1 hold. It follows that

$$|\theta(t)-\phi-t|\le C_1t\epsilon\le C_1T\epsilon,\ \forall t\in[0,T]\tag{5.19}$$

and hence the result (i) follows. From (5.12), using Assumption 5.1, (5.15), 5.19 and the assumption that $r(0) = \Phi(\phi)$ we have that

$$\|r(t)\| \leq C_1 K \epsilon^2 + \int_0^t [C(T)\epsilon \|r(s)\| + C_1 \epsilon^2] \, \mathrm{d}s \quad \forall t \in [0, T]. \tag{5.20}$$

Application of the Gronwall lemma yields

$$\|r(t)\| \leq C_1 (K + T) \epsilon^2 \mathrm{e}^{C(T)\epsilon t}, \quad \forall t \in [0, T]. \tag{5.21}$$

Thus (5.12) gives, using (5.20), (5.21) and Assumption 5.1,

$$\|r(T) - B\Phi(\phi)\| \leq \int_0^T [C(T)\epsilon^3 + C_1 \epsilon^2] \, \mathrm{d}s \leq 2 C_1 T \epsilon^2,$$

for ϵ sufficiently small. This yields (ii).

We also derive a related estimate used in the proof of Corollary 5.6. By an argument similar to that giving (5.20) we may show that

$$\|r(t) - \xi\| \leq \|B(0, t; \phi) - I\| K \epsilon^2 + \int_0^t [C(T)\epsilon \|r(s)\| + C_1 \epsilon^2] \, \mathrm{d}s.$$

Using (5.5) and (5.21) it follows that, for t sufficiently small, there exists $C_3 > 0$ such that

$$\|r(t) - \xi\| \leq C_3 t K \epsilon^2 \Rightarrow \|r(t)\| \leq (1 + C_3 t) K \epsilon^2. \tag{5.22}$$

(iii), (iv) Consider the equations

$$\begin{array}{rclr} r_t^i & = & A(\theta^i) r^i + g(r^i, \theta^i), & r^i(0) = \Phi(\phi^i) \\ \theta_t^i & = & 1 + h(r^i, \theta^i), & \theta^i(0) = \phi^i \end{array} \tag{5.23}$$

for $i = 1, 2$. Define $\delta = \phi^1 - \phi^2$, $\rho = r^1 - r^2, \gamma = \theta^1 - \theta^2$ and

$$\|\rho\|_\infty = \sup_{0 \leq t \leq T} \|\rho(t)\|.$$

Note that, since $\Phi \in \Gamma$, we have that

$$\gamma(0) = \delta, \ \|\rho(0)\| \leq \epsilon |\delta|. \tag{5.24}$$

Now, by the mean value theorem,

$$\gamma_t = \bar{h}(r^1, r^2, \theta^1)\rho + h_\theta(r^2, \beta)\gamma \tag{5.25}$$

where

$$\bar{h}(r^1, r^2, \theta) := \int_0^1 h_r(z r^1 + (1 - z) r^2, \theta) \, \mathrm{d}z$$

and $\beta = \beta(t) := (1 - \zeta)\theta^1 + \zeta\theta^2$ for some $\zeta = \zeta(t) \in [0, 1]$.

Thus, by (5.24),

$$\gamma(t) = \delta + \int_0^t \{\bar{h}(r^1(s), r^2(s), \theta^1(s))\rho(s) + h_\theta(r^2(s), \beta(s))\gamma(s)\} \, \mathrm{d}s. \tag{5.26}$$

Hence, by Assumption 5.1, we obtain

$$|\gamma(t)-\delta| \le C(T)\|\rho\|_\infty + \int_0^t C_1\epsilon|\gamma(s)|\,\mathrm{d}s \quad \forall t\in[0,T] \tag{5.27}$$

and application of the Gronwall lemma yields

$$|\gamma(t)| \le C(T)[|\delta| + \|\rho\|_\infty], \quad \forall t\in[0,T]. \tag{5.28}$$

Also, we have that

$$\rho_t = A(\theta^1)\rho + [A(\theta^1)-A(\theta^2)]r^2 + \bar{g}(r^1,r^2,\theta^1)\rho + g_\theta(r^2,\alpha)\gamma,$$

where

$$\bar{g}(r^1,r^2,\theta) := \int_0^1 g_r(zr^1+(1-z)r^2,\theta^1)\,\mathrm{d}s$$

and $\alpha=\alpha(t) := (1-\zeta)\theta^1 + \zeta\theta^2$ for some $\zeta=\zeta(t)\in[0,1]$.

Hence integration yields

$$\begin{aligned}\rho(t) = \;& B(0,t;\phi^1)\rho(0) + \int_0^t B(s,t;\phi^1)\{[A(\theta^1(s)) - A(\phi^1+s)]\rho(s)\\ &+[A(\theta^1(s)) - A(\theta^2(s))]r^2(s) + \bar{g}(r^1,r^2,\theta^1(s))\rho(s)\\ &+g_\theta(r^2(s),\alpha(s))\gamma(s)\}\,\mathrm{d}s.\end{aligned} \tag{5.29}$$

From 5.19, (5.21), (5.24), (5.28) and Assumption 5.1 we deduce that

$$\begin{aligned}\|\rho(t)\| &\le C_1\epsilon|\delta| + \int_0^t C(T)\epsilon[\|\rho(s)\| + \epsilon|\gamma(s)|]\,\mathrm{d}s\\ &\le C_1\epsilon|\delta| + C(T)\epsilon[\epsilon|\delta| + \|\rho\|_\infty].\end{aligned} \tag{5.30}$$

Hence we deduce that, for ϵ sufficiently small,

$$\|\rho\|_\infty \le C(T)\epsilon|\delta| \quad \forall t\in[0,T]. \tag{5.31}$$

Thus (5.28) gives

$$|\gamma(t)| \le C(T)(1+\epsilon)|\delta| \tag{5.32}$$

and (5.27) yields, for ϵ sufficiently small,

$$|\gamma(T)-\delta| \le \epsilon C_2(T)|\delta|$$

as required for (iii). Now let $B^i := B(0,T;\phi)$ and note that

$$\|B^1 - B^2\| \le c|\delta|.$$

Returning to (5.29) we obtain, by (5.31), (5.32)

$$\|\rho(T) - B^1\rho(0)\| \le \int_0^T C(T)\epsilon[\|\rho(s)\| + \epsilon|\gamma(s)|]\,\mathrm{d}s \le \epsilon^2 C(T)|\delta|.$$

Hence

$$\|\rho(T) - (B^1r^1(0) - B^2r^2(0))\| \;\le\; \|\rho(T) - B^1\rho(0)\| + \|(B^2-B^1)r^2(0)\|$$

$$
\begin{aligned}
&\le \epsilon^2 C(T)|\delta| + CK\epsilon^2\|\delta| \\
&\le \epsilon^2 C_2(T)|\delta|
\end{aligned}
$$

as required for (iv).

Again we derive a related estimate used in the proof of Corollary 5.6. From (5.29) an argument similar to that yielding (5.30) gives

$$
\|\rho(t)-\rho(0)\| \le \|B(0,t;\phi^1)-I\|\|\rho(0)\| + \int_0^t C(T)\epsilon[\|\rho(s)\| + \epsilon|\gamma(s)|]\,\mathrm{d}s.
$$

Thus (5.24), (5.30), (5.31) and (5.32) yield the existence of $C_4 > 0$:

$$
\|\rho(t)\| \le (1 + C_4 t)\epsilon|\delta|, \tag{5.33}
$$

for t and ϵ sufficiently small. Hence, by (5.31) and (5.32), we have from (5.26)

$$
\begin{aligned}
|\delta| &\le |\gamma(t)| + C_1 t\|\rho\|_\infty + \int_0^t C_1\epsilon|\gamma(s)|\,\mathrm{d}s \\
&\le |\gamma(t)| + C_4 t\epsilon|\delta|.
\end{aligned}
$$

Thus

$$
|\delta| \le \frac{|\gamma(t)|}{1 - C_4 t\epsilon}.
$$

Hence (5.33) gives us

$$
\|\rho(t)\| \le (1 + C_3 t)\epsilon|\gamma(t)| \tag{5.34}
$$

for t sufficiently small. (Note that we have chosen the same constant C_3 as appears in (5.22) without loss of generality.)

(v), (vi) Consider the equations

$$
\begin{aligned}
r^i_t = A(\theta^i)r^i + g(r^i,\theta^i), \quad r^i(0) = \Phi^i(\phi) \\
\theta^i_t = 1 + h(r^i,\theta^i), \theta^i(0) = \phi
\end{aligned} \tag{5.35}
$$

for $i = 1, 2$. Define ρ, γ and $\|\rho\|_\infty$ as in the proof of (iii) and (iv). Note that

$$
\|\rho(0)\| \le \|\Phi^1 - \Phi^2\|, \quad \gamma(0) = 0.
$$

As in cases (iii) and (iv), (5.25) holds and, since $\gamma(0) = 0$, it follows from (5.28) that

$$
|\gamma(t)| \le C(T)\|\rho\|_\infty \quad \forall t \in [0,T]. \tag{5.36}
$$

Then (5.29), together with Assumption 5.1, gives

$$
\begin{aligned}
\|\rho(t)\| &\le C_1\|\Phi^1 - \Phi^2\| + \int_0^t C(T)\epsilon[\|\rho(s)\| + \epsilon|\gamma(s)|]\,\mathrm{d}s \\
&\le C_1\|\Phi^1 - \Phi^2\| + \int_0^t C(T)\epsilon\|\rho\|_\infty\,\mathrm{d}s;
\end{aligned}
$$

hence, for ϵ sufficiently small,

$$\|\rho\|_\infty \leq C_2(T)\|\Phi^1 - \Phi^2\| \quad \forall t \in [0,T].$$

Application to (5.36) gives (v) as required; application to (5.29) gives

$$\|\rho(T) - B\rho(0)\| \leq \int_0^T C(T)\epsilon[\|\rho(s)\| + \epsilon|\gamma(s)|]\,ds \leq \epsilon C_2(T)\|\Phi^1 - \Phi^2\|$$

as required for (vi). □

Armed with the bounds on G and H proved in Lemma 5.2 we can proceed to show that the map $\mathcal{T}$ given in (5.18) is well defined. Note that we have already assumed ϵ sufficiently small in the proof of Lemma 5.2. Unlike the case of unstable manifolds considered in Section 3 we will not detail the bounds on ϵ required in the sequel – we simply observe that our arguments hold for ϵ sufficiently small. Without loss of generality we will use the same constant ϵ^* to denote the upper bound on ϵ sufficient for all our arguments to work.

Lemma 5.3 Let Assumption 5.1 hold. If $\Phi \in \Gamma$ then there exists $\epsilon^* > 0$, such that if $\epsilon \in (0, \epsilon^*], \mathcal{T}\Phi$ given by (5.18) is well defined.

Proof. We show that, for every $\theta \in \mathbb{R}$, there is a unique $\phi \in \mathbb{R}$ so that (5.18) is well defined. That is we solve the equation

$$\theta = \phi + T + H(\Phi(\phi), \phi)$$

for ϕ, given $\theta \in \mathbb{R}$ and $\Phi \in \Gamma$. To do this we use the contraction mapping theorem. Consider the iterates

$$\begin{aligned} \phi^{k+1} &= \theta - T - H(\Phi(\phi^k), \phi^k), \\ \psi^{k+1} &= \theta - T - H(\Phi(\psi^k), \psi^k). \end{aligned}$$

Clearly $\phi^k \in \mathbb{R}$ yields $\phi^{k+1} \in \mathbb{R}$ and furthermore, by Lemma 5.2(iii),

$$|\phi^{k+1} - \psi^{k+1}| \leq C_2\epsilon|\phi^k - \psi^k| \leq \tfrac{1}{2}|\phi^k - \psi^k|$$

for ϵ sufficiently small; thus the mapping is a contraction on $\mathbb{R}$ and the existence of ϕ given θ follows. □

We may now show that $\mathcal{T}$ maps Γ to itself. We assume throughout the remainder of this section that K is given by (5.16).

Lemma 5.4 Let Assumption 5.1 hold. Then there exists ϵ^* such that, if $\epsilon \in (0, \epsilon^*]$ then the mapping $\mathcal{T} : \Gamma \mapsto \Gamma$.

Proof. From (5.18), (5.6) and Lemma 5.2(ii) we obtain, for ϵ sufficiently small and $\Phi \in \Gamma$,

$$\|(\mathcal{T}\Phi)(\theta)\| \leq \alpha K\epsilon^2 + 2C_1T\epsilon^2.$$

Noting that K is given by (5.16) and that $2C_1 < 4C_1$ we deduce that

$$\|(\mathcal{T}\Phi)(\theta)\| \le \frac{4\alpha C_1 T\epsilon^2}{1-\alpha} + 4C_1T\epsilon^2 \le \frac{4C_1T\epsilon^2}{1-\alpha} = K\epsilon^2.$$

(This argument has been constructed to be robust under an increase of C_1 by a factor of 2.)

By Lemma 5.2(iv) it follows that

$$\begin{aligned}\|(\mathcal{T}\Phi)(\theta^1) - (\mathcal{T}\Phi)(\theta^2)\| &\le \alpha\|\Phi(\phi^1) - \Phi(\phi^2)\| + \epsilon^2 C_2|\phi^1 - \phi^2| \\ &\le [\alpha + \epsilon C_2]\epsilon|\phi^1 - \phi^2|.\end{aligned}$$

By Lemma 5.2(iii) and (5.18) it follows that

$$|\theta^1 - \theta^2| \ge |\phi^1 - \phi^2| - C_2\epsilon|\phi^1 - \phi^2|.$$

Thus, for ϵ sufficiently small,

$$|\phi^1 - \phi^2| \le \frac{|\theta^1 - \theta^2|}{1 - C_2\epsilon}.$$

Combining we find that

$$\|(\mathcal{T}\Phi)(\theta^1) - (\mathcal{T}\Phi)(\theta^2)\| \le \frac{[\alpha + \epsilon C_2]\epsilon}{1 - \epsilon C_2}|\theta^1 - \theta^2|.$$

We deduce that, for $\epsilon : 2\epsilon C_2 \le 1 - \alpha$,

$$\|(\mathcal{T}\Phi)(\theta^1) - (\mathcal{T}\Phi)(\theta^2)\| \le \epsilon|\theta^1 - \theta^2|$$

as required.

It remains to establish that $(\mathcal{T}\Phi)(\theta)$ is periodic. If we set $\phi \mapsto \phi + T$ in (5.18) we obtain

$$\begin{aligned}(\mathcal{T}\Phi)(\psi) &= B\Phi(\phi + T) + G(\Phi(\phi + T), \phi + T), \\ \psi &= \phi + 2T + H(\Phi(\phi + T), \phi + T).\end{aligned}$$

Since A, g, h and Φ are T-periodic in θ it follows that $\psi = \theta + T$ and $(\mathcal{T}\Phi)(\psi) = (\mathcal{T}\Psi)(\theta)$ and thus periodicity of $(\mathcal{T}\Phi)(\theta)$ follows. □

Finally we may prove:

Theorem 5.5 Let Assumption 5.1 hold. Then there exists $\epsilon^* > 0$ such that, for all $\epsilon \in (0, \epsilon^*]$, equations (5.13) have an invariant manifold $\Phi \in \Gamma$ satisfying (5.17).

Proof. To show the existence of the invariant manifold for (5.13) it is sufficient, by Lemma 5.4, to show that $\mathcal{T} : \Gamma \mapsto \Gamma$ is a contraction. Consider the equations

$$\begin{aligned}(\mathcal{T}\Phi^i)(\theta) &= B\Phi^i(\phi^i) + G(\Phi^i(\phi^i), \phi^i), \\ \theta &= \phi^i + T + H(\Phi^i(\phi^i), \phi^i)\end{aligned} \tag{5.37}$$

for $i = 1, 2$. It follows that

$$\begin{aligned}&\|(\mathcal{T}\Phi^1)(\theta) - (\mathcal{T}\Phi^2)(\theta)\| \\ &\quad \le \|B\Phi^1(\phi^1) - B\Phi^2(\phi^2)\| + \|G(\Phi^1(\phi^1), \phi^1) - G(\Phi^2(\phi^2), \phi^2)\|. \quad (5.38)\end{aligned}$$

Hence

$$\begin{aligned}\|(\mathcal{T}\Phi^1)(\theta) - (\mathcal{T}\Phi^2)(\theta)\| &\le \|B\Phi^1(\phi^1) - B\Phi^2(\phi^1)\| + \|B\Phi^2(\phi^1) - B\Phi^2(\phi^2)\| \\ &+ \|G(\Phi^1(\phi^1), \phi^1) - G(\Phi^2(\phi^1), \phi^1)\| + \|G(\Phi^2(\phi^1), \phi^1) - G(\Phi^2(\phi^2), \phi^2)\|.\end{aligned}$$

Using Assumption 5.1 and Lemma 5.2, it follows that

$$\|(\mathcal{T}\Phi^1)(\theta) - (\mathcal{T}\Phi^2)(\theta)\| \le (\alpha + C_2\epsilon)\|\Phi^1 - \Phi^2\|_P + (\alpha + C_2\epsilon)\epsilon|\phi^1 - \phi^2|.$$

Also, by similar manipulations, it follows that

$$|\phi^1 - \phi^2| \le C_2\|\Phi^1 - \Phi^2\|_P + C_2\epsilon|\phi^1 - \phi^2|.$$

Combining these two estimates we obtain

$$\|(\mathcal{T}\Phi^1)(\theta) - (\mathcal{T}\Phi^2)(\theta)\| \le (\alpha + C_2\epsilon)\|\Phi^1 - \Phi^2\|_P + \frac{(\alpha + C_2\epsilon)}{(1 - C_2\epsilon)}\epsilon C_2\|\Phi^1 - \Phi^2\|_P.$$

Again, choosing ϵ sufficiently small, we obtain

$$\|(\mathcal{T}\Phi^1)(\theta) - (\mathcal{T}\Phi^2)(\theta)\| \le \tfrac{(}{1} + \alpha)\|\Phi^1 - \Phi^2\|_P.$$

Since this holds for all θ it follows that the mapping $\mathcal{T}$ is contractive on Γ with constant $(1+\alpha)/2 < 1$ and existence of an invariant manifold follows. □

Given ϵ^* and Φ from Theorem 5.5, it is a corollary that equations (1.1) have a periodic solution:

Corollary 5.6 (**Periodic solutions**) Assume that there exists a C^3 diffeomorphism $\chi : \mathbb{R}^p \mapsto \mathbb{R}^p$ which renders (1.1) in the form (5.2) under $u = \chi(v)$; assume further that there exists $\epsilon \in (0, \epsilon^*/2]$ such that A, g and h satisfy (5.4) and (5.9) for all $r \in B(0, \epsilon)$, all $\theta \in \mathbb{R}$ and all $t \in \mathbb{R}$. Then (1.1) has a periodic solution $\bar{u}(t)$ comprising the sets of points

$$\{u \in \mathbb{R}^p | u = \chi(v),\ v = (\Phi(\theta)^T, \theta)^T : \theta \in \mathbb{R},\ \Phi \in \Gamma\}.$$

Proof. By virtue of the coordinate tranformation, it is sufficient to show that the invariant manifold Φ constructed in Thoerem 5.5 yields a periodic solution of (5.2). To do this we introduce some notation. Let $\bar{S}(t)$ denote the semigroup for $v(t) = (r(t)^T, \theta(t))^T$ given by the solution of (5.2). Thus $v(t) = \bar{S}(t)(\xi^T, \phi)^T$. Let

$$\mathcal{M} := \{(r^T, \theta)^T \in \mathbb{R}^p : r = \Phi(\theta)\}.$$

Now consider the set

$$\mathcal{M}(t) := \bar{S}(t)\mathcal{M}.$$

Thus $\mathcal{M}(t)$ is obtained by taking every point in $\mathcal{M}$ and evolving it forward t time units under equation (5.2), that is under $\bar{S}(t)$. It is our aim to show that $\mathcal{M}(t) \equiv \mathcal{M}$ for all t sufficiently small; then we deduce that the closed curve given by the invariant manifold Φ of Theorem 5.5, which is invariant under the time T evolution of the differential equation (5.2), is actually invariant under the evolution of (5.2) for any time t sufficiently small. This shows that $\mathcal{M}$ is a periodic solution, since it is a closed curve.

Note that, from (5.12) and (5.14), every point $(r^T, \theta)^T \in \mathcal{M}(t)$ satisfies

$$\begin{aligned} r &= B(0,t;\phi)\Phi(\phi) + G(\Phi(\phi),\phi;t), \\ \theta &= \phi + t + H(\Phi(\phi),\phi;t) \end{aligned} \tag{5.39}$$

for some $\phi \in \mathbb{R}$. Analogously to the proof of Lemma 5.3, we may show that there is a graphical relationship between r and θ so that $r = \Psi(\theta)$ for some $\Psi \in C(\mathbb{R}, \mathbb{R}^{p-1})$. Furthermore, use of (5.22) and (5.34) show that there exists $C_3 > 0$ such that

$$\|r\| = \|\Psi(\theta)\| \leq (1 + C_3 t)K\epsilon^2$$

and

$$\|\Psi(\theta_1) - \Psi(\theta_2)\| \leq (1 + C_3 t)\epsilon|\theta_1 - \theta_2|.$$

Hence, if t is chosen sufficiently small that $(1 + C_3 t) \leq 2$ and ϵ sufficiently small that $\epsilon \in (0, \epsilon^*/2]$, then we deduce that $\Psi \in \Gamma$.

Finally, observe that

$$\bar{S}(T)\mathcal{M}(t) = \bar{S}(T)\bar{S}(t)\mathcal{M} = \bar{S}(t)\bar{S}(T)\mathcal{M}.$$

But, $\mathcal{M}$ is invariant under $\bar{S}(T)$ by construction and hence we have

$$\bar{S}(T)\mathcal{M}(t) = \bar{S}(t)\mathcal{M} = \mathcal{M}(t).$$

Hence $\mathcal{M}(t)$ is invaraint under $\bar{S}(T)$. Since all points in $\mathcal{M}(t)$ may be represented by means of a graph Ψ lying in Γ the uniqueness implied by Theorem 5.5 gives $\Psi = \Phi$. Hence $\mathcal{M}(t) = \mathcal{M}$ and the proof is complete. □

5.2. Periodic solutions

We now modify the analysis of Section 1 to prove the existence and convergence of a closed invariant curve for the numerical method (1.2) which lies close to the periodic solution $\bar{u}(t)$ of (1.1). Let $m\Delta t = T$ and

$$U_m = S^m_{\Delta t}U, \quad u(T) = S(T)U$$

denote the solutions of (1.2), (1.1) respectively subject to the same initial condition U. We know, from Theorem 3.8, that there exists $C_5 = C_5(T)$ such that

$$\begin{aligned} \|S^m_{\Delta t|\Delta t}U - S(T)U\| &\leq C_5\Delta t^r, \quad \forall U \in \mathbb{R}^p, \\ \|\mathrm{d}S^m_{\Delta t}(U) - \mathrm{d}S(U;T)\| &\leq C_5\Delta t^r, \quad \forall U \in \mathbb{R}^p. \end{aligned} \tag{5.40}$$

We define $v(t)$ and V_n by $u(t) = \chi(v(t))$ and $U_n = \chi(V_n)$ where χ is the C^3 diffeomorphism given in Corollary 5.6. We also set

$$V = (\xi^T, \phi)^T \in \mathbb{R}^p, \quad V_m = (R^T, \Theta)^T \in \mathbb{R}^p$$

We may define semigroups appropriate for the variables $v(t)$ and V_n from those in the original variables by

$$\begin{aligned} \bar{S}(t)v &:= \chi^{-1}(S(t)\chi(v)), \\ \bar{S}^n_{\Delta t}v &:= \chi^{-1}(S^n_{\Delta t}\chi(v)). \end{aligned} \tag{5.41}$$

Then, since χ is a C^3 diffeomorphism it follows from (5.40) and (5.41) that there exists $C_6 = C_6(T)$ such that

$$\begin{aligned} \|\bar{S}^m_{\Delta t}V - \bar{S}(T)V\| &\leq C_6\Delta t^r, \quad \forall V \in \mathbb{R}^p : \xi \in B(0,\epsilon) \\ \|\mathrm{d}\bar{S}^m_{\Delta t}(V) - \mathrm{d}\bar{S}(V;T)\| &\leq C_6\Delta t^r, \quad \forall V \in \mathbb{R}^p : \xi \in B(0,\epsilon). \end{aligned} \tag{5.42}$$

(Here the derivatives of the semigroups $\bar{S}^m_{\Delta t}$ and $\bar{S}(t)$ are defined analogously to those for $S^m_{\Delta t}$ and $S(t)$.)

We now exploit the existence theory derived for the periodic solutions of (5.2) to study the numerical method. From (5.13) and (5.42) we obtain

$$\begin{aligned} R &= B\xi + \tilde{G}(\xi,\phi), \\ \Theta &= \phi + T + \tilde{H}(\xi,\phi), \end{aligned} \tag{5.43}$$

where

$$\begin{aligned} \|\tilde{G}(\xi,\phi) - G(\xi,\phi)\| &\leq C_6\Delta t^r, \\ \|\tilde{H}(\xi,\phi) - H(\xi,\phi)\| &\leq C_6\Delta t^r, \\ \|\mathrm{d}\tilde{G}(\xi,\phi) - \mathrm{d}G(\xi,\phi)\| &\leq C_6\Delta t^r, \\ \|\mathrm{d}\tilde{H}(\xi,\phi) - \mathrm{d}H(\xi,\phi)\| &\leq C_6\Delta t^r. \end{aligned} \tag{5.44}$$

Here $\mathrm{d}G$, $\mathrm{d}\tilde{G}$, $\mathrm{d}H$ and $\mathrm{d}\tilde{H}$ denote the Jacobians of G and $\tilde{G}$ with respect to V.

For simplicity we assume that (5.44) holds for all $v \in \mathbb{R}^p$; since all the analysis takes place within the ball $\xi \in B(0,\epsilon)$ the results hold for any numerical approximation satisfying (5.42). Using (5.44), together with Lemma 5.2, it follows that for Δt sufficiently small, $\tilde{G}$ and $\tilde{H}$ satisfy estimates analogous to those appearing in Lemma 5.2 for G and H but with $C_1 \mapsto 2C_1$ and $C_2 \mapsto 2C_2$. Hence, following the proofs of Lemmas 5.3, 5.4 and Theorem 5.5 with $C_1 \mapsto 2C_1$ and $C_2 \mapsto 2C_2$, we can prove the existence of an invariant manifold $\Phi_{\Delta t} \in \Gamma$ for (5.43). Specifically we seek a $\Phi_{\Delta t}$ with the property that

$$\xi = \Phi_{\Delta t}(\phi) \Leftrightarrow R = \Phi_{\Delta t}(\Theta). \tag{5.45}$$

Comparison with (5.39) shows that to do this is equivalent to finding a fixed point of the mapping $\mathcal{T}_{\Delta t}$ defined by

$$\begin{aligned}(\mathcal{T}_{\Delta t}\Phi)(\Theta) &= B\Phi(\phi) + \tilde{G}(\Phi(\phi), \phi),\\ \Theta &= \phi + T + \tilde{H}(\Phi(\phi), \phi).\end{aligned} \tag{5.46}$$

Using this formulation we can prove the following theorem:

Theorem 5.7 Let Assumption 5.1 hold. Then there exists $\Delta t_c > 0$ and $\epsilon^{**} > 0$ such that, for all $\Delta t \in (0, \Delta t_c]$ and $\epsilon \in (0, \epsilon^{**}]$ equations (5.43) have an invariant manifold $\Phi_{\Delta t} \in \Gamma$, satisfying (5.45). Furthermore, $\Phi_{\Delta t}$ is close to the graph Φ constructed in Theorem 5.5 in the sense that there exists $C_7 = C_7(T)$ such that

$$\|\Phi - \Phi_{\Delta t}\| \le C_7 \Delta t^r.$$

Proof. The existence of such a manifold follows precisely as in the proof of Theorem 5.5, except that C_2 is enlarged by a factor of 2. Hence further reduction of ϵ is necessary in the proof.

To prove closeness of the manifolds we again use, essentially, the uniform contraction principle; note that both Φ and $\Phi_{\Delta t}$ lie in Γ. Thus, using the contractivity of $\mathcal{T}_{\Delta t}$ we obtain

$$\begin{aligned}\|\Phi(\theta) - \Phi_{\Delta t}(\theta)\| &= \|\mathcal{T}\Phi(\theta) - \mathcal{T}_{\Delta t}\Phi_{\Delta t}(\theta)\|\\ &\le \|\mathcal{T}\Phi(\theta) - \mathcal{T}\Phi_{\Delta t}(\theta)\| + \|\mathcal{T}\Phi_{\Delta t}(\theta) - \mathcal{T}_{\Delta t}\Phi_{\Delta t}(\theta)\|\\ &\le \tfrac{1}{2}(1+\alpha)\|\Phi - \Phi_{\Delta t}\|_P + \|\mathcal{T}\Phi_{\Delta t}(\theta) - \mathcal{T}_{\Delta t}\Phi_{\Delta t}(\theta)\|.\end{aligned}$$

Now, from (5.18), (5.46) and (5.44) and the Lipschitz properties of $\mathcal{T}_{\Delta t}\Phi \in \Gamma$, we deduce that

$$E := \|\mathcal{T}\Phi_{\Delta t}(\theta) - \mathcal{T}_{\Delta t}\Phi_{\Delta t}(\theta)\|$$

satisfies

$$E \le \|\mathcal{T}\Phi_{\Delta t}(\theta) - \mathcal{T}_{\Delta t}\Phi_{\Delta t}(\Theta)\| + \|\mathcal{T}_{\Delta t}\Phi_{\Delta t}(\theta) - \mathcal{T}_{\Delta t}\Phi_{\Delta t}(\Theta)\|.$$

Hence, since $\mathcal{T}_{\Delta t}\Phi_{\Delta t} \in \Gamma$, we have from (5.18), (5.43) and (5.44) that

$$E \le C_6(1+\epsilon)\Delta t^r.$$

Thus, putting these estimates together we obtain

$$\|\Phi - \Phi_{\Delta t}\|_P \le \frac{2C_6(1+\epsilon)}{1-\alpha}\Delta t^r.$$

This completes the proof. □

As a corollary of Theorem 5.7 we consider the approximation of periodic solutions in (1.1) by closed invariant curves in (1.2). Recall the notions (3.1), (3.2) of distance.

Corollary 5.8 (**Periodic solutions under approximation**) Assume that (1.1) has a hyperbolic, stable periodic solution $\bar{u}(t)$ comprising the set of points $\mathcal{P}$. Then (1.2) has closed invariant curve comprising the set of points $\mathcal{P}_{\Delta t}$ and, furthermore, there exists a constant $C > 0$:

$$d_{\rm H}(\mathcal{P}_{\Delta t}, \mathcal{P}) \leq C\Delta t^r.$$

Proof. If (1.1) has a hyperbolic, stable periodic solution then a transformation χ exists rendering (1.1) in the form to which Corollary 5.6 applies. Application of Theorem 5.7 yields the required closeness result. □

5.3. *Bibliography*

The standard local construction of periodic solutions uses the Hopf bifurcation which facilitates the construction of a periodic solution branching from an equilibrium solution; see Guckenheimer and Holmes (1983) and Drazin (1992). For background material describing global questions concerning the existence of periodic solutions in (1.1) see, for example, Hale (1969), Hartman (1982) and Guckenheimer and Holmes (1983). In particular, Hale (1969) presents the full details of the coordinate transformation $u = \chi(v)$ which is central to the analysis described here.

The first article to study the effect of numerical approximation on periodic solutions in a general context was Braun and Hershenov (1977). They studied stable hyperbolic periodic solutions and employed the time T map to perform the analysis (where T is the period). The next article to address such questions was Brezzi *et al.* (1984) where the existence of Hopf bifurcation points, and the resulting closed invariant curves close to the Hopf point, was studied for numerical methods (1.2) operating in a parameter regime close to that in which a Hopf bifurcation occurs in (1.1). The work of Braun and Hershenov (1977) was generalized in Doan (1985) to encompass multistep methods and the general case of hyperbolic periodic solutions which are not necessarily stable. Neither of the articles (Braun and Hershenov, 1977) nor (Doan, 1985) obtained precise orders of convergence for the approximate invariant curve. A little later Beyn (1987a) generalized the work of Braun and Hershenov (1977) to encompass arbitrary hyperbolic periodic solutions, obtaining precise orders of convergence; his approach is similar to that in Braun and Hershenov (1977) but, rather than employing the time T map he uses the time Δt map of the differential equation. The whole subject area was put in a very clear setting in Eirola (1988; 1989) where results similar to those of Beyn derived but in the stronger C^k topology, the value of k depending on the smoothness of the vector field $f(\cdot)$ in (1.1). In turn the work of Eirola can be viewed in a very general setting concerning the stability of invariant circles of mappings in $\mathbb{R}^p$; see Pugh and Shub (1988). The proof given in this article is closely related to the proofs in Braun and Hershenov (1977) and Beyn (1987a); it is far from optimal in the sense that only stable periodic solutions are considered and the result is in the C^0 topology. However, this presentation has been chosen because it is self-contained, relatively simple and is similar to the method of analysis used to construct unstable manifolds in Section 4. A recent article (Alouges and Debussche, 1993) is concerned with extensions of the work referenced here to partial differential equations. In this context the work of Titi (1991) is also of interest.

We have not described here an analogous theory for the behaviour of invariant tori under numerical approximation. Such a theory has recently been developed in Lorenz (1994) using the approach of Fenichel (1971) to determine an appropriate coordinate system analogous to the coordinate system described in Hale (1969) used to study periodic solutions.

6. Uniform asymptotic stability and attractors

6.1. Background theory

In this section we consider the effect of numerical approximation on general objects which are attracting, in certain senses to be made precise, for solutions of (1.1). In most cases we do not make specific statements about the nature of the dynamics within the attracting object so that our framework will be sufficiently general to include, for example, the strange attractors observed in the Lorenz equations (2.10). Thus our assumptions will be the existence of an arbitrary compact set possessing some form of attractivity. Up to this point we have considered the numerical approximation of equilibrium points, invariant sets in the neighbourhood of equilibrium points and periodic solutions. In all cases our methodology has been the same: we have employed the contraction mapping theorem to develop an existence theory for the object in question and then used the uniform contraction principle to incorporate the effect of numerical approximation. In this section a different approach will be necessary since there is no known existence theory based on the contraction mapping theorem which we can exploit.

We start by defining the type of objects of interest to us, together with certain of their properties which we require. In this section, all of our definitions and theorems concern the continuous semigroup $S(t)$. At the end of the section we detail which definitions and theorems can be generalized from $S(t)$ to $S^n_{\Delta t}$.

Definition 6.1 A set A *attracts* a set B under $S(t)$ if, for any $\epsilon > 0$, there exists $t^* = t^*(\epsilon, B, A)$ such that $S(t)B \subset \mathcal{N}(A, \epsilon)$ $\forall t \geq t^*$. A compact invariant set A is said to be a an *attractor* if A attracts an open neighbourhood of itself. A *global attractor* is an attractor which attracts every bounded set in $\mathbb{R}^p$.

Example Define the set $\mathcal{A} := (x, y) \in [-1, 1] \times \{0\}$ and note that $\mathcal{A}$ is clearly compact; by arguments similar to those used to establish the example following Definition 3.1, we deduce that $\mathcal{A}$ is invariant under the differential equations

$$x_t = x - x^3, \quad x(0) = x_0$$

$$y_t = \lambda y, \quad y(0) = y_0.$$

Denote the semigroup for the x-equation by $S_x(t)$ and for the y-equation by $S_y(t)$. By (3.8), the solution of these equations has the property that $S_x(t)[-a,a] \to [-1,1]$ as $t \to \infty$, for any $a > 0$; furthermore, $S_x(t)0 = 0$. Clearly $S_y(t)y_0 = \mathrm{e}^{\lambda t}y_0$.

It follows that, for any $\lambda \geq 0$, $\mathcal{A}$ attracts any set $(x,y) \in [-a,a] \times \{0\}, a \geq 0$. Furthermore, if $\lambda < 0$ then $\mathcal{A}$ is an attractor – since then $[-a,a] \times [-\epsilon, \epsilon]$ is attracted to $\mathcal{A}$ for any $a, \epsilon \geq 0$. Indeed, since $a, \epsilon \geq 0$ are arbitrary, it is a global attractor. □

Attractors are often constructed by applying the following theorem.

Theorem 6.2 Assume that $B \subset \mathbb{R}^p$ is a bounded open set such that $S(t)\bar{B} \subset B \; \forall t > 0$. Then $\omega(B)$ is an attractor which attracts B.

Proof. Since $S(t)\bar{B} \subset B$ it follows that

$$\omega(B) = \bigcap_{s \geq 0} \overline{\bigcup_{t \geq s} S(t)B} \subset \bigcap_{s \geq 0} \overline{\bigcup_{t \geq s} B} = \bar{B}. \tag{6.1}$$

Thus $\omega(B)$ is bounded. Furthermore, $\omega(B)$ is closed and invariant by Theorem 3.3 and it follows that $\omega(B)$ is a compact invariant set.

We now show that $\mathcal{A} := \omega(B)$ attracts B. Assume that it does not. Then, for all sufficiently small $\epsilon > 0$, there does not exist t^* such that $S(t)B \subset \mathcal{N}(\mathcal{A}, \epsilon) \; \forall \; t \geq t^*$. Thus there exists $x_k \in B$ and $t_k \to \infty$ such that $S(t_k)x_k \not\subset \mathcal{N}(\mathcal{A}, \epsilon)$. But $S(t_k)x_k$ is a bounded sequence contained in B and hence has a convergent subsequence $S(t_{k_i})x_{k_i} \to y \in \overline{B}$. By Definition 3.2 $y \in \mathcal{A}$ and this is a contradiction.

Note that $\omega(B) \subset \bar{B}$ by (6.1). We show that, in fact, $\omega(B) \subset B$. Assume for the purposes of contradiction that $\exists y \in \omega(B) \cap \partial B$ (where $\partial B = \bar{B} \backslash B$). Since $\omega(B)$ is invariant it follows that, for any $t > 0 \; \exists x \in \omega(B) : S(t)x = y$. But, since $\omega(B) \subset \bar{B}$ we have $x \in \bar{B}$ and hence, by assumption, $y = S(t)x \in B$. This is a contradiction and thus no such y exists. Thus $\omega(B) \subset B$.

Now, since $\omega(B) \subset B$ is closed it follows that, for ϵ sufficiently small, $\mathcal{N}(\omega(B), \epsilon) \subset B$. Since $\omega(B)$ attracts B it follows that $\omega(B)$ attracts an open neighbourhood of itself and the proof is complete. □

Example Consider the equation $u_t = u - u^3$. Let $B = (-a, a), a > 1$. Then $\omega(B) = [-1, 1]$ as shown in the second example following Definition 3.2; furthermore, $\omega(B)$ is an attractor: on $|u| = a$ we have $\mathrm{d}|u|^2/\mathrm{d}t < 0$ and hence $S(t)\bar{B} \subset B \; \forall t > 0$. Theorem 6.2 gives the desried result. □

A second attracting object of interest to us is now defined:

Definition 6.3 A compact set Λ is *uniformly stable* if for each $\epsilon > 0$ $\exists \delta = \delta(\epsilon) > 0$ such that $\text{dist}(U, \Lambda) < \delta \Rightarrow \text{dist}(S(t)U, \Lambda) < \epsilon$ $\forall t \geq 0$; a compact set Λ is *asymptotically stable* if there exists $\delta_0 > 0$ and for each ϵ, a $T = T(\epsilon)$ such that $\text{dist}(U, \Lambda) < \delta_0$ implies that $\text{dist}(S(t)U, \Lambda) < \epsilon \quad \forall t \geq T$. A compact set Λ is *uniformly asymptotically stable* if it is uniformly stable and asymptotically stable.

Theorem 6.4 The following properties hold for uniformly asymptotically stable sets:

(i) uniformly stable sets are positively invariant;
(ii) an attractor is uniformly asymptotically stable;
(iii) if Λ is uniformly asymptotically stable then $\mathcal{A} = \omega(\Lambda) \subseteq \Lambda$ is an attractor.

Proof. We first prove (i). For contradiction assume that Λ is uniformly stable and not positively invariant. Then $\exists \tau > 0, \epsilon > 0$ and $U \in \Lambda$ such that $\text{dist}(S(\tau)U, \Lambda) > \epsilon$. But, since the set is uniformly stable it follows that $\exists \delta = \delta(\epsilon)$ such that $\text{dist}(U, \Lambda) < \delta \Rightarrow \text{dist}(S(\tau)U, \Lambda) < \epsilon$. Since $\text{dist}(U, \Lambda) = 0$ this gives the required contradiction.

Now consider (ii). It is automatic that an attractor is asymptotically stable since it attracts a neighbourhood of itself. Thus it suffices to show uniform stability. Assume for contradiction that $\mathcal{A}$ is an attractor, attracting a neighbourhood W, but it is not uniformly stable. Thus, for any $\epsilon > 0$ there exists a sequence of times $\{t_j\}_{j=1}^{\infty}$ and a sequence $\{x_j\}_{j=1}^{\infty}$ with $x_j \in W$ for each j and $x_j \to x \in \mathcal{A}$, such that $S(t)x_j \in \mathcal{N}(A, \epsilon), t \in [0, t_j)$ and $S(t_j)x_j \notin \mathcal{N}(\mathcal{A}, \epsilon)$. Now let $H = \{x, \{x_j\}_{j=1}^{\infty}\}$ and note that, since $x_j \in W$ (a bounded set) and since $x_j \to x$, we have that $H \subset W$ is compact. Since $\mathcal{A}$ attracts W it follows that $\mathcal{A}$ attracts H and hence that $\omega(H) \subset \mathcal{A}$. We deduce that, for any $T > 0$, the sequence $S(t_j - T)x_j \to z \in \omega(H) \subset \mathcal{A}$. But $S(T)S(t_j - T)x_j = S(t_j)x_j \to S(T)z$. By the invariance of ω-limit sets (see Theorem 3.3) it follows that $S(T)z \in \mathcal{A}$. However, $S(T)z \notin \mathcal{N}(\mathcal{A}, \epsilon)$ since $S(t_j)x_j \notin \mathcal{N}(\mathcal{A}, \epsilon)$ and this gives the required contradiction.

Finally we prove (iii). Since Λ is postively invariant it follows that $\mathcal{A} = \omega(\Lambda) \subseteq \Lambda$ and hence $\mathcal{A}$ is bounded; $\mathcal{A}$ is closed by Theorem 3.8 and hence compact. Let $H = \mathcal{N}(\Lambda, \delta_0)$ where δ_0 is given by the definition of asymptotic stability. To show that $\mathcal{A}$ attracts H it is sufficient to show that $\omega(H) \subseteq \mathcal{A}$. Clearly $\omega(H) \subseteq \Lambda$ since Λ is asymptotically stable. Let $y \in \omega(H)$. Since $\omega(H)$ is invariant (by Theorem 3.8) it follows that, for each $t_k > 0$ there exists $x_k \in \omega(H) \subseteq \Lambda$ such that $S(t_k)x_k = y$. Thus the sequence $S(t_k)x_k \to y$ as $k \to \infty$. But, since $x_k \in \Lambda$ it follows that $y \in \omega(\Lambda) = \mathcal{A}$ and the result follows. □

Thus the important fact distinguishing attractors and uniformly asymptotically stable sets is that the former are necessarily invariant whilst the latter need only be positively invariant. The following example illustrates this.

Examples Consider the equation $u_t = -u$, $u(0) = U$. Any interval $[-a, a], a \geq 0$ is a uniformly asymptotically stable set. However, only the point 0 is an attractor since $[-a, a]$ is not invariant for $a > 0$.

It is well known that the existence of stable equilibrium points can be deduced from construction of appropriate Lyapunov functions. Converse results are also available and the following result, deducing the existence of a Lyapunov function from the uniform asymptotic stability of a compact set, will be extremely useful.

Theorem 6.5 Given a compact uniformly asymptotically stable set Λ there exists $R_0 > 0$ and a Lyapunov function $V : \mathcal{N}(\Lambda : R_0) \to \mathbb{R}^+$ satisfying the following three properties:

(i) there exists $L > 0 : |V(x) - V(y)| \leq L\|x - y\| \quad \forall x, y \in \mathcal{N}(\Lambda, R_0)$;
(ii) there exist continuous, strictly increasing functions α, β: $[0, R_0) \to \mathbb{R}^+$ with $\alpha(0) = \beta(0) = 0$ and $\alpha(r) < \beta(r)$ for $r > 0$ such that

$$\alpha(\mathrm{dist}(x, \Lambda)) \leq V(x) \leq \beta(\mathrm{dist}(x, \Lambda));$$

(iii) there exists a constant $C > 0$ such that

$$V(S(t)U) \leq \mathrm{e}^{-Ct} V(U), \; 0 \leq t \leq T$$

provided $S(t)U \in \mathcal{N}(\Lambda, R_0), \; 0 \leq t \leq T$.

Proof. This theorem is proved in Theorem 22.5 of Yoshizawa (1966) with a slightly different conclusion in part (iii); that our point (iii) holds may be deduced from Theorem 4.1 of Yoshizawa (1966). □

Corollary 6.6 Let $\beta(R_1) = \alpha(R_0)$ and assume that $\mathrm{dist}(U, \Lambda) < R_1$. Then

$$S(t)U \in \mathcal{N}(\Lambda, R_0) \quad \text{and} \quad V(S(t)U) \leq \mathrm{e}^{-Ct} V(U) \quad \forall t \in [0, \infty). \tag{6.2}$$

Proof. For the purpose of contraction assume that there exists $T > 0$:

$$\mathrm{dist}(S(T)U, \Lambda) < R_0, \quad t \in [0, T) \quad \text{and} \quad \mathrm{dist}(S(T))U, \Lambda) = R_0.$$

By Theorem 6.5(iii) it follows that $V(S(t)U) \leq \mathrm{e}^{-Ct} V(U), t \in [0, T]$. From Theorem 6.5(ii) we have that

$$\begin{aligned} \alpha(R_0) &= \alpha(\mathrm{dist}(S(T)U, \Lambda)) \\ &\leq V(S(T)U) \leq \mathrm{e}^{-CT} V(U) \\ &\leq \mathrm{e}^{-CT} \beta(\mathrm{dist}(U, \Lambda)) \\ &\leq \mathrm{e}^{-CT} \beta(R_1) = \mathrm{e}^{-CT} \alpha(R_0). \end{aligned}$$

This is a contradiction, hence no such T exists and $S(t)U \in \mathcal{N}(\Lambda, R_0)$ $\forall t \geq 0$. Theorem 6.5(iii) then completes the proof. $\Box$

We now discuss structural assumptions on the vector field $f(\cdot)$ in (1.1) which yield the existence of attractors and uniformly asymptotically stable sets.

First we consider the assumption

$$\exists \epsilon, R > 0 : \langle f(u), u \rangle \leq -\epsilon \quad \forall u : \|u\| = R. \tag{6.3}$$

(Such an assumption holds, for example, for (1.1) under (2.13) for any $R \geq (a+\epsilon)/b$ although (2.13) cannot hold for globally bounded vector fields.) We may now prove:

Theorem 6.7 Consider (1.1) under (6.3) and let $B = \{u \in \mathbb{R}^p : \|u\|^2 < R\}$. Then the semigroup $S(t)$ has an attractor given by $\mathcal{A} = \omega(B)$.

Proof. By Theorem 6.2 it is sufficient to show that $S(t)\bar{B} \subset B$ $\forall t > 0$. From (6.3) we have,

$$\frac{1}{2}\frac{\mathrm{d}}{\mathrm{d}t}\|u\|^2 = \langle f(u), u \rangle, \quad \forall u \in \mathbb{R}^p \tag{6.4}$$

and hence, for $\partial B = \bar{B} \backslash B$,

$$\frac{1}{2}\frac{\mathrm{d}}{\mathrm{d}t}\|u\|^2 = \langle f(u), u \rangle \leq -\epsilon, \quad \|u\| \in \partial B.$$

This shows that trajectories on ∂B point into B and establishes the required property of B; the result follows. $\Box$

A second class of systems of interest to us are gradient systems. Recall the definition $\mathcal{E} = \{v \in \mathbb{R}^p : f(v) = 0\}$ from Section 1.

Definition 6.8 The dynamical system $S(t)$ generated by (1.1) is said to define a *gradient system* if $\exists F \in C(\mathbb{R}^p, \mathbb{R})$, called a *Lyapunov function*, satisfying

(i) $F(u) \geq 0$ for all $u \in \mathbb{R}^p$;
(ii) $F(u) \to \infty$ as $\|\mathbf{u}\| \to \infty$;
(iii) for a solution of (1.1) $F(S(t)U)$ is nonincreasing in t;
(iv) if $F(S(t)U) = F(U)$ for $t > 0$ then $U \in \mathcal{E}$.

A particular case where gradient systems arise is when $\mathbf{f}$ is a gradient vector field, so that

$$f(u) = -\nabla F(u). \tag{6.5}$$

With this assumption, taking the inner-product in (1.1) with u_t yields

$$\frac{\mathrm{d}}{\mathrm{d}t}\{F(u(t))\} = -\|u_t(t)\|^2. \tag{6.6}$$

Hence Properties (iii) and (iv) of Definition 6.8 hold; thus, if F satisfies (i) and (ii) then (1.1), (6.5) defines a gradient system. The following theorem shows that the dynamics of a gradient system must be relatively simple.

Theorem 6.9 If (1.1) is a gradient system then $\omega(U) \subseteq \mathcal{E}$. If, furthermore, the zeros of f are isolated and $\omega(U)$ is bounded then $\omega(U) = x$ for some $x \in \mathcal{E}$.

Proof. Now let x, y be two points in $\omega(U)$. Thus, without loss of generality, there exist sequences t_i and τ_i with $\tau_{i-1} < t_i < \tau_i$ such that

$$S(t_i)U \to x, \quad S(\tau_i)U \to y.$$

By Definition 6.8(iii) we have

$$F(S(\tau_i)U) \leq F(S(t_i)U) \leq F(S(\tau_{i-1})U).$$

Hence, by continuity, we deduce that $F(x) = F(y)$. Now, since $\omega(U)$ is positively invariant by Theorem 3.3 we have that for any $x \in \omega(U)$ and $t > 0$, $y = S(t)x \in \omega(U)$. But $F(x) = F(y)$ and, by Definition 6.8(iv), we deduce that $x \in \mathcal{E}$ yielding the first result.

By Theorem 3.3 we know that, if $\omega(U)$ is bounded it is connected. Since $\omega(U) \in \mathcal{E}$ and $\mathcal{E}$ comprises isolated points it follows that $\omega(U)$ is a single point $x \in \mathcal{E}$. □

Theorem 6.10 Consider a gradient dynamical system $S(t)$ generated by (1.1), 6.5. Assume that $F(u)$ has the property that

$$\exists \xi > 0 : v \in \mathcal{E} \Rightarrow F(v) \leq \xi.$$

Then the set $\overline{B(R)}$, where

$$B(R) := \{u \in \mathbb{R}^p : F(u) < R\},$$

is uniformly asymptotically stable for any $R > \xi$ and, furthermore, $\mathcal{A} = \omega(B(R))$ is a global attractor for any $R > \xi$.

Proof. Let $R > \xi$. First observe that $\overline{B(R)}$ is a closed bounded set because of Definition 6.8(ii); hence it is compact. Note that, for any $\epsilon > 0$, we may choose $r > 0$ such that

$$F(x) < r \Rightarrow \operatorname{dist}(x, B(R)) < \epsilon. \tag{6.7}$$

We may also choose $\delta > 0$ such that

$$\operatorname{dist}(x, B(R)) < \delta \Rightarrow F(x) < r. \tag{6.8}$$

Now, by (6.8), if $\operatorname{dist}(U, B(R)) < \delta$ then $F(U) < r$. By Definition 6.8(iii) we have $F(S(t)U) \leq F(U) < r$. Hence, by (6.7) we have $\operatorname{dist}(S(t)U, B(R)) < \epsilon$ as required to establish uniform stability (see Definition 6.3). Furthermore, choose δ_0 such that

$$\text{dist}(x, B(R)) < \delta_0 \Rightarrow F(x) < R + l.$$

Then let

$$\eta(l) := \inf_{\{x: R \leq F(x) \leq R+l\}} |f(x)| > 0; \tag{6.9}$$

the strict positivity follows since there are no equilibria with $F(x) \geq R$ and the set

$$\{x \in \mathbb{R}^p : R \leq F(x) \leq R + l\}$$

is compact. If $\text{dist}(U, B(R)) < \delta_0$ then $F(S(t)U) < R+l \ \forall t \geq 0$ by Definition 6.8(iii). Assume, for contradiction, that $F(S(t)U) \geq R \ \forall t > 0$. Then, from (6.6), we obtain

$$F(S(t)U) - (R + l) \leq -t\eta(l)^2 \tag{6.10}$$

yielding a contradiction for $t > l/\eta(l)^2$. Hence there exists $T \in (0, l/\eta(l)^2]$ such that $F(S(t)U) \leq R$ for all $t \geq T$. Thus $\text{dist}(S(t)U, B(R)) = 0 \ \forall t \geq T$ and asymptotic stability (see Definition 6.3) is proved, establishing uniform asymptotic stability.

Finally, (6.6) and (6.9) show that

$$F(S(t)U) = R \Rightarrow \frac{\mathrm{d}}{\mathrm{d}t}\{F(S(t)U)\} \leq -\eta(0)^2$$

so that trajectories on the boundary of $B(R)$ point into the set and, hence, by Theorem 6.2, $\omega(B(R))$ is an attractor. Since l is arbitrary and $B(R)$ is attracted to $\mathcal{A}$, we deduce from (6.9) and (6.10) that the bounded set $B(R+l)$ is attracted to $B(R)$ in finite time $T \leq l/\eta(l)^2$. Since l is arbitrary this gives the required global attraction. □

Recall Definition 4.1 of the unstable manifold given in Section 3. The following theorem elucidates the structure of the attractor for gradient systems.

Theorem 6.11 If $\mathcal{A}$ is a compact global attractor then it comprises all solutions of (1.1) which exist and are uniformly bounded for all $t \in \mathbb{R}$ Under the same assumptions as Theorem 6.10, $S(t)$ has a global attractor $\mathcal{A}$ given by

$$\mathcal{A} = \{U \in \mathbb{R}^p : \text{dist}(u(t), \mathcal{E}) \to 0 \quad \text{as} \quad t \to -\infty\}.$$

Furthermore, if $\mathcal{E}$ comprises only hyperbolic equilibrium points then

$$\mathcal{A} = \bigcup_{v \in \mathcal{E}} W^u(v).$$

Proof. The proof of the first part of this result may be found in Hale *et al.* (1984) and the remainder in Hale (1988). □

Example Consider the equation $u_t = u - u^3$. This is a gradient system with Lyapunov function $F(u) = (1 - u^2)^2/4$. The equilibria are the points $\{0, +1, -1\}$ and all are hyperbolic since $f'(0) = 1$, $f'(\pm 1) = -2$. Furthermore, the points ± 1 are stable whilst the point 0 is unstable; by (3.8) it follows that the unstable manifold of 0 is $(-1, 1)$. Thus, by Theorem 6.11, the global attractor is the set $\mathcal{A} = [-1, 1]$. This result is an agreement with the construction of $\mathcal{A}$ in the second example following Definition 3.2. □

We will show in Section 6.3 that if $S(t)$ has an attractor $\mathcal{A}$ then $S^n_{\Delta t}$ has an attractor $\mathcal{A}_{\Delta t}$ satisfying

$$\text{dist}(\mathcal{A}_{\Delta t}, \mathcal{A}) \to 0 \quad \text{as} \quad \Delta t \to 0.$$

This shows that, in the limit as $\Delta t \to 0$, every point on the numerical attractor is close to a point on the true attractor and is known as *upper semicontinuity*. We will also show that, in general, the converse is not true – we will only be able to prove *lower-semicontinuity*, namely that

$$\text{dist}(\mathcal{A}, \mathcal{A}_{\Delta t}) \to 0 \quad \text{as} \quad \Delta t \to 0,$$

under very special assumptions on the nature of the flow on the attractor; these essentially amount to the system being in gradient form, or something closely related to it – see Sections 6.4 and 6.5. The following example illustrates the essential difficulty in trying to derive lower-semicontinuity results:

Example Consider equation (1.1) in dimension $p = 1$. It follows that, for $F(u) : \mathbb{R} \mapsto \mathbb{R}$ defined so that $F'(u) = -f(u)$, we have that (6.6) holds and the system is in gradient form provided F satisfies (i) and (ii) of Definition 6.8. In this case, all solutions will have their ω-limit sets contained in the set of equilibrium points.

Now consider (1.1) with $f(u) \mapsto f_\epsilon(u)$ given by

$$f_\epsilon(u) = \left\{ \begin{array}{cc} -(u+1)^3 + \epsilon, & u \leq -1, \\ \epsilon(u^3/2 - 3u/2), & -1 < u < 1, \\ -(u-1)^3 - \epsilon, & u \geq 1, \end{array} \right\} \tag{6.11}$$

This vector field is $C^1(\mathbb{R}, \mathbb{R})$ and satisfies (i) and (ii) of Definition 6.8 for each $\epsilon \geq 0$. Hence, by Theorem 6.10 the system has a global attractor $\mathcal{A}_\epsilon$, say, for each $\epsilon \geq 0$.

The potential $F_\epsilon(u)$ satisfying $F'_\epsilon(u) = -f_\epsilon(u)$ and $F_\epsilon(0) = 0$ is shown in Figure 3 for $\epsilon > 0$ and $\epsilon = 0$ respectively.

Examination of these figures, together with application of Theorem 6.11, shows that, for every $\epsilon > 0$, the attractor of (1.1), (6.11) is given by

$$\mathcal{A}_\epsilon = \{0\}, \quad \epsilon > 0,$$

a single point; a similar analysis for $\epsilon = 0$ shows that

$$\mathcal{A}_0 = [-1, 1],$$

an entire interval. Thus the perturbed attractors with $\epsilon > 0$ are contained in the unperturbed attractor at $\epsilon = 0$ but not the other way around. This shows that the attractor $\mathcal{A}_0$ is *upper semicontinuous* with respect to $\epsilon > 0$ but it is not *lower semicontinuous.* Although the perturbation induced by ϵ in this example is not directly analogous to a numerical approximation, it nonetheless indicates an important point – without strong assumptions it may be difficult to prove lower semicontinuity of attractors with respect to perturbations of any kind, including those induced by numerical approximation.

The difficulty observed in the example is a consequence of the fact that certain portions of the attractor may attract very slowly – specifically slower than exponentially – and hence disappear under perturbation. In order to get around this difficulty it is natural to consider a slightly enlarged object which does have a form of exponential attraction. This is one motivation for the consideration of the weaker concept of a uniformly asymptotically stable set for which it is possible to prove both upper and lower semicontinuity.

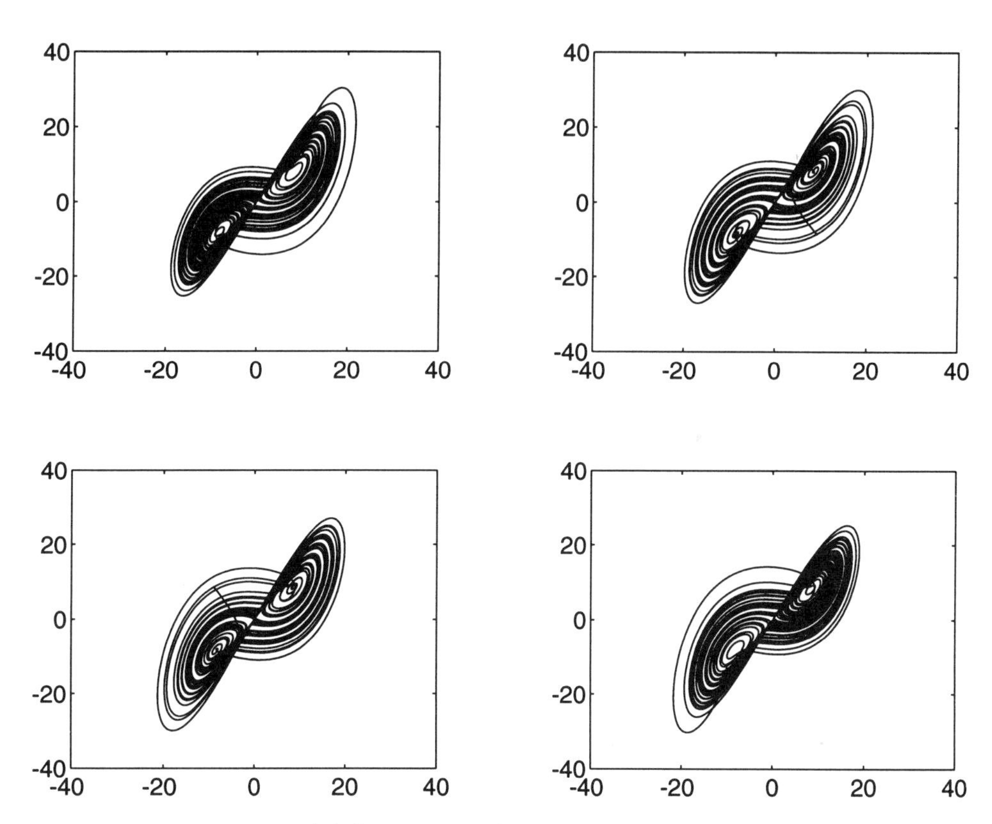

Fig. 3. The potential $F_\epsilon(u)$ (vertical axis) against u for $\epsilon = 0.01$ (top) and $\epsilon = 0.0$ (bottom).

Important remark All our definitions and theorems concern the continuous semigroup $S(t)$. However, the Definitions 6.1,6.3 and 6.8 and Theorems 6.2, 6.4 and 6.11 can all be extended to a discrete semigroup $S^n_{\Delta t}$ simply by replacing t by n in the definitions and using the continuity of $S^n_{\Delta t}$. Theorem 6.9 can also be extended but the extension is slightly less trivial – see Humphries and Stuart (1994).

6.2. Continuity of uniformly asymptotically stable attracting sets

We now consider (1.1) under the assumption that there exists a compact set Λ which is uniformly asymptotically stable.

Theorem 6.12 (**Uniformly asymptotically stable sets under approximation**) Assume that the semigroup $S(t)$ for (1.1) has a compact, uniformly asymptotically stable set Λ. Then there exists $\Delta t_c > 0$ such that, for all $\Delta t \in (0, \Delta t_c]$, the approximating semigroup $S^n_{\Delta t}$ for (1.2) has a compact, uniformly asymptotically stable set $\Lambda_{\Delta t} \supset \Lambda$ which satisfies

$$d_{\mathrm{H}}(\Lambda_{\Delta t}, \Lambda) \to 0 \quad \text{as} \quad \Delta t \to 0.$$

Furthermore there exists a set $B \supset \Lambda_{\Delta t}$ and $T = T(\Delta t)$ with the property that

$$S^n_{\Delta t} B \subset \Lambda_{\Delta t} \ \ \forall n : n\Delta t \geq T.$$

The proof will be performed through a sequence of lemmas. Recall the definition of R_1 from the proof of Corollary 6.6.

Lemma 6.13 There exists $r_0 \in (0, R_1)$ and $\Delta t_0 > 0$ such that

$$U \in \mathcal{N}(\Lambda, r_0) \Rightarrow S^1_{\Delta t} U \in \mathcal{N}(\Lambda, R_0) \quad \forall \Delta t \in (0, \Delta t_0].$$

Furthermore it then follows that

$$V(S^1_{\Delta t} U) \leq \mathrm{e}^{-C\Delta t} V(U) + KL\Delta t^{r+1}.$$

Proof. Define r_0 and Δt_0 by

$$\beta(r_0) = \alpha(R_0/2) \quad \text{and} \quad K\Delta t_0^{r+1} = R_0/2. \tag{6.12}$$

Note that $r_0 < R_1 < R_0$, so that, by Corollary 6.6,

$$U \in \mathcal{N}(\Lambda, r_0) \Rightarrow S(t)U \in \mathcal{N}(\Lambda, R_0) \quad \forall t > 0 \quad .$$

Theorem 6.5(iii) shows that we have

$$V(S(\Delta t)U) \leq \mathrm{e}^{-C\Delta t} V(U) < V(U). \tag{6.13}$$

But

$$\alpha(\mathrm{dist}(S(\Delta t)U, \Lambda)) \leq V(S(\Delta t)U) \tag{6.14}$$

and

$$V(U) \leq \beta(\text{dist}(U, \Lambda)) < \beta(r_0) = \alpha(R_0/2). \tag{6.15}$$

Equations (6.13)–(6.15) imply that

$$\alpha(\text{dist}(S(\Delta t)U, \Lambda)) \leq \alpha(R_0/2) \Rightarrow \text{dist}(S(\Delta t)U, \Lambda) \leq R_0/2 \; . \tag{6.16}$$

Now,

$$\begin{aligned} \text{dist}(S^1_{\Delta t}U, \Lambda) &= \inf_{x\in\Lambda} \|S^1_{\Delta t}U - x\| \\ &\leq \inf_{x\in\Lambda} \|S(t)U - x\| + \|S(t)U - S^1_{\Delta t}U\| \\ &\leq \text{dist}(S(\Delta t)U, \Lambda) + \|S(\Delta t)U - S^1_{\Delta t}U\|. \end{aligned} \tag{6.17}$$

Thus (6.16), (6.18), (6.12) and Assumption 3.7(iii) gives

$$\text{dist}(S^1_{\Delta t}U, \Lambda) \leq \tfrac{1}{2}R_0 + K\Delta t^{r+1} \leq R_0 \forall \Delta t \in (0, \Delta t_0].$$

Now, by Theorem 6.5(i) and Assumption 3.7(iii) we have

$$|V(S^1_{\Delta t}U) - V(S(\Delta t)U)| \leq L\|S^1_{\Delta t}U - S(\Delta t)U\| \leq KL\Delta t^{r+1}.$$

Hence, by Theorem 6.5(iii),

$$\begin{aligned} V(S^1_{\Delta t}U) &\leq V(S(\Delta t)U) + KL\Delta t^{r+1} \\ &\leq \mathrm{e}^{-C\Delta t}V(U) + KL\Delta t^{r+1}, \end{aligned}$$

by (6.13). □

The previous lemma gives a bound on U which ensures that $V(S^1_{\Delta t}U)$ is defined. Using this fact we now construct a positively invariant set B for $S^1_{\Delta t}$.

Lemma 6.14 The set $B = \{x \in \mathcal{N}(\Lambda, R_0) : V(x) < \alpha(r_0)\}$ is open and $\Lambda \subset B \subset \mathcal{N}(\Lambda, r_0)$. Furthermore, there exists Δt_1 such that B is positively invariant under $S^1_{\Delta t}$ for all $\Delta t \in (0, \Delta t_1]$.

Proof. The set B is open since V is continuous and $V(x) < \alpha(r_0)$ for all $x \in B$. The set B contains Λ since $V(x) = 0$ if $x \in \Lambda$. Let $U \in B$; then

$$\alpha(\text{dist}(U, \Lambda)) \leq V(U) < \alpha(r_0) \tag{6.18}$$

by the properties of $\alpha(\cdot)$ and B. Thus $\text{dist}(U, \Lambda) < r_0$ which implies that $U \in \mathcal{N}(\Lambda, r_0)$ as required. Hence $B \subset \mathcal{N}(\Lambda, r_0)$. Now we define Δt_1, the largest number in $(0, \Delta t_0]$ such that

$$\frac{KL\Delta t^{p+1}}{1 - \mathrm{e}^{-C\Delta t}} < \tfrac{1}{4}\alpha(r^*) \quad \forall \Delta t \in (0, \Delta t_1], \tag{6.19}$$

where

$$r^* = \tfrac{1}{2}\beta^{-1}(\alpha(r_0)). \tag{6.20}$$

Note that

$$r^* < \tfrac{1}{2}\alpha^{-1}(\alpha(r_0)) = \tfrac{1}{2}r_0 < r_0.$$

Thus

$$\alpha(r^*) < \alpha(r_0). \tag{6.21}$$

Now, if $U \in B \subset \mathcal{N}(\Lambda, r_0)$ then by Lemma 6.13, (6.18) and (6.19) we have

$$\begin{aligned} V(S^1_{\Delta t}U) &\leq e^{-C\Delta t}V(U) + KL\Delta t^{p+1} \\ &< e^{-C\Delta t}\alpha(r_0) + \tfrac{1}{4}\alpha(r^*)(1 - e^{-C\Delta t}) \\ &\leq \tfrac{1}{4}\alpha(r_0) + \tfrac{3}{4}e^{-C\Delta t}\alpha(r_0) < \alpha(r_0). \end{aligned}$$

In addition, $S^1_{\Delta t}U \in \mathcal{N}(\Lambda, R_0)$ by Lemma 6.13 and so we have $S^1_{\Delta t}B \subset B$ as required. □

We now construct the approximate uniformly asymptotically stable set $\Lambda_{\Delta t}$.

Lemma 6.15 Define

$$\eta(\Delta t) = \frac{2KL\Delta t^{r+1}}{(1 - e^{-C\Delta t})}, \quad \Delta t \in (0, \Delta t_1]$$

and define

$$\Lambda_{\Delta t} = \{x \in \mathcal{N}(\Lambda, R_0) : V(x) \leq \eta(\Delta t)\}.$$

Then $\Lambda_{\Delta t}$ is compact, positively invariant, contains Λ in its interior and satisfies $d_{\mathrm{H}}(\Lambda_{\Delta t}, \Lambda) \to 0$ as $\Delta t \to 0$.

Proof. $\Lambda_{\Delta t}$ is bounded since Λ is bounded and V is continuous; $\Lambda_{\Delta t}$ is closed by construction. Hence it is a compact set. Clearly, by (6.19) and (6.21) it follows that

$$\Lambda_{\Delta t} \subset B \subset \mathcal{N}(\Lambda, r_0), \quad \Delta t \in (0, \Delta t_1], \tag{6.22}$$

since $\alpha(r^*) < \alpha(r_0)$. Thus by Lemma 6.13, if $U \in \Lambda_{\Delta t}$ we have

$$\begin{aligned} V(S^1_{\Delta t}U) &\leq e^{-C\Delta t}V(U) + KL\Delta t^{r+1} \\ &\leq e^{-C\Delta t}\eta(\Delta t) + \tfrac{1}{2}\eta(\Delta t)(1 - e^{-C\Delta t}) \leq \eta(\Delta t). \end{aligned}$$

Since $S^1_{\Delta t}U \in \mathcal{N}(\Lambda, R_0)$ by (6.22) and Lemma 6.13, we have $S^1_{\Delta t}U \in \Lambda_{\Delta t}$ as required.

It is clear that Λ is contained in the interior of $\Lambda_{\Delta t}$ since V is continuous and $\eta(\Delta t) > 0$. Thus $\mathrm{dist}(\Lambda, \Lambda_{\Delta t}) = 0$. Also $\mathrm{dist}(\Lambda_{\Delta t}, \Lambda) = \sup_{x \in \Lambda_{\Delta t}} \mathrm{dist}(x, \Lambda)$. But, for every $x \in \Lambda_{\Delta t}$ we have that $\mathrm{dist}(x, \Lambda) \leq \alpha^{-1}(\eta(\Delta t))$. Since $\eta(\Delta t) \to 0$ as $\Delta t \to 0$ it follows that $\mathrm{dist}(\Lambda_{\Delta t}, \Lambda) \to 0$ as $\Delta t \to 0$. □

We now show that iterates starting in B are absorbed into $\Lambda_{\Delta t}$ in a finite number of steps under $S^1_{\Delta t}$, giving asymptotic stability.

Lemma 6.16 There exists $\Delta t_2 \in (0, \Delta t_1]$ such that, for any $U \in B\backslash\Lambda_{\Delta t}$ and any $\Delta t \in (0, \Delta t_2]$,

$$V(S^1_{\Delta t}U) \leq \mathrm{e}^{-C\Delta t/4}V(U).$$

Furthermore, if

$$T(\Delta t) = \frac{4}{C}\ln\{\alpha(r_0)/\eta(\Delta t)\},$$

then there exists a $\delta_0 > 0$ such that, $\mathcal{N}(\Lambda_{\Delta t}, \delta_0) \subset B$ and, furthermore, if $U \in \mathcal{N}(\Lambda_{\Delta t}, \delta_0)$ then $S^n_{\Delta t}U \in \Lambda_{\Delta t}$ $\forall n : n\Delta t \geq T(\Delta t)$.

Proof. Let $U \in B\backslash\Lambda_{\Delta t}$. It follows that

$$\eta(\Delta t) \leq V(U) \leq \alpha(r_0). \tag{6.23}$$

Then, by Lemma 6.13 and (6.23) it follows that, for $\Delta t \in (0, \Delta t_1]$

$$\begin{aligned} V(S^1_{\Delta t}U) &\leq \mathrm{e}^{-C\Delta t}V(U) + KL\Delta t^{r+1} \\ &= \mathrm{e}^{-C\Delta t}V(U) + \tfrac{1}{2}\eta(\Delta t)(1 - \mathrm{e}^{-C\Delta t}) \\ &\leq \tfrac{1}{2}(1 + \mathrm{e}^{-C\Delta t})V(U). \end{aligned}$$

Now define $\Delta t_2 = \min\{\Delta t_1, \gamma\}$ where $\mathrm{e}^{-C\gamma} + 1 = 2\mathrm{e}^{-C\gamma/4}$. Then

$$V(S^1_{\Delta t}U) < \mathrm{e}^{-C\Delta t/4}V(U)$$

as required. Iterating this bound shows that, if we assume that $S^j_{\Delta t} \in B\backslash\Lambda_{\Delta t}$ for $j = 0, \ldots, n-1$ then

$$V(S^n_{\Delta t}U) < \mathrm{e}^{-C\eta\Delta t/4}V(U). \tag{6.24}$$

From (6.23) it follows that

$$V(S^n_{\Delta t}U) < \mathrm{e}^{-C\eta\Delta t/4}\alpha(r_0)\ .$$

If $n\Delta t \geq T(\Delta t)$ then

$$\mathrm{e}^{-Cn\Delta t/4} \leq \eta(\Delta t)/\alpha(r_0)$$

so that

$$V(S^n_{\Delta t}U) < \eta(\Delta t).$$

Hence $S^n_{\Delta t}U \in \Lambda_{\Delta t}$, a contradiction. It follows that $S^n_{\Delta t}$ enters $\Lambda_{\Delta t}$ for some $n > 0 : n\Delta t \leq T(\Delta t)$.

Finally, to exhibit asymptotic stability we need to find an appropriate δ_0 such that $\mathcal{N}(\Lambda_{\Delta t}, \delta_0) \subset B$. Let $\delta_0 = \frac{1}{2}\beta^{-1}(\alpha(r_0)) > 0$. Then, if $U \in \mathcal{N}(\Lambda_{\Delta t}, \delta_0)$ we wish to show that $U \in B$. Now, for any $z \in \mathbb{R}^p$,

$$\mathrm{dist}(U, \Lambda) = \inf_{y\in\Lambda} \|U - y\| \leq \|U - z\| + \inf_{y\in\Lambda} \|z - y\|.$$

Choose z: $\|U - z\| = \inf_{y \in \Lambda_{\Delta t}} \|U - y\|$. Then, by Lemmas 6.15 and (6.19), (6.20) we obtain

$$\begin{aligned}
\operatorname{dist}(U, \Lambda) &\leq \operatorname{dist}(U, \Lambda_{\Delta t}) + \operatorname{dist}(\Lambda_{\Delta t}, \Lambda) \\
&< \delta_0 + \alpha^{-1}(\eta(\Delta t)) \\
&= \tfrac{1}{2}\beta^{-1}(\alpha(r_0)) + \alpha^{-1}\left(\frac{2KL\Delta t^{r+1}}{1 - \mathrm{e}^{-C\Delta t}}\right) \\
&< \tfrac{1}{2}\beta^{-1}(\alpha(r_0)) + \alpha^{-1}(\tfrac{1}{2}\alpha(r_*)) \\
&< \tfrac{1}{2}\beta^{-1}(\alpha(r_0)) + r_* = \beta^{-1}(\alpha(r_0)).
\end{aligned}$$

Hence $\operatorname{dist}(U, \Lambda) < \beta^{-1}(\alpha(r_0))$ which implies that $V(U) \leq \beta(\operatorname{dist}(U, \Lambda)) < \alpha(r_0)$. Hence $U \in B$ and the result follows. □

Finally we show that the stability is uniform.

Lemma 6.17 Let $\Delta t \in (0, \Delta t_2]$. Then, for each $\epsilon > 0$ there exists $\delta = \delta(\epsilon, \Delta t) > 0$ such that

$$U \in \mathcal{N}(\Lambda_{\Delta t}, \delta) \Rightarrow S^n_{\Delta t} U \in \mathcal{N}(\Lambda_{\Delta t}, \epsilon) \quad \forall n \geq 0 \, .$$

Proof. By Assumption 3.7(i)

$$\|S^1_{\Delta t} U - S^1_{\Delta t} V\| \leq (1 + K\Delta t)\|U - V\|.$$

Let

$$\delta = \min\{\delta_0, \tfrac{1}{2}\epsilon(1 + K\Delta t)^{-T(\Delta t)/\Delta t}\} \, .$$

Let $U \in \mathcal{N}(\Lambda_{\Delta t}, \delta)$. If $U \in \Lambda_{\Delta t}$ then positive invariance implies the result automatically, by Lemma 6.15. Hence suppose $U \notin \Lambda_{\Delta t}$. Choose $V \in \Lambda_{\Delta t}$ such that $\operatorname{dist}(U, \Lambda_{\Delta t}) = \|U - V\|$; then $\|S^n_{\Delta t} U - S^n_{\Delta t} V\| \leq (1 + K\Delta t)^n \|U - V\|$. Also, since $V \in \Lambda_{\Delta t}$ we have that $S^n_{\Delta t} V \in \Lambda_{\Delta t}$. Thus

$$\begin{aligned}
\operatorname{dist}(S^n_{\Delta t} U, \Lambda_{\Delta t}) &\leq \inf_{y \in \Lambda_{\Delta t}} \|S^n_{\Delta t} U - y\| \leq \|S^n_{\Delta t} U - S^n_{\Delta t} V\| \\
&\leq (1 + K\Delta t)^n \|U - V\| \\
&= (1 + K\Delta t)^n \operatorname{dist}(U, \Lambda_{\Delta t}) \\
&= (1 + K\Delta t)^n \delta. \qquad (6.25)
\end{aligned}$$

The bound (6.25) and the fact that $\delta \leq \delta_0$ shows that

$$\operatorname{dist}(S^n_{\Delta t} U, \Lambda_{\Delta t}) \leq \tfrac{1}{2}\epsilon \quad \forall n : n\Delta t \leq T(\Delta t).$$

Also, since

$$S^n_{\Delta t} U \in \Lambda_{\Delta t} \quad \forall n : n\Delta t \geq T(\Delta t)$$

we have

$$\operatorname{dist}(S^n_{\Delta t} U, \Lambda_{\Delta t}) = 0 \leq \tfrac{1}{2}\epsilon \quad \forall n \geq 0.$$

□

Proof of Theorem 6.12. Lemma 6.15 establishes existence of a positive invariant set $\Lambda_{\Delta t}$ and its convergence properties. Lemmas 6.16 and 6.17 show uniform asymptotic stability and Lemma 6.16 gives the required absorbtion property. □

In the following section we will give a different construction of a set $\Lambda_{\Delta t}$ which has the same properties as the set $\Lambda_{\Delta t}$ constructed in Theorem 6.12. We conclude this section with a corollary of Theorem 6.12 which relates to the existence of attractors. Note that, whilst uniformly asymptotically stable sets have been proven both upper and lower semicontinuous, the following result establishes only upper semiconinuity of attractors.

Corollary 6.18 (**Attractors under approximation**) Assume that the semigroup $S(t)$ for (1.1) has an attractor $\mathcal{A}$. Then there exists $\Delta t_c > 0$ such that, for all $\Delta t \in (0, \Delta t_c]$, the approximating semigroup $S^n_{\Delta t}$ for (1.2) has a compact, uniformly asymptotically stable set $\Lambda_{\Delta t} \supset \mathcal{A}$ and an attractor $\mathcal{A}_{\Delta t} = \omega(\Lambda_{\Delta t})$ satisfying

$$\operatorname{dist}(\mathcal{A}_{\Delta t}, \mathcal{A}) \to 0 \quad \text{as} \quad \Delta t \to 0.$$

Proof. Note that $\mathcal{A}$ is uniformly asymptotically stable by Theorem 6.4(ii). Hence, by Theorem 6.12 with $\Lambda = \mathcal{A}$, we deduce that (1.2) has a uniformly asymptotically stable set $\Lambda_{\Delta t} \supset \mathcal{A}$. Furthermore, by Theorem 6.4(iii) it follows that $\mathcal{A}_{\Delta t} = \omega(\Lambda_{\Delta t})$ (the limit set being defined through $S^n_{\Delta t}$) is an attractor for $S^n_{\Delta t}$. Note that $\mathcal{A}_{\Delta t} \subseteq \Lambda_{\Delta t}$ and that, by Theorem 6.12,

$$\operatorname{dist}(\Lambda_{\Delta t}, \mathcal{A}) \to 0 \quad \text{as} \quad \Delta t \to 0.$$

Hence

$$\operatorname{dist}(\mathcal{A}_{\Delta t}, \mathcal{A}) \to 0 \quad \text{as} \quad \Delta t \to 0$$

and the proof is complete. □

6.3. Upper semicontinuity of attractors

In this section we consider the numerical approximation of (1.1) satisfying (6.3). Under this assumption it follows from Theorem 6.7 that (1.1) has a local attractor $\mathcal{A} = \omega(B)$ and it is thus natural to study the effect of the numerical approximation of (1.1) on $\mathcal{A}$. It would be possible to use (6.3) to deduce the exsitence of a uniformly asymptotically stable set for (1.1) and then apply the methods of the previous section to deduce all the results in this section. However, we present here an entirely different approach to the problem. Our first result proves that an approximate attractors exists.

Theorem 6.19 (**Existence of an approximate attractor**) Consider the approximation of (1.1) under (6.3) by (1.2) and let $B = \{u \in \mathbb{R}^p : \|u\| < R\}$. Then there exists $\Delta t_c > 0$ such that the semigroup $S^n_{\Delta t}$ has an attractor given by $\mathcal{A}_{\Delta t} = \omega(B)$ for all $\Delta t \in (0, \Delta t_c]$.

Proof. Recall that $\partial B = \bar{B}\backslash B$. By continuity of f and (6.3) it is possible to choose $\delta > 0$ sufficiently small so that

$$\langle f(u), u\rangle \leq -\tfrac{1}{2}\epsilon \quad \forall u : R - \delta \leq \|u\| \leq R. \tag{6.26}$$

Recall also that, by Assumption 3.6, there exists $L > 0$ such that $\|f(u)\| \leq L$ for all $u \in \mathbb{R}^p$. Now choose $\Delta t_c > 0$ so that

$$L\Delta t \leq \tfrac{1}{4}\delta,\ K\Delta t^{r+1} \leq R,\ RK\Delta t^r \leq \tfrac{1}{10}\epsilon,\ K\Delta t^{r+1} \leq \tfrac{1}{8}\delta\ \forall \Delta t \in (0, \Delta t_c]. \tag{6.27}$$

In the remainder of this proof we assume that $\Delta t \in (0, \Delta t_c]$. We deduce that

$$\|u(t+s) - u(t)\| \leq \int_t^{t+s} \|f(u(\tau)\|\, \mathrm{d}\tau \leq L\Delta t < \tfrac{1}{2}\delta \quad \forall s \in (0, \Delta t_c]. \tag{6.28}$$

Thus, if $R - \delta/2 \leq \|U\| \leq R$ we have, from the positive invariance of B under $S(t)$ and (6.28), that

$$R - \delta \leq \|S(t)U\| \leq R, \quad t \in (0, \Delta t_c].$$

It then follows from (6.4) and (6.26) that, for all $\Delta t \in (0, \Delta t_c]$,

$$\|S(\Delta t)U\|^2 - \|U\|^2 = 2\int_0^{\Delta t} \langle f(S(\tau)U), S(\tau)U\rangle\, \mathrm{d}\tau \leq -\epsilon\Delta t.$$

Hence by Assumption 3.7(iii), (6.26) and (6.27) we have

$$\begin{aligned}
\|S^1_{\Delta t}U\|^2 - \|U\|^2 &\leq \|S(\Delta t)U\|^2 - \|U\|^2 \\
&\quad +[2\|S^1_{\Delta t}U\| + \|S(\Delta t)U - S^1_{\Delta t}U\|] \\
&\quad \|S(\Delta t)U - S^1_{\Delta t}U\| \\
&\leq -\epsilon\Delta t + [2\|S(\Delta t)U\| + 3\|S(\Delta t)U - S^1_{\Delta t}U\|] \\
&\quad \times\|S(\Delta t)U - S^1_{\Delta t}U\| \\
&\leq -\epsilon\Delta t + [2R + 3K\Delta t^{r+1}]K\Delta t^{r+1} \\
&\leq -\epsilon\Delta t + 5RK\Delta t^{r+1} \\
&\leq -\tfrac{1}{2}\epsilon\Delta t.
\end{aligned}$$

Thus we have proved that

$$R - \tfrac{1}{2}\delta \leq \|U\| \leq R \quad \Rightarrow \quad \|S^1_{\Delta t}U\|^2 - \|U\|^2 \leq -\tfrac{1}{2}\epsilon\Delta t. \tag{6.29}$$

Now, if $\|U\| < R - \frac{1}{2}\delta$ we have, by (6.28), (6.27) and Assumption 3.7(iii),

$$\begin{aligned}
\|S^1_{\Delta t}U\| &\leq \|S(\Delta t)U\| + \|S(\Delta t) - S^1_{\Delta t}U\| \\
&\leq \|U\| + \|S(\Delta t)U - U\| + \|S(\Delta t) - S^1_{\Delta t}U\| \\
&< R - \tfrac{1}{2}\delta + \tfrac{1}{4}\delta + K\Delta t^{r+1} \\
&\leq R - \tfrac{1}{8}\delta.
\end{aligned}$$

Hence

$$\|U\| < R - \tfrac{1}{2}\delta \quad \Rightarrow \quad \|S^1_{\Delta t}U\| \leq R - \tfrac{1}{8}\delta. \tag{6.30}$$

It is clear that (6.29) and (6.30) imply that

$$S^1_{\Delta t}\bar{B} \subset B \tag{6.31}$$

and hence, by the generalization of Theorem 6.2 from $S(t)$ to $S^n_{\Delta t}$, we deduce that $\mathcal{A}_{\Delta t} = \omega(B)$ is an attractor for $S^n_{\Delta t}$. □

As the example in Section 6.1 illustrates we cannot expect to obtain lower semicontinuity of $\mathcal{A}$ without imposing further conditions on the dynamical system in question. However, it is possible to derive upper semicontinuity results without further assumptions.

Theorem 6.20 (**Upper semicontinuity of attractors**) Consider the approximation of (1.1) under (6.3) by (1.2). Then the attractors of the semigroups $S(t)$ and $S^n_{\Delta t}$, $\mathcal{A}$ and $\mathcal{A}_{\Delta t}$ respectively, satisfy

$$\text{dist}(\mathcal{A}_{\Delta t}, \mathcal{A}) \to 0 \qquad \text{as} \qquad \Delta t \to 0.$$

Proof. Note that, if $\mathcal{A}_{\Delta t} \subseteq \overline{\mathcal{N}}(\mathcal{A}, \epsilon)$, then $\text{dist}(\mathcal{A}_{\Delta t}, \mathcal{A}) \leq \epsilon$; thus, if we show that for any $\epsilon > 0$ there exists $\Delta = \Delta(\epsilon)$ such that

$$\mathcal{A}_{\Delta t} \in \overline{\mathcal{N}}(\mathcal{A}, \epsilon) \quad \forall \Delta t \in (0, \Delta], \tag{6.32}$$

then the result follows. Thus we aim to prove (6.32).

First we estimate the attraction of B to a neighbourhood of $\mathcal{A}$ under $S^n_{\Delta t}$ by using the attractivity of B to $\mathcal{A}$ under $S(t)$ and the truncation error bound. We have that

$$\begin{aligned} \text{dist}(S^n_{\Delta t}U, \mathcal{A}) &= \inf_{x\in\mathcal{A}} \|S^n_{\Delta t}U - x\| \\ &\leq \|S^n_{\Delta t}U - S(n\Delta t)U\| + \inf_{x\in\mathcal{A}} \|S(n\Delta t)U - \mathbf{x}\| \\ &\leq \|S^n_{\Delta t}U - S(n\Delta t)U\| + \text{dist}(S(n\Delta t)U, \mathcal{A}). \end{aligned}$$

Hence we have that

$$\text{dist}(S^n_{\Delta t}U, \mathcal{A}) \leq \text{dist}(S(n\Delta t)U, \mathcal{A}) + \|S^n_{\Delta t}U - S(n\Delta t)U\|. \tag{6.33}$$

Now, since $\mathcal{A}$ attracts B under $S(t)$ there exists $T = T(\epsilon) > 0$ such that, for all $U \in B$,

$$S(t)U \in \mathcal{N}(\mathcal{A}, \epsilon/2) \quad \forall t \geq T. \tag{6.34}$$

Without loss of generality we may choose $T = N\Delta t$ for some integer N. By Theorem 3.8 it follows that, for any $U \in B$, there exists $\Delta = \Delta(\epsilon) > 0$ such that for any $\Delta t \in (0, \Delta]$

$$\|S^n_{\Delta t}U - S(n\Delta t)U\| \leq \epsilon/2 \quad \forall n, \Delta t : 0 \leq n\Delta t \leq 2T. \tag{6.35}$$

Hence it follows from (6.33), (6.34) and (6.35) that, for all $\Delta t \in (0, \Delta]$

$$\text{dist}(S^n_{\Delta t}B, \mathcal{A}) < \epsilon \quad \forall n, \Delta t : T \leq n\Delta t \leq 2T. \tag{6.36}$$

We now proceed to use induction. Suppose that, for some integer $k \geq 2$

$$\text{dist}(S^n_{\Delta t}B, \mathcal{A}) < \epsilon \quad \forall n, \Delta t : T \leq n\Delta t \leq kT. \tag{6.37}$$

Note that this has been proved true for $k = 2$. Now consider integer n such that $kT \leq n\Delta t \leq (k+1)T$. Choose m and p such that $n = m + p$, $T \leq m\Delta t \leq 2T$ and $p\Delta t = (k-1)T$; thus $p = (k-1)N$. Then $S^n_{\Delta t}B = S^m_{\Delta t}S^p_{\Delta t}B$ and by (6.31) it follows that $S^p_{\Delta t}B \subset B$. This implies that

$$S^n_{\Delta t}B = S^m_{\Delta t}S^p_{\Delta t}B \subset S^m_{\Delta t}B;$$

since $T \leq mh \leq 2T$ it may be shown, from (6.36) with $n \mapsto m$, that

$$S^m_{\Delta t}B \subset \mathcal{N}(\mathcal{A}, \epsilon)$$

and hence that

$$S^n_{\Delta t}B \subset \mathcal{N}(\mathcal{A}, \epsilon) \quad \forall n, \Delta t : kT \leq n \leq (k+1)T.$$

This, together with (6.37), completes the inductive step and we deduce that

$$\text{dist}(S^n_{\Delta t}B, \mathcal{A}) < \epsilon \quad \forall n, \Delta t : T \leq n\Delta t < \infty. \tag{6.38}$$

Finally recall that

$$\mathcal{A}_{\Delta t} = \bigcap_{m \geq 0} \overline{\bigcup_{n \geq m} S^n_{\Delta t}B}$$

and since (6.38) holds it follows that

$$\overline{\bigcup_{n \geq N} S^n_{\Delta t}B} \subseteq \overline{\mathcal{N}}(\mathcal{A}, \epsilon),$$

where $N\Delta t = T$, and hence $\mathcal{A}_{\Delta t} \subseteq \overline{\mathcal{N}}(\mathcal{A}, \epsilon)$ as required. □

Note that Theorem 6.20 does not give a rate of convergence for the quantity $\text{dist}(\mathcal{A}_{\Delta t}, \mathcal{A})$. This is since nothing is assumed about the *rate of attraction* of the attractor. If the rate is assumed exponential then a stronger result can be proved and we obtain the error bound given in Theorem 6.21 below. Note that the bound is less than the rate of convergence of individual trajectories and reflects the competition between the exponential attraction to $\mathcal{A}$ (which determines α) and the exponential divergence of trajectories on $\mathcal{A}$ (which determines K).

Theorem 6.21 (Rate of convergence of attractors) Consider the approximation of (1.1) satisfying (6.3) by (1.2) and assume that the attractor of the semigroup $\mathcal{A}$ is exponentially attracting in the sense that there exist $C_1 > 0, \alpha > 0$ such that

$$\text{dist}(S(t)B, \mathcal{A}) \leq C_1 e^{-\alpha t}. \tag{6.39}$$

Then there exist $\Delta t_c, C_2 > 0$ such that the the global attractors $\mathcal{A}$ and $\mathcal{A}_{\Delta t}$ of $S(t)$ and $S^n_{\Delta t}$ respectively, satisfy

$$\operatorname{dist}(\mathcal{A}_{\Delta t}, \mathcal{A}) \leq C_2 \Delta t^\beta,$$

where $\beta = \alpha r/(K+\alpha)$, for all $\Delta t \in (0, \Delta t_c]$.

Proof. Using (6.33) and the arguments following it in the proof of Theorems 6.20, 6.39 and using Theorem 3.8, we obtain, for any $U \in B$,

$$\operatorname{dist}(S^n_{\Delta t}U, \mathcal{A}) \leq C_1 \mathrm{e}^{-\alpha T} + \mathrm{e}^{KT}\Delta t^r \quad \text{for} \quad n, \Delta t : T \leq n\Delta t \leq 2T. \tag{6.40}$$

We can balance the contributions in (6.40) to find the relationship between T and Δt which optimizes the error. We find that

$$\Delta t \propto \mathrm{e}^{-(\alpha+K)T/r}$$

is the appropriate choice. This shows that there exists $C_2 > 0$ such that, for any $U \in B$,

$$\operatorname{dist}(S^n_{\Delta t}U, \mathcal{A}) \leq C_2 \Delta t^\beta \quad \text{for} \quad n, \Delta t : T \leq n\Delta t \leq 2T.$$

Proceeding with an induction argument as in Theorem 6.20 we obtain the required result. □

We can use the construction of Theorem 6.20 to prove a result closely related to Theorem 6.12.

Theorem 6.22 (**Uniformly asymptotically stable sets under approximation**) Consider the semigroup $S(t)$ for (1.1) under (6.3) with attractor $\mathcal{A}$. Then there exists $\Delta t_c > 0$ such that, for all $\Delta t \in (0, \Delta t_c]$, the approximating semigroup $S^n_{\Delta t}$ for (1.2) has a compact, uniformly asymptotically stable set $\Lambda_{\Delta t} \supseteq \mathcal{A}$ which satisfies

$$d_{\mathrm{H}}(\Lambda_{\Delta t}, \mathcal{A}) \to 0 \quad \text{as} \quad \Delta t \to 0.$$

Proof. Let $\Delta t \in (0, \Delta t_c]$ where Δt_c is given by Theorem 6.20. To prove this result define

$$\Lambda_{\Delta t} = \mathcal{A}_{\Delta t} \cup \bigcup_{n=0}^{\infty} S^n_{\Delta t}\mathcal{A}$$

where $\mathcal{A} = \omega(B)$ and, by Theorem 6.19, $\mathcal{A}_{\Delta t} = \omega(B)$ (the limit sets being defined through $S(t)$ and $S^n_{\Delta t}$ respectively.) Note that, by Theorem 6.19, both $\mathcal{A}$ and $\mathcal{A}_{\Delta t}$ are contained in B so that $\Lambda_{\Delta t} \subseteq B$. Since $\Lambda_{\Delta t}$ contains $\mathcal{A}_{\Delta t}$ which is an attractor for $S^n_{\Delta t}$, it follows that $\Lambda_{\Delta t}$ is asymptotically stable for $S^n_{\Delta t}$. Note also that $\Lambda_{\Delta t}$ is positively invariant under $S^n_{\Delta t}$ since $\mathcal{A}_{\Delta t}$ is invariant by Theorem 3.3 and $\cup_{n=0}^{\infty} S^n_{\Delta t}\mathcal{A}$ is positively invariant by construction; thus it follows by using a similar argument to that used in establishing Theorem 6.4(ii), that $\Lambda_{\Delta t}$ is uniformly stable.

It remains to establish the error bound. Since $\Lambda_{\Delta t} \supseteq \mathcal{A}$ we have

$$\text{dist}(\mathcal{A}, \Lambda_{\Delta t}) = 0 \quad \forall \Delta t \in (0, \Delta t_c].$$

Note that, since $\mathcal{A} \in B$ we have from (6.38) that, for any $\epsilon > 0$, there exists $T = T(\epsilon) > 0$ such that

$$\text{dist}(S^n_{\Delta t}\mathcal{A}, \mathcal{A}) < \epsilon \quad \forall n, \Delta t : T \leq n\Delta t < \infty.$$

Furthermore, since $S(t)\mathcal{A} = \mathcal{A}$ it follows from (6.35) that there exists $\Delta = \Delta(\epsilon) > 0$ such that, if $\Delta t \in (0, \Delta]$ then

$$\text{dist}(S^n_{\Delta t}\mathcal{A}, \mathcal{A}) \leq \tfrac{1}{2}\epsilon \quad \forall n, \Delta t : 0 \leq n\Delta t \leq 2T.$$

Also Theorem 6.20 yields

$$\text{dist}(\mathcal{A}_{\Delta t}, \mathcal{A}) < \epsilon$$

for $\Delta < \Delta(\epsilon)$. Combining these estimates shows that

$$\text{dist}(\Lambda_{\Delta t}, \mathcal{A}) < \epsilon$$

for $\Delta < \Delta(\epsilon)$ as required. □

6.4. Lower semicontinuity of attractors

As we have seen, lower semicontinuity results are not true in general. However, if we make assumptions about the nature of the flow on the attractor $\mathcal{A}$ then it is possible to prove lower semicontinuity with respect to numerical perturbations. One important case where this is possible is when the dynamical system $S(t)$ is in gradient form and the set $\mathcal{E}$ is a bounded set containing only hyperbolic equilibria. We assume this henceforth. The method of proof is to decompose the attractor $\mathcal{A}$ according to the value of $F(\cdot)$ and build up the nearby approximate attractor $\mathcal{A}_{\Delta t}$ starting from the smallest value of F on the attractor.

We make the following assumption throughout this section:

Assumption 6.23 The dynamical system (1.1) has vector field f given by (6.5) and is a gradient system; furthermore, the set $\mathcal{E}$ of equilibrium points is bounded and comprises only hyperbolic equilibria.

Thus we may enumerate the set of equilibrium points of $S(t)$ as

$$\mathcal{E} = \{x_1, \ldots, x_M\}. \tag{6.41}$$

Let

$$v_1 > v_2 > \ldots > v_N$$

be the distinct points of $\{F(x_1), \ldots, F(x_M)\}$. Since $\mathcal{E}$ is bounded the assumptions of Theorem 6.10 hold. Thus the set $B = B(R)$ given in that theorem is uniformly asymptotically stable for any $R > \xi$. Hence we may define

$$E^k = \{x \in \mathcal{E}: F(x) = v_k\}, \quad U^k = \{x \in B: F(x) < v_k\},$$
$$W^k = \bigcup_{x \in E^k} W^u(x), \quad \mathcal{A}^k = \bigcup_{j=k}^N W^k. \tag{6.42}$$

We will require the following lemma concerning these sets:

Lemma 6.24 Consider a dynamical system (1.1) under Assumption 6.23. Then $S(t) : U^k \mapsto U^k$, $k = 1, \ldots, N$ and $\mathcal{A}^k$ attracts all compact sets in U^{k-1}.

Proof. See Hale (1988), Theorem 3.8.7. □

Since $\mathcal{A}^1$ contains the unstable manifolds of all equilibria, it follows from Theorem 6.11, noting that all equilibria are assumed hyperbolic, that $\mathcal{A}^1 = \mathcal{A}$. On the other hand, since all points in E^N must be stable as there are no equilibria with lower values of F than those in E^N and (6.6) holds, we deduce that $\mathcal{A}^N = E^N$. It is straightforward to show that E^N is close to its discrete counterpart; we use this as the basis of an inductive proof to build up the properties of the approximate attractor. We require the following lemma – recall the definition $\mathcal{E}_{\Delta t}$ of the fixed points of $S^1_{\Delta t}$.

Lemma 6.25 Under Assumption 6.23 there exist $C, \Delta t_c > 0$ such that, for all $\Delta t \in (0, \Delta t_c]$, the semigroup $S^1_{\Delta t}1$ has M fixed points, $X_j \in \mathcal{E}_{\Delta t} \subset B$ $j = 1, \ldots, M$, all of which are hyperbolic and satisfy $\|x_j - X_j\| \leq C\Delta t^r$. Furthermore, for any $\epsilon > 0$ there exists $\epsilon' > 0$ such that

$$\operatorname{dist}(W^{u,\epsilon}(x_j), W^{u,\epsilon'}_{\Delta t}(X_j)) \leq C\Delta t^r.$$

Proof. Let Δt_c be given by the minimum Δt_c found in Theorems 4.10 and 4.11 and Corollary 4.13. By Theorem 4.10 the existence and closeness of the approximate fixed points follows and, by Theorem 4.11 we deduce that no others exist. Corollary 4.13 gives the required bound for the local unstable manifolds. □

Thus we may set

$$\mathcal{E}_{\Delta t} = \{X_1, \ldots, X_M\}. \tag{6.43}$$

and let

$$V_1 > V_2 > \ldots > V_N$$

be the distinct value of $F(\cdot)$ on the members of $\mathcal{E}_{\Delta t}$. Definitions analogous to (6.42) can be made for the dynamical system generated by (1.2). Thus we define

$$E^k_{\Delta t} = \{x \in \mathcal{E}_{\Delta t}: F(x) = V_k\} \text{and} U^k_{\Delta t} = \{x \in B: F(x) < V_k\},$$
$$W^k_{\Delta t} = \bigcup_{x \in E^k_{\Delta t}} W^u_{\Delta t}(x) \text{and} \mathcal{A}^k_{\Delta t} = \bigcup_{j=k}^N W^k_{\Delta t}. \tag{6.44}$$

Notice that since the global attractor must include the union of all unstable manifolds of fixed points by Theorem 6.11, we have

$$\mathcal{A}^k_{\Delta t} \subseteq \mathcal{A}_{\Delta t}, k = 1, \ldots, N \tag{6.45}$$

and this is the only property required of the $\mathcal{A}^k_{\Delta t}$. We now use the decomposition to prove lower semicontinuity for the numerical method.

Theorem 6.26 **(Lower semicontinuity of attractors)** Consider (1.1) under Assumption 6.23 with attractor $\mathcal{A}$. Then there exists $\Delta t_c > 0$ such that for $\Delta t \in (0, \Delta t_c]$ the numerical solution (1.2) possesses an attractor $\mathcal{A}_{\Delta t}$ which satisfies

$$d_{\mathrm{H}}(\mathcal{A}_{\Delta t}, \mathcal{A}) \to 0 \qquad \text{as} \qquad \Delta t \to 0.$$

We postpone the proof of the theorem until after the following lemma, which is fundamental in the proof.

Lemma 6.27 Consider (1.1) under Assumption 6.23. Assume that there exists Δ_k such that

$$\mathrm{dist}(\mathcal{A}^k, \mathcal{A}^k_{\Delta t}) \leq \frac{\epsilon}{2^k} \quad \forall \Delta t \in (0, \Delta_k]. \tag{6.46}$$

Then there exists Δ_{k-1} such that

$$\mathrm{dist}(\mathcal{A}^{k-1}, \mathcal{A}^{k-1}_{\Delta t}) \leq \frac{\epsilon}{2^{k-1}} \quad \forall \Delta t \in (0, \Delta_{k-1}]. \tag{6.47}$$

Proof. In the proof it is useful to observe from (6.42) and (6.44) that

$$\begin{aligned} \mathcal{A}^{k-1} &= W^{k-1} \cup \mathcal{A}^k \\ \mathcal{A}^{k-1}_{\Delta t} &= W^{k-1}_{\Delta t} \cup \mathcal{A}^k_{\Delta t}. \end{aligned} \tag{6.48}$$

Suppose that (6.46) holds. Now, since $\mathcal{A}^{k-1}_{\Delta t} \supseteq \mathcal{A}^k_{\Delta t}$ by (6.48), it follows that

$$\mathrm{dist}(\mathcal{A}^k, \mathcal{A}^{k-1}_{\Delta t}) \leq \mathrm{dist}(\mathcal{A}^k, \mathcal{A}^k_{\Delta t}) = \frac{\epsilon}{2^k}.$$

Furthermore, by (6.48) we deduce that

$$\mathrm{dist}(\mathcal{A}^{k-1}, \mathcal{A}^{k-1}_{\Delta t}) = \max(\mathrm{dist}(W^{k-1}, \mathcal{A}^{k-1}_{\Delta t}), \mathrm{dist}(\mathcal{A}^k, \mathcal{A}^{k-1}_{\Delta t})).$$

Hence, to establish (6.47), it is sufficient to show that

$$\mathrm{dist}(W^{k-1}, \mathcal{A}^{k-1}_{\Delta t}) \leq \frac{\epsilon}{2^{k-1}}. \tag{6.49}$$

Recall the notation (3.3) and let

$$\Gamma^{k-1} = \bigcup_{x \in E^{k-1}} \left(W^{u,\delta}(x) \cap \partial B(x, \delta) \right),$$

for some $\delta > 0$. The set Γ^{k-1} is compact and, by Lemma 4.2, we have

$$W^{k-1} = \bigcup_{x \in E^{k-1}} \left\{ W^{u,\delta}(x) \right\} \cup \bigcup_{t>0} S(t)\Gamma^{k-1}. \tag{6.50}$$

To establish that (6.49) holds we consider three separate cases corresponding to a breakdown of W^{k-1} into three different subsets.

(a) Note that $\Gamma^{k-1} \subset U^{k-1}$ by (6.6). By Lemma 6.24 $\mathcal{A}^k$ attracts all compact subsets of U^{k-1} and so there exists t_{k-1} such that

$$\mathrm{dist}(S(t)\Gamma^{k-1}, \mathcal{A}^k) \leq \frac{\epsilon}{2^k} \quad \forall t \geq t_{k-1}.$$

But by the inductive hypothesis (6.46) and (6.48) we have

$$\begin{aligned} \mathrm{dist}\Big(\bigcup_{t \geq t_{k-1}} S(t)\Gamma^{k-1}, \mathcal{A}^{k-1}_{\Delta t} \Big) &\leq \mathrm{dist}\Big(\bigcup_{t \geq t_{k-1}} S(t)\Gamma^{k-1}, \mathcal{A}^{k}_{\Delta t} \Big) \\ &\leq \mathrm{dist}\Big(\bigcup_{t \geq t_{k-1}} S(t)\Gamma^{k-1}, \mathcal{A}^{k} \Big) + \mathrm{dist}(\mathcal{A}^k, \mathcal{A}^k_{\Delta t}) \\ &\leq 2\epsilon/2^k = \epsilon/2^{k-1}. \end{aligned} \tag{6.51}$$

(b) Recall the time t_{k-1} given in (a), the constant $\Delta_k > 0$ from (6.46) and the constant K from Assumption 3.7.

Now, given $x \in E^{k-1}$, let $X \in E^{k-1}_{\Delta t}$ be the approximate fixed point given by Lemma 6.25. By Lemma 6.25 it follows that there exist $\Delta^1 > 0$ and $\delta' > 0$ such that

$$\mathrm{dist}(W^u_\delta(x), W^u_{\delta',\Delta t}(X)) < \frac{\epsilon}{2^k e^{K(t_{k-1}+\Delta t_k)}} < \frac{\epsilon}{2^{k-1}}, \quad \forall \Delta t \in (0, \Delta^1]. \tag{6.52}$$

Hence

$$\mathrm{dist}(W^u_\delta(x), \mathcal{A}^{k-1}_{\Delta t}) < \frac{\epsilon}{2^{k-1}}, \tag{6.53}$$

since $W^u_{\delta',\Delta t}(x) \subseteq \mathcal{A}^{k-1}_{\Delta t}$.

(c) Now we show that $\mathrm{dist}(S(t)\Gamma^{k-1}, \mathcal{A}^{k-1}_{\Delta t}) \leq \epsilon/2^{k-1}$ for $t \in (0, t_{k-1}]$. By Theorem 4.11 we can choose $\Delta t^2 > 0$ such that for $\Delta t \in (0, \Delta t^2]$ and any $v, w \in \mathbb{R}^p$ satisfying

$$\|v - w\| \leq \frac{\epsilon}{2^k e^{K(t_{k-1}+\Delta_k)}} \tag{6.54}$$

we have

$$\|S(n\Delta t)v - S^n_{\Delta t}w\| \leq \epsilon/2^{k-1} \quad \text{for} \quad n\Delta t \leq t_{k-1} + \Delta_k. \tag{6.55}$$

Now let $\Delta_{k-1} = \min(\Delta t^1, \Delta t^2, \Delta_k)$. Suppose $\Delta t < \Delta_{k-1}$ and that $u \in S(t)\Gamma^{k-1}$ for $t \in (0, t_{k-1}]$. Then, by Lemma 4.2, there exists v so that

$$v \in \overline{B}(x, \delta) \cap W^u(x)$$

for some $x \in E^{k-1}$, such that $S(n\Delta t)v = u$ and $n\Delta t \in [0, t_{k-1} + \Delta t_k]$. By (6.52) there exists $w \in W^u_{\Delta t}(X)$ such that (6.54) holds. Now, by (6.55),

$$\|u - S^n_{\Delta t}w\| = \|S(n\Delta t)v - S^n_{\Delta t}w\| \leq \epsilon/2^{k-1} \quad \forall \Delta t \in (0, \Delta_{k-1}].$$

Since the unstable manifold is invariant by Theorem 4.2 and is contained in the attractor by Theorem 6.11, we have $S^n_{\Delta t}w \in \mathcal{A}^{k-1}_{\Delta t}$; hence it follows that

$$\text{dist}(u, \mathcal{A}^{k-1}_{\Delta t}) \leq \epsilon/2^{k-1}.$$

But u is an arbitrary point in $\bigcup_{t\in(0,t_{k-1}]} S(t)\Gamma^{k-1}$ and hence

$$\text{dist}\left(\bigcup_{t\in(0,t_{k-1}]} S(t)\Gamma^{k-1}, \mathcal{A}^{k-1}_{\Delta t}\right) \leq \frac{\epsilon}{2^{k-1}}. \tag{6.56}$$

By the definition (6.50) of W^{k-1}, the estimates (6.51), (6.53) and (6.56) together establish that

$$\text{dist}(W^{k-1}, \mathcal{A}^{k-1}_{\Delta t}) \leq \epsilon/2^{k-1}$$

and complete the proof of the lemma. □

Proof of Theorem 6.26. First note that, since $S(t)$ has an attractor $\mathcal{A}$, it follows from Corollary 6.18 that $S^1_{\Delta t}$ has a nearby uniformly asymptotically stable set $\Lambda_{\Delta t}$ and an attractor $\mathcal{A}_{\Delta t} = \omega(\Lambda_{\Delta t})$ satisfying

$$\text{dist}(\mathcal{A}_{\Delta t}, \mathcal{A}) \to 0 \quad \text{as} \quad \Delta t \to 0.$$

Thus it remains to establish the lower semicontinuity result that

$$\text{dist}(\mathcal{A}, \mathcal{A}_{\Delta t}) \to 0 \quad \text{as} \quad \Delta t \to 0.$$

It is sufficient to prove that given any $\epsilon > 0$ there exists $\Delta = \Delta(\epsilon)$ such that if $\Delta t < \Delta$ then $\text{dist}(\mathcal{A}, \mathcal{A}_{\Delta t}) \leq \epsilon$.

Recall the notation and decomposition of $\mathcal{A}$ and $\mathcal{A}_{\Delta t}$ given in (6.42) and (6.44). Note that, as described above, $\mathcal{A}^N = E^N$ and $\mathcal{A}^N_{\Delta t} = E^N_{\Delta t}$. Applying Lemma 6.25 we deduce that there exists $\Delta_N > 0$ such that

$$\text{dist}(E^N, E^N_{\Delta t}) = \text{dist}(\mathcal{A}^N, \mathcal{A}^N_{\Delta t}) \leq \epsilon/2^N \quad \forall \Delta t \in (0, \Delta_N].$$

By induction, using Lemma 6.27, we deduce that there exists Δ_k such that, for $k = 1, \ldots, N$

$$\text{dist}(\mathcal{A}^k, \mathcal{A}^k_{\Delta t}) \leq \epsilon/2^k \quad \forall \Delta t \in (0, \Delta_k].$$

In particular, since $\mathcal{A}^1 = \mathcal{A}$ and $\mathcal{A}^1_{\Delta t} \subseteq \mathcal{A}_{\Delta t}$ by (6.45) we deduce that

$$\text{dist}(\mathcal{A}, \mathcal{A}_{\Delta t}) \leq \epsilon \quad \forall \Delta t \in (0, \Delta_1].$$

This completes the proof. □

6.5. *Lower semicontinuity of global unstable manifolds*

In this section we examine the lower semicontinuity of the global unstable manifolds of (1.1) with respect to numerical perturbation. Recall that we have already studied a related question for the local unstable manifold in Section 4 – see Corollary 4.13. Since the global unstable manifold of a fixed point is necessarily contained in the global attractor by Theorem 6.11, it is natural to study them in the context of attractors. As a corollary we shall obtain a simpler proof of Theorem 6.26 since, for gradient systems with only hyperbolic equilibria, the attractor comprises unstable manifolds of equilibrium points.

Theorem 6.28 **(Global unstable manifold under approximation)** Assume that (1.1) has an equilibrium point v and that V is the equilibrium point of (1.2) which converges to v as $\Delta t \to 0$. Then if the unstable manifold $W^u(v)$ is bounded it follows that

$$\mathrm{dist}(\overline{W^u(v)}, \overline{W^u_{\Delta t}(V)}) \to 0 \qquad \text{as} \qquad \Delta t \to 0.$$

Proof. It is sufficient to prove that, given any $\epsilon > 0$, there exists $\Delta > 0$ such that for every $y \in W^u(v)$ there exists $Y \in W^u(V)$ with the property that $\|y - Y\| \leq 2\epsilon$ for $\Delta t \in (0, \Delta]$.

Recall $\partial B(v; r)$ and Γ given by (3.3) and (4.7). Now set

$$\mathcal{W} = W^u(v) \backslash W^{u,\epsilon}(v). \tag{6.57}$$

Then, for ϵ sufficiently small,

$$\mathcal{W} = \bigcup_{t>0} S(t)\Gamma.$$

Since $W^u(v)$ is bounded it follows that $\overline{\mathcal{W}}$ is compact. It may be noted that $\{B(x;\epsilon): x \in \mathcal{W}\}$ is an ϵ-cover for $\overline{\mathcal{W}}$ and hence, since $\overline{\mathcal{W}}$ is compact, we may extract a finite subcover. Denote this subcover by $\{B_i(\epsilon)\}_{i=1}^I$ and note that each $B_i(\epsilon)$ contains a point $y_i \in \mathcal{W}$, where $B_i(\epsilon) = B(y_i, \epsilon)$. By construction there exists $x_i \in \Gamma$ and $T_i > 0$ such that $S(T_i)x_i = y_i$ for each $y_i \in \mathcal{W}$. Now, by Corollary 4.13, it follows that there exists $X_i \in W^u(V)$ and $\Delta(i) > 0$ such that

$$\|x_i - X_i\| \leq \epsilon/(2\mathrm{e}^{KT_i}) \quad \forall \Delta t \in (0, \Delta(i)]; \tag{6.58}$$

by the invariance of the unstable manifold (see Lemma 4.2) it follows that $Y_i = S^n_{\Delta t} X_i \in W^u(V)$.

It now follows from Theorems 3.8 and (6.58) that

$$\|y_i - Y_i\| \leq (\mathrm{e}^{KT} - 1)\Delta t^r + \tfrac{1}{2}\epsilon$$

for $\Delta t \in (0, \Delta(i)]$. Thus, by further reduction of $\Delta(i)$ if necessary, we find that

$$\|y_i - Y_i\| \leq \epsilon, \quad \forall \Delta t \in (0, \Delta(i)].$$

Since I is finite, we deduce that there exists $\{Y_i\}_{i=1}^I$ each lying in $W^u(V)$ and $\Delta > 0$ such that

$$\max_{1 \le i \le I} \|y_i - Y_i\| \le \epsilon \quad \forall \Delta t \in (0, \Delta].$$

Thus, since y_i is the centre of $B_i(\epsilon)$, we deduce that for every $y \in B_i(\epsilon)$ and $i : 1 \le i \le I$ there exists $Y_i \in W^u(V)$ such that

$$\|y - Y_i\| \le 2\epsilon \quad \forall \Delta t \in (0, \Delta].$$

Since the $B_i(\epsilon), i = 1, \dots, I$ form a cover of $\overline{\mathcal{W}}$ we deduce that

$$\text{dist}(\overline{\mathcal{W}}, W^u(V)) \le 2\epsilon \quad \forall \Delta t \in (0, \Delta]. \tag{6.59}$$

Now, by Corollary 4.13 there exists $\delta' > 0$ such that

$$\text{dist}(W^{u,\delta}(v), W^{u,\delta'}_{\Delta t}(V)) \le 2\epsilon \quad \forall \Delta t \in (0, H], \tag{6.60}$$

possibly by further reduction of Δ. Putting (6.59) and (6.60) the result follows by (6.57). □

We now use this result to study lower semicontinuity of attractors. We make the following assumption:

Assumption 6.29 The dynamical system (1.1) has a global attractor $\mathcal{A}$ where

$$\mathcal{A} = \bigcup_{x \in \mathcal{E}'} \overline{W^u(x)}$$

and $\mathcal{E}'$ comprises a finite number of hyperbolic equilibrium points of (1.1).

Note that this assumption is a consequence of Assumption 6.23 but that Assumption 6.29 is weaker. For example, Assumption 6.29 admits the equations (3.12) with global attractor given by the disc $x^2 + y^2 \le 1$. Under Assumption 6.29 we may prove lower semicontinuity of the attractor, yielding a simpler proof of Theorem 6.26.

Corollary 6.30 (**Lower semicontinuity of attractors**) Consider (1.1) under Assumption 6.29. Then there exists $\Delta > 0$ such that for $\Delta t \in (0, \Delta]$ the numerical solution (1.2) possesses an attractor $\mathcal{A}_{\Delta t}$ which satisfies

$$d_{\text{H}}(\mathcal{A}_{\Delta t}, \mathcal{A}) \to 0 \qquad \text{as} \qquad \Delta t \to 0.$$

Proof. It follows from Corollary 6.18 that there exists an approximate attractor $\mathcal{A}_{\Delta t}$ satisfying

$$\text{dist}(\mathcal{A}_{\Delta t}, \mathcal{A}) \to 0 \qquad \text{as} \qquad \Delta t \to 0.$$

Thus it remains to establish the lower semicontinuity result that

$$\text{dist}(\mathcal{A}, \mathcal{A}_{\Delta t}) \to 0 \qquad \text{as} \qquad \Delta t \to 0. \tag{6.61}$$

Let $v \in \mathcal{E}'$. By Theorem 6.28 we deduce that there exists a fixed point V of $S^1_{\Delta t}$ such that

$$\operatorname{dist}(\overline{W^u(v)}, \overline{W^u_{\Delta t}(V)}) \to 0 \qquad \text{as} \qquad \Delta t \to 0.$$

But, by Theorem 6.11, $\overline{W^u_{\Delta t}(V)} \subseteq \mathcal{A}_{\Delta t}$ and hence it follows that

$$\operatorname{dist}(\overline{W^u(v)}, \mathcal{A}_{\Delta t}) \to 0 \qquad \text{as} \qquad \Delta t \to 0.$$

Since Assumption 6.29 holds it is clear that (6.61) follows and the proof is complete. □

6.6. Bibliography

The material in Section 6.1 can be found in several books, notably Hale (1988), Bhatia and Szego (1970), Hirsch and Smale (1974) and Yoshizawa (1966). For Theorem 6.2 and related results see Hale (1988). For Theorem 6.5 see Yoshizawa (1966); Bhatia and Szego (1970) also contains similar results on converse theorems for Lyapunov functions. The importance of gradient systems is that they enable an explicit decomposition of the dynamics into equilibrium points together with a Lyapunov function decreasing along all trajectories connecting equilibrium points at $t = \pm\infty$; this idea can be generalized to allow the equilibrium points to be replaced by more general limit sets. For results concerning gradient systems see Hirsch and Smale (1974) and Hale (1988).

The first article concerning a detailed analysis of the effect of numerical approximation on sets possessing some general form of attractivity is Kloeden and Lorenz (1986). They essentially proved Theorem 6.12, which concerns upper and lower semicontinuity of uniformly asymptotically stable sets Corollary 6.18, concerning upper semicontinuity of attractors, is a consequence of their work. The approach of Kloden and Lorenz, using Lyapunov functions, was generalized to partial differential equations in Kloeden and Lorenz (1989); it is extended to multistep methods in (Kloeden and Lorenz, 1990). Theorem 6.20 is due to Hale and Raugel (1989) although their result is more general, concerning an arbitrary attractor in a Banach space. A similar result may also be found in Temam (1988). As can be seen by comparing the work required to prove Corollary 6.18 (via Theorem 6.12) and Theorem 6.20, the approach of Hale *et al.* (1988) is considerably shorter than that of Kloeden and Lorenz (1986) if interest is focused only on attractors. Furthermore, as pointed out in Hill and Suli (1993), results strongly related to those of Kloeden and Lorenz (1986) concerning uniformly asymptotically stable sets can be deduced from the approach of Hale *et al.* (1988) – this is then proved in Theorem 6.22. Theorem 6.21, concerning the rate of convergence of the attractor when it is exponentially attracting, has not appeared in the literature; however, the basic idea for the proof is contained in the study of exponential attractors in Babin and Vishik (1992). The issues concerning the derivation of upper semicontinuity results for partial differential equations are considerably more subtle than for ordinary differential equations since, frequently, the spaces in which the attractors lie are not sufficiently regular to apply standard smooth initial data error bounds; Larsson (1989) contains a self-contained and clear presentation of this issue in the context of the finite-element approximation of the reaction-diffusion equation and that work is generalized to cover the Cahn–Hilliard equation in Elliott and Larsson (1992). In Yin-Yan (1993) similar issues are considered for finite difference approximations of the Navier–Stokes equation and in Lord and Stuart (1994) for finite difference approximations of the Ginzburg–Landau equation. The whole question of the existence of global attractors under approximation is reviewed in Humphries *et al.* (1994).

Section 6.4 contains an exposition of the work of Hale and Raugel (1989) concerning the lower semicontinuity of attractors for gradient systems. The presentation given here is closely related to that given in Humphries and Stuart (1994) where Runge–Kutta methods are studied in this context. As can be seen, the proof is not at all straightforward and for this reason more accessible proofs have been sought. An alternative, more accessible approach is described in Section 6.5, culminating in Corollary 6.30. This approach is due to Humphries (1994); it is interesting to note that, whilst a more general class of problems is considered in Section 6.5 than in Section 6.6, Humphries' proof of lower semicontinuity is more straightforward than that of Hale and Raugel – compare Theorems 6.26 and Corollary 6.30. Furthermore, the approach of Humphries also yields further information about the global unstable manifolds and can be trivially modifed to obtain upper semicontinuity of global unstable manifolds. The approach of Humphries is extended to partial differential equations in Humphries *et al.* (1994).

Note that we have described a variety of results in this section some of which supersede others, at least on a superficial level. However, since it is not clear in which direction these results can be generalized, we feel it worthwhile to document in detail the various approaches to these problems.

7. Conclusions

In this article we have concentrated on the convergence of limit sets and invariant sets of a time continuous semigroup, under numerical approximation of the evolution semigroup by a time discrete semigroup. It should be clear that a fairly full picture of this subject has now been developed and that, furthermore, there are a variety of approaches to some of the questions studied. It is natural to ask at this point what future directions are likely to be of scientific interest in this subject area. We give a purely subjective answer by describing areas likely to be fruitful for future development.

7.1. *Convergence of attractors*

As the work of Section 6 shows, it is not in general possible to prove lower semicontinuity of attractors. In practice this means that numerical computations may 'miss' part of the true attractor. The only situations in which it is currently possible to prove both upper and lower semicontinuity are those in which something is known about the flow on the attractor – specifically, lower semicontinuity has been proved for certain hyperbolic gradient systems (Section 6.4) and for systems whose attractors are the union of the closure of unstable manifolds of equilibria (see Section 6.5). The important point about the assumptions made in Sections 6.4 and 6.5 is that they amount to a form of *hyperbolicity* of the flow on the attractor. There are two directions that the study of attractors can be taken.

The first is to investigate further the hyperbolicity conditions on the flow on the attractor which yield lower semicontinuity results. However, since such questions are by no means fully understood even in the context of smooth perturbations of the vector field in (1.1), this question appears quite difficult. To pursue this avenue will require a parallel development of the general theory of structural stability of attractors. The work of Pliss and Sell (1991) is of interest in this context.

The second is to weaken the concept of attractor to obtain an object which is, for example, exponentially attractive and retains favourable properties under perturbation. In a sense this is what the concept of uniformly asymptotically stable sets (see Section 6.2) does. However, other generalizations are possible. The inertial manifold is an enlargement of the global attractor to obtain an exponentially attracting object and it is possible to prove both upper and lower semicontinuity for the inertial manifold – see Foais *et al.* (1988), Demengel and Ghidaglia (1989) and Jones and Stuart (1993). If this work could be combined with, for example, the approach of Pliss and Sell (1991) it might be possible to make useful deductions about the relationship between the flows on the true and numerical attractors. The concept of *intertial sets* is of also interest (see Eden *et al.* (1990)). The inertial set is an enlargment of the attractor to a positively invariant set which is contained in an inertial manifold and is exponentially attracting. It is plausible that this object is both upper and lower semicontinuous with respect to numerical perturbations of the semigroup.

7.2. *Shadowing*

We have not explicitly described the subject of shadowing at all in this article although it will almost certainly play an increasingly important role in making statements about the meaning of long-time computations. The general area of shadowing is enormous and there is not room in this article to do it justice. We briefly mention some existing literature in this area.

Seminal work in this area can be found in Hammel *et al.* (1987; 1988) where the effect of round-off error is studied on the computer iteration of certain chaotic maps, such as the quadratic map and the Henon map. Subsequently this work was generalized to consider the effect of numerical approximation and the following references in the area are representative of this growing field: Chow and Palmer (1990a,b), Chow and Van-Vleck (1993), Corless (1992), Corless and Pilyugin (1993) and Sauer and Yorke (1991).

The concept of shadowing is closely related to the notion of *backward error analysis*, familiar to numerical analysts. The approach of Beyn (1987b) to the approximation of phase portraits is a form of backward error analysis and the idea has been taken further in Eirola (1993), Corless and Corliss (1991) and Elliott and Stuart (1994).

7.3. Direct numerical approximation of invariant sets

In this article we have been mainly concerned with numerical approximation of the semigroup $S(t)$ generated by (1.1). This approach, where the invariant sets of the differential equation are observed indirectly as corresponding invariant sets in the numerical method, is sometimes termed the *indirect approach* (Beyn, 1992). An alternative is the *direct approach* where a numerical method is constructed to compute a given invariant set directly. To do this it is necessary to set-up defining equations, typically a boundary value problem, for the invariant set of interest. This is an area in which there is some existing work but in which there is much room for further development, especially as increases in computational power mean that computations previously prohibitively expensive, in comparison with standard indirect simulations, are now straightforward. We briefly describe some of the existing literature.

The simplest invariant objects (1.1) are, of course, equilibrium solutions and much literature exists concerning solution of equilibrium problem exists; indeed, the development of the subject is such that excellent packages now exist – for example the package PITCON; see Rheinboldt (1986). It is also true in the case of periodic solutions arising from Hopf bifurcations that excellent packages exist – see for example the package AUTO described in Doedel and Kervenez (1986). The computation of quasi-periodic solutions (invariant tori) has not yet evolved to the extent where automatic software is available. However, considerable advances have been made and the following references summarise the existing literature: Aronson *et al.* (1987), Van Veldhuizen (1988), Dieci *et al.* (1991) and Dieci and Lorenz (1993). The article by Moore (1993) contains a unified treatment of computational techniques for periodic solutions and invariant tori.

It appears that it is not possible to formulate the question of the existence of a general compact invariant set as a boundary value problem. However, there are a number of direct computational methods of relevance to the study of general, possibly strange, invariant sets. In Beyn (1990) the concept of computing *connecting orbits* directly was introduced and analysed. Connecting orbits are solutions of (1.1) which connect together two limit sets as $t \to \pm\infty$; they are of importance in understanding dynamical systems in many different contexts, including the existence of chaos. The work of Beyn has been taken further in Bai *et al.* (1993), Moore (1993) and Liu *et al.* (1993). Connecting orbits may be viewed simply as the intersection of stable and unstable manifolds; the direct computation of stable and unstable manifolds is considered in Homburg *et al.* (1993) and Hubert (1993). In many circumstances the charactersitics of a strange attractor can be well understood by investigating the Lyapunov exponents; an article containing recent developments in this area, together with a survey of the existing literature, is Dieci *et al.* (1993).

7.4. Generalization to partial differential equations

Much of the numerical analysis described in Sections 3, 4 and 5 has only been fully developed for ordinary differential equations and there are many interesting remaining questions concerning extensions to partial differential equations. Some of these have been addressed for specific equations (typically in reaction-diffusion) and specific methods.

Acknowledgements This work was supported by the Office of Naval Research, contract number N00014-92-J-1876 and by the National Science Foundation, contract number DMS-9201727. I am greatly indebted to the people who helped in the task of checking this article: Fengshan Bai, Chris Budd, Adrian Hill, Arieh Iserles, Yunkang Liu, Gerald Moore and Alastair Spence.

REFERENCES

F. Alouges and A. Debussche (1991), 'On the qualitative behaviour of the orbits of a parabolic partial differential equation and its discretization in the neighbourhood of a hyperbolic fixed point', *Numer. Funct. Anal. Opt.* **12**, 253–269.

F. Alouges and A. Debussche (1993), 'On the discretization of a partial differential equation in the neighbourhood of a periodic orbit', *Numer. Math.* **65**, 143–175.

D.G. Aronson, E.J. Doedel and H.G. Othmer (1987), 'An analytical and numerical study of the bifurcations in a system of linearly coupled oscillators', *Physica* D, **25**, 20–104.

D.K. Arrowsmith and C.M. Place (1990), *An Introduction to Dynamical Systems*, Cambridge Univeristy Press (Cambridge).

A. Babin and M.I. Vishik (1992), 'Attractors of evolution equations', in *Studies in Mathematics and its Applications*, North-Holland (Amsterdam).

F. Bai, A. Spence, A.M. Stuart (1993), 'The numerical computation of heteroclinic connections in systems of gradient partial differential equations', *SIAM J. Appl. Math.* **53**, 743–769.

W.-J. Beyn (1987a) 'On invariant closed curves for one-step methods', *Numer. Math.* **51**, 103–122.

W.-J. Beyn (1987b) 'On the numerical approximation of phase portraits near stationary points', *SIAM J. Numer. Anal.* **24**, 1095–1113.

W.-J. Beyn (1990), 'The numerical computation of connecting orbits in dynamical systems', *IMA J. Numer. Anal.* **9**, 379–405.

W.-J. Beyn (1992), 'Numerical methods for dynamical systems', in *Numerical Analyis; Proceedings of the SERC Summer School, Lancaster, 1990* (W.A. Light, ed.), Oxford University Press (Oxford).

W.-J. Beyn and E. Dodel (1981), 'Stability and multiplicity of solutions to discretizations of nonlinear ordinary differential equations', *SIAM J. Sci. Stat. Comput.* **2**, 107–120.

W.-J. Beyn and J. Lorenz (1987), 'Center manifolds of dynamical systems under discretization', *Numer. Func. Anal. Opt.* **9**, 381–414.

N.P. Bhatia and G.P. Szego (1970), *Stability Theory of Dynamical Systems*, Springer (New York).

M. Braun and J. Hershenov (1977), 'Periodic solution of finite difference equations', *Quart. Appl. Maths* **35**, 139–147.

F. Brezzi, S. Ushiki and H. Fujii (1984), 'Real and ghost bifurcation dynamics in difference schemes for ordinary differential equations', in *Numerical Methods for Bifurcation Problems* (T. Kupper, H.D. Mittleman and H. Weber, eds), Birkhauser (Boston).

D. Broomhead and A. Iserles (1992), *Proceedings of the IMA Conference on the Dynamics of Numerics and the Numerics of Dynamics, 1990*, Oxford University Press (Oxford).

C.J. Budd (1990), 'The dynamics of numerics and the numerics of dynamics', Bristol Univerisity Applied Mathematics Report AM-90-13.

C.J. Budd (1991), 'IMA Conference on the dynamics of numerics and the numerics of dynamics. Report of the meeting', *Bull. IMA* **27**, 56–58.

K. Burrage and J. Butcher (1979), 'Stability criteria for implicit Runge–Kutta processes,' *SIAM J. Numer. Anal.* **16**, 46–57.

J.C. Butcher (1975), 'A stability property of implicit Runge–Kutta methods', *BIT* **15**, 358–361.

J.C. Butcher (1987), *The Numerical Analysis of Ordinary Differential Equations: Runge–Kutta Methods and General Linear Methods*, Wiley (Chichester).

M.P. Calvo and J.M. Sanz-Serna (1992), 'Variable steps for symplectic integrators', in *Numerical Analysis, 1991* (D.F. Griffiths and G.A. Watson, eds), Longman (London).

M.P. Calvo and J.M. Sanz-Serna (1993a), 'Reasons for a failure. The integration of the two-body problem with a symplectic Runge–Kutta–Nystrom code with stepchanging facilities', in *Equadiff-91* (C. Perello, C. Simo and J. de Sola-Marales, eds), World Scientific (Singapore).

M.P. Calvo and J.M. Sanz-Serna (1993b), 'The development of variable-step symplectic integrators, with applications to the two-body problem', *SIAM J. Sci. Comput.* **14**, 936–952.

J. Carr (1982), *Applications of Centre Manifold Theory*, Springer (New York).

S.N. Chow and K.J. Palmer (1989), 'The accuracy of numerically computed orbits of dynamical systems', in *Equadiff 1989, Prague.*

S.N. Chow and K.J. Palmer (1990a), 'The accuracy of numerically computed orbits of dynamical systems in $\mathbb{R}^k$', Georgia Inst. Tech. Report CDSNS90–28.

S.N. Chow and K.J. Palmer(1990b), 'On the numerical computation of orbits of dynamical systems: the higher dimensional case', Georgia Inst. Tech. Report CDSNS90–32.

S.N. Chow and E. Van Vleck (1993), 'A shadowing lemma approach to global error analysis for initial value ODEs', to appear in *SIAM J. Sci. Stat. Comput.*

P. Constantin, C. Foias, B. Nicolaenko and R. Temam (1989), *Integral Manifolds and Inertial Manifolds for Dissipative Partial Differential Equations, Applied Mathematical Sciences*, Springer (New York).

R.M. Corless (1992), 'Defect-controlled numerical methods and shadowing for chaotic differential equations', *Physica* D **60**, 323–334.

R.M. Corless and G.F. Corliss (1991), 'Rationale for guaranteed ODE defect control', Argonne Nat. Lab. Preprint MCS-P273–1191.

R.M. Corless and S.Y. Pilyugin (1993), 'Approximate and real trajectories for generic dynamical systems', to appear in *J. Math. Anal. Appl.*

M. Crouziex and J. Rappaz (1990), *Numerical Approximation in Bifurcation Theory*, Masson–Springer (Paris).

G. Dahlquist (1975), 'Error analysis for a class of methods for stiff non-linear initial value problems', in *Numerical Analysis, Dundee 1975* (G.A. Watson, ed.), Springer (New York), 60–74.

G. Dahlquist (1978), 'G-stability is equivalent to A-stability' *BIT* **18**, 384–401.

K. Dekker and J.G. Verwer (1984), *Stability of Runge–Kutta Methods for Stiff Nonlinear Differential Equations*, North-Holland (Amsterdam).

F. Demengel and J.M. Ghidaglia (1989), 'Time-discretization and inertial manifolds', *Math. Mod. Numer. Anal.* **23**, 395-404.

R.L. Devaney (1989), *An Introduction to Chaotic Dynamical Systems*, Addison-Wesley (New York).

L. Dieci and J. Lorenz (1993), 'Computation of invariant tori by the method of characteristics', to be submitted.

L. Dieci, J. Lorenz and R.D. Russell (1991), 'Numerical calculation of invariant tori', *SIAM J. Sci. Stat. Comput.* **12**, 607–647.

L. Dieci, R.D. Russell and E. Van Vleck (1993), 'On the computation of Lyapunov exponents for continuous dynamical systems', submitted to *SIAM J. Numer. Anal.*

H.T. Doan (1985), 'Invariant curves for numerical methods', *Quart. Appl. Math.* **3**, 385–393.

E.J. Doedel and J.P. Kervenez (1986), 'AUTO: Software for continuation of bifurcation problems in ordinary differential equations', Appl. Maths Tech. Rep., California Institute of Technology.

P.G. Drazin (1992), *Nonlinear Dynamics*, Cambridge University Press (Cambridge).

A. Eden, C. Foias, B. Nicolaenko and R. Temam (1990), 'Inertial sets for dissipative evolution equations', IMA Preprint Series, #694.

T. Eirola (1988), 'Invariant curves for one-step methods', *BIT* **28**, 113–122.

T. Eirola (1989), 'Two concepts for numerical periodic solutions of ODEs', *Appl. Math. Comput.* **31**, 121–131.

T. Eirola (1993), 'Aspects of backward error analysis of numerical ODEs', *J. Comput. Appl. Math.* **45**, 65–73.

T. Eirola and O. Nevanlinna (1988), 'What do multistep methods approximate?', *Numer. Math.* **53**, 559–569.

C.M. Elliott and S.Larsson (1992), 'Error estimates with smooth and nonsmooth data for a finite element method for the Cahn–Hilliard equation', *Math. Comput.* **58**, 603–630.

C.M. Elliott and A.M. Stuart (1993), 'Global dynamics of discrete semilinear parabolic equations.' to appear in *SIAM J. Numer. Anal.*

C.M. Elliott and A.M. Stuart (1994), 'Error analysis for gradient groups with applications', in preparation.

B. Enquist (1969), 'On difference equations approximating linear ordinary differential equations,' Report 21, Department of Computer Science, Uppsala University.

N. Fenichel (1971), 'Persistence and smoothness of invariant manifolds for flows', *Indiana Univ. Math. J.* **21**, 193–226.

C. Foias, G. Sell and R. Temam (1988), 'Inertial manifolds for nonlinear evolutionary equations', *J. Diff. Eqns* **73**, 309-353.

B.M. Gacay (1993), 'Discretization and some qualitative properties of ordinary differential equations about equilibria', Budapest University of Technology, Technical Report.

D.F. Griffiths, P.K. Sweby and H.C. Yee (1992), 'On spurious asymptotic numerical solutions of explicit Runge–Kutta methods', *IMA J. Numer. Anal.* **12**, 319–338.

J. Guckenheimer and P. Holmes (1983), *Nonlinear Oscillations, Dynamical Systems and Bifurcations of Vector Fields*, Springer (New York).

E. Hairer, A. Iserles and J.M. Sanz-Serna (1990), 'Equilibria of Runge–Kutta methods', *Numer. Math.* **58**, 243–254.

E. Hairer, S.P. Norsett and G. Wanner (1987), *Solving Ordinary Differential Equations, Parts I & II*, Springer (New York).

J.K. Hale (1969), *Ordinary Differential Equations*, Wiley (New York).

J.K. Hale(1988), *Asymptotic Behaviour of Dissipative Systems*, AMS Mathematical Surveys and Monographs 25, American Mathematical Society (Providence, RI).

J.K. Hale and H. Kŏcak (1991), *Dynamics and Bifurcations*, Springer (New York).

J.K. Hale, L. Magalhaes and W. Oliva (1984), *An Introdcution to Infinite Dimensional Dynamical Systems*, Springer (New-York).

J.K. Hale and G. Raugel (1989), 'Lower semicontinuity of attractors of gradient systems and applications', *Annali di Mat. Pura. Applic.* **CLIV**, 281–326.

J.K. Hale, X.-B. Lin and G. Raugel (1988), 'Upper semicontinuity of attractors for approximations of semigroups and partial differential equations,' *Math. Comput.* **50**, 89–123.
S. Hammel, J.A. Yorke and C. Grebogi (1987), 'Do numerical orbits of chaotic dynamical processes represent true orbits?', *J. Complexity* **3**, 136–145.
S. Hammel, J.A. Yorke and C. Grebogi (1988), 'Numerical orbits of chaotic processes represent true orbits', *Bull. AMS* **19**, 465–470.
P. Hartman (1982), *Ordinary Differential Equations*, Birkhauser (Boston).
D. Henry (1981), *Geometric Theory of Semilinear Parabolic Equations. Lecture Notes in Mathematics*, Springer (New York).
J.G. Heywood and R. Rannacher (1986), 'Finite element approximations of the Navier-Stokes problem. Part II: Stability of solutions and error estimates uniform in time,' *SIAM J. Numer. Anal.* **23**, 750–777.
A.J. Homburg, H.M. Osinga and G. Vegter (1993), 'On the numerical computation of invariant manifolds', University of Groningen Technical Report.
A.T. Hill (1994), 'Global dissipativity for multistep methods,' in preparation.
A.T. Hill and E. Suli (1993), 'Upper semicontinuity of attractors for linear multistep methods approximating sectorial evolution equations,' submitted to *Math. Comput.*
A.T. Hill and E. Suli (1994), 'Set convergence for discretizations of the attractor,' submitted to *Numer. Math.*
M.W. Hirsch and S. Smale (1974), *Differential Equations, Dynamical Systems and Linear Algebra*, Academic Press (London).
E. Hubert (1993), 'Computation of stable manifolds', MSc Thesis, Imperial College (London).
A.R. Humphries (1993), 'Spurious solutions of numerical methods for initial value problems', *IMA J. Numer. Anal.* **13**, 262–290.
A.R. Humphries (1994), 'Approximation of attractors and invariant sets by Runge–Kutta methods', in preparation, 1994.
A.R. Humphries and A.M. Stuart (1994), 'Runge–Kutta methods for dissipative and gradient dynamical systems', to appear in *SIAM J. Numer. Anal.*
A.R. Humphries, D.A. Jones and A.M. Stuart (1994), 'Approximation of dissipative partial differential equations over long-time intervals', in *Numerical Analysis, Dundee, 1993* (D.F. Griffiths and G.A. Watson, eds), Longman (New York).
A. Iserles (1990), 'Stability and dynamics of numerical methods for nonlinear ordinary differential equations', *IMA J. Numer. Anal.* **10**, 1–30.
A. Iserles, A.T. Peplow and A.M. Stuart (1991) 'A unified approach to spurious soluions introduced by time discretization. Part I: basic theory', *SIAM J. Numer. Anal.* **28**, 1723–1751.
D.A. Jones and A.M. Stuart (1993), 'Attractive invariant manifolds under approximation', submitted to *J. Diff. Eqns*
U. Kirchgraber (1986), 'Multi-step methods are essentially one-step methods', *Numer. Math.* **48**, 85–90.
P. Kloeden and J. Lorenz (1986), 'Stable attracting sets in dynamical systems and their one-step discretizations,' *SIAM J. Numer. Anal.* **23**, 986–995.
P. Kloeden and J. Lorenz (1989), 'Liapunov stability and attractors under discretization', in *Differential Equations, Proceedings of the Equadiff Conference*

(C.M. Dafermos, G. Ladas and G. Papanicolaou, eds), Marcel-Dekker (New York).

P. Kloeden and J. Lorenz (1990), 'A note on multistep methods and attracting sets of dynamical systems', *Num. Math.* **56**, 667–673.

P. Kloeden and K.J. Palmer (1994), 'Chaotic numerics: the approximation and computation of complicated dynamics', Proceedings of the Conference on Chaotic Numerics, Geelong, Australia, July 12th–16th 1993.

O. Ladyzhenskaya (1991), *Attractors for Semigroups and Evolution Equations*, Cambridge University Press (Cambridge).

J. Lambert (1991), *Numerical Methods for Ordinary Differential Equations*, Wiley (Chichester).

S. Larsson (1989), 'The long-time behaviour of finite element approximations of solutions to semilinear parabolic problems', *SIAM J. Numer. Anal.* **26**, 348–365.

S. Larsson (1992), 'Non-smooth data error estimates with applications to the study of long-time behaviour of finite element solutions of semilinear parabolic problems', Preprint, Chalmers University (Sweden).

S. Larsson and J.M. Sanz-Serna (1993), 'The behaviour of finite element solutions of semilinear parabolic problems near stationary points', submitted to *SIAM J. Numer. Anal.*

L. Liu, G. Moore and R.D. Russell (1993), 'Computing connecting orbits between steady solutions', Technical Report, Mathematics Department, Simon Fraser University.

G.J. Lord and A.M. Stuart (1994), 'Existence and convergence of attractors and inertial manifolds for a finite difference approximation of the Ginzburg–Landau equation,' in preparation.

E.N. Lorenz (1963), 'Deterministic noperiodic flow', *J. Atmos. Sci.* **20**, 130–141.

J. Lorenz (1994), 'Convergence of invariant tori under numerical approximation', in Proceedings of the Conference on Chaotic Numerics, Geelong, Australia, July 12th–16th 1993.

J. Mallet-Paret and G.R. Sell (1988), 'Inertial manifolds for reaction-diffusion equations in higher space dimensions', *J. Amer. Math. Soc.* **1**, 805–864.

M. Medved (1991), *Fundamentals of Dynamical Systems and Bifurcation Theory*, Adam-Hilger (Bristol).

A.R. Mitchell and D.F. Griffiths (1986), 'Beyond the linearized stability limit in nonlinear problems', in *Numerical Analysis* (D.F. Griffiths and G.A. Watson, eds), Pitman (Boston).

G. Moore (1993), 'Computation and paramterization of connecting and periodic orbits', submitted to *IMA J. Numer. Anal.*

G. Moore (1993), 'Computation and paramterization of invariant curves and tori', in preparation.

T. Murdoch and C.J. Budd (1990), 'Convergent and spurious solutions of nonlinear elliptic equations', *IMA J. Numer. Anal.* **12**, 365–386.

A.C. Newell (1977), 'Finite amplitude instabilities or partial difference schemes', *SIAM J. Appl. Math.* **33**, 133–160.

V. Pliss (1966), *Nonlocal Problems in the Theory of Oscillations*, Academic Press (New York).

V.A. Pliss and G.R. Sell (1991), 'Perturbations of attractors of differential equations,' *J. Diff. Eqns* **92**, 100-124.
C. Pugh and M. Shub (1988), 'C^r stability of periodic solutions and solution schemes', *Appl. Math. Lett.* **1**, 281–285.
W.C. Rheinboldt (1986), *Numerical Analysis of Parameterised Nonlinear Equations*, Wiley (New York).
J.M. Sanz-Serna (1992a), 'Symplectic integrators for Hamiltonian problems: an overview', *Acta Numerica*, Cambridge University Press (Cambridge), 244–286.
J.M. Sanz-Serna (1992b), 'Numerical ordinary differential equations vs. dynamical systems', in *Proceedings of the IMA Conference on the Dynamics of Numerics and the Numerics of Dynamics, 1990* (D. Broomhead and A. Iserles, eds), Cambridge University Press (Cambridge).
J.M. Sanz-Serna and S.Larsson (1993), 'Shadows, chaos and saddles', to appear in *Appl. Numer. Math.*
J.M. Sanz-Serna and A.M. Stuart (1992), 'A note on uniform in time error estimates for approximations to reaction-diffusion equations', *IMA J. Numer. Anal.* **12**, 457–462.
T. Sauer and J.A. Yorke (1991), 'Rigorous verification of trajectories for the computer simulation of dynamical systems', *Nonlinearity* **4**, 961–979.
A.B. Stephens and G.R. Shubin (1987), 'Multiple solutions and bifurcations of finite difference approximations to some steady problems of fluid mechanics.' *SIAM J. Sci. Stat. Comput.* **2**, 404–415.
H. Stetter (1973), '*Analysis of Discretization Methods for Ordinary Differential Equations*, Springer (New York).
D. Stoffer (1993), 'General linear methods: connection to one-step methods and invariant curves', *Numer. Math.* **64**.
D. Stoffer (1994) 'Averaging for almost identical maps and weakly attractive tori', submitted.
A.M. Stuart (1989a) 'Nonlinear instability in dissipative finite difference schemes', *SIAM Rev.* **31**, 191–220.
A.M. Stuart (1989b), 'Linear instability implies spurious periodic solutions', *IMA J. Numer. Anal.* **9**, 465–486.
A.M. Stuart (1991), 'The global attractor under discretization', in *Continuation and Bifurcations: Numerical Techniques and Applications* (D. Roose, B. De Dier and A. Spence, eds), Kluwer (Dordrecht).
A.M. Stuart and A.R. Humphries (1992a), 'Model problems in numerical stability theory for initial value problems', submitted to *SIAM Rev.*
A.M. Stuart and A.R. Humphries (1992b), 'The essential stability of local error control for dynamical systems', submitted to *SIAM J. Numer. Anal.*
R. Temam (1988), *Infinite Dimensional Dynamical Systems in Mechanics and Physics*, Springer (New York).
E.S. Titi (1991) 'Un critère pour l'approximation des solutions périodiques des équations de Navier–Stokes', *C.R. Acad. Sci. Paris*, **312**, 41–43.
M. Van Veldhuizen (1988), 'Convergence results for invariant curve algorithms', *Math. Comput.* **51**, 677–697.

S. Wiggins (1990), *Introduction to Applied Nonlinear Dynamical Systems and Chaos*, Springer (New York).

H. C. Yee, D. F. Griffiths and P. K. Sweby (1991), 'Dynamical approach to the study of spurious steady–state numerical solutions for nonlinear differential equations, I: The dynamics of time discretization and its implications for algorithm development in CFD', *J. Comput. Phys.* **97**, 249–310.

Yin-Yan (1993), 'Attractors and error estimates for discretizations of incompressible Navier–Stokes equations', submitted to *SIAM J. Numer. Anal.*

T. Yoshizawa (1966), *Stability Theory by Lyapunov's Second Method*, Publ. Math. Soc. Japan (Tokyo).